AF469274

Integrated Network Management IV

IFIP – The International Federation for Information Processing

IFIP was founded in 1960 under the auspices of UNESCO, following the First World Computer Congress held in Paris the previous year. An umbrella organization for societies working in information processing, IFIP's aim is two-fold: to support information processing within its member countries and to encourage technology transfer to developing nations. As its mission statement clearly states,

> IFIP's mission is to be the leading, truly international, apolitical organization which encourages and assists in the development, exploitation and application of information technology for the benefit of all people.

IFIP is a non-profitmaking organization, run almost solely by 2500 volunteers. It operates through a number of technical committees, which organize events and publications. IFIP's events range from an international congress to local seminars, but the most important are:

- the IFIP World Computer Congress, held every second year;
- open conferences;
- working conferences.

The flagship event is the IFIP World Computer Congress, at which both invited and contributed papers are presented. Contributed papers are rigorously refereed and the rejection rate is high.

As with the Congress, participation in the open conferences is open to all and papers may be invited or submitted. Again, submitted papers are stringently refereed.

The working conferences are structured differently. They are usually run by a working group and attendance is small and by invitation only. Their purpose is to create an atmosphere conducive to innovation and development. Refereeing is less rigorous and papers are subjected to extensive group discussion.

Publications arising from IFIP events vary. The papers presented at the IFIP World Computer Congress and at open conferences are published as conference proceedings, while the results of the working conferences are often published as collections of selected and edited papers.

Any national society whose primary activity is in information may apply to become a full member of IFIP, although full membership is restricted to one society per country. Full members are entitled to vote at the annual General Assembly, National societies preferring a less committed involvement may apply for associate or corresponding membership. Associate members enjoy the same benefits as full members, but without voting rights. Corresponding members are not represented in IFIP bodies. Affiliated membership is open to non-national societies, and individual and honorary membership schemes are also offered.

Integrated Network Management IV

Proceedings of the fourth international symposium on integrated network management, 1995

Edited by

Adarshpal S. Sethi
University of Delaware
Newark
Delaware
USA

Yves Raynaud and
Fabienne Faure-Vincent
University Paul Sabatier
Institut de Recherche en Informatique de Toulouse (IRIT)
Toulouse
France

SPRINGER-SCIENCE+BUSINESS MEDIA, B.V.

First edition 1995

Originally published by Chapman & Hall in 1995
MyCopy version of the original edition 1995

DOI 10.1007/978-0-387-34890-2

A catalogue record for this book is available from the British Library

∞ Printed on permanent acid-free text paper, manufactured in accordance with ANSI/NISO Z39.48-1992 and ANSI/NISO Z39.48-1984 (Permanence of Paper).
www.springer.com/mycopy

CONTENTS

PREFACE

Continuing the spirit of global cooperation established at our three previous landmark conferences, the *Fourth International Symposium on Integrated Network Management* (ISINM) provides an international forum for the diverse members of the network management community. Vendors and users, researchers and developers, standards planners and implementors, LAN, WAN and MAN specialists, systems and network experts, all must find ways to share and integrate network management knowledge.

The Fourth Symposium, ISINM '95, pursues the successful record of the first three, to build this community of knowledge. It continues the pledge to serve the diverse spectrum of interests of the network management community, bringing together the leaders of the field to cover its most central developments and the state of the art. It continues the commitment to high quality technical programs of great distinction, and to stimulate productive multilogues within the network management community.

The technical papers presented in this volume were selected from among 109 submissions through a most rigorous review process. Each paper was reviewed by 4 referees and carefully evaluated by the program committee, to ensure the highest quality. Continuing the tradition of diverse international participation, authors represent some 17 countries including Belgium, Canada, Denmark, England, Finland, France, Germany, Greece, India, Ireland, Italy, Japan, South Africa, Spain, Switzerland, Turkey, and the U.S.A., as well as papers involving international collaborations. Vast sections of the telecommunications, computer communications and computer industries are represented, as well as leading users, academic and industrial research labs.

The contents of the proceedings includes the 50 selected submissions, keynote papers and abstracts from the plenary sessions presented by leading visionaries of integrated systems management, short descriptions of 5 panels involving some of the best technical experts in the field, and the abstracts of papers presented as posters.

The table of contents is organized following the conference framework (tracks/sessions). Three main topics (tracks), including sub-topics (sessions), have been identified as follows:

- *Distributed Systems Management*
  - Distributed Management
  - Policy-Based Management
  - Application Management
  - Service & Security Management
  - Performance & Accounting Management

❐ *Performance and Fault Management*

- ❍ Enterprise Fault Management
- ❍ Event Management
- ❍ AI Methods in Management
- ❍ ATM Management
- ❍ Telecommunications Management Network

❐ *Practice and Experience*

- ❍ Agent Experiences
- ❍ Platform Experiences
- ❍ Management Databases
- ❍ Managed Objects Relationships
- ❍ Managed Objects Behavior

This organization aims at providing a useful reference book and a text book on current research in the field.

We are honoured to present these proceedings of the Fourth ISINM '95. The work included in this volume represents the collective contributions of authors, dedicated reviewers and a committed program committee. We thank Iyengar Krishnan and Paul Min for coordinating the Panels. Thank also to Branislav Meandzija, Wolfgang Zimmer and Doug Zuckerman for useful and helfpul comments, and to Gabi Dreo for helping with the conference database software. Last but not least, we thank Fabienne Faure-Vincent and Pramod Kalyanasundaram for their help with the handling of paper submissions, conference database maintenance, and many other tasks.

We wish to extend our gratitude to the authors of the technical papers and posters, without whom this symposium would not have been possible, and the members of the Program Committee for their help with paper solicitation and review.

And many thanks to all of you for your interest in the ISINM '95 symposium. We hope you will benefit from the technical program, and that you will capture the spirit of the complete *Integrated Network Management Week.*

Adarshpal S. Sethi and Yves Raynaud
Program Co-Chairs
January 15, 1995

SYMPOSIUM COMMITTEES

ORGANIZING COMMITTEE MEMBERS

Wolfgang Zimmer	GMD-FIRST, Germany, General Co-Chair
Douglas N. Zuckerman	AT&T Bell Laboratories, U.S.A., General Co-Chair
Yves Raynaud	Universite Paul Sabatier, France, Program Co-Chair
Adarshpal S. Sethi	University of Delaware, U.S.A., Program Co-Chair
Fabienne Faure-Vincent	Universite Paul Sabatier, France, Program Coordinator
Branislav Meandzija	MetaAccess Inc., U.S.A., Advisory Board Chair
Iyengar Krishnan	The MITRE Corporation, U.S.A., Tutorial and Special Events Chair
Allan Finkel	Morgan Stanley and Company, U.S.A., Vendor Chair
Tom Stevenson	IEEE Communications Society, U.S.A., IEEE/ComSoc Coordinator
Kenneth J. Lutz	Bellcore, U.S.A., IEEE/CNOM Coordinator
Mary Olson	U.C. Santa Barbara, U.S.A., Local Arrangements Chair and Treasurer
Anne-Marie Lambert	Bolt Beranek and Newman, Inc., U.S.A., OC Secretary

ADVISORY BOARD MEMBERS

Lawrence Bernstein	AT&T Bell Laboratories, U.S.A.
Sholom Bryski	Bankers Trust, U.S.A.
Jeff Case	SNMP Research, U.S.A.
Roberta S. Cohen	AT&T Paradyne, U.S.A.
Andre Danthine	University of Liege, Belgium
Michael Disabato	McDonald's Corporation, U.S.A.
Richard Edmiston	Bolt Beranek and Newman, Inc., U.S.A.
Heinz-Gerd Hegering	University of Munich - LRZ, Germany
Dave Mahler	The Remedy Corporation, U.S.A.
Venkatesh Narayanamurti	UCSB, U.S.A.
Izhak Rubin	UCLA, U.S.A.
Otto Spaniol	RWTH Aachen, Germany
Denis Yaro	Sun Microsystems, U.S.A.
Yechiam Yemini	Columbia University, U.S.A.
Makoto Yoshida	NTT Network Information Systems Labs, Japan

PROGRAM COMMITTEE MEMBERS

Raj Ananthanpillai	I-NET, U.S.A.
Anastasios Bouloutas	IBM Watson Research Center, U.S.A.
Stephen Brady	IBM Watson Research Center, U.S.A.
Walter Buga	AT&T Bell Laboratories, U.S.A.
Seraphin B. Calo	IBM Research, U.S.A.
William Donnelly	Broadcom Eireann Research, Ireland
Janusz Filipiak	University of Cracow, Poland
Ivan Frisch	Polytechnic University, U.S.A.
Kurt Geihs	Johann Wolfgang Goethe-Univ., Germany
Joerg Gonschorek	Siemens Nixdorf Inf. AG, Germany
Rodney M. Goodman	California Inst. of Technology, U.S.A.
Shri Goyal	GTE Laboratories, U.S.A.
Varoozh Harikian	IBM International Educational Centre, Belgium
Satoshi Hasagawa	NEC Corporation, Japan
Frank Kaplan	Consilium Inc., U.S.A.
Gautam Kar	Advantis, U.S.A.
Aurel A. Lazar	Columbia University, U.S.A.
Keith McCloghrie	Cisco Systems, Inc., U.S.A.
Paul S. Min	Washington University, U.S.A.
George V. Mouradian	AT&T Bell Laboratories, U.S.A.
Yoichi Muraoka	Waseda University, Japan
Shoichiro Nakai	NEC Corporation, Japan
George Pavlou	University College London, UK
Jan Roos	University of Pretoria, South Africa
Veli Sahin	NEC America, Inc., U.S.A.
Roberto Saracco	CSELT, Italy
David Schwaab	Hewlett-Packard Co., U.S.A.
Morris Sloman	Imperial College London, UK
Einar Stefferud	First Virtual Holdings, Inc., U.S.A.
Colin Strutt	Digital Equipment Corporation, U.S.A.
Liba Svobodova	IBM Research Division, Switzerland
Mark Sylor	Digital Equipment Corporation, U.S.A.
Ole Krog Thomsen	Jydsk Telefon, Denmark
Isabelle Valet-Harper	EWOS/DEC, Belgium
Jill Westcott	BBN Systems and Technology, U.S.A.
Carlos B. Westphall	UFSC-CTC-INE, Brazil

LIST OF REVIEWERS

A. Abdulmalak
S. Aidarous
R. Ananthanpillai
N. Aneroussis
K. Auerbach
L. Auld
K. Bahr
C. Bakker
T.G. Bartz
N. Bauer
A. Benzekri
L. Bernstein
K. Beschoner
M. Besson
S. Brady
H. Braess
R. Brandau
J.M. Bruel
G. Bruno
W. Buga
S. Calo
A. Chandra
K. Chapman
D. Chomel
K.L. Clark
R. Cohen
J. Conrad
A. Danthine
G. Dean
R. de Jager
T. Desprats
M. Disabato
A. Dittrich
W. Donnelly
O. Drobnik
H.P. Eitel
F. Faure-Vincent
M. Feridun
J.M. Ferrandiz
J. Filipiak
A. Finkel
B. Fricke
I. Frisch
S. Fukui
D. Gaiti
J. Galvin
D. Gantenbein
K. Geihs
C. Gerbaud
G. Giandonato
J.P. Golick
J. Gonschorek
R. Goodman
S. Goyal
J. Guion
J. Hall
V. Harikian
M. Hasan
S. Hasegawa
R. Hauser
G. Hayward
H. G. Hegering
R. Hutchins
D. Jaepel
G. Jakobson
A. Johnston
J.F. Jordaan
J.A. George
P. Kalyanasundaram
F. Kaplan
G. Kar
Y. Kiriha
M. Klerer
I. Krishnan
B. Krupczak
A. Lazar
G. Leduc
B. Leon
M. Levilion
J. Li
J.A. Lind
T. K. Lu
L. Lewis
K. Lutz
J.N. Magee
D. Mann
J.L. Marty
S. Mazumdar
K. McCloghrie
B. Meandzija
K. Meyer
P. Min
J. Moffett
P. Moghe
K. Morino
Y. Muraoka
G. Mouradian
S. Nakai
B. Natale
S. Ng
W. Norton
D. O'Mahony
J.J. Pansiot
R. Patton
G. Pavlou
M. Pietschmann
E. Pinnes
A. Pras
T. Preuhs
E. Pring
E.A. Pulsipher
G. Pujolle
R. Purvy
P. Putter
W. Reinhardt
P. Rolin
J. Roos
I. Rubin
H. Saidi
R. Saracco
V. Sahin
A. Sathi
R. Sauerwein
D. Schwaab
R. Schwartzl
N. Scribner
A. Shvartsman
M. Sibilla
S. Siegmann
M. Sloman
O. Spaniol
R. Stadler

LIST OF REVIEWERS (Continued)

D. Subrahmanya
R. Sultan
C. Sunshine
L. Svobodova
M. Sylor
O.K. Thomsen
M. Tobe
S. J. Towers
J. Tsay
J. G. Tsudik
K. Twiddle
A. Valderruten
I. Valet-Harper
F. Venter
M. Wakano
J. Warner
S. Warren
H. Wedde
A. Wedig
R. Weihmayer
C. Westphall
Y. Yemini
M. Yoshida
W. Zimmer
D. Zuckerman

Introduction
Integrating network management and rightsizing in the nineties

Wolfgang ZIMMER, GMD-FIRST, GERMANY

and

Douglas N. ZUCKERMAN, AT&T Bell Laboratories, U.S.A.

1. The Spirit of ISINM

During the two years since our last International Symposium on Integrated Network Management, ISINM '93 in San Francisco, numerous business needs and global competition have ever more strongly firmed themselves as the driving forces to achieve overall *systems* management of the enterprise information infrastructure. The requirement to perform this in the most efficient way is evident. It is well perceived that the high performance computing and communications technology plays a major role in the overall organizational performance.

This has greatly increased the demand for seamless integration of computer applications and communications services into network, systems and technology infrastructures which are robust, flexible and cost-effective to meet very real business challenges. It is this comprehensive provision of the whole information infrastructure mirroring the needs of the enterprise that has emerged as the linchpin of 'rightsizing in the nineties'.

The fourth symposium on Integrated Network Management, ISINM '95, itself has been 'rightsized' to focus on the pivotal role that integrated network management plays in establishing and maintaining an efficient worldwide information infrastructure, needed not only for big customers with worldwide operations.

However, no rightsizing took place in the spirit of the ISINM series: The 1995 symposium continues to provide a world-class program of high-quality technical sessions presented by recognized leaders in their field. They will discuss the critical issues that surround 'Managing Networked Information Services: The Business Challenge for the Nineties', and other related topics of high relevance to you and your colleagues.

2. ISINM History

Beginning with our first symposium in 1989, each ISINM program and its related theme has reflected the historic events in integrated network management, indeed has helped shape them.

- 1989: Improving Global Communication Through Network Management -
When we held the first ISINM in Boston in 1989, the need for comprehensive network management capabilities was apparent after major disasters had occurred in the telecommunications industries in the years before. Standards for enabling integrated network management across multiple vendor networking resources were in the heat of development in international and regional arenas. While some thought that developing these standards was the most difficult path on the road to integrated management solutions, many realized a few years later that standards were the beginning of a long journey. Integrated network management emerged to be one of the most complex and hard to solve problems of our heterogeneous communications community.

- 1991: Worldwide Advances in Integrated Network Management -
After two years, when we held the second ISINM in Washington, D.C., the need for enterprise-oriented management across data and telecommunications applications and distributed systems became increasingly apparent. Principle problems related to incorporating standards into products aimed at providing coherent, integrated network management solutions across future, standards-based, multi-vendor components as well as existing proprietary components. Multi-vendor demonstrations in North America, Europe and Japan seemed to indicate that the time had come when users could competitively procure network management products in any of several countries and be confident that they would interoperate with comparable products in other world regions. That wasn't so.

- 1993: Strategies For The Nineties -
We have learned. We are not at the end of the road - we are not even in the middle. We are only at the beginning and will remain there probably for the greater part of the nineties. Worldwide coordinated strategies are needed to evolve integrated network management in the best way. The beginning of the nineties was characterized by big political, ecological and technical changes in all areas worldwide. The exponential growth of internetworking in general and new multimedia applications based on broadband and mobile network technology will remain the driving forces of the communications area.

However, the element of uncertainty plays a dominant role in all environments. Down-sizing and up-sizing in volume and time requires flexibility to change. These problems are intensified by economic and regulatory constraints, problem complexity, technology advances, standards development, product introductions, market requirements, user demands and other factors which change unpredictably over time.

A paradigm shift took place during these phases: network management systems used for crisis situations in the past evolved to powerful tools for the day-to-day management of systems, services, applications and, of course networks. This brings us up to 1995 and 'Rightsizing in the Nineties.'

3. Rightsizing in the Nineties

During this sometimes turbulent period of rightsizing in all areas, the need for management systems is greater than ever before. Management is a fundamental part of a reliable information infrastructure. It assures the correct, efficient and mission-directed behavior of the hardware, software, procedures and people that use and provide all the information services. Effective management of the information infrastructure is becoming as essential as marketing and selling products. In addition, it helps to raise customer satisfaction. Integrated network management belongs to the enabling technologies of a worldwide information infrastructure.

The path to synergistically using this information infrastructure and the correlated management system faces a number of challenges:

- *Administrational*:
  Administrations need to take better account of the management technology and benefits, with its functions forming an integral part of the total enterprise. Unfortunately, budgets for new networked information services often did or do not adequately address the management part, leading to increased costs after systems crashes, degraded quality of service, etc. When the utilized information backbone is impacted, so is the whole enterprise, with potentially major financial repercussions. Issues such as proactive versus reactive management must be resolved throughout the enterprise to achieve to improved competitiveness.

- *Organizational*:
  The overall organization performance depends upon a high quality information infrastructure. Management systems are currently not considered as a primary life-function within it. Nor is it given full recognition for its intrinsic value to the organizational productivity. All this makes it very difficult to realize the cost-effective and timely use of management systems as the foundation for realizing the full enterprise-wide benefits of the newly re-engineered business processes. Further re-engineering of business processes will be needed and must take the benefits of management systems into account.

- *Bureaucratic*:
  Information technology managers perceive management systems as too expensive for the perceived benefits, and so are inclined to underfund or eliminate them. And in some long-established organizations, 'keepers' of the legacy infrastructure may intentionally or unintentionally get in the way of change. Rightsizing requires not only a flexibility in change of systems, but also the attitude of (some) people needs to change.

- *Security*:
  There is always the need for appropriate privacy and security protection; not only for the financial community, but also for individuals. Powerful expressions of constraints, policies, goals, etc. are required to guarantee this in a flexible and straightforward way. In addition, the public awareness of associated system risks, and related additional features to further minimizing these risks, will lead to a more careful usage, and a higher acceptance at lower overall system prices.

- *Reliability*:
  Our information infrastructure is not considered to be a prominent global safety-critical computerized system. Though it is a very large, globally distributed system, only parts of it fail completely. We know from experience that it will be up and running again after a certain period of time. It is mostly not the hardware, but the software that has been identified as the critical component. There are always risks, we have learned to live with them, but reliable and dependable software (and hardware) is one of the major challenges.

- *Flexibility*:
  If software is the solution, it is also the problem. It must be extensible, meet high performance requirements, and be highly reliable. There are also the haunting issues of how to replace, or, in the interim, adapt legacy systems to meet rapidly changing business and customer requirements. The communication infrastructure is also challenged to incorporate new transport/switching technologies such as SONET/ATM, and to take maximum advantage of promising high-performance computing technologies for integration such as multimedia applications.

- *Scalability*:
  Information systems and applications are continuously evolving at enormously increasing rates. Scalability in volume, performance and price for up to some hundred millions of users has to be addressed in the appropriate way. Initial investments should be kept as low as possible, to allow everyone to be part of the future global village. A subscription to products and to an associated definite product migration plan might be much better suited for the future, than the buy once and get a revision from time to time procedure of the past. Major efforts should be directed towards ensuring that we meet the current needs with low initial investments, and enable smooth migration (upwards scalability) afterward.

So, how do we overcome these and other challenges? Most of the problems outlined above are addressed in many of the papers included in these proceedings. We are certain that you will find viable solution approaches to most of today's problems and future challenges. To be most viable, our integrated network management solutions must: 1) be simple, and 2) impact 'the bottom line' without losing the overall picture of the future. Required are overall management solutions across computer and communication systems, and being part of a collaborative effort within the whole enterprise.

By-and-by the affordable and instant access to any information, independent of geographical location of client and server worldwide, will be as common as using a phone today. Many coordinated activities are needed to ensure it for the benefit of all of us.

4. Future Events

Examination of the papers has shown that we will have a very-high quality program with an excellent mix of topics, organizations and international contributions that we believe will be of high benefit to you.

As the management world continues evolving, this ongoing series of international symposia will continue to foster and promote cooperation among individuals of diverse and complementary backgrounds, and to encourage international information exchange on all aspects of network and distributed systems management.

To broaden the scope of these symposia, the International Federation for Information Processing (IFIP) Working Group (WG) 6.6 on Network Management for Communication Networks, as the main organizer of ISINM events, has been successfully collaborating with the Institute of Electrical and Electronics Engineers (IEEE) Communications Society's (COMSOC) Committee on Network Operations and Management (CNOM). ISINM and the Network Operations and Management Symposium (NOMS) are the premier technical conferences in the area of network and systems management, operations and control. ISINM is held in odd-numbered years, and NOMS is held in even-numbered years. CNOM and IFIP WG 6.6 have been working together as a team to develop both these symposia.

NOMS '96 will take place in Kyoto, Japan, April 16-19, 1996. The next International Symposium on Integrated Network Management (ISINM '97) will be held in the Spring of 1997, in North America on the East Coast or vicinity.

Starting in 1990, IFIP WG 6.6 together with IEEE CNOM has also been organizing the International Workshops on Distributed Systems: Operations and Management (DSOM) which takes place in October of every year and alternates in location internationally. DSOM '95 will be held at the University of Ottawa, Canada, October 16-18, 1995 and will be hosted by Bell-Northern Research (BNR).

For more information on future ISINM, NOMS, DSOM events and other related activities please get in touch with us.

5. Acknowledgements

ISINM '95 is the result of a great coordinated effort of a number of volunteers and organizations. First of all, we would like to thank our main sponsors, IFIP TC 6 and IEEE COMSOC CNOM for the financial support, the College of Engineering, University of California at Santa Barbara for hosting this event, GMD-FIRST and AT&T Bell Laboratories and all other organizations for their continued support.

Following the very huge success of ISINM '93, an intense discussion took place on how to follow it with an even better event. We owe a debt of gratitude to Branislav Meandzija and Mary Olson who both worked with us so hard in the beginning to form the vision of an ISINM '95 in Santa Barbara that would most effectively meet the needs of the network management community in 1995.

The organizing committee of ISINM '95 was formed in September 1993 and has been the main force behind the symposium. We would like to thank (in alphabetical order):

Fabienne Faure, Allan Finkel, Kris Krishnan, Anne-Marie Lambert, Kenneth Lutz, Branislav Meandzija, Mary Olson, Yves Raynaud, Adarshpal Sethi, and Tom Stevenson for enduring with us in this 18-month marathon towards ISINM '95.

The program committee under the tireless leadership of Adarshpal Sethi and Yves Raynaud has once again defined the standard for conferences and proceedings in network management. Its creative work, represented through this book, has clearly selected the main problem areas of integrated network management and the most promising solutions to those problem areas. Our deepest thanks go to Seraphin B. Calo, Janusz Filipiak, Heinz-Gerd Hegering, Frank Kaplan, Gautam Kar, George Pavlou, Jan Roos, Veli Sahin, Morris Sloman, Michelle Sibilla, Mark Sylor and Ole Krog Thomsen, who attended the program committee meeting in Toulouse, all other members of the program committee, and all the additional reviewers who created the outstanding program. Also, special thanks are due Martine De Peretti for her invaluable help with the logistics for the program committee meeting at DSOM '94.

Finally, we would like to thank Clark DesSoye for producing our main symposium brochures such as the advance and final programs, Steve Adler for his enthusiastic pursuit of vendor patrons, and last but not least all vendor patrons for their key role in the vendor program and showcase.

PART ONE

Distributed Systems Management

SECTION ONE

Distributed Management

1

Decentralizing Control and Intelligence in Network Management[1]

Kraig Meyer, Mike Erlinger, Joe Betser, Carl Sunshine
The Aerospace Corporation
P.O. Box 92957, Los Angeles, CA, 90009, USA. Phone: +1 310-336-8114. Email: kmeyer@aero.org

Germán Goldszmidt, Yechiam Yemini
Computer Science Department, Columbia University
450 Computer Science Building, Columbia University, New York, NY, 10027, USA.
Phone: +1 212-939-7123. Email: german@cs.columbia.edu

Abstract

Device failures, performance inefficiencies, and security compromises are some of the problems associated with the operations of networked systems. Effective management requires monitoring, interpreting, and controlling the behavior of the distributed resources. Current management systems pursue a platform-centered paradigm, where agents monitor the system and collect data, which can be accessed by applications via management protocols. We contrast this centralized paradigm with a decentralized paradigm, in which some or all intelligence and control is distributed among the network entities. Network management examples show that the centralized paradigm has some fundamental limitations. We explain that centralized and decentralized paradigms can and should coexist, and define characteristics that can be used to determine the degree of decentralization that is appropriate for a given network management application.

Keywords

Network Architecture and Design, Management Model, Distributed Processing, Client-Server.

1 INTRODUCTION

Some experts in the field of network management have asserted that most, if not all, network management problems can be solved with the Simple Network Management Protocol (SNMP) [3]. This stems in part from the belief that it is nearly always appropriate to centralize control and intelligence in network management, and that SNMP provides a good mechanism to manage networks using a fully centralized management paradigm.

[1]This work was sponsored in part by ARPA Projects A661 and A662. The views expressed are those of the authors and do not represent the position of ARPA or the U.S. Government. This paper approved for public release–distribution unlimited.

In this paper, we explore a number of different applications currently being used or developed for network management. We show that there are real network management problems that cannot be adequately addressed by a fully centralized approach. In many cases, a decentralized approach is more appropriate or even necessary to meet application requirements. We describe such an approach and start to build a taxonomy for network management applications. We specifically identify those characteristics that can be used to determine whether an application is more suitably realized in a centralized or decentralized network management paradigm. From the outset, it should be noted that many, if not most, network management applications can be realized in either paradigm. However, each application has characteristics that make it more suitable to one of the two approaches, or in some cases to a combination of both.

The remainder of this paper briefly lists what these characteristics are, discusses several categories of applications that have these differing characteristics, and analyzes some example applications. The next section describes two contrasting paradigms for network management: centralized and decentralized. Section 3 describes application characteristics that can be used to determine which paradigm is appropriate, along with some typical applications. Section 4 looks at four examples of decentralized applications in more depth. Finally, section 5 provides a conclusion and discussion of future work.

2 NETWORK MANAGEMENT MODELS

Basically, a network management system contains four types of components: Network Management Stations (NMSs), agents running on managed nodes, management protocols, and management information. An NMS uses the management protocol to communicate with agents running on the managed nodes. The information communicated between the NMS and agents is defined by a Management Information Base (MIB).

2.1 Centralized SNMP Management

The Internet-standard Network Management Framework is defined by four documents ([3], [6], [8], [9]). In the Internet community, SNMP has become the standard network management protocol. In fact, SNMP has become the accepted acronym for the entire Internet-standard Network Management Framework. Despite this, it should be noted that SNMP itself need not be bound to the paradigm that has developed around it. SNMP can be used as a reasonably general and extensible data-moving protocol.

To encourage the widespread implementation and use of network management, a minimalist approach has driven SNMP based network management. As noted in [10], "The impact of adding network management to managed nodes must be minimal, reflecting a lowest common denominator." Adherence to this "axiom" has resulted in a network management paradigm that is centralized, usually around a single NMS. Agents tend to be simple and normally only communicate when responding to queries for MIB information.

The centralized SNMP paradigm evolved for several reasons. First, the most essential functions of network management are well-realized in this paradigm. Agents are not capable of performing self-management when global knowledge is required. Second, all network entities need to be managed through a common interface. When many of these entities have limited computation power, it is necessary to pursue the "least common denominator" strategy mentioned above. Unfortunately, in many cases this strategy does not allow for data to be processed where and when it is most efficient to do so.

Even when management data is brought to an NMS platform, it is frequently not processed by applications in a meaningful way. Network management protocols unify the syntax of managed data access, but leave semantic interpretation to applications. Since the semantic heterogeneity of managed data has grown explosively in recent years, the task of developing meaningful management applications has grown more onerous. In the absence of such applications, platform-centered management often provides little more than *MIB browsers*, which display large amounts of cryptic device data on user screens. As first noted in the introduction to [7], it is still the case that "most network management systems are passive and offer little more than interfaces to raw or partly aggregated and/or correlated data in MIBs."

The rapid growth in the size of networks has also brought into question the scalability of any centralized model. At the same time, the computational power of the managed entities has grown, making it possible to perform significant management functions in a distributed fashion.

Contemporary management systems, based on the platform-centered paradigm, hinder users from realizing the full potential of the network infrastructure on which their applications run. This paradigm needs to be augmented to allow for decentralized control and intelligence, distributed processing, and local interpretation of data semantics.

2.2 Decentralized Management by Delegation

Management by Delegation (MBD) [13] utilizes a decentralized paradigm that takes advantage of the increased computational power in network agents and decreases pressure on centralized NMSs and network bandwidth. MBD supports both temporal distribution (distribution over time) and spatial distribution (distribution over different network devices). In this paradigm, agents that are capable of performing sophisticated management functions locally can take computing pressure off of centralized NMSs, and reduce the network overhead of management messages.

At the highest level of abstraction, the Decentralized MBD paradigm and Centralized SNMP paradigm appear the same, as both have an NMS communicating with agents via a protocol. But the MBD model supports a more distributed management environment by increasing the management autonomy of agents. MBD defines a type of distributed process, Elastic Process [4], that supports execution time extension and contraction of functionality. During its execution, an elastic process can absorb new functions that are delegated by other processes. Those functions can then be invoked by remote clients as either remote procedures or independent threads in the scope of the elastic process.

MBD provides for efficient and scalable management systems by using delegation to elastic agents. Instead of moving data from the agent to the NMS where it is processed by applications, MBD moves the applications to the agents where they are delegated to an elastic process. Thus, management responsibilities can be shifted to the devices themselves when it makes sense to do so.

Decentralization makes sense for those types of management applications that require or can take advantage of spatial distribution. For example, spatial distribution may be used to minimize overhead and delay. There is also an entire class of management computations, particularly those that evaluate and react to transient events, that must be distributed to the devices, as they can not be effectively computed in an NMS. Decentralization also allows one to more effectively manage a network as performance changes over time. The ability to download functions to agents and then access those functions during stressed network conditions reduces the network bandwidth that would be consumed by a centralized paradigm.

3 DISTRIBUTING NETWORK MANAGEMENT APPLICATIONS

The two paradigms of network management presented in the previous section might be viewed as contrasting, competing, possibly even incompatible models. The reality is that the SNMP (or centralized) paradigm and the MBD (or decentralized) paradigm are really just two points on a variety of continuous scales. An ideal network management system should be able to handle a full range of network management functions, for example using MBD's elastic processes to distribute management functionality in those cases where distribution is more efficient, but using SNMP's centralized computation and decision making when required. In this way, MBD should be seen as augmenting, rather than competing with, SNMP efforts. In fact, the SNMP community has already recognized the value of distributable management, with a manager-to-manager MIB [2] and some preliminary work on NMS-to-agent communications via scripts.

As previously mentioned, most of the early network management applications were well-suited to centralized control, which explains the success that the centralized SNMP paradigm has had to date. Some newer and evolving applications require a decentralized approach. A good example of an application that requires decentralization is the use of RMON (remote monitoring) probes [12]. RMON probes collect large amounts of information from their local Ethernet segment, and provide an NMS with detailed information about traffic activity on the segment. These probes perform extensive sorting and processing locally, and provide summary and table information via SNMP through a specially formatted MIB. Although this application uses SNMP for data transfer, in actuality, RMON is a realization of an application in the decentralized paradigm.

The question remains, how does one characterize network management applications in such a way that one can determine whether they should be distributed? There are a number of metrics that can be used to judge whether a network management application is more appropriately realized in a centralized or decentralized paradigm. These metrics are illustrated in figure 1 and include the following:

- **Need for distributed intelligence, control and processing.** This scale runs from a low need for distribution (corresponding with centralized intelligence) to a high need for distribution, or decentralized intelligence. An application that requires fast decisions based on local information will need decentralized control and intelligence. Applications that utilize

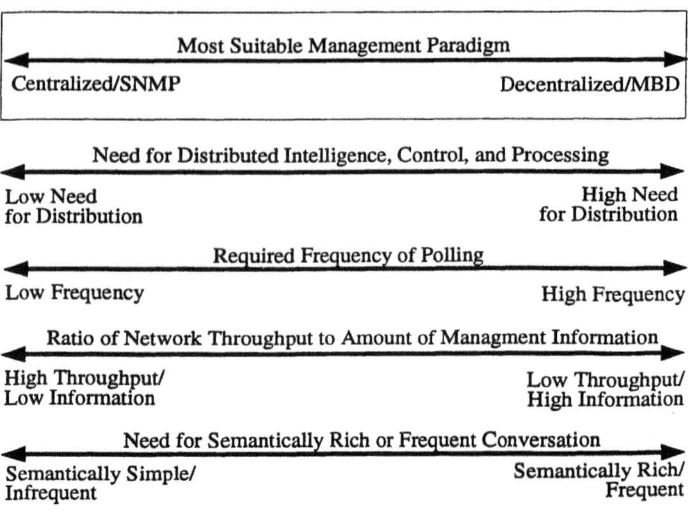

Figure 1: Metrics used to determine decentralization

large amounts of data may find it advantageous, though not always necessary, to perform decentralized processing. A specific example of this is an application that may need to use many pieces of data that can only be obtained by computing database views over large numbers of MIB variables. In this case, the application output may be very small, but the input to it may be an entire MIB.

- **Required frequency of polling.** The need for proximity to information and frequency of polling may dictate that computations be performed in local agents. This scale runs from a low frequency of polling to a high frequency of polling. An example of an application that requires a high frequency of polling is a health function that depends on an ability to detect high frequency deltas on variables.

- **Ratio of network throughput to the amount of management information.** At one end of this scale, the network in question has plenty of capacity relative to the amount of management information that needs to be sent through it. At the other end of the scale, there is a large amount of management information–so much that it conceivably could saturate the lower throughput network. An example of an application with a low throughput/high information ratio is the management of a large remote site via a low bandwidth link. Note that network throughput is affected not only by the amount of bandwidth available but also by the reliability of that bandwidth.

- **Need for a semantically rich and/or frequent conversation between manager and agent.** One end of this scale represents those applications that require only semantically simple and infrequent conversations, meaning that access to data is infrequent and simple data types are all that need to be accessed. At the other end of this scale are applications that require frequent conversations and/or semantically rich interactions, meaning that complex data structures, scripts, or actual executables need to be passed to a remote server. An application that needs to download diagnostic code to agents on demand is an example of one that would require a semantically rich and frequent conversation.

3.1 Centralized Applications

From the discussion of these metrics, we can see that centralization is generally appropriate for those applications that have little inherent need for distributed control, do not require frequent polling or high frequency computation of MIB deltas, have high throughput resources connecting the manager and agent, pass around a small amount of information, and do not have a need for frequent and semantically rich conversations between the manager and agent.

Most network management applications that are currently being used fall into this category. One may argue that this is because the centralized (SNMP) paradigm is the only one that is realized in most commercial products, but in actuality this centralized paradigm was built because the most important network management needs fit these characteristics. The classic example of this is the display of simple MIB variables. Monitoring a router's interface status, or a link's up/down status, involves querying and displaying the value of a single or small number of (MIB) variables, and is well suited to centralized management.

The NMS network map is another example of a tool that requires input from a number of devices to establish current connectivity. Thus a decentralized approach would not provide the connectivity map that a centralized approach can quickly establish via an activity like ping.

3.2 Partially Decentralized Applications

"Partial Decentralization" is appropriate for applications that are bandwidth-constrained, but still require some degree of centralized administrative control. An example of a bandwidth-constrained application is the management of a west coast network by an east coast manager. If the networks are linked by a relatively low bandwidth link, it is desirable for all information about the west coast network to be collected locally by an agent on the west coast, and only summary information be passed back to the east coast. Another case of a "partially decentralized" application is when local networks are autonomous. A department administrator may manage a local network, passing only summary information up to the higher level network manager.

This category of applications also includes those that can be decentralized for the purpose of bandwidth and processor conservation. It may be possible to greatly reduce the amount of bandwidth or centralized processing required by having an agent perform a local calculation over a large amount of data, then reporting the result–a small amount of data–back to the centralized manager. This algorithm may be repeated on each subnet of a large network, effectively breaking one large calculation into many small calculations. Some applications of RMON and health functions fit this profile. Some applications for the management of stressed networks also fit this profile.

Some degree of decentralization is highly desirable for the applications in this category. This may be accomplished by building a midlevel SNMP manager local to the variables being monitored, or by using elastic processes in the MBD paradigm. The SNMP solution is less general in that each midlevel manager must include both agent and NMS capabilities.

3.3 Decentralized Applications

Further analysis of the aforementioned metrics shows that decentralization is most appropriate for those applications that have an inherent need for distributed control, may require frequent polling or computation of high frequency MIB deltas, include networks with throughput constraints, perform computations over large amounts of information, or have a need for semantically rich conversations between manager and agent.

An example in this class is a health function that requires an ability to detect high frequency deltas on a set of MIB variables. A second example may be the management of a satellite or disconnected subnet, where a subnet manager is required to obtain data, make decisions, and change application or network characteristics even when that manager is isolated from the central, controlling manager. Finally, an application may have a need to download diagnostics and control information into a network element dynamically, in an attempt to isolate a problem.

Depending on the generality required, the SNMP manager-to-manager MIB may not be sufficiently general to allow for adequate delegated control for these applications. If frequent reprogrammability is a requirement, decentralization is the logical choice.

4 EXAMPLES OF DECENTRALIZED APPLICATIONS

We have identified four examples of network management applications that should be realized in a decentralized network management paradigm. These include Distributed Intrusion Detection, Subnet Remote Monitoring, Subnet Health Management, and Stressed Domain Management. What is presented below is a description of the activity and an analysis of its requirement for a decentralized approach. Current research efforts are involved in determining quantitative values for centralized and decentralized approaches to these applications.

4.1 Management of Distributed Intrusion Detection

Intrusion detection refers to the ability of a computer system to automatically determine that a security breach is in the process of occurring, or has occurred at some time in the past. It is built upon the premise that an attack consists of some number of detectable security-relevant system events, such as attempted logons, file accesses, and so forth, and that these events can be collected and analyzed to reach meaningful conclusions. These events are typically collected in an audit log, which is processed either in real time or off-line at a later time.

Intrusion detection requires that many potentially security-relevant events be recorded, and thus enormous amounts of audit data are a necessary prerequisite to successful detection. Simply recording all of the audit records results in a large amount of Input/Output (I/O) and storage overhead.

For example, if all audit events are enabled on a Sun Microsystems workstation running Multilevel Secure SunOS, it is possible for a single machine to generate as much as 20 megabytes of raw data per hour, although 1-3 megabytes is more typical [11]. Once the audit records are recorded, they must all be read and analyzed, increasing I/O overhead further and requiring a large amount of CPU processing. Audit data generally scales linearly with the number of users. As a consequence, expanding intrusion detection to a distributed system is likely to result in network congestion if all audit data must be sent to a central location. The CPU requirements scale in a worse than linear fashion: Not only must analysis be performed on each machine's local audit log, but correlation analysis must be performed on events in different machines' local logs. As a result, there is a high motivation to keep processing distributed as much as possible, and to keep the audit record format as standardized as possible.

Historically, the management of distributed intrusion detection has not been addressed in any standardized way. Banning [1] suggests that a list of an audit agent's managed objects should be stored in a MIB, and an audit agent should be managed using a standardized protocol such as CMIP [5]. However, to date, no intrusion detection systems have been widely fielded that perform this function.

Intrusion detection is an excellent candidate application for decentralized management. There is a high motivation for decentralized intelligence and processing because it is very clear that centralized processing won't scale, and that network bandwidth won't accommodate all audit data being sent to a centralized point. Further, there may be a need for a semantically rich conversation between distributed monitors, as they may need to pass relatively complicated structures that are hard to predefine in a MIB.

4.2 Subnet Remote Monitoring (RMON)

As previously mentioned, RMON [12] provides a framework in which remote monitoring probes collect information from local Ethernet segments, and provide this data to NMSs. RMON has in fact taken a hybrid centralized/decentralized approach to management. The RMON agent is responsible for collecting data from the local segment and performing calculations over that data (e.g., determining which stations are generating the largest amount of traffic). On a busy network, this may include maintaining a station table of over 3000 nodes along with packet counts. It is impractical, and inefficient, to download this entire station table to the management station for centralized processing. The entire transaction could easily take minutes, which is likely too slow to be meaningful.

In the RMON MIB a form of distributed processing was used in the creation of the Host Top N function. The Host Top N MIB group provides sorted host statistics, such as the top 20 nodes sending packets, or an ordered list of all hosts according to the number of errors they sent over the last 24 hours. Both the data selected and the duration of the study is defined by the user via the NMS. Once the requested function is setup in the agent, the NMS then only queries for the requested statistics.

Using a pure centralized approach for the Top N transmitting stations,[2] the NMS would have to

[2] Assume that a sort will be performed based on the number of packets transmitted by each station.

request statistics for all the hosts that have been seen on that subnet. Two such sets of requests would have to be made to determine the Top N: one to get a baseline count for each station and one to get the count for each station after a time, t. The difference between the two sets of requests would then be sorted by the NMS for the Top N display.

Assuming that statistics for only one station can be requested in each SNMP message, the total number of SNMP messages is 2 times the number of stations (ns) with a total SNMP cost of: 2 * ns * SC, where SC is the cost of an SNMP message.

If instead, the RMON approach is taken, the Top N function is distributed to the agent and the costs are greatly decreased. In this situation there are two costs. The first cost corresponds to the request that a Top N function be performed for some number of stations N < ns over some period t; the second is the cost of gathering the sorted statistics. Assuming that the set up costs (selection criteria and time period) can be established in two SNMP messages, the cost for a distributed top N function is: 2 * SC + N * SC. In the worst case, N = ns, decentralization costs (2 + ns) * SC. Thus whenever NS > 2, the decentralized approach of RMON is superior–costs less–than the usual centralized approach.

4.3 Management of Subnet Health Applications

Subnet health management is another application that requires some degree of decentralization. One of the difficult problems in a large network is the determination of the health of a subnet, where health is a dynamic function of a number of network traffic parameters. RMON is designed to provide data for the management of subnets. In a network of many subnets, e.g., a corporate network, the SNMP centralized paradigm puts a processing burden on the NMS and a data transfer burden on the network.

Subnet health can be determined using either the centralized or distributed paradigm. In a lightly loaded network, it is acceptable for the NMS to query all the subnets for information. The returned information can then be filtered by the management station to determine subnet health. The problem with this centralized paradigm arises in a loaded or congested network, especially when the amount of information being returned is large. When the network is loaded, the additional traffic generated by querying the subnets for large volumes of data can be significant. Thus the decentralized approach becomes necessary. This is a case where a large amount of information is needed relative to the throughput or bandwidth available on the network.

In the centralized approach the management station has the requirement to make some evaluation of subnet health by first gathering data and second, correlating that data. The decentralized approach localizes the gathering and correlation activities, so the local subnet then has the responsibility only to report its health based on some known health function.

The determination of whether subnet health is a centralized or decentralized activity is made not by the activity itself, but by variables affecting that activity. Thus, it is not the activity of gathering data and evaluating health that determines centralization. Rather, the effects of the network traffic on such gathering and the effects of such gathering on network traffic determine the choice between centralized and decentralized paradigms. This determination should be made dynamically by the NMS, which is able to determine and modify the balance of centralized versus decentralized activity.

The following steps might be taken:

- Using ping or a predefined health function, the NMS determines whether a centralized or decentralized approach should be used.
- If conditions favor a centralized approach, the NMS would request from the RMON agent all data that might be needed for various application tools. This is essentially the current approach.
- If a decentralized approach is determined to be needed, the NMS would request results from predefined RMON agent health functions.
- Based on these health functions, additional health data may be requested and/or new health functions downloaded to the agent. Each health function would put additional emphasis on agent health evaluation.

In some ways the above is a dynamic escalation from the centralized paradigm to the decentralized paradigm based on health functions. The goal of the NMS is to determine subnet health with minimal impact on the network as a whole.

4.4 Management of Stressed Networks

An additional application that is well-suited towards distributed management is the management of stressed networks. Networks in stressed conditions have a number of properties that require different management strategies from unstressed networks. For the purpose of this paper, network stress is defined as sustained operation at high utilization, and includes highly saturated network segments or devices. Related characteristics of such networks include longer delays, reduced effective connectivity, and less predictable responses. Network stress may be caused by failure of network components, causing phenomena such as loss of connectivity, increased packet traffic, and unexpected routing. A common characteristic of stress is that if left unattended, problems tend to escalate, and network resources become less available. The unstable stress phenomena are the most critical to address. Algorithms used for stressed region management must have the following characteristics:

- **Local Autonomy of Algorithm.** The algorithm must have good distributivity, provide most information locally, and only require low management bandwidth outside of the local domain.
- **Stress Containment using Routing.** Routing must be able to bypass problematic regions. Routing algorithms must be very distributed, with routing tables at each domain, and must react to changes in traffic patterns. In stress, there should be alternate routes known locally, but remote verification of reachability is required.
- **Local Network Domain Stabilization.** If the source of a problem is local, the local domain should be able to make decisions to contain and correct problems locally. If a stress source is external, outside consultation is required.

- **Gradual and Graceful Degradation.** Management algorithms should function and network services should continue—albeit with worse performance—as network stress grows. This typically requires a distributed architecture, with low dependency on remote resources and high dependence on local autonomy.
- **Stress Prediction.** Distributed health monitoring allows for local domains to anticipate stress conditions before they actually occur. Countermeasures may be taken locally or may require interaction between domains.

A basic technique for stress monitoring involves the correlation of MIB variables reflecting local stress (such as retransmissions, packet lengths, and timeouts). These correlations should be done on a domain-by-domain basis, for efficient collection of data from neighboring nodes, and thus computations would be distributed. This may also naturally lead to distributed control and decentralization. Local managers would conduct cross-correlations on a regular basis, and patterns of stress could be established and trigger stress alarms for that domain. Similarly, higher level managers would conduct cross-correlations of domain-manager information, to establish "regional" stress propagation, and devise policies and strategies to combat escalating stress. All these activities are very likely to be distributed in a hierarchical fashion among network domains.

A need for distributed control, bandwidth limitations, and other characteristics of stress management indicate that decentralization may provide significant benefits in effectively managing network and system stress.

5 CONCLUSIONS AND FUTURE WORK

We have described two network management paradigms, SNMP and MBD, that have historically represented conflicting views of how networks should be managed. We have shown that the centralized approach associated with SNMP and the decentralized approach of MBD are actually just two points on a continuous scale of network management approaches. We have started building a taxonomy for network management applications and identified a number of characteristics that can help to determine whether a given network management application should be realized in a centralized paradigm, a decentralized paradigm, or some hybrid of the two. Finally, we have focused on four specific examples of network applications and explained why none of them is best realized in a strict, fully-centralized network management paradigm.

We plan to continue to investigate network management approaches through a series of experiments directed at quantifying the choice of network management paradigm. We believe that the costs associated with the various paradigms can be used by applications to dynamically choose among centralized, decentralized, or hybrid approaches to network management. The experiments should also provide additional input to extend the list of characteristics that effect the choice of network management paradigm.

References

[1] D. Banning, et. al. *Auditing of Distributed Systems.* Proceedings of the 14th National Computer Security Conference, pages 59–68, Washington, D.C., October 1991.

[2] J. Case, K. McCloghrie, M. Rose, and S. Waldbusser. *Manager-to-Manager Management Information Base.* Request for Comments 1451, April 1993.

[3] J. Case, M. Fedor, M. Schoffstall, and J. Davin. *A Simple Network Management Protocol (SNMP).* Request for Comments 1157, May 1990.

[4] G. Goldszmidt. *Distributed System Management via Elastic Servers.* Proceedings of the IEEE First International Workshop on Systems Management, pages 31–35, Los Angeles, California, April 1993.

[5] International Standards Organization (ISO). *9596 Information Technology, Open Systems Interconnection, Common Management Information Protocol Specification*, May 1990.

[6] K. McCloghrie and M. Rose. *Management Information Base for Network Management of TCP/IP-based internets: MIB-II.* Request for Comments 1213, March 1991.

[7] B.N. Meandzija, K.W. Kappel, and P.J. Brusil. Introduction to *Proceedings of the Second International Symposium on Integrated Network Management*, Iyengar Krishnan and Wolfgang Zimmer, editors. Washington, DC, April 1991.

[8] M. Rose and K. McCloghrie. *Structure and Identification of Management Information for TCP/IP-based Internets.* Request for Comments 1155, May 1990.

[9] M. Rose and K. McCloghrie. *Concise MIB Definitions.* Request for Comments 1212, March 1991.

[10] M. Rose. *The Simple Book, An Introduction to Management of TCP/IP-based Internets.* Prentice Hall, 1991.

[11] O. Sibert. *Auditing in a Distributed System: SunOS MLS Audit Trails.* Proceedings of the 11th National Computer Security Conference, Baltimore, MD, October 1988.

[12] S. Waldbusser. *Remote Network Monitoring Management Information Base.* Request for Comments 1271, November 1991.

[13] Y. Yemini, G. Goldszmidt, and S. Yemini. *Network Management by Delegation.* Second International Symposium on Integrated Network Management, pages 95–107, Washington, DC, April 1991.

Kraig Meyer is a Member of the Technical Staff at The Aerospace Corporation in El Segundo, CA. He has previously worked as a lecturer and research assistant at the University of Southern California, and as a Systems Research Programmer on the NSFNET project at the Merit Computer Network. His research interests include computer network security, protocols, and management. Kraig holds a BSE in Computer Engineering from the University of Michigan and an MS in Computer Science from the University of Southern California.

Mike Erlinger is a Professor of CS at Harvey Mudd College, and a member of the technical staff at The Aerospace Corporation. Mike has founded and chaired the CS department at Mudd, and has technical program support responsibilities at Aerospace, as well as a lead role in several of the research efforts, such as the Southern California ATM Network. He has also founded and chaired the RMON MIB WG within the IETF. Mike has worked for Micro Technology as Director of Network Products and previously for the Hughes Corporation. His interests are in the areas of network management, software engineering, system administration, and high speed networking.

Joe Betser is the founder and head of the Network and System Management Laboratory at The Aerospace Corporation. Dr. Betser provides the national space programs with ongoing technical guidance and also serves as an ARPA PI. Joe established research collaborations with Columbia University and several California centers active in high speed networking and ATM. His new work focuses on QOS for tele-medicine, tele-multi-media, and other imaging applications. Joe served on the program and organizing committees for NOMS, ISINM, MilCom, and other computer communications events, and in particular, has chaired the vendor program at ISINM'93. Joe holds a PhD and MS in CS from UCLA, and a BS with Honors from Technion, Israel Inst. of Tech.

Carl Sunshine has been involved in computer network research from the early development at Stanford University of the Internet protocols. He subsequently worked at The Rand Corporation, USC Information Sciences Institute, Sytek (now Hughes LAN Systems), and System Development Corporation (now Unisys). Dr. Sunshine's work encompassed a range of topics including network protocol design, formal specification and verification, network management, and computer security. Since 1988 he has been with The Aerospace Corporation, managing computer system research and development for a variety of space programs.

German Goldszmidt is a PhD candidate in Computer Science at Columbia University, where he is completing his dissertation, entitled "Distributed Management by Delegation". He received his BA and MS degrees in Computer Science from the Technion. His Master's thesis topic was the design and implementation of an environment for debugging distributed programs. Since 1988 he worked at IBM Research, where he designs and develops software technologies for distributed applications. His current research interests include distributed programming technologies for heterogeneous systems, and network and distributed system management.

Yechiam Yemini (YY) is a Professor of CS and the Director of the Distributed Computing and Communications Laboratory at Columbia University. YY is the Founder, Director, and Chief Scientific Advisor of Comverse Technologies, a public NY Company producing multimedia store-and-forward message computers. YY is also the Founder and Chief Scientific Advisor of System Management Arts (SMARTS), a NY startup specializing in novel management technologies for enterprise systems. YY is frequently invited to speak in the areas of computing, networks, distributed systems, and the interplay among these areas, and is the author of over 100 publications.

2

Models and Support Mechanisms for Distributed Management[1]

J.-Ch. Grégoire[2]
INRS-Télécommunications
16, pl. du Commerce, Ile des Soeurs,
Verdun, Qc, CANADA H3E 1H6
gregoire@inrs-telecom.uquebec.ca

Abstract

We describe here an experimental environment for distributed network and system administration based on the integration of a small number of simple efficient conceptual models which support a variety of management paradigms. They are implemented in turn by a couple of simple, but powerful mechanisms and a customizable runtime environment. We describe how this environment has been realized around a small and efficient language.

Keywords: distributed systems management, delegation, worm, conceptual models, implementation support architecture.

1 Introduction

Network management has received a lot of attention from standardization bodies, network and computer equipment manufacturers, and has inspired various consortiums. In most cases, network management has been handled, to a large extent, as a distributed database problem, where the management information is acquired remotely then transferred to a central location to be processed [11, 2]. The data is organized as a hierarchical distributed, potentially *object-oriented model* [3, 4]. However, even when the model is object-oriented, it nevertheless supports direct data manipulation as well as a notion of *operation*[3]. In other words, the notion of object provides inheritance of properties and granularity of concepts, but not necessarily encapsulation. It is worth noting, in this case, that the database model is not explicitly recognized as the basis for the management *mechanisms*, and little effort has been made to integrate the results of developments in distributed database technology in standards and platform alike.

The major alternative offered to the database model is a distributed object-oriented *application*. The importance of this model appears to be increasing, even though it has been pushed forward mainly by consortiums [16, 15] rather than official standardization bodies, although the conceptual influence of Open Distributed Processing (ODP) [5] must

[1]parts of this work were submitted to DSOM'94

[2]this work was partially funded by the Chaire Cyrille Duquet en Logiciels de Télécommunications

[3]note that, in this document, operation may mean an action on an object or the operations of the distributed system/network

be acknowledged. This model supports cooperative forms of management and appears to be quite well suited for higher levels of management. Because objects tend to be large grained, and their manipulation through trading and/or brokerage mechanisms may incur a significant operational overhead, this model is not really considered for low-end operations such as data acquisition at this stage.

System management has been more the focus of individual computer systems manufacturers as well as third party suppliers. This form of management is typically aimed at system configuration and information sharing. As such, its problems are different from network management as it focuses more on *dynamic* configuration through information distribution and sharing (e.g. with Hesiod or NIS) whereas configuration in network management tends to be more static. Yet, the support mechanisms used can be related to decentralized and/or hierarchical databases. Monitoring and performance oriented operations are typically done locally.

More recent developments, mainly from private companies, have introduced distributed platforms for operations management, again using a distributed object-oriented model. These new developments bring system management more in line with the concerns of network management, and we thus feel that it is legitimate to try and unify the two notions.

Each form of management uses a unique mechanism, either a distributed database or distributed objects, to support all management tasks. This mechanism is either limited in functionality for efficiency reasons (e.g. SNMP), or turns out to be a rather heavyweight generic tool (e.g. Tivoli, ANSAware). The lack of flexibility in mechanisms leads to inflexible solutions. There are indeed few tradeoffs available in computing power and bandwidth requirements between either mechanisms.

The focus of our study is the identification of a *basic set* of conceptual mechanisms and models (paradigms) necessary and sufficient to support management tasks. Using several mechanisms, as opposed to a single, general one, allows us to have minimal structural overhead for different operations. We can also mix different levels of support for different classes of devices. Overhead indeed increases dramatically as platforms increase in complexity, with a single general mechanism.

A toolkit supporting these conceptual mechanisms allows us to fine tune the *quality of service* for different operations. Performance, availability, integrity and safety are all factors that can be taken into consideration in the selection process. These toolkit consists of a programming language and its runtime environment, which supports remote execution and dynamic interactions.

The structure of this paper is as follows. We first give some general background on network management and its terminology. We then discuss different computational structures used in, or of potential interest for network management. We then discuss another dimension of management, that is the nature of the operations that must be performed. This allows us to introduce our set of mechanisms and show how it can support the functionality required. We show how it can be used, and describe a prototype implementation. We close with a discussion and some conclusions.

2 Background

In this paper, we will be using the "standard" network management framework.

2.1 General notions

A *manager* communicates with *network elements* running *agents.* An agent interacts with the physical (or logical) process to create and maintain *managed object* abstractions. An agent can also act as a *proxy,* that is, hide and create a management compatible abstraction for parts of the network that use a different protocol.

Management is the realization of various functional categories, such as Operations, Administration, Maintenance and Provisioning (OAM&P) in the TelCo tradition, or Accounting, Fault, Configuration, Performance and Security in the OSI perspective.

Network management solutions address the problem of *network element* (or device) management. They incorporate important decisions wrt issues such as

- in band vs. out of band,
- connection based or connectionless,
- protocol efficiency and performance,
- agent resources requirements,
- manager resources requirements,
- complexity of access and manipulation of the information structure.

Network management protocols reflect a conceptual structure of managed information. The *database model* is the underlying structure in international network management standards such as SNMP or CMIS. Basically, the managed resources are treated as a collection of managed objects whose state can be queried and modified from a number of remote managers. The database model naturally suggests itself as long as one views the network as a collection of information sources to browse, and possibly to change.

SNMP, for example, is a connectionless protocol, suitable for small scale networks. Its use of polling to update the database information also generates a volume of traffic which can consume too much bandwidth as networks grow in size: a form of the so-called *probing effect* [13]. Its agents are however rather simple. Its data access paradigm is also quite simple, and consists mainly of variable manipulation. CMIP on the other hand is connection based. Its information model is richer than SNMP's and require more support from the agent. It is meant to be scalable to large networks, but it lacks, as does SNMP, a hierarchy of higher level, inter-manager, information exchange and cooperation structure.

2.2 Problems with current models

There are a number of problem with the current database approach to network management. First, the mechanism actually implemented in protocols is a restricted form of the database model, however, for efficiency reasons.

Atomicity of access is restricted to some operations when it is available at all. Operations can only be performed on a unique network element at a time. Consistency of information retrieval across several network elements cannot therefore be guaranteed, i.e. we cannot manipulate distributed relations.

The complexity of the management work rests on one or several management station(s) which must be capable of browsing the information structure of the managed objects and recover, or modify, specific objects. Managed objects may however spontaneously notify a manager of some change in their status with *traps* or *notifications* (a notion similar to triggers in the database world).

The database model lacks notions of cooperation and grouping. There is no provision in the basic model for cooperation between managers, although the underlying mechanisms can be used to communicate information to another manager. There is also no way of grouping agents into a single element to give it a collective presentation.

In the case of in-band management, when agents have to be polled for updates, the database model may incur a significant load on the network which can be detrimental to normal operations. Scalability then becomes an important issue. Spontaneous notification mechanisms may somewhat alleviate the problem, however.

Finally, the different database models used in administration are non-hierarchical and another mechanism is required to integrate managers for domains that outgrow the model quantitatively of geographically.

2.3 Evolution

More recently, there has been a growing interest in using emerging "standard"[4] distributed OO platforms as a basis for object management or, in another case, at least to support inter-manager communication, acting as an integration platform.

In the first case, a managed entity is defined, accessed and manipulated like an object. Unlike the OSI management object model, operations are the only way to manipulate the state of an object. It is part of an object hierarchy, has an interface that defines the operations that can be performed on it and provides full encapsulation.

In the latter case, a "bridging manager" must provide a bridge between a lower level protocol's data model and the object model, and integrate their operations. The object model is used to allow cooperation between peer managers, rather than developing a manager/agent model.

Because their purpose typically is to be a general purpose communication and computation infrastructure, distributed object oriented platforms tend to carry with them unnecessary luggage in the form of features of marginal use, whose implementation, however, can negatively impact performance overhead. They provide highly flexible, dynamic communication structures whereas most of management's communication patterns tend to be fixed.

2.4 Functional Categories

Network and system management are characterized by *functional categories*, that is, a classification of the various operations which can be performed in the context of management [1]. The functionality is important to us, as it gives us indication on the respective computation

[4]standard here denotes consortium activities, or platforms inspired by ODP

and communication requirements of these classes of functions. We have thus identified four classes of support operations required to implement the functions:

- data copy (e.g. configuration),
- data retrieval (e.g. logging, accounting),
- action (e.g. diagnostic, operation),
- notification (e.g. asynchronous event reporting).

Little is new here. However, we must make and additional distinction on the nature of the communication patterns, which may be between peers or organized hierarchically. Our notion of action is also dynamic, as its effect can be modified to reflect the changing nature of the network. Similarly, notifications, as they result from actions, can also be added dynamically to a system.

This perspective allows us to look more closely at the nature of structural support that is required for different functional categories. Of course, orthogonal to these classes, we have further parameters to take into account such as volume of information, atomicity or distributed actions, but we should not forget that the use of mechanisms becomes more marginal as they get more sophisticated. Furthermore, as is already done in some cases, separate, dedicated, protocols can be used to support very specific, demanding management operations, such as, say, bulk transfer. We shall refine this classification in the next section.

3 A new approach to distributed management

We are building a management environment for networks and applications based on a collection of conceptual mechanisms, such as:

- basic access,
- delegation,
- worm,
- cooperation,
- notification.

These conceptual mechanisms are supported by a *remote execution* and a *local interaction* mechanism.

3.1 Conceptual mechanisms

3.1.1 Basic access

We call *basic access* the simplest general support mechanism. It enables the configuration of the device, as well as accounting operations. It allows the reading, retrieving and modifying chunks or pieces of information. This is the major functionality provided by database-like mechanisms.

3.1.2 Delegation

Delegation is operation and diagnostic oriented. Delegation allows us to dynamically expand the functionality of the network element by transferring executable code to it [8, 17]. This code can either execute a function locally and report back its results, or create a higher level object which can be queried by other mechanisms. Delegation helps to regroup a set of operations on several objects into a single action.

Delegation has several benefits. Delegated management operations are executed locally one the network element, but in a flexible way as the operation can be modified dynamically at any time. It contributes to reducing the bandwidth required, as well as decreasing the latency in the discovery of potential problems and the execution of remedial actions.

3.1.3 Worm

The worm is a recursive form of delegation. In the pursuit of the root of a problem, it can be necessary to trace its symptoms across different machines. When the diagnostic is performed by browsing from machine to machine, a worm can be used to implement the procedure.

A worm can also be used for configuration and accounting style operations for a range of machines. It can also implement features such as topology discovery.

3.1.4 Cooperation

Cooperation is the interaction of several managed objects to achieve a collective modification. It is a peer to peer model, as opposed to the hierarchical function/library model.

The activities of the program are the result of the cooperation of several programs, rather than a single one.

3.1.5 Notification

A notification is an asynchronous, or rather unsolicited, message sent to signal an important change in the NE.

A notification can be sent to a manager, or to another NE.

3.2 Support mechanisms

3.2.1 Remote execution

The technique of remote execution simply means to transfer a program to a machine where it can be repeatedly executed. The transfer process must take care of architectural differences and manage an output channel to a manager.

Remote execution depends on the availability of a core functionality, such as access to management information, on the target platform. It requires an execution mechanism, remotely accessible which must also be reflected in the management information model. It requires a support language in which the management functions can be expressed and also has a type system rich enough to capture the details of the conceptual model.

Run-time safety is a prime concern. We want to guarantee that a program will not fail at run time. For most operations, this can be achieved with a type-safe language, with functional, rather than imperative, characteristics. Type-safe compilation and linking should guarantee that the data is available in the NE interface, represented as a library. A functional language has simple recursive data structures which are safer to manipulate than pointer-based structures.

Remote execution implements basic access, delegation and worm. It supports notification. A program is the largest grain of atomicity provided in the model.

3.2.2 Interaction

Interactions exist at two different levels: either between co-resident or between remote (e.g. on a manager station) programs.

Co-resident interaction can be handled through a simple type message passing interface. An interface must be defined for every type of communication. Two partners exchanging information can exchange some form of token to guarantee that they are using the right interface, as is done in presentation layer negotiation schemes. Remote interaction can be treated as a combination of remote execution and co-resident interaction.

Interaction implements cooperation and support notification in its remote form. The managers must have an interface to capture the interactions.

3.3 Management environment

Since our work is experimental in nature, we are aiming at simplicity and flexibility in the construction of the management support environment. The major complexity of implementing administration with our mechanisms is that, since they are at a lower conceptual level, i.e. they act as enabling mechanisms, and their access is language-based, operations may require some programming. One should note, however, that our mechanism can be enabled by management platform technology similar to what is in use in the industry. Graphical browsers and mouse-based operations activation can hide the assembly, compiling and transfer of a program. By using a lightweight, efficient interpreter environment, the compilation and linking overhead can be kept to a minimum and close to performance levels similar to marshalling/unmarshalling operation times. The information structure can be mapped from a conceptual object oriented structure to the type system of the programming language.

3.4 Complementary mechanisms

Other mechanisms that we have to consider to expand our capability are a mass transfer mechanism, and a multi-way communication structure.

The first one is definitely useful to retrieve, typically, logging or accounting information. In the telecommunication industry, this is done with a different file oriented transfer protocol, such as FTAM.

A multi-way communication structure is a simple way to share information between different parties. Combined with a causal communication structure [6], we can build globally consistent information updates and built consistent views of parts of the network. Such an infrastructure has proved useful to implement distributed monitoring [13], but it has a significant overhead, however, and would be best done by a dedicated, separate structure, installed only as required.

4 Implementation

We have built an experimental delegation/worm environment at INRS-Télécommunications [9, 7]. It is a lightweight environment, flexible and quite suitable for experimentation. It is smaller in size of code and runtime image than the SNMP libraries and SNMP agents we have studied[5].

The environment was built around the CAML language and the CAML-LIGHT virtual machine [12]. This pragmatic, (mostly) functional language has most of the features we required, namely strong, polymorphic typing, separate compilation, an exception mechanism and a rich data model. Its implementation gives us ease of extension, portability, architecture-dependent conversions postponed to linkage time, and a compiler/virtual machine implementation. We have added to it multithreading, that is, the capacity of executing several CAML programs concurrently with preemption, remote loading of compiled code, remote control and monitoring of the threads, inter-thread communications, remote linking and a worm mechanism. The data model of the language is rich, dynamic and flexible and it has been proved to be capable of emulating OO structures.

The interface to managed objects is done through an encapsulated, typed interface (an abstract data type). An interface defines the structure of the information and the operations which can manipulate it. The virtual machine is responsible for retrieving the information relevant to all managed objects and updates the corresponding data structures at regular intervals, as required by the applications. The virtual machine also supports atomicity of access and manipulation to managed objects. It is possible to write different interfaces to the same objects, for different access rights. The interface one uses thus limits the manipulation of the data. The management of access rights is done entirely out of our model. If necessary, the communications between platforms could be encoded, although we have not implemented it.

Interaction between threads is done through type-safe interfaces, implemented using techniques similar to marshalling. Unfortunately, because it uses compilation, the CAML-LIGHT environment does not keep type information at run-time, and we had to introduce our own mechanism. These interfaces are available only locally. For two threads running on different machines to interact, a intermediate, interaction thread must be transferred to the machine where the interaction will occur. The use of such intermediate interaction threads is hidden in communications libraries.

Any administrative task is implemented by a piece of code. This code is compiled from the administration environment, transferred to the target machine where it is linked and executed as a thread. Libraries of executable threads can be managed on the target machines, if the memory is available. Similarly, libraries of precompiled tasks can be stored in the administration environment and transmitted as required. More importantly, each virtual

[5]typically the ISODE snmp and the CMU packages

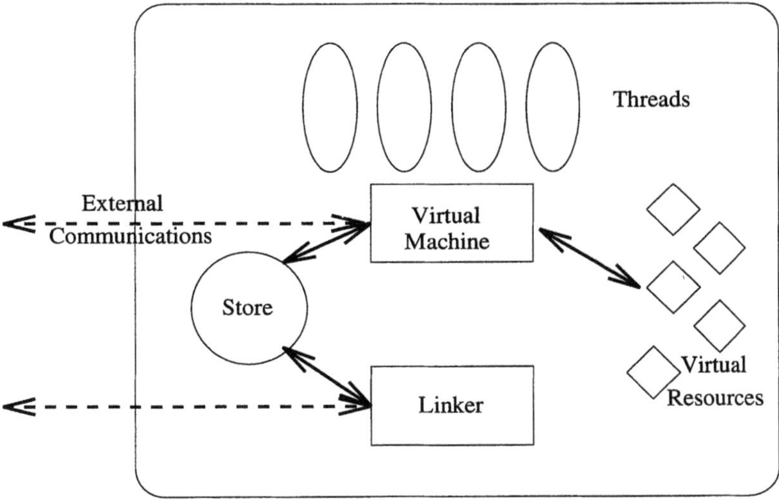

Figure 1: The elements of the management environment

machine stores the libraries which give access to the managed objects abstractions, with which threads have to be linked. Only the interfaces definitions for these libraries need to be available to compile a thread on a manager's site.

Each thread can be activated with specific information, in the form of run-time arguments. Each thread has a "log" channel to recover error information. Another channel recovers normal output. These channels are set up dynamically for each thread and, similarly, each thread can report to a different manager.

Figure 1 illustrates the general structure of the management environment.

We use this environment to remotely manage our distributed heterogeneous workstation environment. We have built an interface to the Unix kernel for system monitoring and we use Unix commands to carry operations. We also have integrated a SNMP access mechanism. Worms have been used to track users, implement load balancing and experiment with several distributed algorithms. We have also replaced the distributed configuration environment of our workstations by a local control managed by delegation. Several forms of resources management are also executed locally in this environment. In this context, distributed management follows the "think globally, act locally" philosophy.

5 Discussion

In our vision of distributed management, all NE's should support remote execution. Local interaction would come second on our list. The resources of the devices could limit the number of resident and active programs, with the possible effect of increasing latency in the response to some operations.

Our perspective, however, is that not all NE's need to support the whole set of mechanisms. By keeping them orthogonal, we can limit the impact of their combination on performance, memory and CPU requirements. It is even possible to cross compile for a machine which doesn't have enough memory for a linker, or enough storage for libraries. There are in fact a wide range of quality of service factors which can be tuned *independently of our conceptual mechanisms.* Different transport protocols can be used, storage can be offloaded to another machine; it is even possible to fully compile a (then) static application where performance can be critical.

One interesting feature of our mechanisms is that they can be used recursively across hierarchical administration domains. The remote execution and interaction mechanisms make no provision on the peer to peer or master/servant nature of the communication relationship. It is straightforward for an agent station to become a server provided it can store the code for the tasks under its control. The remote execution mechanism is directly accessible to the threads.

In our experience, the overhead of the transmission of threads is low. For simple operations, a thread fits in a single packet. The main cost of the execution of threads on the virtual machine is in memory management. We have had to tune a garbage collector to optimize the management of the memory of on-shot threads vs recurring threads. In fact, specific algorithms can be used depending on the degree of sophistication of the operations performed by the agent (i.e. hub vs workstation vs management platform).

Further mechanisms such as file transfer would be left out of the virtual machine. In that respect, it is worth pointing out the fact that distributed network applications coexist with distributed management. Typical examples are distributed network reconfiguration and, more generally routing. Although such mechanisms can be implemented using our mechanisms, they tend to be either integrated in a low level protocol, or are realized with dedicated links. What is at stake here is a tradeoff between flexibility and efficiency. Dedicated mechanisms potentially avoid information extraction and conversion overhead, at the cost of flexibility. Traditionally, real-time applications (e.g. routing) have used dedicated mechanisms whereas less time critical applications (e.g. on-line diagnostic) have used more generic mechanisms, and also potentially more computationally intensive (e.g. AI search techniques) techniques.

6 Comparison with other work

The techniques we have described here have been pursued in various guises, but, to our knowledge, never in a similar integrated context of a toolkit of complementary mechanisms, supported by a programmable environment.

Delegation of duties has been studied both from a operations and an administrative point of view [14]. One application of delegation in a standard's framework is the definition and implementation of higher level managed objects, which compute some chosen function based on the values of other objects. Programmable area managers are another example of delegation of operations. An area manager is responsible for a small network of - say - SNMP managed devices. The area manager is programmable and can perform tasks delegated from a higher level manager, through a suitable, but different protocol. MINERVA [10] is such an environment, where local changes of interest, monitored through SNMP, are reflected into events which in turn trigger the execution of *scripts*, written in a custom language. *Empirical Tools and Technologies*[6] is a commercial company which sells a manager which can execute SCHEME programs which can be remotely downloaded. Let us notice here that SCHEME is not as safe a language as CAML and the risk of occurrence of run-time errors is significantly higher.

The AI notion of *agents* is also similar to our concepts of recursive remote execution, as used by worms. The use of AI agents for distributed network management has been suggested recently by different researchers. In that work, the use of agents to study and improve routing is described. Although such work is usually done by more efficient mechanisms, worms could be programmed to realize such a task.

In spite of the similarities in concepts however, we haven't found anywhere an attempt at providing scalable mechanisms, and to provide a uniform view and uniform support for the manager/NE universe.

7 Conclusions

We have described a perspective of enabling management through a set of simple conceptual mechanisms, rather than a single high level one, and described a management environment based on remote execution and interaction. These mechanisms support a number of paradigms well suited for network and distributed application management. These mechanisms have been implemented in a programming language-based environment. Complementary mechanisms such as file transfer can be done efficiently using a dedicated protocol outside of this environment.

In practice, there seem to already exist a few commercial tools which follow our philosophy of combining several mechanisms, including a form of remote execution, in their management environment. However, they all tend to support a single layer in the management hierarchy and do not share our vision of the recursive application of similar concepts with tradeoffs with regard to the quality of service.

The major benefit that we see in using remote execution as opposed to a database mechanism is the integration of a computational and a data models which allow us to uniformly manipulate the data as well as retrieving it.

Since our focus was on low level enabling mechanisms, there are a large number of concerns that we haven't covered in this short presentation, such as higher level of management coordination, domains and policies, etc. We are currently studying the requirements of the management platform with these considerations in mind.

Acknowledgments.
The development of the distributed platform has been done by F. Gagnon.
N. Greene and F. Gagnon have provided helpful feedback on various drafts of this paper.

[6]this information is based on an exchange with K. Auerbach

References

[1] CCITT Recommendation X.700— ISO/IEC 7498-4: 1992, *Information Technology - Open Systems Interconnection - Management Framework for Open System Interconnection.*

[2] CCITT Recommendation X.711— ISO/IEC 9596-1: 1992, *Information Technology - Open Systems Interconnection - Common Management Information Protocol, part 1: Specification.*

[3] CCITT Recommendation X.720— ISO/IEC 10165-1: 1992, *Information Technology - Open Systems Interconnection - Structure of management information, part 1: Management information model.*

[4] CCITT Recommendation X.722— ISO/IEC 10165-4: 1992, *Information Technology - Open Systems Interconnection - Structure of management information, part 5: Guidelines for the definition of managed objects.*

[5] CCITT Recommendation X.901—ISO/IEC 10746-1 *Basic Reference Model for Open Distributed Processing- Part 1: Overview and guide to use*, 1993

[6] O. Babaoglu and K. Marzullo, *Consistent Global States of Distributed Systems: Fundamental Concepts and Mechanisms*, in "Distributed Systems", E. Müllender, Ed., 2nd Edition, Addison Wesley, 1993.

[7] J-Ch. Grégoire, *Delegation: Uniformity in Heterogeneous Distributed Administration*, LISA VII, Monterey, California, 1993.

[8] J-Ch. Grégoire, *Management with Delegation*, IFIP'93, AIPs Techniques for LAN and MAN Management, Paris, France, 1993.

[9] J-Ch. Grégoire, F. Gagnon, *Implementation of Delegation in Distributed Network Administration*, Canadian Conference on Electrical and Computer Engineering, Vancouver, Canada, 1993.

[10] D.J. Hughes, Z.D. Wu, *Minerva: An Event Based Model for Extensible Network Management*, Proceedings of INET'93, pp. CEC-1—CEC-6.

[11] Internet RFC 1157, *A Simple Network Management Protocol (SNMP)*, 1990.

[12] X. Leroy, "The Caml Light system documentation and user's manual", version 0.6, INRIA, 1993.

[13] M. Mansouri-Samani, M. Sloman, *Monitoring Distributed Systems*, Chap. 12 in *Network and Distributed Systems Management* M. Sloman, Ed., Addison Wesley, 1994.

[14] J.D. Moffett, M.S. Sloman, *Delegation of Authority*, I. Krishnan & W. Zimmer (eds), *Integrated Network Management II*, North Holland (1991), pp 595-606.

[15] Object Management Group, *Common Object Request Broker*, 1992.

[16] Open Software Foundation, *Distributed Management Environment*, 1991.

[17] Y. Yemini, G. Goldszmidt and S. Yemini, *Network management by delegation*, Integrated Network Management II, Elsevier Science Publishers, pp. 95–107, 1991.

3
Configuration Management For Distributed Software Services

S. Crane, N. Dulay, H. Fosså, J. Kramer, J. Magee, M. Sloman, K. Twidle
Imperial College, Department of Computing, London SW7 2BZ.
E-mail: mss@doc.ic.ac.uk

Abstract

The paper describes the SysMan approach to interactive configuration management of distributed software components (objects). Domains are used to group objects to apply policy and for convenient naming of objects. Configuration Management involves using a domain browser to locate relevant objects within the domain service; creating new objects which form a distributed service; allocating these objects to physical nodes in the system and binding the interfaces of the objects to each other and to existing services. Dynamic reconfiguration of the objects forming a service can be accomplished using this tool. Authorisation policies specify which domains are accessible by which managers and which interfaces can be bound together.

Keywords

Domains, object creation, object binding, object allocation, graphical management interface.

1 INTRODUCTION

The object-oriented approach brings considerable benefits to the design and implementation of software for distributed systems (Kramer 1992). Configuring object-structured software into distributed applications or services entails specifying the required object instances, bindings between their interfaces, bindings to external required services, and allocating objects to physical nodes. Large distributed systems (e.g., telecommunications, multi-media or banking applications) introduce additional configuration management problems. These systems cannot be completely shut down for reconfiguration but must be dynamically reconfigured while the system is in operation. There is a further need to access and reconfigure resources and services controlled by different organisations. These systems are too large and complex to be managed by a single human manager. Consequently, we require the ability not only to partition configuration responsibility within an organisation's managers but also to permit controlled access to limited configuration capabilities by managers in different organisations.

This paper describes the SysMan configuration management facilities for open distributed software services. We use the Darwin notation to define the *structure* of a distributed service or application as a *composite object* type which defines internal primitive or composite object instances and interface bindings (Magee 1994). The external view of a service is in terms of interfaces *required* by clients and *provided* by servers. Managed objects implement one or more management interfaces providing management services and event notifications to managers. In the following we use the terms 'object reference' interchangeably with 'interface reference' since an object is uniquely identified by one of its interface references.

A domain-based infrastructure is used to group object references. This can be used to partition management responsibility by grouping those objects for which a manager is responsible. Furthermore, domains provide naming contexts in which interfaces are registered. (An interface can be included in more than one domain.) The domain service thus performs two functions: it associates management policy with groups of objects and it permits managers to associate convenient names or icons with interface references.

A graphical user interface permits a human manager to locate managed objects by browsing through the domain hierarchy. Once located, composite objects may be inspected and their internal configuration of interconnected object instances modified. New applications can be constructed by interactively creating object instances and binding their interfaces to those already registered in the domain service. Figure 1.1 shows the overall environment. A manager locates interfaces in the domain service via a configuration manager object (CM) and invokes operations on these interfaces to create or delete objects, bind interfaces or perform application-specific management.

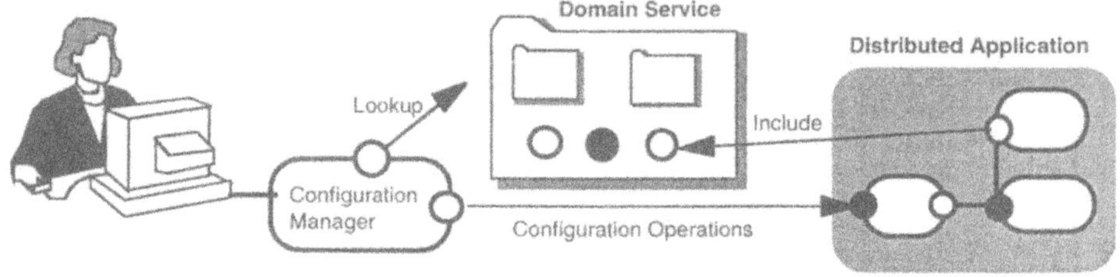

Figure 1.1 Interactive configuration management.

The term 'configuration management' often connotes those activities concerned with setting internal object state, for example: updating routing tables, adjusting numbers of buffers and specifying device addresses. We assume that these functions are performed by invoking operations on objects and use the term to describe the management of the structure of objects constituting a distributed service.

In section 2 we give an overview of the use of domains in the SysMan management environment and then in section 3 we use the Active Badge Location Service as an example to describe the configuration facilities of the Darwin Language. In section 4, we discuss issues relating to creating objects followed by binding of interfaces in section 5. The user interface for configuration management is described in section 6 and is followed by related work and conclusions.

2 MANAGEMENT ENVIRONMENT

2.1 Domains and Policies

Domains provide a means of grouping object interface references and specifying a common policy which applies to the objects in the domain (Sloman 1989, 1994, Moffett 1993, Twidle 1993). A reference is given a local name within a domain and an icon may also be associated with it. If a domain holds a reference to an object, the object is said to be a *direct member* of that domain and the domain is said to be its *parent.* A domain may be a member of another domain and is then said to be a *subdomain*. Policies which apply to a parent domain normally propagate to subdomains under it.

An object (or subdomain) can be included in multiple domains (with different local names in each domain) and so can have multiple parents. The domain hierarchy is not a tree but an arbitrary graph. An object's direct and indirect parents form an *ancestor* hierarchy and a domain's direct and indirect subdomains form a *descendant* hierarchy (Figure 2.1). The domain service supports operations to create and delete domains, include and remove objects, list domain members, query objects' parent sets and translate between path names and object references (Becker 1993).

An authorisation policy is specified by an *access rule* which defines a relationship between managers (in a subject domain) and managed objects (in a target domain) in terms of the management operations permitted on objects of a specific type (Moffett 1993, 1994). Policies applying to a user or manager are defined in terms of a *User Representation Domain* (URD), a persistent representation of that person in the domain system. When they log into the system, a CM object is created and included in the URD. Policies specified for their URD then apply to their CM object.

2.2 Domain Browser

The *Domain Browser* is a graphical interface common to all management applications (Sloman 1993). It permits a human manager to navigate the domain structure; select objects and include or remove them from domains and invoke operations on selected objects. The browser displays tree diagrams with ancestors in the left window, the current domain in the middle and descendants to the right, Figure 2.1. The current domain, `hal`, has two direct parents: `/home` and `/users/staff`, and itself contains two domains: `mydomain` and `tmp`. It is possible to indicate cycles and collapse parts of the tree (not shown in Figure 2.1). A displayed domain can be selected to become the current domain in the window or a new window can be opened.

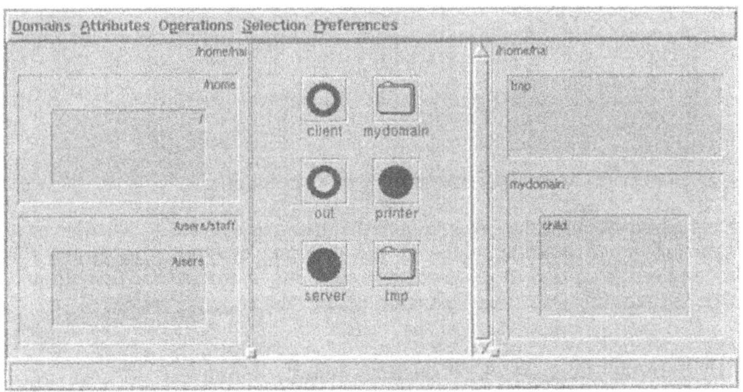

Figure 2.1 Domain window with hierarchy views.

Directories in the UNIX file system can also be displayed as domains via an adapter object included in a domain. (However, it is not possible to include files into domains or object references into a UNIX directory.) The domain browser is used to navigate the file system and select an object template (stored as a program file) which can then be used to create object instances (described further in section 4).

2.3 Operation Invocation

The following attributes are associated with an interface reference present in a domain:

Local name: a textual name which uniquely identifies the interface within the domain.

Object identifier: a unique identifier used to invoke operations on the interface.

Icon reference: specifies its appearance in the graphical interface.

Type reference: used to query a type store for the interface's operation signatures.

The type information associated with an object specifies the operations which can be invoked on the object and the parameters they require. Operations are invoked on an object from the Domain Browser by selecting the object icon in the current domain window then selecting an operation from a pull down menu which lists the names of the operations supported by the object's interface. The Domain Browser uses the operation name and associated type information to generate a dialogue box for the user to supply required arguments, Figure 2.2.

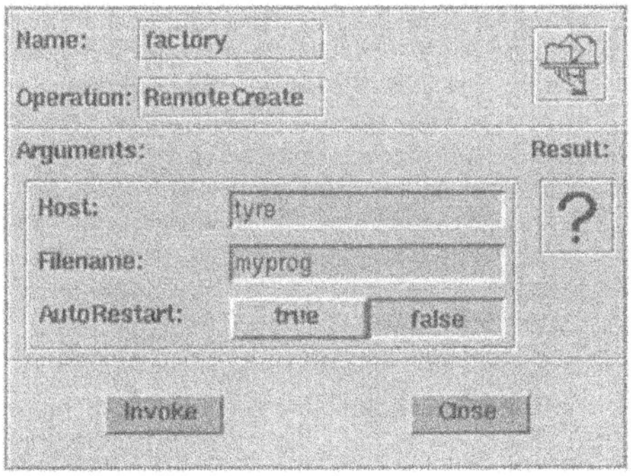

Figure 2.2 Dialogue box to invoke operation with parameters.

The user enters parameters for the invocation in the dialogue box and presses Invoke. The user interface performs the invocation, updating the dialogue box with the result. The domain browser also supports *drag-and-drop* invocation; selecting an icon in one domain and dropping it onto another invokes the `include` operation on the destination domain.

3 ACTIVE BADGE LOCATION SERVICE

Examples in this paper are taken from an Active Badge system implemented using the SysMan environment. Active Badges (Harter 1994) emit and receive infrared signals which are received and transmitted by a network of infrared sensors connected to workstations. Badges can be worn by people or attached to equipment. The system permits the location and paging of badges within range of a sensor.

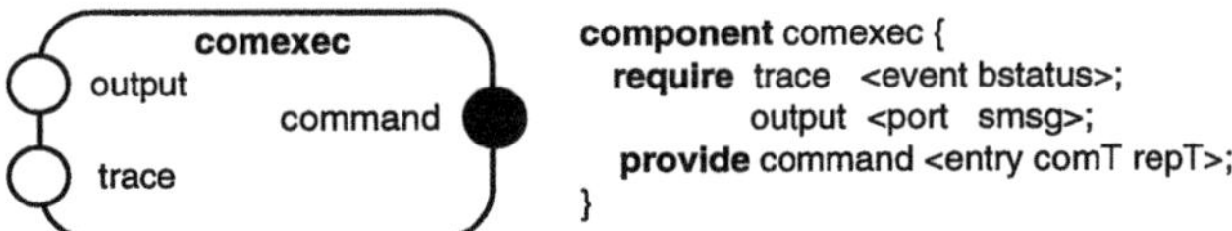

Figure 3.1 Component type.

The object in Figure 3.1 provides a service via an interface (depicted by a filled circle) but requires two external services (empty circles). It executes badge commands to set off a badge's internal beeper or to illuminate status LED's. By convention, the first word of the type specification (in angle brackets) is the ***interaction mechanism*** class. For example, `command` accepts 'entry' calls with a request of type `comT` and a reply of `repT`. To execute a command, it is first necessary to locate a badge. Consequently, `comexec` requires the `trace` service which gets location events of type `bstatus` from an event service. The component sends a message to the sensor network to transmit the command to the badge, once found via `output` which possesses 'port' semantics.

Composite distributed services are constructed by composing object instances, Figure 3.2. The `sensornet` component controls access to the sensor network. Each requirement (empty circle) in this example is for a port (`output`) to which messages are sent, and each provision (filled circle) is a port (`input`) on which messages are received. Internal interfaces can be made visible at a higher level by binding them to the composite component interface, e.g. `M.output` is bound to `sensout` and `sensin` to `D.input`.

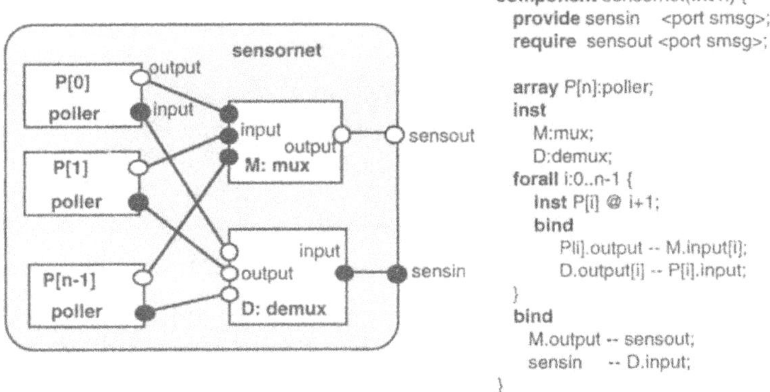

Figure 3.2 Composite component type.

Each poller component is located on a different workstation and controls a multidrop RS232 line of sensors. It requires a service to output badge location sightings and provides a service on which it transmits commands. In general, many requirements may be bound to a single provided interface; however, in this case, each poller instance's `output` is bound to a separate `input` port to allow the multiplexor `M` to identify the particular poller `P[i]` from which a message is received. Pollers are distributed by the expression `inst P[i]@ i+1` to locate each instance (`P[i]`) on a separate machine (`i+1`). Machine identifiers are mapped to physical machines at run time which permits a configuration specification to be reused in different environments.

The `sensornet` component of Figure 3.2 forms a subcomponent of the badge manager, `badgeman`, Figure 3.3. This server provides the following interfaces:

> `where` to query the locations of all badges,
>
> `location` to receive all location-change events,
>
> `trace` to receive location change events for a particular badge,
>
> `command` to execute a command on a badge.

When `badgeman` is created, it registers these interfaces in the domain 'badge' (which is assumed to exist). Darwin's `export` statement indicates that the reference to a provided service interface should be registered externally. Conversely, an `import` statement allows required services to be found in the domain service.

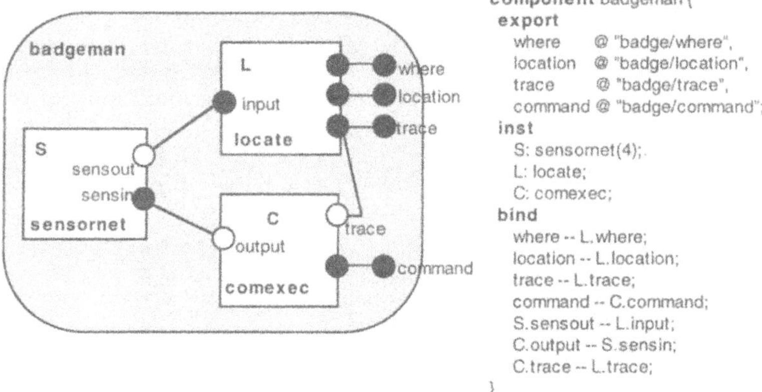

Figure 3.3 Exporting services to the domain service.

In practice, on-line configuration of the badge system is desirable (for example to add new pollers as the sensor network is extended). In the following, we demonstrate how the composite service of Figure 3.2 may be represented in Darwin to permit dynamic configuration.

4 OBJECT CREATION

As we have seen, an object can contain multiple composite or primitive objects, distributed over many nodes. It can export multiple service interfaces which can be included in domains to permit binding. In this section we describe management facilities supporting object creation.

Local Creation Service (LCS)

This is provided by the operating system. For example, the badge server can be created simply by executing a command from a UNIX shell. Once executing, its interfaces appear in the domain 'badge'. (This implies that the operating system, which is outside of the domain service context, must be able to include interfaces in a domain.)

Remote Creation Service (RCS)

The example in Figure 3.2 requires a remote creation service to instantiate a poller at a node different to that of the multiplexor and demultiplexor. This service creates distributed objects by providing access to the LCS on a remote node, (Crane 1994).

Internal (Darwin) Creation Service

A Darwin program may create objects statically at the time the composite is instantiated or dynamically using the keyword `dyn`. New objects may be instantiated entirely within an existing object or they may make use of the LCS or RCS to create composite objects on new nodes. In Figure 4.1, `master` dynamically creates a badge proxy to handle each request for command execution.

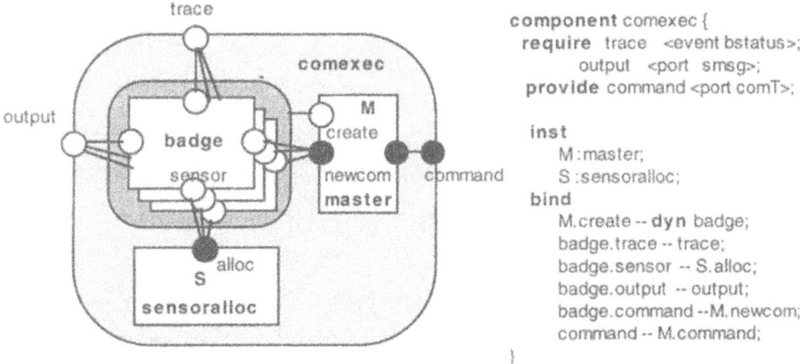

Figure 4.1 Dynamic object instantiation.

Application-Provided Creation Service

An application interface may provide a specific operation to create objects in the context of the composite object. For example, Figure 4.2 depicts a simplified version of Figure 3.2 in which poller objects can be added by invoking the `newpoll` service. It uses a different poller object taking a single parameter which determines its location (c.f. Figure 3.2).

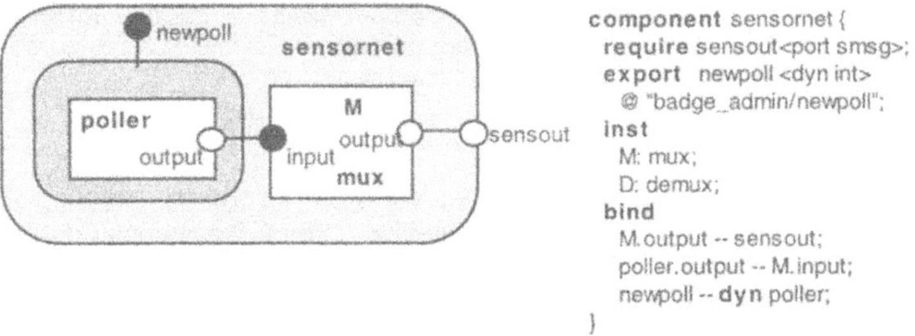

Figure 4.2 Dynamic object instantiation service.

Interactive Object Creation

The configuration manager permits a human manager to access all creation services via a graphical interface which is described in section 6. Figure 4.3 indicates how this service uses the other creation services.

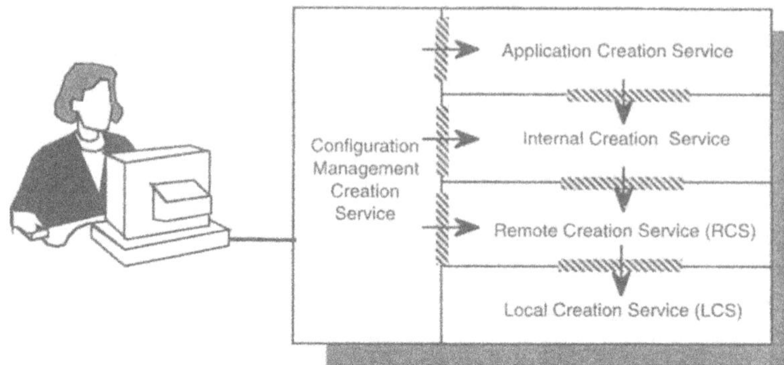

Figure 4.3 Creation mechanism relationships.

5 OBJECT BINDING

A required interface must be bound to a provided interface before a client can invoke operations on a server. There are two fundamental binding operations:

Binding create a link between a required interface on a client and a provided interface on a server using an external 'third-party'.

Unbinding destroy an existing binding.

Rebinding is performed by first unbinding and then binding. Destroying a running object instance will generally require its interfaces to be first unbound.

Whereas it may be assumed that unbound program components are in a consistent state prior to binding, this is certainly not always the case before unbinding. Therefore a protocol is needed for 'safe' unbinding and rebinding. It will be explained in section 5.4.

5.1 Third-Party Binding

In the examples of sections 3 and 4, bindings are performed by an external third party (the manager) or they are defined by a Darwin configuration. Objects being bound do not play an active part in binding; they are unaware of the interfaces to which they are bound. The advantage of this approach is that structure is defined explicitly rather than being hidden in an object's internal state. Figure 5.1 shows the stages in the interaction of a configuration manager with the domain service to locate and bind interfaces.

This example requires certain access rules to be present: `manager` requires 'lookup', 'bind from' and 'bind to' permission on `badge`. An additional access rule specifies the operations the client can invoke at the server interface.

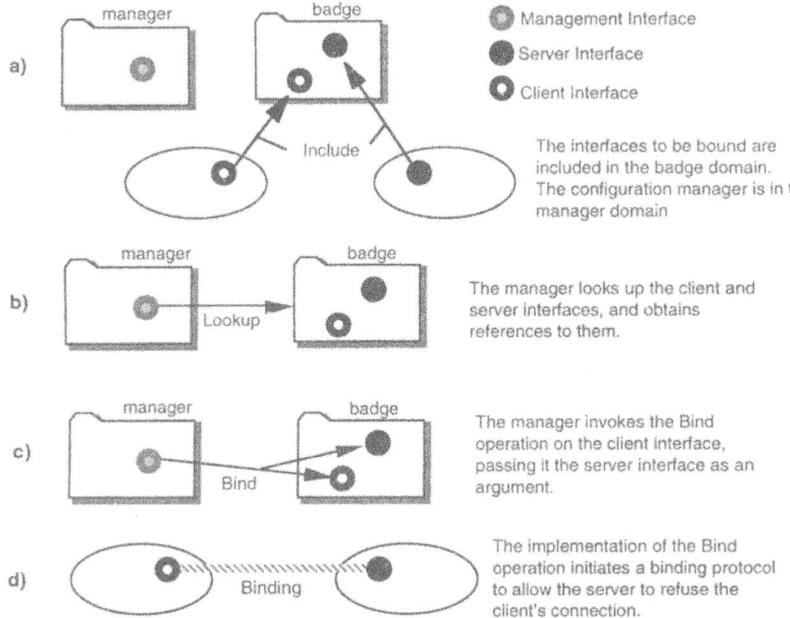

Figure 5.1 Third party binding interactions.

5.2 First-Party Binding

Many distributed systems, e.g. (ANSAware 1993), make use of first-party bindings in which a client locates a server using a name server and establishes the binding itself. This type of binding is very common in open systems: a client can locate the services it requires with no intervention by a manager. It assumes that the information which enables the client to find the required service (a name or service description) is compiled into the client or passed as an instantiation parameter.

The Darwin language also permits first-party bindings to be specified for a composite object:

```
component view (int dt) {
      require  locations<entry int statT>;
}

component where (int dt=0) {
      import locations @ "badges/where";
      inst v: view (dt);
      bind v.locations -- locations;

}
```

This shows a client of the badge manager which polls the latter's `where` service periodically (or once if no parameter is given). It queries the badge domain for the `where` service, gives it an internal name (`locations`) and binds its internal interface to this service. The client requires an access rule permitting 'lookup' and the server requires an access rule permitting 'include' on the domain `badges/where`.

5.3 Dynamic Invocation Bindings

A third type of binding arises when a reference to a *provided* interface is passed in a message to a client, which implicitly assigns it to a *required* interface, and uses it to invoke operations on the provided interface. This mechanism is suitable for dynamic environments which cannot afford the overhead of either first- or third-party bindings.

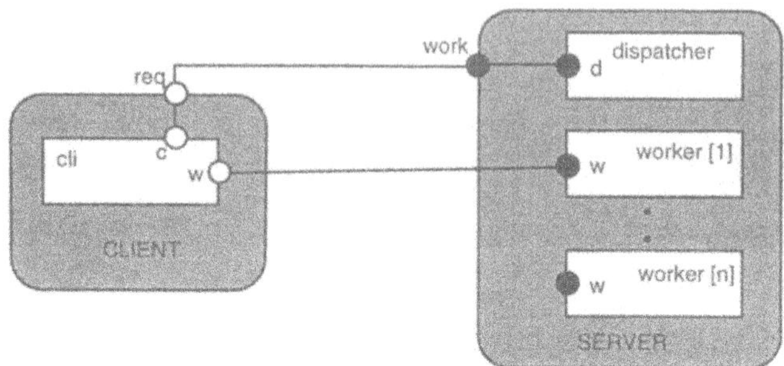

Figure 5.2 Dynamic binding example.

In Figure 5.2, the client's `req` interface reference is initially bound to the server's `work` interface. To access one of the server's worker processes, the client sends a request and receives a reply containing a reference to a worker's interfaces (at `dispatcher`'s discretion) which it assigns to w. Communication between `cli` and `worker` then proceeds independently of dispatcher.

An access rule is required to permit the binding between `cli` and `worker,` but this can only be checked when `cli` invokes an operation on `worker` (unless a bind protocol has previously been executed).

5.4 Safe vs. Unsafe Unbinding

Destroying a binding is more complicated than creating one, because it might be in use at the time of removal. Bindings are part of an application's overall state, and applications normally require *safe* unbinding, requiring a consistent state to be reached before bindings are destroyed (Kramer 1990). A request for immediate unbinding is usually unsafe (but perhaps desirable in certain circumstances).

In general, safe unbinding entails the co-operation of the programmer. Our approach requires programmers to mark bindings *critical* in sections of code where unbinding would cause inconsistency. When bindings may be safely removed, they are marked safe. If an unbind request arrives when an interface is critical, it is blocked until the binding becomes safe.

In the Regis system (Magee et.al. 1994, Crane 1994), many communication styles are available. The simplest and most flexible of these is the *message port*, but programmers must explicitly render them safe for unbinding. Regis also provides objects similar to Ada's *entries* which have semantics similar to RPCs. These are safe to reconfigure as long as no calls are outstanding on them, which can be determined by the support system. Another communication object, providing an even more rigid style of communication, is the *event distributor* used in the badge system. For these objects, safety is synonymous with the desire to receive event notifications; when enabled, they are *critical*, when disabled they may be safely rebound. (An attempt to transmit on an unbound interface will block the transmitting process until binding occurs.)

6 INTERACTIVE CONFIGURATION MANAGEMENT

As mentioned in section 1, the configuration manager (CM) supports Domain browser facilities to locate interfaces, and functions to display composite object structure and invoke operations on interfaces from within a 'configuration window', as will be explained in section 6.2. The CM permits a user to associate an invocation signature with an interface and to specify an icon to represent it. For example, to create an object at a particular node, the CM is used to locate an object type in the file system which is then dropped onto the required node icon in the RCS domain, Figure 6.1.

6.1 Configurable Composite Objects

The configuration management view of a distributed application is an extension of the domain browser view. A user employs the domain browser to navigate to a composite object. A composite object with visible structure is represented by a *Configuration Domain* which displays internal interfaces as icons (Figure 6.1). This *domain view* is similar to an ordinary domain but it is not possible to include external objects *into* a configuration domain although objects can be included *from* a configuration domain *into* other domains. A configuration domain can optionally display a *structural view* showing bindings between internal interfaces, permitting a manager to monitor the system structure and make changes to it. There may not be a complete view of all internal interfaces, but only the rebindable ones on which configuration operations are possible. The configuration domain is effectively a management interface to a composite object and is included in a domain when the object is created. Objects visible in a configuration domain may themselves be configurable composite objects.

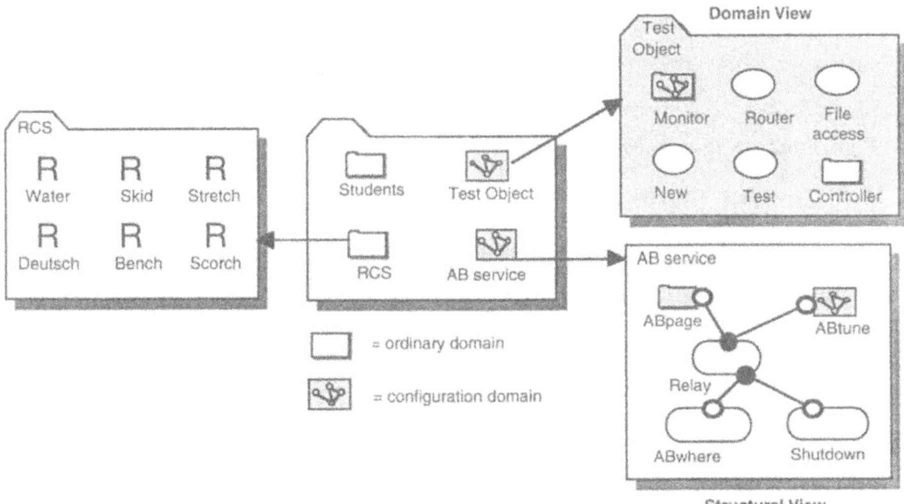

Figure 6.1 Domains and services with special and default icons.

6.2 Configuration Window

A user of the CM performs interactive binding in a *configuration window* which displays a structural view of a configuration.

Figure 6.2 shows how `Relay` in the `AB Service` configuration window can be bound to `New Relay` in domain `Test` by a drag-and-drop operation. The drop invokes the Bind operation on the target, and results in the configuration window being updated to show the new binding.

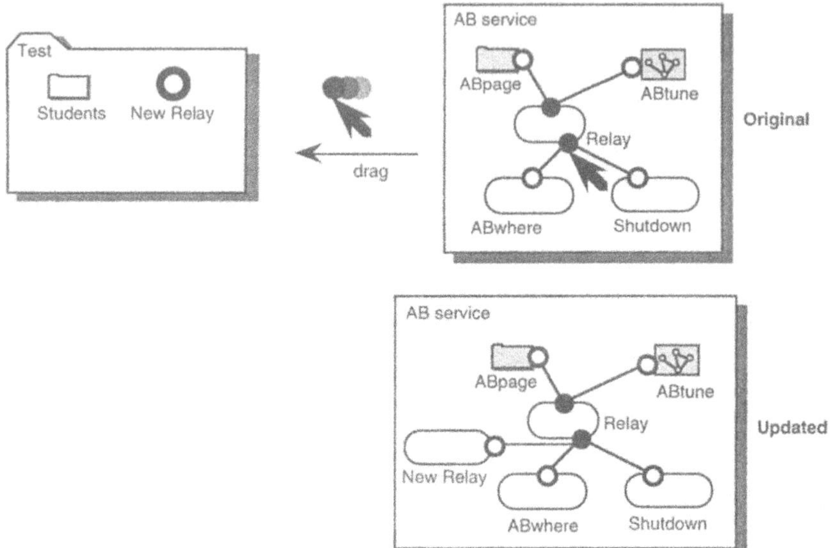

Figure 6.2 Drag-and-drop binding.

6.3 Current Status

The domain browser, object invocation via dialogue windows and structural views of configuration domains have been implemented and drag-and-drop interactions are being implemented. The Darwin compiler works in the Regis programming environment and has been modified to support ANSAware objects. The RCS allows creation of distributed objects defined by a Darwin program.

7 CONCLUSIONS AND RELATED WORK

This paper has shown how a graphical interactive configuration management facility can be used to manage software objects comprising a distributed application or service. Our approach has evolved over many years of experience with Conic, REX, and Darwin which have been used by industrial and academic institutions.

The use of directories in name servers to hold references to objects is common in distributed systems (Leser 1993), but domains extend this concept to applying policies to contained objects. The naming provided by domain path names is for user convenience rather than to provide a unique name for an object. DEC also use the concept of domains to group objects for management purposes (Strutt 1991) and the Ansa Trader uses domains as a trading context (ANSAware 1993). Our approach goes further than trading in that it shows how to use domains for interactive configuration management.

A number of other systems provide a configuration language (Agnew 1994, Zimmermann 1994, Barbacci 1993), which have some similarities to Darwin but our approach is the only one to combine *static* initial configurations, *dynamic preplanned* reconfigurations and *evolutionary* or unplanned dynamic reconfigurations. We cater for configuration management of both 'closed' systems (i.e. single applications) and 'open' systems consisting of multiple applications bound using the domain service. However, further work is needed to gain more experience with the current models for safe reconfiguration (Kramer 1990) and some of the more restrictive but practical proposals (Agnew 1994).

Key concepts in our approach are:

- *Explicit structure*. Both the Darwin notation and the graphical configuration view explicitly identify software structure in terms of object instances and interface bindings. A graphical tool, capable of generating Darwin code, allows design of composite components by stepwise refinement (Kramer 1993).
- *First- and third-party binding*. First-party binding is useful in some circumstances. However object-oriented systems which only support first-party binding often require binding information to be embedded in clients. This makes reuse difficult. Third-party binding permits structural information to be defined at the configuration level, resulting in configurations which are 'cleaner' and easier to understand.
- *Hierarchical Composition*. The ability to create composite services interactively *and* within the Darwin language provides a very powerful way to generate new services from existing services either statically or dynamically.
- *Evolution* is supported at two levels: pre-programmed change can be incorporated in composite objects using the dynamic facilities of the Darwin language, while interactive configuration management facilities are used to introduce new types and replace existing ones with minimal interruption of service.
- *Domains* provide a means to group interfaces and partition the overall management of the system by representing organisational or physical structure. Combined with access rules they provide scope for specifying policies relating managers to managed objects.
- The *domain browser* allows management interfaces to be located and draws upon the experience of file system user interfaces found in many operating systems.

8 ACKNOWLEDGEMENTS

The authors acknowledge the support of the Commission of the European Union through Esprit project 7026 (SysMan) and DTI support of Eureka project IED 4/410/36/002 (ESF). We acknowledge the contribution of our colleague Keng Ng to the concepts described in this paper.

9 REFERENCES

Agnew B., Hofmeister C., Purtilo J. (1994) Planning for Change: a Reconfiguration Language for Distributed Systems, In IOP/IEE/BCS *Distributed Systems Engineering*, **1:5**, 313–322.

ANSAware (1993) Application Programming in ANSAware – Document RM.102.02. APM, Poseidon House, Castle Park, Cambridge CM3 0RD, UK.

Barbacci M., Weinstock C., Doubleday D., Gardner M., Lichota R. (1993) Durra: a Structure Description Language for Developing Distributed Applications, *IEE Software Eng. Journal*, **8:2**, 83–94.

Becker K., Raabe U., Sloman M., Twidle K. (eds.) (1993) Domain and Policy Service Specification. IDSM Deliverable D6, SysMan Deliverable MA2V2. Available by FTP from dse.doc.ic.ac.uk.

Crane S., Twidle, K. (1994) Constructing Distributed UNIX Utilities in Regis. In *Proc. Second Int. Workshop on Configurable Distributed Systems*, IEEE Computer Society Press, 183–189.

Harter A., Hopper A. (1994) A Distributed Location System for the Active Office, *IEEE Network*, Jan./Feb. 1994, 62–70.

Kramer J., Magee J. (1990) The Evolving Philosophers Problem: Dynamic Change Management. *IEEE Trans. Software Eng.*, **SE-16:11**, 1293–1306.

Kramer J., Magee J., Sloman M., Dulay N. (1992) Configuring Object-based distributed programs in REX, *IEE Software Eng. Journal*, **7:2**, 139–140.

Kramer J., Magee J., Ng K., Sloman M. (1993) The System Architect's Assistant for Design and Construction of Distributed Systems. In *Proc. 4th IEEE Workshop on Future Trends of Distributed Computing Systems*, 284-290.

Leser N. (1993) The Distributed Computing Environment Naming Architecture. In IEE/IOP/BCS *Distributed Systems Engineering*, **1:1**, 19–28.

Magee J., Dulay N., Kramer J. (1994) REGIS: A Constructive Development Environment for Distributed Programs. In IOP/IEE/BCS *Distributed Systems Engineering*, **1:5**, 304–312.

Magee J. (1994) Configuration of Distributed Systems, Chapter 18 of *Network and Distributed Systems Management* (ed. Sloman M.), Addison Wesley, 483–497.

Moffett J., Sloman M. (1993) User and Mechanism Views of Distributed System Management. IEE/IOP/BCS *Distributed Systems Engineering*, **1:1**, 37–47.

Moffett J. (1994) Specification of Management Policy and Discretionary Access Control. Chapter 17 of *Network and Distributed Systems Management* (ed. Sloman M.), Addison Wesley, 455–480.

Sloman M., Moffett J. (1989) Domain Management for Distributed Systems. *Integrated Network Management I*, (eds. Meandzija B., Westcott J.), North Holland, 505–516.

Sloman M., Magee J., Twidle K., Kramer J. (1993) An Architecture for Managing Distributed Systems. In *Proc. 4th IEEE Workshop on Future Trends of Distributed Computing Systems*, 40–46.

Sloman M., Twidle K. (1994) Domains: A Framework for Structuring Management Policy. Chapter 16 of *Network and Distributed Systems Management* (ed. Sloman M.), Addison Wesley, 433–453.

Strutt C. (1991) Dealing with Scale in an Enterprise Management Director. *Integrated Network Management II* (eds. Krishnan I., Zimmer W.), North Holland, 577–593.

Twidle K. (1993) Domain Services for Distributed Systems Management, PhD Thesis, Department of Computing, Imperial College.

Zimmermann M., Drobnik O. (1994) Specification and Implementation of Reconfigurable Distributed Applications. In *Proc. Second Int. Workshop on Configurable Distributed Systems*, IEEE Computer Society Press, 23–35.

SECTION TWO

Policy-Based Management

4

Using a Classification of Management Policies for Policy Specification and Policy Transformation

René Wies

Munich Network Management Team
University of Munich, Department of Computer Science
Leopoldstr. 11b, 80802 Munich, Germany
Phone: +49-89-2180-3139
Email: wies@informatik.uni-muenchen.de

Abstract

Policies are derived from management goals and define the desired behavior of distributed heterogeneous systems, applications, and networks. To apply and deal with this idea, a number of concepts have been defined. Numerous policy definitions, policy hierarchies and policy models have evolved which are all very different, as they were developed from diverse points of view and without a common policy classification.

This paper presents and structures the characteristics of policies by introducing a general classification for policies and showing how this classification leads to and aids in the specification of policies. Furthermore, we outline the ideas of a policy life cycle, and that of policy transformation. Policy transformation is a refinement process with conflict resolution which converts policies to become applicable within a management system using management services, such as systems management functions, distributed services, etc.

The paper further looks at aspects to be considered when defining policy templates and concludes with a number of open issues still to be looked at in this field of management policies.

Keyword Codes: K.6.4; C.2.4
Keywords: Network and Systems Management, Management Policy, Policy Classification, Policy Transformation, Policy Hierarchy, Policy Templates

1 Introduction and Motivation

The primary and overriding objective of network and systems management is to maintain network and system availability and aid in extending the network and systems, enhance the performance, provide security, reduce operating overhead (repetitive tasks), and decrease the cost of running the information technology infrastructure. Despite the fact, that providers of network and system services are aware of these objectives, the problem of translating these goals into actions remains. A possible approach to tackle this problem are management policies, which provide a (semi-) formal concept to record and structure the objectives, to refine them depending on the infrastructure, and apply them through the use of management systems.

Policies, as we define them, are derived from management goals and *define the desired behavior of distributed heterogeneous systems, applications, and networks.* It is important to recognize that policies specify only the *information* aspects of this desired behavior, i.e. *what* behavior is desired; they do not describe the precise actions to be taken, i.e. *how* the behaviour can be achieved and maintained.

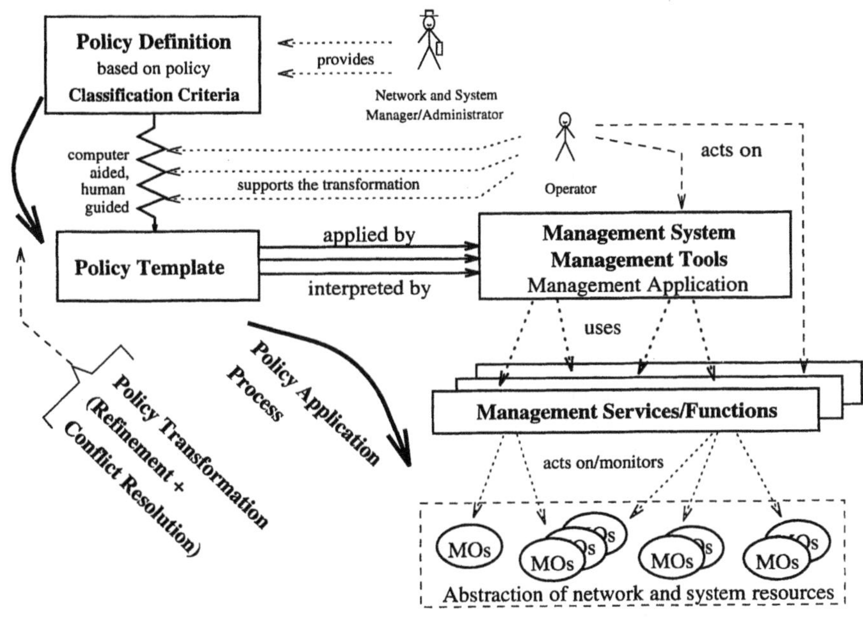

Figure 1: Transformation and application of policies

Low level, technical policies may be wrongly seen similar to the `behaviour` attribute in GDMO ([ISO 10165-4]) templates for managed object classes (MOCs). However, whereas the behavior template for use in a managed object classes (MOCs) defines the *possible or available* behavior of the resource it represents, a policy defines the *desired* behavior, i.e. it is a restriction on the possible behavior. For high level policies this analogy cannot be drawn and we will deal with the policy hierarchy in Section 3.1.

In contrast to the ongoing standardizing work on domains and policies ([ISO 10040/2], [ISO 10164-19], [ISO 10746-1]), our definition indicates that policies are primarily independent from the concept of domains, yet policies can be either applied to or used to define domains of managed objects.

As will be described in Section 3.1, policies may range from high level i.e. abstract non-technical policies to low level technical policies, depending on how the desired behavior of the managed resources is specified. However, unlike [MACA 93] we do not see policies to cover the wide spectrum from business goals and strategies (societal and economic policies) to the executable policies (procedural policies), nor are our policies necessarily executable by some *unsophisticated program* as suggested in [BEHO 93]. The level of abstraction in terms of the desired behavior of distributed heterogeneous systems, applications, and networks, depends on the degree of detail contained in the policy definition and the *ratio* of business related aspects to technological aspects within the policy (see Figure 4 and Section 3 for a detailed discussion). Thus, policies as we will deal with them for the remainder of this paper, do not describe business goals but are derived from them, nor are they executable management scripts, even though management scripts could be generated from low level policies ([WIES 94]).

Only once we know exactly what aspects characterize policies and how policies can be processed, we can start to embed the concept of *management policies* into a suitable management architecture, or possibly extend existing or develop new management architectures. It is our goal to combine for example the numerous formal concepts for the definition of technical security policies (e.g. [MARR 93], [WARE 94]) and the abstract architectures for the application of business or corporate policies (e.g. [IDSM 93]) into *one comprehensive concept* which can deal with policies of all levels in the hierarchy. Furthermore, to avoid the task of having to define implementation specific extensions as is the case for existing Managed Object definitions [HABO 91], the structure and components of a policy object definition must take their future realization (implementation and application) into account. On the grounds of the issues presented throughout the following sections, we will briefly discuss examples of commercial systems and network management tools near the end of the paper in Section 4.2.

Figure 1 illustrates the main ideas described in this paper. The classification presents valuable input for both, the definition of new policies and the realization of policies from existing policy catalogues. The classification criteria are the aspects to be exactly looked at when defining policies. The transformation process is the most difficult part, as it must convert generally abstract and informal policies into low level policies which can be applied to the environment. The transformation process primarily consists of refinement steps and possibly some conflict resolution ([MOFF 94]). *The end products of this transformation process are not policies that act directly on managed resources but rather specifications on how to apply management tools and how to utilize management functions or management services offered by a management system.* Besides the concepts of policy classification, transformation, and application, two other concepts, policy hierarchy and policy life cycle, are introduced to complete the picture of management policies. Yet, the policy classification builds a common basis for all following issues, as it summarizes and organizes the important characteristics of policies.

2 Policy Classification

The large number of policies calls for a classification, i.e. a well-defined set of (as far as possible orthogonal) grouping criteria. The main goal of such a classification of policies is:

1. to get a better grasp of what is meant by management policies and what can be achieved through their use.

Other goals are for example:

2. to identify differences and commonalities between policies in order to specify different classes of policies;

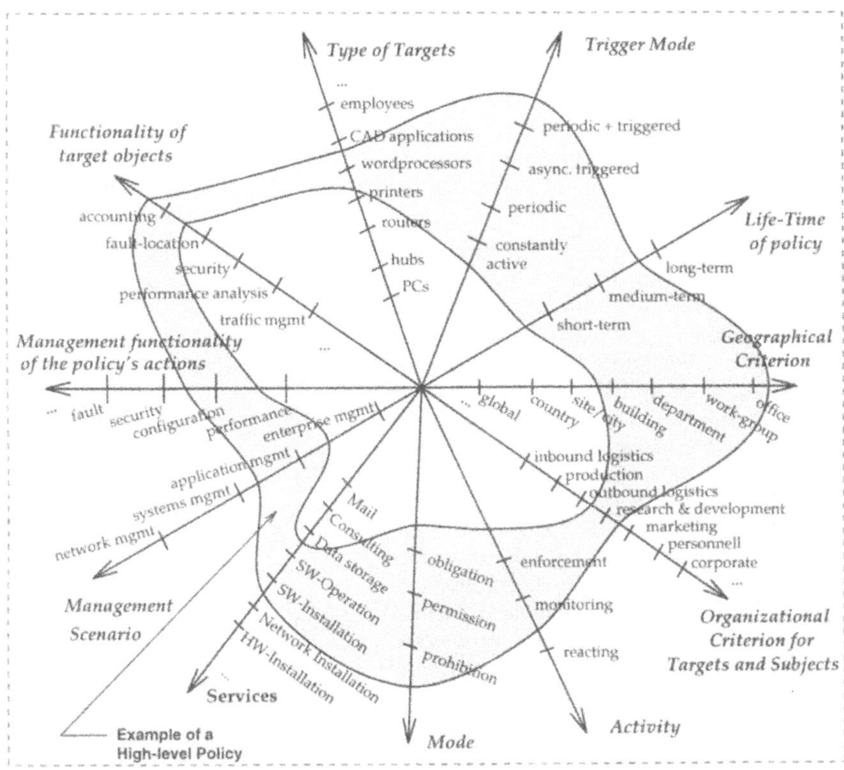

Figure 2: Criteria for Policy Classification

3. to derive a policy hierarchy for the process of policy transformation; and
4. to derive and verify the components of a formal definition of policies.

Several network and system service providers (e.g. FidoNet, VirNet) have gathered their policies in *policy catalogues* which are written in an informal way. A structure, if at all present, is given by the services the company offers to its customers, e.g. policies specific to mail services, data storage, data processing, consulting services, software installation. A thorough analysis of these policy catalogues from numerous network and system service providers and talks with network and system managers, administrators, and operators (e.g. at debis, BMW, LRZ) have allowed us to collect a list of criteria for the classification of policies which are illustrated in Figure 2 in form of a multi-dimensional diagram.

Most of these dimensions can be associated with one or more of the ODP viewpoints[1]

[1] A list of the dimensions and their associated ODP viewpoints would be beyond the scope of this paper, especially as this association is very vague and hence of little use.

([ISO 10746-1]). However, using only the five ODP viewpoints would again group different characteristic properties of policies together and would thus not help to explain and organize management policies.

The precise labels of the axes, i.e. the different categories for each criterion inevitably depend on the level of abstraction, i.e. the policy's position in the policy hierarchy, which will be discussed later. However, the *names* of the dimensions (e.g. trigger mode, life time) will remain the same, whichever level of abstraction we look at; only the labels are refined within each dimension.

For the sake of brevity, we will not explain the different dimensions further, as most of them are self-explanatory. In this Figure, a high level policy (drawn in light grey) covering several categories per dimension is indicated. It describes a real-life example for the following scenario: *The computer science faculty consists of several departments, each of which owns the same number of floating licenses for the word processing system as there are full-time researchers at each department. There are also a limited number of spare licenses for part-time researchers which are distributed on demand. The policy, stating that the departmental licenses must be used before the spare licenses are allocated and distributed, must be enforced and its enforcement monitored.* Using the above classification can only supply a very simplified representation of the policy, as the refinement of the categories and dimensions as well as the notion of domains are neglected. However, this example illustrates, that every dimension must be considered when defining a policy.

As the above stated goals show, this classification is not designed to describe a policy completely, or even to cover all aspects of a policy. For example the axis *type of targets, functionality of targets, geographical criterion* and *organizational criterion* may appear as *one* attribute called *target domain* in a policy template. These domains may either be resolved during the transformation process or remain as is in the policy template to be resolved later by some other domain-resolver in the management system. This issue will be discussed further in Section 3.2 when we take a look at the transformation process.

The classification provides a basis for the derivation and structuring of policies as well as possible hints towards their transformation. In addition, the refinement of the axes in combination with the application of a policy hierarchy will lead to one policy template definition suitable for all levels of the hier-

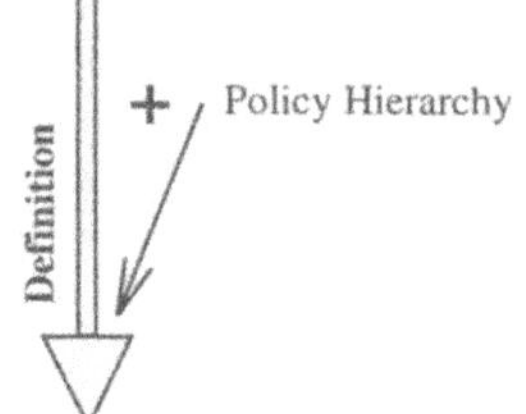

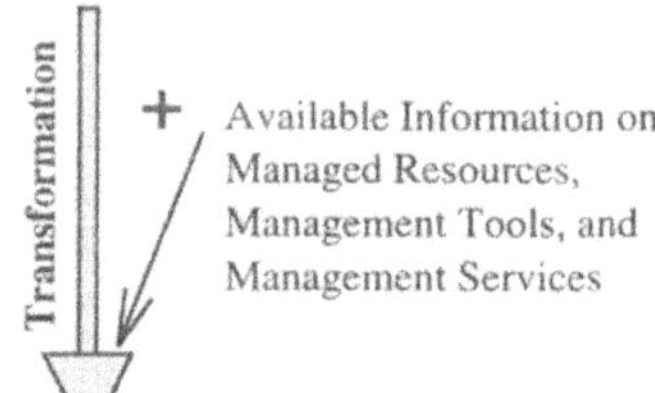

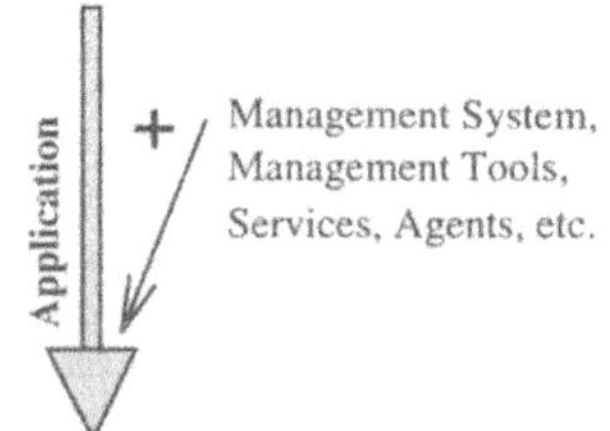

Figure 3: The simplified path from policy classification to policy application

archy. A policy, no matter from which level of the hierarchy, can be analyzed and structured along the above criteria. This process of defining, analyzing, and structuring policies is the starting point for the processing and application of management policies. This approach is illustrated in Figure 3.

3 Policy Hierarchy and Transformation Process

3.1 Policy Hierarchy

When analyzing catalogues of policies from various network and system service providers, it becomes apparent how different policies can be. Different in terms of their values for the above classification dimensions and in their level of abstraction or degree of detail. Security policies specifying the precise format of the allowed password structure or the IP addresses of systems to be protected by firewalls are *mixed* with abstract policies describing the required availability and accessability of printers or policies documenting the precautions to be taken when using a specific management tool.

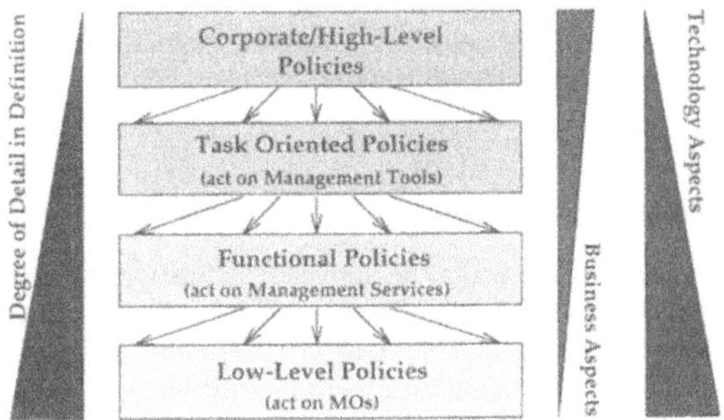

Figure 4: The Policy Hierarchy

To guarantee that all policies are applied to their targets (provided they are not in conflict with each other), it is essential to structure these policies. Thus, a policy hierarchy is a way of splitting the vast number of policies into smaller groups of different levels of abstraction, which can be further processed in distinct steps and transformed into applicable low-level policies. Examples of policy hierarchies can also be found in [MACA 93] and [NGUY 93].

The levels of the hierarchy also represent different *views* on policies. Examples of such views are: the view of a corporate network manager who only sees and only specifies corporate/high level policies; or the view of a network operator, who sees functional policies and realizes them through the use of a management system which in turn may use specific management functions or management services.

Thus, a policy hierarchy defines the levels within the management environment at which policies are applied. As Figure 4 illustrates, the policy hierarchy distinguishes between the following:

- Corporate policies or high level policies: These are directly derived from corporate goals and thus embody aspects of strategic business management rather than aspects of technology oriented management. To allow their application within the management environment, they have to be refined to one of the three policy types below.
- Task oriented policies: Their field of action is sometimes referred to as task or process management, where they define the way how management tools are to be applied and used to achieve the desired behavior of the resources.
- Functional policies: These policies operate at the level of and define the usage of management functions, such as the OSI *systems management functions* ([ISO 10164-X]), the OSF/DME *distributed services* ([DME 92]), or OMG's *object services* ([OMG 92a, OMG 92b]); and
- Low level policies: They operate at the level of managed objects (MOs). MOs in this context refer to simple abstractions of managed network and system resources, and not MOs for e.g. systems management functions.

If a policy can be implemented by use of management functions or management services the last level of the hierarchy may not be reached during the refinement process, nor may it be necessary to define policies at this low level. Furthermore, certain policies can be assigned to exactly one level of the hierarchy, yet other (less well defined) policies may be assigned to different levels and thus must be split into separate policies before the transformation process can be applied.

3.2 Transformation Process

Following the definition of policies using the above classification, each characteristic property can be further detailed to allow a stepwise refinement of the policy. In other words, the lower the level of abstraction, the more precise and detailed will the definition become, i.e. the granularity of the criteria increases. However, in addition to the refinement of a policy, the transformation process can also be used to identify the targets, subjects, and necessary monitor objects. For example, a high level policy calling for a weekly backup of all the company's data may be refined to specify the backup media for different workstation clusters and also identify the system administrators, that are responsible for operating stackers or changing tapes.

To illustrate the refinement of policies, the axis labeled *management functionality* for example can be further subdivided to describe the policy's actions more precisely. As illustrated in Figure 5, an unspecific high-level security policy may be further refined into two separate policies, one responsible for assuring the confidentiality of data and the other for assuring the confidentiality of traffic flow ([ISO 7498-2]).

The number of stages in this transformation process may vary depending on the axes and their labels. Thus, for some axes a derived value may not be refined further while for other dimensions the process of refinement must carry on until the final value is determined.

The questions that arise now are:

- How is this transformation achieved ?
- When does this process of refinement end ?

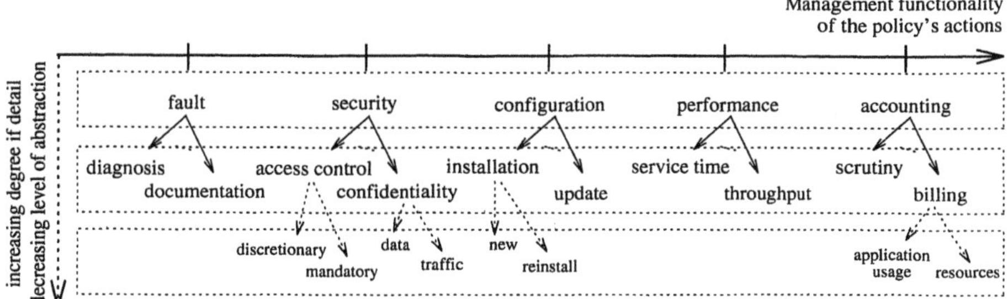

Figure 5: Refinement of classification criteria

To answer the first question, in some cases this process may be automated, yet generally we expect to apply the idea of *computer aided - intuition guided* processing [BBBD 85], i.e. with the helping hand of an expert operator. The question whether this transformation process can be automated or to which degree an automation can be achieved cannot be answered at this stage. However, to interpret the semantics of policies and for any automation of this process (fully computerized or human guided), extensive management information on the managed environment, the management capabilities of the involved systems, and information on available tools, platforms, etc., is essential.

A completely different approach, could be to limit this transformation process to a syntactical transformation which could make concepts like *skolem reduction* applicable. Yet, neglecting semantical interdependencies of policies is not satisfactory as these will probably cause the majority of conflicts.

The transformation will end, when the reached degree of detail cannot be refined further or when a mapping between the value (object, action, etc.) to managed objects or management functions of the management system is possible. Thus, it is a process of merging the results from a *top-down* approach (i.e. the refinement of policies) with the results from a *bottom-up* approach (i.e. the analysis of available management functionality). For example, if the derived targets or monitor objects can be related to existing MOs or if the management actions to be performed can be mapped to management functions or services, the process of refinement will end. However, if a transformation is not possible, for whichever reason (lack of information, conflicts, etc.) the policy may need to be re-defined or taken care of by a human operator. This process was summarized in Figure 1.

The example of floating licenses introduced in Section 2 would need to be refined for example to identify the license servers which are to be configured or to identify the clients which need to be monitored to verify the policy's enforcement. Furthermore, the trigger mode (asynchronous triggering by license requests) and life time of the policy would need to be further detailed during the transformation process.

In Section 2 we already mentioned the problem of resolving domains. This can be either done during the transformation process or resolved later by a domain-resolver within the management system. The latter approach has the advantage of a more simple policy transformation process but allows no (or very limited) conflict resolution until the policy is actually applied. It merely shifts the complexity of resolving conflics from the transformation process to the management system which applies the policy. The former approach (conflict resolution during transformation) leads to a more complex transformation process with e.g. backtracking methods, but it also causes severe problems when it comes to dynamically changing domain-members. For example, newly added devices/objects to a target domain must be dynamically added to the policy's targets, which may result in new conflicts, possibly new monitoring strategies, or even a complete new transformation of the policies concerned. However, to deal with or even answer the question of which alternative for conflict resolution is more practical and sensible, is beyond the scope of this paper.

3.3 Policy Life Cycle

Before we move on to the derivation of policy templates, the policy life cycle is introduced, as it provides valuable information on aspects to be incorporated in template definitions. The life cycle does not only influence the attributes of a policy template (e.g. trigger mode, lifetime), it also hints towards the actions and notifications to be specified within a template and the management functions and services required to implement a policy. The life cycle is characterized by the fact that a policy can divided into an enforcement and a monitoring part ([WIES 94]). The policy enforcement part can be activated by trigger objects (asynchronous events) or through the use of monitor objects.

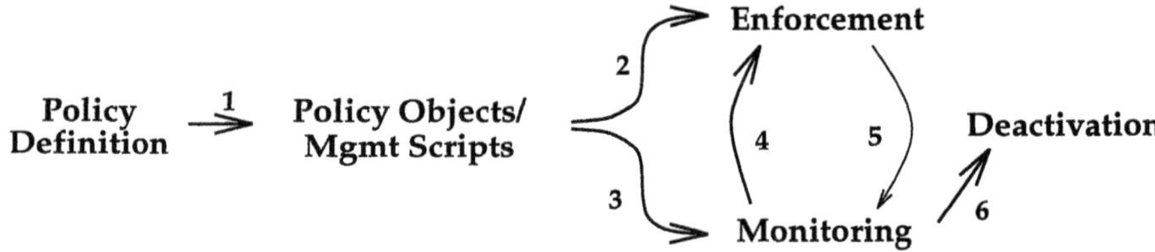

Figure 6: The policy life cycle

The numbers in Figure 6 are to be interpreted as follows:

1. policy transformation: as described in Section 3.2, high level policies are refined and transformed into low-level policies and further processed to become applicable within the management environment.
2. application of *active* policies: policies are activated through the management system or specific management applications. *Active* policies first carry out certain actions and later monitor changes upon which the policy may again react, provided such actions are specified in the policy. (Suspending and resuming policies will be treated as part of policy enforcement.)
3. application of *monitoring* policies: *Monitoring* policies have no initial enforcement part and only monitor certain MOs and possibly react if necessary. The monitoring may also be done using for example a monitoring systems management function.

4. policy adaption or change: The reaction upon changes during the lifetime of a policy can be treated just like the initial enforcement actions. This is because changes in the managed environment may lead to a change in the overall enforcement of the policy, for example additions to a target domain may require a completely new configuration of all other domain members.
5. changes leading to new requirements on monitoring, triggering or enforcement actions: As for the above situation, a change in the enforcement actions may require a new monitoring strategy. For example the deletion of one domain member may no longer require the monitoring of this resource.
6. deletion of policies: short and medium term policies will become obsolete at some point in time. For example when they are replaced by new policies or their domain of targets is removed from the environment.

From the above life cycle, certain characteristics concerning the functionality of necessary underlying management services can be derived. For example, a policy must be able to emit notifications concerning the change in a target's characteristics, or actions must be carried out on the policy if a domain is changed. Furthermore, functions to activate, pause, resume, delete or change a policy must be specified to allow an effective implementation and application of policies.

4 Policy Templates and Commercial Products

4.1 Derivation of Policy Templates

In this section we will use the term policy template rather than policy object, because the notion of objects tends to call for a formal definition of objects which can be implemented using some object oriented language, compiled with some sophisticated compiler, and simply applied to the environment. However, research in this field is far from this stage, even though standards are already being defined.

A policy template must be defined for policies of all levels of the hierarchy. Thus we need a template which suits abstract high level policies as well as technical low level policies. Furthermore, the syntax must allow an effective and efficient use by both managers and operators, for specifying and refining policies. In [SLOM 93] a policy structure consisting of five object attributes (modality, subject scope, target scope, activity, and constraints) is proposed. This may be sufficient for a very abstract representation of policies that involve human operators, but even for high level and abstract technical policies, more information must be held in the template in a structured format. Based on the classification criteria of Figure 2, a policy template could have the format shown in Figure 7. However, not all dimensions are represented by components in the proposed template because for example, the modality is of little use when trying to refine policies or apply them to management services.

Actions, or methods as they are called in the object oriented world, could be grouped into a set of administrative actions for creating, deleting, pausing, resuming a policy application, and sets of operational actions for example for adding or deleting targets, subjects or monitors; or sets of actions for changing other characteristics of a policy such as the trigger mode or time mode. The ultimate goal is of course the mapping of these templates to management functions and services as discussed earlier.

```
POLICY TEMPLATE
        Author(s):
        CreationDate: (mm/dd/yy)
        StatusOfRefinement: (pending, completed/applicable, stopped
                due to conflicts, stopped due to lack of information, etc.)
        DerivedFromParentPolicy:
        GoalAndActivity: (free-text, detailed and semi-formal description
                of what is to be enforced and monitored;
                and how to react to changes)
        ManagementScenario: (network management, systems management,
                application management, enterprise managementt)
        ManagementFunctionality: (fault, accounting, configuration,
                performance, security management)
        Service: (services involved or effected by the policy)
        LifeTime: (duration of application)
        SubjectCharacteristics/Domain: (tools, mgmt. functions, etc.)
        TargetCharacteristics/Domain: (functionality, site, type, etc.)
        TriggerMode: (async.Triggered, syncronous, asyncMonitoring,
                periodicMonitoring, etc.)
        TriggerCharacteristics/Domain: (monitoring objects, triggering events, etc.)
        PolicyProcessOrScript: (formal description of the management
                script or management process/steps to be executed to
                enforce the policy)
        Notifications: (emitted notifications due to policy
                violations, enforcement/monitoring failures, etc.)
REGISTERED AS { ... }
```

Figure 7: Example of a policy template

4.2 The Scope of Commercial Products

Almost all major vendors of management systems and platforms ([JAND 94]) have recognized the advantages of using policies and domains in integrated management. Products such as HP's Dolphin ([PGMM 93]), Cabletron's MaestroVision ([MAES 93]), and Tivoli's TME offer some rudimentary functionality to define domains and apply policies. Yet, all concepts are proprietary and very limited in their functionality.

For example, the systems management system HP-Dolphin uses an object-oriented Prolog-like language for the specification of policies The transformation of a policy is not automated nor system supported, but must be done manually by the programmer. Furthermore, these low level policies are applied to system-specific, non-generic, and non-standardized objects.

Tivoli's Management Environment ([WELL 94]) defines *Core Object Services* and *Common Management Services*, including services for the definition of *Policy Regions* (commonly known as domains) and *Policy Objects*. These Policy Objects consist of a set of customizable programs usually written as shell or Perl scripts. They are not designed to use other management services to enforce policies nor is the concept open for the integration of global and system-independent policies.

5 Conclusions and Future Work

Policies are a powerful concept for the management of distributed heterogeneous systems, networks and applications. In this paper we did not present finished work, but rather ongoing research.

Our classification criteria and the template generated from them have proven to allow a fairly complete description of policies and their characteristics. Thus, new policies can be defined and existing *raw* policies may be detailed using the classification criteria. The refinement process followed by mapping the enforcement and monitoring activities to management functions can be done manually, provided of course management functions are available at the lowest level. Yet, the automation, or at least a systematical approach, supported by a *policy specification and application tool* to achieve this refinement is essential to apply (large numbers of) policies. Therefore, the refinement process marks the main focus of our future research in this field. Thus, while our definition, concepts and classification criteria appear to be powerful as well as natural, in our next step we will prove that they are also practical by refining policies from different management scenarios. However, the goal is not to design a fully automated "policy-system" but rather to develop a structured approach for a generally manual but computer aided, tool supported transformation and application of management policies.

Acknowledgements

The author wishes to thank the members of the Munich Network Management Team for fruitful discussions and valuable comments to preliminary versions of this paper. The MNM Team directed by Prof. Dr. Heinz–Gerd Hegering is a group of researchers of the University of Munich, the Technical University of Munich, and the Leibniz Supercomputing Center of the Bavarian Academy of Sciences.

References

[BBBD 85] F.L. Bauer, R. Berghammer, M. Broy, W. Dosch, F. Geiselbrechtinger, R. Gnatz, E. Hangel, W. Hesse and B. Krieg-Brückner, *The Munich Project CIP, Vol 1: The Wide Spectrum Language CIP-L.*, volume 183 of *Lecture Notes in Computer Science*, Springer, 1985.

[BEHO 93] Karsten Becker and David Holden, "Specifying the Dynamic Behavior of Management Systems", In Manu Malek, editor, *Journal of Network and Systems Management*, volume 1, pages 281 – 298, Plenum Publishing Corporation, September 1993.

[DME 92] Open Software Foundation, *OSF Distributed Management Environment (DME) Architecture*, 1992.

[DSOM 91] IFIP, *Proceedings of the IFIP/IEEE International Workshop on Distributed Systems: Operations & Management*, October 1991.

[DSOM 93] IFIP, *Proceedings of the IFIP/IEEE International Workshop on Distributed Systems: Operations & Management*, October 1993.

[HABO 91] H.-G. Hegering, S. Abeck and Th. Böhnke, "Converting MIB-Descriptions into MIB-Implementations", In [DSOM 91].

[IDSM 93] "Domain and Policy Service Specification", IDSM Deliverable D6 / SysMan Deliverable MA2V2, IDSM Project (ESPRIT III EP 6311) and SysMan Project (ESPRIT III EP 7026), October 1993.

[ISO 10040/2] "Information Technology – Open Systems Interconnection – Systems Management Overview – Amendment 2: Management Domains Architecture", PDAM 10040/2, ISO/IEC, November 1992.

[ISO 10164-19] "Information Technology – Open Systems Interconnection – Systems Management – Part 19: Management Domain and Management Policy Management Function", CD 10164-19, ISO/IEC, January 1994.

[ISO 10164-X] "Information Technology – Open Systems Interconnection – Systems Management – Management Functions", IS 10164-X, ISO/IEC.

[ISO 10165-4] "Information Technology – Open Systems Interconnection – Structure of Management Information – Part 4: Guidelines for the Definition of Managed Objects", IS 10165-4, ISO/IEC, August 1991.

[ISO 10746-1] "Basic Reference Model of Open Distributed Processing – Part 1: Overview and Guide to Use", WD 10746-1, ISO/IEC, November 1993.

[ISO 7498-2] "Information Processing Systems – Open Systems Interconnection – Basic Reference Model – Part 2: Security Architecture", IS 7498-2, ISO/IEC, 1988.

[IWSM-I 93] Wesley W. Chu and Allan Finkel, editors, *Proceedings of the IEEE First International Workshop On Systems Management, Los Angeles*, IEEE, April 1993.

[JAND 94] Mary Jander, "Management Frameworks", *Data Communications International*, February 1994.

[MACA 93] M. Masullo and S. Calo, "Policy Management: An Architecture and Approach", In [IWSM-I 93].

[MAES 93] Calypso Software Systems, "MaestroVision 2.0 beta 1", Release Notes, Calypso Software Systems, Inc., 1993.

[MARR 93] Randy Marchany, "Writing a Site Security Policy: RFC 1244", In [IWSM-I 93].

[MOFF 94] Jonathan D. Moffett, *Specification of Management Policies and Discretionary Access Control*, chapter 17, pages 455–481, In [SLOM 94], June 1994.

[NGUY 93] Thang Nguyen, "Linking Business Strategies and IT Operations for Systems Management Problem Solving", In [IWSM-I 93].

[OMG 92a] "Object Management Architecture Guide", Document 92-11-1, Object Management Group, September 1992.

[OMG 92b] "Object Services Architecture", Document 92-8-4, Object Management Group, August 1992.

[PGMM 93] Adrian Pell, Chen Goh, Paul Mellor, Jean-Jacques Moreau and Simon Towers, "Data + Understanding = Management", In [IWSM-I 93].

[SLOM 93] Morris Sloman, "Specifying Policy for Management of Distributed Systems", In [DSOM 93].

[SLOM 94] Morris Sloman, *Network and Distributed Systems Management*, Addison-Wesley, June 1994.

[WARE 94] Willis H. Ware, "Policy Considerations for Data Networks", *Computing Systems, The USENIX Association*, 7(1):1–44, 1994.

[WELL 94] Caroline Wells, "Tivoli Systems, Inc., Tivoli Management Environment (TME)", *Datapro Integrated Network Management*, January 1994.

[WIES 94] René Wies, "Policies in Network and Systems Management – Formal Definition and Architecture –", In Manu Malek, editor, *Journal of Network and Systems Management*, volume 2, pages 63 – 83, Plenum Publishing Corporation, March 1994.

Biography

René Wies received his diploma (Diplom-Informatiker, M.Sc.) in computer science from the Technical University of Munich, Germany, and a MBA-MDP degree from the Graduate School of Management, Boston University in Japan. Currently he is a Ph.D. student at the University of Munich and a member of the Munich Network Management Team, directed by Prof. Dr. Heinz-Gerd Hegering. He does research on integrated network and systems management, emphasizing on management policies. He is a member of the IEEE and GI.

5

Concepts and Application of Policy-Based Management

B. Alpers, H. Plansky
Siemens AG
Otto-Hahn-Ring 6, 81739 München, Germany
{Burkhard.Alpers,Herbert.Plansky}@zfe.siemens.de

Keywords
Domain, policy, policy hierarchy, policy formalisation

Abstract
Due to downsising, deregulation and tremendous growth, computing and telecommunications systems have become heterogeneous and complex environments with multiple players involved. For managing such systems more powerful methods than those used for network element management are needed. It must be possible to structure and partition management according to responsibilities, to provide managers with higher-level abstractions and to enable them to flexibly adapt the way of management to their specific needs. To fulfil these requirements, we introduce *domain-based management policies*. Policies allow the specification of management intention on different levels of abstraction. We discuss policy classification, formalisation and hierarchies. Then we present an architecture for policy enforcement services. Finally, we outline two application scenarios in the areas of distributed systems and telecommunication.

1. Introduction

Computing and telecommunication systems and services are of vital importance for enterprises as well as for organisations. Two current trends are leading to a rapid change in the structure of these systems and services: "downsising" and "deregulation". Downsising means the substitution of mainframe-systems by smaller networked systems leading to more flexible and extensible systems which form a complex and heterogeneous environment. Deregulation in the area of telecommunications results in a variety of new services, like Virtual Private Network (VPN) offered by independent service providers.

Management in such a heterogeneous, complex and dynamic multi-manager scenario requires efficient management methods which offer more functionality than the traditional network element-oriented management:

- It must be possible to *structure and partition* management responsibilities amongst several managers with different roles.
- Management systems must allow an *abstraction* of management to prevent the managers from becoming flooded with low-level network alarms and details.

- Management systems should be *flexible* and reconfigurable to adapt to the introduction of new services, systems, applications or users.
- Management tasks should be largely *automated.*

The *domain-based policy concept* we introduce in this paper serves to fulfil these requirements. Policies enable the specification of management roles on a higher level of abstraction and hence deliver a powerful method for structuring management of large systems. Management policies need to be formalised without building them inflexibly into manager components .

In chapter 2 of this paper we elaborate the policy concept. After defining and classifying policies we discuss formalisation and hierarchies. Then, we relate our concepts to other work in this area. Chapter 3 deals with realisation aspects. We present an implementation architecture and describe an engineering model for implementing policies. In chapter 4 we present two application scenarios in the areas of distributed systems and telecommunication. Chapter 5 contains conclusions and describes future work.

The general domains and policy concepts described in this paper, the implementation architecture and the application to X.400 management are based on results of the IDSM (Integrated Distributed Systems Management, 6311) and SysMan (7026) ESPRIT projects.

2. Concepts of Policy-based Management

2.1 Principles and Classification of Policies

Management of networks, systems, and applications involves monitoring the activities and states of these systems, making management decisions, and performing management control actions. This management process, particularly the process of decision making is guided by *management policies.* Policies describe management activities rangeing from lower-level management operations to higher level objectives which are determined by economical or social requirements. In our approach we will use the concept of management policies to specify and formalise the *semantics* of management, whereas the state-of-the-art management techniques concentrate on the *syntax* of management, e.g. management protocols and information models, etc.

We define policy in general as *"information which influences the behaviour of managers and managed objects".* A policy determines the management objectives, the roles, and the responsibilities between a subject set containing managers (human managers or applications) and a target set containing managed objects. Policy specifications can be interpreted by human managers. In order to automatically enforce management policies, they must be formalised and represented in the management system. Then, manager applications can be made responsible for the enforcement of policies, i.e. they have to interpret policies and to align their behaviour with the policies. This separation of policies and manager objects enables the dynamic change of policies without changing the managers.

The variety of management tasks results in a multitude of policies. There are many criteria for classifying policies (functional area, application area etc., see e.g. [Wies94]). The criteria we present in the sequel are particularly important for identifying generic classes which are interesting for many application areas. These are therefore very promising candidates for further formalisation, as discussed in the next section:

Criterion: Relationship between managers and managed objects

Managers have obligations and rights with respect to managed objects:

Authorisation policies define what management activities a manager is *allowed* to do in terms of the operations he is authorised to perform on a set of managed objects. An authorisation policy may be positive (permitting) or negative (forbidding).

Obligation policies specify what management activities a manager *must* or must not do.

Criterion: Influence on the managed system

Policy can describe the activity of managers (*manager action policies*), or the desired behaviour of the influenced system resulting in *state achievement policies* or *state change restriction policies*:

- manager *action* policy: this is a (conditioned) sequence of actions to be performed by managers on managed objects;
- *state achievement* policy: this specifies the state of the system to be achieved in terms of existence of objects, attribute value ranges, relations; this could also include the optimisation of certain attributes to reach an optimal state;
- *state change restriction* policy: description of allowed or undesired state changes: this specifies state changes of the managed system which are allowed or forbidden.

Criterion: Abstraction-level of the policy

The concept of policies supports the abstraction of management. Policies with lower-level of abstraction are directly linked to the infrastructure of the managed system, whereas higher-level policies need further interpretation to lead to concrete management actions. Abstract policies can be used to form policy hierarchies, where an abstract policy is mapped on other less abstract policies (see 2.3).

2.2 Policy Formalisation

For analysing policies and automating their enforcement, policies must be formally represented. The complexity and variety of management policies makes it impossible to find just *one* "policy description language" covering all aspects of policy. In the IDSM project a generic policy object class is formalised which serves as a reference for defining more specific policy object classes. The generic policy object class describes a relation between manager objects and managed objects.

This object class was specified according to ISO's "Guidelines for the Definition of Managed Objects" (GDMO, see [GDMO92]) in the following way:

```
idsmPolicy      MANAGED OBJECT CLASS
                DERIVED FROM idsmManagedObject;
                CHARACTERISED BY
                      mandatoryPolicyPackage    PACKAGE
                      BEHAVIOUR idsmPolicyMandatoryAttributes BEHAVIOUR DEFINED AS
                      "This package contains the set of attributes which must be
                      specified for any policy object class";;
                      ATTRIBUTES
                      adminPolicyState    GET-REPLACE,
                      operationalState    GET,
                      subject             GET-REPLACE,
                      target              GET-REPLACE;;
                CONDITIONAL PACKAGES
                      globalConstraintsPackage  PRESENT IF ...
                      propagationModePackage    PRESENT IF ...
                      companionPoliciesPackage  PRESENT IF ...
```

In addition to the attributes for state management the `mandatoryPolicyPackage` contains attributes to determine the subject and the target of management. For these attributes we use domains which are arbitrary groupings of objects (see [DSOM94]). This way the subject and target scope can be easily specified without defining policy for each single object. The optional `globalConstraintsPackage` allows to impose restrictions on policies concerning the location of manager and managed objects or the time when the policy is valid. The `propagationModePackage` can be used to specify that from the domains used for scoping subject and target not only the direct members are considered but also the members of sub-domains (i.e. domains contained as members in other domains). In the `companionPolicyPackage` one can list associated policies which should be activated together with this policy (see next section on refinement).

This generic class serves as a starting point: Subclasses must be built in order to specify exactly what the managers grouped as "subject" should or are allowed to do with the managed objects grouped in the "target". This must be expressed in additional specific attributes. In the IDSM project we formalised authorisation policies and reporting policies (for exact specifications see [IDSM-D7]). The `idsmAuthorisationPolicy` class has additionally a package which allows the specification of permitted or forbidden management operations (Get, Set, Create, Delete, Action) including parameter values. As a specific obligation policy we defined the `idsmReportingPolicy` class which has a package where the events to be reported and the destinations can be specified.

Whereas the formalisation of authorisation policies is relatively straight forward, for obligation policies this is far more complex since what a manager is responsible for can vary considerably. This is usually fixed in job descriptions or functional specifications of automated managers. In order to abstract from concrete tasks one has to look for generic patterns in such descriptions. As a starting point we use the policy classes we identified in the last section according to the classification criterion "influence on the system". In order to formalise these classes one needs an underlying model of the system to be managed. ISO provides a description language for such a model with its GDMO. Thus, actions, states and - to some extent - state changes can be formalised wrt. such a model. The more semantics such a model covers the more fine-grained can the description of management obligation be: If the model covers only a few control variables, a management specification can only be very coarse. In other words: The management description can be only as detailed as the object model it relates to. Having a model, the following ways of formalising the "influence classes" are conceivable:

- *manager actions*: a scripting language could be used to formalise such (potentially conditioned) sequences of action. The examples in [Wies94] and [Moffett93] can be considered as written in a pseudo scripting language.
- *system state*: starting point for a state description language are to be found in the literature on monitoring distributed systems (see [Mansouri93]). In the DOMAINS project, language constructs for specifying management goals were investigated (see [Becker94]). These goals include the description of desired states.
- *state changes*: Note first that for capturing the dynamics of the system to be managed a very rich model is necessary. Since in GDMO the dynamic behaviour cannot be specified formally (except for notifications), this description language is not powerful enough. In [Bean93] a petri-net model is suggested for modelling the dynamics. Control can then be specified by disabling controllable transitions. If such a model exists, an exact specification of the desired behaviour is possible.

There is obviously a trade-off between model complexity and the power of model-based management formalisation. Therefore, there will likely be different models and hence multiple model-dependent management formalisations.

2.3 Policy Hierarchy and Refinement

A management policy is often linked with other policies. This relationships can be peer-to-peer interactions with possible conflict situations or hierarchical dependencies. A policy hierarchy promises additional abstraction and automation of management activities: Ideally, the operator defines and changes higher-level policies whereby these changes propagate automatically to lower-level policies. This automation can only be reached, if the relations and dependencies between policies are formalised. An example for management using policy hierarchies can be found in the telecommunication environment. In TMN (Telecommunication Management Network) management consists of several layers: the lowest level, the network element management layer, deals with network elements, the middle layer deals with network management, the upper layer with service management. In the service layer an obligation policy deals with end-to-end communication links and quality-of-service (QoS) parameters and could be: "The bandwidth of communication links should not be more than 10% under the prescribed value". The target domain of this policy consists of the communication link managed objects; the subject domain comprises the service managers. This policy would be translated into lower-level policies on the network and network element management layer.

Policy refinement can be performed in several ways (see also [Moffett93]):

- The obligation of a policy can be refined by mapping it onto sub-policies; e.g. in Fig. 2.1., the availability policy is translated into test and monitoring policies. The task of finding appropriate sub-policies requires knowledge about the system, the configuration, etc. and cannot be automated in general.
- Delegation of management tasks: The target set, i.e. the objects to which the policy is applied, can be split into targets of other policies with the same obligation, but with different subject sets.

Since a policy consists of policy attributes, their values must also be translated to attribute values of lower level policies. Figure 2.1 shows an example where a certain availability of the system should be reached. The availability is guaranteed by an availability policy which is translated to a *test policy* and a *monitoring policy* on the system resources (here: storage disks). The several degrees of availability result in several test and monitoring policy attribute values.

availability (av) policy	90%< av <92% of scheduled operating time	92% < av < 94% of scheduled operating time	94% < av < 96% of scheduled operating time
test policy	every 3 h short test of all important components, every 24 h extended test	every 30 min. short test of important components, every 6 h extended tests of all components	every 5 min. short test of all important components, every 30 min. extended tests of all components
monitoring policy	alarm if disks are used to 90 %	alarm if disks are used to 80 %	alarm if disks are used to 70 %

Figure 2.1 Example for the translation of policy attributes

As we have seen in section 2.2, policies are formalised by policy objects and must be interpreted by managers. A policy hierarchy can be used to check the adherence of managers to policies. A way to detect policy violation is objective-driven monitoring [Mazumdar91], where monitoring and reporting policies are derived from higher-level policy obligations.

2.4 Comparison with other work

The concepts of domains and policies have recently received considerable attention in research and standardisation.

ISO

ISO considers these concepts in two drafts: The System Management Overview (Proposed Draft Amendment [IS0-10040]) defines the basic concepts, a specific Systems Management Function (Committee Draft 10164-19: [ISO-10164-19]) describes objects, relations and services for the management of domains and policies. In ISO a domain is a group of managed objects, which is determined by a grouping criterion, whereas in our approach a domain is an explicit enumeration of domain members. These two approaches can be combined allowing two kinds of domains, one with explicit and one with implicit members. The user can decide which membership is appropriate for his application. In ISO policies are defined in a more narrow sense than in our approach. Policies are defined as a set of rules, which restrict the behaviour of managed objects. A system management rule is one of the following:

- a constraint on the allowed operations, including permissible parameters and their values,
- an assertion on the allowed attribute values,
- an assertion on the emissions of notifications, including permissible parameters and their values,
- an assertion on the replies of operations, including permissible parameters and their values.

Our notion of policy is sufficiently broad to include the ISO approach as a special policy class.

X/Open

Work on policy specification and support is also in progress in the X/Open consortium. A preliminary specification for support of policies in a CORBA environment (see [CORBA91]) is expected to appear in 1995. In the current working document policies seem to be restricted to the definition of default and allowed values of managed objects. This type of policies is a special case of our general policy definition. Managed objects can be grouped in policy regions, which are similar to domains. In contrary to our management domains, policy regions are not allowed to overlap. This restriction leads to a less flexible concept, because in many cases policy domains will have to overlap.

Research

In the research area several approaches deal with policy hierarchies. [Calo93] develops concepts and supporting tools for formalisation and enforcement of policies. A policy architecture is presented which is based on policy hierarchies. Several policy layers are defined: (1)societal policies, (2)directional policies, (3)organisational policies, (4)functional policies, (5)process policies, and (6)procedural policies. Layers (1)-(3) are abstract, whereas layers (4)-(6) are subject to formalisation and automated interpretation.

[Moffett93] investigates policy hierarchies and the formalisation of policy refinement. The concept of hierarchy is the prerequisite for the refinement of policies, where policies are transformed to lower level policies and actions. Policy hierarchy concepts are supported in our approach by policy classes and a policy refinement service (see 2.3 and 3.1).

The concepts in this paper are partially based on results of the DOMAINS ([Alpers93], [Becker94]) and Domino [Domino92] projects. DOMAINS defines domains as areas of authorisation including a manager whom can be given goals. These goals are roughly comparable to our state-based policies. From Domino we adopted the concept of domains as object groups and the representation of policies as relationships between subject and target sets and the class of manager action policies.

3. Realisation of Policy Concepts

3.1 Implementation Architecture

In this section we describe the services for supporting the domain-based policy concepts and an implementation architecture which is the basis for our implementation in the IDSM project [IDSM-D6]. The centrepiece of our implementation environment is a management platform which gives access to managed objects offered by OSI agents or by management applications acting as agents residing on the same or on a remote platform. Figure 3.1 shows the architecture with supporting services. These services are offered as Service Managed Objects to make them accessible for applications or other services. The Domain Service and Domain Service User Interface contained in the figure support the flexible definition of management domains. These components are described in [DSOM94].

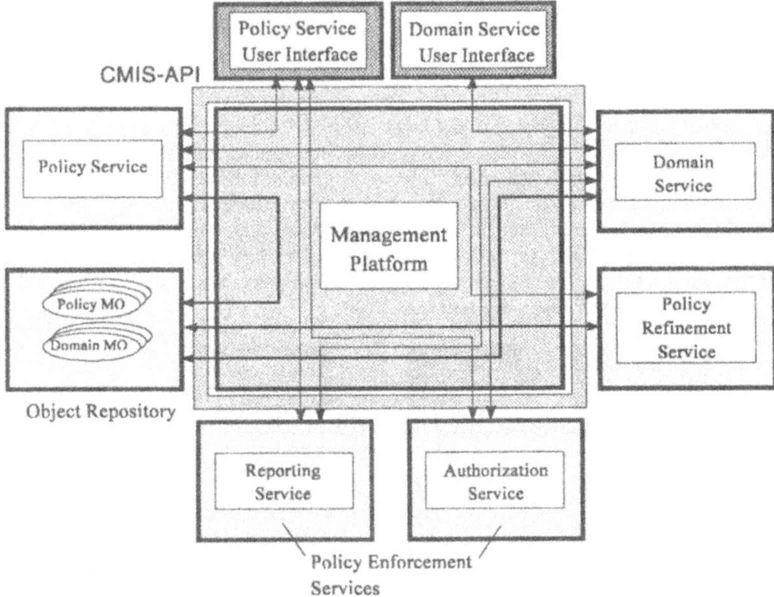

Figure 3.1: Platform-based Implementation Architecture

In this approach the policy concept is supported by a *Policy Service, Policy Service User-Interface, Policy Enforcement Services,* and *Policy Refinement Services.*

- **Policy Service (PS):** The PS supports storage, retrieval and analysis of policy objects belonging to certain policy object classes. It allows to create, delete, query and modify instances of these classes. It is important to note that the Policy Service *does not enforce policy.* Dedicated enforcement services for certain classes of policies interpret instances and enforce them using the mechanisms provided by the underlying infrastructure.
- **Policy Service User Interface (PS-UI):** The PS-UI offers the operations of the PS in a user friendly manner. It allows to look up policy objects and change attributes, to create new policy objects from scratch or to copy and modify existing policies. Moreover, the user can activate and deactivate policies; this does not lead to an operation on the PS but on the respective policy enforcement service.
- **Policy Enforcement Services (PES):** A policy enforcement service is an application, which is able to interpret policy objects and map the object information onto the mechanisms of the infrastructure. Therefore, an enforcement service hides the details and peculiarities of the available mechanisms allowing the user to concern himself with the higher-level abstraction provided by the respective policy object class (POC). Since an enforcement service has to deal with available mechanisms, there might be several different enforcement services for one POC. Moreover, there will be different enforcement services for different POCs. We envisage an increasing set of formalised policy classes and correspondingly an increasing set of enforcement services such that policy-based management will more and more determine the structure and semantics of the management system.

  In the IDSM project [IDSM-D7] we build policy enforcement services for the instantiable *authorisation* and *reporting policy* classes we formalised, i.e. the authorisation resp. reporting service for enforcing authorisation resp. reporting policy objects. These services offer operations to activate and deactivate policies.

  - Authorisation Service: Authorisation policy objects contain information on the subject domain, the target domain, and the rights members of the subject domain should have on members of the target domain. The underlying mechanisms supported by the infrastructure in our environment are access control lists which are used by the platform or by services to perform access control when invocations occur. Thus, the authorisation service transforms domain-based authorisation into information used by the environment for doing the actual access control but is not involved in the latter itself.

- Reporting Service: Reporting policy objects contain information on the subject domain, the target domain, event discrimination and destination of reports. The underlying mechanisms supported by the infrastructure in our environment are OSI event forwarding discriminator objects which can be created, manipulated and deleted in OSI agents via a management platform. The reporting service transforms domain-based reporting policies into information used by the underlying platform and the agents to perform the actual event reporting but is not involved in the latter itself.

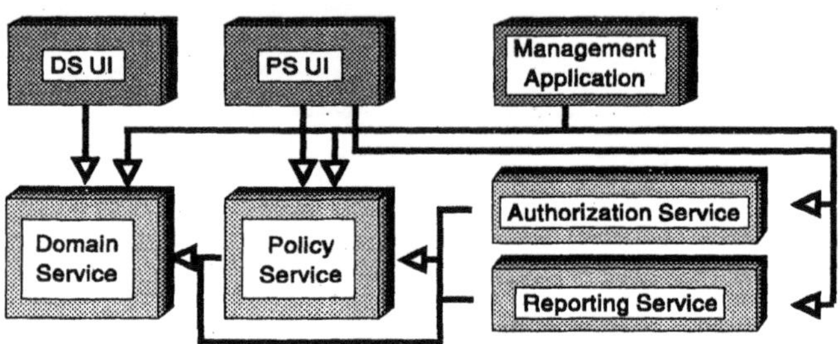

Figure 3.2: Functional Architecture

Figure 3.2 shows the usage relationship between the supporting components and management applications. The Domain Service (DS), the Policy Service (PS), and the policy enforcement services can be accessed by human managers via the respective user interfaces. The PS uses the DS to realise its analysis functionality (detection of potential authority conflicts). When the enforcement services (authorisation and reporting) are requested to activate or deactivate a policy object, they invoke the PS to look up the content of the policy (subject, target etc.). Then they use the DS to determine the scope of the policy to be activated or deactivated. Having this information they invoke the appropriate available infrastructure mechanisms. Other management applications use the DS to determine their scope, they use the PS to store and retrieve policy objects and the available enforcement services to activate and deactivate policies.

- **Policy Refinement Service (PRS):** For the interpretation of policy objects we propose in addition to the policy enforcement services a policy refinement service. A policy enforcement service maps a policy object to the infrastructure mechanisms of the system, whereas a refinement service should be independent from the infrastructure. The main purpose of a PRS is to enable policy hierarchies. A PRS should be able to map a policy object to other policy objects. Since a policy object consists of policy attributes, the refinement process depends on the policy attributes. In contrast to the enforcement services, where a specific service is responsible for a certain policy class, there will be only one type of refinement service. This service will refine all policy objects which cannot be directly enforced by PESs.

3.2 Policy Engineering Model

Policy influences the behaviour of manager and managed objects. It is the centrepiece of each management task. If such a task is to be automated based on policy concepts, one has to perform several activities which we describe in the sequel:

1. Structured Informal Policy Description
In this step, policies must be informally described as objects and their attributes. The description comprises the policy subjects and targets, the desired/permitted behaviour and applicable constraints.

2. Formalising the Policy using GDMO
In section 2.2 we defined a generic policy object class which contains attributes for subject, target, and constraints, from which specific policy classes can inherit. So, the main task consists of formalising the desired behaviour. Here, we have to determine a formal description of the actions and procedures of a manager or of the desired managed object behaviour. For the reporting policies which we formalised in IDSM, this results in attributes for specifying the events to be reported and the destination of event reports.

3. Extending the Policy Service by the new Policy Object Class
In order to make instances of the new class available for documentation and analysis by the Policy Service, the new class must be known in the Policy Service. It depends on the implementation of the latter whether or not adaptations need to be made.

4. Implementation of a Policy Enforcement Service (PES) for the new Class
Once a policy object class has been formalised, one has to investigate how the abstraction provided by the policy can be mapped onto management mechanisms of the underlying platform and systems or onto functionality of already existing management applications. If this is possible, then the enforcement of the policy can be fully automated. In this case, a PES specific to the policy class under consideration is to be designed and implemented. This enforcement service is a special management application which is able to interpret and enforce instances of the newly created policy object class. Enforcement is usually not restricted to a one-time or periodically performed sequence of actions. Does, for example, the policy specify a desired system state, then it is not sufficient to set up this state once. One has also to monitor the state and to take measures in case of deviations. For monitoring purposes the enforcement service to be implemented can ideally use a Reporting PES by creating and activating a reporting policy object. So, Policy Enforcement Services can have usage relationships.

5. Creation of policy object instances and activation
In the operational phase a policy, is introduced into the management system by creating an instance of the policy class. The policy is enforced by invoking the "activate" operation of the PES for the class. The latter enforces and monitors the policy until the policy is deactivated. Finally, deletion terminates the life-cycle of a policy object.

4. Application

We apply the policy concepts to two scenarios:

- In the IDSM project domains and policies are specified for managing the X.400 service over interconnected LANs, i.e. in the area of distributed system and service management.
- In the telecommunications area we identify domains and policies for specifying the interactions between managers in customer network management.

We give an outline of these applications in the sequel of this chapter (for a broader treatment of the first resp. second scenario see [Veldkamp94] resp. [Alpers95]).

4.1 Managing X.400 over interconnected LANs

In the IDSM project we use a pilot site consisting of several Local Area Networks (LANs) which are interconnected by a Wide Area Network (WAN). On top of this network an X.400 message handling application is provided. The local networks contain PCs and workstations. Those stations which are attached to the mail system have an X.400 User Agent and dedicated workstations serve as X.400 Message Transfer Agent (MTA). For this pilot a management system is being built based on industrial platforms providing access to OSI or SNMP managed objects. This system consists of the Domain Service, Policy Service, Authorisation Service, Reporting Service and specialised management applications which use the services. Moreover, the applications can be used for realising higher-level policies not yet formalised (see Figure 3.1).

We specify authorisation policies to separate the areas of authority for the managers involved in managing the whole system. For each local site we define a domain of local managers and a domain of local managed objects and create an *authorisation policy* object which gives the manager objects rights (GET, SET, ACTION, CREATE, DELETE) on the members of the managed domain. Local managers can delegate rights to sub-managers which are in charge of managing specific services like the local X.400 service. For this, the managed objects relevant for X.400 are grouped in a sub-domain, a sub-domain of the local managers is created, and the domains are related to each other by another authorisation policy object. The target domain should also include objects like the PCs and workstations running X.400 software but the X.400 managers should only have read rights thus allowing them to see for example whether the configuration is adequate. Furthermore, we construct technology-specific domains for which we create reporting policy objects. It is for example useful to create a reporting policy which states that managers should be informed if the file systems in the workstations running MTA software are filled by more than 95%.

The fault management application which is being developed deals with fault diagnosis in interconnected LANs. It will help to analyse problems within the pilot site by providing status information of network and logical components for the human manager. The application observes errors and establishes correlations between them to detect the faulty component. Errors are observed by retrieving management information from MIBs, by polling attribute values, or by receiving event notifications. Additionally, diagnostic tests must be performed. The application uses domains and obligation policies to collect the management information. The obligation policies are:

- *reporting policy* in order to receive event notifications;
- *polling policies* for reading attributes of managed objects;
- *testing policies* for obtaining information on state and connectivity of managed objects.

The reporting policy will be enforced by the IDSM Reporting Service (see 3.2). The polling and testing policies are specific to the fault management application. The targets of the policies are specified by domains.

4.2 Customer Network Management

As a consequence of deregulation in the telecommunication area several new players have entered the scene. Beside the network operator providing bearer services there are providers of value-added services like virtual private network (VPN). Moreover, Customer Network Management (CNM) allows customers to manage subscribed services which are offered by service providers who in turn are customers for network services offered by public network providers. Figure 4.1 shows an example of the hierarchy of services and the corresponding hierarchy of CNMs.

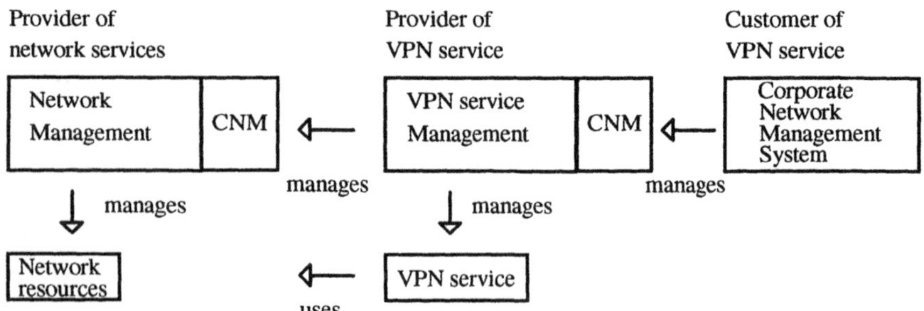

Figure 4.1: Roles in CNM-VPN environment

In this scenario, the service provider has to control the interaction with the customer managers and has to decide which customer requests can and should be fulfilled. In the sequel we show how this problem can be solved by using domains and policies for authorisation and reporting.
In order to organise access of different customers we introduce *authorisation policies*. For each customer we create a customer manager domain where customer manager objects, i.e. object representations for customer managers, are to be included. Moreover, we create a domain of managed objects for which the customer managers get access rights. In this domain we group all managed objects which are of concern for a specific customer, e.g. the VPN object representing the customer's VPN, link objects representing the links within the VPN, objects representing usage and accounting information and so on. Note that these objects are partly specific to CNM and partly overlap with managed objects the service manager uses also for service management purposes; i.e. the respective domain might contain references to objects which are also members of domains used for service management. Given these domains, we specify an authorisation policy which gives the members of the customer manager domain (possibly restricted) access to the objects in the domain of managed objects. The access rights are specified in terms of allowed operations. This way areas of authority can be easily specified along the borderlines of customer concern thus providing a clear authorisation structure in the management system.
Giving customers access to managed objects is not sufficient. Besides this, the customer must be informed in case the service cannot be provided as fixed in a service level agreement or in case of other situations (e.g. excessive usage) the customer might be interested in. For specifying information to be reported, we create or reuse domains and specify *reporting policies*. If, for example, we have already created a managed object domain for a customer, then we can reuse this domain as target for a reporting policy unless we want to reduce the scope and specify different policies for subsets which we then have to include in sub-domains. E.g., we can group all link objects for a VPN in a sub-domain and specify a reporting policy for these objects. The subject domain consists of managers who are responsible for the reporting policy and who will use the Reporting Service for setting up the infrastructure mechanisms accordingly.

5. Conclusions and Future Work

Management of complex networked systems with many parties involved needs concepts and services for organisation and semantic specification of management activity and intention. In this paper we presented the domain-based policy concept for this purpose. We formalised a generic policy object class using GDMO from which specific classes can be derived. Two such classes for authorisation and reporting policies have also been formalised. Formalisation is the prerequisite for implementing enforcement services which realise policies using underlying mechanism. For reporting and authorisation policies such enforcement services are being built on top of OSI-based platforms. The application of our concepts to management of X.400 over interconnected LANs as well as to customer network management shows their value for structuring and abstraction.
In our future work we plan to extend the area of policy-based management by formalising more policy classes and constructing the respective enforcement services. Candidates for formalisation are in particular generic policies which are relevant in many application areas. Here, we think of reporting policies for more complex situations and model-based policies to specify the desired system state. Moreover, technologies to support the enforcement of such policies have to be investigated.
Other conceptual areas where more work is required are policy hierarchies and formal refinement as well as policy analysis where conflict definition, detection and resolution must be investigated.

Acknowledgements

We gratefully acknowledge contributions of our partners in the IDSM and SysMan projects from Bull, Imperial College London, IITB Fraunhofer Institute, MARI, NTUA, AEG, Alcatel Austria, and ICL.

6. References

[Alpers93] Alpers, B., Becker, K., Raabe, U.: DOMAINS: Concepts for Networked Systems Management and their Realisation, Proc. GLOBECOM 93, Houston 1993

[Alpers95] Alpers, B., Plansky, H.: Applying Domains and Policy Concepts to Customer Network Management, to appear in proceedings of ISS '95

[Becker94] Becker, K., Holden, D.: Specifying the Dynamic Behaviour of Management Systems, J. Network Systems Management, vol. 1, no. 3,1994

[Calo93] Calo, S. B., Masullo, M. J., "Policy Management: An Architecture and Approach", Proc. IEEE Workshop on Systems Management, UCLA, Calif., 14.-16.April 1993

[CORBA91] Object Management Group: The Common Object Request Broker: Architecture and Specification, Doc. No. 91.12.1, 1991

[Domino92] Sloman, M., Moffett, J., Twidle, K.: Domino Domains and Policies: An Introduction to the Project Results", Domino Report Arch/IC/4, February 1992

[DSOM94] Alpers, B., Plansky, H., "Domain and Policy-Based Management: Concepts and Implementation Architecture", 5th IFIP/IEEE International Workshop on Distributed Systems: Operations and Management, Toulouse, 10.-12. October 1994

[IDSM-D6] IDSM Deliverable D6/SysMan Deliverable MA2V2: "Domain and Policy Service Specification", October 1993

[ISO-10040] ISO/IEC IS 10040/PDAM 2: Systems Management Overview - Amendment 2: Management Domains Architecture, 2.11.1993

[ISO-10164-19]ISO/IEC JTC1/SC21/WG4: Management Domain and Management Policy Management Function, Committee Draft, 21.1.1994

[Mazumdar91] Mazumdar, S., Lazar, A.A., "Objective-Driven Monitoring", Integrated Network Management II, Ed. I. Krishnan, 1991, p. 653-678

[Moffett93] Moffett, J. D., Sloman, M. S., "Policy Hierarchies for Distributed Systems Management", IEEE Journal on Selected Areas in Communications, Vol. 11, No. 9, December 1993, S.1404-1414

[Veldkamp94] Veldkamp, W., Mitropoulos, S.: Integrated Distributed Management in LANs, Proc. NOMS 94, Orlando 1994

[Wies94] Wies, R.: Policies in Network and Systems Management - Formal Definition and Architecture, J. Network Systems Management, vol. 2, no. 1, 1994)

7. Biography

Burkhard Alpers received his Ph.D in mathematics from the University of Hamburg in 1988, where he specialised in geometry and algebra. Since 1989 he has been working with the research and development laboratories at Siemens, Munich. His research interests are in the field of network and system management.

Herbert Plansky received his Ph.D in electrical engineering from the Technical University of Munich in 1993, where his area of study was picture coding algorithms, digital signal processing, and VLSI architectures. Currently, he is a member of the research and development laboratories at Siemens, Munich. His research interests are in the field of network and system management of data and telecommunication networks. He is member of VDE/ITG (German association of electrical engineers).

6
Towards policy driven systems management

Phillip Putter[1], Judy Bishop[2], Jan Roos[2]
pputter, jbishop, jroos@cs.up.ac.za

Abstract

As the size and complexity of information systems increase organisations need to rely on integrated management systems to an even greater extent to ensure that users' requirements of information systems are met. *Management policies* have been identified as a mechanism by which changing user requirements can be captured and introduced into management systems in order to modify the behaviour of managed systems. This paper refers to systems which allow policies to be stated explicitly with the purpose of modification of management system structure and behaviour as *policy driven* management systems. This paper sets the scene and shows the way towards policy driven systems management.

Keywords

Management policy, systems management, meta-objects

1. POLICY DRIVEN SYSTEMS MANAGEMENT

1.1. Systems management

Systems management can be defined as the activities required to ensure that information systems function in accordance with user requirements and objectives [MOF94]. Figure 1 shows a common way in which management activity is frequently viewed [OSI91]. The Figure shows a manager interacting with a managed object (MO) via a management interface. MOs represent abstractions of managed resources. Managed resources have functional responsibilities which they fulfil through interactions via

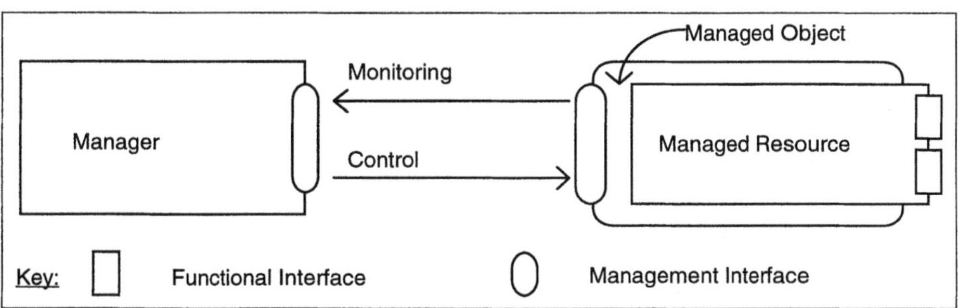

Figure 1 Management Activity

a functional interface.

Managers are regarded as objects with management responsibilities, and may themselves be the subject of higher level management. Managers manage MOs by monitoring their behaviour, making

[1]Emerging Technologies Group, Momentum Life, P.O. Box 7400, Hennopsmeer 0046, SOUTH AFRICA
[2]Computer Science Department, University of Pretoria, Pretoria 0002, SOUTH AFRICA

management decisions based on the monitored behaviour, and modifying MOs' behaviour through management operations [MOF94].

Systems management concerns itself with activities which attempt to ensure that the functional service levels required by the systems' users are met. Systems management does not concern itself directly with the functional activities of managed systems, and could therefore be seen as a *meta* activity. This paper argues that the meta nature of systems management can be exploited explicitly achieve systems that are easier to extend and modify, something which is discussed in more detail in Section 3.

1.2. Complexities of large scale systems management

The nature of the management activities shown in Figure 1 is, of course, much simplified. Large scale management systems can consist of large numbers of managed- and managing objects. Very large management systems introduce a number of problems [SLO94]:

- Large scale management systems often cross organisational boundaries, making it difficult to manage such systems from a central point.
- Large management systems usually require multiple managers, and more often than not, hierarchies of managers, introducing problems with the delegation of authority and responsibility.
- Managed objects could fall in the responsibility domain of more than one manager, introducing the possibility of conflicting management requirements of different managers being brought to bear on MOs.
- The large numbers of MOs make it impossible to manage MOs individually, introducing requirements for grouping mechanisms.
- As the size and complexity of management systems increase it becomes more important to automate as much as possible of the management process to assist human managers with systems management.

Successful management of large systems requires mechanisms which simplify the management process to [MAS93]:

- Raise the level of abstraction at which interactions occur, allowing managers to interact with groups of MOs instead of individual MOs.
- Deal with systems in terms of management policies instead of controls, allowing users of information systems to specify required service levels instead of specifying how these required services levels can be achieved.
- Automate the process by which management policy is captured and transformed to control operations required to achieve policy goals.

1.3. Pressures on organisations

Organisations experience tremendous economic, social and technological pressures to survive in changing business environments [SLO94]. Large portions of information systems are dedicated to meeting business needs that will change. Modifications to dedicated information systems are costly and time consuming [POO91]. The dynamic nature of business requirements places pressure on shorter delivery times to exploit business opportunities as and when they present themselves.

Changing business requirements combined with ever increasing complexity and scale of mission critical information systems place additional pressures on systems designers and developers. Successful information systems need to be delivered rapidly and in such a way that it can adapt to changing business requirements with ease.

Policy driven management systems, by their very nature, need to be able to adapt to policy changes. Later sections of this paper will focus in detail on the mechanisms required to achieve policy driven management systems. Many of these mechanisms should also make a positive contribution to fulfilling organisational requirements for modifiable and extendible information systems.

2. MANAGEMENT POLICY

2.1. Introduction

Management policy can be defined as the relationship between a set of managers and a set of managed objects which obligates and authorises managers to perform management activities on managed objects. Management policy serves as guidelines which influences the decision making process in the light of given constraints [NGU92, SLO94].

This paper is based on the point of view that all entities in managed and managing systems should be modelled and implemented as objects. These include:

- managers,
- managed objects,
- policies,
- functional systems and subsystems, and
- services provided by support environments or management platforms.

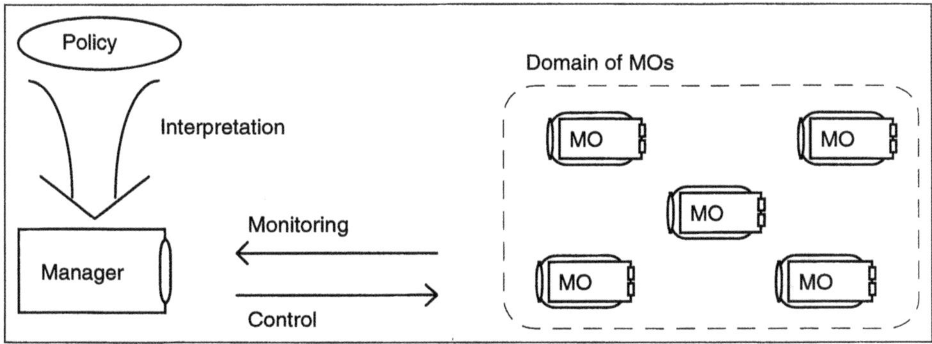

Figure 2 Policy in the management process

Managed and managing systems are viewed as groups of objects which collaborate to fulfil a common goal [ROO93]. Management policy can be used as a mechanism to capture the goals and requirements of the end users of information systems. Captured user requirements and goals need to be transformed into management operations which can be used to influence the behaviour of managed systems to ensure that user requirements are met.

Policy statements transformed into management operations in such a way influence both the behaviour of the objects of which the managed and managing system consist and system structure [SLO94, MAS93]. Managers interpret policy and uses it to guide decision making in the management process, as shown in Figure 2.

2.2. Policy as a separate concern

Policy exists everywhere in organisations' management systems and can be seen in the functioning and operational rules of organisations. Policy statements are usually represented implicitly by burying them deep within the code of managed and managing systems, and are seldom articulated [POO91].

The close linkage between policy and the systems introduces a number of problems [MOF94, MAS93, NGU92]:

- It makes it difficult to capture, store, query and modify policies. This causes policy to be managed in various ad hoc ways.
- Managers cannot interpret policy in a consistent way, making it difficult to implement re-usable managers.
- It leads to inconsistencies, confusion and conflict.
- It makes it very difficult to modify policy dynamically and to forecast the effect of policy changes.
- It makes it difficult and time consuming to modify policies because changes need to be represented in terms of changes to implemented systems.

Research shows that it is becoming well accepted that policy should be modelled and implemented as a separate concern [MOF94, NGU93, MCB91]. Modelling policies as a separate concern:

- recognise them as deliberate,
- forces them to be well defined,
- makes it easier to verify their correctness, and
- makes it easier to manage policies.

A separation between managers and the policies which influence their behaviour allows managers to be re-used in different contexts and permits management policies to be modified and interpreted by managers [MOF94].

2.3. Policy management

Management systems where policy is regarded as a separate concern require policy management services to succeed. Policy management services should provide mechanisms to [SLO94, NGU93]:

- Create, modify and delete policies.
- Represent and interpret policies.
- Store and retrieve policies.
- Negotiate the resolution of conflicting policies.
- Communicate new and modified policies to affected managers.

A consistent model for the representation of policies is required to effectively manage policies [NGU92, MOF94]. The policy model should capture policy on as high a level of abstraction as possible and should allow high level policies to be transformed to into concrete plans and procedures to achieve the required goals [MAS93].

The size and complexity of large management systems require automation of various aspects of systems management [MOF94, SLO94]. One of the main goals of policy driven management systems is to automate as much as possible of the management process. Higher level policies are usually more abstract, and require specific attention to ensure that sufficient information is gathered to allow them to be transformed to the management operations required to fulfil them.

The policy model should provide mechanisms to refine abstract policies to detailed operational plans and should be automated as far as possible [MCB91, SLO94]. Automated processes should exploit human expertise on all levels of transformation. Some of the aspects which might be automated include:

- Capturing policy statements from end users and bridging the gap between policy and the operations required to support them.
- The detection of problems in captured policies, e.g. insufficient detail to allow policies to be transformed to plans and procedures.
- Diagnosis of policies to detect which managers and MOs are affected by them, and the distribution of policies to affected managers.

- The detection and resolution of policy conflicts.
- The transformation of policy statements into management operations.

Representation of the relationships between policies is required to allow human managers to determine that stated policies have been satisfied [MOF94]. Policy relationships can be represented as networks of policy nodes, and should form part of the policy model. Relationships between policies should provide a controllable linkage between policies and plans and procedures [MCB91, NGU92].

2.4. Policies and domains

Because large management systems consist of large numbers of managers and MOs, policy cannot be applied to individual MOs. Instead, a structuring mechanism can be used to simplify large management systems by grouping MOs to which a common policy applies. Domains form a framework for partitioning of large systems and simplify large systems by acting as a naming construct for objects. Domains can also be used for the specification of viewpoints with specific focus on systems [SLO94].

Domains are objects which can be used to group other objects together. The reason for grouping the objects together is unimportant. The remainder of this paper distinguishes between domains grouping objects for the application of a common management policy (referred to as *management domains*) and domains grouping objects to structure functional systems (referred to as *functional domains*).

2.5. Relationships between policies

The relationship between policies closely resemble the organisational structure. Most organisations are structured in a hierarchical fashion in which authority is delegated downward, leading to the definition of hierarchies of policies. Policy hierarchies are characterised by partitioned targets, refined goals and objectives, and delegated responsibilities. A single high level policy may lead to many lower level policies [MOF94, MAS93].

In the case where an organisation follow a management style which is less hierarchical in nature, overlaps in management responsibility are more likely to occur. As soon as multiple managers manage the same MOs or a manager fulfils more than one management responsibility the possibility of policy conflicts exist.

Different approaches can be taken to manage conflicting policies: conflict can be avoided or resolved when it occurs. Some authors propose a combination of resolution and avoidance of policy conflicts for optimal results. Another approach might be to allow a degree of policy conflict. A detailed discussion of the detection and resolution of policy conflict is beyond the scope of this paper. Interested readers are referred to [MOF94].

3. IN PURSUIT OF POLICY DRIVEN MANAGEMENT SYSTEMS

3.1. Introduction

This paper argues that policy driven systems management requires a new way of looking at existing systems management solutions. Without the separate consideration of a number of concerns, the implementation of policy driven systems management would be extremely difficult, if not impossible. Aspects requiring separate consideration include:

- management policy,
- the use of objects as building blocks for the construction of management systems,
- object grouping mechanisms, and
- the system's structure and behaviour, also referred to as its self representation.

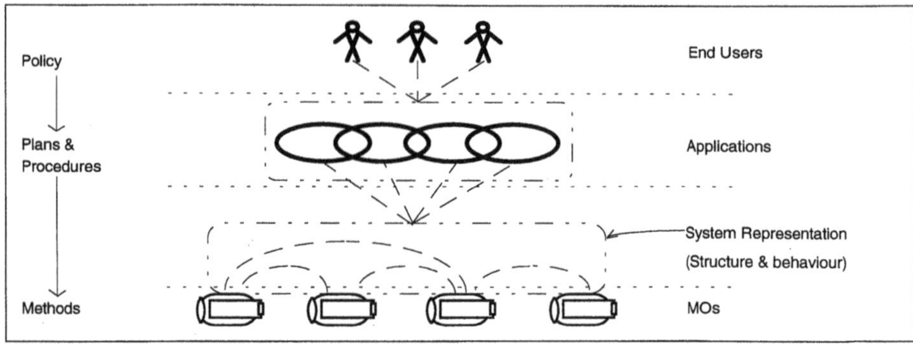

Figure 3 Management Systems

Figure 3 gives a graphical representation of a management system in which the above mentioned concerns have been modelled separately. The following sub-sections discuss the modelling and implementation requirements of these separate concerns in more detail.

3.2. Modelling requirements

Objects

Policy driven management systems fit elegantly into the object-oriented paradigm [ROO93, BEK93]. All policy driven system management entities should be modelled as objects. Objects encapsulate state and behaviour and are defined in terms of the attributes visible at its boundaries and its behaviour.

Objects can represent abstractions of physical equipment, logical components, or collections of information. An object's behaviour is defined by the operations which may be applied to it and its reaction to environmental stimuli. An object's state can only be manipulated and queried via operations exported to the environment by its interface.

This paper argues that objects should be viewed as a set of attributes and a set of operations which, when combined, realise a specific abstraction of a real world entity. Different abstractions of the same real world entity can be formed by combining different sets of attributes and operations. In this way all objects can realise different abstractions of themselves. For the purpose of this paper different abstractions of an object will be referred to as *viewpoints* on the object.

Viewpoints

The ability of an object to represent different viewpoints of itself is regarded as an useful abstraction mechanism which can be used to model and implement policy driven management systems. Figure 1 shows an example of a management viewpoint and a functional viewpoint of the same object. Viewpoints can also be used effectively to modify interfaces offered to particular objects in the environment: in this way it is possible to allow objects to add and remove operations to extend or restrict its interface.

This paper views all systems as groups of objects related to fulfil a common purpose. In this way management systems consist of one or more manager objects related to a number of managed objects with a common purpose of fulfilling the management requirements imposed by the policy to which the system conforms.

Functional systems, on the other hand, consist of a group of related objects with a common functional goal. Management systems can be seen as a different viewpoint of the functional system. The different viewpoints focus on specific aspects and allow users to abstract away from detail which is not necessary in its specific requirements. Some objects may form part of more than one grouping, e.g. a management grouping to apply a specific management policy and a functional grouping to achieve a specific functional goal.

Support for different management viewpoints are also required. Examples of these different viewpoints can be the different day-to-day operational management requirements and the requirements of strategic managers. A more detailed discussion of different management viewpoints can be found in [PUT93].

Domains are discussed as a grouping mechanism in Section 2.4. Although most authors agree that domains should be used only to apply management policy, this paper agrees with the argument that domains can also prove to be extremely valuable for the formation of functional groupings of objects [SLO94, MAS93].

Domains can be seen as different implementation viewpoints on the same objects: in one instance objects may be grouped for the application of management policy, in the other to achieve a common functional goal.

System representation

All systems consisting of groups of related objects have a distinct *structure* and *behaviour*. A system's structure describes the way in which objects are related to fulfil the system's goal, i.e. in a hierarchical fashion or as a network of co-operating objects. A system's behaviour describes how the related objects interact with each other and with the environment. Behaviour can focus on either the functional- or the management behaviour of systems.

Both the structure and behaviour of a management system are influenced by the management policy to which it conforms [PUT93]. Management policy serves as the baseline requirement of management systems, dictating which objects should be used to fulfil the requirements (the system structure), and how they should interact (the system behaviour).

Policy changes influence the behaviour of the managing system directly and the behaviour of managed systems indirectly: management policy guides a manager's decision making process and the manager's interaction with the managed system modifies the managed system's behaviour.

This paper argues that, as policy influences system structure and behaviour, both the structural and behavioural aspects of management systems need to be represented and implemented separately to create open ended systems that can become policy driven. The structural and behavioural aspects of systems can be referred to as the system's representation and can be implemented using meta-objects [BEK93], as discussed in the next sub-section.

Meta-objects

While objects define computation, meta-objects describe and monitor this computation. A meta-object contains information about an object's structure (e.g. relationships with other objects) and about its behaviour (e.g. the way messages are handled, objects are printed, created and initialised) [MAE87].

Meta-objects can be attached to functional objects to enrich the base object's behaviour [BEK93]. It is possible, for instance, to add management functionality to a functional object to allow the functional object to act as a managed object, as shown in Figure 4. Figure 4 shows an object which represents a router that allows packets to be passed between LANs using different communication protocols. This fictitious component will be used to illustrate the use of meta-objects to enrich functional objects, and will also be used in the example in Section 4.

Figure 4 shows two user viewpoints and a mechanism viewpoint of the router object. The first user viewpoint represents the functional viewpoint of the router, the second user view represents the management viewpoint of the router. In order to keep the example simple the router is assumed to have very simple functionality: it inputs a communication protocol A data packet, converts it to communication protocol B, and outputs the converted packed. In order to provide this server the object has three functional operations: read, convert, write. The functional object has only two attributes which can be used as object handles to the packets read and written: *inPacket* and *outPacket*. Each of the functional operations need to be mapped to the real resources' API instructions to perform the relevant

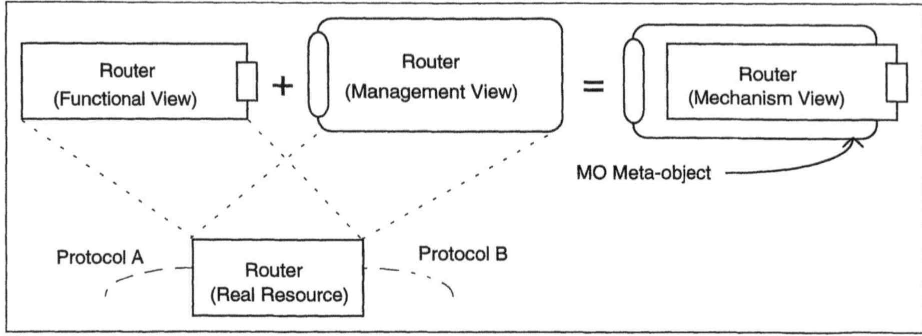

Figure 4 Object enrichment through meta-object

operations.

The management view of the router allows a manager to query the router state, and to reset the *packetsConverted* attribute. In order to fulfil its management responsibilities the router managed object offers three operations: *get_throughput*, *reset_throughput* and *get_state*. The managed object has two attributes: *packetsConverted* and *state*. The packetsConverted attribute keeps track of the number of packets converted since initialisation or the last reset. Each of the managed object operations needs to be mapped to the real resources' API instructions to perform the relevant operation.

The mechanism view of the router of Figure 4 shows how the two user views can be combined to implement a manageable router. The functional router object is enriched with a meta-object containing

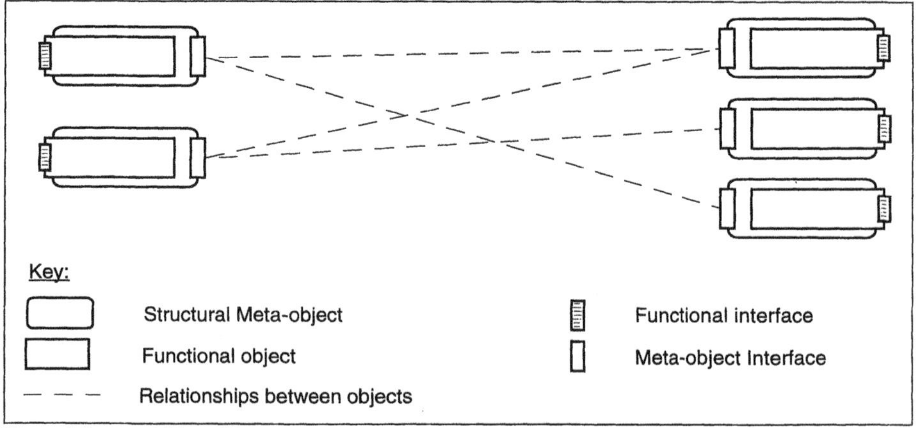

Figure 5 Structural Meta-objects

the management attributes and operations.

In the same way that the functional router object can be enriched with management functionality meta-objects can also be used for the systems' structural representation. Figure 5 shows a number of functional objects related to fulfil a common purpose. Each object's relationships with other objects can be represented by structural meta-objects. The structural meta-objects encapsulate the behaviour related to system structure and are manipulated to modify and traverse these relationships. This clear separation between the functional objects and their structural aspects allows systems' structure to be changed without affecting functional behaviour.

Objects interact with their meta-objects by passing messages to them. All messages not understood by objects are passed to attached meta-objects. In this way levels of meta-objects can be built, referred to as a reflective tower. Examples of possible uses for meta-objects include [BEK93]:

- Explicit system representation (structure and behaviour).
- Enriching objects with managed object and / or managing object behaviour.
- The implementation of distribution transparencies.
- The implementation of exception handlers.

A detailed discussion on the mechanisms which can be used to attach meta-objects to objects and the use of meta-objects in the implementation of distributed management systems can be found in [BEK93].

3.3. Implementation requirements

Policy driven management systems require distributed object-oriented support environments. Apart from support for distribution of objects the support environment should provide:

- Policy services which allow policies to be captured, represented and managed.
- The representation of system structure and behaviour using meta-objects.
- Support for the implementation of domains.

A detailed discussion of the support environments for policy driven management systems is beyond the scope of this paper. Interested readers are referred to [PUT93].

The effective implementation of policy driven management systems require support for sufficient abstraction mechanisms to allow developers to manage the complexity of large management systems as well as the efficient implementation of higher level abstractions. Issues arising from the efficient implementation of policy driven management systems are beyond the scope of this paper, but are discussed in detail in [SLO94].

4. EXAMPLE

4.1. Introduction

The purpose of this section is to present an example of a policy driven management system in order to clarify the concepts presented by this paper. This example, although simple, will highlight the essence of policy driven management systems, as size limitations prohibit the inclusion of a detailed example.

4.2. Problem statement

Consider an organisation that has a large number of routers that form an essential part of its mission critical business processes. Because of the importance of these components a strategic management decision was made that a manager should assume the management responsibility over all protocol conversion services. A simple policy statement was formulated to guide the manager in his decision making: Ensure that all protocol conversion equipment remain operational at all times.

In order to allow the protocol conversion management system to be integrated with the organisation's larger systems management solutions the protocol conversion manager should report to a higher level manager.

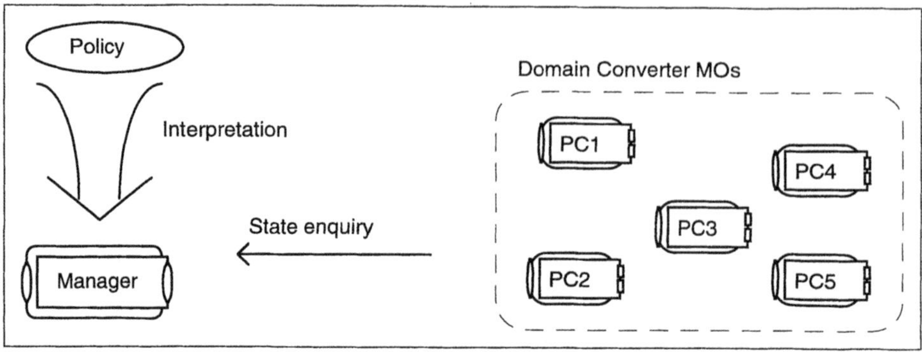

Figure 6 Router Management System

4.3. The system

The structure of the management systems required to fulfil the management policy requirements is clear from the policy statement. At least one manager object (the policy subject) should be associated with managed objects representing all the routers in the organisation (the policy targets).

The goal of the management system is to ensure that all routers remain operational at all times. The goal's requirements dictate the required behaviour of the management system: all routers should be polled at regular time intervals to determine their current operational state. If any routers are found to be in a state other than OK a trouble ticket is generated which is handled by the organisation's systems management solution.

Figure 6 gives a graphical representation of the management system required to fulfil the stated management policy. The figure shows the manager object interacting with a domain of managed objects representing router objects. The manager object and the functional router objects present managed object interfaces via managed object meta-objects which have been attached to it.

Because of the separation between the policy to which the management system conforms and the manager interpreting the policy it is possible to change the management policy, to for instance, relax the restriction that policy routers need to be operational all the time to a availability requirement of 90%. Such a relaxation could, for instance, lead to the modification of the manager's behaviour in that the manager might increase the time between state queries directed at router managed objects, with the required effect on the management system.

4.4. Implementation

Experiments have shown that it is possible to attach meta-objects to Smalltalk objects by capturing the Smalltalk message passing behaviour. It is possible to attach meta-objects to any Smalltalk objects in this way. Meta-objects are Smalltalk objects that encapsulate state and implement behaviour in terms of operations[3]. Communication between objects and their meta-objects take place by passing messages between the object and the attached meta-object. Any messages not understood by the functional object are automatically passed on to attached meta-objects via extensions to the Smalltalk messaging infrastructure. A detailed discussion about capturing the Smalltalk message passing behaviour is beyond the scope of this document, interested readers are referred to [BEK93].

---

[3]Object operations are called *methods* in Smalltalk. The term operation will be used for the remainder of this section to avoid unnecessary confusion between the use of the terms *method* and *operation*.

In this way it is possible, for instance, to implement a router MO with the attributes *packetsConverted* and *state* and the operations *get_throughput*, *reset_throughput* and *get_state*, which can be attached to a router functional object. Attaching the MO meta-object to the router functional object the enriches the functional object's behaviour, as it now responds to the MO operations as well as the functional object operations *read*, *convert* and *write*. The passing of control between the object and the attached meta-object is handled transparently. To the object invoking the operation it seems as if the operation was performed by the functional object. Invoking any of the MO operations on a router object with an attached MO meta-object will cause the MO operation to be performed by the meta-object and the result of the operation to be.

The experimentation also focused on the explicit representation of the management system structure by constructing management systems from collections of functional objects. It was shows that it is possible to separate the relationships between objects from the objects themselves. No attempt was made to modify these relationships dynamically as result of policy modifications, relationships were identified and explicitly constructed in the experimentation.

The results of the experimentation, showed that:

- that it is possible to modify object behaviour by attaching meta-objects.
- the separation between the objects which make up management systems and the relationships between these objects can be achieved.

The authors feel that more research is required to determine the extent to which the modification of behaviour and structure can be automated.

5. CONCLUSIONS

This paper presented an approach to policy driven systems management. Research covered by this paper attempts to exploit leading edge object-oriented principles in management systems. The research offers a fresh look at the managed and managing systems involved in systems management.

The pursuit of policy driven systems addresses a number of concepts which should have a positive effect on the extendibility of functional systems. These include:

- The separate modelling, implementation and management of policies.
- Separation of structural and behavioural concerns of object-oriented systems.
- Using meta-objects to enrich object behaviour and modify object interfaces.
- Exploiting the implementation of viewpoints on objects to realise more than one abstraction of an object.

The exploitation of object-oriented principles to the degree presented by this paper has not yet found wide acceptance in the area of systems management. This paper argues that, without these principles, it will become increasingly difficult for management systems to provide the users of information systems with the assurance that end-user requirements have been met, and to adapt to changes in user requirements.

Although experimentation with the concepts presented are still in its early stages, current prototyping results are very encouraging. No claims are made that the work presented solves all the complexities faced in the area of policy driven management systems, but it is argued that the research points the way to some solutions of existing management problems.

6. ACKNOWLEDGEMENT

All glory to the Lord our God, originator of all knowledge, without whom none of the work presented here would have been possible.

7. REFERENCES

BEK93 Reflective Architectures: Requirements for Future Distributed Environments. Bekker C, Putter P. Proceedings of the Fourth Workshop on Future Trends of Distributed Computing Systems, Lisbon Portugal (1993).

MAE87 Concepts and Experiments in Computational Reflection. Maes P. Proceedings of Object-Oriented Programming: Systems, Languages, and Applications, ACM SIGPLAN Notices, Vol. 22, Nr 12 (December 1987).

MAS93 Policy Management: An Architecture and Approach. Masullo MJ, Calo SB. Proceedings of the IEEE First International Workshop on Systems Management Los Angeles (1993).

MCB91 A Rule Language to Capture and Model Business Policy Specifications. McBrian P, Niezette M, Pantaziz D, Selviet AH, Sundin U, Theodoulis B, Tziallas G, Wohed R. Proceedings of the Third International Conference on CAiSE, Norway (1991).

MOF94 Specification of Management Policy and Discretionary Access Control. Moffett JD. Chapter 28 of Network and Distributed Systems Management. Sloman MS, Kappel K. Addison Wesley (1994).

NGU93 Incorporating Business Management Policy into Information Technology. Nguyen TN. Proceedings of the Second International Symposium on Network Management, Halfway House, South Africa (1993).

OSI91 Open Systems Interconnection: Management Overview. IS10040 (1991)

POO91 Representing Business Policies in the Jackson System Development Method. Poo CD. The Computer Journal Vol 34 no 2 (1991).

PUT93 A general building block for distributed systems management. Putter P. Masters Dissertation, University of Pretoria (1993).

ROO93 Modeling Management Policy using Enriched Managed Objects. Roos J, Putter P, Bekker C. Proceedings of the Third IFIP/IEEE International Symposium on Integrated Network Management, San Francisco (1993).

SLO94 Domains: A Framework for Structuring Management Policy. Sloman MS, Twidle K. Chapter 17 of Network and Distributed Systems Management. Sloman MS, Kappel K. Addison Wesley (1994)

YEM91 Network Management by Delegation. Yemini Y, Goldsmidt G, Yemini S. Proceedings of the Second IFIP Symposium on Integrated Network Management, Washington USA (1991)

ABOUT THE AUTHORS

Phillip Putter is responsible for the development of networks, distributed systems infrastructures and systems management at Momentum Life, South Africa's fourth largest life assurer. Phillip completed an MSc at the University of Pretoria in 1993, and is currently busy with a PhD with research into policy driven systems management.

Judy Bishop is professor of computer science at the University of Pretoria, where she specialises in distributed systems and programming languages. Judy obtained her PhD in computer science from the University of Southampton in 1977. Judy has been a member of IFIP's Working Group 2.4 (System Programming Languages) since 1987, has served on several international programme committees, and is the author of seven books on programming.

Jan Roos teaches Computer Networking at the Department of Computer Science of the University of Pretoria. Jan obtained an MSc in Computer Science from the University of South Africa, and has extensive research and development experience in the areas network design and network management. In 1988 Jan spent a year in Germany researching high speed networking and network management.

SECTION THREE

Panel

7

Distributed Management Environment (DME): Dead or Alive ?

Moderator: *Allan FINKEL,* Morgan Stanley and Company, U.S.A.

Panelists: *J. Scott MARCUS,* Bolt Beranek and Newman, Inc., U.S.A.
Lance TRAVIS, Open Software Foundation, U.S.A.
Nguyen HIEN, IBM Watson Research Center, U.S.A.

The Open Software Foundation's Distributed Management Environment (DME) effort was undertaken several years ago amidst great expectations and much fanfare. DME promised a standardized management framework, a consistent set of management applications, and a common user interface technology. Software Management, License Management, and Print Services were identified as key applications. A Data Engine was designed to allow for a shared object model, with objects distributed by Object Servers. Management Request Brokers were designed to provide connectivity between applications and objects. Event Management Services were designed to allow applications and administrators to be notified of problems and changes.

For all its promise, DME is not yet pervasive in customer environments. The challenges it faced highlight the difficulty of designing standards for fields dominated by niche players. Lance Travis will review the original design of OSF/DME, its current status, and future outlook. J. S. Marcus in "Icaros, Alice and the OSF DME" will comment on the disparity between the "management technology we would like to have, and the technology we are capable of successfully delivering". Nguyen Hien will discuss the OSF Data Engine and the technical challenges its designers and developers faced. To assess the successes and failures of DME, the panelists will discuss the difficulties of developing a shared management information model, the value and pitfalls of object-oriented technology, and the difficulties of integrating software developed by disparate organizations into a coherent offering.

8

Icaros, Alice and the OSF DME

J. Scott Marcus
BBN Internet Services Corp.
150 Cambridge Park Drive, Room 20/321
Cambridge, MA 02140, U.S.A.
(617) 873-3075 phone, (617) 873-5620 FAX
smarcus@bbn.com

Abstract

In 1990, the Open Software Foundation (OSF) set out an ambitious quest: to create an integrated management platform that would unify systems management and network management within the context of an object-oriented management framework. The totality of this task proved to be over-ambitious. This paper explores a few of the factors that made the project more difficult than had been assumed.

1 Introduction

> "Icaros was delighted. He flew steadily at first, but when they had got clear out of sight of the haven, he became excited... higher and higher he went, until he came too close to the sun, and the sun melted the wax, and the feathers fell out of his wings, and he dropped into the sea like a stone."
>
> W.H.D. Rouse, *Gods, Heroes, and Men of Ancient Greece*

The disparity between the management technology we would like to have, and the technology that we are capable of successfully delivering today, is daunting. Small wonder, then, that so many network and system management development projects fail!

When the Open Software Foundation (OSF) set out to create a Distributed Management Environment (DME), they did so with high hopes -- and indeed, why should they not have? The OSF represented the technical and marketing might of some of the greatest computer companies in the world. The task before them was the most ambitious that OSF had attempted, but also the most urgently needed: a stable, widely deployed platform that would offer modern software development tools and standards-compliant Application Programming Interfaces (APIs) for both system management and network management. The objectives seemed to many to be challenging, but achievable. The DME was to "unify the worlds of systems and network management" by ushering in a new era of object-oriented distributed management **[OSF91]**.

The OSF has been unable to deliver on these promises. They have delivered OSF software, and this software may in time achieve wide deployment **[OSF94]**; however, the OSF as delivered is a wan shadow of the comprehensive, integrated functionality that was initially intended. The first DME software that the OSF delivered, under the moniker DME 1.0, consisted of the distributed applications that were originally intended to serve as a mere proof of concept for the DME management framework. The conventional network management portions of the DME, the Network Management Option (NMO), do not appear to represent a major advance over the technology that was available to the OSF when it began the DME acquisition process in August of 1990 **[OSFRFT]**. The Object Management Framework (OMF) with its integral Object Request Broker (ORB), the heart of the modern object-oriented portions of the DME, has been delayed to the point where it was no longer necessary or beneficial for OSF to deliver its own ORB. Perhaps most significant, the vaunted integration of system and network management is still simply nowhere in sight.

The trade press has been quick to fasten on simplistic explanations for this failure, ranging from Machiavellian machinations on the part of OSF's sponsors to gross incompetence on the part of OSF. The reality is, of course, more complex.

It is not the author's intent to cast stones. I have driven my share of management projects into the ground, and am convinced that there is more that we can learn from our failures than from our successes. The DME represents an important case study – it has a story to tell. It can tell us a great deal about the probable future of integrated, distributed system and network management, if only we're willing to listen.

2 False assumptions

To a significant degree, the failure of the OSF to deliver DME reflects false assumptions and internal contradictions underlying the DME, in the author's opinion. A few examples should suffice:

- There is no fundamental difference between network management and system management.
- Distributed management of objects is the same as management of distributed objects.
- An object-oriented system so simplifies software development that every problem is easy to solve.
- If one Management Information Model is good, then two must be even better, and three, better still!
- A production system is the merely the last release of a research prototype.

The balance of this paper explores these assumptions, one by one.

3 Conflicting paradigms of management

> "...a paradigm is a 'reality model' in which I express my thoughts in language which reflects the semantics of my reality. Choose a different paradigm, and you will be using different semantics in your language, and be operating in a different reality...

> ...[Some people] live in the *Computing Paradigm.* [They] solve networking problems by making them go away; those are somebody else's problems. [They] spend lots of time talking about APIs as though APIs solve communication network problems...
>
> Networkers... live in the *Network Paradigm.* They assume that they own it all, ... or that what they don't own is out of scope.
>
> Internauts live in the *Internetworking Paradigm*; knowing, designing, and building networks of networks where one must always assume that the other end of any connection/association will be owned and controlled by someone else."
>
> Einar Stefferud, "Paradigms Lost"

As Einar Stefferud has pointed out, it may not be obvious that people are arriving at radically different conclusions because they are starting from radically different premises **[STEF94]**. The Computing Paradigm focuses on APIs, the Network Paradigm on private, closed protocols, and the Internetworking Paradigm on open protocols as a means of achieving interoperability. To a significant degree, the OSF confused itself by never consciously recognizing the internal inconsistencies among these paradigms, nor consciously choosing the most appropriate paradigm for the task at hand.

The OSF operated primarily in the Computing Paradigm, and secondarily in the Network Paradigm. This left them at a serious disadvantage in comparison to their most capable competitors, who operated for the most part in the Internetworking Paradigm. The OSF was primarily oriented toward system management rather than network management. In consequence, they focused on management APIs, such as the X/Open Management Protocol (XMP) – which, in fact, is not a protocol at all. They also placed a great deal of emphasis on the use of OSF Distributed Computing Environment (DCE) Remote Procedure Calls.

Figure 1, which follows, may help to clarify the implications of the Computing Paradigm for the DME. The central DME "cloud" represents a number of DME-capable workstations. They interact with one another by means of OSF DCE RPC. They communicate with external communication devices – such as routers and bridges – by means of the public, standard protocols, SNMP and CMIP.

In the context of the DME, DCE RPC must be viewed as an internal ***proprietary*** distribution mechanism, not as an open protocol for communication among heterogeneous systems. The OSF chose not to create a lightweight, agent-only version of the DME; therefore, there was no practical possibility that vendors of communication devices such as bridges and routers would implement DME. This, in turn, guaranteed that DME could not use DCE RPC to communicate with any management platform other than another copy of the DME. We discuss this point further in the next section.

Open APIs and essentially closed use of RPC protocols took center stage in the design and planning of DME. This left scant room for the Internetworking Paradigm, for the use of open protocols in support of heterogeneity. Open protocols were relegated to a relatively minor role – the Network Management Option – on the periphery of the DME.

There is a subtle point here, one that bears repeating. The fundamental architecture of the OSF DME was felt by OSF to support *heterogeneity,* in that it supported hardware and software platforms from multiple vendors; however, it was perceived by many in industry as being restricted to *homogeneity*, in the sense that it supported communications only with other realizations of itself, the OSF DME. Moreover, DME was realized in a way that made it unsuitable for implementation into communications gear. This resulted in a deep and fundamental schism between classical network management and DME-based systems management – a schism that ran directly counter to OSF's stated intent of unification of systems management and network management. The next section of this paper further elaborates on this theme.

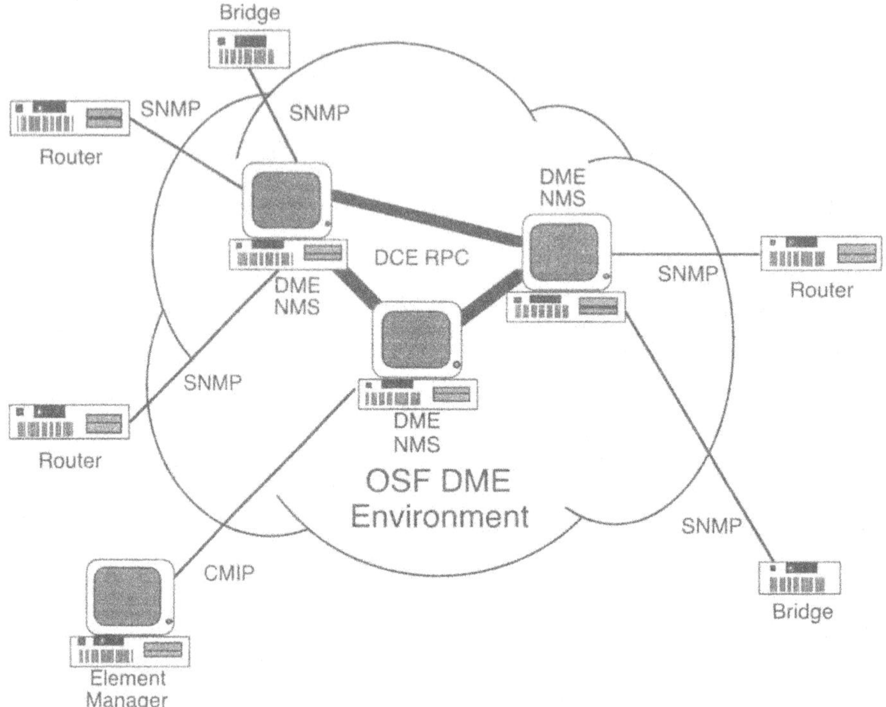

Figure 1. Management protocol environment of the OSF DME.

4 Distributed management of objects, or management of distributed objects?

> "You might as well say," added the Dormouse, which seemed to be talking in its sleep, "that 'I breathe when I sleep' is the same thing as 'I sleep when I breathe'!"
>
> "It *is* the same thing with you," said the Hatter.
>
> Charles Lutwidge Dodgson (Lewis Carroll), *Alice in Wonderland*

The DME was intended from the first to support distributed management. But... eh... *what* exactly was supposed to be distributed? The platform, or the things it was managing? Were they different?

The very ubiquity that was sought for the DME served in many instances to cloud the issue. DME was expected to be present on workstations from a very wide variety of workstation vendors. If DME were truly to be everywhere, then what need to distinguish between the manager and the agent -- between the ruler and the ruled? DME systems distributed everywhere could communicate happily and simply using peer protocols, in the form OSF's Distributed Communications Environment (DCE).

The reality is, of course, that we live in a pluralistic and heterogeneous world. Even in the most optimistic scenario, the DME would need to coexist with many other management solutions and technologies, just as it would need to coexist with routers and bridges in the scenario presented in the previous section. OSF's difficulty in recognizing these realities flowed naturally from the paradigms within which they operated.

The DME designers attempted to use DCE RPC and its associated mechanisms to solve a myriad of problems for the DME, from distribution to naming to security. This was a natural enough choice from point of view of engineering economy, but it severely limited the potential ability of the DME to interact with other management systems.

Consider the distribution mechanisms, for example. The OSF DCE protocols are documented, but OSF never had any concept of standardizing DME's *use* of DCE RPC in order to establish these protocol operations as an open protocol interface to other platforms. In consequence, DME's use of DCE had to be viewed as a closed and private mechanism for internal distribution of the DME management platform, rather than an open strategy for distributed management. It operated only within the "cloud".

Analogous inconsistencies appear in the security strategy for the DME. DME was intended to capitalize on the Kerberos-based security framework of DCE RPC, in order to achieve authentication and access control. However, this security strategy could only be relevant between copies of the DME -- elsewhere, a different strategy would have to be used for authentication and access control.

Outside of the DME distributed application, DME could presumably interact with other systems using SNMP or CMIP management protocols. The harmonization of DME's DCE-based security model, however, with those inherent in the new security features in SNMP Version 2.0 and in the GULS-based security (a generic upper-layer OSI security model) emerging for CMIP is a profoundly difficult problem. In each case, the semantics of the security model for these network management protocols differ somewhat from that of Kerberos. In general, it is difficult or impossible to map from one security model to another without losing the ability to verify the correct operation of the system.

In sum, the DME security model, and many other aspects of DME operation, are inherently applicable only to a homogeneous DME environment. They are not open, general solutions.

5 Object orientation as a panacea

There are many who would argue that object orientation is without a doubt the answer. I, for one, would like to better understand the question.

Object oriented programming is a very promising technique for the development of management applications. Object orientation appears to be a natural model for objects under management. Nonetheless, experience to date with true object-oriented management is somewhat limited.

The management framework used for SNMP, the most popular management protocol, can not be said to be object-oriented. The Management Information Model of OSI network management uses object-oriented modeling techniques **[OSIMIM]**, but not all management platforms that implement OSI network management use object-oriented techniques. Overall, the existing network management platforms can not be said to represent a compelling proof of the practicality of object-oriented management.

The ability of object-oriented solutions to scale to very large networks, to continue to operate in the face of network outages and partitions, and to accommodate changes and enhancements over time to MIBs have not been demonstrated to the author's satisfaction. It seems premature to assume that object-oriented management platforms will painlessly solve any conceivable problem.

In retrospect, it seems clear that OSF moved too quickly to embrace the Common Object Request Broker Architecture (CORBA) **[CORBA]** as a panacea. CORBA did not provide a defined means of interoperability among systems, nor were its interactions with traditional network management protocols specified. Immature and unstable CORBA specifications appear to have significantly delayed OSF's delivery of the Object Management Framework (OMF), the heart of the DME.

6 The Management Information Model

> "When *I* use a word," Humpty Dumpty said, in rather a scornful tone, "it means just what I choose it to mean -- neither more nor less."
>
> "The question is," said Alice, "whether you *can* make words mean so many different things."
>
> Charles Lutwidge Dodgson (Lewis Carroll),
> *Through the Looking-Glass*

The recent work of the Internet Interoperable Management Committee (IIMC), sponsored by the Network Management Forum (NMF), has demonstrated that mapping from SNMP protocols to CMIP, or vice versa, is workable (**[IIMC1]** and **[IIMC2]**). Actually, many vendors have implemented similar mappings between SNMP and CMIP over the years.

As the number of models of management information increase, however, the combinatorial problem of mapping from one to another becomes less and less tractable.

The DME supports at least three management information models:

1. The SNMP Structure of Management Information (SMI), based on RFCs 1155 **[RFC1155]** and 1212 **[RFC1212]**, expressed in Concise MIB Definition notation (with a variant required in the near future, in the form of the SNMP Version 2.0 SMI **[RFC1442]**);
2. the OSI CMIP Management Information Model (MIM), based on ISO IS 10165-1 **[OSIMIM]** and -2 (the DMI) **[OSIDMI]**, expressed in accordance with the Guidelines for the Definition of Managed Objects (GDMO, ISO 10165-4) **[OSIGDMO]**; and
3. a CORBA-based DME object model, expressed in I4DL.

The multiple object models confused and complicated many aspects of the DME. Consider, for instance, the DME Graphical User Interface (GUI). The DME GUI was initially tied to the

Tivoli-derived Object Management Framework (OMF) of the DME. The GUI could not act on the attributes of SNMP or CMIP objects in the absence of an *adapter object* that would map the object to an equivalent I4DL definition. Initially, OSF did not intend to provide any adapter objects. This would have resulted in a GUI with no cognizance at all of SNMP objects! OSF, under prodding from its user group, implemented an adapter object for SNMP MIB-I.

Mapping among these three management information models could be cumbersome or even impractical. There are two possibilities: provide any-to-any translation, or translate everything into a common model (which, in the case of the DME, would have to be I4DL).

Consider an example. Suppose that you speak German, and I speak English. We can solve our communication problems by learning one another's language, and learning to translate from one to the other. This is the IIMC approach. Granted, there may be some loss of meaning when going from a language with richer semantics to a language with more meager semantics, but we should still be able to communicate after a fashion. This works well enough.

Add a French speaker to our little clique. We now have three possible translations. Add a Spaniard, and we have six. It quickly gets out of hand.

An alternative approach is for both of us to agree to speak some third language -- an Esperanto. This avoids the proliferation of required translations. If we are only trying to translate between German and English, however, this is less satisfactory than direct translation, because the likelihood of loss of meaning is far greater when we translate everything *twice*. The Esperanto approach becomes more attractive if everyone can be induced to use it, yet one has to wonder: Esperanto could be viewed as a technical success, but a marketing failure. Does I4DL really make sense as the *lingua franca* of the DME?

Moreover, the translation of SNMP MIB definitions of GDMO MO specifications into CORBA IDL must still be viewed today as a research project, not as a trivial matter of software engineering. The Joint Inter-Domain Task Force (JIDM) of X/Open and the NM Forum has made a good start in this area (**[JIDM1]** and **[JIDM2]**), but it is only a start.

The OSF developed a methodology for translating SNMP MIBs into I4DL. They considered serveral possible approaches to this translation. One possibility was to perform a very direct and literal translation of the SNMP definition; a second was to create a highly abstract and user-oriented I4DL specification based on the SNMP MIB definitions; and a third was to create a definition that follwed the original SNMP MIB closely, but with some adaptations in the interest of neatness and comprehensibility. After some thought, they settled on the last of these three options, and translated SNMP MIB I as a proof of concept.

The problem with this third approach is that it implies manual translation of SNMP MIBs, rather than fully automated translation. This is impractical. New SNMP MIBs are proliferating rapidly today. A typical SNMP management product simply compiles the SNMP MIB definitions in order to provide at least a primitive access to management information – neither software development nor human intelligence are required, ideally, in order to access a new device with a new MIB. In the absence of an automated translation capability, OSF could have no hope of catching up with the flood of new MIB definitions.

The NMF-sponsored IIMC project, by contrast, chose a mapping from SNMP MIBs into GDMO that is largely mechanical. It can be done. In fact, an object-oriented infrastructure like that of the DME lends itself to the most primitive possible translation from external managament information models to internal ones – more complex and user-friendly views can always be layered on top of simple and direct translations.

7 Software integration

"Why should we be in such desperate haste to succeed, and in such desperate enterprises?"

Henry David Thoreau, *Walden*

Back in September of 1991, when the OSF first announced the technologies it had selected from those submitted in response to the DME RFT (Request For Technology), many knowledgeable observers concluded that the OSF had bitten off more than it could chew. OSF's initial strategy was to merge HP OpenView's SNMP and CMIP protocol implementations with Bull's XMP API software, to integrate the Wang / Banyan event services in with these, and to overlay all of these with Tivoli's object-oriented management platform and IBM's data engine, as depicted in Figure 2 below **[OSF91]**. This management framework would then be integrated with a variety of management applications, which would serve to demonstrate the capabilities of the management framework.

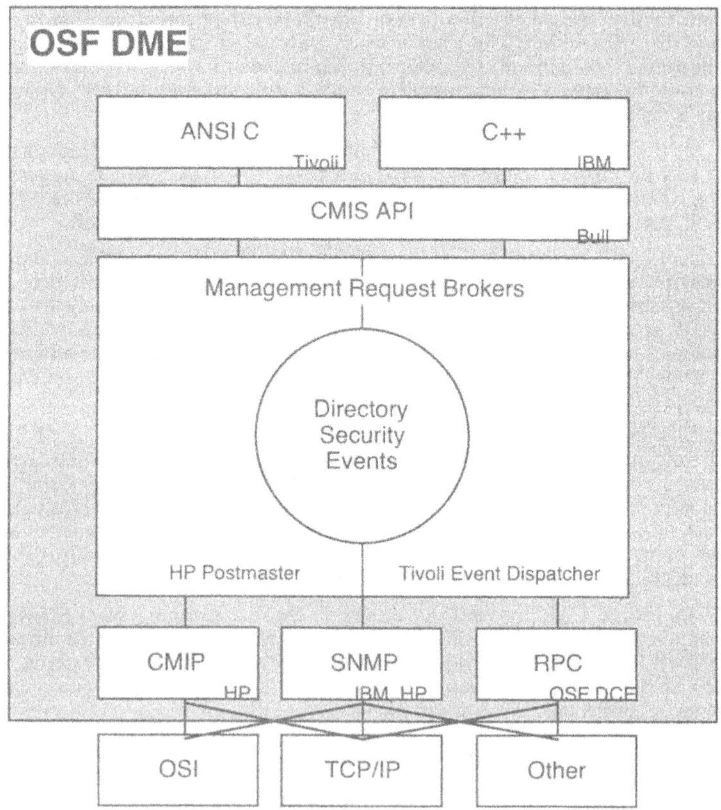

Figure 2. Implementation architecture of the OSF DME.

The notion that a single coherent system could quickly be created from these widely disparate piece parts was simply naive. Furthermore, the software components were not mature enough to be used as OSF intended.

It should also be clear from the foregoing discussion that many aspects of the DME involved advanced technologies that were barely beyond the stage of research. They were not yet ripe for production deployment.

8 Concluding remarks

> "Those of us who succeed, and fail to push on to a still greater failure, are the true spiritual middle-classers."
>
> Eugene O'Neill

Clearly the OSF did not realize their original DME objectives. Were they foolish to have tried?

In the author's opinion, they were not. OSF realized at the outset that the world of the Nineties did not need the OSF to create yet another SNMP MIB browser. To do so would intrude on existing products that were already successful in the marketplace, without offering software of greater utility to end users.

Instead, OSF attempted to leapfrog the existing state of the art. This may have been a risky approach, but it is not clear, from a business perspective, that they had better alternatives. More to the point, the problem that the OSF was trying to solve was the right problem. The integration of systems management and network management may be as elusive today as it was in 1990, but potentially it is just as valuable today as it appeared to be in 1990.

We should all strive to keep the lessons of the DME in mind as we move forward with other management projects. The underlying technology was subtly complex. It was not yet ready for "prime time". The most fundamental shortcoming of the DME is that it over-reached itself – it flew too high, too far above the comfortable and the mundane, too close to the sun.

At the same time, we should strive to remember that Icaros did not fly alone. His father, the master technologist Daedalus, learned from his son's mistakes and ultimately succeeded where Icaros failed. If we wish to ultimately rise above the limitations of today's management systems, we must be willing not only to take risks, but also to learn from the errors of others.

References

[OSFRFT] Open Software Foundation, "Request for Technology: Distributed Management Environment", August, 1990.

[OSF91] Open Software Foundation, "OSF Distributed Management Environment Rationale", September, 1991.

[OSF94] Open Software Foundation, "Distributed Management Environment: An Overview", March, 1994.

Open Software Foundation, "Using DCE and DME to Manage Software in DCE-Based Environments", March, 1994.

Open Software Foundation, "OSF Distributed Management Environment: The DME Network Management Option" (brochure), April, 1994.

Bruce Papazian and J. Scott Marcus, "Issues for a Graphical User Interface for the DME RFT", June, 1991.

W.H.D. Rouse, *Gods, Heroes and Men of Ancient Greece: Mythology's Great Tales of Valor and Romance,* Mentor, 1957.

[STEF94] Einar Stefferud, "Paradigms Lost", in *ConneXions*, Volume 8, Number 1, January, 1994.

Charles Lutwidge Dodgson (Lewis Carroll), *Alice's Adventures in Wonderland*, Random House, 1946. Originally published in Great Britain in 1865.

Charles Lutwidge Dodgson (Lewis Carroll), *Through the Looking-Glass*, Random House, 1946. Originally published in Great Britain in 1872.

[RFC1155] RFC 1155, M. Rose and K. McCloghrie, "Structure and Identification of Management Information for TCP/IP based internets", May 1990.

[RFC1212] RFC 1212, M. Rose, K. McCloghrie - Editors, "Concise MIB Definitions", March 1991.

[RFC1442] RFC 1442, PS, J. Case, K. McCloghrie, M. Rose, S. Waldbusser, "Structure of Management Information for version 2 of the Simple Network Management Protocol (SNMPv2)", May, 1993.

[OSIMIM] *ISO/IEC 10165-1, Information Technology - Open Systems Interconnection - Structure of Management Information - Part 1: Management Information Model*, 1991.

[OSIDMI] *ISO/IEC 10165-2, Information Technology - Open Systems Interconnection - Structure of Management Information - Part 2: Definition of Management Information,* 1992.

[OSIGDMO] *ISO/IEC 10165-4, Information Technology - Open Systems Interconnection - Structure of Management Information - Part 4: Guidelines for the Definition of Managed Objects*, 1991.

[IIMC1] *ISO/CCITT and Internet Management Coexistence (IIMC): Translation of Internet MIBs to ISO/CCITT GDMO MIBs*, Draft 3, August 1993.

ISO/CCITT and Internet Management Coexistence (IIMC): ISO/CCITT to Internet Management Proxy, Draft 3, August 1993.

ISO/CCITT and Internet Management Coexistence (IIMC): ISO/CCITT to Internet Management Security, Draft 3, August 1993.

[IIMC2] *ISO/CCITT and Internet Management Coexistence (IIMC): Translation of ISO/CCITT GDMO MIBs to Internet MIBs, Draft 3*, August 1993.

[CORBA] Object Management Group, *The Common Object Request Broker: Architecture and Specification*, OMG Document Number 91.12.1, December, 1991.

[JIDM1] Subrata Mazumdar, "Translation of SNMPv2 MIB Specification into CORBA-IDL: A Report of the Joint Xopen/NM Forum Inter-Domain Taskforce", July, 1993.

[JIDM2] Tom Rutt, "Comparison of the OSI management, OMG and Internet management Object Models: A Report of the Joint XOpen/NM Forum Inter-Domain Management Task force," March, 1994.

SECTION FOUR

Application Management

9
Managing in a distributed world

A. R. Pell, K. Eshghi, J-J. Moreau, S. J. Towers
Hewlett-Packard Laboratories
Filton Road, Stoke Gifford, Bristol, BS12 6QZ, United Kingdom
E-mail: {arp,ke,jjm,sijt}@hplb.hpl.hp.com
Telephone: +44 117 922 8762
Fax: +44 117 922 8920

Abstract

The task of networked systems management has become increasingly complex in recent years. Reducing this complexity and permitting easy management are major challenges to the acceptance of networked systems and applications. This paper introduces a language for describing these systems and applications and gives an example of its use.

Keywords

Networked systems management, application management, distributed management, model description, print spooling, management protocols

1 INTRODUCTION

The task of systems management has changed radically in the past 10 years. This has been caused in part by the explosive growth in computing power available in most organisations and also by the widespread distribution of these systems throughout organisations. As a consequence, it is no longer a simple matter for the MIS department to manage the available computing resources – even keeping track of where those resources are is quite taxing!

Allied to this growth in system power has been a major paradigm shift in the construction of large software systems, typified by the move towards client-server applications. Although the number of such applications in common use is, as yet, relatively small, the problems that they bring to the system manager are immense.

This paper describes a research project, Dolphin, which seeks to provide some responses to these challenges. In the next section, we describe in detail some of the problems facing system managers. We then introduce a language for describing the systems and applications that must be managed, and we give a practical example of its use. Finally, we describe our experience with this system, and outline some interesting research problems that still remain to be solved.

2 THE MANAGEMENT DILEMMA

One does not need to look far to observe recent trends in computing that are having far-reaching effects on the way many businesses function. The move away from a small number of large mainframe computers operated by the MIS department towards many physically smaller, yet often equally powerful, workstations is happening throughout most industries. With this decentralisation of computing have come a number of other moves – to distributed client-server applications on the one hand, and to mobile computing with its consequent change of user expectations on the other. These moves have commonly been portrayed as downsizing, partly because of the physical size reduction, but also, perhaps, because of hoped-for consequent reductions in the size and influence of the MIS department.

Looked at from a management standpoint, however, the picture is rather different. No longer is it possible to look in a single place to determine the health of a particular application – its vital signs may be spread across many machines, and these may be located over a wide area. No longer is it possible to reliably ensure the correct operation of all systems at all times – personal workstations and especially mobile workstations may come and go at a whim. Perhaps, above all, it is no longer possible to easily identify exactly *who* is managing particular systems and applications. To some extent, every person sitting at every workstation may be acting as a manager for some part of the networked systems and applications. It doesn't take much arithmetic in many organisations to recognise that, far from downsizing networked systems management, we have in fact seen an upsizing in this area.

Redressing this balance and ensuring that networked systems, and especially the business applications for which they are used, continue to serve the business most effectively are the major challenges facing businesses intent on rightsizing in the nineties.

3 DESCRIBING DISTRIBUTED SYSTEMS

In order to effectively manage any system, it is necessary to construct a description of it. This might be an informal sketch on paper or in a manual from which the experienced system manager can work, or might be embodied in the code of an application tailored to managing a particular system. Furthermore, it is necessary to have some idea of the task for which the description will be used – will it be for system installation, configuration, fault diagnosis and so on? Often this results in management applications that only perform one of these tasks, and which may not interwork with those performing other tasks, even on the same system or for the same application. There are, of course, some description languages, such as the Guidelines for the Definition of Managed Objects (GDMO) (CCITT, 1992), which describe the information that may be examined or changed in a system or applications, but often stop short of putting real semantics on those operations, except in comments.

The Dolphin description language takes a different slant. It is declarative in nature. and makes no assumptions about the purpose for which the description will be used. Rather, it concentrates on giving a precise description of the *correct* functioning of the system. Since the number of ways in which a system can fail may be very large, and may be dependent on parts of the system outside its direct scope (e.g. network stacks), describing the correct configuration is a significantly easier task, especially for the system designer.

The management tasks to be performed, such as configuration, diagnosis and monitoring, may now be described in a generic manner making use of the descriptions. Since these tasks are not dependent on the particular system to be managed, they may form part of the core of the management system. Descriptions of other systems and applications may now easily be added and the same management tasks performed. Of course, the interpretation and use of the description depends on the task being performed. For configuration, it can be regarded as a specification of things to install and change; for diagnosis, it is a list of things to check.

This separation of description and function in management reduces considerably the complexity of developing management tools, and ensures consistent behaviour for administrators through use of the same system description.

3.1 Models and objects

The fundamental component of the Dolphin language is the model. At one level, a model is simply a structuring component that refers to a collection of objects. It is, however, the unit by which the management system knows about systems and applications and it is conventional that a model should describe a whole system or a well-defined part of one. We can rely on a model being present in its entirety or totally absent. Models may themselves import other models, permitting the decomposition or specialisation of descriptions.

The Dolphin language is object-oriented in nature. Object definitions describe the fundamental components of the system to be managed with inheritance being used in the usual way. Instances of each of these object definitions will represent objects that are discovered in the real world. Figure 1 shows examples of some of the objects that might occur in the specification of a machine running the UNIX® operating system.

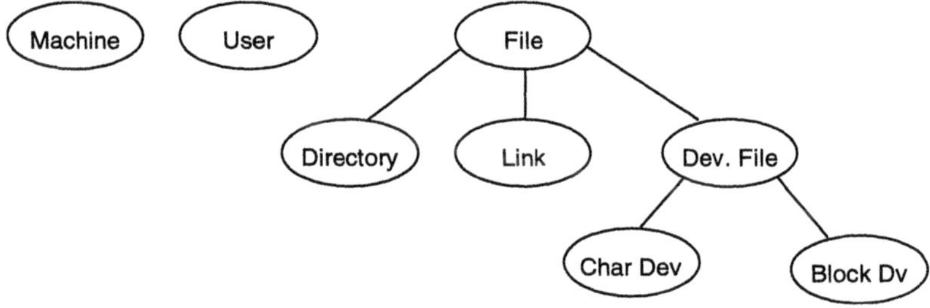

Figure 1 Object hierarchy

3.2 Location

It is insufficient, however, to only consider applications that run on a single machine. Whilst the management of such applications is not trivial, it would be short-sighted to consider that these applications are typical of those being deployed by many organisations, either now or in the future. It is necessary, therefore, to introduce a concept of the location of objects.

Some objects are fundamentally self-locating. For example, an HP-UX machine or a network printer can be reached simply by looking up its network address in a name server. Other objects, such as users and files, are not directly network addressable and information about them must be obtained via some other object, such as a computer system. Furthermore, unique identification of such objects is typically only assured within the confines of a single system – two users with the same name on different machines are not required to be the same

---

® UNIX is a registered trademark in the United States and other countries, licensed exclusively through X/Open Company Limited.

user. Thus, the concept of the location of an object provides us with the ability to both locate and uniquely identify all objects being managed. This will be especially important when an application is distributed across multiple machines.

Here are Dolphin definitions for some of the objects in Figure 1. The ISA keyword indicates inheritance.

```
OBJECT UnixMachine
OBJECT File LOCATION UnixMachine
OBJECT Link ISA File
```

3.3 Attributes

The characteristics of a particular object are described by its attributes. There are two principal types of attribute in the Dolphin language:

- basic attributes represent information that can be obtained from the real world, for example, the name of a user or the owner id of a file.
- derived attributes, also called rules, represent higher level information and depend on the status and value of other attributes. For example, a particular user may read a given file if the user's id and the file owner id match, and the appropriate permissions are set on the file.

Here are Dolphin definitions for some attributes of the objects in Figure 1.

```
[User u] name [String s]
[File f] ownerId [Integer id]
[User m:u] canRead [File m:f]
IF
    [u] id [id] &
    [f] ownerId [id] &
    [f] ownerMode ['read']
```

The notation "m:" in the last attribute is used to refer to the location of the users and files. In this case, since the same name is used, the rule can only be true if the user and file are located on the same machine. As can be seen, it is only necessary to specify this location once for each variable in a rule.

3.4 Connecting to the real world

Without further adornment, the language presented here could describe many things. In order to be able to obtain information from the real world, it is necessary to ground the management system to some access mechanisms. This is done through request and action definitions.

A *request definition* describes how some particular information source in the real world may be interpreted. For example, a request about users on an HP-UX system would relate the /etc/passwd file to a collection of users with their names, user ids, home directory name and so forth. Alternatively, the online status of a network printer might be obtained by a request using some known SNMP variable.

By contrast, an *action definition* describes how a particular change may be made to the configuration of a system in the real world. For example, an action to disable a user on an HP-UX system might make a change to the /etc/passwd file to put a * in the password field. The action definition must also express any side-effects that may occur whilst performing the action.

The exact mechanism for doing this is not described here, but the interested reader is referred to Pell (1993).

4 PRACTICAL EXPERIENCE - A PRINT SPOOLER

In this section, we give a practical example of the use of the Dolphin language to model a distributed application. The application that we have chosen is the print spooler of the UNIX® System V operating system, and its derivatives. We first give a very brief tutorial on the relevant internal workings of the spooler. For more details, the reader is referred to the manuals (Hewlett-Packard Company, 1993).

4.1 Background

For clarity, we distinguish between a *physical printer* that produces paper and a *logical printer* that is the representation of a physical printer internal to the print spooler. One physical printer is likely to have a logical printer representing it on many computer systems. Indeed, it is possible for one physical printer to have multiple logical printers representing it on a *single* computer system, each providing the appearance of a separate personality for the printer. We shall not, however, deal further with this latter case.

Every computer system that has the print spooler installed will have a number of configured *logical printers*. These are the printers to which a user of the system may send print jobs. They may be of three types:

- A *local printer* is one for which the corresponding physical printer is directly connected to this computer system. Information about the physical local printer will also be obtained from this computer system.
- A *remote printer* is one for which another computer system acts as a server. Jobs destined for this printer will be sent to the server for further handling. Note that it is possible for a job to pass through many servers before reaching one that can actually print the job.
- A *networked printer* is one for which this computer system has final responsibility for printing jobs, but where the corresponding physical printer is connected directly to the network. Information about the physical printer can, therefore, be obtained directly from the printer.

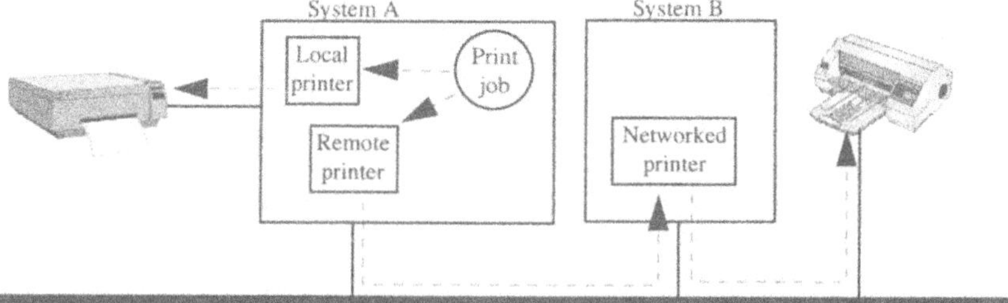

Figure 2 print spooling system

Figure 2 shows how these logical printers interact in a typical spooling system. At each node within the printing system, there are additionally a number of requirements of the local configuration, such as provision of adequate disc space, permission checking and so forth.

4.2 Object structure

Our brief description of the spooling system has identified some of the principal objects that must be modelled. Besides these "visible" components of the application, however, we must also model some hidden components – the application components that do the real work. We identify two:

- The print spooler application itself. This represents the application that the end-user would recognise. A user complaint might refer to some part of this application, e.g. printing a file, obtaining a queue listing, deleting a spooled file. The way that this object is modelled must, therefore, reflect in part its use in the real world.
- The scheduler. This represents the permanent "live" part of the print spooler. It must be running at all times and receives requests from the application mentioned above and acts on them.

We can now construct definitions for all these objects as follows:

```
OBJECT LogicalPrinter LOCATION Machine.
OBJECT LocalPrinter ISA LogicalPrinter.
OBJECT RemotePrinter ISA LogicalPrinter
OBJECT NetworkedPrinter ISA LogicalPrinter
OBJECT Lp LOCATION Machine
OBJECT Scheduler LOCATION Machine
```

4.3 Rule definitions

In order to describe the attributes of the objects that we have now defined, we apply a process of stepwise refinement, always keeping in mind the structure of distribution in the application. Note that, as we refine and develop our rules, we encounter the need for information that we will later request from the real world. This will be the basis for our basic attributes.

The print spooler application

At this stage, we must decide how to present the functionality of the spooler to the administrator. Our spooler does not permit restriction of the use of printers to particular users, so we only model whether it is possible for *anybody* to print on a given printer. The specification of this rule looks as follows:

```
[Lp m:_] canPrintOn [LogicalPrinter m:p]
```

This attribute will be true for a certain logical printer (p) if it is possible for any user to print on that printer. Note that this is an attribute of the print spooler application (Lp), since this is likely to be the source of a user complaint. The notation "_" is used to refer to an arbitrary object of the given type. Often, as in this case, there will only be one object instance, e.g. the print spooling application, on a given machine.

In the body of this rule, we express the conditions that must hold in order for it to be possible to print on the given printer. These are that we must be able to determine the name of the machine on which the application resides and that of the desired printer, and that the local scheduler must be capable of printing to the named printer on behalf of the local machine (the scheduler's client). Thus, the full rule becomes:

```
[Lp m:_] canPrintOn [LogicalPrinter m:p]
IF
    [m] name [clientName] &
    [p] name [printerName] &
    [Scheduler m:_] canPrintOn [printerName] for [clientName]
```

The scheduler

In considering the management view of the scheduler – the active component of the print spooler – we must be aware that it will be accessed remotely when a print job originates on a machine other than the server for the corresponding physical printer. In this case, the machine which originates the print job must determine both the name of the remote server, and the corresponding printer name on that server. There is no requirement that the given printer name actually exists on the remote server, however! From a management viewpoint, therefore, we model the ability of the scheduler to print on a named printer for a particular named client. The type DomainName is a special form of string whose content is restricted to the form of Internet domain names.

```
[Scheduler m:s] canPrintOn [String printerName] for [DomainName clientName]
```

In order that the scheduler can perform as indicated, the following conditions must be satisfied. The scheduler must itself be running and there must exist a logical printer of the given name, which must be accepting jobs and be able to print for the named client. This is expressed in the full form of the rule:

```
[Scheduler m:s] canPrintOn [String printerName] for [DomainName clientName]
IF
    [s] running &
    [LogicalPrinter m:p] name [printerName] &
    [p] acceptingJobs &
    [p] canPrintJobFrom [clientName]
```

Logical printers

So far, we have made no distinction between the various types of logical printer introduced earlier. Now it is time to do this. Recall earlier that we defined an object LogicalPrinter to represent any type of logical printer. There will not, however, exist any "logical printers" in a real system – only its subclasses will be instantiated. So we can define the canprintJobFrom attributes of these subclasses of logical printer.

Successful printing on a locally connected printer requires that the logical printer be enabled, and that its device file be accessible to the user whose name is 'lp' – the owner of the print spooling system. In this case, we choose to ignore the name of the particular client requesting service. We could, however, enhance this rule at a later stage to check this against some security policy.

```
[LocalPrinter m:p] canPrintJobFrom [DomainName _]
IF
    [p] enabled &
    [p] devFile [m:devFile] &
    [User m:lp] name ['lp'] &
    [lp] hasReadWriteAccessTo [devFile]
```

Similar constraints occur in the case of the networked printer, except that explicit checking of the ability to print is deferred to the (directly accessible) printer itself. This is embodied in the use of the isOk rule, which is defined appropriately for each type of network printer. It need not be of concern, however, to the designer of the print spooler manager.

```
[NetworkedPrinter m:p] canPrintJobFrom [DomainName _]
IF
    [p] enabled &
    [p] networkPrinter [np] &
    [np] isOk
```

The final case, that of the remote logical printer, is the most interesting since it is in this description that the fundamental requirement for managing a *distributed* application is embodied.

Here, in addition to determining whether the logical printer is enabled locally, we determine the remote server for this printer, and the name of the associated logical printer on that server. Finally, we determine whether the scheduler *on the remote node* (server) can print to the named printer on behalf of the given client.

```
[RemotePrinter m:p] canPrintJobFrom [DomainName clientName]
IF
    [p] enabled &
    [p] remotePrinterName [rpName] &
    [p] remoteServer [server]
    [Scheduler server:_] canPrintOn [rpName] for [clientName]
```

4.4 Basic attributes

Having defined all of the rules that are required, we have also determined the basic information that is required to manage the spooling service. All of this information is readily available through normal UNIX® commands or configuration files.

```
[Scheduler s] running
[LogicalPrinter p] name [String s]
[LogicalPrinter p] enabled
[LogicalPrinter p] acceptingJobs
[LocalPrinter m:p] devFile [File m:df]
[NetworkedPrinter np] networkPrinterName [String npName]
[RemotePrinter rp] remoteServer [Machine m]
[RemotePrinter rp] remotePrinterName [String rpName]
```

Each of these attributes must be represented by a (part of a) request to the real world. For example, the status of the scheduler can be checked by using the 'lpstat -r' command. Similarly, information about the various logical printers can be gleaned from a configuration file. In each case, the return from the request (typically a string or an SNMP variable) is transformed by the request processor into some information about basic attributes, such as the running state of the scheduler.

4.5 Using the model

Having defined our model of the print spooling application, let us consider how it would be used in order to diagnose a potential fault. The symptom that we wish to resolve is that a user cannot print on the printer called laser1 whilst working on the machine called machine1.

We start with the attribute that we defined earlier for the print spooling application itself:

[Lp m:_] canPrintOn [LogicalPrinter m:p]

We must instantiate the variables represented here. In this case, m will be a machine whose name is machine1, and p will be a logical printer whose name is laser1. Throughout the remainder of this section, we will represent this by the notation:

[Lp m:_] canPrintOn [LogicalPrinter m:p] (m.name='machine1', p.name='laser1')

Note that, in order to do this, we must determine the actual type of the logical printer whose name is laser1. This we do by performing a request to the machine m for information on its available logical printers. This will tell us the precise type of each. Let us suppose, for this example, that laser1 is a remote logical printer.

Now, we can expand the rule above and, by using the known information about m and p, we reach the next major goal:

[Scheduler m:_] canPrintOn ['laser1'] for ['machine1'] (m.name='machine1')

In order to verify this, we must first check whether the scheduler is running. For this, we can perform a straightforward request to the machine m. We must then determine whether a logical printer called laser1 exists, and whether it is accepting jobs. We know the former condition to be satisfied because of the way that we reached this point, but we must perform another request to determine that the printer is, in fact, accepting jobs. Finally, we reach the next goal:

[RemotePrinter m:p] canPrintJobFrom ['machine1']
(m.name='machine1', p.name='laser1')

Now, we must determine whether this remote printer is enabled and what are its remote server (suppose its name is printserv) and remote printer name (suppose it is attic). Assuming all of these can be successfully found using requests, we reach the goal:

[Scheduler server:_] printOn ['attic'] for ['machine1'] (server.name='printserv')

At this point, we continue to check and expand the various rules, but this time using information from the printserv machine. Finally, either all conditions will have been found to be satisfied, or some condition will not succeed. In the latter case, this information may be used to identify a possible fault, although further work may be required to obtain a full diagnosis.

5 EXPERIENCE AND FUTURE WORK

The Dolphin management system presently operates in a centralised manner, that is, there is a single management station which makes requests to agents on a number of managed nodes for information. This has proved to be quite satisfactory so far in managing a relatively small number of systems – roughly a large workgroup or a small site. Furthermore, the approach of separating system descriptions and task specifications has proved to be a good way to build management systems since the descriptions are easily written and understood, and the quality of the management provided (diagnosis, etc.) is more than adequate. There remain, however, a number of interesting research issues to be solved.

The centralised nature of the system causes some problems of scale. We are limited principally by the ability of the management station to hold and manipulate large quantities of information from many systems. In order to progress further, to true enterprise management for example, we require the ability for multiple management stations to coexist and to cooperate in the management tasks. For example, in our print spooling example, it would be possible to configure the system to print on a printer on the other side of the world. This might mean that the remote server would be outside the scope of the local management station. So, the assistance of a remote management station must be sought to obtain and check certain information on behalf of the local manager. Determining which remote manager to use, and arranging for the diagnosis or fix to be split between the management stations is a challenging problem.

Agents are presently assumed to be passive. That is, they do not generate asynchronous events when some piece of information changes. This means that, whilst performing a management task, fresh information will be gathered from managed systems even if nothing has changed. The introduction of asynchronous events, which may be regarded as something like an unsolicited request for information, together with persistent storage of information, would give a more responsive management system, whilst not sacrificing accuracy.

There is also no notion of time within the management system. That is, only immediately available information is used when performing management tasks. Introducing such a concept would permit more expressive modelling of applications that themselves have a notion of time, or would allow historical views of the systems being managed.

6 RELATED WORK

We divide related work into two parts – the provision of information for management, and the use of that information.

6.1 Management protocols

There are a number of emerging standards for information retrieval from managed systems. In the Internet world, and increasingly wider, the SNMP standard (Schoffstall, 1990) is used to specify data that may be retrieved from systems and devices. Historically, this has been used for the *monitoring* of network devices, although recent work on the Host Resources MIB (Grillo, 1993) has extended this monitoring to the lower levels of computer system activity. Application monitoring still remains a missing area. There has, until now, been no widespread use of SNMP for actually changing the configuration of systems and devices. This is largely because of SNMP's reliance on a single password, or community name, per device for write permission. With the emergence of SNMPv2 (Case, 1993), this objection to the use of SNMP for active management may go away. It remains to be seen, however, just how widespread is the adoption of SNMPv2.

The lack of information about applications on the desktop is being addressed by the Desktop Management Task Force (DMTF) – a industry grouping of the major players in the PC marketplace. They have recently produced a first release of their Desktop Management interface (DMI) (DMTF, 1994a), together with an initial set of component descriptions (DMTF, 1994b). The DMI provides access to an extensible set of components which may describe the desktop operating environment as well as installed applications. It is hoped that software and hardware manufacturers will move to producing components for the DMI as part of their products, thus evolving towards full management of desktop systems. It is too early to tell how well this will be realised, although one shortcoming that will need to be addressed is that the DMI is intended as a local interface only. A mapping has been defined from the component descriptions to SNMP MIBs, but this hides some of the benefits of the DMI specification, such as easy access to the component definitions.

In the telecommunications world, the CMIS/CMIP standard (CCITT, 1991 and ISO, 1991) is in widespread use. This provides a broader scope for definition of managed objects through the Guidelines for the Definition of Managed Objects (GDMO) (CCITT, 1992). However, the required implementation of these protocols is perceived by many to be much more heavyweight than SNMP or DMI, and it looks unlikely to become established in any field other than telecommunications.

The Dolphin management system adopts a liberal attitude to this diversity of management protocols, permitting the use of information from many different sources to be used in the management of systems and applications. In addition, it is possible to take these definitions as a starting point for Dolphin object definitions, and then to build higher level semantics on top.

6.2 Management products

In most cases, the above management protocols simply convey data about the systems being managed. With the exception of some limited textual comments, there is little attempt to express the semantics of the systems which are, of course, essential for effective management. There are a number of management products which introduce these semantics.

The TME management environment from Tivoli Systems Inc. (Kramer, 1993) presents to the system manager a similar view to the Dolphin system. That is, it is an object-oriented management framework which, for the applications supported, has a similar model structure. There, however, the similarity seems to end since TME requires separate underlying programs for its supported application areas such as user management, security management and so on. This contrasts with our approach of using a single management application augmented by *descriptions* of the systems and applications to be managed, which are more easily adaptable to changing requirements.

Unicenter, a product of Computer Associates, addresses the needs of large system installations including tools for routine tasks such as storage management, problem management, help desk setup and so forth (Ricciuti, 1992). It is not clear to us how this is constructed, although we suspect that it again uses a number of separate underlying programs, with the difficulties of consistency outlined earlier.

There are, in addition, emerging products from various major computer manufacturers which provide some of the functionality available in the Dolphin system, whilst not giving the same flexibility of specification.

7 SUMMARY

In this paper we have presented a language for describing systems and applications to be managed, and have shown how this may be done, even in a distributed environment. This language and technology are embodied in the HP OpenView AdminCenter product from Hewlett-Packard which supports configuration and change management in an enterprise.

The principal benefits from using this approach to building a management system lie in the ready capture of the necessary management understanding through a rich descriptive language, and the uniform application of this knowledge to the various facets of system management. Although the initial work in constructing a comprehensive model might appear to be somewhat more than comparable approaches, we believe that the long-term gains in productivity for system implementors and managers far outweigh this initial investment.

8 REFERENCES

Case, J. *et al* (1993) Introduction to version 2 of the Internet-standard network management framework, *Internet request for comments 1441*.

CCITT (1991) Recommendation X.710 (1991). Common management information service definition for CCITT applications.

CCITT (1992), Recommendation X.722 (1992) | ISO/IEC 10165-4 (1992). Information technology – open systems interconnection – structure of management information: guidelines for the definition of managed objects.

Desktop management task force (1994a) Desktop management interface specification, version 1.0.

Desktop management task force (1994b) PC standard groups, version 1.0.

Grillo, P. and Waldbusser S. (1993) Host Resources MIB, *Internet request for comments 1514*.

Hewlett-Packard Company (1993), System administration tasks, in *HP-UX manuals release 9.0*.

ISO (1991) ISO/IEC 9595 (1991). Information technology – open systems interconnection – common management information service definition.

Kramer, M.I. (1993) Enterprise system management: the quest for industrial-strength management for distributed systems. *Patricia Seybold's Distributed Computing Monitor* **8**(6), 3–23.

Pell, A.R. *et al* (1993), Data + understanding = management," *IEEE first international workshop on systems management*, Los Angeles, California.

Ricciuti, M. (1992) Industrial strength UNIX management tools *Datamation* **38**(10) 73–74.

Schoffstall, M. *et al* (1990) A simple network management protocol (SNMP), *Internet request for comments 1157*.

ABOUT THE AUTHORS

Adrian Pell received the B.Sc. degree in mathematics from the University of Reading in 1978. He joined Hewlett-Packard Laboratories in 1987 where he is now a senior engineer in the networks and communications laboratory. His research interests are in the development and management of networked computer systems and applications.

Kave Eshghi received the Ph.D. degree in computer science from Imperial College of Science and Technology in 1986 and joined Hewlett-Packard Laboratories in 1988. In 1991, he went to Stanford University to work on a collaborative project with Professor Yoav Shoham on application of agent oriented programming to the design and development of software agents. In 1993, he returned to Bristol where he has been working on model based system management.

Jean-Jacques Moreau received the Diplôme d'Ingénieur (M.S.) degree in telecommunications and computer science from Telecom Bretagne University, France in 1989. Since then, he has been a member of technical staff at Hewlett-Packard Laboratories, Bristol, UK. His research interests include distributed systems, computer languages, file systems and operating systems, and telecommunications.

Simon Towers received the B.Sc. degree in physics from Birmingham University in 1982 followed by the D.Phil. degree from Oxford University in 1985. He is presently a senior project manager at Hewlett-Packard's European research laboratories at Bristol in the UK. His research interests are in the management of networked systems and distributed applications.

10

POLYCENTER License System: Enabling Electronic License Distribution and Management

Timothy P. Collins
Digital Equipment Corporation
153 Taylor St. , Littleton, MA 01460-1407 USA,
collins@nac.enet.dec.com

Abstract

License management is a neglected area of systems management. First-generation license systems have focused on preventing unauthorized software use. The POLYCENTER License System reaches out to provide an infrastructure for an electronic distribution chain, from software publishers and distributors to the end user. Novel security and customization features along with support for industry standard APIs make PLS convenient and safe to use.

Keywords

Software licenses, software asset management, LSAPI, public-key cryptography

1. INTRODUCTION

Software is usually licensed, not sold. The actual title or ownership of the software remains with its producer or publisher. The end user buys an agreement with the publisher to use the software and the media on which the software is delivered. This agreement, or software license, describes the conditions under which the user may run the publisher's software. A typical PC license agreement might state that the program may be installed on as many computer systems as is convenient for the end user so long as there is no possibility that the software could be used in two places at one time. This allows for the case where a user has both a home and a work machine. However, it forbids installing the software on a network for anyone to use.

Software license constraints are seldom reasonable for the network administrator who needs the freedom to install software in locations which make the most sense for performance and storage management reasons. This can mean using a large and slow server for seldom used software packages, or redundant installations on several user workstations to facilitate legitimate (if temporary) transfers of licenses. End users frequently move their organizations into legal violation of software licenses. Most see little wrong with borrowing their neighbor's software if they need it to complete some task. The end result is widespread software piracy, occasional lawsuits (Didio, 1993), and the loss of billions of dollars of revenue worldwide for publishers.

First-generation license systems, such as Digital Equipment Corporation's License Management Facility (LMF) give useful assistance to the task of enforcing license agreements. Unless the user loads a unique license key for a program into a system, the program will not run. A variety of policies encoded into LMF help enforce those portions of the paper contract for which LMF can collect information. This can have dramatic effects on profitability. Recently Digital added software licensing to one of its PC products and increased revenues by tens of millions of dollars the first year.

Several years ago Digital embarked on a second generation license system project. The result of this effort is the POLYCENTER License System (henceforth PLS). PLS was developed to extend the reach of automated software licensing activities beyond simple enforcement of program use. Some of the key requirements were to:

- Support distributors: licenses are issued mostly by distributors, not publishers.
- Distribute software licenses across the network.
- Support centralized administration from a single management console.
- Support license policies which change faster than the software does.
- Recognize that sometimes it is just as important to allow unlicensed software to run as it is to stop it from running.
- Decryption-based security—where there is, in effect a secret password hidden in the system—was considered unwieldy and unsafe. Find something better.

The resulting PLS system met these requirements in the following way:

- Two kinds of licenses are supplied: *license agreements* and *issuer agreements* The latter is a license to create agreements, thus providing a distribution chain.
- Extraordinary levels of customization are possible.
- Public key cryptography, using the RSA algorithm, is used extensively to block forgery attempts. PLS is immune to reverse-engineering attacks.
- Usage logs provide a mechanism for system administrators to understand actual usage and *overdraft* events. Software purchases can be based on hard data.
- PLS conforms to the "License Service Application Programming Interface" (LSAPI, 1993) and will conform to proposed OMG and X/Open standards.

2. DISTRIBUTED LICENSE MANAGEMENT

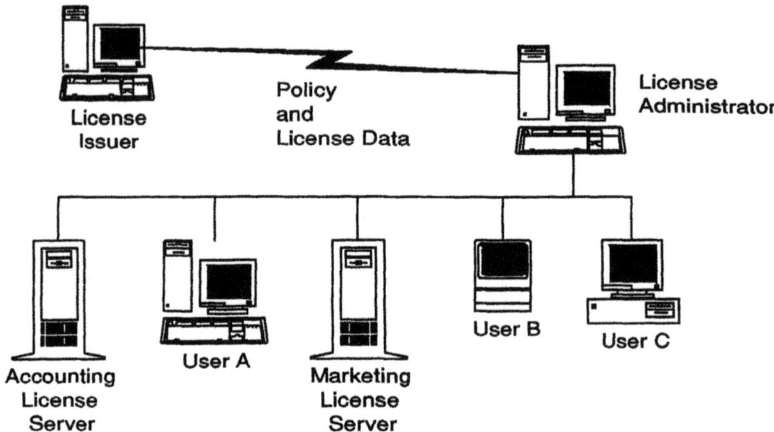

Figure 1 Network License Administration.

Figure 1 shows how a license is created and used. A license issuer creates licenses on their workstation using the PLS software. The licenses are held in the issuer's PLS database from which they may be extracted into one or more small ASCII files. These may then be copied to magnetic media or sent electronically (see Fenkel, 1993) to their destination. Along with supplying a GUI to create licenses interactively, PLS supplies an API through which the issuer's own order fulfillment system can create licenses automatically.

A license administrator may then load or "import" the licenses, can monitor usage of those licenses, and may set permissions on them. This includes entering user names or node names (or other features used for licensing) and "activating" the licenses. In the example in Figure 1 the licenses have been partitioned into those for the "accounting" and those for the "marketing" departments.

End user computers are configured with a list of license servers. These license servers will be searched for licenses when a licensed program is run. If the first server does not have the necessary resources then others on the list will be searched. This permits the license administrator to move licenses around the network or to re-configure the network without adversely affecting the end users. Further, multiple servers help improve performance and mitigates any inconvenience resulting from unavailable servers.

Each license comes with a number of **license units**. A license unit is an arbitrary measure of value. One unit might represent one user. Alternatively, a license might come with 100 units but require that 10 units be used for a PC and 100 units for a mainframe. This is commonly termed a "capacity license." License units resemble currency printed by the license issuer. The value of the units from one issuer is unlikely to match the value of units from another. PLS ensures that units of unequal value are not mixed or combined.

Some portion of the license units supplied by a license are temporarily held reserved while a program is running and then released when the program exits. Should an application make a request, and all the license units are tied up in prior requests, then a failure reply will result. The application developer has the choice of determining how their program will respond. The program may exit back to the operating system, it may run but with reduced features, or it may ignore the result of the request. Consumptive licenses are also supported. For consumptive licenses the units are permanently deducted from the license rather than being released for use by others. This is good for trial licenses and for controlled pay-as-you-go metering styles.

2.1 "License Service Application Programming Interface" (LSAPI)

The LSAPI defines a small number of calls which are to be added to an application program to be licensed (LSAPI, 1993). The benefit for the application programmer is that this job need be done only once. The specification was written by an assemblage of over 24 system and software vendors to serve as a standard license interface for application software. At the heart of the LSAPI are three calls:

Table 1 LSAPI Calls.

LSRequest()	Ask permission to run.
LSUpdate()	Program is still in use. Optionally, adjust license units used.
LSRelease()	Program exiting, release held resources.

The **LSRequest** call is called by the application to receive initial permission to run. LSRequest supplies the publisher name, the product name, and the version number. These must uniquely identify a program.

The program may also request a particular license system to use, a suggested number of license units to use, a comment for whatever logging system might be present, and a challenge value. The LSAPI allows the client-side code to request either a specific license system or to try all of the license systems available using the reserved name "LS_ANY". This allows multiple license systems to coexist on the same network without requiring any intervention on the part of the end user.

The **LSUpdate** call is used to allow the application to check in with the license system to make sure that the original request is still valid. For example, if the license system were re-started, and the original request information lost, then too many users might be able to run. The update guards against this. It also represents an opportunity for the application to claim more license units as circumstances warrant. The final reason for an update request is to inform the license system that the application is still running and that the automatic time-out check should be re-started.

The **LSRelease** call releases any resources held by PLS for others to use. It optionally supplies a number of license units to consume and a comment for whatever logging system which might be present. It also turns off the automatic time-out mechanism.

An optional **challenge mechanism** provides a way for the server and the application to believe that they are interacting with valid or authentic license system components. The challenge algorithm is based on the notion of shared secrets: a list of numbers both the client program and the license system both have but do not reveal to each other. The challenge mechanism can only be circumvented by a competent programmer who can examine the running code and determine how to extract the secrets either from the application or the server. These secrets could then be used for forge any number of perfectly acceptable licenses. For better security, PLS uses an additional technique—digital signatures using the RSA algorithm—described below in section 3.2 "Security Data".

2.2 The Extended LSAPI

The LSAPI does not supply a mechanism whereby the user's name, for example, can be collected and passed into the receiving license system. PLS supplies an optional **field list** mechanism to pass data in and out of a the normal LSAPI request, update or release calls. A field list is a list of paired name strings and value strings. A handful of routines are supplied which allow creating, adding to, modifying and destroying field lists. The LSAttach call tells the PLS software that a particular field list is to accompany a particular LSAPI request. Thus only the normal LSAPI interface is needed to communicate field lists with the license system.

An important use of the field list mechanism is to pass in the public key of the application's license data for verification purpose. This will be explored below in section 4. "Securing the Distribution Chain".

2.3 Usage Enforcement Policies

The first release of PLS on Windows NT comes with a small number of powerful enforcement policies. While more enforcement policies will be released for other platforms as well as for Windows NT over the next few months, these basic enforcement policies include:

- **Concurrent Use.** One unit held or deducted per request. Subscriber lists are provided for user and node names, but are ignored if not used.
- **Node Lock.** One unit held or deducted per request per node. A subscriber list is provided for node names. Change duration set by issuer.
- **Personal Use.** One unit held or deducted per request per user name. A subscriber list is provided for user names. Change duration set by issuer.

Other features included in the first release are support for the Dallas Semiconductor model DS1425 hardware security device, overdraft licenses, amendment licenses, capacity licenses, and embedded licenses. Any or all of these may be used with any of the three above enforcement policies, making the effective number of enforcement policies much greater.

An **overdraft** license is one which allows the creation of a 0 unit license to satisfy a request or an update. This allows the end user to always succeed in an attempt to use the software, and an overdraft event is added to the usage log. Allowing forgiving access to the license agreements is a distinct improvement over the normal pattern, which results in users quietly "borrowing" what they feel they need to do their jobs.

An **amendment** license agreement is a license agreement which seeks out another license agreement and replaces it Amendment license agreements let issuers correct mistakes to already deployed license agreements without having to issue new (and valuable) corrected license agreements.

Capacity license agreements compute the number of license units required for a request or an update by matching the requester's hardware in a list of possible hardware types. Thus, the a mainframe product can cost less if it runs on a PC.

3. LICENSE DATA

The PLS server database holds the objects which comprise both licenses and licenses to issue licenses. This section starts by describing how PLS objects may have their behavior customized (by adding rules and data fields) and secured against forgery (by receiving a digital signature.) Then the **license agreement** and **issuer agreement** objects are described.

3.1 Policy Data

First-generation license systems come equipped with a fixed repertoire of enforcement policies. They are typically coded in a third-generation language (C usually.) The license system code is typically split between the application program which makes the request and the license system which fulfills that request. The application may query the license system for information, interpret that information, and compute what to do next. The license system itself may be bound into the computer's operating system. For a description of a proposal to do this, see (Hauser, 1994).

The consequence of this is a very long lead time to make changes in the terms and conditions. Re-built software is usually only distributed to end users when there is a new release of that software. This bottleneck is felt keenly by distributors of electronic licenses, whether they are participants in the distribution chain or if they are trusted end users making their own electronic licenses. They simply don't have the source code and can never get it. If business, competitive, or legal circumstances require rapid changes in license policy they must wait for changes to be made by the application producer, the license system vendor, or both. This assumes they can persuade either party to change their code.

PLS removes most of the policy computations for controlling use and issuing from all program binaries. Neither application programs nor the various management clients contain policy code. Further, PLS removes much of the policy computations from the PLS executables as well. Rather than encode key portions of the terms and conditions in C, a more "fourth-generation" approach was taken:

- An interpreted Pascal-like rule language allows distributors to author their own terms and conditions. These rules are invoked at key moments in processing usage requests and may control the success or failure of any operation. They may change field values to affect subsequent computations, and they may issue log events. Because these rules are interpreted and not compiled they work on any PLS system without modification.
- An extensible database which lets issuers define their own data fields on agreements, set them to initial values, and manipulate them through rules.

The data which comprises the rules and data field information is loosely termed **policy data.** Like all data in PLS it may be moved to a customer's computer via ASCII text files. The consequence of this is that new kinds of authorizations may be developed and installed in days or even hours. Contrast this with the worst case of using a license system bundled with an operating system released every 18 months.

3.2 Security Data

PLS uses the RSA public key cryptosystem algorithm for forgery protection and authentication. RSA was invented in 1977 by Ron Rivest, Adi Shamir, and Leonard Adleman (Rivest, Shamir, Adleman 1977). The RSA algorithm is used to compute a digital signature for PLS objects which protects them against forgery. See also (Tardo and Alagappan, 1990) and (ACM 1992).

The digital signature is based on a matched pair of keys (very, very large prime numbers.) One is termed the **public key** and accompanies the data to be signed where ever it goes. The other is termed the **private key** and is closely guarded. The RSA digital signature technique exploits a crucial property of the keys: they are 2-way ciphers. This means that data encrypted with one key can only be decoded with the other key, and vice-versa. The algorithm is such that it is impossible (for all practical purposes) for an attacker to deduce one key from the other. Here's how a digital signature works.

First a digest of the data to be protected is computed. This is a large number (but much smaller than the message) which is a function of the value of the message. The digest has the property that a small change in the data causes a massive change in the number. Also, the message digest algorithm is designed so that it is impossible to create a reasonable-looking data stream which matches a given digest number. It is this digest value which is actually encrypted and decrypted. This reduces the amount of space consumed by the digital signature, as well as the amount of computation to encrypt and decrypt it.

Next the digest is encrypted with the private key and packaged with the data. We now have a **digital signature**. It gives us two valuable pieces of information about the object: who made the object (or changes to an object) and whether or not it was changed.

To see if someone changed the object we re-compute for ourselves the message digest. We then compare this new message digest with the old message digest. The old message digest is computed by decrypting the signature with the public key. If the old and new digests fail to match then the object has been changed. If they do match then only the holder of the private key could possibly have created the data and it is authentic.

3.3 License Agreement

License Agreement is the name given to the type of object that plays the role of an electronic license. A license agreement embodies all the constraints which determine whether or not a program may be run by a particular user. Once a user has actually been given permission to run under a license agreement then a **license** or **grant** is said to exist for a product. There can be no guarantee that a user has a license just because they have a license agreement. PLS must determine this.

All license agreements have at least these features:

- Start and end dates, plus an optional life span duration that begins after the license agreement is enabled for the first time.
- License units, plus a style field for allocative or consumptive accounting.
- A pair of license unit values to hold the number of units required for request or update calls respectively. A 0 value for either or both is permitted.
- An indication of the version(s) of the product.
- A selection weight value to be set by the license administrator. The license administrator might prefer to force the more selective licenses (such as the already assigned user license agreements, or individual product license agreements) to be used before the more general licenses (such as a concurrent use license agreement or a group license agreement.)
- A variety of subscriber lists for user or node names..
- Various title and comment fields allowing the license agreement to be self-documenting to a large degree.

To satisfy an LSRequest or LSUpdate call, PLS first locates all the license agreements which apply to that version. PLS then tests them to see if the user passes their usage constraints. These remaining candidate license agreements are then sorted in selection weight order. As many license agreements as are needed to get enough license units to satisfy the request are combined and their license units deducted. This information is held on the **license** data structure which keeps track of the user process which made the request, the product involved, and from which license agreements the units came.

3.4 Issuer Agreement

Software developers seldom sell their software to end users directly. Most sell through a network of distributors. (See Figure 2.) For most PC software, far more than two thirds of the retail cost of software is a consequence of distribution chain markups. Finding ways to make software distribution channels both shorter and more efficient translates directly into cost savings for software buyers.

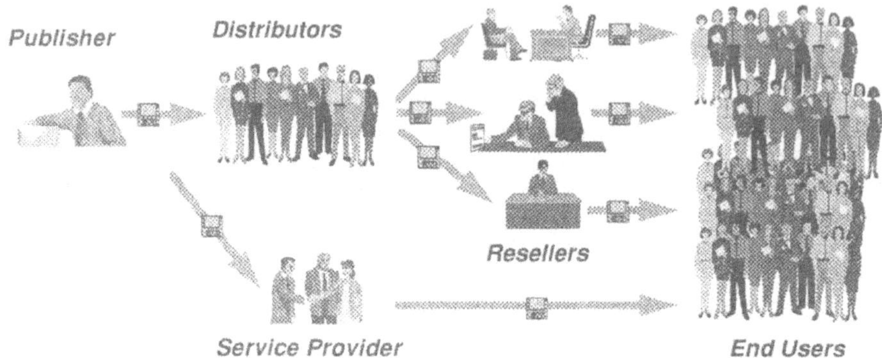

Figure 2 Software Distribution Chain.

The **issuer agreement** is a special sort of license agreement controlling the creation of either license agreements or subsequently issued issuer agreements. The issuer agreement specifies the kinds of usage constraints to follow, how many license units are available, and who may use those units. This other party may either make license agreements (and sell them to end users) or issuer agreements (and sell these to other issuers further down the distribution chain) . PLS keeps a record of precisely which issuer agreement was the source of units for any subsequent license or issuer agreement. This constitutes an audit trail from the original producer, through any and all distributors or end-user issuers, to the final electronic license.

An interesting new channel for software distribution is the situation where the end user makes their own license agreements as they need them. The end user is trusted by their distributor to pay them fairly for what they take. Experiments by Digital with this model actually resulted in higher software revenues for the Digital and increased satisfaction on the part of the buyer. This reflects the sentiment that large corporate buyers want to be treated as partners and not adversaries, and will usually work faithfully to this end.

The first issuer agreement is termed the **root** issuer agreement and may specify an unlimited supply of license units. Issuer agreements created from other issuer agreements are termed sub-issued issuer agreements.

4. SECURING THE DISTRIBUTION CHAIN

An object may have as an attribute the public key of another party from whom it is willing to accept as providing data to fulfill some role. Presuming no objects have been forged (which can be reliably proven) then the entire chain of objects which comprise the distribution chain can be authenticated using digital signatures.

- A producer can create a Product object providing the public key used to validate all license agreements and issuer agreements on the distribution chain.
- A producer can encode and protect "root" issuer agreements signed using the same key as the "product" object. They are, of course, entitled to issue agreements for their own products.
- An issuer agreement can specify the identity (public key) of another party on the next step of the distribution chain. This chain can be any length.
- An issuer agreement can specify the identity (public key) of another party entitled to make license agreements from their issuer agreement.

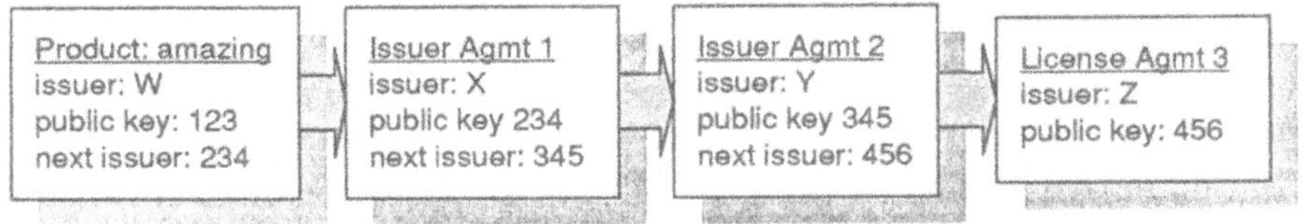

Figure 3 Securing the Distribution Chain.

We can now examine a sequence of digitally signed objects comprising the distribution chain for a single license agreement in Figure 3:

- Party W makes a Product indicating that the next issuer's public key for all "root" issuer agreements must be "234."
- Party X receives the Product and issues Issuer Agreement 1. Party X is the only party possessing the private key portion which corresponds to the public key "234".
- Party Y receives Issuer Agreement 1 and issues Issuer Agreement 2. Party Y is the only party possessing the private key portion which corresponds to the public key of "345".
- Party Z receives Issuer Agreement 2 and issues License Agreement 1. Party Z is the only party possessing the private key portion which corresponds to the public key of "456."

Thus the entire distribution chain is protected by unbreakable RSA signatures: product to issuer agreement to optional sub-issuer agreement(s) to license agreements. The public keys prove that each object was created under authority of the one before. The series begins with the "product" entry in the license system.

How then is the public key for the "product" object to be trusted? What if some hostile issuer creates their own otherwise identical product object, and issues their own internally consistent chain of issuer and license agreements? The key is to find a trusted source for that public key from some source outside the license system.

Privacy enhanced mail systems (such as Apple Computer's AOCE, or RSA Laboratory's TIPEM package) use an RSA certificate chain rooted at a TLCA ("top level certifying authority") to prove the authenticity of public keys. A RSA certificate is a signed document containing the names and public keys of parties whose public keys need to be verified. The TLCA provides a master RSA certificate containing the public keys needed to verify all the RSA certificates. This provides a single key value which, if unaltered, vouches for the integrity of all the other public keys. Such certificates are crucial for supporting public commerce using electronic means. For a cogent discussion of the value of digital signatures and the need for verifiable certificates nationwide see (Chokhani, 1994).

An alternate mechanism to verify public keys can eliminate the need for users to go to an outside party to have their public keys certified. The application program can pass the public key for the product object as part of the extended LSAPI LSRequest. The public key must match the public key on the product in the repository for the request to succeed. The rationale is that anyone wanting to substitute this key with one of their own could more easily patch around the LSRequest call instead. One way to look at a modification attack is to consider it equivalent to a viral invasion. Software licensing services do not protect against viruses which might damage license calls. Such services should be part of the underlying operating system instead.

5. CONCLUSION

Electronic licensing will, over the next few years, add more complexity to the task of managing resources in a network. Easy to use management tools will be necessary to ease the pain of introducing this emerging technology. At the same time licensing offers many benefits to the network administrator.

Software buyers can save money by developing a history of software use and from this knowing exactly how much software they need to buy. Perhaps every PC does not need a high-end word processor if a concurrent use license for only one-half of the PCs will get the job done. This should help cut down on the amount of "shelfware" sitting in people's bookcases. Buying quantities of licenses should garner discounts. This should be possible if the buyer is confident of the amount really needed. Finally, licensing should help reduce the expense of sudden, unexpected shifts of resources within a company. Rather than pay higher, retail prices in an emergency (such as a new hire, or a department started or shut down) the cheaper, bulk licenses can be rapidly deployed.

PLS raises the bar for all future license systems. Software publishers cannot be content with systems which do not provide a secure distribution channel into the marketplace. RSA digital signatures and PLS customization features supply the kind of enabling technology that electronic licensing needs. The advent of totally electronic distribution of software will cause major changes in the ways software is bought and sold. System managers need to start thinking now about what those changes will mean to them in the years to come.

6. REFERENCES

ACM (1992). Special section on encryption standards and the proposed digital signature standard. *Communications of the ACM*, **35**(7), pp. 36-54.

Chokhani, S. (1994) Toward a National Public Key Infrastructure. *IEEE Communications Magazine*, September, 1994,

Didio, L. (1993) Crackdown on Software Bootleggers Hits Home, *LAN Times*, **10**(22).

Frenkel, G. (1993), Software Distribution: The Electronic Way, *PC Magazine*, September 28, pp. 11-16.

Hauser, R.C. (1994) Does Licensing Require New Access Control Techniques? Comunications of the ACM, **37**(11), pp. 48-54.

LSAPI (1993), *License Service Application Programming Interface*, Version 1.1, December 14, 1993. Available by sending Internet mail to *lsapi@microsoft.com*, or by sending mail to Dave Berry, Microsoft Developer Relations, 1 Microsoft Way, 4/2, Redmond WA 98052-6339, or from CompuServe, or from the author.

Rivest, R.L., Shamir, A., and Adleman, L. (1978) A method for obtaining digital signatures and public-key cryptosystems. *Communications of the ACM*, **21**(2), 120--126.

Tardo, J. and Alagappan, K. (1990) *Sphyinx: Global Authentication using Public Key Certificates*. Digital Equipment Corporation.

7. BIOGRAPHY

Tim Collins received his BS in Zoology from the University of Massachusetts in Amherst in 1977. Over the last 17 years he worked as a scientific programmer, helped build CASE tools for structured analysis, tool integration, and configuration management, and was architect for PLS. Current interests include autonomous intelligent agents, OMG standards, object-oriented programming, and next-generation user interfaces.

11

A Resource Management System Based on the ODP Trader Concepts and X.500 *

A Warren Pratten, James W. Hong, Michael A. Bauer
J. Michael Bennett and Hanan Lutfiyya
Department of Computer Science
University of Western Ontario
{warren,jwkhong,bauer,mike,hanan}@csd.uwo.ca

Abstract

Distributed computing systems are composed of various types of hardware and software resources. Providing a reliable and efficient distributed computing environment largely depends on the effective management of these resources and the services that they provide. ISO has begun work on a proposed standard for Open Distributed Processing (ODP). The ODP framework includes a mechanism called the Trader which provides a framework for exchanging services in an open distributed computing environment. This paper presents a design of a resource information management system which employs and extends the ODP Trader concepts to facilitate the management and use of resources, information about resources and the services provided by the resources. We describe the architecture, information model, and user interface of the resource management system. We also describe a prototype implementation which uses the X.500 Directory Service as its repository for resource information and report on our experience with it to date.

[**Keywords:** distributed resource management system, ODP Trader, X.500 Directory Service, information repository, distributed computing resources]

1 Introduction

The trend of computing in the 90's is towards distributed computing. Computing systems, which are geographically dispersed, are interconnected through communications networks and cooperate to achieve intended tasks. Such computing systems are composed of a variety of hardware and software resources. Some of these resources are static, such as devices and others are dynamic, like servers which may come and go as demand dictates. As the size and heterogeneity of these computing systems increase, so too will the number and type of resources. Since users of these systems

*This research work is supported by the IBM Center for Advanced Studies and the Natural Sciences and Engineering Research Council of Canada.

depend on these resources, the effective and efficient use of these resources will be critical. An essential prerequisite of such use and sharing is the management of the various distributed resources, including keeping track of what resources are available, where they are located, what their properties are, what their statuses are, etc. Management of resources also includes maintaining similar information about the services that the resources provide. This is especially important in a distributed environment where systems come and go, servers are migrated or replicated, etc.

Resource management has always been a primary concern in centralized computing environments and operating systems. However, managing resources is much simpler in centralized systems than in distributed systems, since the resources are confined to a single location and, in general, the operating system has full control of them. In distributed computing systems, these resources are scattered throughout the distributed computing environment and no single entity has full control of these resources. Thus, the management of resources and there associated services in a distributed computing environment is inherently more difficult. As part of our work into services and tools to help manage a distributed computing environment [1, 7], we have looked into problems associated with the management of resources, information about the resources and their services.

ISO has begun work on a proposed standard for Open Distributed Processing (ODP) [10]. Included in this proposed standard is a mechanism called the Trader, which provides a framework for "trading" services in an open distributed computing environment [11]. "Trading" is an ODP term that is defined as the sharing of services between ODP entities (or objects). The ODP framework (including the Trader) has been continuously going through design and refinement stages and no implementation of the ODP environment currently exists. Although there has been some work on the refinement of the Trader [3, 9] and investigation of the potential uses in distributed computing environments [12, 15], more work is required for it to become an acceptable international standard.

Our interest in the ODP Trader is motivated by several goals. First, we required a resource information management facility as part of our work investigating distributed systems management services[8]. We feel that the ODP Trader can be a good candidate to support such a management facility to maintain and provide information about resources and their services. Second, we believe that a functional component such as the Trader will be an essential component in a distributed computing environment and thus requires further research in its role, use, and interoperability with other components. Ultimately, our aim is to communicate our experiences (both design and implementation) with the Trader to the developers and users of the ODP framework.

In this paper, we present a design of a resource information management system. The aim of the system is to help manage and facilitate use of resources, information about resources and their services in a distributed computing environment. Our motivation is to use such a system to support a variety of management activities, but it can also be used to support applications and users in general. The design of the information management system is based on the ODP Trader and, hence, we refer to it as the Trader-Based Resource Management System (TBRMS). We present an architecture of TBRMS and its major components. We also describe a prototype implementation of TBRMS, which uses the X.500 Directory Service [5, 6] as its repository for resource information.

The rest of the paper is organized as follows. In Section 2, we provide a brief overview of ODP Trader. Section 3 discusses general requirements for a resource management system in a distributed computing environment. Section 4 presents a design for the Trader-Based Resource Management System. Section 5 describes our implementation effort of a TBRMS prototype using the X.500 Directory Service. Our experience with it to date is provided in Section 6. We summarize our work and discuss possible future work in Section 7.

2 Overview of the ODP Trader

The ODP is a set of draft standard documents [10, 11] that are aimed at a variety of architectures, networks, and operating systems to provide an open distributed processing environment. The ODP Trader is one component of the ODP environment. The Trader's purpose is to provide a matchmaking facility between ODP objects.

The real advantage of the ODP Trader is in large distributed environments where objects need to be made aware of the services available. The Trader allows ODP objects to be configured into an ODP environment without prior knowledge of the services or service providers within that environment. The Trader allows this by acting as a third party that enables the dynamic service selection and the linking of clients and servers.

The ODP Trader document [11] discusses a large number of components that will comprise the Trader. Some of these deal specifically with trading policies, security requirements, accounting requirements, transfer requirements, quality of service, and federation. However, for our purposes we are looking at starting with a minimal set of functions that can later be extended to handle other concerns.

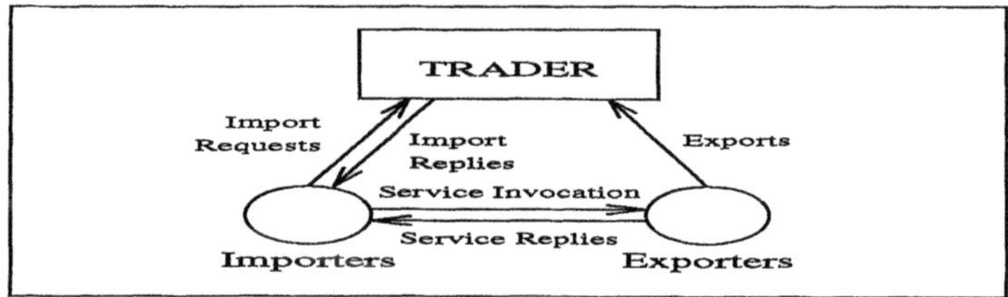

Figure 1: ODP Trader and its Clients

At the core of the ODP Trader system are the interactions among four different types of objects: traders, importers, exporters, and services (see Figure 1). An *exporter* is an ODP term for a service provider. It is an object with a service that it wishes to make available to other objects. Providing a service is accomplished by *exporting* the service to the Trader. An exporter is also able to later *withdraw* (*e.g.*, make unavailable) the service. In ODP terminology, a requester of services is known as an *importer*. The expectation is that importers in the ODP environment can operate without any prior knowledge of where the required services are or which object provides them. To find these services the importer must make a service request to the Trader. The Trader then returns to the importer the details of the services matching the service request if any exist. A *service* is a function provided by an exporter for use by other ODP objects. A service may be one of the following types: an atomic operation (*e.g.*, write), a sequence of operations (*e.g.*, open, write, close), or a set of operations (*e.g.*, read, write, open, close).

A service is exported in the form of a *service offer* which describes the service being made available. An importer discovers services by sending import requests to the Trader. The main component of an import request is the *service request*, which is a set of assertions that describes the desired service. The import request also provides information describing the method and scope of the search to be used by the Trader.

It is the purpose of the Trader to match the *service requests* of the importers with the *service offers* of the exporters. This is done by matching the assertions in the *service request* with the assertions that compose the *service properties* of the offered services. The Trader sends to an importer the details of the services (including location) that match its service requirements.

3 TBRMS Requirements

We have based our TBRMS design upon three primary requirements: providing a functional architecture for the TBRMS, providing a simple set of service interfaces, and employing a repository for storing resource information.

3.1 Functional Architecture

A certain level of functionality will be needed to exist within the TBRMS to adequately respond to client requests. The TBRMS will need components to communicate with the clients, parse their requests, and provide a means through which resource information may be stored, retrieved, updated, and deleted. The TBRMS should also offer some means of assuring the status of resources for which it is responsible and a method of controlling client access to the resource information.

The TBRMS architecture should be clean, extensible, and modularized. It should allow the delegation of tasks to various subcomponents in such a way that the requests of the clients are dealt with in a logical, coordinated, and timely fashion.

3.2 Service Interfaces

The TBRMS service interfaces should provide simple access to the TBRMS for the TBRMS clients. Three types of interactions will be involved between the clients (or users) and the TBRMS and thus appropriate interfaces should be provided to support them.

General Client Interactions: All clients will need a method of establishing an association with the TBRMS and later breaking that association when the TBRMS service is no longer required.

Exporter Interactions: The TBRMS must provide interfaces through which an exporter may add, change, or remove any description of resources it wishes to make available to other TBRMS clients.

Importer Interactions: The TBRMS must provide a method which allows an importer to do resource discovery based on its resource requirements and the resource descriptions maintained by the TBRMS.

3.3 Resource Information Repository

The very nature of the TBRMS requires that a resource information repository that stores resource information form a crucial element of our TBRMS design. Some of the necessary characteristics of the resource information repository are: extensible data modeling capabilities, general naming scheme, distributed service, heterogeneous data sources, good performance, and security [7].

4 Design of TBRMS

In this section, we present a design for the Trader-Based Resource Management System. We describe the architecture of TBRMS as well as its service interfaces.

4.1 TBRMS Architecture

Our TBRMS architecture defines the major components that interact to function as the TBRMS. These components are TBRMS Coordinator, Request Parser, Access Control, Inventory Control, Matcher, Resource Information Maintainer, and Federator. Figure 2 illustrates the TBRMS architecture.

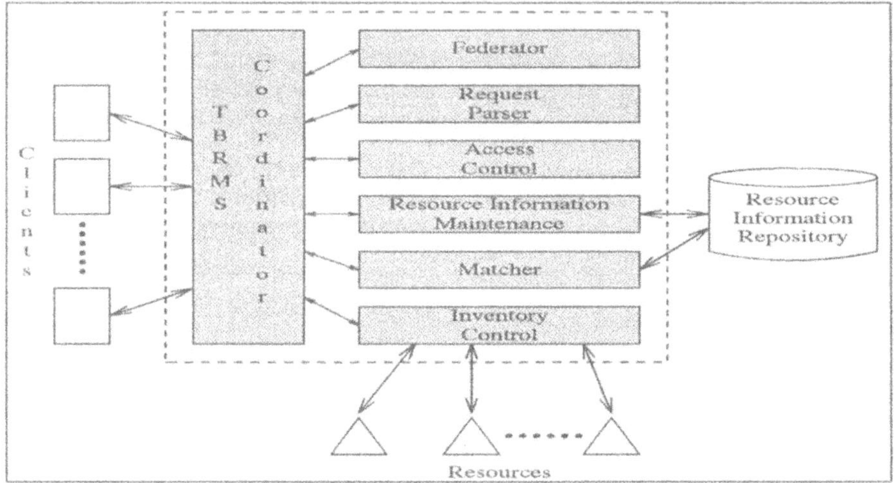

Figure 2: TBRMS Architecture

TBRMS Coordinator: This component coordinates activities within the TBRMS and acts as a front end to the TBRMS. As client requests are received by the TBRMS, the Coordinator acts upon them by interacting with the other TBRMS components. It coordinates the activities within the TBRMS to produce timely responses to client requests.

Request Parser: This component takes the client requests and translates them into an internal format which will later be translated into requests of the type understood by the Resource Information Repository.

Access Control: This component is used to determine the extent to which clients may make use of the TBRMS. For example, an importer must be registered with the TBRMS before it may request resources, and a client must be the owner (exporter) of a resource to modify or withdraw it.

Inventory Control: This component is used to interact with resources to enquire about their status, including determining whether a resource is still up and running.

Resource Information Maintenance: This component exists to provide an interface to the Resource Information Repository. It provides the functionality that allows the TBRMS to

- add new information on resources
- delete information on resources
- modify information on resources
- list available resources
- search for specific resources

Matcher: This component queries the Resource Information Repository for resources. The queries are generated by the Request Parser component based on the resource requests of a client. The Matcher returns all resources matching the original request.

Federator: To be effective in a distributed environment the TBRMS should not be a centralized service but should instead be distributed in some manner. The Federator component provides the means by which two TBRMSs could communicate to share the resources each manages with the other. The Federator component in part determines which resources may be shared with another TBRMS. The ODP Trader document [11] describes the federation (or interworking) of Traders which other work has examined [2, 13, 17].

4.2 TBRMS Service Interfaces

The service interfaces of the TBRMS system represent points of interactions between the TBRMS and its clients. These interfaces have been grouped by function, namely client, importer, and exporter. The details of the interface specifications can be found in [14].

4.2.1 Client

Before any client (importer or exporter) may make use of the TBRMS we require that the client first register with the TBRMS. Accordingly, when a client is finished making use of the TBRMS, we require that the client deregister itself. Although strictly speaking this set of interfaces is not necessary for a working TBRMS, we felt that there should exist some method by which the TBRMS could keep track of its clients. Forcing clients to register before using the TBRMS allows the TBRMS to have knowledge of its clients. This will become more important with security extensions to the TBRMS.

register: The operation called *register* allows a client to register itself with a TBRMS. Since a client may use the TBRMS to both import and export resources there is no need for the client to state what use it will make of the TBRMS.

deregister: The operation called *deregister* allows a client to deregister itself from a TBRMS.

4.2.2 Importer

Importers are TBRMS clients which have resource requirements that need to be fulfilled. The set of importer operations provide a method that allows a client to do some resource discovery and eventually provide the information necessary to reference a particular resource.

search: The operation called *search* can be used by an importer to discover the resources matching a set of resource requirements. The matching criteria is an expression using attribute-based matching to represent the resource requirements of the importer. The TBRMS returns to the client references for those resources matching its stated requirements.

list: The operation called *list* is used by an importer to retrieve the details of a particular resource. A client may use the list operation on a variety of resources to select the most appropriate resource to fulfill its resource needs. An importer client uses the previously acquired resource identifier for the resource of interest.

select: The operation called *select* is used by an importer client to retrieve the interface to a resource. The client must supply a previously obtained resource identifier.

4.2.3 Exporter

Exporters are TBRMS clients which have resources they are willing to make available to other clients in the distributed system. Although the exporter allows other processes to use its resources, the exporter maintains control of the resource and may change or withdraw the resource at its convenience.

export: The operation called *export* is used by an exporter wishing to make a resource available through the TBRMS. The exporting client supplies to the TBRMS the resource properties for a resource. The resource properties are expressed as a list of assertions about the resource.

withdraw: The operation called *withdraw* is used by an exporter which, after previously exporting a resource, now wishes to remove the reference of the resource from the TBRMS. Note that withdrawing a resource is not necessarily equivalent to deleting or killing the resource. It simply removes the resource from the TBRMS, restricting any new usage by other clients.

update: The operation called *update* is used by an exporter which, after previously exporting a resource, now wishes to update some or all values associated with that resource; for example an exporter may want to change the values associated with the attributes queueLength and costPerPage for an exported printer resource. Strictly speaking this operation could be accomplished by the sequence of withdrawing the resource and then exporting the resource with the updated information, but one advantage of allowing updates is that the resource retains its resource identifier.

4.2.4 Status Responses

It is a basic assumption of the TBRMS system that the clients may rely on the TBRMS being in good working order. This is true because clients might depend on the TBRMS to provide essential services. Therefore it is important that the clients receive from the TBRMS messages indicating the status of their operations on the TBRMS interfaces. Examples of status responses would be Ok, clientUnknown and resourceNotFound.

5 TBRMS Prototype Implementation

A prototype Trader-Based Resource Management System has been developed to demonstrate that the TBRMS provides a viable means where by resources may be managed in a distributed computing environment. In this section, we present the details behind the TBRMS prototype implementation.

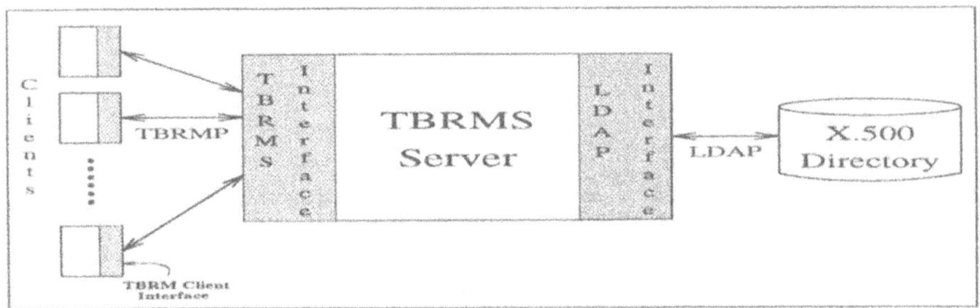

Figure 3: Prototype Implementation of TBRMS Architecture Using X.500

Figure 3 illustrates the architecture of the TBRMS prototype which is based on the TBRMS architecture described in Section 2. Work with the prototype has taken place within the UWOCSD Systems Lab. This lab is comprised of a network of heterogeneous computers consisting of Sun Sparc, Sun 3, IBM RS6000 and MIPS workstations as well as a 10-processor Sequent Symmetry.

The prototype TBRMS server runs on one of the Sun Sparc workstations. Clients running on all system lab machines have successfully interacted with the prototype TBRMS server. The client-TBRMS communication is provided by the Trader-Based Resource Management Protocol [14] which was implemented using the Open Network Consortium's (ONC) Remote Procedure Call mechanism [4]. The TBRMS Service Interfaces described in Section 4.2 are mapped onto the operations offered by the TBRMS.

The prototype relies on the X.500 Directory Service [5, 6] as its resource information repository. The X.500 Directory Service possesses some essential properties that satisfy the requirements of our resource information repository, in particular its powerful *information modelling capability*, *global naming scheme*, *distributed service*, and *simple access interface* [7, 18]. The X.500 Directory contains entries (or objects) which describe information about entities (*e.g.*, resources). An object-oriented approach is used for modelling directory information objects and allows the users to define any information object class by either extending existing classes or defining entirely new classes.

The prototype TBRMS uses the ISODE Quipu 8.0 implementation of X.500 [16] and a directory service agent (DSA) running on a second Sun Sparc workstation within the lab. The TBRMS accesses the DSA through the light-weight directory access protocol (LDAP) [19].

At present, the prototype TBRMS only does a weak form of access control. Each client and resource is assigned a unique identifier which is used in any subsequent interaction with the TBRMS. Authentication is performed using this identifier to ensure a client has the ability to perform its requested actions. For example a check is made before a client is allowed to *update* or *withdraw* a resource. Currently all authentication is carried out by performing search and read operations on the X.500 directory information. That is, when a client makes a request the TBRMS uses the identifier provided by the client to search the directory. If an entry with a matching identifier is found the client is assumed to be valid. Similarly if the request involves either withdrawing or updating a resource then the operation is allowed only if the directory entry contains both the client's and resource's identifiers.

The actual *resource types* were implemented using X.500's object classes [14]. This provides a good method of ensuring type checking on resource definitions. When a resource is exported one of its attributes must be a resourceType. The value associated with the resource type is used as part of the X.500 object class.

6 Experience

To demonstrate the functionality of the TBRMS we show how a sample client-server application has been modified to use the TBRMS. The application is a locally developed password maintenance system. The password maintenance system consists of one password server (or daemon) program (passwdd) and multiple password client programs (passwd) running throughout the distributed computing environment in the Department of Computer Science. This password maintenance system provides the ability for users to change their passwords from remote machines. Typically one machine acts as the server for a domain and access to the server is limited to password clients within that domain. Whenever a user within the domain wishes to change his/her password, they use the local password client program which connects with the password daemon and changes the user's password on their behalf. Figure 4 illustrates this password maintenance system.

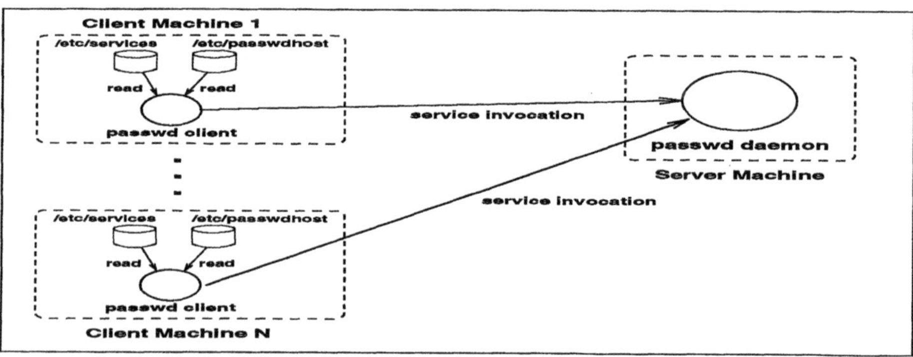

Figure 4: The Password Maintenance System

In order for the client to contact the password daemon, the client must have some way of locating the daemon. The original version of the password client reads two different files to locate the daemon. The first file (`/etc/passwdhost`) tells the client which machine is running the daemon. The other file (`/etc/services`) tells the client which port on that machine the daemon is listening to. Both these files remain relatively static, meaning that if the daemon is moved to a new machine the `/etc/passwdhost` and `/etc/services` files on all client machines would need to be updated by hand.

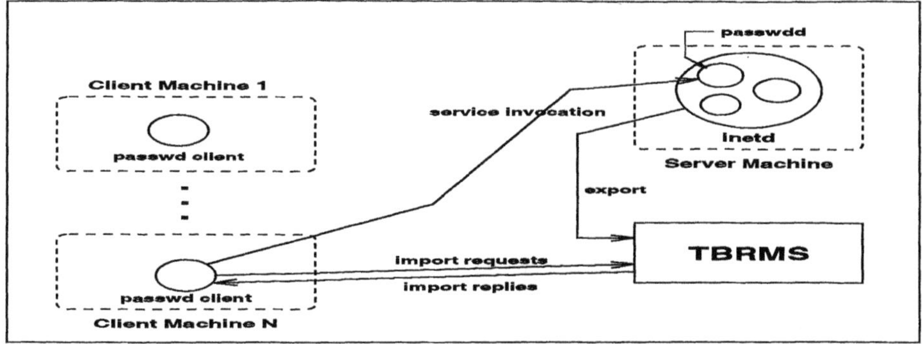

Figure 5: The Password Maintenance System using TBRMS

Using the TBRMS simplifies locating the password maintenance service in the network. Figure 5 illustrates the new password maintenance system using TBRMS. For the purposes of our previous discussion we could view the password daemon program as being an independent server. In actual fact access to the password daemon is controlled by an Internet services daemon called inetd which is responsible for invoking the password daemon when a client contacts the appropriate service port.

Another way of viewing inetd is as a service provider and passwdd is one of the services it offers. The resource type tbrmInetdService was defined to describe the services offered by inetd. Since the services offered by inetd are a resource sub-type of the more general tbrmGeneralResource type we can specify the tbrmGeneralResource in our definition for tbrmInetdService and then only specify the new attributes that define the new resource type.

Using the TBRMS with the password maintenance program meant making modifications to the resource provider (in this case inetd) and the resource requester (passwd client). The inetd had to be modified to *export* the resources it offered, which in this case meant exporting passwd. Since many programs rely on inetd it was potentially risky to modify it. Instead a program inetd.init was developed which essentially performed the *register* and exports that inetd would if it had been modified. When inetd.init is killed it *withdraws* the inetd services and *deregisters* before dying as inetd would.

The inetd.init program exports the passwdd program by providing the passwdd's properties to the TBRMS. One of the essential properties inetd.init provides is the resourceInterface for the password daemon. The resourceInterface includes information about where passwdd is running, which port it is associated with, and what protocol it is expecting to use with the password client program.

The password client program had to be modified to use the TBRMS for locating the password server program. To find a suitable password server the client provides the TBRMS with its resource requirements. In the case of the password client it was important to find a passwdd program that served the right domain and used the same protocol. The password clients resource requirements were: resourceName = passwdd and protocol = uwocsdTwistedEudora and serviceDomain = syslab.csd.uwo.ca. When a resource matching was found the password client was able to use the resourceInterface to successfully interact with the password server program.

The success of the TBRMS prototype helps show that the TBRMS design is a feasible mechanism for managing resources in a general heterogeneous computing environment.

7 Concluding Remarks

This paper was motivated by the need and importance of managing resources in distributed computing systems. We examined the requirements for resource management, particularly using the Trader concepts proposed by the ODP standards. We presented a design of a Trader-Based Resource Management System, consisting of an architecture and resource management service interfaces.

Our prototype implementation of a Trader-Based Resource Management System using the X.500 Directory as its information repository has been completed and we have just started using it for managing a variety of distributed system resources. The performance measurement on the current prototype show that the time between a client's request and the TBRMS's reply is on the order of a couple of seconds. While work can be done to optimize this time, it does show that using the TBRMS does not add a significant overhead to the client's performance.

We are also in the process of instrumenting the client resource management service interface onto distributed applications and services that may utilize the TBRMS. As we reported earlier in this paper, the X.500 Directory possesses many characteristics that are quite desirable for supporting the operation of the resource management system as well as for the modelling of the resources that are to be managed by it.

For future work, it has been suggested that X.500 might serve a useful purpose in facilitating the federation of Traders [13]. We plan to examine federating our TBRMSs using X.500. This is natural since we are already using X.500 in our TBRMS implementation. A main use of the TBRMS is being planned in the area of distributed systems management. We plan to integrated the TBRMS into the distributed systems management testbed being currently developed here at the University of Western Ontario [8].

Our hope is that our current and future work with the ODP Trader can be beneficial to the refinement of the Trader standard itself as well as to the potential users of the Trader in various computing environments.

References

[1] M. Bauer, P. Finnigan, J. Hong, J. Rolia, T. Teorey, and G. Winters. Reference Architecture for Distributed Systems Management. *IBM Systems Journal*, 33(3):426–444, September 1994.

[2] M. Bearman and K. Raymond. Federating Traders: An ODP Adventure. *Proc. of the IFIP Workshop on Open Distributed Processing*, Berlin, Germany, 1991.

[3] M. Bearman and K. Raymond. Contexts, Views and Rules: An Integrated Approach to Trader Contexts. *Proc. of the International Conference on Open Distributed Processing*, pages 153–163, Berlin, Germany, September 1993.

[4] John Bloomer. *Power Programming with RPC.* O'Reilly & Associates, Inc,, Sebastopol, CA, 1992.

[5] CCITT. *The Directory - Overview of Concepts, Models and Services, CCITT X.500 Series Recommendations.* CCITT, December 1988.

[6] CCITT. *The Directory - Overview of Concepts, Models and Services, Draft CCITT X.500 Series Recommendations.* CCITT, December 1991.

[7] J. W. Hong, M. A. Bauer, and J. M. Bennett. Integration of the Directory Service in the Network Management Framework. *Proc. of the Third International Symposium on Integrated Network Management*, pages 149–160, San Francisco CA, April 1993.

[8] J. W. Hong, M. A. Bauer, and H. L. Lutfiyya. Design of the Distributed Systems Management Testbed. *Technical Report, in preparation*, Dept. of Computer Science, University of Western Ontario, 1994.

[9] J. Indulska, K. Raymond, and M. Bearman. A Type Management System for an ODP Trader. *Proc. of the International Conference on Open Distributed Processing*, pages 141–152, Berlin, Germany, September 1993.

[10] ITU-TS. *Basic Reference Model of Open Distributed Processing Part 1: Overview and Guide to the Use of the Reference Model.* ITU-TS Rec X.901, ISO/IEC 10746-1, July 1992.

[11] ITU-TS. *Draft ODP Trading Function.* ITU-TS SG7.Q16 Draft Recommendation, July 1994.

[12] C. Popien and B. Hager. The ODP Trader Functionality Applied to the Integrated Road Transport Environment. *Proc. of the Globecom'93*, pages 1202–1206, Houston, TX, November 1993.

[13] C. Popien and B. Meyer. Federating ODP Traders: An X.500 Approach. *Proc. of the ICC'93*, Geneva, Switzerland, May 1993.

[14] A W. Pratten. Resource Management in a Distributed Computing Environment. *MSc. Thesis*, Dept. of Computer Science, University of Western Ontario, London, Ontario, Canada, September 1994.

[15] P. Putter and J. D. Roos. Relationships: Implementing Transparency in Distributed Management Systems. *Proc. of the IEEE First International Workshop on Systems Management*, pages 118–124, Los Angeles, CA, April 1993.

[16] C. J. Robbins and S. E. Kille. *The ISO Development Environment: User's Manual Version 8.0.* X-Tel Services Ltd., June 1992.

[17] A. Vogel, M. Bearman, and A. Beitz. Enabling Interworking of Traders. *Proc. of the IFIP International Conference on Open Distributed Processing*, Brisbane, Australia, February 1995.

[18] C. Weider, R. Wright, and E. Feinler. *A Survey of Advanced Usages of X.500.* Internet Draft, IETF DISI Working Group, October 1992.

[19] W. Yeong, T. Howes, and S. Hardcastle-Kille. *Lightweight Directory Access Protocol.* Internet Engineering Task Force OSI-DS Working Document 26, August 1992.

About the Authors

A Warren Pratten received his BA in Geography, his BSc in Computer Science and his MSc in Computer Science from the University of Western Ontario in 1989, 1992 and 1994 respectively. He is currently working as a systems administrator in the department. His research interests include distributed computing and systems management. He can be reached via electronic mail at `warren@csd.uwo.ca`.

James W. Hong is a research associate and adjunct professor in the Department of Computer Science at the University of Western Ontario. He received his BSc and MSc from the University of Western Ontario in 1983 and 1985 respectively and his doctorate from the University of Waterloo in 1991. He is a member of the ACM and IEEE. His research interests include distributed computing, software engineering, systems and network management. He can be reached via electronic mail at `jwkhong@csd.uwo.ca`.

Michael A. Bauer is Chairman of the Department of Computer Science at the University of Western Ontario. He received his doctorate from the University of Toronto in 1978. He has been active in the Canadian and International groups working on the X.500 Standard. He is a member of the ACM and IEEE and is a member of the ACM Special Interest Group Board. His research interests include distributed computing, software engineering and computer system performance. He can be reached via electronic mail at `bauer@csd.uwo.ca`.

J. Michael Bennett is an associate professor in the Department of Computer Science at the University of Western Ontario. He received his doctorate from the University of Western Ontario in 1972. He has been active in the Canadian and International groups working on the X.500 standard. He is a member of the ACM and IEEE. His research interests include distributed computing, network management, computer system performance, communications and computer architecture.

Hanan L. Lutfiyya is an assistant professor of Computer Science at the University of Western Ontario. She received her B.S. in computer science from Yarmouk University, Irbid, Jordan in 1985, her M.S. from the University of Iowa in 1987, and her doctorate from the University of Missouri-Rolla in 1992. She is a member of the ACM and IEEE. Her research interests include distributed computing, formal methods in software engineering and fault tolerance. She can be reached via electronic mail at `hanan@csd.uwo.ca`.

SECTION FIVE

Service and Security Management

12

Standards for Integrated Services and Networks

J P Chester & K R Dickerson

RACE Consensus Management Office
165 boulevard du SOUVERAIN
B-1160 Brussels

Tel: +32 2 674 85 22
Fax: +32 2 674 85 38
Email: joe@ric.iihe.ac.be

Abstract

This paper discusses the technical requirements and the standards that are required before global services can be implemented across multiple network operator and service provider domains in Europe. Two advanced service scenarios are described to illustrate the sort of global services that are required, and the problems of implementing these using current technology are discussed. The most important standards bodies for solutions to these problems are then identified.

Keywords

IN, Multimedia, Network Architecture, Personal Mobility, Private and Public Networks, Services, Signalling, Standards, Terminal Mobility, TMN, VPN.

1 Introduction

The telecommunications market in Europe is seeing a proliferation of service providers and network operators. In order to provide common services to customers across the range of network operator and service provider domains it is necessary to provide standards that will allow interconnection and interoperability between networks and services across Europe.

This paper describes the problems that currently prevent the provision of global services, and introduces the standards that are required to allow interconnection of networks and interoperability between services.

2 Scenarios for Advanced Service Provision

Scenarios are a useful way to present advanced applications and services in order to estimate the likely demand. Otherwise it may be difficult to visualise the capabilities of applications and services which it is impossible to personally experience. Scenarios are also a useful way to stimulate discussion on a wide range of issues associated with the application or service, ranging from usage and usability issues through to the implications for service creation, management and services platforms.

The scenarios described here cover PSCS (Personal Services Communications Space) and Hypermedia. They represent ends of a spectrum of service opportunities that provide on the one hand capabilities that will allow users to communicate with each other independent of their physical location, and on the other the ability to easily access a wide range of information sources at a wide range of bit rates. The combination of these two scenarios with a user-friendly interface would provide the holy grail where instant access can be obtained to people or information anywhere in the world.

2.1 PSCS

This scenario was developed by the MOBILISE project [3] and is based on a development of the UPT concept as defined by ETSI [4]. It is based around the concept of personal mobility. - the user can move between geographical locations and can still be contacted on a pre-specified number.

Key concepts in this scenario are personal numbering, number portability, and personalisation and customisation of services. Personal communication offers the ability to communicate in different roles and to organise communication according to the user's preferences. Users can play different roles and set up different routings for calls depending on the caller, the time of day and other requirements. The link with mobile services is extremely important because customers will want to access these services via mobile as well as fixed terminals.

2.2 Hypermedia

The second scenario is based on the concept of a global village, sometimes referred to as cyberspace, a space full of information objects. Multimedia is already bringing the ability to see, hear (and eventually smell) your colleagues remotely, as well as to view and point at shared objects on screen. This concept is extended through the use of explicit links between multimedia objects to become hypermedia. This provides the ability to sit at a terminal and set up instant video connections to colleagues and experts and to access all the world's knowledge in a variety of media.

The key to this scenario is high quality video, voice and data communications with fast response times. It requires high bit rates and generally makes greater use of multimedia and multipoint services than the PSCS scenario.

3 Barriers to Implementation

Today there are a number of barriers to providing these sorts of services, especially where they must be provided globally across a range of network operator and service provider domains. These problems were investigated by the ETSI DASH Task Group which reported in May 1994 [1]. Problems identified include:

- The difficulty of interworking between public and private networks and services. The provision of services such as VPN depend on interworking capabilities between public and private domains. Regulatory developments are also likely to lead to the breaking down of the traditional barriers between public and private domains, and will heighten the need for convergence between the two sectors.

- The difficulty of interworking between fixed and mobile services. Different architectures are currently used for fixed and mobile areas. This may prevent similar services being offered across the two environments.

- The difficulty of creating and managing distributed services in an IN structured environment. There may also be a problem of interoperability between similar services created using different service creation environments.

The problems currently prevent supplementary services available in a private domain (such as a PBX) from being extended transparently over a public network or to a mobile terminal. This will be even more the case in the future with a greater range of IN-supported services.

4 Standards Required for Implementation

The layered model in Figure 1 can be used to illustrate the type of standards issues. There are important issues at all layers in the model, from the application layer down through the service infrastructure, transport and network access layers, down to the physical transmission layer. All of these must be correct and interwork satisfactorily in different environments in order to offer effective services to customers.

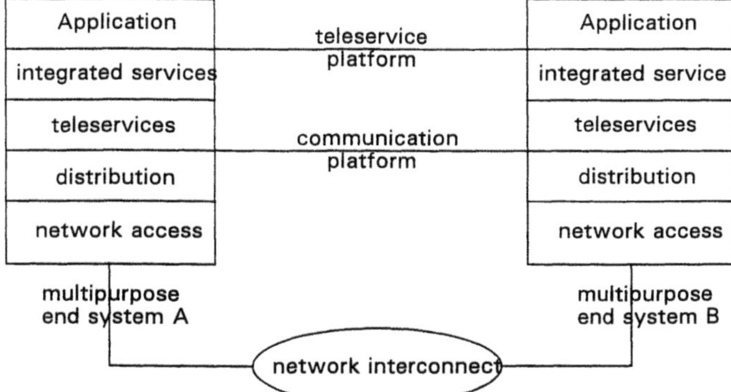

Figure 1 Layered model of telecommunications service provision.

Functions required for service implementation can be provided in either the terminal or the network and must be complementary.

The priority work areas that will need standards to be developed to overcome these barriers are described in the remainder of this section.

4.1 Service and network capability description

Service and network architecture description methods are required to ensure the consistent description of services specified for different platforms, and to allow services to be described in a network independent fashion. This requires:

- The revision of the I.130 3-stage method [7] to provide network-independent service descriptions at Stage 1, and sufficiency flexibility to cover services requiring broadband, mobile and multimedia capabilities.
- A movement away from rigid service descriptions, as provided by CCITT for ISDN services, and towards the reduced level of specification associated with the IN approach. This will allow a larger range of more flexible services to be provided to customers, based on agreed sets of common network capabilities.
- The use of a common state model as a basis for all service descriptions. This will provide a greater degree of interoperability between services.

Network capabilities to support IN services are being defined in three phases, known as capability sets 1-3. The current schedule for these is as follows: CS1 (1994), CS2 (1996) and CS3 (>1996). The following issues must be addressed:

- It is still to be decided which IN service features will be included in CS2 and CS3. It is important that the necessary capabilities are provided to allow the scenarios described in Section 2 to be implemented.
- How will these services be created and deployed effectively?

- The evolution of IN towards the distributed platform approach based on ODP that will be required for CS3.

The last two issues are addressed further in the following section.

4.2 The relationship between TMN and IN

It is essential that the services required for the implementation of the scenarios described in Section 2 can be created and managed effectively. Key issues relate to the interactions between the Service Creation Environment, the IN Platform, and the TMN. The interactions defined between these entities must allow for market requirements, including both bilateral and multilateral relationships between the various actors. The DASH report [1] represents a useful starting point for defining the requirements of these interactions. The DASH model for the relationship between the TMN and IN is shown in Figure 2. This is oriented towards IN CS1 type services, and highlights the interfaces to the basic call processing state machine.

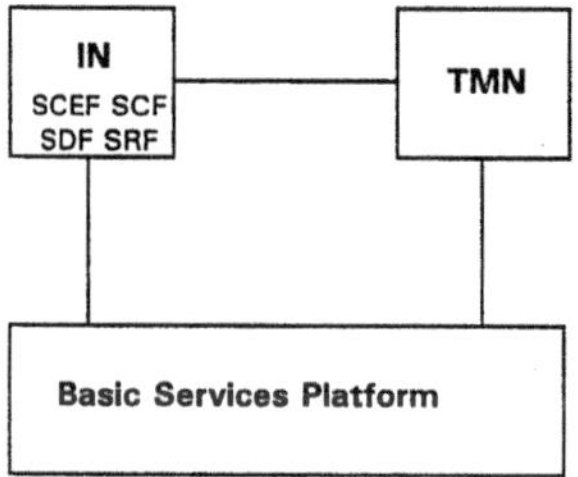

Figure 2 DASH model for the integration of TMN and IN.

The requirements for IN CS2 type services will involve a high degree of distribution of management and control, as well as enhancement of the basic call model to include more advanced services such as network and non-call related services. A refinement of the DASH model, known as the SMP Model, is shown in Figure 3. The main advantage of this model is that it highlights a more important set of interactions for further study, in the context of IN CS2 and CS3 and to meet the objectives of TINA-C. These interactions focus more on the information systems viewpoint, and show clearly the need for detailed study of a number of major issues in the telecommunications services environment as a whole.

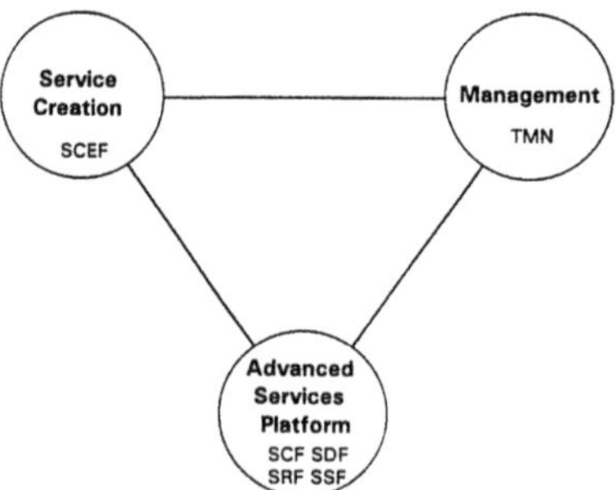

Figure 3 SMP Model for the integration of TMN and IN.

A more detailed analysis of the use of the SMP model to derive requirements for R&D and for standardisation activity is given in [5,6]. Some of the key results of the use of the model are presented below.

4.2.1 Interaction between Service Creation and Management

Service creation will take place in a non-real time Service Creation Environment, off-line from the services execution platform. As a consequence, there is a need to establish processes and procedures for:

- The effective interaction between the Management entity of the telecommunications services environment and the Service Creation entity. This interaction will govern processes and procedures, as well as deal with the two-way flow of information, and the transfer of service logic, service data, results and performance information.

- The deployment (i.e. the transfer of service logic) via the TMN to the Execution Platform. It is a principle of quality management of the telecommunications services environment that upgrading of the Execution Platform is under Management control. The service data always involves changes to TMN data.

It is considered important for some types of services that subscribers have a limited ability to customise certain features of the services in accordance with their preferences. Separation of service logic and service data is an important principle. Such service customisation must be under management control, and the functionality of the Management entity will need to make specific provision for this.

4.2.2 The use of Transaction Processing Technology

The interactions between the three entities in Figure 3 involve high levels of information transfer and processing. Research is needed on the data consistency aspects of distributed systems control, for current services as well as for distributed IN and TMN type services. The use of Transaction Processing technology seems to be a promising means to achieve this.

4.2.3 Distributed Control of new classes of services

The introduction of more advanced services will require distribution of the control and management functions. In addition, the requirement with each change in services for rapid upgrading of the SSP functionality, and of the capabilities of the SCP - SSP and SCP - SCP protocols needs further study. The use of DCE type architectures may provide solutions to some of these issues.

4.2.4 Building Blocks

In the Service Creation Environment, to enable rapid and flexible service provision there is a need to construct new services from building blocks. These building blocks should not be standardised, but the interfaces to the building blocks should be clearly specified by manufacturers (e.g. with template descriptions) in order to allow service providers to put together the building blocks in the correct way. This applies to both IN services and TMN applications.

4.2.5 *Emerging IT technologies for IN services and TMN applications*

De facto standards, such as X/Open, OMG and OSF, and emerging IT technologies such as DCE, OLTP, CORBA and Motif, are currently a major influence in the management area. Additional effort is needed on the integration of these into the environment for telecommunications services.

4.2.6 *Implications of inter-domain interactions via X*

For inter-domain interaction between telecommunications services environments under the control of different administrations (however organised) the only interaction will take place via the X reference point. The requirements of the telecommunications services environment on this reference point need to be considered, and must include:

- distribution of service logic between domains, or
- distribution of service descriptions to allow interoperable services to be created in different SCEs,
- transfer of service customisation configuration data, and control of the resulting changes.

4.2.7 *The need for Multi-vendor services environments*

In order to ensure that services can be deployed across multi-vendor management and service execution platforms, these platforms need agreed APIs. The output of the service creation activity should be standardised in accordance with these APIs.

4.3 *Personal mobility*

In order to meet the needs of the PSCS scenario, a number of issues need to be resolved and new standards put in place.

PSCS is a distributed service, and as such will reply on an early implementation of IN CS2. One of the key issues is how the distribution aspects are implemented in the Advanced Services Platform.

Another requirement of the PSCS scenario involves the maintenance and updating of customer profile information. This is a management function, and means are needed to implement this requirement through the TMN. In addition, PSCS implies that the management and control of the services are also distributed. An interesting issue is whether there is a need for two different approaches to the distribution issues in both the advanced services platform and the management platform.

Distributed management of resources in different administrative domains has been studied by a number of projects, including RACE PREPARE and EURESCOM P226 and P230. There has been two very useful workshops on UPT management organised by R2083. A Co-operative Model is seen as satisfying the requirements of the current operators for integrity of control over their own resources. Implementation of distributed management of PSCS type services may require implementation of transaction processing schemes between the various domains.

4.4 Fixed vs. mobile service interworking

It is important that services can interwork and can originate and terminate on either fixed or mobile networks. In ETSI UMTS is currently being specified to provide an infrastructure that will support the use of mobile terminals. In ITU this is known as FPLMTS.

4.5 User interface (service configuration) issues

The usability of single services has been well investigated during the RACE I and II programmes. The focus is now moving towards the integration and configuration of services to meet user requirements. The concurrent execution of tasks, perhaps using different media and requiring the sharing of screens of multimedia information between participants, must be made simpler for users. Even professional users have problems interacting with more than two tasks simultaneously. Foreground and background tasks must also be seamlessly integrated to help minimise some of the inevitable performance limitations of current systems and to best exploit future, more capable systems. For example, the effective implementation of the hypermedia scenario described in Section 2 requires background tasks to be constantly operating to search for information and to update indexes.

5 Standards Bodies

The most important working groups producing standards for IS&N are associated with ITU-T, ETSI and ISO/IEC. The flow of information between these groups is shown in Figure 4.

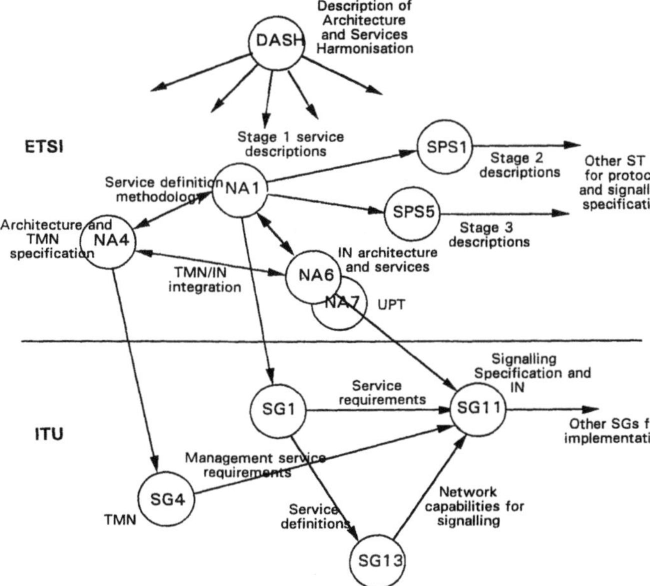

Figure 4 The relationship between the standards groups most important to IS&N.

There are also important standards produced by ISO/IEC and de-facto standards.

5.1 European Telecommunications Standards Institute (ETSI)

ETSI is European regional standards body and it the primary focus of work from RACE. ETSI produces Technical Standards (ETSs) and Technical Reports (ETRs). Neither of these are mandatory in their own right, but they can be referenced by the CEC in European Directives, which then become mandatory on all European equipment suppliers. Standards in ETSI are drafted by 11 Technical Committees (TCs), each divided into a number of Sub-Technical Committees (STCs). The most important TCs to IS&N are:

NA Network Aspects are the core of the work in RACE towards IBC. It is necessary to contribute to NA to ensure that the current networks can evolve towards the seamless integrated broadband network(s) of the future. NA1 covers services, NA2 covers numbering and addressing, NA4 covers architectures and TMN, NA5 covers broadband, NA6 covers IN and NA7 covers UPT. All STCs are important to IS&N.

SPS Signalling, Protocols & Switching. Service and network capability requirements should be provided by NA to SPS as shown in Fig 4, so that the signalling and protocol specifications can evolve to meet the requirements of future services. However, signalling capabilities are often defined in advance of the services for which they are provided, and this will increasingly be the case for IN services where, to enable maximum flexibility of service provision, a full set of services are not defined before the signalling capabilities are implemented. All STCs within SPS are important, especially SPS3 which covers digital switching.

SMG Special Mobile Group. Mobile access is becoming increasingly important to users and must be integrated seamlessly into IBC. The most important STC is SMG5 which is currently specifying UMTS. SMG1 and SMG3 are also important.

5.2 International Telecommunications Union (ITU)

The ITU produces recommendations that are applicable worldwide and so must be addressed to enable global services to be provided. Work in the telecommunications standards sector of ITU is carried out in 15 Study Groups (SGs). The most important for ISN are:

SG1 Service Definition. A wide range of services are being defined including multimedia and multipoint conferencing services.

SG4 Network Maintenance. This includes all recommendations on the TMN.

SG11 Switching and Signalling. This includes all recommendations on IN.

SG13 General Network Aspects. This includes the specification of B-ISDN and the specification of the network capabilities required to support multimedia services.

5.3 International Organisation for Standardisation (ISO)

ISO standards cover all fields except for electrical and electronic engineering which is covered by IEC, and telecommunications which is covered by ITU. The technical work of ISO is done in technical committees (TCs) and their subcommittees (SCs) and working groups (WGs).

The work relevant to ISN is covered in joint groups with IEC. The most important of these are:

- ISO TP
- ISO CMIP/CMISE

5.4 Other standards

De facto standards such as X/Open, OMG and OSF, and emerging IT technologies such as DCE, OLTP, CORBA and Motif are a major influence in the TMN area. The Internet community has also been very successful in establishing de-facto standards for such things as routers and messaging systems. These were available before and are operating in competition to internationally recognised standards for systems with similar functionality.

6 Conclusions

This paper has listed the high priority areas in which work is needed. It is not suggested that all projects can or need to contribute to all the above areas. However, it is important that some means be found through the current management activities to better coordinate this joint effort. In particular, while efforts in TINA, EURESCOM and the CEC funded RACE/ACTS Programmes are essentially independent activities, it is important that there be a means to coordinate the effort in the interest not only of harmonised solutions, but also of more cost effective R&D effort for the companies involved (both industry and operator).

7 Glossary

CCITT International Telegraph and Telephone Consultative Committee. The part of the ITU responsible for (non-mandatory) recommendations on public telecommunications services. CCITT publishes telecommunications recommendations in the form of books; the most recent is the Blue Book (1988).

ETSI European Telecommunications Standards Institute. A non-profit making organisation setting telecommunications standards in Europe. ETSI has 12 technical committees (TCs) dealing with telecommunications, IT (in cooperation with CEN) and broadcasting (in cooperation with the EBU).

ISO The International Organisation for Standardisation. A federation of national standards bodies. ISO sets worldwide standards in any field not covered by a specialist standards body.

ITU The International Telecommunications Union. An agency of the United Nations based in Geneva. It is responsible for telecommunications standards worldwide and has 5 parts including CCITT and CCIR. On 1 March 1993 CCITT and CCIR were merged into a single part of ITU responsible for telecommunications standards.

PSCS Personal Services Communications Space.

UMTS Universal Mobile Telecommunications System.

UPT Universal Personal Telecommunications.

8 References

[1] ETSI TCR-TR/NA-003001 "Recommendations towards the Harmonisation of Architecture and Service Description Methodologies".

[2] *Standardisation in Information Technology and Telecommunications*, Commission of the European Communities DG XIII: 200 rue de la Loi, B-1049 Brussels, Belgium.

[3] MOBILISE PSCS Concept: Definition and CFS - Draft Version. Deliverable 4, RACE Project R2003, June 1993.

[4] ETSI DTR/NA-10100 "UPT Phase 1 - Service Description".

[5] Report of workshop on UPT Management, Bonn, May 1994.

[6] Report of joint STG meeting of IS&N STGs, STG JOI(94)1/R, Brussels, 18 May 1994.

[7] CCITT Recommendation I.130 "Method for the Characterisation of Telecommunication Services supported by an ISDN and Network Capabilities of an ISDN".

ISO standards can be obtained from the ISO Central Secretariat, 1, rue de Varembé, Case postale 56, CH-1211 Geneva 20, Switzerland.

CCITT Recommendations can be obtained from ITU Headquarters, Place des Nations, CH-1211, Geneva 20, Switzerland.

ETSI Technical Reports and ETSI Technical Standards can be obtained from the ETSI Secretariat, 06921 Sophia Antipolis Cedex, France.

13

Customer requirements on teleservice management

J. Hall, I. Schieferdecker, M. Tschichholz
GMD-FOKUS
Hardenbergplatz 2, 10623 Berlin, Germany
Phone: +49-30-25499200, E-mail: hall|ina|tschichholz@fokus.gmd.de

Abstract

This paper examines some of the issues arising from customer requirements concerning the management of end-to-end services and guaranteed end-to-end quality of service. The implications of supporting the desired management capabilities both horizontally (inter-domain cooperative management) and vertically (from the service to the network elements) are discussed using as examples work currently being undertaken in two complementary projects.

Keywords

TMN, user requirements, inter-domain management, quality of service (QoS)

1 CONTEXT

The developments in the telecommunications world that are leading to an integrated broadband environment will result in an open service market where a variety of advanced multimedia services will be on offer in a competitive arena. These developments have been initiated by two main thrusts - liberalisation and advances in technology. Liberalisation implies greater consideration of customers' needs as teleservice providers will only be successful if their services are viable in the market place. The ability to meet customers' requirements will therefore play an increasingly important role not only in the national arena but also internationally as improved fast speed communication links promote the emergence of a global market place. Liberalisation in offering services is being accompanied by an evolution, if not revolution, in networking and information technology. High speed integrated broadband communications (IBC) over ATM, together with faster LANs, can support the transmission of multimedia streams, including audio, video, and text, over the same digital infrastructure. Highly sophisticated multimedia services will be able to provide support for cooperative working in a variety of areas and as users become familiar with the availability and flexibility of such services they will make greater demands on the services being provided.

The trend is therefore towards a service-driven market where services are offered on a competitive basis in response to specific customer needs. This increase in services being offered is creating new challenges for management, both in managing advanced multimedia services with different characteristics and quality of service requirements as well as in meeting customer demands for more control over the services that are offered.

These issues are being investigated in the European research and development programme RACE, which is promoting research into advanced technologies enabling the next generation of services to be created and openly available in an IBC environment. This paper discusses work from two RACE projects investigating how customer requirements affect, and are being supported by, management. Section two discusses customer and end-user requirements vis-à-vis the management of the telecommunication services they purchase and use. Sections three and four introduce two examples of how these requirements can be met by management, first for managing an organisation's virtual private network (VPN) as part of its corporate telecommunications network (CTN), and then for ensuring end-to-end quality of service in a multimedia collaboration service. Conclusions are presented in section five.

2 REQUIREMENTS

Corporate customers in an integrated teleservice environment are expected in the first instance to be organisations operating in a distributed, increasingly global, market where communications and the distributed handling of information are essential to success in their core business. Such customers are becoming more demanding and sophisticated; they must see the benefit from subscribing to a new service or to new features in an existing service, and this must be at a price that they are prepared to pay. Services are judged not only according to cost, but also on quality of service, which is defined as user satisfaction with service performance as it is perceived at the user interface, including service availability, reliability, and flexibility. Customers expect high levels of connectivity, bandwidth on demand, convenience, and teleservices tailored to their specific requirements, and they will select the services that most closely meet their requirements.

The impact of corporate customer and end-user requirements on the management of teleservices is being investigated here with respect to two groups of teleservice which have been selected as representative of the type of service that customers will be purchasing: VPN data services, and multimedia teleservices. VPN data services offer a more efficient and flexible alternative to leased lines for organisations wishing to connect geographically distributed sites. End-to-end multimedia teleservices will be offered by value-added service providers in a variety of areas, with multimedia collaboration services and multimedia mail among those currently being developed. In many cases multimedia services will be used in connection with a VPN service providing the underlying data communication service, and possibly offered together with a VPN service by the same service provider as part of a one-stop-shopping package.

In a competitive environment, customer requirements concerning management of the services purchased will determine what is offered. Customers using VPNs instead of leased lines for connecting their various sites will regard the VPN together with their local networks as forming one CTN (ETSI, 1993). Customers will therefore request a certain quality of service and control over their network and may wish automated access to the service provider's management system via standardised management interfaces that can interoperate with their own management systems to provide end-to-end CTN management. The customer management services required include requesting end-to-end bandwidth dynamically according to the application being used over the VPN; modifying on-line the configuration of their VPN, for example by reallocating addresses; changing user profiles and access controls; obtaining information about the status and usage of the VPN; receiving notifications about faults or alarm thresholds being exceeded; or modifying QoS choices dynamically. The customer management system and the service provider management system have to cooperate to support such management functionality. For example, if the customer management system needs information about its VPN or wants to change the VPN configuration parameters, the service provider must offer the required functionality to the customer over a well-defined interface, and must also interact with the network management systems across which the VPN is offered in order to pass down the modifications requested by the customer to the network service operators.

As users become more familiar with multimedia teleservices, quality of service must be approached from the users' point of view and not from the network-oriented view as has traditionally been the case (Seitz, et. al., 1994). QoS attributes are user-specific and should be determined and controllable by users before and during service use respectively. Customers will not be interested in how a service is provided and over which networks it runs, but they will be interested in comparing services in terms of QoS user-oriented attributes that the service provider can support to suit their particular needs. The choice and trade-offs between the various options available must be comprehensible and meaningful to both customer and end user. As bandwidth alone will no longer be the determining factor, customers will wish to make their own decisions regarding quality of service and price, determining the kind of quality they wish to receive for the price that they are prepared to pay. When using a multimedia teleservice, users will want to specify their performance requirements and obtain assurance that these requirements can be met. If they cannot be met users may prefer to modify their original requirements. If during the course of using a service the QoS guarantee fails, the user should be informed and given a choice of alternatives, including terminating the service, or dropping one of the media streams (Ferrari, 1990).

The high performance necessary for real-time multimedia services encompasses not only speed, but also reliability and availability, integrity, operability, delay, and accuracy, including synchronisation accuracy between media streams. Different kinds of media stream require different levels of latency, bandwidth, jitter, and error resilience. Advanced multimedia services are time-critical and need management support for ensuring agreed qualities of service. In the case of multimedia collaboration services, management has to promote end-to-end service guarantees for the different time scales of a service. The overall quality of service depends on the management of both the service and all the networks on which it runs from end to end, including the customer's own networks. The capabilities of each end system also need to be considered. The end-to-end management therefore has to include management capabilities for each layer of the communication stack, i.e., for the multimedia service itself, for the end system, and for the various networks used for interconnecting the conference partners. Consequently, the end-to-end management has to adopt a vertical management approach. Specifically, it needs more elaborate network management interfaces which will be the basis for building the service management.

The above requirements from customers and end users highlight specific management challenges in an advanced service environment. In addition, teleservices such as VPN and multimedia collaboration are end-to-end services crossing several networks and administrative domains. Management of such services therefore involves both security and inter-domain management issues. Customers will only use the teleservices being offered if these services can support customer security requirements concerning confidentiality and integrity of the data being handled both by the teleservices themselves as well as by the management services offered to the customer (O'Connell and Donnelly, 1994). Inter-domain management issues concern particularly the question of how several autonomous domains, including the customer premises network, can cooperate to provide end-to-end QoS management, and how network operators can make available the required functionality to service providers over the network/service management boundary. It also requires an understanding of the intra-domain enterprise network capabilities in order to integrate them with the end-to-end inter-domain capabilities (Tschichholz, et. al., 1995).

3 INTER-DOMAIN VPN SERVICE MANAGEMENT FOR CORPORATE NETWORKS

The RACE project PREPARE (PREPilot in Advanced REsource management) is investigating the issues arising from managing end-to-end IBC services spanning heterogeneous networks belonging to public and private organisations. IBC VPN services providing corporate networking facilities between customer premises networks (CPNs) have been selected as examples of end-to-end services requiring inter-domain cooperative management. The CPNs are part of the CTN and are therefore also involved in providing end-to-end services to users. PREPARE has been developing support for cooperative end-to-end management, focusing on performance and configuration management, including management support for guaranteeing quality of service. Customer management services are a specific subject of investigation in the project, particularly the interactions between the customer's management system and the service provider's management system. Advanced management services are being produced for the customer management system that enable it to obtain information about its virtual networking facilities and to make configuration and other changes to these facilities.

PREPARE has adopted the Telecommunications Management Network (TMN) framework for designing and implementing its inter-domain management system, including the relevant management services and management information model. However, in the many cases where standards did not exist, PREPARE took a pragmatic approach about where to place the management functionality. An inter-domain management architecture based on the TMN functional architecture (CCITT, 1992) was defined for the PREPARE testbed*. It consists of four subnetwork management domains and one or more end-to-end service management domains, or TMNs, which incorporate the customer management systems and the service provider management system and which interface to each subnetwork TMN (see Figure 1). The end-to-end services are managed by cooperating service operations systems that interact over the x reference point. The service provider is responsible for providing on-line service management capabilities to the customer at the x reference point (via the X interface at the service level) to allow, for example, the customer to reconfigure the VPN or to change QoS parameters. The customer management system is not concerned with the underlying networks and their management. It sees only the end-to-end service with the management services

* The PREPARE testbed consisted in the first phase of an ATM WAN and DQDB MAN providing the public network which connects customer premises networks comprising token ring LANs and ATM MUXs. This allows for both connectionless and isochronous VPN services. In the second phase (1995) ATM LANs are being connected via the European pilot ATM network.

offered to the customer being mapped to management actions across the service/network management boundary (Schneider and Donnelly, 1993).

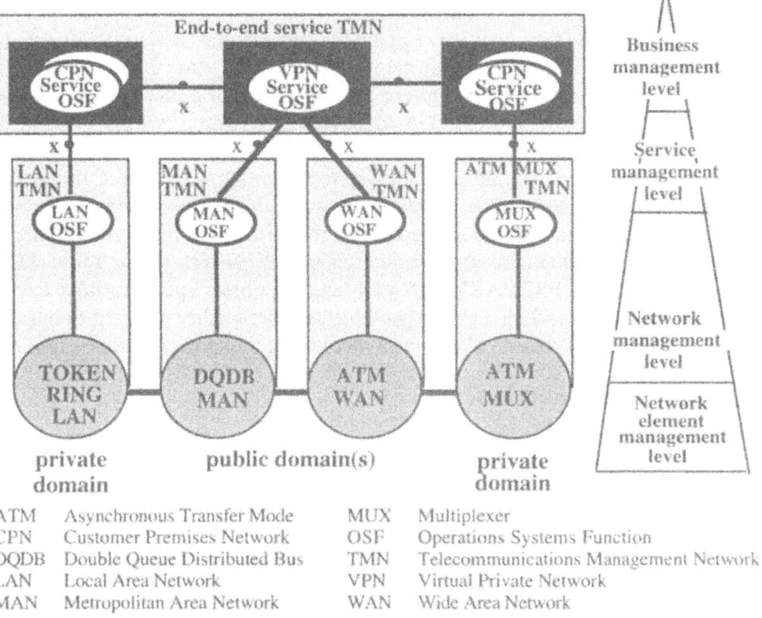

Figure 1 The PREPARE Management Architecture.

In order to manage an end-to-end service there must be an integrated view of management between the CPN management system and the service provider's management system at the service level. This implies not only standardised interfaces but also a shared understanding of the management information being manipulated. Inter-domain and intra-domain management information models have been specified according to (ISO, 1992) using where possible existing managed object definitions from standards bodies and other organisations, such as the Network Management Forum. Internal intra-domain subnetwork information models have been defined for the subnetworks that enable the service-level functionality available to the customer management system to be supported within the individual TMNs of the subnetworks involved and made available over an open X interface from each TMN. The inter-domain information model designed for end-to-end management of the VPN services across the testbed ensures that all TMNs involved in managing the end-to-end service share the same understanding of the management information and of the management functionality in the form of the operations that can be performed on the managed objects, i.e., shared management knowledge (PREPARE, 1994). As the service-level information must be available on an end-to-end basis, a globally consistent view is maintained by an Inter-Domain Management Information service which enables the management information to be uniquely identified throughout the testbed and retrieved in a uniform manner wherever it is located (Tschichholz and Donnelly, 1993).

Some examples of the types of managed object are given as an illustration. At the service level, information about the end-to-end service is supported by the operations systems (OSFs) within the end-to-end service TMN, i.e., the service provider's domain and the CPN domains at the customer's locations (see Figure 2). In the CPN OSF, *cpn* gives the service provider details of the customer network that are relevant to operating the service at that site. In the same OSF are *endUser*, *te* (terminal equipment), *vdl* (virtual direct line), and *userStream*, representing a communication path between two or more end points which has specific QoS requirements and which can be booked in advance with a certain priority level. In the service provider's VPN OSF are held details about customers and their VPNs, for example, *pVLL* (public virtual leased line) which is a logical representation of public network resources available between customer access points in the same VPN and which is specified using a profile containing information about bandwidth, start time, end time, and quality of service, *userStream* which represents a communication path between two or more TEs, and *cap* (customer access point) (PREPARE, 1994). Managed object specifications have also been produced for each subnetwork in order to provide the functionality required to manage the end-to-end service within the subnetwork, such as monitoring the QoS parameters of a connection.

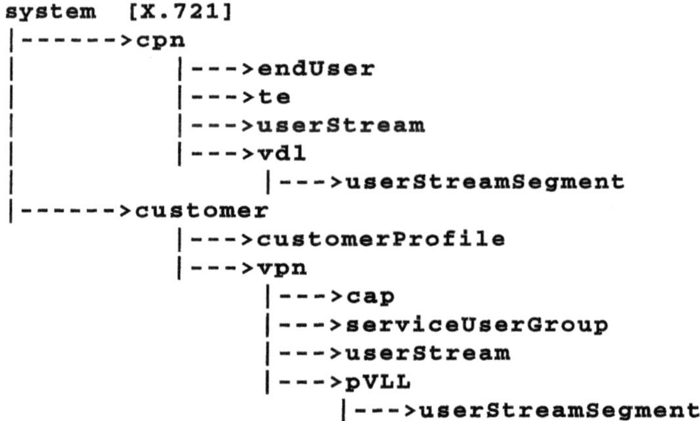

Figure 2 The End-to-End VPN Service Containment Tree.

Customers can be provided with a set of management services allowing them to manage the VPN service according to their particular needs. Management service functions supporting both end-to-end management across the testbed as well as allowing value-added service providers access to the bearer services of the subnetworks were specified. The service functions granting the customer organisation management capabilities concerning its VPN service and made available by the service provider at the X interface were classified into four groups. The first group is concerned with customer administration, providing a means of managing customer service information, such as the customer profile, list of permitted end users, customer access ports. A customer management system can retrieve information and also modify parts of the customer data using the services in this group. The second group consists of the end-user access management service functions which enable the customer to retrieve information about end users and to modify this information, for example, to add and remove end users both of the VPN service itself and of the VPN management service. The traffic and switching management service functions in the third group enable connections between service end users to be created, modified and deleted. Information on the bandwidth for a connection can be retrieved and dynamically modified by the customer management system. The fourth group of service functions are concerned with service performance and quality of service, enabling service performance information to be retrieved and analysed and end-to-end connections to be tested. Notifications when the performance of the VPN goes beyond specified threshold values can also be transmitted to the customer management system. Management service functions have been defined for each subnetwork, so ensuring that customer requests can either be met at the service level or can be mapped to the network management level, for example, by allocating virtual paths through the ATM network (PREPARE, 1993).

Scenarios for two VPN services were designed in order to demonstrate the various activities that can take place during the provision and operation of the VPN services and which focus on different aspects of inter-domain end-to-end service management. They were specified in terms of the CMIS operations that the relevant management service functions are mapped down to. The scenarios developed for the connectionless bearer service show, for example, how information relating to the customer's VPN can be retrieved, how the customer can modify information, such as that relating to the customer's access bandwidth, in order to improve the quality of service, or add and remove allowed VPN connections between customer access points. The scenarios specified for the connection-oriented isochronous bearer service in conjunction with a multimedia teleconferencing application running over it enable the initial teleconference to be set up and bandwidth to be renegotiated during the session, and also provide a situation where QoS degradation during the conference has to be handled. For example, when reacting to the QoS degradation, the customer management system can invoke service functions such as *getPVLLStatus*, and can request changes to the quality of service with a *modifySession* request to the service provider's management system (PREPARE, 1993).

The scenarios were realised for the PREPARE demonstrator, which enabled the PREPARE TMN architecture to be validated and which showed how a self-contained design for a specific testbed could be used to investigate the requirements of customers and end users for active participation in end-to-end service management. Many of the issues are not adequately covered in the standards work, if at all, and so the work in PREPARE can, by testing many of the ideas in a real demonstrator, show how customer requirements on network management need to be met by value-added service providers and network operators and how the relevant management architecture and information model should be designed to support this from the service level down to the network element level.

The PREPARE work has demonstrated the need for an inter-domain architecture with clearly defined interfaces between different domains in order to provide cooperative end-to-end service management. An end-to-end service management information model allows the management operations available across the X interface to be clearly specified and made available to external TMNs. Current work is extending this architecture in order to investigate, among other things, customer management services for multimedia teleservices in an environment composed of multiple value added service providers and network operators.

4 QoS SUPPORT FOR MULTIMEDIA TELESERVICES

The RACE project TOMQAT (Total Management of End-to-End Service Quality for Multimedia Applications in IBC Networks) is investigating support for end-to-end QoS management for multimedia services in an interconnected broadband environment and, in particular, how the customer perception of quality of service can be translated into network performance parameters and mapped to specific requirements on the network provider. It is designing and implementing QoS management, including management services based on appropriate information models and performance models. The project is using the TMN framework for its management architecture and is testing its approach over an ATM testbed* using the multimedia application Joint Viewing and Teleoperation System which has high QoS requirements as it combines video conferencing with (multimedia) application sharing between several participants in a wide area environment [Dermler, et. al., 1993].

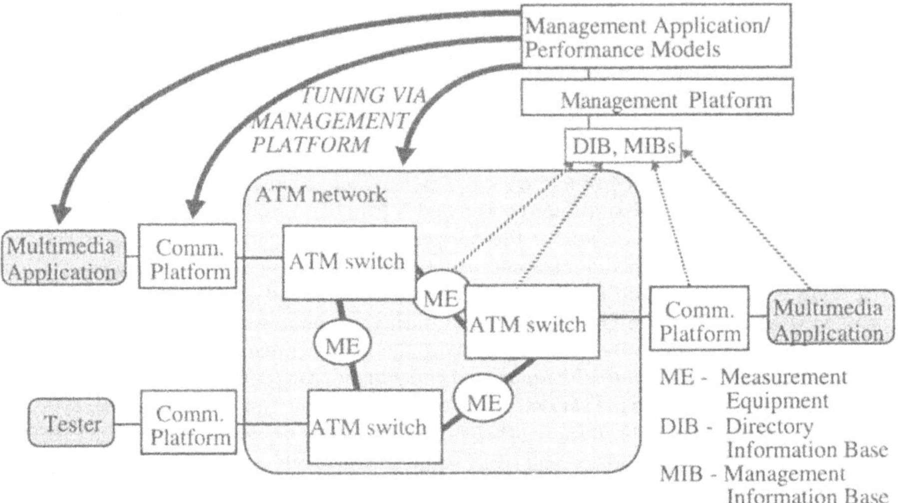

Figure 3 The Quality of Service Management Configuration.

The QoS requirements from customers and users need to be supported by a management system that can monitor the performance of the teleservice being used and make the appropriate modifications to ensure that the performance can be maintained. QoS parameters are subjective and so it is necessary to look at the QoS requested by users and map this to network performance parameters which are objective and can be measured. Performance measurements will provide the management system with information about the status of the network. If performance is not sufficient to support the required QoS, corrective action can be taken. There currently exists a lack of guidelines for end-to-end management of the QoS perceived by users, including the management of all factors affecting this quality. Total quality of service management is concerned not only with the network element and network TMN layers but also with the service layer and necessitates both static planning of network and component resources as well as the dynamic real-time management of QoS attributes of running multimedia services.

* The testbed is based on BALI (Berlin LAN Interconnection Net), an ATM infrastructure connecting ATM LANs in Berlin that is connected to the German and European pilot ATM networks.

Total management of end-to-end QoS means that the end-to-end system performance is relevant and that the global, end-to-end system has to be considered when making local choices about the best communication system to use. QoS management at the end system therefore includes the end-user communication stack as well as the cumulative effect of the performance of each subnetwork constituting the whole end-to-end network. Management is used to tune the network performance by observing network element performance, with performance models aiding the decision-making process. This configuration is shown in Figure 3.

In general, a multimedia application is realised by a number of different service elements supporting, for instance, the audio and video communication between end users, the coordination of joint processing on shared documents or the session management to establish/renegotiate or release connections. Some of these service elements carry functionalities of local influence (for example, the control of local peripheral devices such as cameras, monitors, microphones), whereas others provide functionalities of global significance (such as joint editing). It is not sufficient to rely exclusively on the information and management capabilities of the used service alone but on the whole end-to-end network. More information about the data flow within the network is needed so that suggestions can be made, based on long-term end-to-end observations, for new network parameters/thresholds to be used within the network control strategies. It should also be possible to tune the network using global information, as opposed to local information available to the network. This implies a more elaborate management interface between network operator and service provider to allow for network monitoring down to network element monitoring and network tuning.

A management architecture has been designed to meet the management requirements for total quality of service (see Figure 4). It has adopted the domain approach as a means of structuring the environment and has designated QoS managers with two kinds of roles: a Service QoS Manager (SQM) for each domain and a local End-System QoS Manager (ESQM) for each end system. A master SQM will be provided by the value-added service provider of the multimedia collaboration service (TOMQAT, 1994).

The ESQMs realise QoS management functionalities in the user's local environment and rely on network element (NE) functions offered in this environment. Basic network element functions in an ATM network are, for instance:

- *ATM switch*: set-up thresholds for traffic management strategies, change routing tables.
- *Measurement equipment and mediation devices*: collect data, alarm on performance degradation.
- *Multimedia application*: change compression algorithm for audio or video, adjust frame size of a video at the source, adjust playout buffer.

An ESQM will support QoS management functions such as: QoS negotiation; monitoring of NEs to collect performance-oriented data; QoS analysis and evaluation to derive network performance parameters and to forecast future or to determine current QoS bottlenecks; QoS decision-making based on appropriate performance models to decide on better/optimal parameter settings for NEs; and local QoS configuration by tuning NEs with the above identified parameters to remove performance bottlenecks. The SQMs realise QoS management functionalities that are responsible for coordinating the multimedia application within a specific domain, and for maintaining required global end-to-end user QoS characteristics between different domains. Typical functions offered are: QoS negotiation coordination; global monitoring of network performance and evaluation of end-to-end QoS; global decisions on NE parameters settings in the whole network using end-to-end performance models; and global network tuning in order to guarantee end-to-end QoS. The QoS management strategy refers to the way of finding better/optimal parameter settings for the required QoS. After predicting/observing a QoS degradation the manager will either tune the network or renegotiate QoS requirements with the end user.

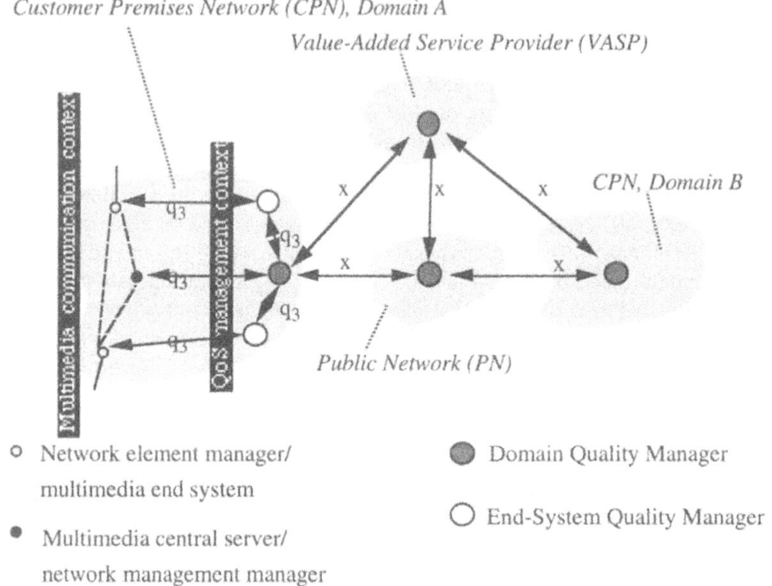

Figure 4 The Quality of Service Management Architecture.

One example - QoS influenced routing - is considered here. During connection set-up QoS requirements are specified by the end user. In the case of the video stream, its frame refresh rate and frame size are given. Due to costly communication resources, the multimedia collaboration service uses a variable bit rate (VBR) service with sustainable cell rate (SCR) and peak cell rate (PCR) suitable for video transmission. A virtual channel with corresponding SCR and PCR will be established. The QoS requirements for the multimedia collaboration service are forwarded to the ESQM responsible for maintaining the contracted QoS.

The throughput and cell error rate of a specific virtual channel is measured using ATM measurement equipment which is connected to a physical link of the ATM network. Due to overload and congestion in the ATM network, the throughput at some communication link may drop below the specified SCR, which implies an unacceptable QoS degradation for the end user – the video might simply stop. This causes the generation of a notification which will be forwarded to the ESQM via the operations system of the measurement equipment. In order to re-establish the contracted QoS the communication link can be tuned. The ESQM must have knowledge about alternative links and their mean load observed in the past. Based on this knowledge, the ESQM can choose a new route for the virtual channel that is used for transmitting the video stream. It will modify the routing tables of some switches to avoid the congested link. If it is not possible to achieve the required bandwidth through alternative links, renegotiation of end-user QoS requirements will be necessary. The user will be informed that only smaller video frames or a lower frame rate can be supported. The user can decide whether to reduce the video in size or the frame rate. Of course, another possibility for the user will be simply to terminate the connection.

TOMQAT is producing a series of similar scenarios such as QoS based call admission control, playout delay adaptation, and end-to-end application control, showing that QoS can be maintained by using the QoS management service but will degrade crucially otherwise. Special emphasis is being placed on the mapping from end users' QoS requirements down to the network element performance. Based on this, appropriate monitoring functionalities to observe the local and global network state as well as network tuning strategies at each vertical layer of the communication stack will be derived. As TOMQAT aims at developing a running QoS management environment, deeper insights into the layered management architecture are expected. As far as the management interfaces of public network providers are concerned, it has emerged that more management capabilities are needed at the boundary between network and service provider in order to offer QoS management to the end users.

5 CONCLUSIONS

The work discussed here is investigating how customer and user requirements can be met by management in an advanced service environment. The result of the work so far shows that any management capability offered to customers and end users must be underpinned by an architecture and corresponding information model that structures the essential management functionality and responsibility at all levels of this environment.

The development of management services for the customer management system in a CTN has emphasised the significance of a standardised open interface over which management functionality can be offered by the service provider so that it can be used with the customer's existing management system. The work has shown that such functionality, although offered at the service level, must be mapped to operations offered by each of the underlying networks supporting the end-to-end service. A modelling approach is required that can structure the management functionality in an architectural framework, specify the management information model that is needed in each TMN, and designate the operations available at each interface in the architecture. Without this underlying support, customer management requirements cannot be met. End-to-end management of teleservices enables end-to-end services to be managed efficiently and their quality of service to be supported. The work in designing QoS management for such teleservices has shown that the high requirements from demanding multimedia applications necessitate a well thought-out performance management and quality control system in order to guarantee end-to-end QoS to the end user.

In a heterogeneous environment cooperative management between the customer, service provider and network operator entails both an end-to-end inter-domain (horizontal) approach providing cooperation between customers and service providers, as well as a vertical layered approach ensuring management capabilities at each communication level in order to guarantee end-to-end QoS. The boundary between service and network management is proving to be crucial as it is here that the interaction between network operator and service provider takes place. The conclusion from the work discussed in this paper is that the current functionality offered by network management is not sufficient to support end-to-end service management for quality of service requirements and that work on the end-to-end management support and its interoperability with intra-domain management is essential before customer and end-user requirements can be satisfied. The two projects discussed here are contributing to this work.

ACKNOWLEDGEMENTS

The authors wish to thank their colleagues at GMD-FOKUS for many fruitful discussions and the partners of the PREPARE and TOMQAT consortia for their contributions to the ideas presented here. This work was partially supported by the Commission of the European Communities (CEC) under projects R2004 PREPARE and R2116 TOMQAT of the RACE II programme. This paper does not necessarily reflect the views of the PREPARE and TOMQAT consortia.

REFERENCES

CCITT (1992) Principles for a Telecommunications Management Network, CCITT Recommendation M.3010, ITU, Geneva.

Dermler, G., et. al. (1993) Constructing a Distributed Multimedia Joint Viewing and Tele-Operation Service for Heterogeneous Workstation Environments, in *Proceedings Fourth IEEE Workshop on Future Trends of Distributed Computing, Lisbon.*

ETSI (1993) Strategic Review Committee on Corporate Telecommunications Networks. Report to the Technical Assembly, SRC5 Final Report.

Ferrari, D. (1990) Client Requirements for Real-Time Communication Services, *IEEE Communications Magazine*, **28**, 65–72.

ISO (1992) Information Technology – Open Systems Interconnection – Structure of Management Information, Part 4: Guidelines for the Definition of Managed Objects, ISO/IEC International Standard 10165-4.

O'Connell, S. and Donnelly, W. (1994), Security Requirements of the TMN X-Interface within End-to-End Service Management of Virtual Private Networks, in *Proceedings of the RACE International Conference on Intelligence in Broadband Services and Networks, Aachen, September 1994,* 207–217.

PREPARE (1993) CNM and PNM Specification Based on MIS, PREPARE deliverable R2004/LME/WP6/DS/I7011/b1.

PREPARE (1994) Final TMN Information Model Specification, PREPARE deliverable R2004/BRI/WP2/DS/P/017/b1.

Schneider, J.M. and Donnelly, W. (1993) An Open Architecture for Inter-Domain Communications Management in the PREPARE Testbed, in *Proceedings of the 2nd International Conference on Broadband Islands, Athens, June 1993*, 77–88.

Seitz, N.B, et. al. (1994), User-Oriented Measures of Telecommunication Quality, *IEEE Communications Magazine*, **32**, 56–66.

TOMQAT (1994) Architecture of the TOMQAT System and Definition of the Net Infrastructure, TOMQAT Deliverable R2116/TUB/WP2/DS/P/006/b1.

Tschichholz, M. and Donnelly, W. (1993) The PREPARE Management Information Service, in *Proceedings of the RACE International Conference on Intelligence in Broadband Services and Networks, Paris, November 1993,* IV/3/1–12.

Tschichholz, M. et. al. (1995) Information Aspects and Future Directions in an Integrated Telecommunications and Enterprise Management Environment, to be published in *Journal of Network and Systems Management.*

BIOGRAPHIES

Jane Hall is a senior scientist at the Research Institute for Open Communication Systems (FOKUS) in the Management in open Systems (MinoS) group at GMD Berlin. She has been working in several European projects (COST11, ESPRIT, RACE) in the area of network and service management. Her current research interests are quality of service management and management of teleworking environments.

Ina Schieferdecker graduated in mathematics at the Humboldt University in Berlin in 1990 and received her PhD in computer science from the Technical University of Berlin in 1994. She is currently working at the Research Institute for Open Communication Systems (FOKUS) at GMD Berlin in the Performance Laboratory. Her current research interests are formal specification, verification and performance evaluation of communication networks and management systems.

Michael Tschichholz received his diploma in computer science from the Technical University of Berlin in 1982. He has been working in the area of open communication systems (E-mail, Directory, OSI Management, TMN) since 1980, and is actively contributing to international standardisation work. He is working at GMD-FOKUS and is the head of the Management in open Systems (MinoS) group. He has participated in several national and international management related projects. His current research interests are related to multi-domain management based on TMN and ODP.

14

Secure remote management

S. N. Bhatti, G. Knight
Department of Computer Science, University College London, Gower Street, London WC1E 6BT, England, UK
saleem@cs.ucl.ac.uk, knight@cs.ucl.ac.uk
and
D. Gurle, P. Rodier
CNET, France Telecom, 905 Rue Albert Einstein, 06921 Sophia Antipolis Cedex, France
gurle@sophia.cnet.fr, rodier@sophia.cnet.fr

Abstract

Much of the network management technology today still centres around a remote monitoring approach. One would like to have a more intrusive management capability but in a large distributed system one must have confidence that management activities can not be subverted, whether by accident or by malicious intent. To achieve this goal, one requires the management applications to have security mechanisms that will prevent unprivileged users from altering the system accidentally but also, more importantly, to prevent possible attacks from a third party who may disrupt or misuse services. This paper describes some services and mechanisms with which the authors have experimented to allow secure remote management of a distributed system in a real service environment. Although there are many standards documents describing various security mechanisms, some aspects of these documents are not stable and in other cases we can not apply the mechanisms they describe due to restrictions in our development and deployment environment. In such cases we have had to make some adaptations.

Keywords

Network Management, Security Management, Distributed Systems Management.

1 INTRODUCTION

The provision of secure management facilities for distributed applications is very important if the applications operate in an environment that is geographically widely dispersed, operating over a mixture of private and public networks. In general, one must assume that such underlying networks are insecure; that management information may be destroyed or stolen; that malicious third-parties may be able to gain access to the networks and disrupt management activities in a variety of ways. In such cases, the management and security facilities we require must be placed in the parts of the system we can trust – in the applications themselves.

Part of the motivation in the development of the security services described in this document is that they will be deployed in a real service environment, namely in the management of a large X.400(84) [X.400, 1984] mail network.

The mail network also uses a X.500(88) [X.500, 1988] directory service. There are two prime considerations:

- **Protecting existing security mechanisms** If applications are to be controlled remotely then there is a danger that any security services native to the application may be undermined by a third party through subversion of the management exchange. In the particular case of X.400, for example, a range of services including confidentiality, non-repudiation etc, are already provided, so subversion of management control could allow, for example, a third-party to cause mail for one user to be delivered to another. In fact, the native security services will be only as strong as the security services that are provided for management.
- **Deployment in a real service environment** We must develop technology which can be deployed commercially in the short to medium term. Consequently we have sought security solutions which can readily be introduced into today's OSI environment.

1.1 Contents of this paper

The aim of this paper is to give a detailed description of security services and mechanisms we have implemented and the APIs which have been introduced. The analysis behind our choices of security services and mechanisms is summarised in Section 2; a much fuller description of them is given in [Knight et al, 1994]. Section 3 discusses the mechanisms we used and some of the restrictions that affected our design decisions. The implementation of the mechanisms is described in Section 4 with a short summary in Section 5.

2 STATE OF THE ART

The issues of providing security service have received much attention recently, and the the activities of various standards bodies and consortia have resulted in a number of relevant documents. There has been work from ISO and the ITU [X.509, 1988] [X.511, 1988] [X.800, 1991] [CD10183.2, 1992] which includes specific access control features for the purposes of management [CD10164-9, 1992]. This is of particular relevance to the work described in this document which is to be applied in an OSI environment. The ISO mechanisms rely on an infrastructure that uses the RSA system of Public Keys [Rivest et al, 1978] distributed by use of the X.500 directory service [X.500, 1988].

A set of standards that are still maturing but may be extremely important are the Generic Upper Layers Security (GULS) documents [GULS, 1992].

The X.500 documents are stable and use of the infrastructures they describe has been successfully demonstrated in the PASSWORD project [Kirstein et al, 1992]. The ISO/ITU work documents on access control are not fully stable (in the final draft stages). However, they contain mechanisms that were considered to be useful and will probably be present in the final standards, hence it was decided to proceed with their use. The GULS work was not considered sufficiently stable to implement.

A discussion with respect to security **threats** that an application may be subject to, the requirements of the security **services** that are to be implemented, and the **mechanisms** that can be used to realise the services can be found in [Knight et al, 1994].

In this paper, we identify **managed** (server or agent) and **managing** (client or manager) roles for our applications, and we consider a third party that tries to subvert that exchange of information between remote entities in managed and managing roles. Management applications use the OSI Common Management Information Service (CMIS) [CMIS, 1990] [CMIP, 1990].

3 SECURITY MECHANISMS

Many of the mechanisms which implement the security services we require are based on encryption techniques. In choosing mechanisms we have tried to follow the pattern which prevails in the OSI world but, at the same time, to borrow from the other work (such as that for SNMPv2 [Case et al, 1993]) which is geared particularly to the needs of management. The principle difference between OSI and SNMPv2 management services is that the OSI one establishes a long-term, reliable association whilst SNMPv2 does not. This has some impact when security mechanisms are considered for use:

- Confidentiality and integrity mechanisms typically require the two communicating parties to have shared knowledge of a secret value. Without an association it is usual to expect this secret to be known to the two parties *a priori* and it must be stored securely by each of them ready for use – this is what happens in SNMPv2. With an association it is natural to negotiate a new secret value when an association is established thus eliminating the need for secure storage.
- The OSI protocols employed maintain an association that guarantees sequenced delivery of PDUs with very high probability. Further, each PDU has an `invokeID` field – an integer which we can insist must take values from a known sequence. This greatly simplifies the design of a stream integrity mechanism. To achieve the same with SNMPv2 requires a rather complex shared clock mechanism.
- Once an association has been established it will normally be held for a comparatively long period. This makes it reasonable to implement quite complex security mechanisms in the association establishment phase in the knowledge that they will be used only rarely. It is feasible, for example, to use Public Key encryption in association establishment.

With these considerations in mind, the following mechanisms and services were chosen:

- **authenticated associations** To authenticate associations through the use of Public Key encryption using the RSA algorithm. This is the mechanism described in [X.509, 1988] and [X.511, 1988].
- **integrity checks** To add cryptographic checksums to all management PDUs calculated according to the MD5 [Rivest, 1992] algorithm.
- **sequence numbers** To use a well-known sequence for the values of the `invokeID` in ROS PDUs [ROS, 1989].
- **confidentiality** To use secret key encryption in the form of the Data Encryption Standard (DES) [DES, 1988] for protecting confidential data.
- **access control** To implement access control as per [CD10164-9, 1992].

Initial investigations of these mechanisms are reported in [Knight et al, 1994].

3.1 Authenticated associations

For providing mutual peer authentication, the communicating parties exchange credentials based on the X.509 authentication framework and syntaxes defined in the X.511 Directory protocol. When an application (initiator) wishes to establish an association with a peer (responder), it first constructs credentials which consist of the following information:

- **user certificate** This certificate contains the initiators identity cryptographically signed by a Certification Authority (CA) in accordance with X.509.

- **recipient identity** The responder may be able to assume many identities so the initiator provides a Distinguished Name (DN) informing the responder which identity it expects.
- **session key** A secret value that will be used by the mechanism for protecting the PDUs sent on the association. The session key is encrypted using the recipients Public Key before transmission.

Both the encrypted session key value and the recipient DN value are signed by the sender to ensure that they can not be tampered with by a third party. It would have been more convenient to carry this information in the user certificate, but this is not possible due to the certificate's syntax.

An ASN.1 syntax called `SessionCredential` is used to carry the information listed above. We use the `StrongCredentials` syntax [X.511, 1988] for the user certificate and our own `Session-Key` syntax. (At moment, we use the `SessionKey` value for protecting PDUs only – see note on the implementation of confidentiality later). We also use the ASN.1 macros `SIGNED` and `ENCRYPTED` [X.509, 1988].

The `SessionCredential` is sent in the `userInfo` parameter of `CMIPUserInfo` syntax (which is in turn passed to the peer as the `user-information` parameter of the AARQ PDU [ACSE, 1992]).

Authenticating an association is a comparatively expensive operation since the RSA algorithm is complex and we must implement it in software; therefore we perform this authentication just once at association set-up time. Integrity checks – which are relatively cheap – are then applied to all subsequent PDUs sent on the association. In this way we obtain a strong assurance about the origin of PDUs on an association. The authenticated identity can also be used as input to access control decisions.

In fact, we restrict the identity to be a X.500 Distinguished Name (DN); ie. a distinguished name of an entry in the global X.500 Directory. This guarantees that identities are globally unique and is well-suited to authenticating entities such as applications, people, etc.

Before any authentication can take place, an entity must establish its right to assume a particular identity and obtain the corresponding secret key. How this is done is a purely local matter; for example, use of **smart-card** technology or simply the use of a UNIX **filestore** where the secret key information is associated to UNIX **userids** on the client system.

When the responder replies, it sends its own certificate.

3.2 Integrity checks and sequence numbers

These two mechanisms are presented together as their realisation is closely linked. For each PDU sent, the integrity check is evaluated as follows:

1. **Encode PDU using DER to produce byte stream** The CMIP PDU is encoded using the Distinguished Encoding Rules (DER) [X.509, 1988] for ASN.1 to produce a byte stream B. DER ensures that the 'shortest' BER encoding is always used.
2. **Evaluate MD5 checksum for byte stream and session key** The byte stream, B, has the session key value, k, appended to it and the MD5 value for the resulting byte stream B_s is used as the input to the MD5 algorithm, resulting in an 128bit checksum, c:

$$B_s = B * k \tag{1}$$

$$c = MD5(B_s) \tag{2}$$

where the $*$ operator results in a byte stream that is the concatenation of its byte stream arguments.

We carry the value of the checksum outside the CMIP PDU itself as the value of the ROS **invokeID**. However, before the checksum can be used as the **invokeID** value, we must try and ensure that any outstanding **invokeID** values ona an association are unique. As there is no mandatory parameter in CMIP PDUs that will guarantee that during a single association the PDUs (and so the DER byte stream) will be unique, the same checksum will be produced for those PDUs that are the same even if the generation of the CMIP PDUs is separated in time; this provides a potential attacker the opportunity to replay that PDU. Uniqueness is achieved by using a generated sequence of numbers which are combined with the checksum value. The sequence of numbers is generated by using a seed from the session key, so only the initiator and responder can be aware of the sequence. The sequence numbers are used to form the **invokeID** as show by equation 3 and decoded by the receiver using equation 4.

$$i = f(n, c) \tag{3}$$

$$c = g(i, n) \tag{4}$$

where:

- i is the eventual value of the ROS **invokeID**
- n is one of the numbers in our generated sequence of numbers
- c is the checksum value for the CMIP PDU
- f is used by the sender of the CMIP PDU
- g is used by the receiver of the CMIP PDU

Here, the values of n form a known sequence. CMIP and the OSI upper layers provide ordered delivery of PDUs, so we can use such a sequence number mechanism with confidence. As the receiver of the PDU knows n, it can evaluate c for the received CMIP PDU locally and compare it with the received value of the **invokeID** of the ROS PDU.

As there may be many n outstanding, all replies to an initiator request will have to be checked against all these values of n using g. Therefore, the functions f and g should not be computationally expensive, in order to maintain performance.

We have chosen to use the exclusive-OR function for f and its inverse for g. For n, we use the sequence of numbers from a pseudo random number generator. The seed for the generator is taken from the session key value exchanged. As clients and agents communicate asynchronously (in general), there are actually two sequences, one directed from an application that is the initiator of the association (the **initiator-sequence**) and one directed in the opposite direction (the **responder-sequence**). An important property of the number generator we decide to use is that we know its period, so we can ensure that we can always uniquely identify PDUs.

There is a possible weakness in this method which is explained below.

Possible method of attack to the integrity check mechanism

The method of using sequence numbers generated from the session key relies on the fact that the sequence numbers can not be evaluated without knowledge of the session key (which was transmitted encrypted using the recipients Public Key and signed by the sender). However, as the PDUs maybe sent in plaintext in the absence of a confidentiality mechanism, a potential attacker can see when identical PDUs are sent and knows that the only difference in the ROS **invokeID** is that there is a different sequence number. So, it would theoretically be possible (though very computationally expensive) for an attacker who has knowledge of the function f, to work out the sequence and so possibly deduce the value of the session key. While no analysis has been conducted to evaluate the potential for such an attack, we feel that it would be difficult to perform. Further, we advocate the use of the optional `currentTime` field in CMIP reply PDUs to further deter an attacker. We feel that the use of the `currentTime` field with a granularity of 0.001 seconds (or greater, if possible) in most practical cases would deter such an attack, as this would result in different byte streams for PDUs that might otherwise be identical.

However, as the CMIP `currentTime` field is only available in replies and is optional, we can not insist on or guarantee its use in all cases. Therefore, another solution to evaluating the checksum value from the PDU byte stream may be as follows:

$$B_{sn} = B * k * n_t \tag{5}$$

$$c = MD5(B_{sn}) \tag{6}$$

$$i = c \tag{7}$$

where n_t is the two's complement representation of the number n in the least whole number of bytes.

In this case it would be sufficient for n to be part of a monotonically increasing sequence. The drawback with this solution is, however, that it may require the receiver of replies to perform many calculations of B_{sn} and c if there are many outstanding n, which would affect performance greatly.

3.3 Confidentiality

To prevent unauthorised persons inspecting the contents of a PDU, we can encrypt the bytes that make up the CMIP PDU or the ROS PDU. One method for this would be to negotiate a new transfer syntax for the encrypted encoding of PDUs. However, this may not be possible, for instance if we have bought a stack from a vendor that does not support our encryption transfer syntax. Indeed, this is the case in our service environment and so this solution is not desirable. Instead, we require some application level mechanism rather than a presentation layer mechanism to allow us to send encrypted data.

Moreover, encryption (unless supported by hardware) can be quite computationally expensive and we are sensitive to the general requirement that management operations should not noticeably affect the performance of the systems they are managing. Also, we may not require the encryption of the whole PDU, just certain fields that carry sensitive information. For instance, for a CMIS M-Get request, we may not care that a third part is able to inspect the replies and determine that they are indeed replies to a M-Get request, but we would like to prevent disclosure of the attribute values.

The design of our confidentiality mechanism revolves around the use of ASN.1 macros that embody the encryption and decryption process, converting between an encrypted byte stream and 'wrapper' syntax that carries the encrypted byte stream. The wrapper syntax allows us to selectively encrypt certain fields of a CMIP PDU without modifying the syntax of the PDU. The use of an encryption algorithm and any data associated with the use of the algorithm is notified at association set-up, and indeed the session key value could be used.

When the first experiments were conducted on implementation of the described mechanism, it was decided to use DES. As there was no hardware available to us, we had to rely on software implementations of DES. Our own software implementation achieved an approximate throughput of 0.75 Mb/s (on a Sun4 IPC) and introduced noticeable additional load on the host machine. These constraints were considered unacceptable to allow deployment of the mechanism in our service environment.

Given the fact that confidentiality was not identified as a high-priority security service for our demonstrator and the likely impact on performance of software-based encryption we have not yet proceeded with a full implementation. In the remainder of this paper we concentrate on the services which have been implemented; authentication, integrity and access control.

3.4 Access control

Initially, the access control work at UCL used a mechanism based on security labels and is described in [Knight et al, 1994]. This relied on the use of DNs rather than labels, and has now been supplemented by an access control list scheme implemented by CNET.

ISO has standardized in [CD10183.2, 1992], generic access control models and procedures and in [CD10164-9, 1992] has applied these principles to OSI management. We use these documents and draw from [OMNIPoint016, 1992] to specify and implement our access control service based on an access control list (ACL) scheme. To provide an access control service, we adopt the scenario that a user of a client management application is trying to operate upon some information held at the agent application, and that the identity of this user has already been authenticated during association set-up, and that this authenticated identity information is available to the access control mechanism.

In the context of OSI systems management, control of access to management information may be required in each of the following cases [OMNIPoint016, 1992]:

- association establishment
- request to perform a management operation
- the forwarding of notifications as EventReports

We describe hereafter access control as applied only to the first of these three. An access control mechanism *is* applied to management requests, but it is not based on the use of any additional information received with the request, relying only on the fact that request was received on a secure, authenticated association. Access control for EventReport emission is not applied for the reasons exposed in [OMNIPoint016, 1992]. Also, we separate the activities required to offer access control into **OSI management** activities involving operations upon management information that relate to access control, and **operational** activities which are specific to the access control mechanism which acts upon the access control information. To explain; access control is achieved by passing access control information (ACI) to an access control decision function (ADF). The management activities required to initialise or maintain access control information for each of the items listed above would be very similar, but the operational activities that take place to apply this information will be different when considering each item, i.e. the access control decision function will behave differently.

Access control policy representation

The basic building block of OSI management is the Managed Object (MO). The MO is an abstract representation of a resource that is to be managed. OSI management essentially revolves around the remote manipulation of the attributes, actions and notifications of MOs. Our ACL based scheme is modeled according to the managed object classes defined in [CD10164-9, 1992]. As the policies are themselves realised as managed objects, the same CMIP operations can be used to control and manage the user selected policies. ACI is part of a policy and is expressed as the attributes of a MO. The policies are represented as a set of **rules**. For each policy there may be a set of **global** and **default** rules which can **deny** or **grant** access. The access control information that is used as input to the access control decision function is defined as:

- **access control rules** According to the access control policy representation in use, we use attributes of managed object classes `accessControlRules`, `globalRules` and `default-Rule` to represent access control rules.
- **initiator-bound ACI** This is the access control information provided by the initiator of a management request. In our case, we rely on the distinguished name that was authenticated at association set-up.

- **target-bound ACI** This access control information identifies the management information on which operations are to be performed. In our case, this is the given by the distinguished name of the user and CMIP parameters that identify the MO instances that are to be operated upon, e.g. `managedObjectIntstance`, `scope`, etc.

Both the initiator-bound ACI and the target-bound ACI could also use information that is sent on a per request basis in the `accessControl` field of a CMIP PDU. For our identity based ACL scheme, however, it is sufficient to use the DN of the user: we have confidence that the DN is genuine as it has been authenticated and we are also aware that the request PDU itself has been verified by the integrity mechanisms.

When expressing access control policy the user must state precisely the level of granularity that is required. Although [CD10164-9, 1992] allows very fine granularity we limit the access control to the coarsest granularity, applying it effectively to the association. The reasons for this are mainly concerned with performance and are discussed in [Knight et al, 1994].

Use of access control information

We use the managed object class `aclInitiators` to store target-bound ACI. We use a DN that can be matched against the DN that was authenticated during association set-up. A particular `aclInitiators` instance is identified by a `globalRules` object instance that contains a list of `aclInitiators` and the permissions which apply. The `globalRules` object can grant or deny permissions so with the granularity we have chosen for our access control, we have effectively an all-or-none approach to allowing access to the managed system.

4 DESCRIPTION OF IMPLEMENTATION

Authentication services are provided by the OSISEC package [OSISEC, 1993]. The CMIP implementation is provided by the MSAP library that is part of the OSIMIS [OSIMIS, 1993] management platform. Upper layer OSI services are provided by the ISODE [ISODE, 1991] package. All these packages are implemented in C and C++, running under UNIX type operating systems. Our CMIS/P library is called MSAP.

4.1 Authenticated associations

The authentication information is passed to the peer in the CMIP `userInfo` field. This is passed in the ASN.1 type `EXTERNAL` represented by the MSAP type `External` to the MSAP library calls.

Although there is a complex exchange of information between applications, the information that MSAP user may need to supply in order to set up a secure association is fairly simple. A MSAP user is provided with the following interface to pass information about the secure association to the MSAP library:

```
typedef struct AuthAssocIntegrityInfo_s {
    char *name;          /* String DN of user/application         */
    char *peer;          /* String DN of peer user/application    */
    char *ca;            /* String DN of Certification Authority  */
    char *dsa;           /* Name or ISODE-format address of DSA   */
    char sessionKey[8];  /* MD5 Key - zeros if no key for this assoc */
} AuthAssocIntegrityInfo;
```

The elements `name`, `peer` and `ca` take the form of a human readable DN to identify the user (or user application), the peer (or the peer application) and the Certification Authority (CA) that has signed their credentials, respectively. An example is of a human readable DN is:

```
"c=GB@o=University College London@ou=Computer Science@cn=Saleem Bhatti"
```

The **dsa** is the name of your local X.500 DSA. The **sessionKey** element is the shared secret that will be used to create unforgeable MD5 checksums, as the seed for the random number sequences for the PDUs and also as the DES key. Mapping between this data structure and the ASN.1 **EXTERNAL** representation is provided by the following simple API:

```
int makeAcseUserInfo(AuthAssocIntegrityInfo *info, External **external);
int getAcseUserInfo(External *external, AuthAssocIntegrityInfo **info);
```

4.2 Integrity checks and sequence numbers for CMIP PDUs

There are three aspects to the implementation of the integrity check:

- **Managing the session key values** A new session key must be generated for each association. Knowledge of the session key is required to generate the integrity checks for the CMIP PDUs. The session key information must be accessed by a separate 'sessionKey-manager' function.
- **The MD5 algorithm and checksum generation** The implementation of the MD5 algorithm is taken from RFC1321 [Rivest, 1992].
- **Generating sequence numbers for an association** Each of the two sequence number flows is generated by an 'ID-manager' function.

Managing the session keys

A new session key must be generated for each new association that is set up. To allow this, the following API allows access to a sessionKey-manager function:

```
char *makeMd5Key();
int setMd5Key(const int fd, const char *key);
char *getMd5Key(const int fd);
```

The function **makeMd5Key()** generates a random key value which can be copied to the **sessionKey** element of the **AuthAssocIntegrityInfo** structure. A call to **setMd5Key()** registers the key for use. Both the MSAP library and the MSAP user may then use **getMd5Key()** to access the key value for the association with file descriptor **fd**.

Making the MD5 checksum for a PDU

The API for obtaining the MD5 checksum value for a CMIP PDU is as follows:

```
int makeMd5Value(const int fd, PE pdu, MD5Value *check);
```

makeMd5Value() effectively implements equations 1 and 2. The session key value for the association identified by **fd** is found by **makeMd5Value()** by interrogating the sessionKey-manager function. In our implementation, the checksum, c, does not have to be evaluated by the user of MSAP for CMIP PDUs being sent; this is automatically done in the MSAP library when the PDU is generated before being passed down to ROSE.

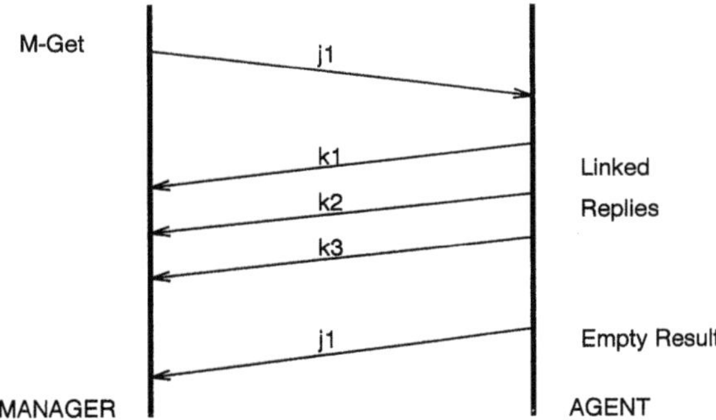

Figure 1 Example showing use of sequence numbers for CMIP PDUs

Generating sequence numbers for the PDUs

For our integrity check mechanism, we are using the mechanism identified in equations 3 and 4. So far we have described the implementation of the functions f and g as well as the generation of c. It remains to describe the generation of n. The sequence is generated by the use of the pseudo random number generator **random(3)** with a state size of 32 bytes. **random(3)** implements a non-linear additive feedback random number generator which has good randomness properties. A user of the MSAP library must generate the next number in the sequence and then pass this as one of the parameters to the MSAP function calls. Should a user wish to generate part of the sequence and then return to a previous point, P, in the sequence, the user is provided with functions to save the state of the random number generator at point P and then restore this state so restarting the random number generator at P.

Use of n and i for applications in a manager role

In the following text the subscript m identifies values associated with a manager application and the subscript a identifies values associated with an agent application. For a request identified by the pair $\{n_m, i_m\}$, the reply PDUs will use values for n as follows: a single CMIP reply PDU, or an EmptyResult PDU will use the n_m value from the original request, and any other PDU will use n_a, part of the responder-sequence. When a manager wishes to issue a M-CancelGet, i_m is used to identify the M-Get to be cancelled.

Where a CMIP PDU is not created, the MSAP library creates the bytes stream B from the DER encoding of n values. In the case of an 'empty' PDU, where the PDU is an EmptyResult, the byte stream B is the DER encoding of n_m and in other cases it is formed from the DER encoding of the next value in the agent sequence n_a.

This is best illustrated with an example. We show a M-Get request that results in the generation of linked replies in Figure 1. Each arrow represents a PDU. The number on top of an arrow is the value of n used to evaluate i for that PDU. The j numbers are successive numbers in the n_m sequence, and the k numbers are successive numbers in the n_a sequence.

The numbers in the sequences are generated by a simple API:

```
void setIRStatus(const int fd, const IRStatus s);
int makeId(const int fd);
int makePeerId(const int fd);
```

The association file descriptor, **fd**, a call to **setIRStatus()** with registers whether the applications is the initiator or responder for that association. This will allow generation of the two number sequences using makeId() and makePeerId().

4.3 Access control

An initial set of aclInitiators objects are created and initialised at agent start-up. These objects are always present and can be used to control not only access to the management information held at the agent, but also control the modification of access control information, e.g. creation of new access control objects or modification of existing ones.

Location of the access control operational activities

The standards leave open the location of access control elements that perform the access control decision function. They can be treated as being logically distributed between the manager and agent systems. However, in OSIMIS the access control decision function is built into the agent software (i.e. local and centralised), and this is in keeping with the way in which the mechanism with which the identity of communicating parties is authenticated within our model. The manager system must always authenticate itself to the agent, and the agent will always confirm the association by sending its own credentials, but the manager application does not have to authenticate the agent if it so desires.

The access control decision function is represented by an instance of an accessControlRules managed object, and there is only one instance of this object in the management information tree. It represents the access control decision function. The following access control information is located in the agent system:

- **policy-ACI** This is represented by the defaultAccess and denialGranularity attributes of the accessControlRules object class and the globalRules object class. Instances of the globalRules objects exists to deny and grant access. Each one identifies, by the value of its initiatorList attribute, a list of other managed objects to whom access is granted or denied. The deny lists are checked before the grant lists and there is a defaultAccess attribute used when an identity is not matched. The security manager can allow or deny access to CMIS services by modifying this attribute for the unknown users. Thus we effectively have **privileged** users, **normal** users and **restricted** users.
- **target-ACI** This is represented by instances of the aclInitiators managed object class which identifies a set of peers by their DN. There are two instances of this class; one which is associated with the list of restricted users (the deny list) and the other that is associated with the list of privileged users (the grant list).

The object instances are part of the management information tree of the agent, and so can be operated upon just like other managed object instances using the CMIP primitives to perform management activities.

Performance considerations

To improve the performance of the access control decision function there is a simple (volatile) cache mechanism which caches all the identities of peers that frequently access the agent during the time it is active (**retained ACI**). When a new access request arrives, if the access control information has not been modified, then a simple table look-up for the initiator in this cache speeds up association set-up. This avoids scoping and filtering operations needed to interrogate the information in the object instances to find the permission for the authenticated identity and reduces the time for evaluating the access decision by approximately 50%.

5 SUMMARY AND CONCLUSIONS

The following security mechanisms have now been implemented within the OSIMIS management platform:

- **authentication** using the RSA based Public Key method.
- **sequence numbers** and MD5 **integrity checks** to provide data origin authentication and stream integrity.
- **access control** based on the use of access control lists to provide protection against unauthorized access to management applications and management information.

These are now being evaluated within MIDAS (ESPRIT Project 6331) as part of the final demonstrator.

For computationally expensive encryption there may be additional load introduced on the host machine and so we must seek to use it only where necessary and preferably with the aid of hardware. We find that we are prepared to pay the price of software RSA encryption for authentication at association set-up but providing confidentiality using software for DES is not practical for our service environment. Experiments continue at UCL with our confidentiality mechanism.

For the integrity check and stream integrity mechanism, MD5 is relatively cheap, and we are again prepared to accept this to be implemented in software. We feel that the integrity mechanism described in this paper for the CMIP protocol could be applied to any ROS based protocol.

An access control list scheme is well suited for implementing the access control service we require for managing our X.400 system, providing protection against deliberate attack and accidental misuse. However, implementing partially the `itemRules` and `targets` managed object classes would allow the agent finer granularity for control. However, we do not see a reasonable way of implementing `itemRules` completely within OSIMIS and still maintain performance. The use of the identity based ACL scheme coupled with the integrity schemes for the CMIP PDUs allows us to make access control decisions without requiring further information. This means that we do not incur the additional overhead of processing any per request access control information. Also, the increase in performance is considerable with the introduction of caching.

6 ACKNOWLEDGEMENTS

The work conducted at University College London was partially financed by the MIDAS project under the ESPRIT funding initiative.

7 REFERENCES

[X.400, 1984] CCITT Recommendation X.400, Message Handling Systems: System Model Service Elements, Geneva, 1984.

[Rivest et al, 1978] R. L. Rivest, A. Shamir, L. A. Adleman, A Method for Obtaining Digital Signatures and Public Key Cryptosystems, Communications of the ACM, number 21, volume 2, pages 120 - 126, February 1978

[X.500, 1988] CCITT Recommendation X.500, The Directory – Overview of Concepts, Models and Services, Geneva, March 1988.

[X.509, 1988] CCITT Recommendation X.509, The Directory – Authentication Framework, Geneva, March 1988.

[X.511, 1988] CCITT Recommendation X.511, The Directory – Abstract Service Definition, Geneva, March 1988.

[X.800, 1991] CCITT Recommendation X.800, Security Architecture for Open Systems Interconnection for CCITT Applications, Geneva, 1991

[CD10183.2, 1992] ISO/IEC CD 10183.2, Information Technology – Open Systems Interconnection – Security Frameworks in Open Systems – Part 3: Access Control, 16 June 1992.

[CD10164-9, 1992] ISO/IEC CD 10164-9.3, Information Technology – Open Systems Interconnection – Systems Management – Part 9: Objects and attributes for Access Control, Borehamwood, UK, December 1992.

[GULS, 1992] ISO/IEC CD 11586, Information Technology – Open Systems Interconnection – Generic Upper Layers Security, December 1992.

[Kirstein et al, 1992] P. T. Kirstein, P. Williams, Piloting Authentication and Security Services Within OSI applications for R&D information (PASSWORD), UCL Department of Computer Science, April 1992.

[Case et al, 1993] J. Case, K. McCloghrie, M. Rose, S. Waldbusser, Introduction to version 2 of the Internet-standard Network Management Framework, Internet RFC 1441, April 1993.

[OMNIPoint016, 1992] Network Management Forum, Application Services: Security of Management, OMNIPoint/NM-Forum 016, Bernardsville, NJ, August 1992.

[Rivest, 1992] R. Rivest, The MD5 Message-Digest Algorithm, Internet RFC 1321, 16 March 1992.

[ROS, 1989] ISO/IEC 9072, Information processing systems – Text Communication – Remote Operations, 1989.

[DES, 1988] National Institute of Standards and Technology, Data Encryption Standard, FIPS Publication 46-1, January 1988.

[Knight et al, 1994] G. Knight, S. Bhatti, L. Deri, Secure Remote Management in the ESPRIT MIDAS project, Proceedings of IFIP WG 6.5 International working Conference on Upper Layer Protocols, Architectures and Applications, Barcelona, June 1994

[ACSE, 1992] CCITT Recommendation X.227, Connection Oriented Protocol Specification for the Association Control Service Element, September 1992.

[CMIS, 1990] ISO/IEC 9595, Information technology – Open Systems Interconnection – Common management information service definition, May 1990.

[CMIP, 1990] ISO/IEC 9596, Information technology – Open Systems Interconnection – Common management information protocol specification, May 1990.

[OSISEC, 1993] UCL Department of Computer Science, The OSI Security Package OSISEC User's Manual, May 1993.

[OSIMIS, 1993] UCL Department of Computer Science, The OSI Management Information Service User's Manual, Version 1.0 for system version 3.0, February 1993.

[ISODE, 1991] UCL Department of Computer Science, The ISODE User's Manual, Version 7.0, July 1991.

8 BIOGRAPHIES

Saleem N. Bhatti received a B.Eng.(Hons) in *Electronic and Electrical Engineering* in 1990 and a M.Sc. in *Data Communication Networks and Distributed Systems* in 1991, both from

University College London. Since October 1991 he has been a member of the Research Staff in the Department of Computer Science, involved in various communications related projects. He has worked particularly on Network and Distributed Systems management.

Graham Knight received his M.Sc. from UCL in 1980 and has since worked in the Computer Science department as a researcher and teacher. He is now a Senior Lecturer and has led a number of research efforts in the department. These have been concerned mainly with two areas; network management and ISDN. These interests have been pursued through three ESPRIT projects; INCA, PROOF and MIDAS. The network management activities have led ultimately to the OSIMIS management platform whilst the ISDN activities have resulted in the design, production and ultimate deployment of the UCL Primary Rate ISDN gateway.

David Gurle received his M.Sc. in *Computer Science and Telecommunications* in 1992 from Ecole Superieure d'Ingenieurs en Genie des Telecommunications et en Informatique (Paris – Fontainebleau). He worked for one year in Digital on CORBA and Intelligent Networks before joining CNET in 1993. Since then, he has worked on network and distributed systems management.

Philippe Rodier received his engineering degree in *Mechanical Sciences* in 1978 from Insitut National des Sciences Appliquees (Lyon). He worked for four years in Thomson CSF. He than worked for five years in Texas Instruments. He received his M.Sc. in *Computer Science* in 1988 from Cerics. Since 1988 he has worked in CNET and since 1992 he leads a group which focuses on applications of computing to network management.

SECTION SIX

Panel

15

Security and Management: The Ubiquitous Mix

Moderator: *Lee LaBARRE,* The MITRE Corporation, U.S.A.

Standards based management capabilities are becoming widely available in many network and distributed applications products; but unsecured access to the control capabilities they offer could allow accidental or deliberate damage to the network transmission and application services. Also, standards based security capabilities for such products are emerging that will require remote management of their security mechanisms, and security auditing.

The panelists will discuss the concepts relating security and management, the status and relationship of management and security standards, and issues related to their use in the secure management of resources in the data, telecommunications, and client server environments.

SECTION SEVEN

Performance and Accounting Management

16

An architecture for performance management of multimedia networks

Giovanni Pacifici and Rolf Stadler
Center for Telecommunications Research – Columbia University
Room 801 Schapiro Research Building
New York, NY 10027–6699
giovanni@ctr.columbia.edu, rolf@ctr.columbia.edu

Abstract

A principal requirement for multimedia networks is the ability to allocate resources to network services with different quality–of–service demands. The objectives of achieving efficient resource utilization, providing quality–of–service guarantees, and adapting to changes in traffic statistics make performance management for multimedia networks a challenging endeavor. In this paper, we address the following questions: what is the respective role of the real–time control system, the performance management system, and the network operator, and how do they interact in order to achieve performance management objectives? We introduce an architecture for performance management, which is based on the idea of controlling network performance by tuning the resource control tasks in the traffic control system. The architecture is built around the L–E model, a generic system–level abstraction of a resource control task. We use a cockpit metaphor to explain how a network operator interacts with the management system while pursuing management objectives.

Keywords

Multimedia networks, performance management, quality–of–service, resource control, network architectures

1 INTRODUCTION

Future multimedia networks will carry traffic of different classes, such as video, voice, and data. Each one of these has its own set of traffic characteristics and performance requirements. Sufficient resources, such as link bandwidth and buffer space, must be allocated to each call of a traffic class in order to guarantee the required quality–of–service (QOS).

As opposed to data networks, which perform best–effort data delivery, the concepts of time and resource are crucial to multimedia networks. Since multimedia networks provide QOS guarantees to user traffic, they contain real–time control functions as part of their traffic control systems. A typical service requirement for a data network is error correction, which is achieved by an end–to–end protocol; a typical requirement for a multimedia network is the guarantee of maximum end–to–end delay on a virtual circuit, which is based on the cooperation of distributed real–time control tasks. Therefore, *the tasks of controlling and allocating resources under QOS constraints are central in multimedia networks*. Note that resources are allocated on various levels of abstraction or granularity, such as per cell, call, or traffic class.

In a multimedia network environment, three entities are involved in the task of controlling and allocating resources — namely, the traffic control system, the performance management system, and the network operator. So far, little work has been done to define the role of these entities and to specify their interactions.

In this paper, we define the task of performance management for multimedia networks and provide an architecture for achieving this task. Specifically, we describe the role of the traffic control system, the performance management system, and the network operator, as well as their interactions. Further, we show how such an architecture relates to a standard management framework like that of ISO/CCITT (ISO, 1991). Two main directions of research activity concentrate on performance management. One direction deals with developing algorithms for resource control tasks that are designed to operate in real–time and make efficient use of resources in a dynamic environment. Usually, these efforts focus on improving the performance of a specific resource control task such as scheduling, buffer management, or admission control. The work described in (Lee and Ray, 1993) is an example of research in this field. The second direction involves activities within the standardized frameworks for network management, such as these developed jointly by the ISO and CCITT committees (ISO, 1991), or by the Internet community (Case et al., 1990; Rose and McCloghrie, 1990). These frameworks provide models to define the structure of management information, and they specify protocols for exchanging this data between functional entities known as managers and agents. Unified modeling of performance–related management information (Neumair, 1993) and the definition of generic interfaces for monitoring (Hayes, 1993) fall into this category.

While recognizing the importance and necessity of the above activities, we follow a third avenue of investigation in this paper, which is essential to meeting the challenges presented by the comprehensive performance management of future multimedia networks. First, our direction focuses on *managing the complete set of resource control tasks* in the traffic control system, by defining a generic abstraction of these tasks. This allows us, from a resource control perspective, to perceive the traffic control system as a collection of resource control subsystems with identical structures and control interfaces. This approach reduces the complexity of the performance management system which controls those subsystems, thus simplifying the design of a performance management framework. Second, having recognized that *performance management attempts to pursue potentially conflicting objectives*, such as the guarantee of QOS versus the obtaining of a high degree of multiplexing, we believe that a system which supports a human operator in implementing the desired strategy is crucial to a performance management framework.

We study *functional descriptions* of a performance management architecture in the form of data flow diagrams. We argue that this kind of description is necessary, in addition to the structural description supported by the standard management frameworks.

The paper is structured as follows. In Sec. 2, we discuss the task of performance management for multimedia networks and outline an architecture to perform this task. Specifically, we define the roles of traffic control and management systems, as well as that of the human operator. In Sec. 3, we refine the architecture by presenting a generic model for resource control tasks and by describing the interaction between the entities involved in the performance management task. Also, we discuss how our architecture relates to the ISO/CCITT management framework. Finally, in Sec. 4, important results of this work are summarized and a few remaining issues are discussed.

2 PERFORMANCE MANAGEMENT FOR MULTIMEDIA NETWORKS

We define the task of performance management for multimedia networks as that of pursuing (high–level) *management objectives*. These objectives can be grouped into two classes. The first class deals with providing network services that meet the needs of customer applications, such as service reliability and QOS guarantees. The second class deals with defining resource allocation strategies that provide benefits for the service provider. Controlling end–to–end packet delays and call blocking rates fall into the first class of management objectives, whereas pursuing high resource utilization and favoring one type of traffic (service) over others fall into the second category. The first class of management objectives favors increasing the resources allocated to each call, while the second class focuses on achieving a high level of resource utilization. These are conflicting requirements, which have to be balanced.

In multimedia networks two different subsystems operate on network resources — namely, a management system and a real–time traffic control system (Lazar, 1991). The following questions arise: What is the role of these systems in the performance management task? How do they interact to achieve high–level management objectives? What is the role of the network operator?

To address these questions, we introduce the architecture outlined in Fig. 1, which contains two subsystems and assumes the presence of an operator. The traffic control system directly regulates the competition for network resources and operates in real–time. The performance management system controls the operations of the traffic control system, while the network operator supervises these activities, pursuing management objectives. The different subsystems in Fig. 1 interact asynchronously and run on different time scales. In order to cope with the high–speed and dynamic nature of user traffic, the real–time traffic control system works on a time scale of μs to ms, while the performance management system and the network operator act on a time scale of seconds or minutes.

In the remainder of this paper, the term "performance management" will refer to the combined activity of all entities in the architecture shown in Fig. 1, whereas the term "performance management system" will be used only for a subsystem of this architecture, and may be thought of as a system structured according to the ISO/CCITT management framework.

2.1 The role of real–time control and performance management

Given the dynamic nature of traffic patterns in a multimedia network, a real–time *traffic control system* is required to regulate the competition for resources among the different traffic classes. The task of this system is to provide the QOS to network users, by utilizing network resources in an efficient way. The traffic control system can be seen as a collection of mechanisms, each of which operates asynchronously and solves a specific resource control problem. Examples of real–time control mechanisms are buffer management and scheduling, flow control, routing and admission control (Gilbert et al., 1991). The operations of the traffic control system can be tuned by changing *control parameters* associated with each mechanism. Changing the parameters of a single controller results in a different resource control policy for that controller and, in turn, may result in a different operating point for all other controllers. The *network state* is the result of the interaction of these real–time control mechanisms.

The task of the *performance management system* is to provide the functionality for pursuing management objectives. The performance management system executes its task by interacting with the real–time control system, following the monitor/control paradigm. This means that it monitors the network state and takes control actions in order to influence this state. Control actions result in changing specific parameters in the real–time control system. The interaction of the performance

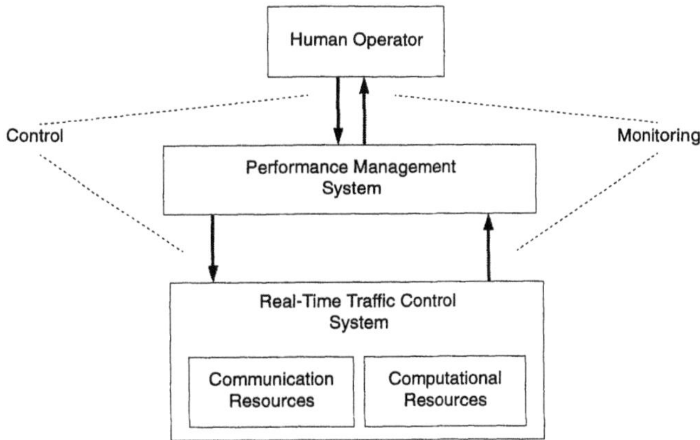

Figure 1: Performance management for multimedia networks

management system with the real–time control system is asynchronous, due to the different time scales on which the functional components in both systems run (Lazar and Stadler, 1993).

The management system is controlled by a human *operator*. Network operators perform actions to influence the network state, and are responsible for achieving management objectives. They monitor the network state represented as dynamic visual abstractions on a graphical interface, and perform operations by acting upon *management parameters*. A detailed example, describing the management parameters used for controlling the traffic mix in a multimedia network, is presented in (Pacifici and Stadler, 1995).

From the above discussion, we gather that the focus of performance management for future multimedia networks is different from that of classical approaches proposed for data networks. Influenced by the OSI Reference Model, performance management is often understood as monitoring and controlling protocol entities and associated service access points (Neumair, 1993; Cellary and Stroinski, 1989). While this is certainly valid for data networks, we argue that, for the case of multimedia networks, the focus should be different — namely, that of managing resource control tasks. In our approach, the performance management system interacts with the real–time control system, which, in turn, operates on protocol engine parameters and network resources. Executing performance management functions means operating management parameters that tune resource control tasks. We justify our point of view by the fact that multimedia networks provide real–time services, and resource control plays a central and critical role. Data networks, such as the existing Internet, do not guarantee QOS, and, as a result, their resource control tasks are much less complex.

2.2 The role of the network operator — the cockpit metaphor

Since the network is the heart of every distributed service, the failure of large parts of a network can result in a disaster for customers, and, as a consequence, for the service provider. Therefore, experienced operators supervise the operation of a network to prevent such scenarios from occurring. As we explained in the last section, supervision for future multimedia networks may be even more important than for today's networks, due to the complex interactions inside the traffic control systems. To explain the role of human operators and the way they interact with the management system while pursuing management objectives, we use the metaphor of a pilot flying an airplane.

A pilot operates the aircraft in reaction to and in anticipation of environmental conditions, as expressed by wind, visibility, air pressure, etc. The pilot has no influence on the environment and on how it evolves. In a similar way, a network operator performs actions to handle the current and anticipated load pattern of the network traffic, while guaranteeing the required QOS to network services and allowing a high utilization of network resources. The traffic load pattern changes over time and cannot be influenced by the operator. However, operators are responsible for maintaining the network state within a stability region that allows reliable operations. When the traffic pattern changes, so does the network state, and the operator "navigates" the network state back into the stability region, if necessary.

A pilot operates on high–level controls such as yoke, handles, and control sticks, the positions of which relate to specific settings of the airplane's control surfaces such as elevators, ailerons, rudders, and flap positions. Similarly, the network operator sets management parameters. Modifications to these parameters are translated by the management system into control parameters that influence the way network control mechanisms operate, thereby affecting the network state. Operators observe the reaction of the system in response to control actions in the same way a pilot observes the flight instruments changing to adjustments of the flight controls. The relationships between an aircraft's speed and vertical velocity, on the one hand, and elevators and throttle, on the other, are complex, and a pilot understands them through practice. Likewise, we think that understanding certain relationships between management parameters and the network state in large multimedia networks will be based in large part on experience and expertise.

While steady–state conditions hold, an autopilot system can control the aircraft and perform automated functions. In difficult situations or during unprecedented events, however, the pilot takes control. Such situations might include a sudden change in the weather or the occurrence of turbulences. Also, the takeoff and landing procedures are normally executed by the pilots themselves. We believe that, in an analogous way, performance management functions can be automated when the network operates in a stability region subject to minor fluctuations in the traffic load patterns. Operators, however, will always be needed to handle difficult situations. In such conditions, they will decide which functions should be executed and when they should be run assisted perhaps by an expert system. Aircraft takeoff and landing operations can be compared to adding or removing parts of the network during operation — tasks that have to be performed in every network on a regular basis and need human supervision.

3 AN ARCHITECTURE FOR PERFORMANCE MANAGEMENT

In this section, we develop a performance management architecture that integrates the network subsystems that participate in the resource management task. We present an abstraction of the traffic control system with respect to resource control and utilize this model to define a framework that allows management operations to influence the behavior of traffic control mechanisms.

3.1 Modeling resource control tasks — the L–E model

The traffic control system of multimedia networks contains a collection of resource control subsystems, each of which implements a specific task, such as admission control or routing. Each of these subsystem regulates access to a specific resource, by responding to requests that are generated by functions external to the resource control subsystem. The behavior of the resource control task (i.e., the way it responses to service requests) can be influenced by changing a set of control parameters associated with the subsystem.

The main functional components, of a resource control subsystem, together with the interactions among components and with the outside world, are identified in the *L–E model* shown in Fig. 2. We use a functional model in Fig. 2, in order to focus on functional components as well as the data exchanged and accessed by them (Rumbaugh et al., 1991).

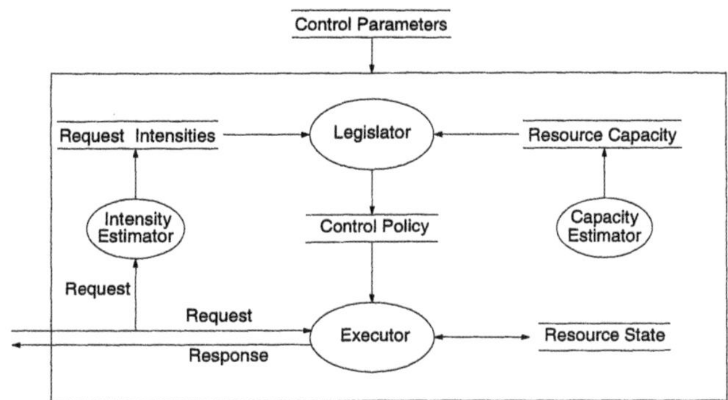

Figure 2: Data–flow diagram of the L–E model

The main idea behind the L–E model is that the task of computing a control policy for allocating a resource in a dynamic environment is separated from the task of binding this resource to a particular communication service. Following this separation, the model contains two types of mechanisms, the *legislator* and the *executor* (see Fig. 2). A pair of these mechanisms, one of each type, interact to perform a specific resource control task, e.g., controlling access to a physical network link.

The *legislator* generates a set of rules, which must be observed when allocating a resource. This set of rules is called the *control policy*. The *executor* regulates access to the communication resource while observing the current control policy. In other words, the executor implements the control policy computed by the legislator.

The executor is driven by external stimuli. Its task is to serve requests that are initiated by functions external to the resource control subsystem. The legislator, in contrast, is either invoked by the executor or runs on its own and periodically recomputes the control policy. It performs its operation usually on a time scale much slower than that of the executor, since the computational complexity of a resource control subsystem resides in the legislator part.

Legislator and executor interact by sharing a data object — the control policy — which is written by the legislator and read by the executor. The interaction between legislator and executor can be either synchronous or asynchronous. In the synchronous case, the legislator invokes the executor, e.g., in the form of a function call. The routing scheme in the plaNET traffic control system (Gopal and Guerin, 1994) works in this way. In the case of asynchronous interaction, legislator and executor form a loosely coupled subsystem. Each mechanism runs on its own time scale, and they communicate asynchronously via the shared policy object. This approach can be found in the adaptive routing schemes of today's long distance telephone networks (Girard, 1990). Note that asynchronous interaction between legislator and executor allows them to run independently and on different time scales. Therefore, they can be optimized according to different requirements: the executor guarantees response times, while the legislator optimizes the utilization of the resource, e.g., by minimizing a given cost function.

The L–E model allows for a wide range of possible implementation decisions. It covers single threaded, distributed, as well as parallel implementations of resource control subsystems, depending on whether the mechanisms are intended to run on the same or different machines and whether their interaction is designed to be synchronous or asynchronous. Further, the model supports the case where several executors share the same legislator.

In order to manage resources in an efficient way, the resource control system of multimedia networks must be able to adapt dynamically to changes in the network state and traffic statistics. In the L–E model this is achieved by the legislator, which periodically recomputes the control policy, taking into account the latest value of the request intensities and the resource capacity.

Our model contains two mechanisms that generate the dynamic abstractions needed by the legislator to recompute the control policy. The *intensity estimator* calculates the request intensities, by filtering the stream of service requests, and the *capacity estimator* computes the resource capacity, based on traffic statistics and configuration data. Note that the capacity of a network link (expressed in cell/sec) can be seen as a constant configuration parameter, while the capacity of a high–level abstraction of the same link (i.e, the maximum number of video, voice and data calls that can be multiplexed at any given time on that link) varies continuously, following changes in traffic characteristics. Examples of capacity estimation techniques that provide high–level abstractions of link resources can be found in (Ferrari and Verma, 1990; Hyman et al., 1991). Both the intensity and capacity estimators run on the same time-scale as the legislator and generate new estimates for each new computation of the control policy.

The L–E model provides the framework for dynamically influencing the resource control task, by associating *control parameters* with each mechanism, i.e., with legislator, executor, intensity estimator, and capacity estimator. Control parameters of a legislator include the QOS constraints for handling requests and the utility generated for granting access to the resource, as well as the time interval between two consecutive recomputations of the control policy. The length of the estimation interval, which reflects the capability of the system to respond to changes in the traffic statistics, is a typical control parameter for the intensity estimator. The robustness of the capacity estimation processes is a parameter associated with conflicting objectives. In the case of link admission control, it relates to the trade–off between using the link bandwidth efficiently and providing cell–level QOS guarantees (Pacifici and Stadler, 1995).

All these control parameters provide the fundamental capability to influence how a resource control system works, namely, by affecting the QOS constraints under which it operates, its adaptivity related to changes in the environment, and its robustness in guaranteeing the QOS under varying traffic loads and conditions.

The L–E model is based on our experience with designing and implementing traffic control mechanisms for multiclass networks. Tab. 1 identifies some elements of the L–E model for the most important resource control tasks in a multimedia system. For example, the TCP/IP flow control task (Jacobson, 1988) can be modeled as an end–to–end protocol entity (executor) that performs transport operations according to a maximum window size (control policy). The window size is determined by the flow controller (legislator), which computes the size of the window using the estimated link bandwidth available to a specific user (capacity estimation) and the transmission rate (request intensities) of the specific user source (Jacobson, 1988). The system state is defined by the number of transmitted cells not yet acknowledged.

The tasks of scheduling and buffer management — to give another example — can be modeled in the same fashion. Here, the policy is defined by time sharing (scheduling) and space partitioning

Task	Control Policy	Resource State	Resource Capacity	Request Intensities
Admission Control	State transition matrix	Number of active calls	Schedulable Region	Call arrival rates and call holding times
VC Routing	Set of routes	Number of active calls per link	Collection of Schedulable Regions	Call arrival rates and call holding times
Flow Control	Window size	Number of cells in system	Available link bandwidth	Cell arrival rates
Buffer Mngt.	Buffer partitions	Number of cells in buffer	Buffer space	Cell arrival rates
Scheduling	Link partitions	Number of cells in buffer	Link bandwidth	Cell arrival rates

Table 1: Modeling resource management tasks in a multimedia network

(buffer management) of the resources among each traffic class. The system state is determined by the number of cells in the buffer, while the request intensities are given by the cell arrival and departure rates. The link speed and the buffer size define the resource capacities, which are available as configuration parameters. The admission control task and its functional model are discussed in (Pacifici and Stadler, 1995).

With the above discussion we want to illustrate that that our model is truly generic in the sense that it is not restricted to a particular resource control task. Note that Table 1 is based on specific control algorithms. The choice of different algorithms can result in different table entries for control policy, resource state, etc.

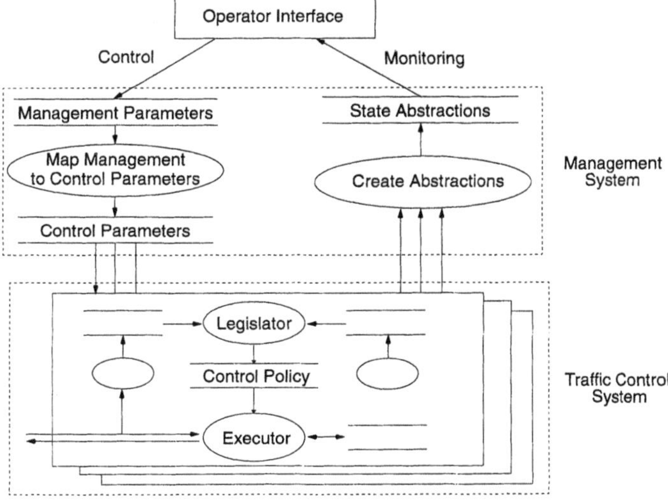

Figure 3: Interaction between the operator, the management system, and traffic control tasks

3.2 Integrating resource control and performance management

From the point of view of performance management, the traffic control system can be seen as a set of subsystems, each performing a specific resource control task. As described in Sec. 3.1, a set of control parameters can be associated with each resource control subsystem. These parameters define the control interface between the management and traffic control systems. The management system writes them while the traffic control system reads them. This scheme allows for asynchronous interaction between functional components of both systems, thus enabling these components to run on different time scales and at different locations. By modifying control parameters, the management system influences the behavior of a resource control subsystem, and, therefore, changes the way resources are allocated.

There are two main reasons for including the L–E model in a framework for performance management. First, in order to tune resource allocation, specific knowledge about the algorithms involved in resource control and the way the resource control subsystem is implemented is not required in the management system. This allows a clear split between performance management and the traffic control system, with the set of control parameters defining the control interface. Second, the L–E model provides generic classes of control parameters that can be made accessible to the management system.

The management system presents a high–level view of the network state to the operator in the form of dynamic visual abstractions. The operator manipulates a set of management parameters. Changes in these parameters are translated into modifications to control parameters that influence the behavior of traffic control components (see Fig. 3).

A straightforward way to support an operator with management capabilities is to make each control parameter directly available at the operator interface. For example, a control parameter that is defined within a certain interval can be associated with a management parameter (both values may be related by a monotonic mapping, such as with a linear or logarithmic function), which can be presented to the operator by means of the visual abstraction of a slider. Changing the position of the slider will result in a change of the control parameter, which, in turn, will affect the corresponding resource control subsystem.

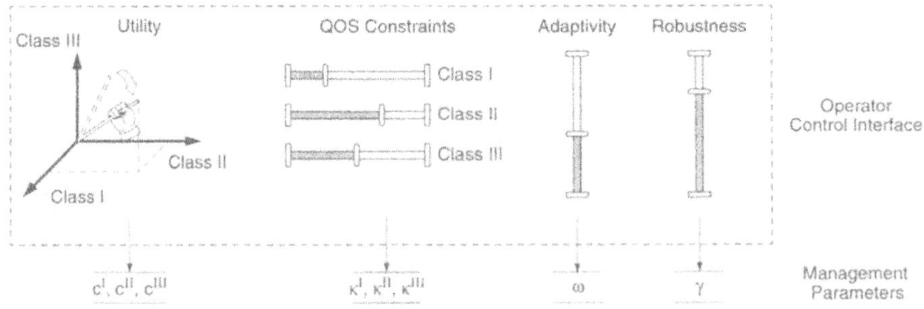

Figure 4: Visual abstractions and management parameters associated with the task of managing the communication resources of a multimedia network

Figure 4 introduces a sample set of management parameters associated with the task of managing the communication resources of a multimedia network, and shows the visual abstractions that allow an operator to change the management parameters, thus affecting the performance of the network. In this example, the management parameters relate to network utility, QOS constraints, as well as adaptivity and robustness of the resource control system. In (Pacifici and Stadler, 1995) it is shown how the task of link admission control can be managed, by using these four different types of management parameters.

Obviously, a network operator needs the capability to tune not only each single controller in the traffic control system, but also sets of controllers simultaneously, for example, all controllers on a specific route or inside a certain network region. Therefore, the operator interface provides *selection capabilities* that allow an operator to choose a set of objects (e.g., links, nodes, network regions, or the whole network) that determine the domain of controllers on which a management operation is to be executed. A management operation thus involves a selection operation and the setting of a management parameter. The management system then maps this data onto both the settings for control parameters and the domain of controllers affected by the operation, and distributes the settings to the traffic control system.

Note that a single management parameter can be associated with several classes of controllers. A management parameter related to robustness, for example, can be associated with control parameters in resource control systems that implement call routing, call admission control, and cell scheduling. Again, the mapping from the management to the various control parameters is performed by the management system.

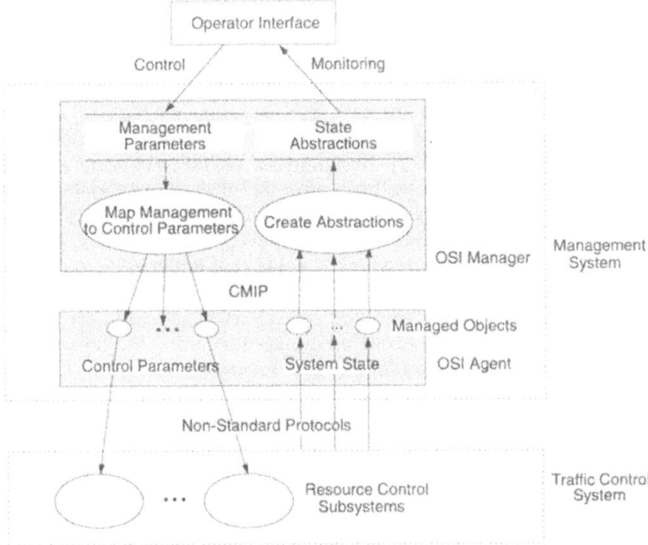

Figure 5: Performance management within the OSI framework

Having described the concepts of our architecture, the question arises, how do they relate to a management framework, such as the one standardized by ISO (ISO, 1991)? In that framework, the system to be managed is conceptualized as a global database, the Management Information Base (MIB). The MIB contains a set of managed objects, which represent network entities. Managed objects are implemented on OSI agents, and can be accessed and manipulated by OSI managers by a standard protocol called CMIP. Therefore, monitoring and controlling a system means reading and changing managed objects in a standardized way.

Figure 5 shows our approach. We propose that the control parameters associated with network mechanisms be modeled and implemented on agents as managed objects, which are part of the management system. Further, network state information should be modeled and implemented in the same way, and thus be accessible for management purposes. The mapping and abstraction functions should be implemented on a manager, because they support network functions that operate on the global space of managed objects, which will be distributed over several agents. While the interaction between the manager and the agents is standardized, there is no standard protocol for the communication between a managed object and a resource control mechanism.

4 DISCUSSION

We believe that the architecture presented in this paper opens the way for building powerful tools for network operators who manage the resources of a multimedia network. The selection functionality allows them to choose a set of objects on the operator interface, so as to define a domain of controllers (such as a link, a path, a network region, or the whole network) on which a management operation is to be executed. Operators can change, for every selected domain, the QOS constraints and the utility generated by the user traffic in this domain, and they can tune the adaptivity and robustness of resource control functions in the same fashion. These tools support network operators in their task of navigating the managed system — here we use a term from the cockpit paradigm — effectively and safely. Operators have at their disposition high–level controls in order to keep the appropriate balance when pursuing different, potentially conflicting objectives. These objectives include providing QOS on the cell–level and call–level, keeping up a high degree of multiplexing, securing network utilization, and maintaining a highly responsive and yet stable system.

We are currently experimenting with the design of our architecture using a network emulator, which runs functional components of a traffic control and management system of a multimedia network. The emulator is implemented on a KSR parallel machine. It emulates a 50 node network, in which traffic statistics can be dynamically changed at every network access point. The operator interface runs on an Indigo2 workstation, which is connected to the KSR via an ATM link. We can demonstrate, for example, how the traffic mix in the network can be influenced by executing management operations that affect link resource controllers in selected network domains. The effect of management operations can be observed in real-time, using the capability of visualizing call blocking rates and network utilization for any selected network domain.

All examples presented in this paper relate to managing communication resources — indeed, one of the classic subjects in traffic control. Since our framework is generic, other resources, such as computational resources, can be included. Because the traffic control system needs resources to operate, these can be abstracted using the L–E model, and, therefore, their performance can be managed according to our framework. For telephone networks, performance management of traffic control systems has been recognized as a crucial issue (Kühn et al., 1994), and we believe that it will play an equally important role in emerging multimedia networks.

Finally, we believe that our framework can be applied to *managing the performance of real–time services*, such as access to a video server or to a multimedia database, since the resource control systems associated with these services can be abstracted using the L–E model. Furthermore, it can be extended to include the computational resources of multimedia workstations, thus leading to a framework for managing and controlling resources in a distributed multimedia application environment. The architecture proposed in (Campbell et al., 1994) can be seen as a step in this direction, though network management aspects are not addressed there. Note that our approach allows the *integration of the network management and service management tasks* — as far as performance is concerned — which opens interesting perspectives for further investigation.

References

Campbell, A., Coulson, G., and Hutchison, D. (1994). A quality of service architecture. *Computer Communication Review*, 24(2):6–27.

Case, J., Fedor, M., Schoffstall, M., and Davin, C. (1990). *A Simple Network Management Protocol (SNMP)*. RFC–1157.

Cellary, W. and Stroinski, M. (1989). A performance management architecture for protocol entity optimization. In Meandzija, I. B. and Westcott, J., editors, *Integrated Network Management, I*, pages 227–234. Elsevier Science (North–Holland), Amsterdam, The Netherlands.

Ferrari, D. and Verma, D. C. (1990). A scheme for real-time channel establishment in wide-area networks. *IEEE Journal on Selected Areas in Communications*, SAC-8(3):368–379.

Gilbert, H., Aboul-Magd, O., and Phung, V. (1991). Developing a cohesive traffic management strategy for ATM networks. *IEEE Communications Magazine*, 20(10):36–45.

Girard, A. (1990). *Routing and Dimensioning in Circuit–Switched Networks.* Addison–Wesley, Reading, MA.

Gopal, I. and Guerin, R. (1994). Network transparency: The plaNET approach. *IEEE/ACM Transactions on Networking*, 2(3):226–239.

Hayes, S. (1993). Analyzing network performance management. *IEEE Communications Magazine*, 31(5):52–58.

Hyman, J. M., Lazar, A. A., and Pacifici, G. (1991). Real-time scheduling with quality of service constraints. *IEEE Journal on Selected Areas in Communications*, 9(7):1052–1063.

ISO (1991). *Information Processing Systems — Open System Interconnection — Systems Management Overview*. ISO/IEC, IS 10040.

Jacobson, V. (1988). Congestion avoidance and control. In *Proceedings of the ACM SIGCOMM*, pages 316–329, Stanford, CA.

Kühn, P. J., Pack, C. D., and Skoog, R. A. (1994). Common channel signaling networks: Past, present, future. *IEEE Journal on Selected Areas in Communications*, 12(3):383–394.

Lazar, A. A. (1991). An architecture for real-time control of broadband networks. In *Proceedings of the IEEE Global Telecommunications Conference*, pages 289–295, Phoenix, AZ.

Lazar, A. A. and Stadler, R. (1993). On reducing the complexity of management and control of broadband networks. In *Proceedings of the Workshop on Distributed Systems: Operations and Management*, Long Beach, NJ.

Lee, S. and Ray, A. (1993). Performance management of multiple access communications networks. *IEEE Journal on Selected Areas in Communications*, 11(9):1426–1437.

Neumair, B. (1993). Modeling resources for integrated performance management. In Hegering, H. and Yemini, Y., editors, *Integrated Network Management, III*, pages 109–121. Elsevier Science (North–Holland), Amsterdam, The Netherlands.

Pacifici, G. and Stadler, R. (1995). Integrating resource control and performance management in multimedia networks. In *Proceedings of the IEEE International Conference on Communications*, Seattle, WA.

Rose, M. and McCloghrie, K. (1990). *Structure and Identification of Management Information for TCP/IP based Internets*. RFC–1155.

Rumbaugh, J., Blaha, M., Premerlani, W., Eddy, F., and Lorensen, W. (1991). *Object–Oriented Modeling and Design*. Prentice–Hall, Englewood-Cliffs, NJ.

Giovanni Pacifici received the *Laurea* and the *Research Doctorate* degrees from the University of Rome "La Sapienza" in 1984 and 1989 respectively. As a student, his main activities were focused on the performance evaluation of access control protocols for local and metropolitan area networks, with an emphasis on the integration of voice and data. In the course of his studies, he was a Visiting Scholar at the Center for Telecommunications Research, Columbia University, where he designed and implemented a monitoring and traffic generation system for MAGNET II, a high speed metropolitan area network. In 1989, he joined the staff of the Center for Telecommunications Research as a Research Scientist. His research interests include resource control, performance management and real-time quality of service estimation for broadband networks. Dr. Pacifici is a member of IEEE and ACM.

Rolf Stadler received a master's degree in mathematics and a Ph.D. degree in computer science from the University of Zurich in 1984 and 1990, respectively. His thesis work focused on the specification of communication systems. During 1991 he was a post-doctoral researcher at the IBM Zurich Research Laboratory, involved in developing a traffic management system for a broadband LAN/WAN environment. From 1992 to 1994 he was a Visiting Scholar at the Center for Telecommunications Research, Columbia University. In 1994 he joined the staff of the Center for Telecommunications Research as a Research Scientist. His current interests include management, control, and services with respect to broadband networks. Dr. Stadler is a member of IEEE and ACM.

17

Network Performance Management Using Realistic Abductive Reasoning Model

G. Prem Kumar and P. Venkataram
Department of Electrical Communication Engineering
Indian Institute of Science
Bangalore - 560 012, INDIA
(Tel : (+91) (080) 3340855; Fax : (+91) (080) 3347991
e-mail : {prem, pallapa}@ece.iisc.ernet.in)

Abstract

Performance degradation in communication networks can be viewed to be caused by a set of faults, called *soft failures*, owing to which the network resources like bandwidth can not be utilized to the expected level. An automated solution to the performance management problem involves identifying these soft failures and use/suggest suitable remedies to tune the network for better performance. Abductive reasoning model is identified as a suitable candidate for the network performance management problem. An approach to solve this problem using the *realistic abductive reasoning model* is proposed. The realistic abductive inference mechanism is based on the parsimonious covering theory with some new features added to the general abductive reasoning model. The network performance management knowledge is assumed to be represented in the most general form of causal chaining, namely, hyper-bipartite network. Ethernet performance management is taken up as a case study. The results obtained by the proposed approach demonstrate its effectiveness in solving the network performance management problem.

Keywords

Network Performance Management, Network Fault Diagnosis, Realistic Abductive Reasoning Model, Parsimonious Covering Theory, Ethernet Performance Management.

1 INTRODUCTION

Communication network management (Cassel,1989), (Sluman, 1989) is drawing a lot of attention as the networks are spreading geographically and the number of heterogeneous devices and services supported by them are increasing exponentially. Network performance management is a complex part of present day network management (Hayes, 1993). The necessity for performance management arises when the network continues to function but in a degraded fashion because of one or more of the reasons such as temporary congestion that causes delayed transmission, failure of higher level protocols and mischevious users (Metcafe, 1976). In this work, the performance degradation is considered as a soft failure since the network is only partially afffected but is still in operation; on the other hand, if some of the devices in the network are not functioning or if the network is not able to run, then it is considered as a hard failure.

There are some specialized problems in the network management, that have to be considered. The entire information required for management may not be available at once and there may be missing information, both of which, the management center needs to confirm with the respective managed nodes. In this paper, we present a two step approach that aids the network performance management. First step involves identification of a set of faults from the given soft failures by using Realistic Abductive Reasoning Model (*Realistic_ARM*) (Prem, 1994) that is modelled as a diagnostic problem solver. In the second step, the system suggests suitable remedies to tune the network for better performance.

The fundamental idea behind abductive reasoning is "reasoning to the best explanation" (Pople, 1973). Based on the given symptoms (or manifestations), initially, it uses forward chaining to anticipate all the possible causes of the symptoms (also called disorders), and then it uses backward chaining to confirm whether the explanation is supported to a required degree of confidence. Ever since parsimonious covering theory (Reggia, 1985), (Peng, 1987), (Peng, 1990) is developed for abductive reasoning with sound mathematical foundation, there has been a shift in attention from deductive reasoning to abductive reasoning. Abductive reasoning generates all the possible explanations which may require further refinement to arrive at appropriate covers (Bylander, 1991). Deductive reasoning, though generates only appropriate covers, will not generate those required covers which it would have generated in the presence of missing information. Both, abductive and deductive reasoning strategies are far from the reality. The proposed approach, which uses Realistic_ARM for solving the network performance management problem is a compromise between the two strategies and attempts to find explanations for a given set of symptoms. The knowledge used by Realistic_ARM is assumed to be represented in the most general form of causal chaining, namely, hyper-bipartite network.

We briefly describe the realistic abductive reasoning model in Section 2. Section 3 discusses the network performance management problem and highlights the applicability of realistic abductive reasoning model in solving the problem. The algorithm is presented in Section 4. A case study, Ethernet performance management is discussed in Section 5. And finally, conclusion follows in Section 6.

2 THE REALISTIC ABDUCTIVE REASONING MODEL

Realistic abductive reasoning model (Prem, 1994) is a modified version of the abductive reasoning model (Peng, 1990) to solve the diagnostic problems effectively in a realistic scenario. This model uses abductive inference mechanism based on the parsimonious covering theory with some new features added to the general model of diagnostic problem solving.

2.1 Notation

Definition 1 : The *diagnostic problem, P,* is a 4-tuple $< M, D, H, L >$ where $M = \{m_1, m_2, \ldots, m_e\}$ is a set of manifestations causing a set of disorders, $D = \{d_1, d_2, \ldots, d_f\}$ either directly or via a set of hypotheses (which could be a manifestation or a disorder), $H = \{h_1, h_2, \ldots, h_r\}$. And, $L = \{l_{i,j} | i \in M \cup H, j \in H \cup D\}$ is a set of causal links joining any two related elements in M, H and D. In a general case, there are many causes to each of the manifestations, many effects to each of the disorders, and both causes and effects to each of the hypotheses.

Definition 2 : *Hyper-bipartite network* is an acyclic graph, G = $< M, D, H, L >$, where M is a set of manifestations (in the bottom most layer), D is a set of disorders (in the top most layer) and H is a set of hypotheses (in one or more intermediate layers). All elements of M, H, and D are represented as nodes in their respective layers. And, L is a set of edges joining any two related nodes in M, H and D. Let the number of layers in the graph be N.

Definition 3 : *Layered network* is an acyclic graph $G^* = < M, D, H^*, L^* >$, constructed from the hyper-bipartite network G, where each node belonging to M, H^* and D are connected only to the nodes in its neighboring layers. The procedure to convert a hyper-bipartite network into a layered network, *Build_Layered_Net*, is discussed in Section 4.

Definition 4 : A *symptom* is an observed manifestation/hypothesis/disorder.

Definition 5 : A *volunteered symptom* is a hypothesis/disorder at layer i $(1 < i \leq N)$ observed to be present.

A hypothesis/disorder *covers* a symptom if there is a causal pathway from the hypothesis/disorder to the symptom.

Definition 6 : A *cover* or an *explanation* is a set of hypotheses/disorders that covers all the given symptoms.

In solving the diagnostic problem, P, where the representation is in the form of a layered network, G^*, jth cover of layer i (1 ≤i <N), $c_j{}^i = \{h_1, h_2, \ldots, h_s\}$, is a set of disorders at layer $(i+1)$, which covers the symptoms at layer i. At each layer, there may be more than one explanation for the given symptoms and they are placed in the *cover set* of that layer, $C_i = \{c_1{}^i, c_2{}^i, \ldots, c_t{}^i\}$. While at the top most layer, a volunteered symptom is simply added to each cover of the cover set if it is not already present.

Definition 7 : *Intermediate cover* (t_k^i), of layer i, is a cover belonging to the the cover set (T_i) being generated, which provides an explanation for the symptoms being explored but may or may not provide explanation for the unexplored symptoms.

Definition 8 : *Direct disorder*, $dd \in D$, of a manifestation/hypothesis is the direct cause of the manifestation/hypothesis mapping on to the top most layer.

Definition 9 : *Irredundancy* is the parsimonious criteria used in Realistic_ARM to refine the cover set by eliminating the redundant covers. A cover c_j^i is redundant if there exists another cover c_k^i, which is a subset of c_j^i.

Definition 10 : The *solution* to a diagnostic problem is the set of all explanations for the given symptoms.

2.2 The Realistic_ARM

Inference process used in the abductive reasoning, based on parsimonious covering theory, is similar to the model of sequential *hypothesis-test* cycle of human diagnostic problem solving (Peng, 1990). The "hypothesis" part *covers* the given symptoms and generates *parsimonious* covers. The "test" part of it is the question-answering process to explore for more symptoms to discriminate the generated covers. This cycle continues, taking one symptom at a time, until all relevant questions are asked and all symptoms are processed.

The diagnostic knowledge in Realistic_ARM is represented in the form of a hyper-bipartite network. In this model, all the manifestations/hypotheses have direct disorders. All the elements belonging to M, D, H^* exist only in their respective layers. Any symptom belonging to any layer may appear at any time during the reasoning process. All the possible manifestations that could be present in a layer because of the existing manifestions through common disorders (the disorder a manifestation causes along with some other manifestations/hypotheses) are queried at once before starting the reasoning process for that layer. The advantage here is two fold : (i) all the covers will be generated with the same set of symptoms, and (ii) especially in the networking environment, queries for the presence of manifestations need a lot of time in collecting the information and it is good to present them at the earliest.

In the rest of this section, we describe the realistic abductive reasoning model approach to solve a general diagnostic problem.

Solution to the diagnostic problem where the knowledge base is represented in the form of a hyper-bipartite network is found by converting it into a layered network and solving it as a series of bipartite networks, moving upwards one layer at a time. A cover for the symptoms in layer $(i-1)$, c_j^{i-1}, becomes symptoms for layer i. (C_0 is initialized to $\{\emptyset\}$.) In addition to these, some more symptoms that are added at layer i by user input (or interactive querying) together form jth *symptom set* at layer i, for which an intermediate cover set T_i is built in the following way : at layer i, starting with a symptom, all its disorders get into different covers since each of them separately provide an explanation for that symptom. For the subsequent symptoms, if a cover is already providing the explanation, the cover will remain unchanged. Otherwise, for an intermediate cover, t_k^i, that is not providing an explanation for a symptom, m_l, append only those disorders of m_l, which are supported by prespecified number of symptoms, one at a time to form new covers and delete t_k^i. If no new covers are generated, then append the direct disorder of m_l to t_k^i. After the covers are built to provide explanation to all the symptoms, the parsimonious criteria, namely, *irredundancy* is applied and a few covers are eliminated. T_i is then appended to the cover set C_i and reinitialized to $\{\emptyset\}$ to take up next symptom set of that layer. When all the symptom sets are explored, C_i is made irredundant. This process repeats for all the layers till the top most layer is reached. At the top most layer, the volunteered symptoms are simply added to each cover of the cover set if they are not already present. After covering the symptoms of the top most layer, if there are any more symptoms left uncovered, the reasoning process repeats from the bottom most layer. The intention here is to *cover* the symptoms only at their respective layer along with the other symptoms of that layer to avoid too much of guess in generating the covers and retain the simple layered network architecture with out additional dummy nodes. For details, refer (Prem, 1994).

3 ADAPTATION OF REALISTIC_ARM TO SOVLE NETWORK PERFORMANCE MANAGEMENT PROBLEM

The fact that the Realistic_ARM is a compromise between the extreme cases of abductive and deductive reasoning models is utilized to solve the network performance management problem. Since in the network scenario, there may be missing information and all the infomation required for fault identification may not be available at the time of diagnosis. If the deductive reasoning mechanism is applied to such a problem, the fault can not be identified since all the symptoms may not be present. At the same time, the abductive reasoning approach will result in too many number of unwanted explanations for a given set of symptoms and subsequently, it will be very difficult to say which is the correct explanation (a set of faults), that caused the degradation in the network performance. The realistic abductive reasoning model discussed in the previous section can be found to satisfy the requirements of the problem.

The prespecified number of symptoms required to support a given symptom before concluding a disorder (fault) is a variable. This can be set based on the incremental step in which the performance needs to be tuned. Intermediate layer of diagnostic knowledge base enables a hypothesis to be given in any form, namely, from the lower layers as a result of reasoning process or as a symptom in the respective layer. The direct disorder to every symptom whether it is in the bottom most layer or in the intermediate layer allows the fault to be concluded very precisely without waiting for the rest of the symptoms to conclude the faults in the topmost layer.

The realistic abductive reasoning model in its original form allows the reasoning mechanism to query back the user (here, the managed nodes) to confirm the missing symptoms before concluding any fault. But, since performance tuning can not be deferred for such a long time before all the required symptoms are obtained, this can be relaxed since the model allows some tolerance on the number of symptoms required to conclude reason for degradation in the network performance.

By suitably constructing the network fault knowledge model required for performance tuning, this model can be found to give very good results for the problem. A case study of Ethernet performance management, discussed in Section 5, illustrates this approach.

4 THE ALGORITHM

The performance management model (described as algorithm, Performance_Mgt) presented in this section accepts a set of symptoms given as soft failures from the monitoring information and identifies remedies for the set of faults concluded using Realistic_ARM. Since the knowledge base which is in the form of hyper-bipartite network is converted into a layered network, the symptoms can be allowed to enter at any stage of the inference process.

Nomenclature

1. *temp_man* is a set of symptoms at layer of inference. (By both, one of the covers of the previous layer and the symptoms of that layer.)
2. *prim_man*, is a set of symptoms available at all the layers, holds the symptoms provided by the user excluding the symptoms explored in all the previous layers (if the manifestation is present in the next layer because of dummy nodes created by *Build_Layered_Net*, they are retained).

3. *sec_man*, is a set of symptoms available at all the layers, holds all the symptoms that are provided by the user.
4. *More_Manifs*, a boolean, is TRUE if there are any more symptoms found to exist at a layer by either input or when asked interactively through common disorders of the existing symptoms. Otherwise it is FALSE.

```
Algorithm Performance_Mgt
{
  var i,j,pre_lay_cov_count : int;
  Call procedure Build_Layered_Net;
  Read the given symptoms into prim_man and sec_man.
  C0 = { ∅ };
  loop :
  for(i = 1; i < N; i++)
  {
    pre_lay_cov_count = |C(i-1)| ; j = 0;
    For all the symptoms of layer i, query the related manifestations
         through common disorder and place them in prim_man.
    do
    {
      temp_man = ∅;
      if(|C(i-1)| >0)
         Get jth cover of layer (i - 1) into temp_man.
      Append symptoms of layer i that are present in prim_man to temp_man.
      Ti = Gen_Covers(temp_man);/*Generate covers -
                 for the symptom(s) present in temp_man.*/
      Ci = append(Ci, Ti);
    }while(--pre_lay_cov_count> 0);
    Delete layer i symptoms from prim_man if they do not exist in layer (i + 1).
    Remove redundant covers from Ci.
  }//end of for(i< N, no. of layers)
  Append the disorders of layer N present in layer prim_man to each of -
       the covers if they do not already exist.
  Remove redundant covers from CN.
  Delete the symptoms of layer N from prim_man.
  if(there some symptoms are still left in prim_man)
    prim_man = ∅; Copy sec_man to prim_man and goto "loop".
  Output the final covers, CN.
  Suggest suitable remedies for CN to improve the network performance.
}//end of algorithm Performance_Mgt
```

```
function Gen_Covers(temp_man)
{
  var k,p,q,u,v : int;
      cov_added : boolean;
  T_i = { ∅ };
  for(k = 0; k < |temp_man|; k++)
  {
    if(k == 0)
    {
      for(u = 0; u < v, no. of disorders of kth symptom; u++)
      {
       if(uth disorder of symptom k is supported by a prespecified number -
            of symptoms)
          t_{|T_i|++}^i = {uth disorder};
      }
      if(|T_i| == 0)
          t_{|T_i|++} = {direct disorder of symptom k};
    }//end of if(k == 0)
    else //if(k ≠ 0)
    {
      q = |T_i|;
      for(p=0; p< q; p++)
      {
        cov_added = FALSE;
        for(u = 0; u < v, no. of disorders of symptom k; u++)
        {
          if(uth disorder of symptom k is supported by a prespecified number -
               of symptoms and ∈ t_p^i) /* t_p^i is already a cover for k */
            goto next_cover;
        }//end of for( u < v)
        for(u = 0; u < v, no. of disorders of symptom k; u++)
        {
          if(uth disorder of symptom k is supported by a prespecified -
               number of symptoms)
            t_{|T_i|++}^i = append(t_p^i, uth disorder); cov_added = TRUE;
        }//end of for( u < v)
        if(cov_added == TRUE)
          Mark t_p^i for deletion; goto next_cover;
        t_{|T_i|++}^i = append(t_p^i, direct disorder of symptom k);
        next_cover: ;
      }//end of for(p < q)
      Delete those covers marked for deletion from T_i and update |T_i|.
    }//end of else if(k ≠ 0)
    T_i = Gen_Irr_Covers(T_i); //Make irredundant after each symptom is explored
  }//end of for(k< |temp_man|)
  return T_i;
}//end of function Gen_Covers
```

```
procedure Build_Layered_Net
{
  Retain the nodes of the hyper-bipartite network.
  For each layer i, ( 1 ≤ i ≤ (N − 2) ), of hyper-bipartite network :
      if there is a link from layer i to layer (i + 1), retain the same in the -
           layered network.
      if there is a link (say l_{hm,hn}) from manifestation/hypothesis at layer i -
           to hypothesis/disorder at layer (i + k), k> 1, replace it by creating a -
           dummy node with the name same as h_m at all the intermediate -
           layers and connect them.
}//end of procedure Build_Layered_Net
```

5 ETHERNET PERFORMANCE MANAGEMENT MODEL - A CASE STUDY

In this section, we consider a restricted Ethernet model to illustrate the ideas presented in this work. We assume that reader is aware of the Ethernet operation (Metcafe, 1976), (Boggs, 1988).

We consider a Ethernet network performance management model with the following assumptions.

- The information that needs to be *monitored* for the purpose of performance tuning is collected from the stations and the channel. And, that information, which is beyond the normal (both above and below the normal limits) are reported as symptoms.
- Some monitoring information like *load is normal* and *collisions are with in the range* are included to support the diagnostic process by eliminating the unnecessary fault sets which otherwise raise false alarms.
- there may be some missing information and the entire information may not be available at the time of diagnosis.

5.1 The Ethernet Performance Management Knowledge Model

The Ethernet performance management knowledge base (Boggs, 1988), (Hansen, 1992), (Feather, 1992), (Feather, 1993) is constructed as a hyper-bipartite network (see Figure 1). This maps the network performance management knowledge onto a model suitable for the Realistic_ARM.

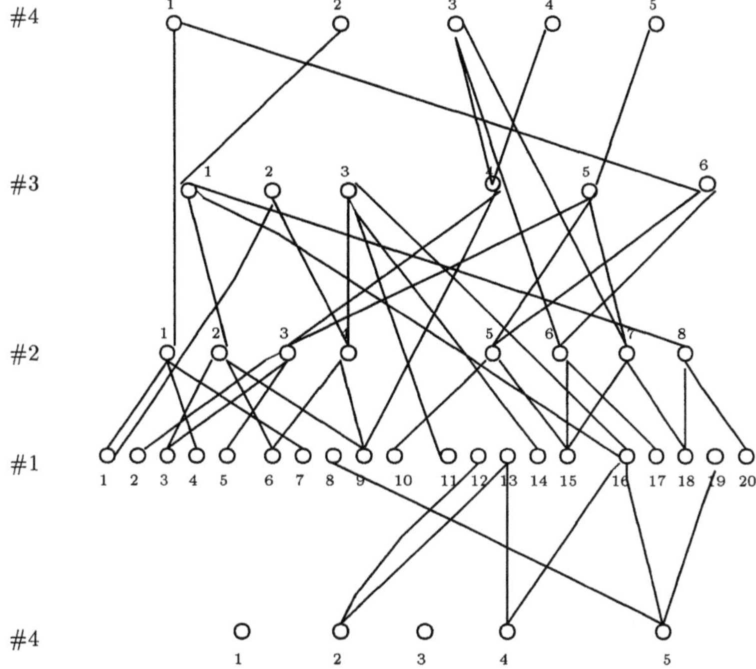

Figure 1 Ethernet Performance Management Knowledge Model. Layer 4 is shown in two places to avoid clumsiness; bottom most one connecting from layer 1 and top most one connecting from layers 2, 3.

Legend:

Layer #1 :

1. Packet loss below normal
2. Packet loss normal
3. Packet loss above normal
4. Load below normal
5. Load normal
6. Load above normal
7. Collisions below normal
8. Collisions normal
9. Collisions above normal
10. Large packets below normal
11. Large packets normal
12. Large packets above normal
13. Small packets below normal
14. Small packets normal
15. Small packets above normal
16. Broadcast packets normal
17. Broadcast packets above normal
18. Packet loss on spine above normal
19. Load on spine normal
20. Load on spine above normal

Layer #2 :

1. Light traffic
2. Heavy traffic
3. Buffers are insufficient
4. Users are many
5. Preambles are many
6. Broadcast packets are many
7. Spine flooded with too many small packets
8. Heavy traffic on spine

Layer #3 :
1. (F1) Babbling node; (Remedy, R1) : Faulty Ethernet card, report to the network manager
2. (F2) Hardware problem; (Remedy, R2) : Request the network manager to initiate Fault Diagnosis measures
3. (F3) Jabbering node; (Remedy, R3) : Ensure many packets are not above the specified size
4. Too many retransmissions
5. Under utilization of channel as many small packets are in use
6. Attempt for too many broadcasts
Layer #4 :
1. (F4) Bridge down; (Remedy, R4) : Report to the network manager
2. (F5) Network paging; (Remedy, R5) : Allocate more primary memory to the required nodes
3. (F6) Broadcast storm; (Remedy, R6) : Selectively control the broadcast packets
4. (F7) Bad tap; (Remedy, R7) : Report to the network manager along with the specified tap
5. (F8) Runt storm; (Remedy, R8) : Ensure many packets are not below the specified size

The fault knowledge base, constructed in the form of a hyper-bipartite network will be transformed into a layered network for a given diagnostic problem. The inference mechanism proceeds from the bottom most layer to the top most layer to find a solution for a given set of symptoms.

Based on a single symptom one should not conclude all its related faults which need some more symptoms to ascertain the validity. In this case, the fault corresponding to the direct disorder should be concluded. At the same time, one should be able to guess the most appropriate explanation even if a few of the symptoms are missing as is generally the case with networks due to the loss of information. Realistic_ARM can be found to solve all these problems related to the network performance management very effectively.

5.2 Results

The algorithm, Realistic_ARM, was run for various sets of symptoms (from Layer 1 of Figure 1) and some of the results are given in Table 1. The prespecified number of symptoms required to support any symptom before concluding a fault is set to 1.

Table 1 Sample results for Ethernet performance model

Sl.No.	Symptoms	Suggested Remedy
1.	3,6,12,18,20	{ R5 }
2.	1,4,10,15,17	{ R4 }
3.	3,9,18,20	{ R1 }
4.	10,15,16,18	{ R8 }

From Table 1, it can be observed that the covers generated by the proposed model contain appropriate explanation for any given symptoms without much of extra guess. Otherwise, generating so many covers is computationally expensive and, further, it requires elimination of inappropriate covers using some heuristic method. The proposed model avoids these problems and still makes an appropriate guess which proves to be useful to solve the performance management problem.

To demonstrate an example, consider the soft failures given as Sl. No. 4 in Table 1. The soft failures, observed as symptoms, are number of large packets below normal (Layer #1, 10), small packets above normal (Layer #1, 15), packet loss on spine above normal (Layer #1, 18) and the number of broadcast packets are with in the normal range (Layer #1, 16; this is a test but not a symptom). The fault concluded is "Runt storm" and the remedy is to ensure by possible means of control that, too many small packets are not injected into the network.

6 CONCLUSION

The abductive reasoning has been shown to be well suited for the specialized problems of network performance management. Realistic Abductive Reasoning Model is then used to solve the network performance management problem. This approach has been illustrated with the help of Ethernet performance management model. The explanation provided by the model is appropriate and shall not have much of extra guess. The results obtained by the proposed model are more appropriate and quite encouraging.

REFERENCES

Boggs D. R., Mogul J. C., and Kent C. A. (1988) Measured Capacity of an Ethernet : Myths and Reality, *Comp. Comm. Reveiw*, 222–234.

Bylander T., Allenmang D., Tanner M. C., and Josephson J. R. (1991) The Computational Complexity of Abduction, *Artificial Intelligence*, 49, 25–60.

Cassel L. N., Patridge C. and Westcott J. (1989) Network Management Architectures and protocols : Problems and Approaches, *IEEE Jl. on Selected Areas in Comm.* 7(7), 1104–1114.

Feather F. E. (1992) Fault Detection in an Ethernet Network via Anomaly Detectors, *Ph.D. thesis*, Dept. Electrical and Computer Engineering, Carniegie Mellon University.
Feather F. E., Slewlorek D. and Maxion R. (1993) Fault Detection in an Ethernet Network Using Anomaly Signature Matching, *Comp. Comm. Reveiw*, 279–288.
Hansen J. P. (1992) The Use of Multi-Dimensional Parametric Behavior of a CSMA/CD Network for Network Diagnosis, *Ph.D. thesis*, Dept. Electrical and Computer Engineering, Carniegie Mellon University.
Hayes S. (1993) Analyzing Network Performance Management, *IEEE Comm. Magazine*, 31(5), 52–59.
Metcafe R. M. and Boggs D. R. (1976) Ethernet : Distributed Packet Switching for Local Computer Networks, *Comm. of ACM*, 19(7), 395–404.
Peng Y. and Reggia J. A. (1987) Diagnostic Problem–Solving with Causal Chaining, *Intl. Jl. of Intelligent Systems*, 2, 395–406.
Peng Y. and Reggia J. A. (1990) Abductive Inference Models for Diagnostic Problem-Solving, Springer-Verlag, New York.
Pople H. (1973) On the Mechanization of Abductive Logic, in *Proc. of Intl. Joint Conf. on Artificial Intelligence*, 147–152.
Prem K. G. and Venkataram P. (1994) A Realistic Model for Diagnostic Problem Solving using Abductive Reasoning Based on Parsimonious Covering Principle, in *3rd Turkish Conf. on Artificial Intelligence and Neural Networks* (TAINN'94), Ankara, Turkey, 1–10.
Reggia J. A., Nau D., Wang P. and Peng Y. (1985) A Formal Model of Diagnostic Inference, *Information Sciences*, 37, 227–285.
Sluman C. (1989) A Tutorial on OSI Management, *Comp. Networks and ISDN Systems*, 17, 270–278.

Prem Kumar Gadey received his B.Tech. (Electronics & Communication Engineering) from Sri Venkateswara University in 1990 and M.Tech. (Artificial Intelligence & Robotics) from University of Hyderabad in 1992. Since then he is a Ph.D. student in department of Electrical Communicaiton Engineering, Indian Institute of Science, Bangalore. His major research interests include Communication Networks, Internetworking, Distributed Computing, Expert Systems and Artificial Neural Networks. Currently he is focussing on applying Artificial Intelligence techniques to the area of Network Management. He is a student member of IEEE Communication Society.

Pallapa Venkataram received his Ph. D. degree from The University of Sheffield, England, in 1986. He is currently an Associate Professor in department of Electrical Communication Engineering, Indian Institute of Science, Bangalore, India. He worked in the areas of Distributed Databases, Communiation Protocols and AI applicaitons in Communication Networks and has published many papers in these areas.

18

Connection Admission Management in ATM Networks Supporting Dynamic Multi-Point Session Constructs

P. Moghe and I. Rubin
Department of Electrical Engineering, UCLA, CA 90024-1594.
{pmoghe,rubin}@ee.ucla.edu

Abstract

A framework for admission management of session-level requests exhibiting space/time heterogeneity is developed. A single sub-threshold based link-level connection admission scheme for a mix of uni-point/static session and multi-point/dynamic session Virtual Channel Link requests (VCLRs) is designed and evaluated under different scenarios. Aside from external blocking, **internal loss** is introduced as an important QOS parameter for multi-point/dynamic session services. Concepts of **service-optimal** and **throughput-optimal sub-thresholds** are formulated. Finally, we outline a network algorithm that designs link-level sub-thresholds in accordance with end-to-end session-level QOS parameters.

Keywords

Multi-point and Multi-party Resource Allocation, Performance Management, QOS Management, Connection Admission Management in ATM Networks.

1 INTRODUCTION

Unlike traditional connection establishment protocols that treat a **call** as a monolithic end-to-end object (used for one service type, using one channel or connection), BISDN signaling needs to be tailored to incorporate an efficient mechanism to service multi-point and multi-media traffic [1][2][3]. In this context, we redefine a **call** as a high-level distributed network object that describes the communication paths connecting the clients. A **View** or a **Session** is the call-context of each client. In the most general case it represents a broadcast tree rooted at a client; its leaves comprising the recipient clients (also called *sink-clients*). Each session is implemented at setup time through end-end Virtual Channel Connection requests (VCCRs). A **VCC**, identified by a unique source VCI, is an end-end directional logical tree between source and sink clients. Each fork represents multicasting of information cells. A VCC itself is established through a sequence of Virtual Channel Link requests (VCLRs). A **VCL** is the basic logical component of our relationship model and represents a logical connection (and a single channel bandwidth allocation) between adjacent switching nodes.

Applications such as multi-media conferencing and information browsing/sharing can be built using the above constructs. As the ATM layer matures, it is our contention that the admission management of these constructs, at the connection layer (above the ATM layer), will pose future challenges. In this work, we formulate appropriate connection-level QOS vectors and design a simple threshold-based admission scheme to handle heterogeneous session constructs.

The paper is organized as follows: In section 2, the problem is motivated and an objective is formulated. In section 3, the single-link (SL) admission model is described, evaluated and tuned for the chosen optimality measures. Section 4 discusses some numerical results of the SL Model. In section 5, we outline a two-tiered network algorithm that uses the SL model to design distributed network-wide sub-thresholds.

2 PROBLEM DEFINITION

We recognize two important resource allocation tradeoff issues related to the bandwidth demand of session requests:

- Spatial Heterogeneity: Multi-point vs. Uni-point Session Requests
  Multi-point requests are susceptible to higher levels of blocking than uni-point requests in networks with limited multi-cast edge switches. The spatial issue thus requires that the multi-point requests be given special care, so that they are not blocked beyond tolerance.

- Temporal Heterogeneity: Static vs. Dynamic Sessions
  In static sessions, the number of member clients is constant and declared by the session request. Dynamic sessions are characterized by a variable number of clients during their life-time. Reservation of optimal number of VCLs for dynamic sessions is a challenging issue. If enough capacity is not reserved for a carried dynamic session, a secondary request for addition of a new user is liable to be blocked. This can adversely impact the carried users of the session. The resulting service degradation can, in certain applications, be severe enough to cause a subset of carried users to abort the session.

In general, session requests are of two types: *primary* (requests that initiate the session) and *secondary* (requests that add on to existing sessions, preferably reusing their resources). We combine the two heterogeneity issues into a single problem by defining two classes of session requests, $\mathcal{A}$ and $\mathcal{B}$. Class $\mathcal{A}$ requests initiate uni-point/static size sessions. Class $\mathcal{B}$ requests set up a multi-point session through a primary request. If admitted, this is followed by uni-point secondary class $\mathcal{B}$ requests for additional client connections. If secondary requests are blocked, a fraction r of the sink-clients are assumed to abort(*internal loss*). Class $\mathcal{A}$ and $\mathcal{B}$ session-requests generate lower-layer class $\mathcal{A}$ and $\mathcal{B}$ VCLRs at the link level. We assume that the required service quality is specified through session-level QOS vectors for both classes. For instance, class $\mathcal{A}$ and $\mathcal{B}$ applications declare worst-case session-level and link-level(VCLR) external blocking probabilities as Θ_{ex}^{max} and Φ_{ex}^{max} respectively. In addition, the worst-case internal loss probability $\Theta_{in-loss}^{B,max}$ (and corresponding link-level $\Phi_{in-loss}^{B,max}$)

defines the maximal acceptable probability with which a carried class $\mathcal{B}$ client aborts due to secondary blocking.

The *problem objective* is : Given an arbitrary session-request loading pattern, a network routing topology and multi-cast switch locations/specifications, design a threshold-based VCL-layer admission scheme on each link that can be tuned to satisfy the session-level QOS vectors (and possibly achieve connection-level optimality measures).

Since the network-wide problem is daunting to tackle on an end-to-end session basis, our approach is to build and solve exactly a flexible single-link (SL) model. This model makes natural sense since the admission scheme is on a link basis anyway. A network algorithm then approximates the end-to-end effect through its dependence structure.

3 SINGLE LINK MODEL

The link-level admission scheme is outlined next. It uses a sub-threshold (m_A) to reserve space for class $\mathcal{B}$ VCLRs. The SL analytical model is described in section 3.2. Parameters such as r (session dependence), D (initial session size), λ_s (secondary arrival rate per session) are formulated. Under the assumed traffic and service statistics, the VCL layer is analyzed for steady-state performance in section 3.3. Performance measures such as external blocking, internal loss, and aggregate throughput are computed in section 3.4. Feasibility and optimality sub-thresholds are defined in section 3.5.

3.1 VCL Connection Admission Scheme

Let m be the maximum number of VCLs on a link, capable of supporting cell-layer QOS. We assume m to be a known quantity; various studies such as [4][5] focus on admission at the ATM layer and indirectly compute it. Define a *sub-threshold* m_A $(0 \leq m_A \leq m)$. Let D_{mc} be the maximum multi-cast gain of a switch (i.e. the maximum number of copies supported by the switch copy-network), and D be the instantaneous multi-cast demand of a Primary class $\mathcal{B}$ VCLR. Let N_t^{vcl} represent the aggregate carried VCLs on the link at time t. We employ the following admission policy for a VCLR arriving at time t :

3.2 Analytical Model Description

We treat each directional link as a multi-VCL resource. Under the homogeneity assumptions (i.e. each VCL represents equal bandwidth), the VCL layer can be modeled as a pure blocking

Class of VCLR	Characteristics	VCLR Admission Rule
$\mathcal{A}$ (Primary)	Initiates uni-point/static session	$N_t^{vcl} < m_A$
$\mathcal{B}$ (Primary)	Initiates multi-point/dynamic session	$N_t^{vcl} \leq m - D$, $1 \leq D \leq D_{mc}$
$\mathcal{B}$ (Secondary)	Uni-point VCLR, adds onto created session	$N_t^{vcl} < m$

Table 1: Connection Admission Policy.

system with 'm' maximum VCLs.

The concept of an end-to-end session extended to a link is defined as a Link-session (L-session). All VCLs of an L-session (VCL-members) share a unique L-session-id. Each VCL member normally holds for an exponentially distributed time (parameter μ). An L-session terminates when all VCL-members have terminated. The holding time of an L-session represents the interval from its initiation to its termination.

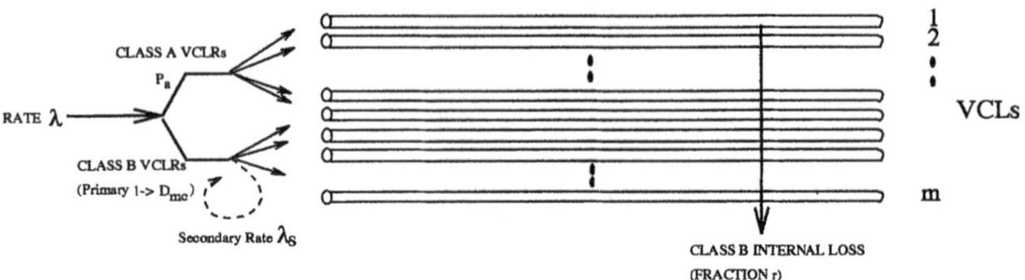

Figure 1: Single Link Model.

Primary VCLRs are assumed to arrive at a node-link User Request Manager, with a Poisson rate λ. A fraction p_a of the VCLRs are class $\mathcal{A}$ VCLRs, the rest class $\mathcal{B}$. Let $\lambda_a = \lambda p_a$, and $\lambda_b = \lambda(1 - p_a)$. Class $\mathcal{A}$ VCLRs represent requests for uni-point, static L-sessions. If admitted, they are allocated a single VCL. A primary class $\mathcal{B}$ VCLR initiates a multi-point, dynamic L-session by first demanding a multi-cast group of D VCLs. D is assumed to be a random number with a distribution $b_i = P\{D = i\}, 1 \leq i \leq D_{mc}$ (Section 4 assumes a uniformly distributed D, so that $b_i = 1.0/D_{mc}$). Each admitted L-session initiated by Primary class $\mathcal{B}$ VCLRs receives additional *secondary* class $\mathcal{B}$ VCLRs at a Poisson rate λ_s. If admitted, the secondary class $\mathcal{B}$ VCLR is allocated a single VCL and the VCL-member set of the corresponding L-session is incremented. Else, a fraction r of its carried VCL-members abort the L-session. Figure 1 illustrates the single-link model. The admission rule has been summarized in Table 1. Our immediate objective is to compute the steady-state VCL-size distribution.

3.3 Analysis

Define the system size process $X = \{X_t, t \geq 0\}$, where $X_t = (X_t^{\mathcal{A}}, X_t^{\mathcal{B}}, X_t^{\mathcal{B}ls}) \triangleq$ Number of $\mathcal{A}$ VCLs, $\mathcal{B}$ VCLs, and $\mathcal{B}$ L-sessions carried at time t. Let $T_n = n$th transition time of X. Define the underlying state sequence $V = \{V_n, n \geq 0\}$, where $V_n \triangleq (V_n^{\mathcal{A}}, V_n^{\mathcal{B}}, V_n^{\mathcal{B}ls}) =$ Number of VCLs carried at time T_n^+. Thus, $X_t = V_n$, for $T_n \leq t < T_{n+1}$, $sup(T_n) = +\infty$.

<u>*THEOREM*</u>: X is a time-homogeneous continuous-time Markov chain over state space $S = \{(i, j, k), \; i = 0, 1, ..m_a; \; 0 \leq (i + j) \leq m; k \in K_j\}$, where $K_j = \{k \,|\, min(1, j) \leq k \leq j\}$, under conditions of session-homogeneity and assumptions of Section 3.2.

We *omit* the proof for brevity. The probability law of X is determined by its transition probability function: $P_t((ijk),(xyz)) \triangleq P\{X_{t+s} = (x,y,z) \mid X_s = (i,j,k),\ s \leq t\} = P\{X_{t+s} = (x,y,z) \mid X_s = (i,j,k)\}$.
Let $S_{loss} = S \cap \{(i,j,k) \mid (i,j,k) \in S, (i+j) = m)\}$ be the state-space subset that represents a full system. The infinitesmal generator rates are derived next:
$\forall (i,j,k) \in S \backslash S_{loss}$,

$$
\begin{aligned}
q_{(ijk),(xyz)} &= \lambda_a, & x = i+1, y = j, z = k, & \quad if\ i+j < m_a \\
&= \lambda_b b_D, & x = i, y = j+D, z = k+1, & \quad if\ D \leq (m-i-j) \\
&= k\lambda_s, & x = i, y = j+1, z = k, & \\
&= i\mu, & x = i-1, y = j, z = k, & \quad if\ i \geq 1, \\
&= \Psi_1(i,j,k), & x = i, y = j-1, z = k, & \quad if\ j \geq 1, \\
&= \Psi_2(i,j,k), & x = i, y = j-1, z = k-1, & \quad if\ j \geq 1, \\
&= 0 & else, &
\end{aligned}
$$

where: $\Psi_1(i,j,k) = j\mu P_{nl}, \Psi_2(i,j,k) = j\mu(1.0 - P_{nl})$ and

$$
\begin{aligned}
P_{nl} &= 1.0 - (\tfrac{k-1}{k})^{j-1}, & for\ j,k > 1 \\
&= 1.0, & for\ j > 1, k = 1, \\
&= 0.0, & for\ k,j = 1
\end{aligned}
$$

$\forall\, (i,j,k) \in S_{loss}$,

$$
\begin{aligned}
q_{(ijk),(xyz)} &= i\mu, & x = i-1, y = j, z = k, & \quad if\ i \geq 1 \\
&= \Psi_1(i,j,k), & x = i, y = j-1, z = k, & \quad if\ j \geq 1 \\
&= \Psi_2(i,j,k), & x = i, y = j-1, z = k-1, & \quad if\ j \geq 1 \\
&= \Psi_3(\alpha_j), & x = i, y = j-\alpha_j, z = k, & \quad if\ \alpha_j \leq \lfloor j\,r \rfloor \\
&= \Psi_4(\alpha_j), & x = i, y = j-\alpha_j, z = k-1, & \quad if\ k \geq 1 \text{ and } \alpha_j \leq \lfloor j\,r \rfloor,
\end{aligned}
$$

where: $\Psi_3(\alpha_j) = k\lambda_s[\{\sum_{l=\lceil \frac{\alpha_j}{r} \rceil}^{min(\lceil \frac{\alpha_j+1}{r} \rceil -1, j)} B(1/k,j,l) I(r < 1.0, k > 1)\} + I(k=1)]$,

$\Psi_4(\alpha_j) = k\lambda_s I(r = 1.0)[\{\sum_{l=\lceil \frac{\alpha_j}{r} \rceil}^{min(\lceil \frac{\alpha_j+1}{r} \rceil -1, j)} B(1/k,j,l) I(k > 1)\} + I(k=1)]$,

$B(p,j,l)$ is the binomial probability of j successes in l trials with success probability p, $I(exp) = 1$ if exp evaluates true, 0 else and $\alpha_j \in \mathcal{Z}^+$.

Assume that under appropriate conditions, steady-state distribution P (of X) and stationary distribution π (of underlying discrete-time Markov chain V) can be computed using balance equations [6].

3.4 Performance Measures

Primary class $\mathcal{A}$ and $\mathcal{B}$ VCLR Blocking Probabilities: Φ_{ex}^{A}, Φ_{ex}^{Bp}, and Φ_{ex}^{Bpg}
These probabilities can be determined by the PASTA property[7].

1. $\Phi_{ex}^{A} \triangleq$ P{Class $\mathcal{A}$ VCLR is blocked} $= \sum_{i=0}^{m_A} \sum_{j=m_A-i}^{m-i} \sum_{k \in K_j} P_{ijk}$

2. $\Phi_{ex}^{Bpg} \triangleq$ P{Primary class $\mathcal{B}$ VCLR (multi-cast group) is blocked}
$= \sum_{l=1}^{D_{mc}} b_l \sum_{i=0}^{m_A} \sum_{j=max(m-l-i+1,0)}^{m-i} \sum_{k \in K_j} P_{ijk}$

3. $\Phi_{ex}^{Bp} \triangleq$ P{Primary class $\mathcal{B}$ (individual) VCL is blocked}
$= \sum_{l=1}^{D_{mc}} \frac{lb_l}{\sum_{k=1}^{D_{mc}} kb_k} \sum_{i=0}^{m_A} \sum_{j=max(m-l-i+1,0)}^{m-i} \sum_{k \in K_j} P_{ijk}$

Secondary class $\mathcal{B}$ VCLR Blocking Probability: Φ_{ex}^{Bs}

$\Phi_{ex}^{Bs} \triangleq$ P{Secondary class $\mathcal{B}$ VCLR is blocked} $= \frac{\sum_{i=0}^{m_A} \sum_{k \in K_j} P_{ijk} k \lambda_s |_{j=m-i}}{\sum_{i=0}^{m_A} \sum_{j=0}^{j=m-i} \sum_{k \in K_j} P_{ijk} k \lambda_s}$

Class $\mathcal{B}$ Internal Loss Probability: $\Phi_{in-loss}^{B}$

$\Phi_{in-loss}^{B} \triangleq$ P{Admitted class $\mathcal{B}$ VCL aborts (is internally lost)}. We derive $\Phi_{in-loss}^{B}$ using busy-cycle arguments. Define the following parameters:

λ_{Bp} (λ_{Bs})	= Offered primary (secondary) class $\mathcal{B}$ VCLR rate
N_{AD}^{tot}	= Aggregate class $\mathcal{B}$ VCLs admitted per busy cycle (primary + secondary)
$N_{IN-LOSS}^{tot}$	= Number of class $\mathcal{B}$ VCLs internally lost per busy cycle
$N_{IN-LOSS}^{ijk}$	= Number of class $\mathcal{B}$ VCLs lost per busy cycle from state $(i,j,k) \in S_{loss}$.

Note that, $\lambda_{Bp} = \lambda_b \sum_{n=1}^{D_{mc}} nb_n$, and $\lambda_{Bs} = \sum_{i=0}^{m_A} \sum_{j=0}^{j=m-i} \sum_{k \in K_j} P_{ijk} k \lambda_s$

Then, N_{AD}^{tot} = Aggregate admission rate of class $\mathcal{B}$ VCLs x Busy Cycle Duration
$= \{\lambda_{Bp}(1-\Phi_{ex}^{Bp}) + \lambda_{Bs}(1-\Phi_{ex}^{Bs})\}(\lambda P_{000})^{-1}$

Also, $\forall (i,j,k) \in S_{loss}$, $N_{IN-LOSS}^{ijk}$ = Number of visits to (i,j,k) per cycle x losses per visit
$= \frac{\pi_{ijk}}{\pi_{000}} \sum_{\alpha=1}^{\lfloor jr \rfloor} \frac{\Psi_3(\alpha)+\Psi_4(\alpha)}{i\mu+\Psi_1(i,j,k)+\Psi_2(i,j,k)+\Psi_3(\alpha)+\Psi_4(\alpha)} \alpha$

Total VCL loss per busy-cycle $N_{IN-LOSS}^{tot} = \sum_{i=0}^{m_A} \sum_{k \in K_j} N_{IN-LOSS}^{ijk}|_{j=m-i}$

Finally, class $\mathcal{B}$ internal loss probability $\Phi_{in-loss}^{B} = \frac{N_{IN-LOSS}^{tot}}{N_{AD}^{tot}}$.

Class $\mathcal{B}$ Loss Prob. Φ_{loss}^{B}, Mean Holding Time HT^B, VCL Throughput TP

1. Class $\mathcal{B}$ (weighted) blocking probability $\Phi_{ex}^{B} = \Phi_{ex}^{Bp}\left(\frac{\lambda_{Bp}}{\lambda_{Bp}+\lambda_{Bs}}\right) + \Phi_{ex}^{Bs}\left(\frac{\lambda_{Bs}}{\lambda_{Bp}+\lambda_{Bs}}\right)$

2. Class $\mathcal{B}$ loss probability Φ_{loss}^{B} is the probability that an arbitrary class $\mathcal{B}$ VCL is externally blocked or internally lost. Then, $\Phi_{loss}^{B} = 1-(1-\Phi_{ex}^{B})(1-\Phi_{in-loss}^{B})$

3. Next, we compute the class $\mathcal{B}$ mean holding time HT^B through Little's law[7]:
HT^B = (Average class $\mathcal{B}$ utilization)/ (Aggregate admission rate of class $\mathcal{B}$ VCLs)
$= \left(\frac{\sum_{i=0}^{m_A} \sum_{j=0}^{m-i} \sum_{k \in K_j} jP_{ijk}}{\lambda_{Bp}(1-\Phi_{ex}^{Bp})+\lambda_{Bs}(1-\Phi_{ex}^{Bs})}\right)$ (used in section 5).

4. Finally, the aggregate VCL throughput (TP) is given by:
$TP = \lambda_a(1-\Phi_{ex}^{A}) + \{\lambda_{Bp}(1-\Phi_{ex}^{Bp}) + \lambda_{Bs}(1-\Phi_{ex}^{Bs})\}(1-\Phi_{in-loss}^{B})$

3.5 QOS and Feasible/Optimal Sub-thresholds

Assume a worst-case VCL QOS vector: $(\Phi_{ex}^{A,max}, \Phi_{ex}^{B,max}, \Phi_{in-loss}^{B,max})$. For simplicity, we combine the worst-case external blocking and internal loss of class $\mathcal{B}$ VCLs into maximum total loss probability $(\Phi_{loss}^{B,max})$ computed as: $\Phi_{loss}^{B,max} = 1 - (1 - \Phi_{ex}^{B,max})(1 - \Phi_{in-loss}^{B,max})$. Further, define $\Phi^{max} = min(\Phi_{ex}^{A,max}, \Phi_{loss}^{B,max})$.

The sub-threshold can be tuned to satisfy feasibility/optimality conditions. The sub-threshold scheme is said to be **feasible** at m_A^* iff $max(\{\Phi_{ex}^{A}\}_{m_A^*}, \{\Phi_{loss}^{B}\}_{m_A^*}) \leq \Phi^{max}$. In Figure 2, the set of feasible sub-thresholds $\mathcal{F}_{m_A}$ is, in general, the set of sub-threshold values bounded by the intersection of Φ_{ex}^{A} and Φ_{loss}^{B} with Φ_{max}.

From the *application* viewpoint, a **service-optimal sub-threshold** $(m_A^*)_S$ is defined such that, if it exists, $(m_A^*)_S \in \mathcal{F}_{m_A}$ and $\{\Phi_{ex}^{A}\}_{(m_A^*)_S} = \{\Phi_{loss}^{B}\}_{(m_A^*)_S}$. To satisfy the integral $(m_A^*)_s$ constraint, we allow for the nearest integer solution to the intersection of Φ_{ex}^{A} and Φ_{loss}^{B}. The sub-threshold $(m_A^*)_s$ defines the operating point at which the network provides the VCLs a service quality (QOS) independent of the higher-layer dependence(class $\mathcal{A}$ or $\mathcal{B}$). Also, note that if $(m_A^*)_s$ cannot be found at an offered load, it follows that there is no feasible solution to the admission scheme!

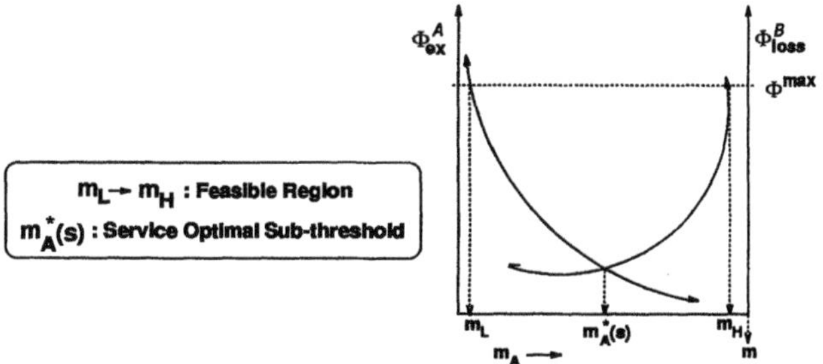

Figure 2: Feasibility and service-optimality issues.

From the *network operator* viewpoint, we select an optimality sub-threshold that maximizes aggregate throughput. Formally, a **throughput-optimal sub-threshold** $(m_A^*)_T \in \mathcal{F}_{m_A}$, such that $\{TP\}_{(m_A^*)_T} \geq \{TP\}_{(m_A^*)}$, $\forall\, m_A^* \in \mathcal{F}_{m_A}$.

4 RESULTS

4.1 Effect of r, m_A on Φ and TP

Note the parameters in the textual legends of Figures 3, 4, and 5. Figure 3 plots class $\mathcal{B}$ primary (individual/group) blocking, secondary blocking and internal loss probability with respect to m_A variation.

Figure 4 compares class $\mathcal{A}$ VCLR external blocking Φ_{ex}^{A} to the class $\mathcal{B}$ total loss probability Φ_{loss}^{B} formulated in Section 3.4. The service optimal point $(m_A^*)_S$ (assuming its feasible) is indicated. Note that r variation at a fixed offered load does not significantly change the performance measures. This is pleasing from the design point of view.

Figure 5 plots aggregate VCL throughput TP over similar conditions. Note that increasing r reduces TP slightly because the batch-loss increase dominates the external blocking reduction. Also, the dynamic variation of TP over m_A is small; increasing m_A increases the $\Phi_{in-loss}$ due to more frequent secondary blocking. This creates more space in the system and consequently reduces class $\mathcal{B}$ external blocking.

Figure 5 also indicates the simulated VCL throughput TP_{sim} for $r = 0.1$. The variation between the analysis and simulation results is no more than 5% (less than 1% for smaller systems). Thus, the session-homogeneity assumption is seen to perform well.

4.2 Throughput-Optimal Sub-threshold Trajectory

Figure 6 illustrates $(m_A^*)_T$ variation with traffic mix parameter p_a. This variation is plotted for two values of initial session size ($D_{mc} = 1, 5$). The secondary arrival rate per L-session is modified at each observation to keep a constant offered load = 0.6.

We observe that as p_a increases, $(m_A^*)_T$ reduces linearly over a significant range. This is equivalent to allocating more resources to class $\mathcal{B}$ VCLRs when the class $\mathcal{A}$ traffic dominates, since *goodput* per admitted class $\mathcal{B}$ VCLR is maximum under this condition.

Also, at a fixed p_a, $(m_A^*)_T$ is larger for larger D_{mc} values (refer to $p_a = 0.5$, where $(m_A^*)_T = 48, 49$ at $D_{mc} = 1, 5$ respectively.). Since secondary arrival rate λ_s is varied to keep offered load constant at both the points, the result offers an important interpretation. Consider the fixed abscissa $p_a = 0.5$. The shift of $(m_A^*)_T$ from 49 to 48 reflects the tradeoff between large initial-size static sessions and small initial-size dynamic sessions. Clearly, at $p_a = 0.5$, the dynamicity of secondary arrivals dominates the initial session size for the overall effect. At an increased value of $p_a = 0.9$, throughput becomes sensitively dependent on every large blocked primary class $\mathcal{B}$ VCLR. Hence, $(m_A^*)_T$ for $D_{mc} = 5$ converges with that for $D_{mc} = 1$. At this point, the initial session size completely counteracts dynamicity due to secondary arrivals.

5 NETWORK ALGORITHM

We present a distributed algorithm that designs network-wide service-optimal sub-thresholds on all the network links. Depending on the location of multi-cast switches and the routing scheme (stochastic routing), it is possible to encode each link (i.e. its offered primary and secondary VCLR traffic pattern, parameters $\lambda, p_a, b_i, D_{mc}, \lambda_s, p_s$) in the SL model format. However, solving independent SL models is inadequate because the offered rates at each link are dependent on the Φ vector of its neighbors.

The network algorithm presented here solves this problem by iteratively modifying the rates through a two-tiered structure. In the first tier, it computes the offered arrival rates

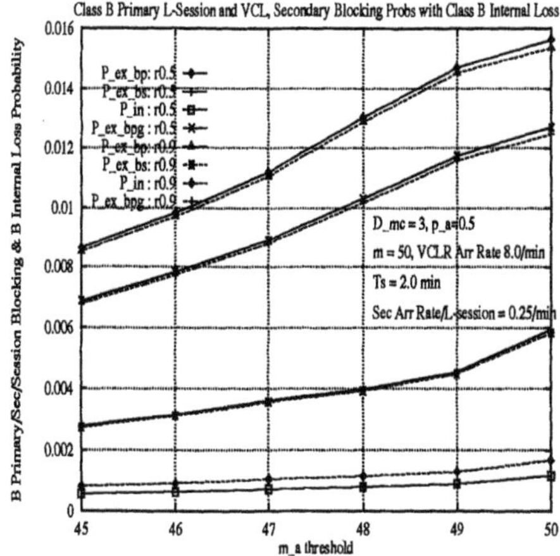

Figure 3: Effect of r, m_A on class $\mathcal{B}$ blocking and internal loss.

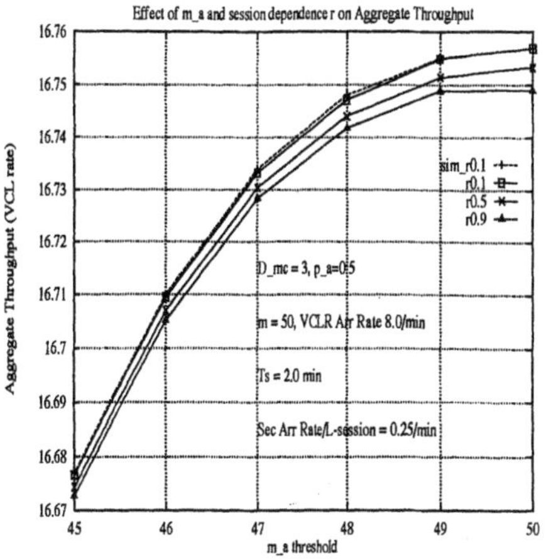

Figure 5: Effect of r, m_A on aggregate VCL throughput.

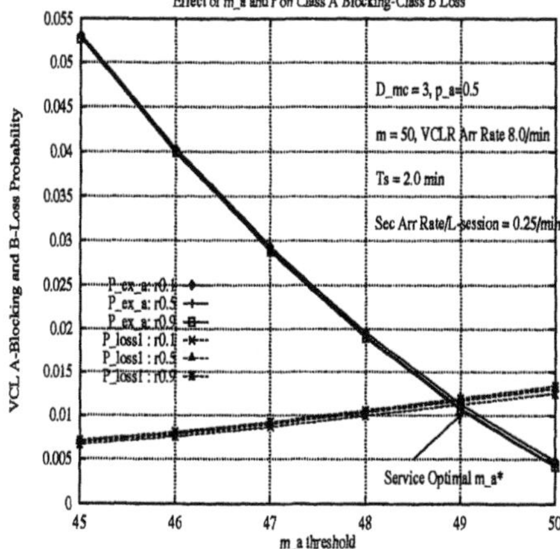

Figure 4: Effect of r, m_A on class $\mathcal{A}$ and $\mathcal{B}$ blocking and loss.

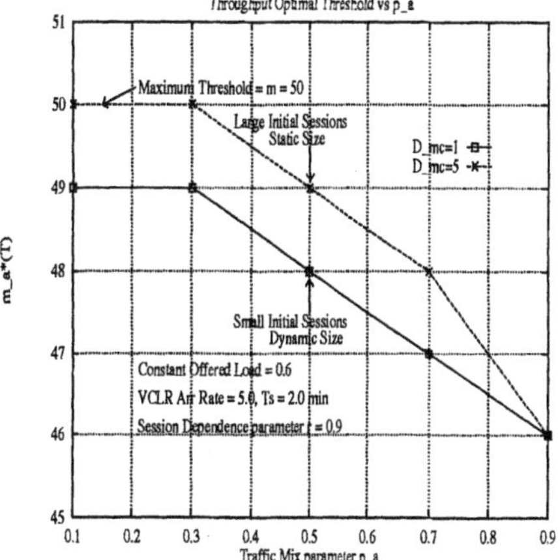

Figure 6: Throughput-Optimal threshold trajectory.

using only the external blocking component. It then calculates the sustained offered rates, as would be seen by the end-to-end connections. In the second tier, it computes the true internal loss on each link by accounting for reflected loss from other links. Finally, the sub-threshold is updated in the instantaneous direction towards service-optimality and the process is repeated.

The basic algorithmic framework follows. We assume for simplicity that sessions are independent of each other. The session QOS vector is $\Theta^{max} = (\Theta_{ex}^{A,max}, \Theta_{ex}^{B,max}, \Theta_{in-loss}^{B,max})$. A session-request is blocked if any component primary VCLR gets blocked. If a secondary VCLR is blocked at any node, a fraction r of the sink-clients of that session, downstream to that request terminate (assuming a topological dependence). $\Theta_{in-loss}^{B,max}$ represents the maximum internal loss probability that the sink-clients can tolerate.

Refer to Figures 7, 8, and 9 for the flow-charts. We qualify these with additional important comments:

1. Φ^{max} vector is derived on each link in the following steps:
   (a) Maximum primary VCLR blocking, $\Phi_{ex}^{Bpg,max} = 1-(1-\Theta_{ex}^{B,max})^{(H)^{-1}}$, and $\Phi_{ex}^{A,max} = 1-(1-\Theta_{ex}^{A,max})^{(H)^{-1}}$ where H = maximum hops traversed by VCCs over all sessions (conservative design).
   (b) $\Phi_{ex}^{Bp,max}$ is related to $\Phi_{ex}^{Bpg,max}$ through a simple bound (given the batch distribution b_i on the specific link): $\frac{\Phi_{ex}^{Bpg,max}}{\sum_{n=1}^{D_{mc}} nb_n} \leq \Phi_{ex}^{Bp,max} < \frac{\sum_{l=1}^{l=D_{mc}} b_l \Upsilon(l)(l-1)}{\sum_{n=1}^{D_{mc}} nb_n} + \frac{\Phi_{ex}^{Bpg,max}}{\sum_{n=1}^{D_{mc}} nb_n}$, where $0 \leq \Upsilon(l) = \sum_{i=0}^{m_A} \sum_{j=max(m-l-i+1,0)}^{m-i} \sum_{k \in K_j} P_{ijk}$. The derivation is omitted for brevity. We conservatively select the lower bound: $\Phi_{ex}^{Bp,max} = \frac{\Phi_{ex}^{Bpg,max}}{\sum_{n=1}^{D_{mc}} nb_n}$.
   (c) Assuming the same bound for secondary blocking, $\Phi_{ex}^{B,max} = \Phi_{ex}^{Bp,max}$. Also, it can be shown that $\Phi_{in-loss}^{B,max} = \Theta_{in-loss}^{Bmax}$ guarantees the sink-clients a feasible internal loss probability.
   (d) As before, $\Phi_{loss}^{B,max} = 1-(1-\Phi_{ex}^{B,max})(1-\Phi_{in-loss}^{B,max})$, $\Phi^{max} = min(\Phi_{ex}^{A,max}, \Phi_{loss}^{B,max})$.
2. In Figure 8, the Dependence Algorithm can be executed in parallel for all links incident on a single node, and sequentially node-wise. The algorithm modifies the holding time of a tagged link by reflecting the holding times of its neighbors on to it. This has the effect of modeling the system-size space effect due to internal loss.
3. The Threshold Guidance algorithm in Figure 9 updates the sub-threshold depending on the current Φ state with respect to the service-optimal threshold (see Figure 2) computed at the given load.
4. If the complexity of the single-link model is $0(SL)$ in an n-node network, the network algorithm can be shown to have a worst-case time-complexity of $O(SL.n^2)$, provided the iterations exhibit constant order. The algorithm has shown promising behavior on the examples tested.

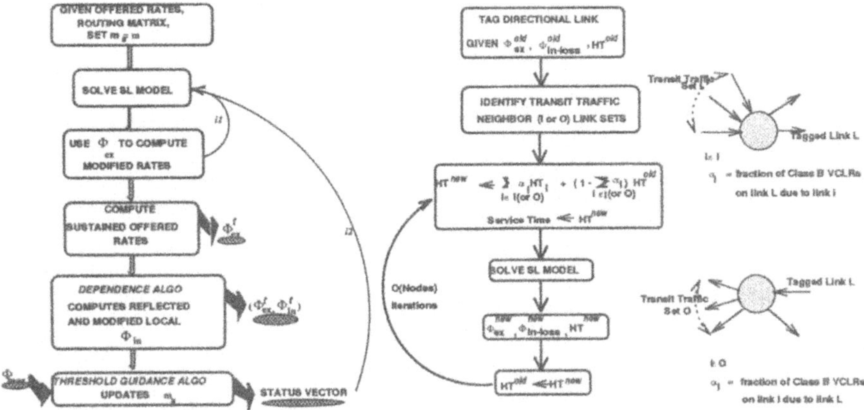

Figure 7: Two-tiered Network Algorithm.

Figure 8: Dependence Algorithm for Internal loss.

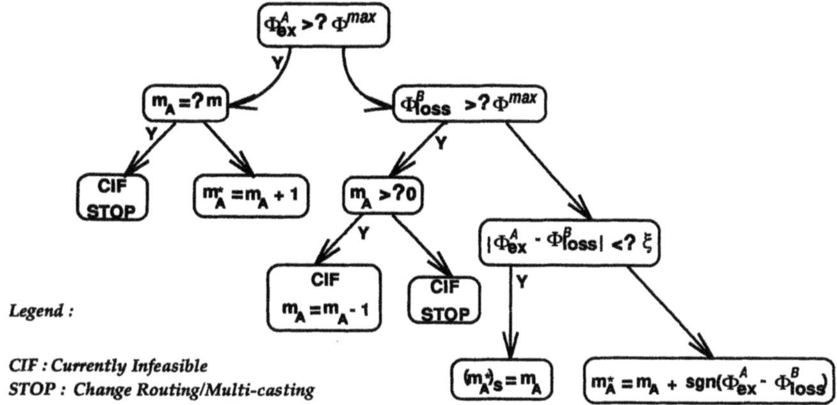

Figure 9: Threshold Guidance Algorithm for sub-threshold updates.

6 CONCLUSIONS

We contend that future multi-media/multi-point applications will require admission management at the connection layer (over and above the ATM layer). In this work, we have formulated a simple threshold-based distributed connection admission scheme for heterogeneous sessions. We have developed appropriate connection-level QOS measures for uni-point/static and multi-point/dynamic sessions. The threshold scheme can be tuned to attain service-optimality. A network algorithm extends this to incorporate end-to-end session requirements.

References

[1] M. Gaddis, R. Bubenik, and J. DeHart, "A Call Model for Multipoint Communication in Switched Networks," *ICC'92*, pages 609-615.

[2] S. Minzer,"A Signaling Protocol for Complex Multimedia Services," *IEEE Journal on Selected Areas in Communications,* 9(9):1383-1394, December 1991.

[3] ANSI T1S1 Technical Sub-Committee, "Broadband Aspects of ISDN Baseline Document," *T1S1.5/90-001*, June 1990.

[4] L. Gun and R. Guerin, "Bandwidth Management-Congestion Control Framework of the Broadband Network Arch.," *Computer Networks and ISDN Systems*, 26(1):61-78, 1993.

[5] H. Saito, "Call Admission Control in an ATM Network Using Upper Bound of Cell Loss Probability," *IEEE Trans. on Comm.*, 40(9):1512-1521, Sept 1992.

[6] E. Cinlar, *Introduction to Stochastic Processes*, Prentice-Hall, Englewood Cliffs, 1975.

[7] L. Kleinrock, *Queueing Systems: Vol I*, Wiley, New York, 1976.

PRATYUSH MOGHE is a graduate student researcher in the department of Electrical Engineering at UCLA. His current research interests focus on admission management of enhanced-calls supporting multi-media/multi-party dynamic applications. He earned the B.E (1988) from the department of Electronics and Telecommunications at College of Engineering, Poona, India, and an M.S (1990) in Electrical Engineering from Clemson University, SC. He was a technical member of the Network Architectures-Services group at GTE Laboratories, Waltham, MA (Summer 1990). He received the Best Student Award (1988 IAF Trophy, Univ. of Poona) and the UCLA Fellowship (1990).

IZHAK RUBIN received the B.Sc. and M.Sc. from the Technion, Israel, and the Ph.D. degree from Princeton University, all in Electrical Engineering. Since 1970, he has been a professor in the UCLA Electrical Engineering Department. He has had extensive research and industrial experience in the design and analysis of telecommunications, computer communications, and C3 networks. He has also been serving as chief-engineer of IRI Computer Communications Corporation. He is an IEEE Fellow, has served as chairman of IEEE conferences, and as an editor of the IEEE Transactions on Communications and of the journal on Wireless Networks.

19

A quota system for fair share of network resources

Çelik C.
Computer Center
Middle East Technical University
Inonu Bulvari, 06531
Ankara, Turkiye
can@knidos.cc.metu.edu.tr

Özgit A.
Dept. of Computer Engineering
Middle East Technical University
Inonu Bulvari, 06531
Ankara, Turkiye
ozgit@metu.edu.tr

Abstract

Interconnected networks of today provide a wide variety of services, which consume widely differing amounts of resources. But unlike other computing resources such as disk space and processing power, the network resource is not that much accounted.

Internet Engineering Task Force (IETF) internet-accounting working group is currently studying this subject. Their approach to the problem is focused on network accounting but does not cover any real-time controls such as quotas or enforcement.

In this paper, a model that increases coordination between accounting mechanisms and access controls is introduced. This model is compatible with the concepts and the architecture introduced by IETF internet-accounting working group. In the proposed model the quota manager is responsible from producing a table of service consumers that have already reached their quotas. This table is formed by using the data accumulated by the accounting system.

Keywords

Network Management, Network Accounting, Quota System, TCP/IP, SNMP.

1 INTRODUCTION

Today computer networks have become a fundamental part of computing. They are used for serving many purposes such as file transferring between computers, cross-login connections, file sharing, distributed computing, electronic mailing, electronic discussion lists, information services, etc. Since the 'network' as a shared physical resource is limited for most cases, it is a reasonable approach to account the usage of network bandwidth. It could also be necessary to impose limitations for the usage, in order to prevent network misuse or even abuse.

This paper is based on the work being carried out by the IETF internet-accounting working group. It describes a system that uses IETF working group's accounting model and adds a quota system to it.

The Internet-accounting architecture model proposes a *meter* that listens on the network to collect information about network usage (Mills, 1991) (Mills, 1992) (Brooks, 1993). A *network manager* tells the meter what kind of information is needed and how much detail the accounting data should contain. This paper introduces a quota system which uses the data collected by the meter and forms a list of hosts that have already reached their quotas. Each service provider such as gateways, file servers, compute servers, etc., may check this list before they serve their users. If a service provider encounters any host that is in the list, it may refuse to provide any service to that host.

After a discussion of the milestones of Internet Accounting Architecture in Section-2, IETF's Internet Accounting Architecture is described in Section-3. The first implementation of the architecture is presented in Section-4. In Section-5, the proposed quota architecture is discussed.

2 HISTORY OF INTERNET ACCOUNTING

IETF Internet Accounting Working Group was formed with the goal to produce standards for the generation of accounting data within the Internet that can be used to support a wide range of management and cost allocation policies. The first publication of the group was titled as 'Internet Accounting Background RFC-1272', published in November 1991 (Mills, 1991).

The milestones of the working group are the following :

- Internet-accounting Background RFC-1272 was published (Mills, 1991).
- SNMP was recommended as the collection protocol.
- Internet-accounting architecture was submitted as Internet-Draft (Mills, 1992).
- Internet-accounting meter MIB was submitted as Internet-Draft (Brooks, 1993).
- The two drafts mentioned above expired 6 months after submission as a draft. And they were then modified several times by the working group.
- Internet-accounting working group was suspended in April 1993, waiting for feedback from implementation experience.
- The first implementation came in October 1993; NeTraMet & NeMaC (Brownlee, 1993).
- The working group started again in March 30 1994. They are planning to publish 'Internet Accounting Architecture' and 'Internet Accounting MIB' RFCs.

3 INTERNET ACCOUNTING ARCHITECTURE

The Internet accounting model, currently a draft of a working group (Mills, 1992), draws from the OSI accounting model. It separates accounting functions into the parts shown in Figure 1.

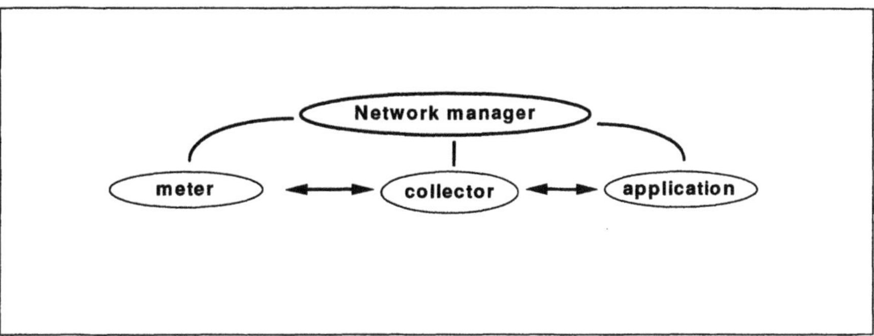

Figure 1 Internet Accounting Functions.

- Network Manager (or simply, Manager) : The network manager is responsible for the control of the meter. It determines and identifies backup collectors and managers as required.
- Meter : The meter performs the measurement of network usage and aggregates the results.
- Collector : The collector is responsible for the integrity and security of data during transport from the meter to the application. This responsibility includes accurate and preferably unforgeable recording of accountable (billable) party identity.
- Application : The application manipulates the usage data in accordance with a policy, and determines the need for information from the metering devices.

The data exchange can be categorized as follows:

- Between Meter and Collector
  The data which travels this path is the usage record itself. The purpose of all the other exchanges is to manage the proper execution of data exchange.
- Between Manager and Meter
  The manager is responsible for controlling the meter. Meter management involves commands which start/stop usage reporting, manage the exchange between meter and collector(s) (to whom do meters report the data they collect), set reporting intervals and timers, and set reporting granularities. Although most of the control information consists of commands to the meter, the meter may need to inform the manager of unanticipated conditions as well as responding to time-critical situations, such as buffer overflows.

- Between Manager and Collector
  Feedback on collection performance and controlling access to the collected traffic statistics are the main reason of this traffic. In most implementations, the manager and thecollector will be the same entity.

Since redundant reporting may be used in order to increase the reliability of usage data, exchanges among multiple entities are also considered, such as multiple meters or multiple collectors or multiple managers.

Internet accounting architecture assumes that there is a "network administrator" or "network administration" to whom network accounting is of interest. The administrator owns and operates some subset of the internet (one or more connected networks) that may be called as "administrative domain". This administrative domain has well defined boundaries. The network administrator is interested in (i) traffic within domain boundaries and (ii) traffic crossing domain boundaries. The network administrator is usually not interested in accounting for end-systems outside his administrative domain (Mills, 1991).

SNMP is the recommended collection protocol. A draft SNMP MIB is already proposed (Brooks, 1993).

The following points are not covered by the IETF working group's proposal :

- User-level reporting is not addressed in this architecture, as it requires the addition of an IP option to identify the user. However, the addition of a user-id as an entity at a later date is not precluded by this architecture.
- The proposal does not cover enforcement of quotas at this time. A complete implementation of quotas may involve realtime distributed interactions between meters, the quota system, and access control.

In the following sections of the paper, a model is introduced which will add a quota system to IETF's proposed architecture.

4 THE FIRST IMPLEMENTATION OF THE PROPOSED INTERNET ACCOUNTING ARCHITECTURE (NeTraMet & NeMaC)

The first implementation of the Internet accounting architecture is NeTraMet (Network Traffic Meter) and NeMaC (NeTraMet Manager/Collector) (Brownlee, 1993).

In this implementation, *network manager* and *collector* are the same entity. *Meter* is another software which can be located on the same host with *manager/collector* or can be located on a different host.

A *traffic flow* is a stream of packets exchanged between two network hosts. *Manager/collector* sends a set of rules to the *meter* which are used for deciding which flows are to be considered and how much detail about the flow will be collected. Rules can be quite detailed so that one can define flow of specific protocols. For example such rules can be stated.

'Count those packets from host X to host Y that are in TCP protocol' or
'Count those packets transferred via telnet connections'

Rules are sent from *manager/collector* to *meter* in SNMP format. Actually, they are variables set in the MIB located in the *meter*.

Figure 2 shows the traffic between *meter* and *manager/collector*.

The *meter* starts collecting data, considering the rules received from *manager/collector*. The flow data collected from the network is also put in the MIB-accounting database that is located in the meter, and the collector gets this data at regular time intervals. All the communication between *manager/collector* and *meter* is done via SNMP.

The MIB for Internet accounting is located in the meter. The structure of this MIB is explained in the following paragraphs.

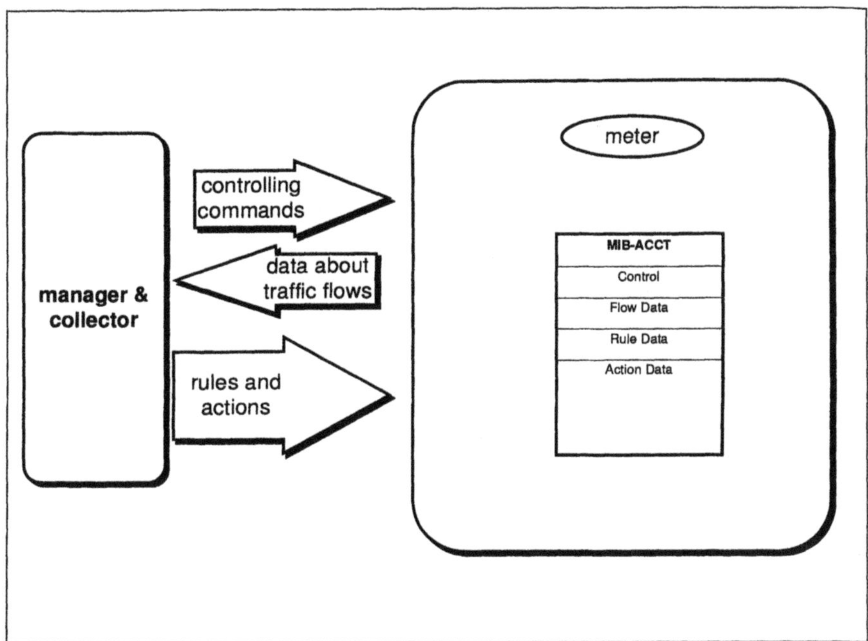

Figure 2 Manager/Collector and Meter.

MIB-acct is composed of four major parts.

- Control : Some parameters to control the meter such as sampling rate, when to send a trap to the manager if the meter is running out of memory, etc.
- Flow data : The counted flows are put here.
- Rule data : Rules for deciding if a flow is to be considered.
- Action data : Action to be performed if the rule's value is matched such as count, tally, aggregate.

'NeTraMet & NeMaC' is the first implementation of internet accounting architecture. NeTraMeT also implements the internet accounting meter services MIB. NeTraMet is available under SunOS and MSDOS operating systems. NeMaC is available under SunOS, SGI-IRIX and HP_UX operating systems. The quota system described in the next section is implemented by using some parts of this software.

5 A QUOTA SYSTEM FOR INTERNET ACCOUNTING ARCHITECTURE

The quota system proposed in this paper is an extension to the IETF's proposed internet accounting architecture.

5.1 Architecture

The accounting system described in 'Internet Accounting Architecture' section collects the accounting data. Quota system processes this data in order to form a list of hosts that have used the system resources beyond their quotas. This list is called the black-list. The algorithm used for deciding which hosts will stay in the black-list and how long will be described in the 'Algorithm' section. The black-list is valid in some domain in the internet. This domain is the mapping of the 'administrative domain' of the 'Internet Accounting Architecture'. More than one copy of the black-list can be located in a domain.

The black-list has been implemented as a MIB entry that can be located on any host running SNMP. It is actually an array of IP addresses. In order for implementing the quota system standard MIB has been modified by adding new variables. The added MIB variables in the ASN.1 notation are shown in Table 1:

Table 1 MIB-quota

blacklist OBJECT IDENTIFIER ::= { experimental 100 }
blacklistTable OBJECT-TYPE SYNTAX SEQUENCE OF blacklistEntry ACCESS read-write STATUS mandatory ::= { blacklist 1 }
blacklistEntry OBJECT-TYPE SYNTAX IpAddress ACCESS read-write STATUS mandatory ::= { blacklistTable 1 }
NoOfEntry OBJECT-TYPE SYNTAX INTEGER ACCESS read-write STATUS mandatory ::= { blacklist 2 }

The first entry 'blacklist' is the highest entry in the MIB-blacklist hierarchy. Its long name is 'iso.org.dod.internet.experimantal.blacklist'. This variable does not hold any value.

The 'blacklistTable' MIB variable defines an array of 'blacklistEntry'. Each 'blacklistEntry' holds an IP address and it is indexed by those addresses. The 'NoOfEntry' shows the number of hosts in the MIB blacklistTable. To set this variable to 0 clears the blacklistTable.

Quota manager has been implemented as a part of *network manager* software. It fills the black-list, the MIB-quota, by using SNMP. MIB quota is a dynamic list and the quota manager decides which IP addresses will enter to and which will exit from the list. Quota manager is also responsible from the consistency of the black-lists, if more than one of them are located in the domain. *Quota manager* does this by updating all of the black-list servers whenever an update is needed.

Service providers (gateways, FTP servers, NFS servers, etc.) in the domain may check the black-list before providing any service and do not give any service if the service requesting host is in the black-list. Each service provider knows which host(s) has up to date black-list in their MIBs and by using SNMP it checks if the service requester is in the black-list or not.

Since the 'Internet Accounting Architecture' allows more than one meter per a Network Manager, the network and quota mangers can use information coming from different networks in the domain.

Figure 3 shows the simplest configuration of the quota system with one meter, one black-list (MIB quota), and one network.

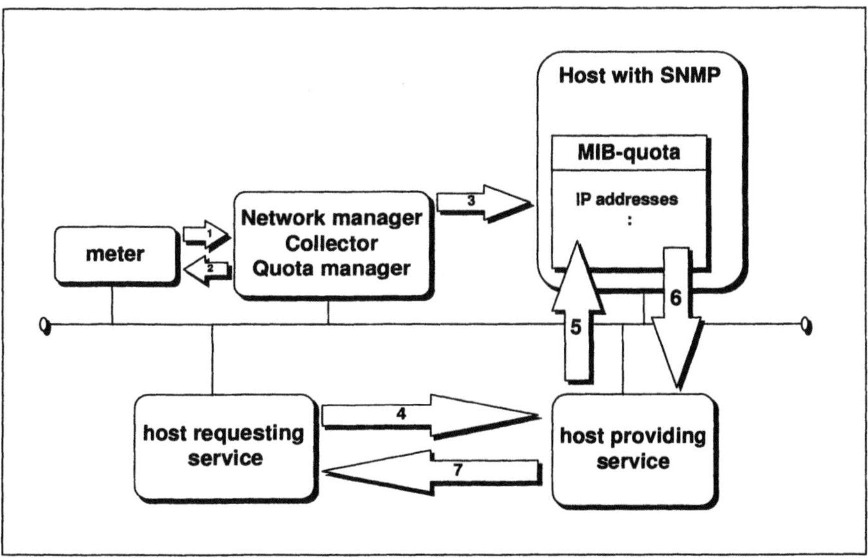

Figure 3 A simple configuration of quota system.

In Figure 3, the arrows denote interactions among various entities. These interactions are explained below.

1,2 : The communication between *meter* and *manager/collector*. This communication is the same as the first implementation of internet-accounting architecture in which *manager* controls the *meter* and *meter* sends the usage reports to the collector.

3 : The *quota manager* fills the MIB-quota on regular time intervals.

4 : A service is requested, such as an ftp request.

5 : The service provider, namely the ftp server, checks if the service requester is in the black-list or not. This is achieved by an SNMP session between *quota manager* and the service provider. It is actually an SNMP-get request of the MIB variable 'iso.org.dod.internet.experimental.backlist.blacklistTable.blacklistEntry.IPaddress made by the service provider.

6 : The answer comes from the host running SNMP agent that maintains MIB-quota. If the IP address of the host requesting the service is in the MIB-quota, it returns that IP address, otherwise it returns something like 'No Such Variable'.

7 : If the IP address of the host requesting service is returned from black-list server, the service provider may not provide the service as the requester is found in the black-list and may return an error. This part is purely implementation dependent. Administrator could implement various alternative models depending on the policy set for that domain.

The proposed quota system can use multiple copies of MIB-quota in a domain. This provides two advantages :

- Availability : If a problem occurs in one of the black-list servers, the alternative one still can be accessed. Of course each service provider knows all of the black-list servers in the domain. They have to know which black-list server to contact first, which one to contact next and so on. Although the updating times of the black-list servers may differ, this won't be a big problem since they are being filled by the same quota-manager.
- Access speed : If the domain is formed of multiple networks, then there will be a performance problem for the service providers to check the black-list through gateways. In such cases, a black-list server can be configured for each network in the domain.

If multiple copies of MIB-quota are desired, the quota manager makes the updates to all of the copies. The updates will be done on regular time intervals. These intervals can be tuned either statically or dynamically by considering the load on the network.

Figure 4 shows a more complicated configuration in which there are three networks, three meters and two black-list servers (MIB-quota).

In the figure the symbols stand for :

R : The router between 3 networks.

Black arrows : The traffic between meters and Manager/Collector. Each of the three networks has a meter in this configuration. Each meter reports the network usage information to the collector. And each of them is controlled by the manager.

X labeled arrows : Since there are two black-list servers in this configuration, the quota manager needs to update both of them on regular time intervals. In order to make necessary additions and deletions to/from MIB-quota, the quota manager makes SNMP-set requests to blacklist-servers. These requests are the same for both of the blacklist-servers.

A labeled arrows : A service is requested from a service provider.

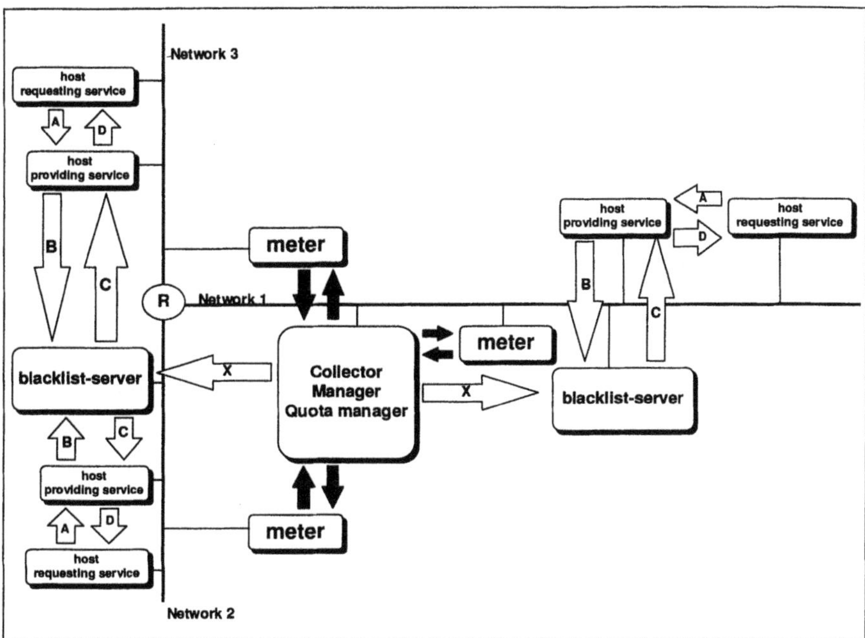

Figure 4 Quota system in multiple network configuration.

B labeled arrows : The service provider checks if the service requester is in the black-list or not. This is implemented by an SNMP-get request. Each service provider makes this request to the nearest blacklist-server. The servers on Networks 1 and 2 makes this request to the blacklist-server on the same network. The one on network 3 makes this request to the blacklist-server on the Network2.

C labeled arrows : This is the answer coming from the blacklist-server. If the IP address of the host requesting the service is in the MIB-quota, blacklist-server returns that IP address, otherwise it returns something like 'No Such Variable'.

D labeled arrows : If the address is in the black-list. The service provider may or may not provide the service. It returns either an error message or the normal response message depending on the specific implementation.

5.2 Algorithm

This algorithm decides which hosts will be put in the black-list and how long they will stay there. Each host starts with U variable assigned to 0 which indicates that no network resources have yet been used. Whenever the host uses the network, the U variable increases proportional to the network usage until a limit *HIGH* is reached. That time the host enters the black-list.

Every night another part of the software decreases the U variable by D. D is the daily increment to the quota of the host. This gives extra network usage chance to the host. A host in the black-list can not use the network resources authenticated by the quota manager. But every night its U variable is decreased. If that comes down to *LOW*, the host is deleted from the black-list. In the current implementation by default U is decreased every night, however this interval can be changed by the network administrator. The network administrator can even give extra usage chance to some of the hosts without considering the algorithm. Another approach can be charging the users for decreasing their U variable and have them to use extra resources.

This is a dynamic quota mechanism, if the host does not use the network its quota is increased, but to some limit. And if it uses the network its quota decreases and enters the black-list if the usage is higher than allowed.

The following figures (Figure 5 and Figure 6) describe the algorithm in flowchart form.

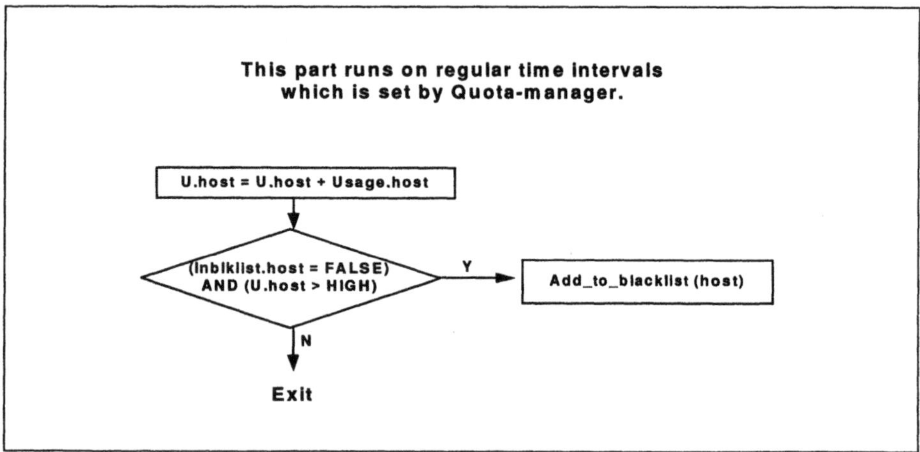

Figure 5

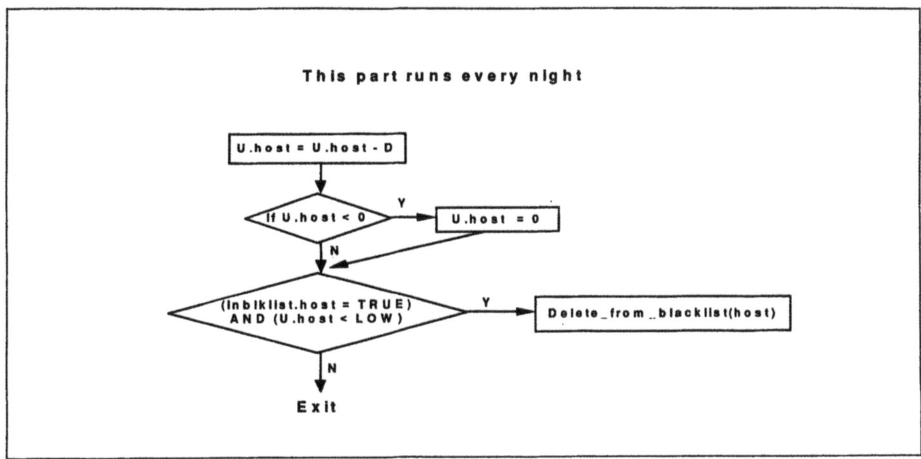

Figure 6

Inblklist.host is TRUE if the host is in the black-list, Add_to_blacklist(host) adds the host to the black-list, Delete_from_blacklist(host) deleted the host from the black-list.

6 SUMMARY

Problems arising from highly loaded networks are not unusual today. Any available resource is consumed by users in a short time. Increasing the available bandwidth does not guarantee to solve this problem permanently. There seems to be a lack of tools that provide a fair share of network resources such as bandwidth. In this study a quota system is proposed to solve this problem in local environments.

Since TCP/IP is the most common networking protocol and SNMP is the most common network management protocol, the study is based on these protocols. As a result of this, it can be ported to many platforms. With the help of this system, network managers may put usage limitations on some of the resources. And this provides a fair share of these resources.

The architecture proposed in this paper could be applied to service usage other than just bandwidth. Meter can collect the traffic for any specific protocol and quota manager can use this data for deciding the usage. A combination of protocols can also be used for deciding the usage.

7 REFERENCES

Brooks, C. (1993) Internet draft. Internet accounting: MIB.

Brownlee, N. (1993) Introductory documentation NeTraMeT & NeMaC (Network Traffic Meter & NeTraMeT Manager/Collector).

Mills, C. Hirsh, D. Ruth, G. (1991) RFC 1272 Internet accounting: background.

Mills, C. Laube, K. Ruth, G. (1992) Internet draft. Internet accounting: Usage Reporting Architecture.

8 BIBLIOGRAPHY

Can Çelik has graduated from Computer Engineering department of Middle East Technical University (METU) in 1991. He is a graduate student in the Computer Engineering Department of METU, he is expected to get M.Sc. degree in Jan. 1995. Mr. Çelik is doing Systems Programming in the Computer Center of METU, specialized in UNIX operating systems.

Attila Özgit is a graduate of Middle East Technical University. He is a faculty member of the Computer Engineering Department and also the Director of Computer Center. His research interests are Operating Systems, Computer Networks and Distributed Systems.

PART TWO

Performance and Fault Management

SECTION ONE

Enterprise Fault Management

20

Towards a Practical Alarm Correlation System

K. Houck, S. Calo, A. Finkel

IBM T. J. Watson Research Center
P.O. Box 704
Yorktown Heights, NY 10598
Email: houck@watson.ibm.com

Abstract

A single fault in a telecommunication network frequently results in a number of alarms being reported to the network operator. This multitude of alarms can easily obscure the real cause of the fault. In addition, when multiple faults occur at approximately the same time, it can be difficult to determine how many faults have occurred, thus creating the possibility that some may be missed. A variety of solution approaches have been proposed in the literature, however, practically deployable, commercial solutions remain elusive. The experiences of the Network Fault and Alarm Correlator and Tester (NetFACT) project, carried out at IBM Research and described in this paper, provide some insight as to why this is the case, and what must be done to overcome the barriers encountered. Our observations are based on experimental use of the NetFACT system to process a live, continuous alarm stream from a portion of the Advantis physical backbone network, one of the largest private telecommunications networks in the world.

The NetFACT software processes the incoming alarm stream and determines the faults from the alarms. It attempts to narrow down the likely root causes of each fault, to the greatest extent possible, given the available information. To accomplish this, NetFACT employs a novel combination of diagnostic techniques supported by an object-oriented model of the network being managed. This model provides an abstract view of the underlying network of heterogeneous devices. A number of issues were explored in the project including the extensibility of the design to other types of networks, and impact of the practical realities that must be addressed if prototype systems such as NetFACT are to lead to commercial products.

1. INTRODUCTION

A single fault in a telecommunication network frequently results in a number of alarms being reported to the network operator. This multitude of alarms can easily obscure the real cause of the fault. This phenomena not only increases the skill and time needed to resolve failures, but also increases the probability that one or more failures will be lost in the confusion caused by others. The resulting increase in "mean time to repair" and support center staffing costs make this problem a frequent source of complaints about current network management systems. In order to solve this problem, we must first understand its origins.

There are a number of reasons why a single fault in a network results in multiple alarms being sent to the network control center. They include:

1. Multiple alarms generated by the same device for a single fault (sometimes known as alarm streaming).
2. The fault is intermittent in nature and each re-occurrence results in the issuance of new alarms.
3. The fault is reported each time a service provided by the failing component is invoked.
4. Multiple components detect (and alarm on) the same condition (e.g., a failing link is detected at both end-points of the link).
5. The fault propagates by causing dependent failures and resultant alarms.

We observe that the first three reasons (above) deal with the same alarm(s) repeated in time, while the last two explain why many different alarms are often triggered by a single fault. With this deeper understanding of the problem, we can now consider solutions.

A variety of solution approaches have been proposed in the literature (Brugnoni(1993), Jordaan(1993), Lor(1993), Sutter(1988)), however, practically deployable, commercial solutions remain elusive. The experiences of the Network Fault and Alarm Correlator and Tester (NetFACT) project, carried out at IBM Research and described in this paper, provide some insight as to why this is the case, and what must be done to overcome the barriers encountered. We divide such barriers into two classes: "basic prerequisites", those things that must be in place before a workable solution can be deployed, and "fundamental technology", the design and algorithms that are needed to solve the problem assuming the basic prerequisites can be put in place. We mention briefly the basic prerequisites and then focus on the fundamental technology issues in the remainder of the paper.

In order for the problem to occur, we can reasonably assume that the most basic of the prerequisites, centralized alarm reporting and storage, is in place. In many cases this may be all the information that is needed to filter out alarms that are repeated in time. Handling different alarms caused by the same fault, however, requires two additional prerequisites: active configuration knowledge (knowledge of the configuration at the time of the failure), and alarm knowledge (knowledge about how the failure condition reported in an alarm from one component relates to other failures in adjacent components of the configuration).

Current technology, such as MAXM(1988), can usually handle the problem of centralized alarm reporting, even from heterogeneous devices using different alarm syntaxes and transport protocols. Standards such as SNMP and CMIP, when fully deployed, will further address the alarm reporting requirements. The problem of acquiring knowledge of the configuration at the time of the failure is somewhat more difficult, but we believe that in most cases this too can be achieved. Active model managers such as RODM (Finkel,1992), that can provide access to sufficiently current representations of the configuration, will help address this need. Alarm knowledge, however, remains an obstacle. We will highlight the requirements in a later section of the paper.

The remainder of the paper discusses the design of the NetFACT system, and our experiences with its development and operation on the Advantis physical backbone network, one of the largest private telecommunications networks in the world. Section 2 provides a overview of the actual algorithms used in the project, section 3 describes the overall system design, and section 4 describes the practical aspects of the problem that had to be accommodated in our design. Section 5 documents some of our observations and conclusions from the project.

2. TECHNICAL OVERVIEW

The approach taken to alarm correlation in NetFACT is to first build a normalized model of the network configuration, normalize the incoming alarms, and then use a generic application to interpret the normalized alarms in the context of the network configuration and prior alarms. This approach stemmed from the observation that three distinct types of knowledge are needed to deduce the underlying faults from the alarms received:

- knowledge about the meaning of the individual alarms,
- knowledge of the network configuration, and
- knowledge of general diagnostic techniques and strategies.

These three types of knowledge would likely come from and, more importantly, be maintained by, separate organizations. Furthermore, alarm knowledge would likely need to be provided and maintained by groups with in depth expert knowledge about the device generating the alarms - this could be many groups, potentially one per type of device. Thus, if the knowledge contained in the system is to be maintainable, it must be partitioned in a way that allows knowledge in any partition to be maintained without awareness or impact to the other partitions. This partitioning is an important and unique aspect of the NetFACT design.

After a brief review of the problem domain in which NetFACT operated, we will describe the diagnostic strategies employed by NetFACT and the representation of the configuration and alarm knowledge required to support those strategies.

2.1 Domain

As background information to aid in understanding the diagnostic strategies and configuration models used by NetFACT, we describe briefly the domain of telecommunications networks, in which NetFACT operated. A telecommunication network multiplexes digitized voice and data circuits onto a smaller number of higher speed backbone circuits that carry data between the multiplexers. These higher speed circuits consist of various sequences of "cable" (e.g., wire, fiber, wireless microwave links) and various pieces of equipment (e.g., CSU's, encryptors, repeaters) that in some way transform, monitor, or amplify the physical or logical representation of the data traveling on the circuit. These high speed circuits can themselves be multiplexed onto even higher speed circuits. When data must be transported over long distances, the "cable" used is actually a telephone carrier provided digital circuit (e.g., DS-1, DS-3). We now consider the abstractions used by NetFACT to model telecommunications networks.

2.2 Configuration Data Model

The diagnostic process makes use of a normalized model of configuration information to obtain the configuration elements and relationships (connectivity, dependency) between these elements. This model is maintained in an object oriented data base, thus allowing it to be shared with other network management applications. In keeping with our objective to make NetFACT sufficiently general to support other types of networks, the normalized model is somewhat more abstract than the description of telecommunications networks given in the previous section. As a result, a given piece of equipment may be represented by more than one object in the model.

In the NetFACT data model, network components are classified as paths, nodes, or shared resources. The normalized data model has one further level of detail (i.e., types of paths, nodes, etc), but this will not be discussed here due to space constraints. A **path** is defined to be a connection between end points, over which the same data flows. A **node** is a network component that in some way processes the data flowing over a path. Paths may contain nodes and other paths. A node with one connection to a path is called an end point of that path. All nodes that are not endpoints have exactly two connections. Nodes may depend on one or more **shared resources**, each of which may also depend on one or more shared resources. A given shared resource may support multiple nodes/shared resources, thus dependency is a many-to-many relationship. To apply this model to telecommunication networks, we use paths to represent both the circuits and "cables" in the network, while nodes are used to model the various pieces of telecommunications equipment on a circuit, including the interface cards in the multiplexors that are the endpoints of the circuits. A complex device with many ports, such as a multiplexor, is modeled as a collection of nodes (representing interfaces) that are dependent upon a common shared resource (representing the common elements of the device such as the power supply, backplane, and control circuitry). More elaborate models can be constructed, if needed.

The normalized relationships modeled by NetFACT include data-flow, composition, and dependency. Data flow and dependency are used to follow the potential propagation of faults, while composition is used to help optimize the diagnostic algorithms by reducing the portion of the network that they must explore in certain situations.

Inheriting from the normalized model are the sub-classes that are unique to each component type. These classes contain any attributes or methods that are needed to convert the alarms for a specific type of device into the normalized form. It is these device type specific classes that are instantiated with the network configuration.

The picture in Figure 1 shows the class instances and relationships that NetFACT uses to model a typical telecommunications circuit. The circuit begins, in the upper left corner, with an IDNX trunk card (N020C050) connected to another IDNX trunk card (N111C000) via a DS1 path (IBM-002003). The DS1 path object does not represent any single physical object but rather the sequence of objects it contains (indicated with the dashed lines). Thus, data flows from the IDNX trunk card (N020C050) through an encryptor (00004553), and then through a CSU (TC006480). At this point the circuit is multiplexed onto a DS3 path (IBM-17958). The use of this DS3 by the DS1 (IBM-002003) is represented by the DS3_channel object (G002); this allows us to follow the original data flow through the DS3 and locate it on the far side. The pair of Network_ports (G006,G003) are used to represent the portion of the circuit provided by a common carrier. Note that the data enters the carrier's network on a DS3 (G006), is demultiplexed by a multiplexor not visible to NetFACT, and exits the carriers network as a DS1 (G003). After exiting the carrier's network, the data flow proceeds thru the CSU (TC0000008), the encryptor (00000004), and finally to the IDNX_Trunk (N111C000) which is the end of the circuit. The multiplexors that are visible to NetFACT are represented by a combination of node objects (e.g., IDNX_Trunk, M13_T1_port) and shared resources (e.g., IDNX_box, M13_box).

In addition to configuration data (i.e., object identity, type, and relationships), the data model also includes real time component status information that is both used and updated in the process of building the normalized alarm representation.

2.3 Diagnostic Strategies

In general, the approach to diagnosis taken in NetFACT is to employ a collection of techniques specialized to the type of topology that encompasses the fault/alarms being considered. Note that we specialize to the type of topology and not to the type of device as is more commonly done in rule based expert systems.

In the current implementation of NetFACT, two diagnostic techniques are used. The first, which we call **path analysis**, handles problems relating to the failure of a path. Typically, one component of the path (either a piece of equipment or a carrier circuit) fails and all communications over the path are stopped. Various components on the path report the failure by generating alarms. Path analysis processes these alarms and determines which

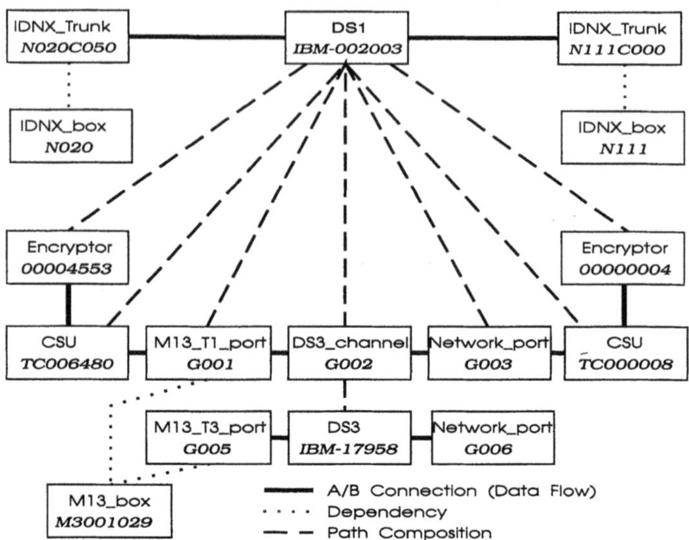

Figure 1. Example of NetFACT Data Model.

components on the path are most likely to be responsible for the failure. Once this determination has been made, the second technique, which we call **tree search**, is used to determine whether the nodes (or sub-paths) identified are responsible for the failure themselves, or whether they are failing because of a problem in components on which they are dependent (e.g., shared resources). Looking at the relationships shown in Figure 1, path analysis locates failures that propagate along horizontal relationships (i.e., data flow), while tree search locates failures that propagate along vertical relationships (i.e., dependency).

Path analysis employs a voting technique to sum all the evidence contained in the alarms received from the nodes on a given path. Each normalized alarm provides an indication of where the cause of the problem might lie, relative to the node reporting the alarm. Possible locations include the node itself, a matching peer device, and somewhere in either direction of data flow. The likelihood that the cause of problem is in any of these possible locations is expressed as an integral number of votes for each possible location. This allows the alarm to express some degree of uncertainty about the precise location of the source of the problem it is reporting. For example, if a CSU detects a problem on its line side, the normalized alarm generated as a result would contain one vote for the device itself (it is always possible that the problem is in the line interface of the CSU) and two votes for all devices on the path in the direction of the line; devices in the direction of the DTE equipment would receive no votes. The votes are summed for each component of the path and the components with the most votes or second most votes are explored using the tree search technique.

Tree search explores the dependency tree for a given component to determine if a lower level problem is causing the component to fail. The exploration process considers both direct evidence (i.e., alarms) and indirect evidence (i.e., how many users of a lower level component are experiencing difficulties.) Indirect evidence must be used because failing components do not always generate alarms. In cases where the failure of a given component or path could be caused by "n" different lower level component failures, for each of which only indirect evidence exists, heuristics are used to choose the component most likely to be the cause of the failure. Diagnostic tests, if available, could also be used to help resolve such ambiguities. If the lower level resource suspected of failing is a path (such as a DS-3), path analysis is invoked recursively.

As the above diagnostic strategies proceed, previously independent problems/alarms are causally related and the overall number of "open" problems is reduced. After a set amount of time, a problem that cannot be related to another is surfaced to the operators through a user interface application. In general, problems in NetFACT are moved through phases (states) of a problem lifecycle. Ignoring some complexities that will be discussed in a later section of the paper, the basic problem lifecycle in NetFACT involves the following states:

Awareness Build internal representation of the alarm and wait briefly for additional related alarms to arrive

Get config Obtains relevant configuration from the configuration model

Diagnosis Use the diagnostic strategies to identify the cause of the alarm

Recovery Await the recovery of the network components impacted by the problem

Closure Mark the problem as closed and direct any further alarms from the components impacted to open new problems

Purge An operator purges the problem from the system

Figure 2, together with the explanation below, shows how the diagnostic techniques are applied to locate the root cause of a problem. The sequence of events is as follows:

1. Components A, B, and E send alarms (The alarm notation shows the number of votes for self inside the circle and the number of votes in each direction of data flow at the ends of the directional arrows.)
2. Path analysis first applies the relative voting information in the alarms to the path configuration
3. Path analysis then sums the votes for each component in the configuration and determines that components C and D are the most likely causes of the path failure; components B and E are second choices
4. Tree search is invoked; only component D is found to have a dependency: it is dependent on component F
5. Components X, Y, and D are all users of component F, but each is on a different path; the paths containing components X and Y are also experiencing failures (not shown)
6. Components X and Y are also prime suspects in their respective path failures (not shown); tree search will identify component F as the most likely cause of failures of the paths containing components X, Y, and D
7. NetFACT will open a single problem with component F as the most likely cause.

2.4 Alarm Normalization

Alarms are received in a variety of syntaxes and must be translated to a normalized form with consistent syntax and semantics. The information contained in the normalized alarm includes:

- The identity of the object to which the alarm refers
- The impact to the behavior of the object (e.g., UP, DOWN)
- Votes representing the likely source of the problem that caused the alarm
- Miscellaneous information such as timestamps and alarm id

When an alarm is received, the corresponding model object is located in the data base and used to determine if the alarm contains new information not reported by previous alarms. Alarms containing new information are normalized and passed to the diagnostic application; the status information in the object model is updated in the process.

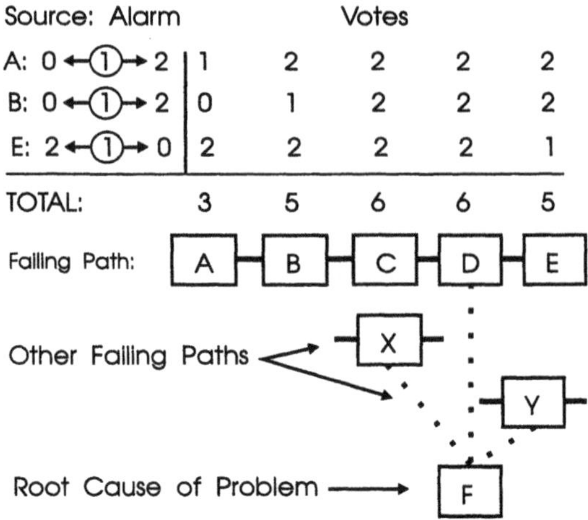

Figure 2. Diagnostic/Correlation Example (see text for explanation).

3. SYSTEM DESIGN

The diagram in Figure 3 shows the components of the NetFACT system and the data flows between them. The system was implemented on an MVS/390 system using IBM's NetView network management system. This allowed the configuration model to be implemented using NetView's Resource Object Data Manager (RODM), a high performance, object oriented data manager (Finkel,1992).

The NetFACT components (Figure 3) are best understood by following the processes in which they participate. NetFACT has a configuration model update process and an alarm handling process. The configuration model update process extracts the current version of the configuration from a number of different tables in an SQL database, and updates the object data model (in RODM) to this version of the configuration. This is accomplished without impacting the availability of the alarm handling process, or other applications that may also be using RODM.

The alarm handling process begins with the receipt of an alarm from the network. NetView's alert automation facilities then select and dispatch the appropriate command procedure (script) to generate the normalized form of the alarm. In the process of doing this, the command procedure locates the corresponding object in RODM and updates its status accordingly. If the alarm contains information that is important to the diagnostic algorithms (and has not been previously reported), it is passed through RODM to the NetFACT application. Here it is operated upon by the diagnostic procedures described in the previous section. If a new problem is identified, an object is created in RODM to represent the problem. The operator interface component can query these objects and display information about the faults they represent to a human operator. In addition, the creation of the problem object can cause a problem record to be opened in a problem management system, such as IBM's INFO/MGT product.

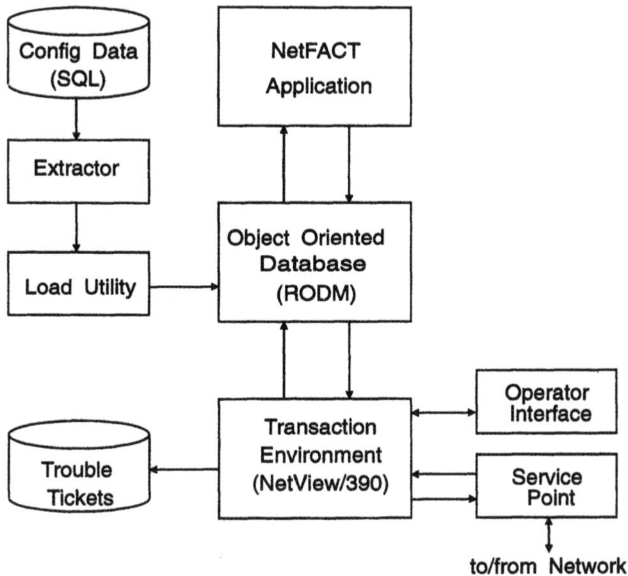

Figure 3. NetFACT System Design.

4. PRACTICAL CONSIDERATIONS

In the process of developing NetFACT and testing it with a real alarm stream from the Advantis physical network, a number of practical problems were encountered. Many of these were solved during the course of the development and we continue to study those that were not. We discuss some of those problems here along with other observations made during the project.

4.1 Noise

The first practical reality that we encountered was "noise". In the ideal case, a problem detected by a component results in one alarm to indicate that the problem has been detected, and another to indicate that the problem has been resolved and correct behavior restored. Some problems do, in fact, result in such clean notifications - unfortunately, many others do not.

We refer to alarms we wish we didn't have to process as "noise" and group them into the six categories shown in Figure 4. The taxonomy is useful because it allows NetFACT to employ different strategies to deal with different kinds of noise.

Alarms that to do not usually indicate a problem with the behavior of the component, although they may help explain a problem reported by other alarms, are classified as insignificant information. The information may optionally be retained in the component's object model, where it can be used in answering specific queries that NetFACT may direct at the object model. Redundant information and streaming alarms can be filtered out with the

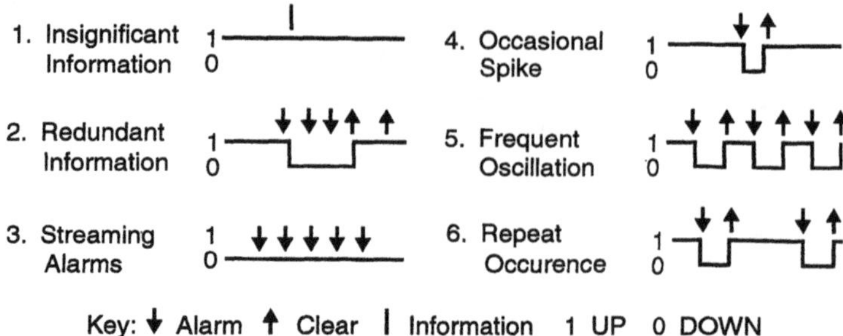

Figure 4. Categories of Noise.

help of the state information kept in the object model. If necessary, streaming alarms can be filtered closer to the source, to avoid the overhead of transmitting them to the central site. The occasional spike is suppressed by extending the problem lifecycle to include a verification stage, where a problem must persist for a specified, but short, period of time before NetFACT will process it. Likewise, frequent oscillations are suppressed by requiring that problems remain resolved for a period of time before they are allowed to enter the closed stage of the lifecycle. Repeat occurrences of a problem occur with sufficient separation that re-diagnosis is appropriate. An automated link from the repeat occurrences back to the original problem was not implemented, although it seems feasible.

4.2 Hidden Dependencies

Many of the systems described in the diagnosis literature, as well as earlier versions of NetFACT, are aimed at the goal of finding a minimum number of faults that will explain the observed symptoms. Since faults occur relatively infrequently, it seems reasonable that the probability of multiple faults occurring simultaneously is extremely low. Thus the heuristic that the correct root cause of a number of related symptoms is likely to be a minimum number of faults. In the real world, this is probably true.

In the world of NetFACT, however, visibility is often limited and not all the dependencies that would point to the single root cause of a set of symptoms are known (Figure 5.) The common cause of a set of path outages may be within a carrier's network. Simply applying the above heuristic may result in incorrectly identifying a common node that is visible to NetFACT as the cause, when really the correct cause is a hidden common element inside the carrier network. To address this problem, NetFACT reports such failures as independent faults. (A possible enhancement would be to group independent failures that occur within the same carrier network at nearly the same time, to suggest a probable correlation.)

This problem helped persuade us to change our basic strategy of how NetFACT should correlate alarms. We changed from using a global optimization strategy based on the "minimum number of faults" heuristic, to a strategy of finding the "best explanation for each symptom". Global knowledge is still used in determining the cause of a symptom, but it is considered weaker evidence than actual alarms, and no overall global optimization is attempted. When a high percentage of users of a resource are experiencing a failure, this suggests that the supporting resource may be responsible. Each user of the resource is

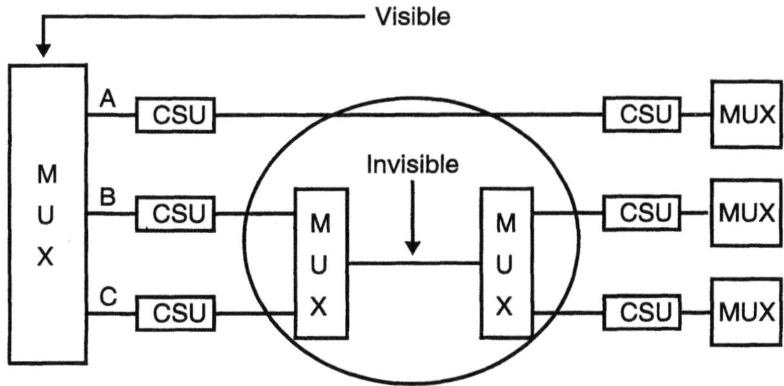

Figure 5. Example Configuration with Hidden Dependencies.

required to believe that the shared resource is the most likely cause of the problem, however, before the association can be made.

4.3 Complex Dependencies

The next problem we encountered involved the nature of the dependency relationships between network components. NetFACT supports only simple dependency relationships, where the dependent resources depend solely on the binary availability of the supporting resources. When a supporting resource fails, all dependents fail with it (and nearly simultaneously). Unfortunately, not all dependencies are simple.

Some resources depend on logical combinations of other resources. For example, the availability of an SNA transmission group depends on at least one of the link in the group being active.

Other resources depend instead, on a quantitative amount of capacity in the shared resource. In these cases, a sudden drop in capacity can cause failures in the dependent resources around the time of the capacity drop. A later failure of a dependent resource is much less likely to be explained by the capacity drop (unless the failing resource has requested additional capacity.) An example of this class of problem is the allocation of capacity on IDNX CPU cards needed to support DS-1 circuits connecting IDNX nodes. When a CPU card fails in a node (multiplexor) without sufficient backup CPU capacity, a number of DS-1 trunks connected to that node are dropped. A related class of problems, involving buffer pools, can be found in packet switches.

4.4 Missing Data

NetFACT relies totally on unsolicited alarms to get its information about what is happening in the network. This limitation was project related and stemmed mainly from the technical difficulties of interconnecting the various network management systems being used at

Advantis, in addition to concerns about the potential of NetFACT interfering with production operations. This limitation proved to be a serious problem. Unsolicited data alone does not always result in a complete or even accurate picture of what is happening in the network. Scenarios involving missing alarms or status updates include:

- Data received from only one end of a path, or one of a pair of matched devices
- No indication that a given device has recovered
- No path to receive data from a remote device

State data derived from alarms and unsolicited status updates must be treated carefully in light of the above. The NetFACT system associates a time stamp with each state of each resource in the state model. This information is very useful when viewing or analyzing resource state information.

It is important to note that if NetFACT were able to solicit status information from the network components, it would be able to use its knowledge of the network status and problems to reduce the number of solicitations needed. Conventional timer driven polling applications would not have such knowledge and therefore would be less efficient at collecting status information. Thus, NetFACT's powerful knowledge base has interesting implications for the overall design of network management systems.

4.5 Implementation

We are often asked about the programming languages and tools used to implement NetFACT. The diagnostic application is written in ANSI C, rather than a rule based language. While there were times when a rule based approach seemed more desirable, we still believe that, overall, the procedural approach resulted in a more robust and maintainable application. C++, had it been available then in the MVS environment, would have resulted in somewhat more maintainable code.

The RODM data store proved quite adequate for our data modeling needs. Both its execution speed and object oriented capabilities greatly facilitated our implementation.

5. OBSERVATIONS, ASSESSMENTS, AND CONCLUSIONS

We now return to the question of what must be done to make systems such as NetFACT into practical, commercial products. Based on our observations, probably the greatest need is to bring together the development of alarm reporting standards, configuration models, and diagnosis/correlation algorithms. By bringing together these currently independent activities, we will be able to insure that the incoming alarms can be understood in terms of the configuration model and that they will contain sufficient information to drive the correlation algorithms. Until adequate alarm reporting standards are in place, systems such as NetFACT will be forced to translate the individual alarms into a suitable normalized form. This translation typically requires knowledge about the semantics of each individual alarm. We believe the discovery and maintenance of this information is both difficult and costly and thus will be a significant impediment to vendors wishing to bring commercial products to market, in anything but a very limited context.

We remain satisfied with the overall system design used in the NetFACT project and we believe it can serve as a model for future implementations. The power of the locally cached object oriented network model can be more fully exploited once solicited diagnostic testing of network components is introduced.

Additional diagnostic/correlation techniques will be needed to support other types of networks such as packet routing networks (IP). How new diagnostic techniques can be easily incorporated into the existing algorithm remains an interesting area for future research. Despite the apparent complexity of such an approach, we remain convinced that a collection of diagnostic/correlation techniques is likely to outperform any single unified approach. In addition, using knowledge of past experience in the diagnostic algorithms offers interesting possibilities. Historical knowledge would be useful in optimizing the search for the cause of a failure, in detecting repeat occurrences of problems, and in recommending a course of action once the cause of a fault has been determined.

Finally we would like to emphasize the importance, benefits, and difficulties of conducting experiments on a real network. In our early work on NetFACT, a small 2-3 node test network was used to collect the alarms resulting from manually induced failures (unplugging various cables). Unfortunately the data collected from these experiments was not indicative of the full magnitude and scope, and especially the interaction, of problems we later encountered running with the full Advantis network. In particular, the problems relating to noise, missing data, and the interaction between multiple faults occurring in the same period of time were not anticipated based on experiences with the "test network" faults. On the down side, the project experienced considerable delays while waiting for various instrumentation and network management connectivity problems to be resolved.

While NetFACT is far from the ideal envisioned by many network managers, it represents an important step toward achieving the goal of developing a global automated system for problem management. The NetFACT project has shown that practical solutions to the problem of alarm correlation are possible, although additional work is necessary, especially in the area of alarm standardization, before such solutions are likely to become commercially available. Furthermore, we believe that our design shows how the vast amount of knowledge used by such a system can be organized and partitioned in a way that will allow it to be easily maintained.

6. REFERENCES

Brugnoni, S., et al. (1993) An Expert System for Real Time Fault Diagnosis of the Italian Telecommunications Network, in ***Proceedings of the International Symposium on Integrated Network Management III*** (ed. H.-G. Hegering and Y. Yemini), IFIP, San Francisco, CA

Finkel, A. and Calo, S.B. (1992) RODM: A Control Information Base. ***IBM Systems Journal*, V31 N2**,252-269

Jordaan, J. F. and Paterok, M. E. (1993) Event Correlation in Heterogeneous Networks Using the OSI Management Framework, in ***Proceedings of the International Symposium on Integrated Network Management III*** (ed. H.-G. Hegering and Y. Yemini), IFIP, San Francisco, CA

Lor, K.-W. E. (1993) A Network Diagnostic Expert System for Acculink(tm) Multiplexers, in ***Proceedings of the International Symposium on Integrated Network Management III*** (ed. H.-G. Hegering and Y. Yemini), IFIP, San Francisco, CA

MAXM Corp. (1992) MAXM System Administrator's Guide, International Telecommunications Management, Inc., Vienna, VA

Sutter, M. T. and Zeldin, P. E. (1988) Designing Expert Systems for Real-Time Diagnosis of Self-Correcting Networks, ***IEEE Network Magazine***,September 1988,43-51

21

Validation and Extension of Fault Management Applications through Environment Simulation

Roberto Manione, Fabio Montanari

CSELT, Centro Studi E Laboratori Telecomunicazioni S.p.A.
Via G. Reiss Romoli 274 - 10148 Torino (ITALY)
Tel. +39-11-2286817, Fax. +39-11-2286862
e-mail: manione@cselt.stet.it

Abstract

Fault management systems are complex applications. Early evaluation of prototypes as well as thorough testing and performance evaluation of the final versions before their deployment are a must.

The present paper presents a simulator of plesiochronous transmission networks, SPRINTER, which has been used to generate test patterns for alarm correlation systems, working on the same kind of networks.

Thanks to the choice of a versatile simulation environment, particularly suited for distributed systems, YES, the implementation of SPRINTER turned out to be elegant and easily extensible.

The approach has been applied to the validation of the alarm correlator SINERGIA; however, the alarm streams generated by SPRINTER could be used to test other correlators working on the same kind of networks.

Furthermore, the proposed simulation approach seems generalizable to other network management applications and areas.

Keywords

Models, Distributed Systems Simulation, Fault Management, Alarm Correlation, Fault Diagnosis, System Testing

1 INTRODUCTION

Fault diagnosis of Telecommunication networks is a fairly complex task, mainly due to the interactions among the different network components along the digital paths; as a consequence of such interactions, a number of equipments across the network emit alarms as a consequence of a single fault.

To cope with this alarms proliferation, correlation techniques are used: their purpose is the isolation and diagnosis of the faults starting from the equipment alarms.

A number of approaches have been proposed; among them are SINERGIA [1] [2], which performs rule based correlation and diagnosis using heuristics taken from the network experts, and IMPACT [3], which uses a model based reasoning approach. In both cases the diagnosis system needs to be verified and validated before its deployment with an extensive number of real cases (i.e. not just test data used in the debugging phase).

On the other hand such real alarm streams are not easy to obtain, particularly during the development phase of the Network Management system; the main disadvantages of the use of real streams are that they span over long time intervals (i.e. weeks), hence they take long times for their collection; furthermore generally they do not contain all the kinds of faults over all the kinds of network equipments in all the network topologies which the diagnosis system claims to deal with

A network simulator can instead be used to generate the test alarm streams; such a simulator is also useful when high volumes of alarms are needed to test the ability of the diagnosis system to stand with given alarm throughputs.

A totally different use of a network simulator is the generation of the diagnostic knowledge to be used by the correlator: new topologies, not known to the correlator, can be simulated and the relative alarm versus fault relations extractedfrom the simulation results.

In the following, the above mutually exclusive usages of a simulator will be called Validation and Extension, respectively; both have been experimented with the SINERGIA alarm correlator.

This paper presents a network simulator, SPRINTER (Simulator of Plesiochronous tRansmIssion NeTworks alaRm handling) built for the validation of the fault diagnosis system implemented at our labs, SINERGIA. The structure and the behaviour of the various network equipments, as far as the alarm handling and propagation is concerned, have been coded into a library of equipment models, usable in the composition of the networks.

A significant number of networks have been built out of the equipment models and extensively simulated. The simulator is able to inject given faults over given equipments and to obtain a timed list of the alarms generated all over the network as a consequence of the faults, either in single or multiple fault contexts; SPRINTER is also able to simulate the ceased alarms stream coming from mending actions over the faulty equipments.

The paper is organised as follows: in section 2 the fault diagnosis system under validation is sketched; section 3 presents the simulation environment, while Section 4 deals with the overall simulator architecture; section 5 reports the validation results on SINERGIA and the first approaches to its extension; finally section 6 draws the conclusions.

2 FAULT DIAGNOSIS IN PLESIOCHRONOUS NETWORKS

The goal of a generic fault diagnosis system is to locate faults in the *digital paths* along the transmission network, which are caused by failures of Lines, Line Terminals (T), Multiplexers/Demultiplexers (M) as well as of the trunk interfaces of Exchanges, Digital Cross Connects (DXCs) and other network devices.

CCITT Recommendations G.704 [4], provide a mechanism for the generation and propagation of fault indication signals (in the following referred to as alarms) in the digital transmission paths, aimed a the easy identification of the faulty equipments; however, almost always, the occurrence of a trouble in one equipment originates alarms from a number of equipments somehow related to it. The main task of the fault diagnosis is to group together all the alarms which are originated from the same physical fault and to find out which equipment needs to be repaired; a more precise diagnosis which identifies the fault diagnosis within the faulty equipment is of course a plus of the diagnosis system. The diagnosis process is not straightforward and sometimes is still carried out by maintenance experts.

In the Italian network the transmission equipments are monitored by proper Mediation Devices, which make the state variables of each monitored equipment available to the diagnosis system. Such variables are in turn driven by the operating status of the equipment and of the digital paths connected to it (as specified in the CCITT recommendations G.704).

Figure 1 pictorially shows of what happens in a real plesiochronous network, e.g. made of Equipments (Multiplexers, M and Line Terminals, L in the picture) and Lines: faults occur from time to time over its components and alarms are generated by the Equipments; in general, different alarms are emitted by a number of equipments in front of any single faults occurring at one equipment of line; alarms are forwarded to a Network Management Center for their correlation, aimed at the isolation of the faulty equipment.

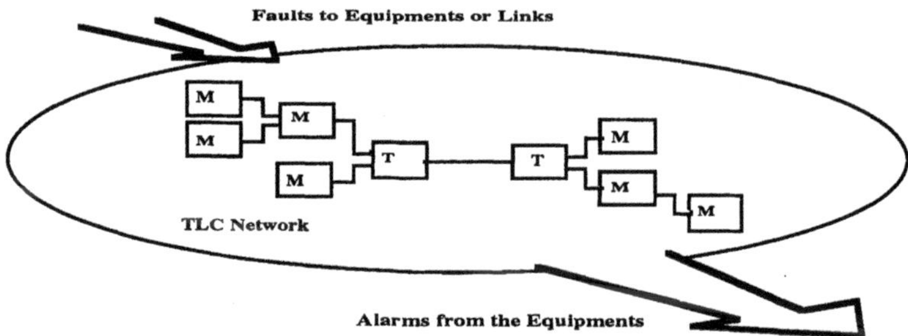

Figure 1 Alarm generation and propagation in Telecommunications Networks

2.1 The SINERGIA alarm correlator and fault diagnostician

The knowledge built in SINERGIA is basically organized as follows: a number of network topologies (e.g. templates) have been selected in a way such as any real network topology can be expressed by instances of such templates; for each template all the feasible alarm patterns have been listed; for each pattern the faulty equipment has been identified, together with the respective fault diagnosis; each template with the associated list of alarm patterns and fault diagnoses has been named Data Sheet; figure 2 shows an example of data sheet.(see [1] for a more formal definition of the templates)

Each fault propagation pattern listed in a Data Sheet holds two different kinds of knowledge: a *Topological Knowledge* of the involved devices, their interconnections and physical characteristics (such as the type of equipment, its bit-rate and its manufacture technique), and an *Expert Knowledge*, derived from the maintenance Experts' experience, regarding the expected alarms for each specific fault. Moreover each pattern embodies the fault diagnosis (i.e. the indication of the faulty device together with the description of the occurred trouble). We can therefore say that the alarm pattern is the fundamental piece of knowledge on which the SINERGIA diagnosis process has been built. Each pattern has been coded as a forward chaining rule; at present about 400 such rules have been encoded into the system.

MPX	TL	TL	MPX	Technique 2Mbit/s CC(N2)	CC(N2C)	FO	PR	Diagnosis
INT	INT	EXT	EXT	X	X			Power Supply
EXT	**EXT**	**EXT**	EXT	X				Line Fault or Regenerator
EXT	**EXT**	·	(-)	X				One way regenerator
(-)	**INT**	EXT	(-)		X			Tx interface or Power supply
EXT	**INT**	EXT	(-)	X				Tx interface or Power supply
EXT	·	·	(-)	X	X	X	X	Rx Line term. Interface or Rx MUX interf
·	**NUR**	·	·	X	X	X		Line regenerator BER > 10^-6
(-)	**EXT**	**EXT**	(-)		X	X		Rx interface of Line Terminals
(-)	**EXT**	·	(-)		X			Rx interface of Line Terminal
INT	URG	(-)	(-)		X	X		Interface
(-)	**URG**	URG	(-)			X		BER > 10^-3 or Regenerator
EXT	**ALIM**	ALLRX	(-)				X	Power
(-)	**ALLTX**	ALLRX	(-)				X	Tx Alignment
(-)	**MIR**	·	(-)				X	No signal received
EXT	**BER**	·	(-)				X	BER > 10^-3
(-)	**ALLRX**	·	(-)				X	No signal received
INT	MIR	·	(-)				X	No signal transmitted
ALIM	MIR	·	(-)				X	Power Supply

Figure 2 An example of Data Sheet

The overall correlation methodology of SINERGIA is built up of two main reasoning steps that implement a sort of *generate and test* paradigm as is depicted in Fig. 3.

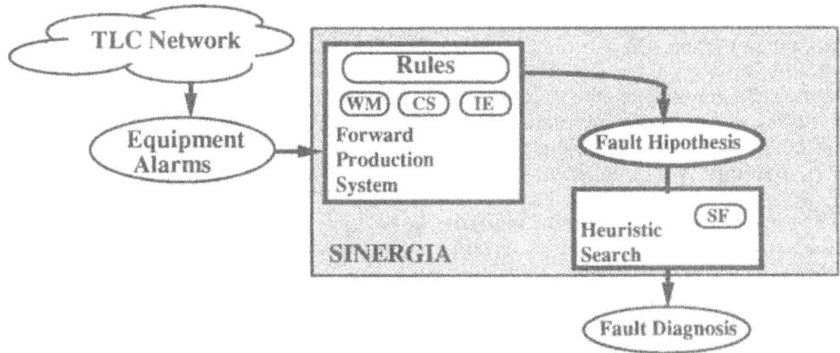

Figure 3 The SINERGIA Architecture

The first step is based on a set of rules (which encode the fault patterns of the Data Sheet) which instantiate fault hypotheses, whilst the second is a heuristic search to determine the best solution among the hypotheses (the fault diagnosis result). In the figure 3 the fundamental blocks of the first step are also depicted. In fact the execution of the rule component relies mainly upon the Working Memory (WM), useful to determine what rules are executable, the conflict set (CS), which contains the executable rules, and the Inference Engine (IE) which governs the whole process.

The rules block works mainly on the alarms collected from the telecommunications network to produce an intermediate result, the Fault Hypotheses set.

The Heuristics Search block selects among the Fault Hypotheses and delivers the Fault Diagnoses, which are the optimal subset of the Fault Hypotheses which best explain, according to a set of criteria, the alarms received from the network; a Scoring Fiunction (SF) is used to rank the Hypotheses Subsets.

Among the more remarkable features of SINERGIA is the ability of its algorithms to exploit the Topological and Heuristic Knowledge, worked out under the hypothesis of single fault, even in case of multiple faults, extending it automatically.

3 SPRINTER MOTIVATIONS AND IMPLEMENTATION CHOICES

A number of needs led us to undertake the implementation of the behaviour of alarm generation across Plesiochronous; they all have been experimented on SINERGIA but are applicable to other diagnosis systems as well:

- to validate the correctness of the knowledge and of the algorithms of Alarm Correlation and Fault Diagnosis systems; in this case the simulation output is the timed stream of alarms coming from the faulty network; such a stream can be sent to the system under test and its diagnosis can be matched against the original faults which was injected in the simulated network;
- to stress Alarm Correlation and Fault Diagnosis systems with heavy load conditions; in this case a low MTBF is specified for the equipment models, in order to obtain time-dense alarm streams from the simulation ;
- to extend the knowledge of rule based Alarm Correlation and Fault Diagnosis systems; in this case the simulation output is the list of all the alarm streams coming from a given (hopefully small) network topology when all the applicable faults are injected, one at a time; this is particularly useful when new alarm correlation rules are to be inserted into the diagnosis system: the simulator can supply the "expert knowledge", once taken from the experts.
- to train the network operators on the alarm correlation task.

With the above needs the main requirement was that the simulator was not "wired" into a monolithic program, but could instead be extended and modified; furthermore, every feasible network topology which could be implemented out of the known equipment types had to be easily described within the simulator.

The main requirements for the simulator were:

- to know about the behaviour of its equipment types; any number of equipment of each type could be used within a given simulated network;
- to allow for the modelling of any network topology which is feasible in the real world;
- to simulate the given network model for a given amount of simulated time: within this interval a number of faults (either given or randomly chosen among the legal ones, for the various equipment types) will be injected; furthermore, after some time (either given or random) the proper ceased alarms will be generated, simulating the mending of the fault;
- to allow for the modelling of new equipment types at any time.

In order to meet the above requirements it was chosen to keep the network models as close as possible to the reality: each equipment was modeled by a Finite State Machine (FSM) and each digital link among the equipment was modeled with a channel. In this way any equipment models has the same interfaces of the respective real equipment and can be interconnected following the same rules.

The chosen simulation environment is YES (Yet another Event driven Simulation environment) [5], developed in CSELT for the functional simulation and for the performance evaluation of generic distributed systems; the simulation language of YES is PROMELA+, a CSELT extended version of the PROMELA language, originally defined at AT&T Bell Labs [6], in turn based on the Hoare's CSP.

The atomic entity of PROMELA+ is the Process, which allows for the modelling of FSMs; processes can communicate either asynchronously or synchronously by means of Channels.

The implementation of SPRINTER within YES turned out to be fairly elegant, since a TLC network model became a distributed system, each equipment became a process and the network topology was represented as a network of channels linking the processes; in this way the behaviour of the equipment could be precisely inserted into the models.

Figure 4 shows the architecture of SPRINTER, based on the library of Equipment models; the fault sequence can be either given or randomly generated among the legal faults for any equipment model.

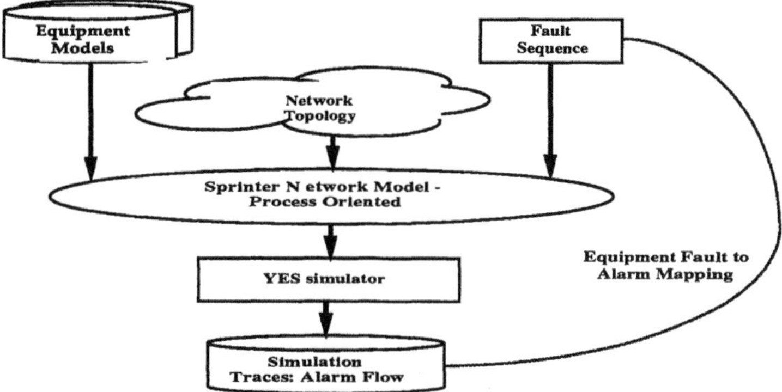

Figure 4 The SPRINTER Architecture

4 THE NETWORK SIMULATOR

The purpose of SPRINTER is to model the behaviour of real generic plesiochronous networks, with respect to the alarms; alarms are generated by individual equipments as a consequence of fault conditions due to internal damages as well as damages occurred over the physical connections among the equipments, as already shown in Figure 1.

4.1 Modelling aspects

SPRINTER models both the structural and the behavioural aspects of the transmission equipments; however, since only the fault management is of concern, only the alarms sent via the alignment and signalling frames are simulated, while the payload streams are not taken into account.

On the structural side, each equipment is further partitioned into its main subfunctions; a subfunction is defined as a module which can be characterised from the fault management point of view by a boolean working state variable (either Working or Out_of_order) and the working state of the module conditions the working state of the whole equipment.

The following functions hold among the internal state **S** of an equipment (e.g. the state of the FSM implemented by the equipment) and all the fault conditions which can occur:

$$\mathbf{W_e} = \mathrm{w_e}(\mathbf{S_e}, \mathbf{M_e}, \mathbf{Li_e}) \quad (1a)$$
$$\mathbf{S_e}(n) = \mathbf{F_e}(\mathbf{S_e}(n-1), \mathbf{M_e}(n), \mathbf{Li_e}(n)) \quad (1b)$$
$$\mathbf{Lo_e} = \mathbf{l_e}(\mathbf{S_e}, \mathbf{M_e}, \mathbf{Li_e}) \quad (1c)$$

where:
S_e is the equipment internal state, boolean vector
W_e is the equipment working state
M_e is the module working state, boolean vector
Li_e is the input link state, boolean vector
Lo_e is the output link state, boolean vector
F_e is future state function, boolean vector

Figure 5 shows, as an example, the structural partitioning of a Line Terminal equipment; the header of the model of such an equipment, with parametric bit-rate, is:

```
proctype TL_8_34_140(chan mux_in, mux_out, lin_in, lin_out, bit_rt, ...)
```

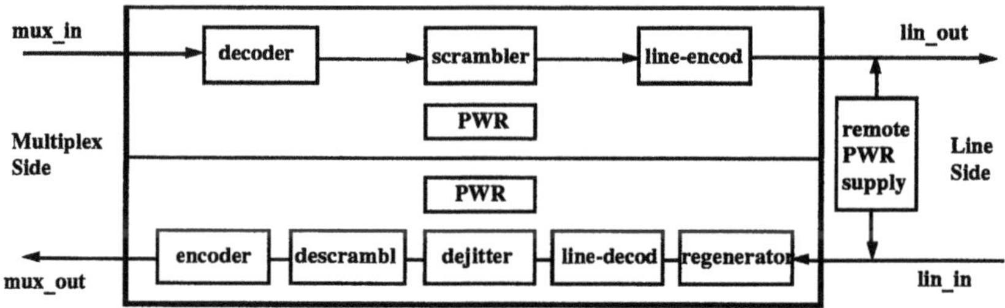

Figure 5 Structural partitioning of a Line Terminal Equipment

On the behavioural side, as stated in (1), each equipment has two main sources of stimuli and two main outputs: figure 6 shows a causal graph of the four entities.

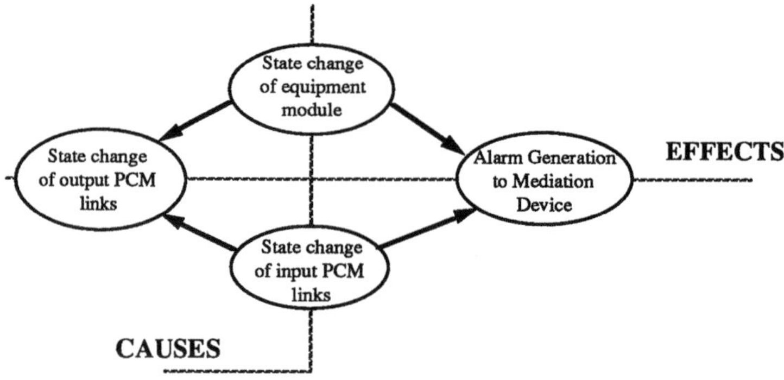

Figure 6 Behaviour model of equipments

The equipment models wait for changes in either the two sources of stimuli; as soon as a new stimulus is received, it updates the internal working state and send the appropriate alarms and signals over the respective outputs, according to what stated in (1).

4.2 Implementation techniques

The characteristics of the YES simulation environment, and particularly of its language, allowed to easily model the transmission networks, keeping models closely adherent to reality.

Every equipment type has been implemented as process template. In the description of a generic network each process model can be instantiated several times to create the different equipments of the same type. Equipments are distinguished one from the other by means of an identification number unique within the network.

PCM links among equipments are modelled with channels; a particular message structure has been defined to represent the relevant information contained in the PCM alignment and signalling frames, according to the CCITT Recommendations (see [4]). Typically they are the AIS (Alarm Indication Signal) and ATL (alarm indication to the remote end) signals.

Due to efficiency reasons, while in the real networks messages are sent continuously within the PCM frames, until ceasing of the cause, in SPRINTER each message is sent only once; no messages are sent across channels until conditions are to be propagated; in this way only the differences among messages are effectively sent.

However, in spite of the reduced message number, message handling in SPRINTER is generally more complex than in reality, since the Plesiochronous transmission technique allows equipments to be transparent to frames of lower hierarchical levels.

For this reason a routing algorithm has been implemented into the equipment models which composes and decomposes the frames for the tributaries, simulating the operations of multi-demultiplexing: in order to allow this every message has been structured into fields containing information about the hierarchy of the message and the tributary number for every possible hierarchy level.

When a multiplexing process receives a transit message from a tributary, it tags it with the tributary number and forwards it to the next higher hierarchical level. Viceversa, during demultiplexing, messages arriving from the composite channel are routed down to the right tributary using the tag contained in the proper field of the message.

The equipment FSM models all the faults and signals processing functions built in the real equipments, as required by the CCITT Recommendations: in particular the generation of alarms to the Mediation Device and signals to the adjacent equipments.

Faults and fault ceasings can be injected to each subfunction of each equipment; when a fault is injected the target equipment updates its state vector accordingly and takes all the appropriate actions; then Grouping and Filtering are performed in order to handle multiple faults: a fault condition may be Masked by the contemporary presence of another fault. Eventually alarms are evaluated and emitted to the outside: a dedicated process (i.e. the "Mediation Device") gathers all the alarms and stores them; Figure 7 shows the above process.

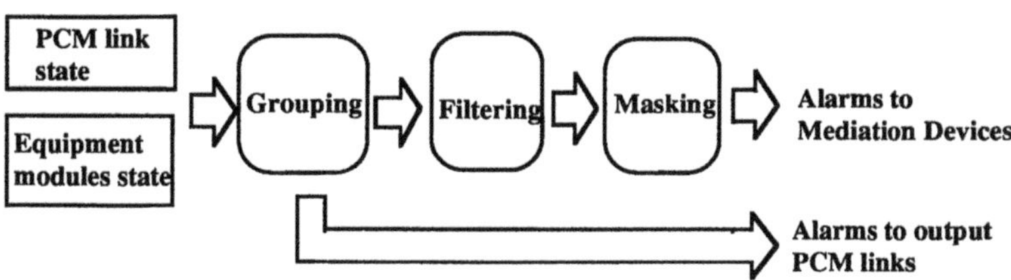

Figure 7 Intermediate steps for alarms generation

4.3 Fault simulation with SPRINTER

The main blocks of knowledge within the simulator is the library of equipment models; however other auxiliary modules are available, which simplify the task of modelling a network: the fault generator, the fault mender and the alarm collector.

The task of modelling a network consists in the instantiation of the equipments and the PCM links connecting them; explicit faults can be specified or in alternative equipments MTBF and repair frequencies can be specified as well.

With the above model the SPRINTER simulates the fault behaviour of the network over a given period of simulated time. A typical structure of a network model is shown in Figure 8; in this case the real network reported in Figure 1 is modeled.

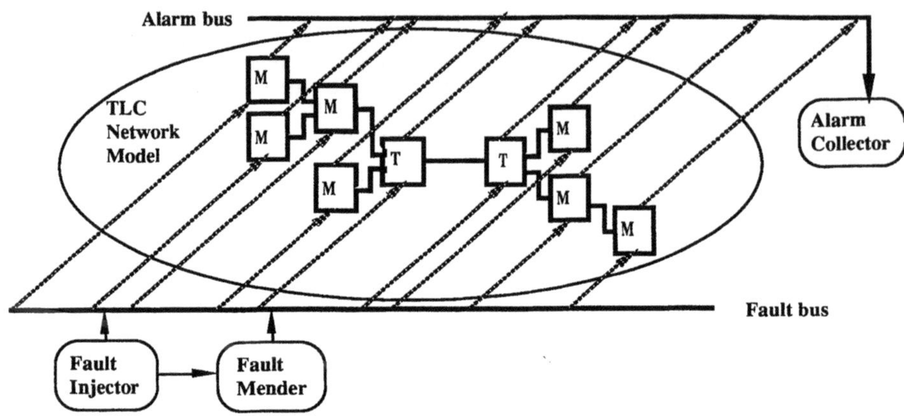

Figure 8 A TLC Network as modeled within SPRINTER

Each equipment shown on the figure 8 is the model of the respective real equipment; all the links wich connect the real equipments are represented in the model, in order to reproduce exaclty the real network topology. Two particular channels classes are highlighted in the picture: the *alarm bus*, by means of which equipment alarms are collected and the *fault bus*, by means of which faults and ceased faults are injected into selected equipments/links; while the former appear also in the reality, although in a less standard form, the latter is obviously used for simulation purposes only: a fault generator process creates faults and injects them through a bus common to all the equipments of the network.

The fault generator works with arbitrary networks; it sends faults not only to the network equipments, but also to a mender process which in turn produces, after a random delay the fault ceasings; working on the distribution of this delay the desired average number of faults present in the network at a time can be obtained; the fault collector simply lists alarms on a file

4.4 Validation of the SPRINTER equipment models

Several tests have been performed on SPRINTER results, at the purpose of validating the model library; the SPRINTER generated alarm traces have been matched against the alarm streams of real network portions; various network topologies have been simulated and large number of faults have been injected on their real equipments; the alarm streams generated by SPRINTER have been compared with the alarms generated by the real test networks. They have shown substantially equal results.

The faults injected in the real equipments were restricted to power supply faults and equipment link faults, because of the lack of controllability of the working state of the real equipments; i.e. the injection of an internal fault could only be done physically acting on the interior of the equipments.

5 VALIDATION AND EXTENSION OF A FAULT MANAGEMENT SYSTEM

5.1 Validation of SINERGIA

SPRINTER has been used to test SINERGIA: a number of networks have been simulated and the simulation results have been submitted to the alarm correlator and diagnostician; SINERGIA outputs have then been matched against the original faults injected into the equipments in the SPRINTER model.

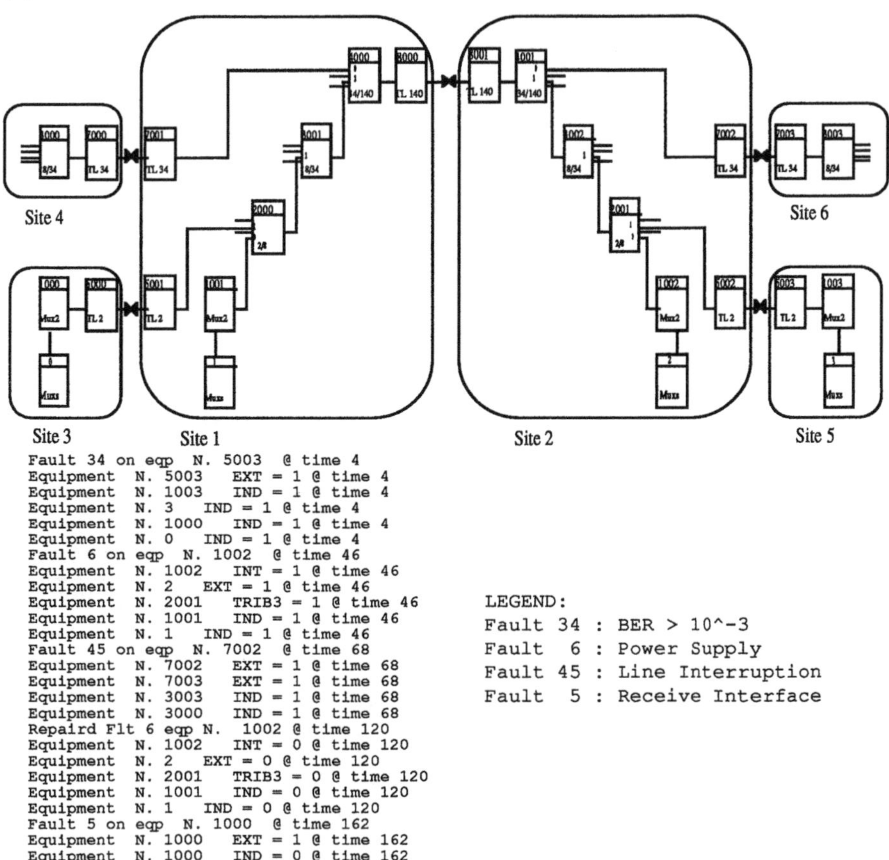

Figure 9 A network example and its SPRINTER log

Figure 9 shows on its top a small network example made of 26 transmission equipments; below a part of the SPRINTER generated Alarm Trace is listed.

The test session run on SINERGIA has shown the substantial correctness of the knowledge about its correlation rules and of the algorithms which exploit it in the diagnosis process.

However, being the Data Sheets produced by human Experts, they could have been somehow wrong and/or incomplete: actually the test session reported one entire missing Data Sheet and 10 missing/wrong rules on known Data Sheet, out of over 400 rules.

5.2 Extension of SINERGIA

SPRINTER has also been used to extend the rule base of an alarm correlator and diagnostician: actually it has been used to generate in advance all the alarm configurations over a given network portion, associating each of them with the fault which caused it.

This process has been exploited in the generation of the missing SINERGIA Data Sheet and in the refinement of the existing ones; in an automated way.

The early results of such experiments are as follows: the TTMT data sheet, the one whose lack was pointed out by the SPRINTER generated test suite has been automatically generated, defining a small network representing the desired topology plus some more equipments at its borders. The TTMT subnetwork, shown of figure 10, has been simulated exhaustively and 36 faults have been injected, in about 15' CPU time on a SUN SPARCSTATION 20; out of the faults 15 different rules have been extracted.

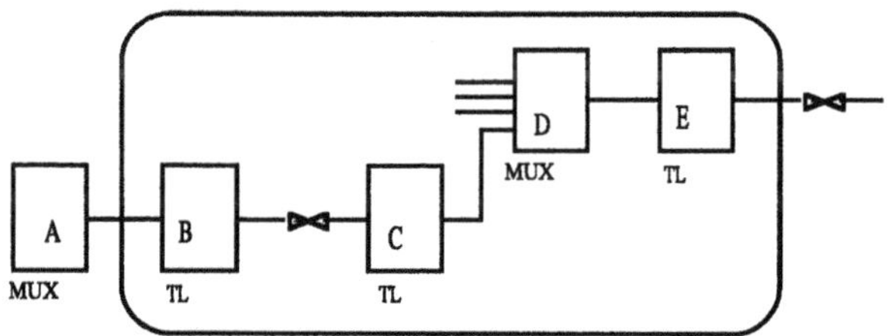

A	B	C	D	E	Diagnosis
(-)	EXT	EXT	-	-	LINE between B and C
(-)	-	EXT	-	-	ERRH,NORXL over C
-	-	NURG	-	-	ERRL over C
(-)	-	INT	-	-	NORXM,DECM,DECL,DESC over C
(-)	INT	INT	-	-	SCR over C
(-)	EXT	INT	-	-	CODL,ALTX over C
(-)	-	INT	TRIBO	-	CODM,ALRX over C
(-)	EXT	INT+ORTAL	-	-	TAL over C
(-)	-	-	EXT	-	FAT,NORX over D
(-)	-	-	TRIBO	-	NORX0,OVTX0,OVRX0 over D
(-)	-	-	INT	-	ALD over D
(-)	-	-	INT+(-)	INT	ALM over D
(-)	-	INT	TRIBO	-	AL0 over D

Figure 10 The TTMT topology analyzed with SPRINTER

The above table reports the TTMT Data Sheet generated by SPRINTER in the case where the E bit rate is 34 Mbit/s.

As a concluding remark about the extension on the knowledge base of an alarm corelator it must be remarked that the new knowledge, derived by network simulation cannot be validated with simulated alarm streams, since possible errors in the equipment models could affect both the knowledge and the test cases, preventing their capture.

6 CONCLUSIONS

Fault management systems and particularly alarm correlators and fault diagnosticians are complex Network Management applications. The creation of a comprehensive functional test suite is not straightforward, since a very deep knowledge of the networks and their equipments is needed; furthermore, even in that case, manually generated test suites cannot guarantee requested coverages.

With the right choice of the simulation environment, fault simulation of networks has proven an effective approach to the test suites generation, for both the functional and performance point of view. Furthermore the same tool has also proven effective in the correction/extension of a fault management application.

The results obtained with SINERGIA confirm the effectiveness of the proposed approach; however SPRINTER could be used virtually without modifications to validate other correlation systems working on plesiochronous networks.

With the encouraging results on the fault management we think that applications belonging to other areas of network management could also take advance of simulation techniques for the modelling of the environment in which they will operate.

REFERENCES

[1] S. Brugnoni, G. s, R. Manione, E. Montariolo, E. Paschetta, L. Sisto, "*An Expert System for Real Time Diagnosis of the Italian Telecommunications Network*", Proc. of ISINM '93, San Francisco, CA, April 1993.

[2] R. Manione, E. Paschetta, "*An Inconsistencies Tolerant approach in the fault diagnosis of telecommunications networks*", Proc. of NOMS '94, Orlando, FA, February 1994.

[3] G. Jakobson, M.D. Weissman, "*Alarm Correlation*", IEEE Network, Nov.93 pg 52-59.

[4] CCITT Recommendations, "*Digital Networks Transmission Systems and Multiplexing Equipment*", G.701-G.941, Yellow Book, Vol. III - Fasc. III.3, Geneva 1981.

[5] E. Chiocchetti, R. Manione, P. Renditore, "*Specification based Performance Evaluation of Distributed Systems for Telecommunications*", (Short Paper), 7th Int. Conf. on Modelling Techniques and Tools for Computer Performance Evaluation, Vienna, 1994.

[6] G.J. Holzmann: "*Design and Validation of Computer Protocols*", Prentice-Hall Int., 1991.

AUTHORS BIOGRAPHY

Roberto Manione graduated in EE in 1983 from Politecnico di Torino. Since then he is working in CSELT; his research interests were formerly in the field of Silicon Compilation, when he was involved in National and European research projects and authored several international publications; since some years he is working in the Network Management field and is project leader in the development of various tools aimed at the functional and performance validation and testing of distributed Network Management systems.

Fabio Montanari graduated in EE in 1994 and has worked in the SPRINTER project for the development of his thesis and after his graduation.

22

Centralized vs Distributed Fault Localization

I. Katzela [1] *CTR-Columbia University* [2]*, New York, NY 10027-6699, USA, tel: (212)854-7378, e-mail: irene@ctr.columbia.edu*

A.T. Bouloutas [3] *First Bank of Boston, Boston MA 0216, USA, tel: (617)434-0534 .*

S.B. Calo IBM T.J. Watson Research Center, Yorktown Heights, NY 10598, USA, tel:(914)784-7514 , e-mail:calo@watson.ibm.com

Abstract

In this paper we compare the performance of fault localization schemes for communication networks. Our model assumes a number of management centers, each responsible for a logically autonomous part of the whole telecommunication network. We briefly present three different fault localization schemes: namely, "Centralized", "Decentralized" and "Distributed" fault localization, and, we compare their performance with respect to the computational effort each requires and the accuracy of the solution that each provides.

1 INTRODUCTION

Usually, a single fault in a large network results in a number of alarms, and it is not always easy to identify the primary source(s) of failure. The problem of fault management becomes even worse when several faults occur coincidentally in the telecommunication network. The fault management process can be divided into three stages: alarm correlation, fault identification, and testing. The first two stages, usually referred to as the fault localization process, correlate the fault indications (alarms) received from the managed objects and propose various fault hypotheses. In the third stage each of the proposed hypotheses is tested in order to localize the fault precisely. The fault localization process is important because the speed and accuracy of the fault management process are heavily dependent on it.

In the past a number of researchers addressed the problem of fault localization in communication networks (Bouloutas, 1992), (Wang, 1993), (Shroff, 1989), (Riese, 1991) [4]. Most of the proposed methods focus on centralized algorithms for fault localization. However, the growth in size and complexity of communication networks may require the partitioning of the management environment into a number of management domains in order to meet organizational and performance requirements. This transition from a centralized management paradigm to a distributed one will require the development of distributed algorithms for fault localization. A distributed fault management approach will be able to shield parts of the network management system from information that is not locally useful, a very impor-

[1] Work done during the author's internship at the IBM T. J. Watson Research Center, NY, Summer 93.
[2] CTR - Center for Telecommunications Research
[3] Work done while the author was with the IBM T. J. Watson Research Center, NY.
[4] Additional references in the area can be found in (Katzela, 1993)

tant function as management centers tend to overflow with information. However, problems that involve objects in more than one domain will have to be resolved collectively by many domain managers in a distributed fashion. This introduces a number of problems that make the design of distributed fault management solutions a challenging task.

This paper is organized as follows: in Section 2 we define the problem of distributed fault localization and present a suitable model for the system; in Section 3 we present three different approaches for distributed fault localization; in section 4 we compare the proposed approaches with respect to the computational effort each requires and the accuracy of the solution each provides; and finally, section 5 concludes the paper. with a summary of the results.

2 MODEL OF THE SYSTEM

We assume a distributed approach to managing communication networks (Katzela, 1993). Each communication network is partitioned into a number of static, disjoint, logically autonomous management domains. Each domain is managed by a single management process [5] which is responsible for it. The managed objects in a domain may or may not be visible to managers of other domains. Hence, each manager has a limited view of the status of other management domains; and, has partial and incomplete information about the state of the network. However, managers from different domains can communicate and exchange information about the status of their domains using a peer to peer type of communication. Each domain manager has adequate knowledge about adjacent managers and domains so that communication can be established and messages passed.

Faults manifest themselves as alarms and alerts that are emitted by the managed objects affected by the fault(s). Alarms are communicated to the managed object's domain manager, which is responsible for identifying the primary source(s) of failure and eventually correcting the fault(s). Each received alarm represents the fault from the point of view of the object that emitted the alarm, and therefore corresponds to partial information about the fault. It is the responsibility of the fault localization system to collect all these partial views of a fault, correlate the information, and infer the real cause(s) of the fault. It is not unusual for an alarm to appear in a managed object belonging to a particular management domain and to indicate a fault in another managed object in a different domain. Since alarms cross management domains, management centers have to collaborate in order to infer the real state of the system.

Note that throughout the paper we assume that faults affect only managed objects but do not affect managers, information transfer processes of management systems, or other parts of the Telecommunication Management Network (TMN). This is a reasonable assumption and it stems from the fact that usually TMN systems have much stricter reliability requirements than the rest of the network.

Before we proceed it is essential to examine the structure of the alarms. Each alarm is characterized by the *domain of the alarm*, which is defined as the set of all independent

[5] In this paper we will use the terms management process, manager and management center interchangeably.

managed objects that could have caused the alarm - in other words, all the managed objects that might be at fault. Note that the domain of an alarm should not be confused with the domain of a management center. The domain of an alarm depends both on the semantics of the alarm and the topology of the communication network. It is the management centers' responsibility to find the domain of a received alarm before proceeding to the fault localization process.

3 FAULT LOCALIZATION APPROACHES

Assume that at a given moment in time a number of alarms appear in a communication network. The objective is to design algorithms that are able to find the *"best"* explanation of the received alarms, i.e., the managed object or the set of managed objects that could have been at fault and caused the alarms. In principle all the managed objects that appear in the domain of a received alarm constitute a possible explanation of it. If two or more alarms share an intersection, these alarms should be examined together because it is more probable that they are caused by the same set of faults. That is the reason why we introduce the notion of a *cluster of alarms.* A *cluster of alarms* is defined as a set of alarms that have intersecting domains. Note that a cluster may span more than one management domain as the alarms that comprise the cluster span more than one domain (Katzela, 1993).

Each cluster of alarms may have a number of explanations. The fault localization algorithm should be able to choose the *"best"* (most probable) among the possible ones. One way to find the most probable explanation would be to associate a probability of failure with each managed object. Then the *"best"* explanation would be the set of managed objects whose combined probability of failure is maximum. Instead of assigning a probability of failure to each managed object we can associate an *"information cost"* which is defined as the negative of the logarithm of the probability of failure for the managed object. For independent faults, the information costs are equivalent to probabilities, but working with them has certain advantages. If we choose to work with information costs then the *"best"* explanation (most probable) will be the set of managed objects whose sum of information costs is minimum (Bouloutas, 1992).

Before we proceed to examine distributed fault localization approaches, we assume that there exists a centralized algorithm that is able to find the most likely errors in a set of managed objects, given a set of alarms and the information costs associated with each managed object. As was shown in (Katzela, 1994) the fault localization problem is NP-Complete. Thus, in general there is no polynomial algorithm that gives the exact solution. One could then either construct a polynomial algorithm that gives an approximate solution, or a polynomial algorithm that gives the correct solution with some probability. In (Katzela, 1993), we present a possible probabilistic algorithm which finds an exact solution if the number of faults in the system is less than k, a parameter. We represent this algorithm by $G_p(A, N, k)$ where A is the received alarm cluster, N is the set of managed objects associated with A, and k is the maximum number of concurrent failures that can be identified by the algorithm. The probabilistic algorithm fails to give a solution with probability Q which is equal to the probability that there are more than k concurrent faults in the system. Hence:

$$Q = P_r(\text{algorithm fails}) \quad = \quad P_r(> k \text{ faults in the network}) \leq 1 - \sum_{i=0}^{k} b(i; N, p) \qquad (1)$$

Each managed object has a probability of failure assigned to it, p is the maximum of all such probabilities for the managed objects associated with the received alarm cluster A, and $b(i; N, p)$ is the probability of N Bernoulli trials, with probability p of success and i successes.

Since each management center can resolve problems that affect managed objects in its domain we will only examine the case where faults affect objects in many management domains. For simplicity we assume that the entire network consists of two domains. A generalization of the results for multiple domains is discussed in (Katzela, 1993). We also, without loss of generality, assume that there is only one cluster of alarms that crosses the boundary between domains.

3.1 Centralized Localization

The first approach, namely *Centralized Localization*, assumes the existence of a central manager that oversees all the domain managers and has a global view of the network. Problems that affect more than one domain could be resolved directly by the central manager as if there were no domain managers. In other words, if the received alarm cluster spans more than one domain, then the domain managers take no action and the central manager uses a centralized algorithm like $G_p(A, N, k)$ to identify the failure. The *Centralized Localization* approach always guarantees to output the optimum explanation of the received alarms. It fails to output an explanation of the received alarms with probability Q given by (1).

3.2 Decentralized Localization

The second approach, namely *Decentralized Localization*, assumes the existence of a central manager that oversees all the domain managers. Problems that affect more than one domain could be resolved in a collaborative way between the central manager and the domain managers. Unlike the first approach, the second one does not require extensive involvement of the central manager.

Assume that m of the L received alarms cross the boundary between the domains. These m alarms might have been produced by a fault in either domain and there is no a-priori information as to whether an alarm that crosses the boundary is explained by a fault in the first domain or by a fault in the second domain, or both. There are 2^m possible explanations for these m alarms depending on whether faults in domain one or domain two explain the alarms. Each domain manager calculates, using perhaps the $G_p(A, N, k)$ algorithm, 2^m optimum solutions, one for each of the possible explanations of the m alarms that cross the boundary.The central manager receives the 2^m optimum solutions from the two domain managers and finds the compatible ones. Two partial solutions are compatible if all the alarms received by both domains are explained in the final solution. Then the central manager selects the compatible global solution of minimum information cost. As one can easily verify, the above described procedure is able to identify the optimum solution given that the two management domains can find the optimum solution for managed objects and alarms in their respective domains (Katzela, 1993).

Finally, since each domain manager uses the probabilistic algorithm for fault identification there is a probability Q' (which will be calculated in section 4.1) that the *Decentralized Localization* fails to output a solution.

3.3 Distributed Localization

The third approach, namely *Distributed Localization* , does not assume the existence of a central manager. The area of the network is divided into two domains, each managed solely by a single domain manager. This strategy tries to find the faults from the point of view of each of the two domain managers without the use of a central manager. Let us examine the problem from the point of view of domain manager one. For each alarm that crosses the boundary, domain manager one would like to associate a probability that this alarm is explained by *Domain Two.* One way to do that would be to represent all the managed objects that belong to *Domain Two* and are associated with the alarm by a *proxy node.* Failure of the proxy node would indicate that some managed objects in *Domain Two* have failed. If one could associate a probability of failure with the proxy node, then the management proces of *Domain One* could use the probabilistic algorithm $G_p(A, N, k)$ to solve a centralized problem in a new expanded domain that includes the managed objects in its domain plus m proxy nodes, one for each alarm that crosses the boundary between the two domains. Here all the alarms that cross the boundary are treated as regular alarms. Once the algorithm has output the optimum solution there will be some alarms that are explained by regular nodes in *Domain One* and some alarms that are explained by *proxy nodes.* The alarms that are explained by *proxy nodes* are the alarms that are not explained by *Domain One* and are hopefully explained by *Domain Two.* The global solution is the one that includes all the regular nodes that appear in the optimum solutions of the two domain managers.

The exact probability of failure for a proxy node is difficult to find since it depends on all the managed objects and all the alarms in the cluster. Thus, the exact calculation of the probability of a proxy node is an NP-complete problem (Katzela, 1993). The best we can achieve is an estimation of the probability of failure for a *proxy node.* The estimated probability for a *proxy node* differs from the exact value by an estimation error. As a result of the estimation errors the distributed localization approach does not always guarantee an optimum global solution (Katzela, 1993).

4 PERFORMANCE COMPARISON

The objective of any fault management process is to minimize the time to localize a fault. The time to localize a fault is the sum of the time to propose possible hypotheses of the fault and the time to do testing in order to verify these hypotheses. Thus, we should minimize the time to perform fault identification and the time to perform testing. The time to identify the fault is affected by the identification algorithm. Hence, the first objective is to minimize the time complexity of the identification algorithm. The second objective is to minimize the time of testing. The time of testing is affected by the number of managed objects that need to be tested which is equal to the number of proposed hypotheses by the identification algorithm. If the management process is able to identify correctly the source of failure, the minimum number of tests is required. Thus, minimizing the number of tests is equivalent to minimizing the number of fault hypotheses, or equivalently, maximizing the accuracy of the fault identification algorithm. Thus, the performance measures of interest are the *time complexity* of the identification algorithm and the *accuracy* of the solution that the identification algorithm provides.

The complexity of the identification algorithm for each of the proposed approaches is a function of the number of nodes associated with the received alarm cluster, the number of alarms that cross domains, and the parameter k of the probabilistic algorithm that is the base of all the proposed localization schemes. On the other hand, the accuracy of each approach depends on the error in the estimations of the probabilities of failure for the managed objects associated with the received cluster of alarms.

4.1 Comparison between Centralized and Decentralized Fault Localization

Assume that the received cluster of alarms, A, is associated with N managed objects that may fail, each with probability p. We would like to compare the performance of the centralized versus the decentralized approach for this network setting. For the centralized approach, the central manager process should use the probabilistic algorithm $G_p(A, N, k)$. For the decentralized approach we assume that we partition the managed system in two domains namely *Domain One* (D_1) and *Domain Two* (D_2). Also we assume that there are m alarms that cross the boundaries between the two domains, and the N managed objects associated with the received cluster of alarms A are partitioned into N_1 objects in D_1 and N_2 in D_2, such that $N = N_1 + N_2$. According to the decentralized approach, each domain manager uses the probabilistic algorithm 2^m times for its area in order to find 2^m optimum solutions, one for every possible interpretation of the alarms that cross the boundary. In each case, D_1 will use $G_p(A_1, N_1, k_1)$ and D_2 will use $G_p(A_2, N_2, k_2)$ to identify the optimum solution. A_1 is the set of alarms that domain manager one takes into account in this instance, A_2 is the set of alarms that domain manager two takes into account in this instance, k_1 is the number of faults that manager one must localize, and k_2 is the number of faults that manager two must localize.

The selected performance measures for comparing these approaches are *accuracy* and *time complexity*. The *accuracy* performance measure has two aspects: The difference between the information cost of the proposed solution and the optimum one; and, the probability that the approach fails to give a solution. By design, both the centralized and the decentralized approaches give the optimum cost solution whenever they give a solution. Thus, we need to discuss only the second aspect of *accuracy*. The centralized approach fails to find a solution with probability Q, which is given by (1). The decentralized approach fails to find a solution with probability Q' which is:

$$\begin{aligned} Q' &= P_r(\text{Decentralized approach fails}) \overset{(1)}{=} Pr(> k_1 \quad \text{faults in } D_1) + \\ &\quad Pr(> k_2 \quad \text{faults in } D_2) - Pr(> k_1 \quad \text{faults in } D_1) \cdot Pr(> k_2 \quad \text{faults in } D_2) \quad (2) \end{aligned}$$

Regarding *time complexity*, the centralized approach has complexity which is bounded by $C_{cen} = O(N^k)$ and the decentralized approach by $C_{dec} = O(2^m \max(N_1^{k_1}, N_2^{k_2}) + 2^{2m})$.

It is obvious that the accuracy of the decentralized approach increases with an increase in the values of k_1 and k_2, which also leads to an undesireable increase in the time complexity of the approach. Suppose that we fix the accuracy of the two approaches and then compare them with respect to time complexity. For a given k (number of faults that the centralized approach can localize) the accuracy of the centralized approach is fixed. We need to calculate the values of k_1 and k_2 such that the two probabilities are equal - the probability that the decentralized approach fails and the probability of failure for the centralized approach, in order to achieve the same accuracy for both approaches. In addition the decentralized approach should be able to identify at least as many faults as the centralized one. Hence:

$$\begin{aligned} P_r(\text{Decentralized approach fails}) &\leq P_r(\text{Centralized approach fails}) \\ \text{and } k_1 + k_2 &\geq k \end{aligned} \tag{3}$$

The unknowns in (3) are the parameters k_1 and k_2. Typically it is difficult to solve such a set of inequalities. In order to simplify our analysis we will propose an approximation for calculating the parameters k_1 and k_2.

Approximate Calculation for k_1 and k_2

As an approximation we assume that the number of faults each domain manager should localize is proportional to the number of managed objects in the alarm cluster that belong to its domain. This assumption is valid and stems from the fact that the managed objects in the system fail independently. Hence :

$$\frac{k_1}{k_2} = \frac{N_1}{N_2} \overset{(3)}{\Rightarrow} k_1 + k_1\frac{N_2}{N_1} \geq k \Rightarrow k_1 \geq \lceil k\frac{N_1}{N} \rceil \tag{4}$$

We approximate the constraint that the probability of failure for the decentralized approach should be less than or equal to the probability of failure for the centralized approach with the following two requirements:

$$\begin{aligned} P_r(> k_1 \text{ faults in } D_1) &\leq \frac{N_1}{N} P_r(\text{Centralized approach fails}) \\ P_r(> k_2 \text{faults in} D_2) &\leq \frac{N_2}{N} P_r(\text{Centralized approach fails}) \end{aligned} \tag{5}$$

The proposed approximation has decomposed the original complex problem in (3) into two simpler problems, one for each domain. Without loss of generality it is sufficient to solve the problem only for D_1. The results are equivalent for D_2.

The new problem for domain one can be stated as follows: *Given a probability of failure for the decentralized approach, what is the value of the parameter k_1 that domain manager one should use in the application of the probabilistic algorithm, so that the following two inequalities hold?*

$$\begin{aligned} P_r(> k_1 \text{ faults in } D_1) &\leq \frac{N_1}{N} P_r(\text{Centralized approach fails}) \\ k_1 &\geq \lceil k\frac{N_1}{N} \rceil \end{aligned} \tag{6}$$

Which is equivalent to the system:

$$\stackrel{(1)}{\Rightarrow} \sum_{i=k_1+1}^{N_1} b(i;N_1,p) \leq \frac{N_1}{N}\sum_{i=k+1}^{N} b(i;N,p)$$

$$k_1 \geq \lceil k\frac{N_1}{N}\rceil \tag{7}$$

The system of inequalities in (7) is still difficult to solve. We would like to simplify it and find a closed form solution for k_1. In order to simplify (7) we should find a simpler expression for $\sum_{i=k_1+1}^{N_1} b(i;N_1,p)$. The form of the expression depends on whether k_1 is $\lesseqgtr \lceil (N_1+1)p\rceil$ (Katzela, 1993). Table 1 summarizes the formulas for estimating k_1 in each case.

Table 1 Formulas for estimating the parameter k_1

$\lceil (N_1+1)p\rceil < \lceil k\frac{N_1}{N}\rceil$	$k_1 > \max\left[\lceil\log_\beta\left(\frac{\frac{N_1}{N}\sum_{i=k+1}^{N} b(i;N,p)}{\Theta}\right)\rceil, \lceil k\frac{N_1}{N}\rceil\right]$
$\lceil (N_1+1)p\rceil > \lceil k\frac{N_1}{N}\rceil$, N_1 large	$k_1 > \max\left[\lceil(\eta - \frac{\frac{N_1}{N}\sum_{i=k+1}^{N} b(i;N,p)}{b(\lceil (N_1+1)p\rceil;N_1,p)}\rceil, \lceil k\frac{N_1}{N}\rceil\right]$
$\lceil (N_1+1)p\rceil > \lceil k\frac{N_1}{N}\rceil$, N_1 small	$k_1 > \max\left[\lceil 2\eta - \frac{\frac{N_1}{N}\sum_{i=k+1}^{N} b(i;N,p)}{b(\lceil (N_1+1)p\rceil;N_1,p)}\rceil, \lceil k\frac{N_1}{N}\rceil\right]$
$\beta = \frac{p}{1-p}$, $\eta = (N+1)p+1$	$\Theta = \binom{N}{N/2}(1-p)^{N+1}\eta$

The formulas in Table 1 provide an overestimation of the value of k_1. Similarly we can calculate k_2. We simply substitute, in the appropriate formulas in Table 1, N_2 for N_1 and k_2 for k_1.

As an example, consider a network scenario where the received alarm cluster A is associated with $N = 100$ managed objects. Each of these objects has a probability of failure $p = 0.01$. We also assume that these N objects are partitioned into the two domains so that $\frac{N_1}{N_2} = \frac{3}{2}$. As is shown in Figure 1, by using the formulas in Table 1, we get k_1 and k_2 close to their exact values. Thus, the overestimation of k_1 and k_2 is very small and the approximation works satisfactorily. A similar behavior, small overestimation, is observed for different values of p. Finally as we stated earlier, the decentralized approach should be able to localize, in total, an equal or larger number of faults. As we can observe from the curves in Figure 1, sometimes the decentralized approach has to localize as much as twice the number of faults as the centralized approach in order to achieve the same probability of failure. Such an increase in the parameters k_1 and k_2 results in increased complexity for the decentralized approach.

A specific problem instance, N_1 managed objects in domain one and N_2 managed objects in domain two, could occur with probability $P_r(N;N_1,N_2)$. It is easy to show that the

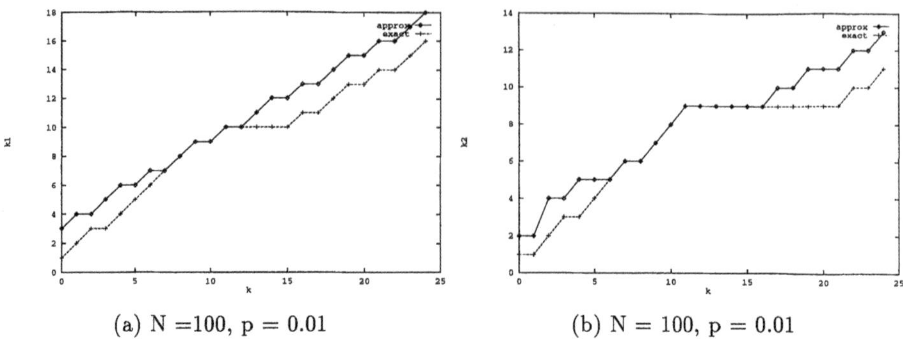

(a) N =100, p = 0.01 (b) N = 100, p = 0.01

Figure 1: Estimation and exact values of k_1, k_2 for different values of k

average complexity per domain manager per problem instance is: $\frac{C_{dec}^{aver}}{problem} = \sum_{i=1}^{N} P_r(N;i;N-i)i^{k_i}$, where k_i is the number of faults that the domain manager can localize in this specific instance. For fixed probability of failure for the decentralized approach, k_i can be calculated by use of Table 1. In principle, the probability of a specific problem scenario, $P_r(N;N_1,N_2)$, could follow any distribution. The value of $P_r(N;N_1,N_2)$ for a specific problem instance indicates how *likely* it is that this problem will be encountered. The concept of *likely* is related to the probability distribution that adequately represents the set of problems to be encountered. Such a probability distribution is difficult to define and analyze. Let us assume for simplicity that the partitioning of N managed objects associated with the received alarm cluster between the two domains is done randomly. Then the probability distribution of the problems to be encountered is a generalized Bernoulli distribution. Thus $P_r(N;N_1,N_2) = \frac{N!}{N_1!N_2!} = \frac{N!}{N_1!(N-N_1)!}$ and the average time complexity of the algorithm will be:

$$C_{dec}^{aver} = 2^m \sum_{i=1}^{N} \frac{N!}{i!(N-i)!} \max\left(i^{k_{i1}}, (N-i)^{k_{i2}}\right) + 2^{2m} = 2^{2m} \sum_{i=0}^{\lceil \frac{N}{2} \rceil} \frac{N!}{i!(N-i)!} i^{k_i} + 2^{2m} \quad (8)$$

We are interested in investigating the conditions under which $C_{dec}^{aver} < C_{cen}$. The average complexity of the decentralized approach depends on the values of N, k, p and m. The parameter m is the one which has the greatest effect on the complexity of the decentralized approach. As we can easily observe from Figure 2, for fixed k and p the time complexity of the decentralized approach remains less than that of the centralized approach up to a certain value of m. For example in Figure 2(a) , for $k = 3$ the complexity of the decentralized approach remains strictly less than that of the centralized approach when $m \leq 4$. Similarly in Figure 2(b) for $k = 5$, the same behavior is observed for $m \leq 7$. Finally, as we could see in Figure 2(b) for the same value of k the allowed number of alarms m that can cross domains so that the complexity of the decentralized approach is less than that of the centralized one increases with a decrease in the probability p. For example for $k = 5$ and $p = 0.1$ in Figure 2(b) the allowed maximum m is $m = 7$, for $p = 0.01$ the maximum m is $m = 5$ in Figure 3(a), for $p = 0.001$ in Figure 3(b) $m = 2$.

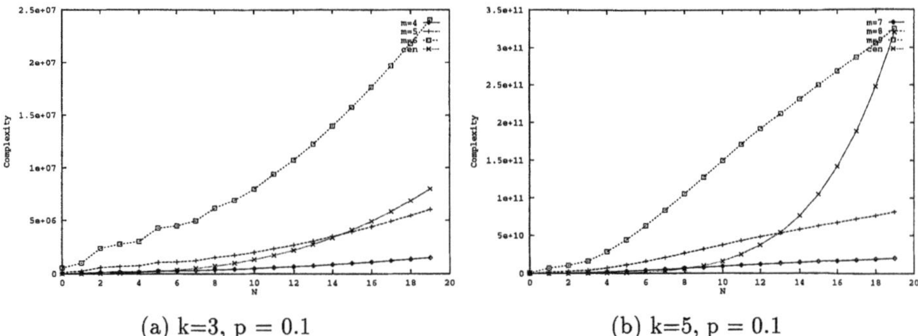

(a) k=3, p = 0.1 (b) k=5, p = 0.1

Figure 2: Average Complexity of Decentralized Approach versus N, for different m and k and the same p. The curve *cen* represents the complexity of the Centralized Approach vs N.

4.2 Comparison between Centralized and Distributed Fault Identification

It is easy to show that the complexity of the distributed approach is always considerably less than the complexity of the centralized approach ($max(N_1^{k_1}, N_2^{k_2}) << N^k$). Hence, it remains to compare the approaches with respect to accuracy. The first aspect of accuracy is the probability of failure of the localization schemes to output a solution. Again we can select k_1 and k_2 for the two domains of the distributed approach so that the probabilities of failure for the centralized and distributed approaches are the same. The corresponding values of k_1, k_2 can be calculated by the use of Table 1.

A possible solution for a received alarm cluster is characterized by its information cost which is the sum of the information costs of the managed objects that are included in the solution. Most of the time there are more than one possible solutions. All of the localization approaches discussed in the previous sections select among all the possible solutions the one which has the minimum information cost. The deviation of any solution from the optimum one is characterized by the difference in the information cost of the solution from the information cost of the optimum (minimum cost) solution. Unlike the centralized and the decentralized localization approaches, the distributed localization does not always guarantee that it can find the optimum solution. This deviation from the optimum solution stems from errors in the approximation of the probability of failure for the proxy nodes.

The exact probability of failure for a proxy node (and thus the exact information cost for the node) is difficult to find since it depends on all the managed objects and all the alarms in the cluster. The best we can achieve is an estimation of the information cost of a proxy node. Thus, the information cost of the proxy node differs from its exact value by an estimation error. The introduction of estimation errors might cause a difference between the solution proposed by the distributed identification algorithm and the optimum one which is given by the decentralized and the centralized approaches. It is of interest to find a bound for the difference between the information cost of the distributed solution and the optimum one. Also it is of interest to investigate how sensitive the distributed solution is to changes

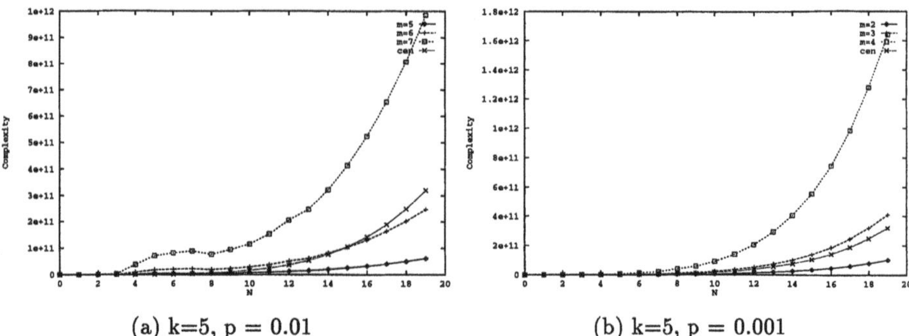

(a) k=5, p = 0.01 (b) k=5, p = 0.001

Figure 3: Average Complexity of Decentralized Approach versus N, for different p, m and the same k. The curve *cen* represents the complexity of the Centralized Approach versus N.

in the information cost of the proxy nodes.

For a given network let I^* be the information cost of the optimum solution; I_{cen}, I_{dec} and I_{dis} the information costs of the solutions that the centralized, decentralized, and distributed localization schemes give, respectively. As we know, $I_{cen} = I_{dec} = I^*$. The objective is to find how close the solution that the distributed approach gives is to the optimum one. In other words we would like to find a bound for $\mid I^* - I_{dis} \mid$. As it was analyzed in (Katzela, 1993), the $\mid I^* - I_{dis} \mid$ is bounded as follows:

$$2 \cdot m \cdot e_{min} \leq \mid I^* - I_{dis} \mid \leq 2 \cdot m \cdot e_{max} \tag{9}$$

Where m is the number of alarms that cross the boundary between the two domains and e_{max}, e_{min} are the minimum and maximum probability estimation errors for all the proxy nodes in both domains. The difference between the information costs in the two approaches depends on the number of alarms m that cross the boundary between domains and the minimum and maximum probability estimation errors for all the proxy nodes in both domains. As it was shown in (Katzela, 1993) the estimation error for the weights of a proxy node is expected to be small. Hence, for small values of m , which is the expected case, the difference $\mid I^* - I_{dis} \mid$ will be also small. Thus, although the decentralized approach does not always guarantee to output the optimum solution, it provides a solution which has information cost close to the information cost of the optimum one.

5 CONCLUSIONS

In this paper we compare the performance between a number of fault localization approaches suitable for a distributed fault management environment. The three proposed methods are namely the *Centralized*, *Decentralized* and *Distributed* Fault Localization approaches. As measures of comparison we used the *accuracy* of the solution and the *complexity* of the identification process that each approach employs. Our comparison proved that the decentralized approach generally has considerably less complexity than the centralized approach, and can provide the same or better solution accuracy. Also the distributed localization approach was proved to have the least complexity of all three schemes in all network settings, but it can not always guarantee an optimum solution. However, as was shown in the previous section, it provides a solution which is almost as accurate as the solution provided by the other two approaches.

5 REFERENCES

Bouloutas, A., Calo, S. and Finkel A. (1992) Alarm Correlation and Fault Identification in Communication Networks. *IBM Technical Report*, **RC 17967**.

Katzela, I., Bouloutas, A. and Calo, S. (1993) Comparison of Distributed Fault Identification Schemes in Communication Networks. *IBM Technical Report*, **RC 19656**.

Katzela, I. and Schwartz, M. (1994) Schemes for Fault Identification in Communication Networks. *CTR Technical Report*, **CU/CTR/TR 362-49-09**.

Riese, M. (1991) Model Based Diagnosis of Networks: Problem Characterization and Survey. *OEGAI-91 Workshop on Model Based Reasoning.*

Shroff, N. and Schwartz, M. (1989) Fault Detection/Identification in the Linear Lightwave Networks *CTR Technical Report*, **CU/CTR/TR 243-91-24**.

Wang, C. and Schwartz, M. (1993) Identification of Faulty Links in Dynamic-Routed Networks *IEEE JSAC*, **11**, 1449-60 .

5 BIOGRAPHY

Seraphin B. Calo received the M.S., M.A., and PhD. degrees from Princeton University, Princeton, New Jersey, in 1971, 1975, and 1976, respectively. Since 1977 he has been a Research Staff Member in the IBM Research Division at the Thomas J. Watson Research Center, Yorktown Heights, New York. He has worked and published in the areas of queueing theory, data communication networks, multi-access protocols, satellite communications, expert systems, and complex systems management. Dr. Calo joined the Systems Analysis Department in 1987, and is currently Manager of Systems Applications. This research group is involved in studies of architectural issues in the design of complex software systems, and the application of advanced technologies to systems management problems. Dr. Calo is involved with IEEE symposia related to networks and computer systems, and was instrumental in establishing the IEEE International Workshop on Systems Management.

Irene Katzela received the Diploma in Electrical Engineering from the National Technical University of Athens, Greece, in 1990 and the M.S. and MPhil degree from Columbia University, New York in 1993 and 1994 respectively. Currently she is working towards her PhD degree, in the area of fault management, at Columbia University. Since 1991 she is a Graduate Research Assistant at the Center for Telecommunication Research at Columbia University. Her other research interests include network management, design and verification of protocols and wireless networking. She is a student member of IEEE and a member of the National Technical Chambers of Greece.

SECTION TWO

Panel

23

Management Technology Convergence

Moderator: *Einar STEFFERUD,* First Virtual Holdings, Inc., U.S.A.

Historically, much of our network management technology has been characterized by a sharp focus on local enclave management, with an unstated assumption that someone owns the entire enclave of interest, so that whatever happens outside that enclave can be seen as being Someone Else's Problem (SEP). It is not surprising that our current environment is loaded with divergent technologies.

For the future, we can see that all these enclaves need to be interconnected into some kind of "internet" where no one owns the whole thing, but where the whole thing still needs to be "managed". Thus the future might be seen as full of conflicts to be resolved in some kind of convergence process.

The panelists will describe their ideas about how convergence might occur. Will it occur by itself in a free and open market ? Will some benevolent vendor resolve everything for us by supplying a truly winning proprietary technology ? Will some single benevolent institutional authority arise to define convergence for everyone ? Or, will convergence simply not happen ?

SECTION THREE

Event Management

24

A Coding Approach to Event Correlation

S. Kliger, S.Yemini
System Management Arts (SMARTS),
199 Main St., White Plains. NY 10601
kliger, yemini@smarts.com

Y. Yemini,[1] *D. Ohsie,*[2] *S. Stolfo*
450 Computer Science Building,
Columbia University, NY 10027
yemini, ohsie, sal@cs.columbia.edu

Abstract

This paper describes a novel approach to event correlation in networks based on coding techniques. Observable symptom events are viewed as a code that identifies the problems that caused them; correlation is performed by decoding the set of observed symptoms. The coding approach has been implemented in SMARTS Event Management System (SEMS), as server running under Sun Solaris 2.3. Preliminary benchmarks of the SEMS demonstrate that the coding approach provides a speedup at least two orders of magnitude over other published correlation systems. In addition, it is resilient to high rates of symptom loss and false alarms. Finally, the coding approach scales well to very large domains involving thousands of problems.

1 INTRODUCTION

Detecting and handling exceptional *events (alarms)*[3] play a central role in network management (Leinwand and Fang 1993, Stallings 1993, Lewis 1993, Dupuy et. al. 1989, Feldkuhn and Erickson 1989). Alarms indicate exceptional states or behaviors, for example, component failures, congestion, errors, or intrusion attempts. Often, a single problem will be manifested through a large number of alarms. These alarms must be correlated to pinpoint their causes so that problems can be handled effectively.

Effective correlation can lead to great improvements in the quality and costs of network operations management. For example, in a recent report on AT&T's Event Correlation Expert (ECXpert™), Nygate and Sterling (1993) report, "..labor savings at a typical US network operations center are between $500,000 and $1,000,000 a year. In addition, at least this amount is saved due to decreased network downtime." The alarm correlation problem has thus attracted increasing interest in recent years as described in a recent survey (Ohsie and Kliger 1993).

A generic alarm correlation system is depicted in Figure 1. *Monitors* typically collect managed data at network elements and detect out of tolerance conditions, generating appropriate alarms. The *correlator* uses an *event model* to analyze these alarms. The event model represents

[1] Work performed while the author was on sabbatical leave at Systems Management Arts.

[2] This author's research was supported in part by NSF grant IRI-94-13847

[3] Henceforth we use the terms *problem events* to indicate events requiring handling and *symptom events* (also *symptoms* or *alarms*) to indicate observable events. The terms *event-correlation* or *alarm-correlation* are used interchangeably to indicate a process where observed symptoms are analyzed to identify their common causes.

knowledge of various events and their causal relationships. The correlator determines the common problems that caused the observed alarms.

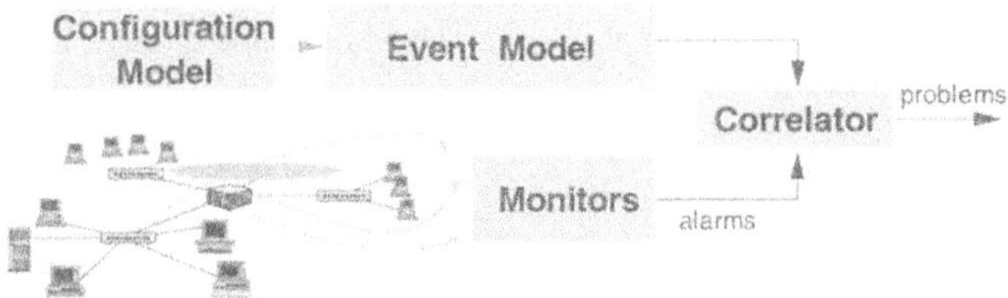

Figure 1: Generic Architecture of an Event Correlation System

An alarm correlation system must address a few technical challenges. First, it must be sufficiently *general* to handle a rapidly changing and increasing range of network systems and scenarios. Second, it must be *scalable* to large networks involving increasingly complex elements. As elements become more complex, the number of problems associated with their operations as well as the number of symptoms that they can cause increases rapidly. Furthermore, propagation of events among related elements can cause dramatic increase in the number of symptoms caused by a single problem. Finally, an alarm correlation system must be resilient to *"noise"* in the inputs to the correlator. This is because alarms may be lost or spuriously generated forming *observation noise* in the alarms input stream. The event-model may also be inconsistent with the actual network, due to insufficient or incorrect knowledge of the configuration model. These inconsistencies form *model noise* in the event model input to the correlator. An alarm correlation system must be robust with respect to both observation and model noise.

Current alarm correlation systems typically fall short of meeting the goals described above (Ohsie and Kliger 1993). Alarms are typically correlated through searches over the event model knowledge base. The complexity of the search seriously limits scalability. To control the search complexity, often the event model knowledge base is carefully designed to take advantage of specific specialized domain characteristics. This limits generality. There are no techniques to select an optimum set of symptoms to monitor or to determine whether observed symptoms provide sufficient information to determine problems. Finally, search techniques derive their computations from the data stored in the knowledge base and arriving alarms. Noise in this data can guide the search in the wrong direction. A more detailed analysis of current correlation systems is pursued in (Ohsie and Kliger 1993).

This paper describes a novel approach to correlation based on coding techniques (Kliger et. al. 1994a). The underlying idea of the coding technique is simple. Problem events are viewed as messages generated by the system and "encoded" in sets of alarms that they cause. The problem of correlation is viewed as decoding these alarms to identify the message. The coding technique proceeds in two phases. In the *codebook selection* phase, an optimal subset of alarms, the *codebook* is selected to be monitored. This codebook is selected to optimally pinpoint the problems of interest and ensure a required level of noise insensitivity. In the *decoding* phase, observed alarms are analyzed to identify the problems that caused them. The coding approach thus reduces the complexity of real-time correlation analysis through preprocessing of the event knowledge model. The codebook selection dramatically reduces the number of alarms that must

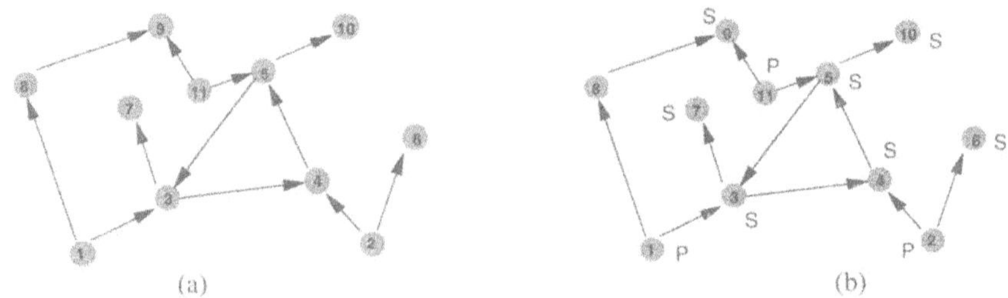

Figure 2: A Causality graph (a) and its labeling (b)

be monitored. It also establishes the relations among these alarms and their causes in a manner that reduces the complexity of the decoding phase.

In what follows we describe the mathematical basis of the coding approach (section 2), develop the technique and establish its properties (section 3), describe a commercial implementation of the coding techniques and a benchmarking of the implementation (section 4) and conclude (section 5).

2 THE MATHEMATICAL BASIS OF EVENT CORRELATION

2.1 Causality Graph Models

Correlation is concerned with analysis of causal relations among events. We use the notation e→f to denote causality of the event f by the event e. Causality is a partial order relation between events. The relation → may be described by a *causality graph* whose nodes represent events and whose directed edges represent causality. Figure 2(a) depicts a causality graph on a set of 11 events.

To proceed with correlation analysis, it is necessary to identify the nodes in the causality graph corresponding to symptoms and those corresponding to problems. A *problem* is an event that may require handling while a *symptom* (alarm) is an event that is observable. Nodes of a causality graph may be marked as problems (P) or symptoms (S) as in Figure 2(b). Note that some events may be neither problems nor symptoms (e.g., event 8) while some other events are both symptoms and problems.

The causality graph may include information that does not contribute to correlation analysis. For example, a cycle (such as events 3,4,5) represents causal equivalence. A cycle of events may thus be aggregated into a single event. Similarly, certain symptoms are not directly caused by any problem (e.g., symptoms 7,10) but only by other symptoms. They do not contribute any information about problems that is not already provided by these other symptoms that cause them. These *indirect symptoms* may be eliminated without loss of information. Henceforth, we will assume that a cauality graph has been appropriately pruned.

2.2 Modeling Causal Likelihood

The causality graphs described so far do not include a model of the likelihood (strength) of causality. The causal implication e→f can be considered as a representation of a proposition "e may-cause f." Often, richer information is available describing the likelihood of such causality. Various approaches and measures have been pursued to model such likelihood. A probabilistic model, for example, associates a conditional probability with a causal implication while fuzzy logic associates a fuzzy measure. Each of these models includes operations to compute the strength of a causal chain between two events or to combine the strength of multiple chains between two events. It is useful to have a general model of likelihood that captures these various techniques as special cases. This model must include a set of causal likelihood measures and operations to compute strength of chains and combine them. We proceed to define and demonstrate such a general model of likelihood.

Define a *semi-ring* as a partially ordered set **L** with an order $\leq$ and two operations * (catenation) and + (combination) such that:

(i) $<\mathbf{L}, *>$ is a semi-group with a unit **1** (a monoid)

(ii) $<\mathbf{L}, +>$ is a commutative semi-group with a unit **0**

(iii) $\forall a,b \in \mathbf{L},\ a*b \leq a,b \quad a,b \leq a+b$

(iv) $\forall a \in L,\ \mathbf{0} \leq a \leq \mathbf{1}$

A semi-ring is used to provide a measure of causality. Elements of **L** provide measures of causal strength with **1** indicating the strongest causality and **0** the weakest. The ordering of likelihood measures is used to compare relative strength of likelihood. The catenation operation is used to compute the strength of causal chains. The combination operation is used to compute the strength of multiple causal chains leading from one event to another.

We give a few examples of semi-rings used to model causal likelihood. The *deterministic model,* uses $\mathbf{L}=\mathbf{D}=\{0,1\}$ with the order $0\leq 1$. The catenation operation is the Boolean **and** $\wedge$, with the unit 1, while the combination operation is the Boolean **or** $\vee$ with the unit 0. Consider now a causality graph whose edges are all labeled with elements from **D**. An edge marked 0 represents a highly unlikely causality while an edge marked 1 represents a sure causality. For simplicity assume that all edges marked 0 have been eliminated. The semi-ring structure permits us to assign likelihood to causal chains between two events. The deterministic likelihood of a causal chain such as $1\rightarrow 8\rightarrow 9$ in Figure 2 is obtained by catenation (**and**) and is trivially 1. Now consider the set of causal chains between two events. The likelihood of this set is obtained by applying the combination operation to the likelihood of all causal chains in the set.

The deterministic model is a simple and commonly used likelihood model. We now introduce another semi-ring, denoted **P**, to model probabilistic causality. **P** consists of the set [0,1] with an ordinary numerical order. The label q on $e\rightarrow f$ models the conditional probability of the event f when e occurs. The catenation operation is the product of probabilities while the combination operation is defined as $q_1+q_2=1-(1-q_1)(1-q_2)$.

The temporal model is denoted **T**. The elements of **T** are non-negative real numbers representing the expected duration for the respective causality to happen. For example, a label of 8.5 on $1\rightarrow 3$ indicates that this causal implication is expected to occur within 8.5 time units (e.g., seconds). The catenation operator * is addition of times (along a causal chain) while the combination operator + is the min operator on real numbers. 0 is the unit with respect to catenation, and ∞ the unit with respect to addition, where ∞ indicates that the causality is unlikely to happen (in any finite time). We use the inverted numerical order as the order on **T**, modeling "sooner" occurrence of events in time. For example, 6.3≥8.5 should be read as "6.3 happens sooner than 8.5".

Similarly, one can establish fuzzy logic models of causal likelihood or other calculus of uncertainty measures such as the Shafer-Dempster model. Furthermore, by combining various models, more complex likelihood measures may be obtained. For example, the semi-ring defined by **PxT** ascribes to a causal edge both probability and expected time of occurrence.

We are now ready to define a *causal likelihood model* as a triplet <**N**, **L**,ϕ> where **N** is a normal form causality graph, **L** is a semi-ring describing a likelihood model and ϕ is a mapping from the edge-set of **N** to **L** assigning a likelihood measure to each causal implication. By varying the semi-ring **L**, a spectrum of models is obtained.

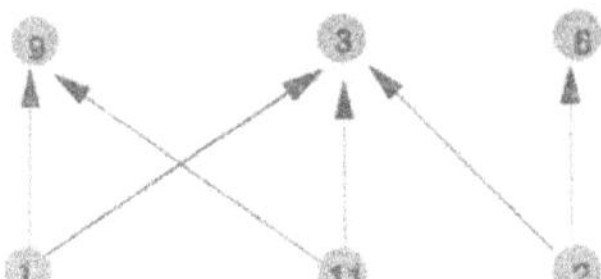

Figure 3: A Correlation Graph

2.3 The Correlation Problem

Correlation analysis is concerned with the relationships among problems and the symptoms that they may cause. Consider the *correlation relation* among problems and symptoms, defined as the closure of the relation → and denoted by ⇒. A correlation p⇒s means that problem p can cause a chain of events leading to the symptom s. This correlation relation may be represented in terms of a bipartite *correlation graph.* Figure 3 depicts the correlation graph corresponding to the causality graph of Figure 2 after pruning indirect symptoms and aggregating cycles.

For a given causal likelihood model <**N**,**L**,ϕ> one can derive a correlation graph **N*** corresponding to the causality graph **N**. Using the catenation operation one can associate a likelihood measure with every causal chain leading from a problem p to a symptom s. The likelihoods of various chains leading from p to s may be combined using the combination operator to provide a likelihood measure of the correlation p⇒s. Thus, for a given causal likelihood model <**N**,**L**,ϕ> there is a corresponding *correlation likelihood model* <**N***,**L**,ϕ > over the correlation graph.

3 THE CODING APPROACH TO ALARM CORRELATION

3.1 Problems, Codes and Correlation

The problem of alarm correlation may be now described in terms of the correlation likelihood model. For each problem p, the correlation graph provides a vector of correlation likelihood measures associated with the various symptoms. We denote this likelihood vector as p̲ and call it the *code* of the problem p. Codes summarize the information available about correlation among symptoms and problems. Code vectors can be best considered as points in an |S|-dimensional space associated with the set of symptoms S, which we call the *symptom space*. Alarms too may be described as *alarm vectors* in symptom space assigning likelihood measures **1** and **0** to observed and unobserved symptoms respectively. A very useful reference for coding theory and techniques is provided by [Roman 1992].

The alarm correlation problem is that of finding problems whose codes optimally match an observed alarm vector. We illustrate these considerations using the example of Figure 3. Figure 4(a) depicts a deterministic correlation likelihood model and Figure 4(b) depicts a probabilistic model. Code vectors correspond to the likelihood of the symptoms 3,6,9 in this order. They are given by 1=(1,0,1), 2=(1,1,0) and 11= (1,0,1) for the deterministic model and by 1=(0.8,0,0.3),

(a) Deterministic model (b) Probabilistic model

Figure 4: Correlation Likelihood Models

2=(0.4, 0.9,0) and 11=(0.5,0,0.9) for the probabilistic model.

Suppose that alarms consisting of symptoms 3 and 9 have been observed. This may be described by an *alarm vector* a=(1,0,1). In the deterministic model either 1 or 11 match the observation a and one would infer that the two alarms are correlated with either problem 1 or 11. Note that these two problems have identical codes and are indistinguishable. Similarly, an alarm vector a=(1,1,0) would match the code of problem 2. How should an alarm vector a=(0,1,0) be interpreted? One possibility is that this is just a spurious false alarm. Another possibility is that problem 2 occurred but the symptom 3 was lost. The choice of interpretation depends on whether loss is more likely than spurious generation of alarms. There are, of course, other more remote possibilities.

Now, suppose that spurious or lost symptoms are unlikely. The information provided by symptom 9 is redundant. If only symptoms 3 and 6 are observed the respective projections of the codes 1=11=(1,0) and 2=(1,1) are sufficient to distinguish and correlate alarm vectors. Since real alarm correlation problems typically involve significant redundancy. The number of symptoms associated with a single problem may be very large. A much smaller set of symptoms can be selected to accomplish a desired level of distinction among problems. We call such a subset of symptoms a *codebook*. The complexity of correlation is a function of the number of symptoms in the codebook. An optimal codebook can thus reduce the complexity of correlation substantially.

To illustrate this consider an example of 6 problems and 20 symptoms depicted in Figure 5(a). The correlation likelihood model is compactly described in terms of a matrix. Matrix elements represent the correlation likelihood parameters of respective problem-symptom pairs.

	p_1	p_2	p_3	p_4	p_5	p_6
1	1	0	0	1	0	1
2	1	1	1	1	0	0
3	1	1	0	1	0	0
4	1	0	1	0	1	0
5	1	0	1	1	1	0
6	1	1	1	0	0	1
7	1	0	1	0	0	0
8	1	0	0	1	1	1
9	0	1	0	0	1	1
10	0	1	1	1	0	0
11	0	0	0	1	1	0
12	0	1	0	1	0	0
13	0	1	0	1	1	1
14	0	0	0	0	0	1
15	0	0	1	0	1	1
16	0	1	1	0	0	1
17	0	1	0	1	1	0
18	0	1	1	1	0	0
19	0	1	1	0	1	0
20	0	0	0	0	1	1

(a) Correlation Matrix

	p_1	p_2	p_3	p_4	p_5	p_6
1	1	0	0	1	0	1
2	1	1	1	1	0	0
4	1	0	1	0	1	0

(b) A Codebook of Radius 0.5

	p_1	p_2	p_3	p_4	p_5	p_6
1	1	0	0	1	0	1
3	1	1	0	1	0	0
4	1	0	1	0	1	0
6	1	1	1	0	0	1
9	0	1	0	0	1	1
18	0	1	1	1	0	0

(c) A Codebook of Radius 1.5

Figure 5: A deterministic correlation matrix and codebooks

Figure 5(b) depicts a codebook consisting of 3 symptoms {1,2,4}. This codebook distinguishes among all 6 problems. However, it can only guarantee distinction by a single symptom. For example, problems p_2 and p_3 are distinguished by symptom 4. A loss or a spurious generation of this symptom will result in potential decoding error. Distinction among problems is measured by the Hamming distance between their codes. The *radius* of a codebook is one half of the minimal Hamming distance among codes. When the radius is 0.5, the code provides distinction among problems but is not resilient to noise. To illustrate resiliency to noise consider the codebook of Figure 5(c) where 6 symptoms are used to produce a codebook of radius 1.5. This means that a loss or a spurious generation of any two symptoms can be detected and any single-symptom error can be corrected.

We illustrate the error-correction capabilities of the codebook of Figure 5(c). A minimal-distance decoder will decode as p_1 all alarms that contain a single-symptom perturbation of p_1. The alarm vectors {011100, 101100, 110100, 111000} will be decoded as a single symptom loss in p_1, while {111110, 111101} will be interpreted as occurrence of a spurious symptom. The total number of alarms that can be generated due to a single symptom perturbation (loss or spurious one) in the 6 problems codes + the null problem p_0=000000 is 42. Therefore, a total of 48 alarm vectors (out of possible 63) will be correctly decoded despite single-symptom observation errors. When two symptom errors occur a minimal distance decoder can detect that errors have occurred but may not decode the alarm vector uniquely.

The considerations above generalize simply to correct observation errors in k symptoms and detect 2k errors as long as k is smaller than the radius of the codebook. Consider now the problem of model errors. That is, what happens when the correlation model itself is incorrect?

For example, suppose problem p_4 in Figure 5 can actually cause symptom 6 even though the model fails to reflect this. This will cause a single symptom error with respect to the code of p_4. Symptom 6 will appear as a spurious symptom whenever p_4 occurs. In other words, *an error in the correlation model is entirely equivalent to an observation error.* In contrast to random observation errors, model errors would appear as persistent observation noise. This persistence may be automatically detected by analyzing correlation logs and then used to correct the correlation model.

In summary, one seeks to design minimal codebooks that accomplish a desired level of insensitivity to observation and model errors. This insensitivity to observation errors is measured by the degree to which codes are distinct. In the case of the deterministic model, distinction among codes is measured by the Hamming distance among code vectors. We will soon see that similar measures of distinction may be used to select optimum codebooks in the case of other likelihood models.

3.2 Coding and Decoding

The coding technique accomplishes significant correlation speeds. Most of the complexity of correlation computations is handled during the pre-processing of codebook selection. The decoding of alarms in real-time can be very fast. Precise complexity evaluation is beyond the scope of this paper and is left for future publications. However, even crude estimates can usefully illustrate the speed gains. The complexity of decoding is logarithmic in the number of direct decodes (alarm vectors whose errors with respect to codes are less than half the radius of the codebook). The number of direct decodes is bounded by $\delta(p,c,k)= (p+1)\sum_{m=0}^{k}\binom{c}{m}$ where p is the number of problems, c is the codebook size (number of symptoms in the codebook) and k is the number of error symptoms to be corrected ($\lceil$radius$\rceil$ - 1). The complexity of decoding is bounded by $\lambda(p,c,k)=\lg[(p+1)\sum_{m=0}^{k}\binom{c}{m}]$. For k<<p, this is of order (k+1)lgp.

In the example of Figure 5(c) p=6, c=6, k=1, the decoding complexity is $\lambda(6,6,1)=\lg[7(1+6)]=\lg 49\sim 6$ search operations. When p=100 and k=1, c may be of the order 10-30 and the complexity of decoding is of the order of 10-12 search operations. Even when $p=10^6$, decoding complexity is of a manageable order of 20(k+1) search operations. In contrast, other knowledge based approaches typically requires an exponential, or even doubly exponential, search in the total number of problems p and symptoms s (s>>c). For p=100 the search complexity may be practically infeasible. For example, Nygate and Sterling (1993) report alarm correlation speeds of ECXpert™ at 0.25 alarms per second for a model involving 10 problems.

We proceed to complete the details of codebook design and decoding for a general correlation likelihood model. The point of departure in codebook design is to define a metric of distinction among codes, generalizing the Hamming distance. This is accomplished by using a distance measure on the likelihood semi-ring **L**. We call a real function d(a,b) on **L** a *distance measure* on **L** if it is symmetric, non-negative and satisfies d(a,a)=0 and $d(a,c)\leq d(a,b)+d(b,c)$ for all $a\leq b\leq c$ in **L**. Given a distance measure d(a,b) on **L**, one can extend it to a measure of distinction among code vectors. Define the distance between two code vectors $\underline{a}=(a_1,a_2,...a_n)$ and $\underline{b}=(b_1,b_2,...b_n)$ as $\mathbf{d}(\underline{a},\underline{b})=\Sigma_{k=1,n}\, d(a_k,b_k)$. For example, in the case of the deterministic model define d(1,1)=d(0,0)=0 and d(1,0)=1 to obtain the Hamming distance.

For the probabilistic model P, a distance measure is given by the log-likelihood measure $d(a,b)=|\lg(a/b)|$ (with lg(0/0)=0 and lg(0/a)=1 for $a\neq 0$). For example, in the probabilistic model of

Figure 5(b), $\underline{1}$=(0.8,0,0.3) , $\underline{11}$=(0.5,0,0.9) and thus **d**($\underline{1}$,$\underline{11}$)= lg(8/5)+lg(9/3)=lg(24/5). Therefore, in the probabilistic model problems 1 and 11 are distinct, in contrast to the deterministic model. Note that the log-likelihood distance measure generalizes the Hamming distance. When all probabilities are 1 or 0 the two measures yield the same distance.

The *radius* of a set of codes P is defined as the one half of the minimal distance among pairs of codes. The radius provides worst case measure of distinction among code vectors. A codebook C is a subset of the set of symptoms S. The code space defined by C is the respective projection of code vectors in the symptom space defined by S. The *radius of the codebook*, $r_C(P)$ is the respective radius among the projections of codes. Clearly, $r_C(P) \leq r_S(P)$. Given a desired level of distinction $d \leq r_S(P)$, the *codebook design problem* is that of finding a minimal codebook C for which $d \leq r_C(P)$. Such codebook provides a guaranteed distinction of at least d among the codes of different problems.

The codebook design problem may be solved by a variety of algorithms. A pruning algorithm, for example, can start with the correlation matrix model and eliminate symptoms until an optimal codebook has been established. The algorithm may be designed independently of the specific likelihood model (semi-ring) and distance measure. This may be used to construct a correlator of great generality.

Given a codebook C, consider now an alarm vector $\underline{a}$ describing observed symptoms. The problem of decoding, as discussed above, is to find problem codes that maximally match $\underline{a}$. It is useful to utilize a correlation measure $\mu(\underline{a},\underline{p})$ for decoding that is, in general, different from the measure of distinction. We illustrate this through the example of Figure 5(a). Suppose the codebook consists of the symptoms C={3,6}. The codes are $\underline{1}$=$\underline{11}$=(1,0) and $\underline{2}$=(1,1). Now consider an alarm vector $\underline{a}$=(0,1). The Hamming distances are **d**($\underline{a}$,$\underline{1}$)=**d**($\underline{a}$,$\underline{0}$)=1 and **d**($\underline{a}$,$\underline{2}$)=2, where $\underline{0}$ =(0,0) is the null problem. Thus the Hamming distance does not distinguish between a lost symptom (correlating $\underline{a}$ with $\underline{1}$) and a spurious symptom (correlating $\underline{a}$ with $\underline{0}$).

In general, a symmetric measure does not distinguish lost from spurious symptoms. We thus permit the correlation measure to be asymmetric. A *correlation measure* on **L** consists of two non-negative functions, $\mu(\mathbf{1},a)$ and $\mu(\mathbf{0},a)$ defined for all $a \in \mathbf{L}$ such that $\mu(\mathbf{1},\mathbf{1})= \mu(\mathbf{0},\mathbf{0})=0$ and if $a \leq b$, $\mu(\mathbf{1},a) \geq \mu(\mathbf{1},b)$ while $\mu(\mathbf{0},a) \leq \mu(\mathbf{0},b)$. For example, in the deterministic case define $\mu(0,1)=\alpha$ as the correlation level between an observation of no symptom when the codebook predicts its occurrence, i.e., a lost symptom. Define $\mu(1,0)=\beta$ as the correlation measure between an observation of a symptom that is not included in a code, i.e., a spurious symptom. A correlation measure on **L** may be easily extended to a correlation measure $\mu(\underline{a},\underline{p})$ between alarm vectors and code vectors. The *problem of decoding* is to find for a given alarm vector $\underline{a}$ the problem codes that minimize the correlation measure $\mu(\underline{a},\underline{p})$ -- best match problems.

In the probabilistic case let $\mu(\mathbf{1},a)= |\lg a|$ (correlation of occurrence) $\mu(\mathbf{0},a)=|\lg(1-a)|$ (correlation of non-occurrence). The correlation measure $\mu(\underline{a},\underline{p})$ is given by the logarithm of the product of probabilities assigned by $\underline{p}$ to events occurring in $\underline{a}$ and the complements of the probabilities assigned to events that do not occur in $\underline{a}$.

To illustrate the use of correlation measure in decoding consider again the codebook of Figure 5(c). The correlation measure for the deterministic model is given by $\mu(0,1)=\alpha$ (loss) and $\mu(1,0)=\beta$ (spurious). The codebook provides guaranteed error-correction for all single symptom errors.

Consider the case when two possible symptom errors occurred. For example, let the alarm vector observed be a=101000. The respective values of the correlation measure for the six problems are 2α, $2\alpha+4\beta$, $2\alpha+\beta$, $2\alpha+\beta$, $\alpha+\beta$, $2\alpha+\beta$. Under all choices of α,β the two candidates decodes are p_1 (two lost symptoms) and p_5 (one lost symptom and one spurious). If $\alpha<\beta$ (loss is more likely) problem p_1 will be decoded and if spurious symptoms are more likely, p_5 will be decoded. If both observation errors are equally likely ($\alpha=\beta$) both problems will be decoded.

Decoding can be accomplished through very fast algorithms. A range of fast decoding algorithms is provided by coding theory. See (Roman 1992) for several possible algorithms with varying tradeoffs. For example, block-decoding techniques aggregate symptoms over a time window and then decode them to find minimal distance codes.

4 IMPLEMENTATION AND BENCHMARKS

The coding technique has been implemented in SMARTS Event Management System (SEMS). In this section we briefly describe the SEMS and the benchmarks used to test its performance. A full description of the benchmarks can be found in (Kliger et. al. 1994b)

SEMS is organized as an event management server and its current implementation runs on Sun Sparcstations under Solaris 2.3. The SEMS server presents interfaces allowing clients to subscribe to problem events of interest, and provides clients with notification upon the detection of problems. Typical SEMS clients include various networked systems managers. For example, a fault manager may subscribe to various fault events, a performance manager may need to handle various congestion events or excessive delay events, while a security manager may need to detect intrusion events. Clients may also include applications running under umbrella management systems such as HP OpenView, Sunnet Manager, or IBM Netview/6000.

A model of a satellite based communications network was used to benchmark the performance of the SEMS. The modelled domain includes close to 4000 managed objects involving some 9500 problems and 6000 symptoms. Random scenarios were created by selecting random subsets of the model. For example, experiments involving 100 problems proceeded by selecting a random choice of 100 problems (out of 9500) to be monitored. All symptoms irrelevant to these problems were discarded, leaving a random number of symptoms from which a codebook was selected.

The model makes two conservative assumptions *unfavorable* to codebook correlation. It assumes an *under-instrumented* system where the number of observed symptoms is much smaller

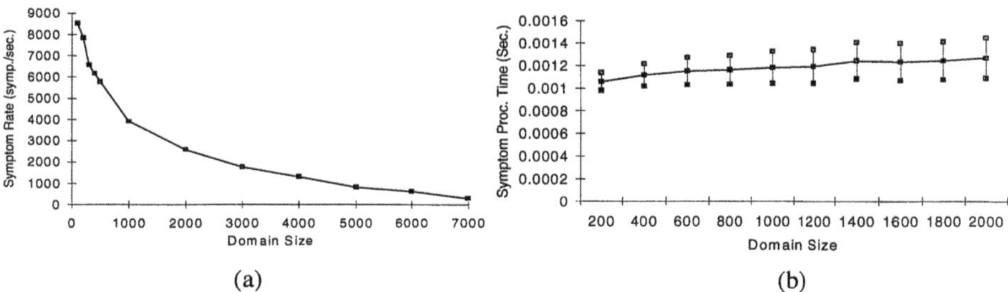

Figure 6: (a) Symptom Processing Rate (b) Symptom Processing Time with Standard Deviation

than the number of problems. Typical systems are over-instrumented. It assumes a *sparse propagation model* where only a small number of symptoms is caused by a typical problem. In a system with complex dependencies, problems can propagate very widely. Real-world situations typically monitor many more symptoms, yielding smaller codebooks, a larger reduction in the number of symptoms to monitor, and faster correlation.

The most important measure of the effectiveness of the coding approach is correlation speed. Figure 6(a) shows the effective event correlation rate measured in symptoms per second of actual elapsed time (the effective event correlation rate includes symptoms which were generated by a problem but not processed by the correlator because codebook reduction removed them from the codebook). In domains with fewer than 4000 problems, symptom processing was measured in thousands of symptoms per second. This is 2-4 orders of magnitude faster than the published figures of 0.25 events per second for ECXPERT (Nygate 1993) and 15 symptoms per second for IMPACT (Jakobson and Weissman 1993).

The fundamental measurement underlying the curve of Figure 6(a) is the elapsed time for processing symptoms. Figure 6(b) depicts these time measurements and the intervals defined by the standard deviation of the measurements. The figures shows that the average speed measures provide a fairly accurate estimate of the actual correlation rates.

Another important aspect of the coding approach its resilience to symptom loss. Figure 7(a) shows the correlation error rates when the probability of symptom loss ranges up to 20%. Even substantial loss or spurious symptoms cause only minimal error probability, falling under 5% when the codebook radius exceeds 1.5.

Our final measure of codebook performance is what reduction is accomplished in the number of symptoms that must be monitored, compared with the total number of relevant symptoms available. The compression factor represents the ratio of the two numbers. This compression is an important feature of the coding approach as it reduces the amount of monitoring and real-time processing of events needed. Figure 7(b) depicts the behavior of the compression factor as the domain size grows. The figure shows that substantial compression is achieved by the codebook.

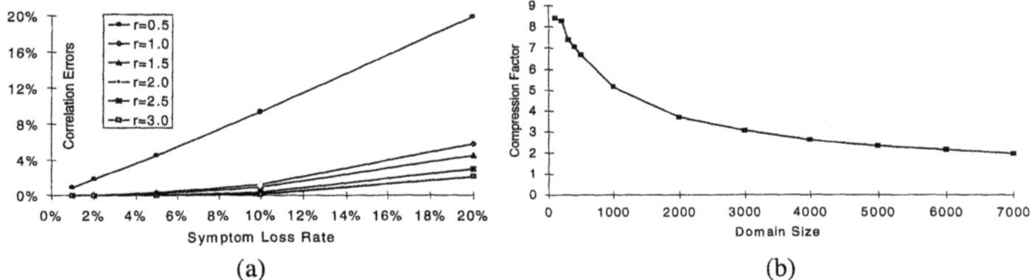

Figure 7: (a) Correlation Error Rate (b) Codebook Compression

5 CONCLUSIONS

This paper provides an overview of the coding approach to event correlation and its mathematical foundations. The coding approach accomplishes the three goals described in the introduction: generality, scalability and resilience to noise. Generality is accomplished through the use of an abstract mathematical formulation of the event correlation process. Scalability is accomplished through a substantial reduction in real time correlation processing due to optimizing symptom sets and fast decoding mechanisms. The complex searches through causality models are performed during the pre-processing phase of codebook design. Resilience to noise is accomplished by selecting codebook symptoms to provide a desired level of guaranteed noise insensitivity.

The coding approach has been implemented in SMARTS Event Management System. The current implementation runs as a server under Sun Solaris 2.3. Preliminary benchmarks confirm the advantages promised by the theoretical analysis.

6 REFERENCES

Dupuy, A., Schwartz, J., Yemini, Y., Barzilai, G. and Cahana, A. (1989) Network Fault Management: A User's View, in *Proc. IFIP Symposium on Integrated Network Management*, North Holland.

Feldkuhn, L. and Erickson, J. (1989) Event Management as a Common Functional Area of Open Systems Management, in *Proc. IFIP Symposium on Integrated Network Management*, North Holland.

Jakobson, G., Weissman, M. (1993) Alarm Correlation, *IEEE Network,* Vol. 7, No. 6.

Kliger, S., Yemini, Y. and Yemini, S. (1994a) Apparatus and Method for Event Correlation and Problem Reporting, Patent Application.

Kilger, S., Ohsie, D., Yemini, Y., Hwang W. (1994b) Decs Performance Benchmarks Summary, *SMARTS Technical Report.*

Leinwand, A., Fang, K. (1993) Network Management : A Practical Perspective Addison Wesley.

Lewis, L. (1993) A Case Base Reasoning Approach to The Resolution of Faults in Communications Networks, in Proceedings *Third International Symposium on Intégrated Network Management.*

Nygate, Yossi and Sterling, Leon (1993) ASPEN - Designing Complex Knowledge Based Systems in *Proceedings of the 10th Israeli Symposium on Artificial Intelligence, Computer Vision, and Neural Networks,* pp. 51-60.

Ohsie, D. and S. Kliger (1993) Network Event Management Survey, *SMARTS Technical Report.*

Roman, Steve (1992) *Coding and Information Theory,* Springer Verlag.

Stallings, W. (1993) *SNMP, SNMPv2, and CMIP The Practical Guide to Network-Management Standards,* Addison Wesley.

Yemini, Y., Dupuy, A., Kliger, S., Yemini, S. (1993) Semantic Modeling of Managed Information in *Second IEEE Workshop on Network Management and Control,* Tarrytown, NY.

7 BIOGRAPHY

Professor Yechiam Yemini is the director of the Distributed Computing and Communications Lab at Columbia University and a co-founder of SMARTS. His interests include broad areas distributed networked systems technologies; he has published over 100 articles and edited 3 books in these areas. **Dr. Shaula Alexander Yemini** is president and co-founder SMARTS. Her past work includes the design of the Hermes Distributed Programming Language, the Concert high level language system and the co-invention (with Rob Strom) of Optimistic Recovery, a technique for transparent fault tolerance in distributed systems. **Dr. Shmuel Kliger** leads the development of SEMS at SMARTS. His research experience includes designing and implementing distributed concurrent logic programming languages and environments. **Professor Salvatore Stolfo** heads the Parallel and Distributed Intelligent Sytems Laboratory at Columbia University, where he led the development of the PARADISER parallel and distributed database rule processing system. **David Ohsie** is a Phd. candidate at Columbia University, where he is currently pursuing his thesis research in causal analysis.

25

Event Correlation using Rule and Object Based Techniques

Y. A. Nygate
AT&T - Bell Laboratories
6200 E. Broad St. - Rm. 2B253, Columbus, OH 43213, U.S.A.
Tel: 614-860-5976 Fax: 614-868-4021 email: yossi@hercules.cb.att.com

Abstract

Today's competitive market place has forced the telecommunications industry to improve their service and reliability. One step that telecommunications companies have taken to reduce network failures is the installation of operations centers to collect data from network elements. These centers are staffed by network managers who monitor network activity by correlating alarms across various operational disciplines (switch, facility, traffic) and relating them to a common cause. Accurate analysis is often difficult due to the volume of data and complexity of problems.

ECXpert is a product developed recently at AT&T to help network managers monitor and analyze alarms, take corrective actions, and minimize disruptions to the network. Successful implementation of event correlation has increased customer revenue since trouble isolation can be done faster, resulting in quicker restoration of service.

The essence of ECXpert is a high level language with which users can specify network events and their correlation with alarms. The system is written in Prolog and C++, a powerful combination which facilitated development to occur on time and in budget. It has been deployed in network management centers throughout the U.S. and is currently being marketed overseas. ECXpert is a success story for Prolog within AT&T.

Keywords

Network Management, TNM, Event Correlation, C++, Prolog, Meta-Languages

1 INTRODUCTION

Total Network Management (Nerys, 1993), TNM, is a very large product developed by AT&T for domestic and international customers. The primary function of TNM is to facilitate early problem detection and prompt repair of telecommunication network facilities and switches. TNM users monitor line-oriented displays to analyze alarms generated by failures, correlate the alarms with knowledge of problem scenarios, and build up a picture of network events. If necessary, a repair request is generated and users continue monitoring to verify that either the problem cleared up or that further action is needed to solve the problem.

Timely generation of repair requests in response to a large volume of alarms has been notoriously difficult to do well. One minor problem, that by itself may be of little importance, can create a major problem when combined with other minor problems. In contrast, one major problem might generate many additional minor problems. Typically, there are many problems occurring concurrently in the network resulting in hundreds of active alarms intermingled on the displays. Users need to group the alarms corresponding to problems, differentiate between alarms that are the underlying causes and those that are results, generate repair orders, and monitor resolution of the problem. Due to the volume of data and complexity of modern telecommunication networks, this task has been difficult to do successfully. This problem is often called *event correlation* and can be defined as the analysis and classification of multiple messages from one or more sources to determine the underlying cause of a failure. The results of correlating alarms correctly can be used to relate the resultant impact and symptomatic troubles to the underlying causes.

Successful implementation of event correlation has increased customers' revenue because trouble isolation can be done faster, resulting in quicker restoration of service. Relating cause to service impact allows prioritization of repairs so problems that cause service outage and loss of revenue can be assigned high priority.

This paper describes the Event Correlation Expert feature package, ECXpert, that was incorporated into the TNM product family to help network managers speedily resolve the problems of analysis, recognition and resolution of alarms.

2 EVENT CORRELATION IN TELECOMMUNICATIONS

2.1 Monitoring Alarms in Telephone Networks

TNM's primary function is to collect, process, and display messages received from network elements (NEs). A typical TNM center may collect thousands of alarms per hour and depending on the type of message, a minor, major or critical alarm may be generated or a previously received alarm may be cleared. Network managers monitor the alarms using line-oriented displays known as awareness screens (AS) that update every 6 seconds. The user can see up to 16 alarms on each page of the AS and can page forward and backward to see other alarms. Each alarm is displayed in a color corresponding to its severity: red for the most severe, blue for the least severe, and green for cleared alarms. The main responsibility of the network managers is to monitor the alarms, picture the current underlying network problems, generate the necessary repair orders, and continue monitoring the network to verify that their analysis was correct and the problem was resolved.

To explain event correlation, I shall present a running example in which the network includes an AT&T 5ESS® switch connected to two other switches made by AT&T and another manufacturer. Between each NE there are a pair of links that together comprise a path. In our example, three network events occur. At 4:24 a hardware failure (X1) occurs on the link between *clli_a* and *clli_y*, at 4:50 high traffic demand (X2) occurs at *clli_z*, and at 5:22 another hardware failure (X3) occurs on the second link between *clli_a* and *clli_y*. The network and location of the events are shown in Figure 1.

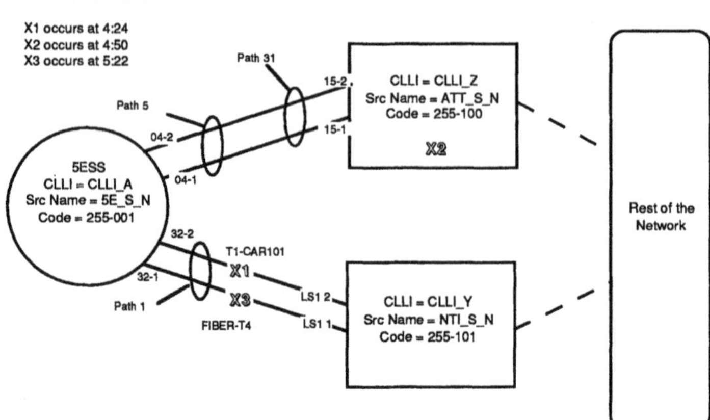

Figure 1 Example Network and Failures.

In telecommunication networks, there are often many more alarms than number of failures as a combination of failures can create additional problems which may result in other alarms being generated. For example, after X1 and X3 occur since all the links between *clli_a* and *clli_y* fail a *path loss* message will be generated. After all three failures occur, since it is impossible for traffic to leave *clli_a* (as *clli_z* and the two links to *clli_y* have failed) a *switch isolation* message will be generated. The set of generated alarms is a function of the nature of the problems, their location and time of occurrence, as well as the configuration of the network. In our example, 19 alarms were generated and displayed on the AS as shown in Figure 2. If the same three events had occurred in different places in the network, possibly only three alarms might have been generated.

DATE	TIME	SYSTEM	A OFFICE	Z OFFICE	TROUBLE INDICATION
09jun	5:30	nti_s_n	CLLI_Y	CLLI_A	LINK FAIL LS1 1
09jun	5:28	5e_s_n	CLLI_A	CLLI_Y	SWITCH ISO 255-001
09jun	5:26	5e_s_n	CLLI_A	CLLI_Y	PATH LOSS 1
09jun	5:24	5e_s_n	CLLI_A	CLLI_Y	LINK FAIL 32-1
09jun	5:22	FIBER-T4	CLLI_E	CLLI_R	HARDWARE FAIL
09jun	5:06	att_s_n	CLLI_Z	CLLI_A	PATH LOSS 31
09jun	5:04	att_s_n	CLLI_Z	CLLI_A	LINK FAIL 15-1
09jun	5:02	att_s_n	CLLI_Z	CLLI_A	LINK FAIL 15-2
09jun	5:00	5e_s_n	CLLI_A	CLLI_Z	PATH LOSS 5
09jun	4:58	5e_s_n	CLLI_A	CLLI_Z	LINK FAIL 04-1
09jun	4:56	5e_s_n	CLLI_A	CLLI_Z	LINK FAIL 04-2
09jun	4:56	att_s_n	CLLI_Z	CLLI_A	CNGSTN RESTART 15-1
09jun	4:55	5e_s_n	CLLI_A	CLLI_Z	OVERLOAD FAIL 04-1
09jun	4:54	5e_s_n	CLLI_A	CLLI_Z	OVERLOAD FAIL 04-2
09jun	4:52	att_s_n	CLLI_Z	CLLI_A	CNGSTN RESTART 15-2
09jun	4:50	att_s_n	CLLI_Z		HI TRAFFIC DMND
09jun	4:29	nti_s_n	CLLI_Y	CLLI_A	LINK FAIL LS1 2
09jun	4:27	5e_s_n	CLLI_A	CLLI_Y	LINK FAIL 32-2
09jun	4:24	T1-CAR101	CLLI_D	CLLI_Q	HARDWARE FAIL

Figure 2 Alarms Generated by Network Failures.

In large networks, many problems occur concurrently resulting in thousands of active alarms. Network managers would be overwhelmed if all these alarms were displayed on every users awareness screen. TNM allows users to specify viewing options that restrict which alarms are displayed (such as specific switch types, or regions) and sorting options (such as by time or severity) These options have to be used prudently. When their view is too restrictive it is often difficult to see the 'big picture' of network problems. Whereas if they do not make enough restrictions, they are unable to read the alarms fast enough to keep up with the flow of information across their screens.

Although useful, viewing options do not utilize the underlying cause and effect relationship that exists between events and alarms. Consequently, an AS will often contain many pages of alarms caused by different problems intermingled on the screen. Because these alarms are not grouped together, it is very difficult to construct an accurate picture of the network problems and differentiate between alarms that are the underlying causes and those that are results.

2.2 Correlation Trees and Correlation Groups

Many network problems can be depicted as a combination of alarms having a specific cause and effect relationship. This can be depicted schematically in a *correlation tree skeleton* as shown in Figure 3.

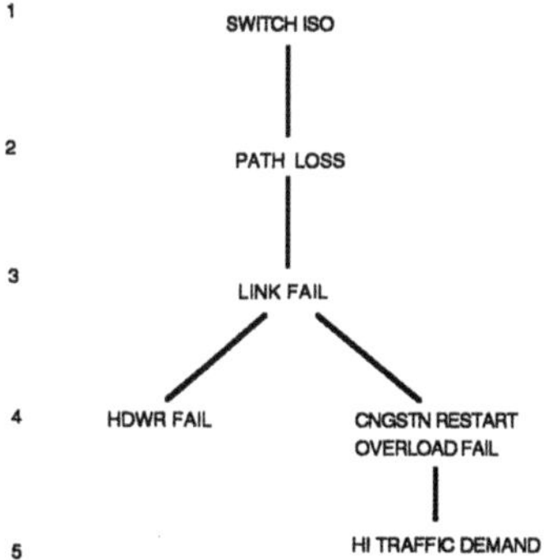

Figure 3 Correlation Tree Skeleton.

```
cor_group = 1 time window = 60 minutes

if new_msg.Trouble[1-2] = path loss precedence = 2 or
if new_msg.Trouble[1-2]=overload fail precedence =4 or
if new_msg.Trouble[1-2]=cngstn restart precedence =4
    new_msg correlates old_msg when
    case old_msg.Trouble[1-3] = hi traffic dmnd
        new_msg.A_Office = old_msg.A_Office or
        new_msg.Z_Office = old_msg.A_Office
    case old_msg.Trouble = anything_else
        (new_msg.A_Office =old_msg.A_Office and
        new_msg.Z_Office = old_msg.Z_Office) or
        (new_msg.A_Office = old_msg.Z_Office and
        new_msg.Z_Office = old_msg.A_Office)
```

Figure 4 Correlation Group.

In these trees, the child/parent link is equivalent to a cause and effect relationship between messages. Equivalent messages (such as *cngstn restart* or *overload fail*) are on the same node. Alternative children to a parent are similar to 'or ' branches, that is a *link fail* can cause a *path loss* whereas either a *hardware fail* or *cngstn restart* /*overload fail* can cause a *link fail.* However, it does not imply that every *link fail* will always cause a *path loss.* For example, one needs all the links between NEs to fail before a path loss occurs. This representation was chosen because users found it intuitive as they often discussed problems in terms of cause and effect. Using these skeletons, the 19 alarms that were generated by the three failures can be represented as a *correlation tree instances* as shown in Figure 5 below.

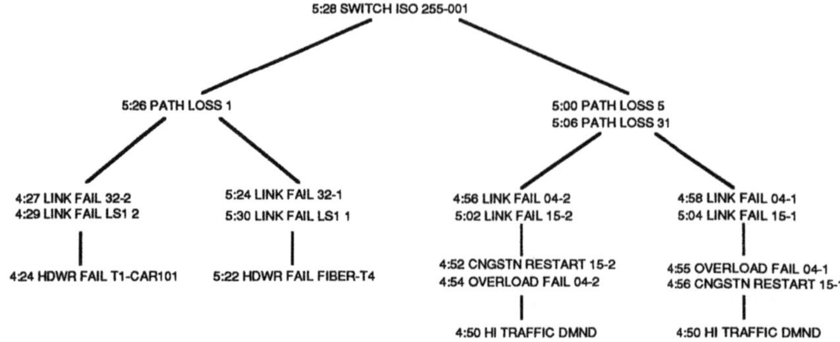

Figure 5 Correlation Tree Instance.

Each node in the correlation tree is a group of one or more equivalent messages (e.g. the *overload fail* at 4:55 and the *cngstn restart* at 4:56) and two nodes are connected if there is a cause and effect relationship between them (e.g. the *link fail* at 4:58 was one of the causes of the *path loss* at 5:00). Each branch of the correlation tree corresponds to a branch in the correlation tree skeleton in Figure 3 with the leaves being the underlying causes of a current network problem and the root being the result. In our example the skeleton contains a *path loss* causing a *switch iso*. Since both *path loss*es correlate the *switch iso* and did not correlate each other they became separate children of the *switch iso*. In our example the 4 leaves (two *hdwr fail*s and two *hi traffic dmnd*s are ultimately the cause of the *switch iso*.

3 ECXPERT

3.1 Correlation Grammar

The primary role of ECXpert is to receive alarms and to dynamically create correlation trees based on the correlation tree skeletons. Since TNM is sold to many customers - each having different levels of network management expertise, performance and security constraints - this package needs to configurable in the field by the customer. ECXpert supports an expert system shell that provides a pseudo-English description language in which users define *correlation groups* that correspond to the correlation tree skeletons. Each correlation group can be viewed as a model of a particular network problem. Using this language users specify

- when a new alarm belongs to a correlation group;
- when a new alarm correlates previous alarms that belonged to this group;
- the cause and effect relationship between alarms;
- what actions to take - e.g. automatic generation of a trouble ticket or invoking a reroute;
- a time window - that is only correlate alarms that have occurred within this time window.

The correlation grammar also allows rules to execute database look ups. Administrators can, then, write correlation groups to make use of network configuration data. For example, a rule might correlate the *link fails* occurring at 4:27 and 4:29 only if they are physically the same link. Figure 4 above, shows some of the rules used to correlate the alarms shown in the correlation tree skeleton in Figure 3.

A correlation group is comprised of three parts. The first part specifies the correlation group number used when displaying the correlation trees and a time window. The second part assigns a precedence to each type of message in this group which corresponds to the level in the correlation tree skeleton. The third part defines when a new message correlates an old(er) message in the group. For example, this rule states that if the first two words in the trouble field of a newly received message are *path loss*, *cngstn restart*, or *overload fail*, the message will belong to correlation group 1 with precedences 2, 4, and 4 respectively. Furthermore, they will correlate an older message in this group whose first three words in the trouble field are *hi traffic dmnd* if the new message's *a_office* field or *z_office* field is the same as the old message's *a_office* field and both alarms occurred less than 60 minutes apart. The grammar allows administrator to define macros; use arithmetic and string comparisons including the use of regular expression; and specify correlation conditions using logical 'and's, 'or's, and parentheses.

In a correlation group of *n* types of messages, there are n^2 possible correlation rules. This could be very large and cumbersome to write and maintain. To reduce the amount of typing, the correlation grammar provides constructs to allow many messages of the same type to be grouped together. For example, if we look at the rules in Figure 4, the *path loss, cngstn restart,* and *overload fail* all use the same correlation rules. In addition, there is one specific correlation condition for the *hi traffic dmnd* message, all the other messages in the group are correlated using the *anything_else* (default) correlation rule clause. This is both intuitive to the users and compact. Using this shorthand notation, most correlation groups have O(nlog(n)) lines with respect to the number of message types.

3.2 Dynamic Manipulation of Correlation Trees.

During normal operation, TNM receives messages from NEs and checks whether the message is

- An alarm - indicating a new problem and is then assigned a severity level and is displayed on the AS and sent to ECXpert for processing.
- A clear message - indicating a previous problem that caused an alarm has been corrected and can be removed from the AS and sent to ECXpert for processing.
- An informational message that can be ignored.

As each alarm is received, ECXpert uses the correlation conditions defined by the correlation groups to add the new alarm to all the relevant correlation trees. The algorithm to do this is quite complex and beyond the scope of this paper. A complete description of the algorithm can be found in (Nygate, 1994). In general, as each alarm is processed one or more of the following actions are taken. The new alarm may

- Be added to a tree - indicating that this alarm is part of a larger problem.
- Start a new tree - indicating a new problem.
- Combine a number of trees - indicating that what were a few small problems before were really only part of a larger one.
- Split a tree - indicating that an underlying cause is responsible for two or more problems.

- Clear an old message - indicating that this problem has now been resolved. This causes the tree to begin to decompose and if nothing new is added; it will eventually disappear.

4 HOW ECXPERT WAS DEVELOPED

The use of single knowledge representations and techniques for knowledge based system has been widely used (Abelson and Sussman, 1985). Successful applications have been reported in the literature in diagnostic systems (Shortliffe, 1976), planners (Ambros-Ingerson and Steel, 1988) and heuristic classification (Clancey, 1983).

However, many problems do not suit the problem solving characteristics of any one particular technique and need to be attacked by a variety of methods. Advocates of applying multiple methods in a single system (Fikes and Kehler 1985) contend that just as a carpenter has many tools, each specialized to its purpose, so should there be many tools in the programmer's kit (Bobrow and Stefik, 1986). Trying to solve a problem that does not fit well into a particular technique may result in programs that are buggy, slow, awkward and long. However, integrating multiple methods does incur a cost. For example, modules may be required to transform between different representations of the same information to optimize processing. But, if the cost is small, the benefits are great. Programmers can choose the most applicable problem solving technique to the module in question.

ECXpert integrates C++ and Prolog in a design that utilizes the run-time efficiency and support for object oriented design of C++ with the powerful meta-programming, semantic parsing, and pattern matching features of Prolog. The design and development of ECXpert was based on ASPEN (Nygate and Sterling, 1993), a new multi-paradigm method for developing knowledge based systems. ASPEN draws on the strengths of those that tout the clarity and success of single problem solving techniques with those who advocate the power and flexibility of multiple methods for software development. This compromise is achieved by providing a structured decomposition that allows each module to use different knowledge based techniques while defining a set number of modules with well delimited borders and functionalities. More information on ASPEN can be found in (Nygate, 1994).

ECXpert is comprised of four main modules - a correlation process, a correlation group compiler, a test correlation process, and a user interface.

4.1 Correlation Process

The correlation process is comprised of a C++ object for collecting alarms from TNM, a Prolog object for executing the correlation algorithm, and a C++ object to manipulate the database containing the composite alarm objects, that is the correlation trees.

The Prolog object can be viewed as a forward-chaining correlation engine that takes each alarm, find what rules it matches and fires a set of rules to update the correlation trees. As each alarm is processed, the Prolog object determines with which trees the new message correlates. Then using the algorithm mentioned in section 3.2, it determines the actions the database object must execute to update the correlation trees.

The correlation process contains 5000 line of C++ code, including 1500 lines for data structure conversion between C++ objects and Prolog lists; 1700 lines of static Prolog code to implement the correlation algorithm; and typically 2500 lines of user-supplied correlation rules.

4.2 The Correlation Group Compiler

The correlation group compiler is written in Prolog and converts user supplied correlation rules into Prolog Horn clauses (Kowalski, 1979). These rules are then dynamically linked with the correlation process. The syntax of the correlation grammar can be specified using a Definite Clause Grammar. Prolog's support for DCGs made the code generation straightforward. As mentioned in section 3.1, the correlation grammar provides a compact notation for combining multiple correlation rules by using 'or's. The compiler expands the disjunctions on the left hand side of the correlation rules and convert them into Horn Clauses. More information on the use of Prolog in ECXpert can be found in (Nygate, 1994).

The development of ECXpert is typical of the meta-programming approach to develop knowledge-based systems as advocated by (Sterling, 1990) and described by (Ümit Yalçinalp, 1991) in her Ph.D. thesis. A high-level language is developed for the application which can be easily compiled into Prolog or executed directly with a simple interpreter.

A typical correlation group contains 100 lines which is compiled into about 250 lines of Prolog code in 5 seconds. The tokenizer, code generator, and a user friendly error handling subsystem to help administrators find and fix syntax errors totals 1500 lines of Prolog code.

4.3 User Interface

When a user selects an alarm on the awareness screen, this module retrieves all the correlation trees to which this alarm belongs and displays them in a pop-up window on the AS monitors. For example, suppose the user selected the *link fail* received at 5:24 on the awareness screen, the correlation window that would be displayed is shown in Figure 6 below.

*********** CORRELATION WINDOW **********

PRECEDENCE	DATE	TIME	L	SYSTEM	A OFFICE	Z OFFICE	TROUBLE INDICATION
SELECTED:	09jun	5:24	2	5e_s_n	CLLI_A	CLLI_Y	LINK FAIL 32-1
				***************** CORRELATION GROUP 1 *****************			
1	09jun	5:28	2	5e_s_n	CLLI_A	CLLI_Y	SWITCH ISO 255-001
2	09jun	5:26	2	5e_s_n	CLLI_A	CLLI_Y	PATH LOSS 1
3	09jun	5:30	2	nti_s_n	CLLI_Y	CLLI_A	LINK FAIL LS1 1
+	09jun	5:24	2	5e_s_n	CLLI_A	CLLI_Y	LINK FAIL 32-1
4	09jun	5:22	1	FIBER-T4	CLLI_E	CLLI_R	HARDWARE FAIL
3	09jun	4:29	2	nti_s_n	CLLI_Y	CLLI_A	LINK FAIL LS1 2
+	09jun	4:27	2	5e_s_n	CLLI_A	CLLI_Y	LINK FAIL 32-2
4	09jun	4:24	1	T1-CAR101	CLLI_D	CLLI_Q	HARDWARE FAIL
2	09jun	5:06	2	att_s_n	CLLI_Z	CLLI_A	PATH LOSS 31
+	09jun	5:00	2	5e_s_n	CLLI_A	CLLI_Z	PATH LOSS 5
3	09jun	5:04	2	att_s_n	CLLI_Z	CLLI_A	LINK FAIL 15-1
+	09jun	4:58	2	5e_s_n	CLLI_A	CLLI_Z	LINK FAIL 04-1
4	09jun	4:56	2	5e_s_n	CLLI_A	CLLI_Z	CNGSTN RESTART 15-1
+	09jun	4:55	2	att_s_n	CLLI_Z	CLLI_A	OVERLOAD FAIL 04-1
5	09jun	4:50	2	att_s_n	CLLI_Z		HI TRAFFIC DMND
3	09jun	5:02	2	att_s_n	CLLI_Z	CLLI_A	LINK FAIL 15-2
+	09jun	4:56	2	5e_s_n	CLLI_A	CLLI_Z	LINK FAIL 04-2
4	09jun	4:54	2	att_s_n	CLLI_A	CLLI_Z	OVERLOAD FAIL 04-2
+	09jun	4:52	2	att_s_n	CLLI_Z	CLLI_A	CNGSTN RESTART 15-2
5	09jun	4:50	2	att_s_n	CLLI_Z		HI TRAFFIC DMND

Figure 6 Correlation Window.

The precedence column corresponds to the precedence in the correlation grammar which allows users to reconstruct the correlation tree. The rest of the columns contain relevant data that also appeared on the awareness screen. For example, the *hi traffic dmnd* at 4:50 is a child of the *cngstn restart* at 4:52; and the *link fail* at 4:56 and the *link fail* at 4:58 are both children of the *path loss* at 5:00 which is in turn the child of the *switch iso* at 5:28. The *overload fail* at 4:55 and the *cngstn restart* at 4:56 are equivalent messages with the *cngstn restart* being the primary message.

Alarms in the correlation window are displayed in the same colors as on the AS, red for the most severe, blue for the least, and cleared alarms in green.

Since many groups can be active at once, the selected message can be in more than one group and each group can span more than one page. The user is able to scroll forward and backwards in the correlation screen looking at each group and page.

The correlation algorithm can also handle missing data. If, for example, neither of the *path loss* messages at 4:58 and 5:04 were received, the *overload fail* at 4:55 and *cngstn restart* at 4:56 would have become children of the *path loss* at 5:06. The precedence column of the correlation window would display a minus sign to signify that a message was missing as shown in Figure 7.

2				09jun	5:06	2	att_s_n	CLLI_Z	CLLI_A	PATH LOSS 31
+				09jun	5:00	2	5e_s_n	CLLI_A	CLLI_Z	PATH LOSS 5
	-									
		4		09jun	4:56	2	5e_s_n	CLLI_A	CLLI_Z	CNGSTN RESTART 15-1
		+		09jun	4:55	2	att_s_n	CLLI_Z	CLLI_A	OVERLOAD FAIL 04-1
			5	09jun	4:50	2	att_s_n	CLLI_Z		HI TRAFFIC DMND

Figure 7 Correlation Window with Missing Messages.

4.4 Test Correlation Process

To facilitate the administrator's role in writing correlation groups we provided a grammar that was intuitive, powerful and compact. In addition, once all the syntax errors in the correlation group were fixed, the administrators were able to verify that the semantics of the correlation group were correct using a test correlation process that incorporated the 'how' and 'why' (Sterling and Shapiro, 1986) tools used in Expert Systems. Administrators were able to provide an input file of high level alarms using the same format as displayed on the AS, send them one by one through a test correlation process and find out 'why' certain messages belonged to which correlation group and 'how ' they correlated with other, older, messages in the group. Once they were satisfied with the results, they could then install the correlation group.

5 USING THE CORRELATION TREE

ECXpert and its use of correlation trees provides many powerful ways of enhancing the effectiveness of network managers. The most obvious and direct improvement utilizes the fact that the leaves of the tree are the causes of the network problem with the root being the consequence. Thus in our example, if a user sees a *switch iso*, he/she can bring up the corresponding correlation window and see that the causes (leaves) are the two *hardware fail*s and the *hi traffic dmnd* and dispatch a repair order to fix these problems immediately. Each of the 'leaf' alarms occur frequently and they typically do not have any major network impact. Without correlation the leaf alarm would not have been fixed as quickly as other more obvious alarms. Once one of the leaves is fixed, all the messages in its branch often become cleared as well. If enough leaves are cleared, the root becomes cleared too. This is clearly shown in the correlation window and allows the user to execute retroactive analysis to see what combination of alarms (i.e. leaves) caused a network event, and how it was resolved (i.e. green branches).

A far more sophisticated but extremely useful feature of ECXpert, is to display on the AS only the alarms that correspond to the roots and the leave in the correlation tree while suppressing intermediate nodes in the tree. This has the immediate impact of reducing clutter on the awareness screens while leaving the critical nodes that show the overall network problems with their corresponding underlying causes.

Users can also set the AS restrictions to show a specific class or set of alarms. Whenever the Event Correlation window is invoked, all the alarms that correlate with the selected alarm are shown. This allows users to peruse the high level alarms but still have access on demand to all the low level contributing alarms. Other features include

- Escalating all the alarms in the correlation tree to the severity of the most severe alarm in that tree. For example, if a critical alarm (displayed as red) is added to a tree, all the alarms in the tree would be escalated and be displayed in red.
- Predicting what other problems must occur before a more serious network situation will occur. This is a very powerful feature as it allows users to estimate how far the network is away from a catastrophe and they can then protect/reserve the critical remaining resources.
- Allowing users to define actions in the correlation group such as setting off audible alarms (particularly useful during night shifts!), generating reports, generating new alarms, automatically starting a repair procedure, etc.

6 RESULTS AND EXTENSIONS

ECXpert has been installed in a number of sites in the U.S. and Europe. The initial customers, at NYNEX (NET and NYT), have been using event correlation to manage their SS7 network since 1992. Many other TNM users have since purchased ECXpert including PacBell, Bell South, SNET, and Bell Atlantic, to correlate alarms in various parts of their network. In addition, TNM has managed to attract 3 other large domestic customers that previously used our main competitor's product partially due to the functionality provided by ECXpert.

ECXpert has increased customer revenue by reducing the amount of time to isolate and repair network problems. Current estimates show that due to decreased network down time and reduced labor costs, savings at a typical U.S. network operation center are in the range of $500,000 and $1,000,000 a year.

The current version of ECXpert can correlate about 1000 alarms per hour with 10 correlation groups active. Although this meets customers current use of the feature package, users are becoming more sophisticated and are adding more and more correlation groups. ECXpert is currently running on a Tandem FT computer and is competing for resources with the rest of TNM. A future release will provide increased performance by allowing ECXpert to run on an adjunct processor.

Other enhancements currently under development include adding a graphics correlation window and replacing the relational database that stores the correlation tree with an object database. We are also working on a learning module to derive correlation groups automatically. For example, suppose in a number of 5 minute windows the messages A, B, C, D, and E occurred in sequence and they all had a common A Office or Z Office. The learning module could generate a correlation group with A causing B, B causing C, etc. We denote this as A→B→C→D→E. Now suppose there were other instances that consisted of (F, B, C, D, E). The learning module could now derive a correlation group with (A or F)→B→C→D→E.

This is a valuable feature as many of our customers do not know all the alarm types and how they should be correlated to network events. We do supply a default set of correlation groups, but the customer needs to configure and add rules to match their particular network needs. The search will be directed by meta-correlation rules written by the customer that will allow them to specify time windows, fields of interest, and strength (that is how many times must a pattern repeat before it is more than just a coincidence). The groups will then be generated automatically and presented to the users for modification, installation, and sometimes for deletion as chance patterns of messages can be grouped.

7 CONCLUSION

Some indication of the benefits of Prolog for this project can be gained by comparing it with another knowledge-based, network management feature package developed for TNM. This application analyzed error messages generated by the 5ESS switch and recommended repair procedures. C5 was used to implement rules collected from 2 experts from the New England Telephone Company. One problem limiting general deployment of this package is that the recommended repair procedures vary between customers. Companies often had different recommendations as to what to try first, what procedures are too risky, what actions are considered a breach of security, etc. Thus, although the system had a large amount of expertise, it was very inflexible and narrow. In contrast to Prolog, a meta-programming approach is not supported by C5. Although they used a pseudo-English description language to capture the experts knowledge, no compiler could be written and the hand encoding into C5 led to many misinterpretations and errors.

Meta-programming is a very powerful feature and useful technique that contributed to the success of ECXpert. Each customer can write their own correlation groups or modify the default groups we provide. They can then compile, test and use these groups in the field without having to interact with AT&T and request that we make the changes. Thus, new correlation groups can be added very quickly and the system can be configured to match each customers individual needs. Prolog not only facilitates the use of meta-programming, but it also allows changes to be dynamically linked with running processes. That is, there is no need to recompile the entire correlation process to use the new set of correlation groups. Nor is there even any need to stop the correlation process. Rather, correlation groups can be complied off-line and then linked dynamically with the running process.

Event correlation is not restricted to telecommunications, but is applicable to many other domains where order must be made of a large volume of related messages. I have spoken to people who have worked as air traffic controllers, power station operators, and chemical plant engineers. They all indicated their need to correlate large volumes of data collected from many pieces of equipment. Moreover, the knowledge required to group these messages together can also be represented as correlation tree skeletons. Thus, the correlation tree skeletons and the correlation algorithm used in ECXpert can be reapplied in many other domains, and should be included as a new generic task in problem solving (Chandrasekaran, 1986). Due to the importance of solving event correlation in leading our competitors and in its potential in other fields, a patent application with the specifics of my algorithm has already been filed by AT&T.

In conclusion, the multi-paradigm implementation provided a powerful environment that enabled us to combine the strengths of logic and object oriented programming. The user definability, dynamic linking, and the high level of abstraction provided by the correlation groups have been keys to success. Customers have a syntax that is powerful enough to configure the system the way they want using a language they can understand.

8 REFERENCES

Ambros-Ingerson, J. A. and Steel, S. (1988) Integrating Planning, Execution, and Monitoring, *Proceedings AAAI*, 83–88.

Abelson, H. and Sussman, G. (1985) Structure and Interpretation of Computer Programs, *MIT Press*, Cambridge.

Bobrow, D. and Stefik, M. (1986) Perspectives in Artificial Intelligence Programming, *Readings in AI and Software Engineering*, Morgan Kaufmann, California.

Chandrasekaran, B. (1986) Generic Tasks in Knowledge Based Reasoning: High Level Building Blocks for Expert System Design, *IEEE Expert*, **1(3)**, 23–30.

Clancey, W. (1983) Heuristic Classification, *AI Journal*, **27**.

Fikes, R.; Kehler, T. (1985) *Communications of the ACM*, **28**, 904.

Kowalski, R. (1979) Logic for Problem Solving, North-Holland, Amsterdam.

Nerys, C. (1993) The Complete Diagnostic Tool: Total Network Management, *Network Edge - AT&T*, 18–22.

Nygate, Y. (1994) ASPEN - Structuring Design of Complex Knowledge Based Systems, *Ph.D. thesis*, Case Western Reserve University.

Nygate, Y. and Sterling, L. (1993) ASPEN - Designing Complex Knowledge Based Systems, *10th Israeli Symposium on Artificial Intelligence*, 51–60.

Shortliffe, E. H. (1976) MYCIN: Computer-Based Medical Consultations, *Elselvier*, New York.

Sterling, L. (1990) Meta-Programming in Logic Programming Tutorial Notes, *Meta 90*, Leuven, Belgium.

Sterling, L. and Shapiro E. (1986) The Art of Prolog, *MIT Press*, Cambridge MA.

Yalçinalp, L. Ü. (1991) Meta-Programming for Knowledge-Based Systems in Prolog, *Ph.D. thesis*, Case Western Reserve University.

9 BIOGRAPHY

Yossi Nygate has been employed by AT&T Bell Labs for the past ten years. He has been responsible for developing telecommunication network management systems integrating C++, C, and Prolog for domestic and international customers. He received his Ph.D. in computer science from Case Western Reserve University in 1994. The focus of his research was on problem solving systems integrating multiple techniques. He received his M.Sc. in computer science from the Weizmann Institute of Science in 1985 in the area of Expert Systems. His current areas of interest include practical applications of AI, planning, and automated learning.

26

Real-time telecommunication network management: extending event correlation with temporal constraints

G. Jakobson, M. Weissman
GTE Laboratories Incorporated
40 Sylvan Rd, Waltham, MA 02254
tel: 1-617-466-2325, fax: 1-617-466-2960, email: gj00@gte.com

Abstract

Event correlation is becoming one of the most central techniques in managing the high volume of event messages. Practically, no network management system can ignore network surveillance and control procedures which are based on event correlation. The majority of existing network management systems use relatively simple *ad hoc* additions to their software to perform alarm correlation. In these systems, alarm correlation is handled as an aggregation procedure over sets of alarms exhibiting similar attributes. In recent years, several more sophisticated alarm correlation models have been proposed. In this paper, we will expand our knowledge-based event correlation model to capture temporal constraints.

Keywords

Real-time telecommunication network surveillance, temporal reasoning, event correlation, network fault propagation, knowledge-based systems

1 INTRODUCTION

Modern telecommunication networks may produce large numbers of alarms. It is not unusual that a burst of alarms during a major network failure may exhibit 40–50 alarms per second. This leads to serious difficulties in the network management process, particularly as follows:

- The inability to follow the stream of incoming events: alarms may pass unnoticed, or be noticed too late.
- The incorrect interpretation of groups of alarms: decision making and application of network controls is based on a single event rather on a macroscopic, generalized event level.
- The concentration of the operations staff on less important events.

Event correlation is becoming one of the most central techniques in managing the high volume of event messages. Practically, no network management system can ignore network surveillance and control procedures which are based on event correlation. The majority of existing network management systems use relatively simple *ad hoc* additions to their software to perform alarm correlation. In these systems, alarm correlation is handled as an aggregation procedure over sets of alarms exhibiting similar attributes.

In recent years, several more sophisticated alarm correlation models have been proposed. In this paper, we will expand our knowledge-based event correlation model (Jakobson and Weissman, 1993) to capture temporal constraints.

2 EVENT CORRELATION DOMAIN

2.1 Definition

We define the task of event correlation as a conceptual interpretation procedure in the sense that a new meaning is assigned to a set of events that happen within a predefined time interval. As discussed in Section 2.3, the conceptual interpretation procedure could stretch from a trivial task of alarm compression to a complex pattern matching task. A typical event correlation is determined to be a dynamic pattern matching over a stream of network events. In addition, the correlation pattern may include network connectivity information, diagnostic test data, data from external databases, and any other information.

Events in the managed network are very diverse in nature. These events include raw event, status, and clear messages from network elements (NEs); events from mediation devices, subnetwork management systems, test systems and other equipment; user action messages from network operator terminals; and system interrupts. The origin of an event could be easily expanded beyond the "real" real-time events. For example, searching through records in a stored event log file, or any other file, could trigger the creation of an "alarm" if the value of some field in the record does not satisfy predefined constraints.

Applying event correlation rules may yield several results. First, a new event (message) may be sent to the operator's terminal. Second, an action may clear or resolve existing events (see Section 4.3). Third, a diagnostic message may be sent about faults occurring in the network. Fourth, a procedure may be called to access a database, run a diagnostic test procedure, generate a trouble ticket, or perform any other executable external procedure. Fifth, an internal system action to store data, change the system mode of operation, or perform any other internal system procedure may be taken. Any event generated as a result of event correlation will be considered a "regular" event coming from the network and, as such, is a subject for futher event correlation. Formally, nothing prevents us from considering the event correlation procedure as a network element which produces events (messages). The process of building correlations from correlations allows the formation of complex multilevel correlations

Networks function in discrete space and time. Even if some process, e.g., temperature in the network management center, takes its physical value in a continuum, it could be, in this particular domain, quantified and thresholded. The time model that we are using will be discrete time, with two modifications: point time and interval time. In point time, the events take place at time moments represented as integers at a predefined time scale (seconds, minutes, etc.). For example, the use of Universal Time (UT) requires that the date/time stamps attached to the event message be reduced to a numeric value in seconds or minutes. Point time is applied to most actions, such as user commands and system interrupts. In interval time, events are described by two time moments, the time of origination and the time of termination. Network events corresponding to NE faults, changes in the system behavior or changes in the state of the sensor or other control equipment, are usually described in interval time.

2.2 The role of event correlation in network management

Event correlation supports the following network management tasks:

- Reduction of the information load presented to the network operations staff by dynamic focus monitoring and context-sensitive event suppression (filtering).
- Increasing the semantic content of information presented to the network operations staff by aggregation and generalization of events.
- Real-time network fault isolation, causal fault diagnosis, and suggestion of corrective actions.
- Analysis of the ramification of events, prediction of network behavior, and trend analysis.
- Long-term correlation of historic event log files and network behavior trend analysis.

2.3 Event correlation types

Depending on the nature of the operations performed on events, we will consider the following types of event correlation:

1. $[a, a, \ldots\ldots, a] \Rightarrow a$	Compression
2. $[a, p(a) < H] \Rightarrow \varnothing$	Filtering
3. $[a, C)] \Rightarrow \varnothing$	Suppression
4. $[n \times a] \Rightarrow b$	Count
5. $[n \times a, p(a)] \Rightarrow a, p'(a), p' > p$	Escalation
6. $[a, a \subset b] \Rightarrow b$	Generalization
7. $a, a \supset b] \Rightarrow b$	Specialization
8. $[a\ T\ b] \Rightarrow c$	Temporal Relation
9. $[a, b, \ldots T, \wedge, \vee, \neg] \Rightarrow c$	Clustering

Event compression (1) is the task of reducing multiple occurrences of identical events into a single representative of the events. No number of occurrences of the event is taken into account. The meaning of the compression correlation is almost identical to the single event *a*, except that additional contextual information is assigned to the event to indicate that this event happened more than once.

Event (alarm) filtering (2) is the most widely used operation to reduce the number of alarms presented to the operator. If some parameter p*(a)* of alarm *a,* e.g., priority, type, location of the NE, time stamp, etc., does not fall into the set of predefined legitimate values H, then alarm *a* is simply discarded or sent into a log file. The decision to filter alarm *a* out or not is based solely on the specific characteristics of alarm *a.* In more sophisticated cases, set H could be dynamic and depend on user-specified criterion or criterion calculated by the system.

Event suppression (3) is a context-sensitive process in which event *a* is temporarily inhibited depending on the dynamic operational context C of the network management process. The context C is determined by the presence of other event(s), network management resources, management priorities, or other external requirements. The change in the operational context could lead to exhibition of the suppressed event. Temporary suppression of multiple events and control of the order of their exhibition is a basis for dynamic focus monitoring of the network management process.

Another type of correlation (4) results from counting and thresholding the number of repeated arrivals of identical events. Event escalation (5) assigns a higher value to some parameter p'*(a)* of event *a,* usually the severity, depending on the operational context, e.g., the number of occurrences of the event.

Event generalization (6) is a correlation in which event *a* is replaced by its super class *b.* Event generalization has a potentially high utility for network management. It allows one to deviate from a low-level perspective of network events and view situations from a higher level. Event specialization (7) is an opposite procedure to event generalization. It substitutes an event with a more specific subclass of this event.

Temporal relations (8) T between events *a* and *b* allow them to be correlated depending on the order and time of their arrival. Different temporal relations for event correlation will be described in Section 5.

Event clustering (9) allows the creation of complex correlation patterns using logical operators ∧ (and), ∨ (or), and ¬ (not) over component terms. The terms in the pattern could be primary network events, previously defined correlations, or tests of network connectivity.

3 REAL-TIME FAULT DIAGNOSIS USING EVENT CORRELATION

3.1 Network fault propagation

The original sources for most network events are physical faults occurring in the managed NEs. These faults can be causally related or not, i.e., they can be independent. Causal relations between faults can be represented by fault propagation rules (Figure 1).

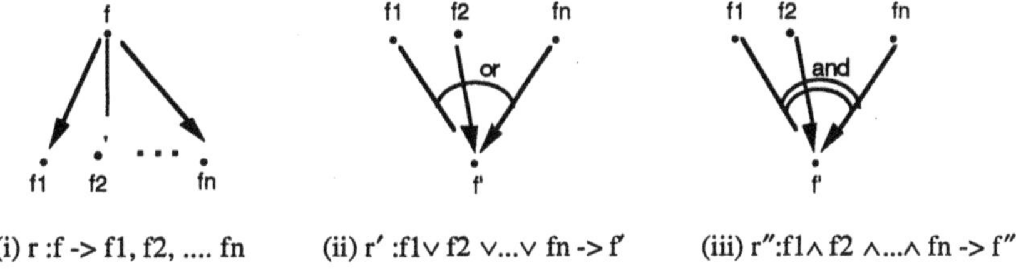

Figure 1 Fault propagation rules.

Rule (i) defines fault f as a root cause for multiple faults f1, f2,..., fn. In rule (ii), fault f′ could be caused by any of the faults f1, f2,..., fn; while in rule (iii), all faults f1, f2,..., fn should be present in order to cause fault f′.

Composition of fault propagation rules forms an acyclic fault propagation graph where f′ from rule (ii) corresponds to the or-node, and f″ from (iii) corresponds to the and-node. A set of independent fault propagation graphs form the fault propagation model of the network.

A fuzzy fault propagation model could be constructed by supplying fault likelihood distribution for faults f1,..., fn in initial rules (i), and defining likelihood calculation algorithms for the logical or (ii) and the logical and (iii) nodes. Different fuzzy reasoning models could be used here, however this topic is beyond the scope of this paper.

Determination of fault propagation rules is a subject of domain knowledge acquisition. It is based on general principles of telecommunication systems, physical construction of NEs, and the behavior of the individual NEs and the whole network. Many fault propagation rules can be derived by examining the network configuration (connectivity) model. For example, knowing the nature of the faults f1–f4, and the fact that NEs NE1 and NE3, and NE2 and NE3 are connected (Figure 2), one may derive causal propagation rules f1 –> f3, f4 and f1 ∨ f2 –> f4.

3.2 Heuristics of fault detection

Not all faults have events associated with them. Such faults can be recognized indirectly by correlating available events. Let's consider a simple fault propagation model, shown in Figure 3.

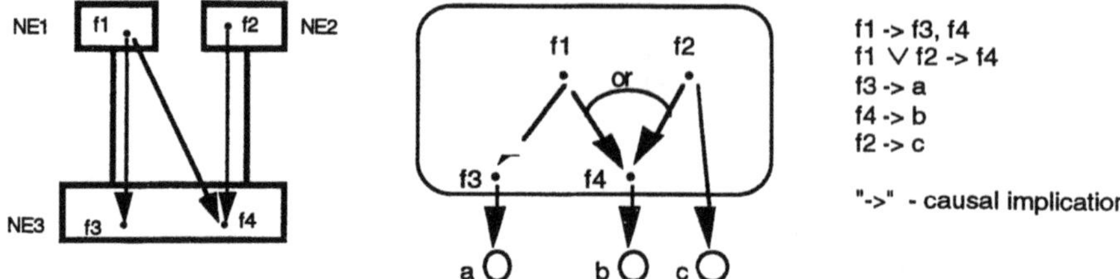

Figure 2 Network fault propagation.

Figure 3 Network diagnostic using alarm correlation.

Fault f3 is caused by fault f1, while f4 could be caused by f1 or f2 or by both of them. Fault f3 is exhibited by alarm a, f4 by alarm b, and f2 by alarm c. Faults f3 and f4 may also happen independently of faults f1 and f2. For example, if alarm *a* was generated, the reason could be directly fault f3 (no presence of the fault f1), or the reason could be fault f1, which consequently caused fault f3. By correlating alarms into simple Boolean patterns, one can construct the following fault diagnostic rules:

Rule 1: if (not a) and b and c then f1 — definitely no
f2 — definitely yes
f3 — definitely no
f4 — unlikely a root cause

The fact that alarm *a* is not present allows us to conclude that fault f3 and, consequently, fault f1 didn't happen. Obviously, the fault should happen, because it is the sole reason for alarm c. Generally, alarm b could be caused either by fault f4 as a consequence of faults f1 or f2, or by fault f4 as an independent root cause. In our example, fault f1 didn't happen, so alarm c could be potentially caused by fault f4 as the root cause or as a fault caused by f2. The presence of fault f2 definitely caused f4, and it is unlikely that f4 happened simultaneously as a root cause and as a fault caused by f2.

Rule 2: if *a* and b and not c then f1 — likely
f2 — definitely no
f3 — unlikely a root cause
f4 — unlikely a root cause

Rule 3: if *a* and b and c then f1 — likely
f2 — definitely yes
f3 — likely
f4 — unlikely a root cause
(f1, f3) — unlikely together

Rule 4: if *a* and not b and c then "Error in alarm message processing"

4 TIME-DEPENDENT EVENT MANAGEMENT

4.1 Events

Formally, an event is a pair (preposition, time quantifier), in which preposition describes the content of the event, and time quantifier is a moment in the point time, or a time interval of duration of the event in the interval time. Without losing generality, we will refer to prepositions as messages. (Strictly speaking, a preposition is a formal representation of a message obtained after parsing the message.) Further in the paper, we will use the following notation:

event = (message, time quantifier)

time quantifier = [t1, t2]
t1 — time of origination
t2 — time of termination

The origination time of the event is issued by the NE or its management system. The event message sent to the event list for display at an operator terminal stays there until it is cleared by the network management system or by the operator. The event will be ultimately eliminated from the event list either by clearing or expiration of the lifespan, whichever comes first.

In addition to the event clearing procedures, an event can "die by a natural cause," i.e., when the event expiration time is over. Event expiration time is determined by the lifespan of the event, *a* potential maximum duration of the event. The lifespan is assigned duration based on event class, and depends on the practices and policies of the particular network management domain.

For many NEs, the events (alarms) are issued pair-wise — the original event message manifesting a beginning of some physical phenomenon, e.g., a fault, and a complimentary clear message manifesting the end of the phenomenon. After origination, these two logically inverse messages may exist together, unless a clear command to remove the first message is issued by the network operator. In network management systems that support logical reasoning and event correlation, the logically inverse messages should be detected and resolved automatically.

4.2 Correlation window

Each event correlation process has an assigned correlation time window (Figure 4), a maximum time interval during which the component events should happen. The correlation process will be started at the time of arrival of the first component event (event *a*) and stopped as the last component event (event *c*) arrives. As any other event, correlation has its time of origination, time of termination, and lifespan. By definition, the time of origination of the correlation is equal to the time of origination of the last component event.

Event correlation is a dynamic process so that the arrival of any component event instantiates a new correlation time window for some correlation. This means that the correlation time window for some correlation slides in time to capture new options to instantiate a correlation. However, if temporal constraints are assigned to the component events, e.g., event *b* should be always after event *a*, no correlation time window is started when event b arrives.

Determining the length of the correlation window and the lifespan of an event (correlation) directly affects the potentials of creating correlations. Widening the correlation window and increasing the lifespans increases the chance of creating a correlation. For very fast processes, e.g., a burst of alarms during T3 trunk failure, the width of the correlation window could be seconds, while for slow processes, such as analyzing a trend of failures from an alarm log file, the correlation window may be several hours, or even several days, long. The same is true for the lifespan: informative events could last several seconds, while the lifespan of critical events should be indefinite, i.e., these events should be always cleared by the operator or by the system.

The right value for the correlation window and the lifespan will emerge from the practice of managing a specific network.

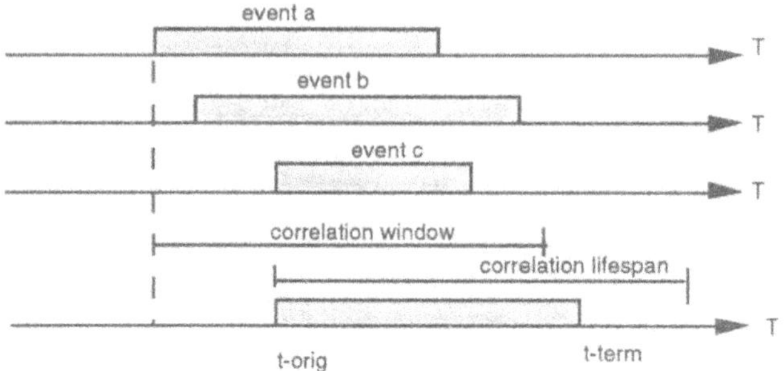

Figure 4 Correlation time window and lifespan.

4.3 Dynamic event memory

Each event, either originated in the managed network or produced by the event correlation process, will be placed into Dynamic Event Memory. Events residing in Dynamic Event Memory are available for the correlation processes. An event is removed from Dynamic Event Memory if one of the following happens:

- Command Resolve is issued by the operator or by the system.
- The lifespan of the event is over.

Command Resolve effectively "kills" the event, preventing its future use. This command should be handled carefully by the operator and the system.

As events are originated, they are also placed into the Event List. The Event List keeps events only for display purposes. An event is removed from the Event List under the following circumstances:

- Command Clear is issued by the operator or by the system.
- Command Resolve is issued by the operator or by the system.
- A new correlation is originated which contains this event as its component.
- The lifespan of the event is over.

The pragmatics of the Clear action is to reduce the number of displayed event messages, while at the same time leaving the events in Dynamic Event Memory for potential future correlations.

4.4 Predictable run-time behavior

Predicting the worst-case run-time behavior is a prerequisite for real-time network event surveillance and fault management systems. The system must guarantee network event collection, parsing, correlation, fault diagnosis, display, and execution of corrective actions within the time limits set by the specific network operational standards. Slow transmission and processing of network messages may result in distortion (masking) of the real sequence of actual physical events and result in incorrect correlations. Two definitions of the predictable execution time exist:

- Event processing must complete before the new event, i.e., $\partial < t$, where ∂ is the maximum event processing time and t is the minimum time interval between synchronous or asynchronous events.
- Event processing must complete during a predefined time interval, i.e., $\partial < D < t$, where D — predefined event processing time.

The value of D could be in the order of tenth or hundredth of seconds during the peak of alarm bursts. For example, a cellular switch supporting a region with 30–40 cell sites produces normally 2–3 alarms per minute. A medium-size wireline network with 3–4 large Class 5 switches and 8–10 digital cross-connects may produce during a major T1/T3 trunk failure tens of alarms per second. Collecting and parsing these alarms should be very fast. It is not unusual that even very fast network management platforms with clock speed of 150 MHz or higher need event buffering for correlating bursts of alarms.

5 TEMPORAL REASONING FOR EVENT CORRELATION

5.1 Issues

Temporal reasoning, reasoning about time, plays a critical role in monitoring network events. The system should be able to reason about the relative and absolute times of occurrence of events, duration of events (or duration of the absence of events), and sequence of events. The time interval between events can be defined on a quantitative time scale or on a qualitative time scale.

5.2 Temporal relations

In this section we will examine temporal relations that will be used to build time-dependent event correlations between events. Temporal relations together with temporal reasoning rules form temporal event calculus (Allen, 1983).

For the specific tasks of network event correlation, we will define the following set of temporal relations. Let e1 and e2 be two events defined on an interval time: e1 = (msg1, [t1, t1′]) and e2 = (msg2, [t2, t2′).

1. Event e2 by an interval h starts after event e1.

$$e2\ AFTER(h)\ e1 \iff t2 > t1 + h$$

2. Event e2 by an interval h follows event e1.

$$e2\ FOLLOWS(h)\ e1 \iff t2 \geq t1' + h$$

From the definitions (1) and (2) follows that

$$\text{if } e2\ FOLLOWS(h)\ e1 \text{ then } e2\ AFTER(d + h)\ e1,$$

where d is the duration of the event e1.

3. Event e2 by an interval h ends before event e1 (ends).

$$e2\ BEFORE(h)\ e1 \iff t1' \geq t2' + h$$

Note that relations (starts) AFTER and (ends) BEFORE are not logically inverse.

4. Event e2 by an interval h precedes event e1.

$$e2\ PRECEDES(h)\ e1 \iff t2 \geq t1' + h$$

The following statement is true:

$$e2\ FOLLOWS(h)\ e1 \iff e1\ PRECEDES(h)\ e2$$

5. Event e2 happens during event e1.

$$e2 \text{ DURING } e1 \Leftrightarrow t2 \geq t1 \text{ and } t1' \geq t2'$$

The following derivation rule holds between DURING, BEFORE, and AFTER:

If e2 DURING e1, then e2 AFTER e1 and e2 BEFORE e1 (and *vice versa*).

6. Event e1 starts at the same time as event e2.

$$e1 \text{ STARTS } e2 \Leftrightarrow t1 = t2$$

Obviously the following rule holds:

If e2 AFTER(h) e1 and e1 AFTER(h) e2, then e1 STARTS e2 (and *vice versa*).

7. Event e1 finishes at the same time as event e2.

$$e1 \text{ FINISHES } e2 \Leftrightarrow t1' = t2'$$

Similarly, as for the previous case, the following rule holds:

If e2 BEFORE(h) e1 and e1 BEFORE(h) e2, then e1 FINISHES e2 (and *vice versa*).

8. Event e1 coincides with event e2.

$$e2 \text{ COINCIDES with } e1 \Leftrightarrow t2 = t1 \text{ and } t1' = t2'$$

As a consequence of the definition of the coinciding events, the following is true:

If e2 COINCIDES with e1, then e2 STARTS e1 and e2 FINISHES e1 (and *vice versa*).

If e2 DURING e1 and e1 DURING e2, then e2 COINCIDES with e1 (and *vice versa*).

9. Event e1 overlaps with event e2.

$$e1 \text{ OVERLAPS } e2 \Leftrightarrow t2' \geq t1' > t2 \geq t1$$

From the definition OVERLAPS, it follows that

If e1 OVERLAPS e2, then e2 AFTER(h) e1 and e1 BEFORE(h) e2.

Regarding the algebraic properties of the temporal relations, we can say that all of them are transitive, except OVERLAPS, while STARTS, FINISHES, and COINCIDES are also symmetric relations.

6 KNOWLEDGE FRAMEWORK FOR EVENT CORRELATION

6.1 Model-based approach

Our approach to event correlation uses the principles on model-based reasoning originally proposed in (Davis, Shrobe, and Hamscher, 1982) for troubleshooting electronic circuit boards. The idea of the model-based approach is to reason about the system from representation of its structure and functional behavior. We will extend this model into real-time event correlation.

The structural representation means the description of the NEs and the topology of the network. Under the topology, we understand not only the connectivity between NEs but also the containment relations between the elements.

The behavioral representation describes the dynamic processes of event propagation and correlation. These processes are described using correlation rules. Each rule activates a new event (correlation), which in its turn may be used in the firing condition of the next correlation rule. In the following sections, we describe the components of the overall event correlation model: the network configuration model, event correlations, and correlation rules.

6.2 Network configuration model

Networks are composed of NEs. Traditional examples of NEs are switches, digital cross-connect systems, channel service units, trunks, routers, bridges, etc.

In a broader sense, a NE is any real or virtual hardware or software entity that composes the telecommunication network or the surrounding environment. The network itself can be considered a NE, e.g., at a certain level of abstraction, a local area network could be considered a NE of a regional network. Following the given definition of the NE, a virtual private network overlaid on a physical public network could be considered a NE, or a cell site of a cellular network is a NE, or an amplifier in a power supply unit is a NE, etc. All NEs working together (whether physically connected or not, contained one in another or not) form the network configuration model.

Each particular NE is described by its model, which is instantiated from the corresponding NE class model. Network element classes (models) form a class-subclass hierarchy. All NE classes, except the terminal classes, are mathematical abstractions of existing "real" NEs, while the terminal classes describe the types of existing NEs.

Following the inheritance paths in the class hierarchy, the constraints, attribute values, and default values of a class (parent) will be passed to its subclasses (children). There are two types of built-in constraint types in the classes: connectivity constraints and containment constraints. On the NE class level, the connectivity constraints will determine the possible connections between the NEs, while the containment constraints define the possible containment relations between the NEs. These constraints, originally defined by the domain expert, will be passed to the terminal classes of the hierarchy, and then enforced during instantiation of a NE model corresponding to the physical NE. For example, if a switch type A can be connected only to a digital cross-connect type B, then this constraint is enforced when a particular network connectivity model is constructed.

6.3 Correlations and correlation rules

On a phenomenological level, a correlation is a statement about the events happening on the network, e.g., Bad-Card-Correlation states that some port contains a faulty circuit card. On a system internal level, a correlation is an object-oriented data structure that contains its component objects and attributes. All correlations are organized into a correlation class hierarchy. The root node of the correlation class hierarchy describes the basic correlation class. The terminal nodes represent correlation classes, which will be instantiated each time particular conditions are met in the stream of incoming events.

A correlation rule defines the conditions under which correlations are asserted, e.g., if there is a red carrier group event (CGA) from a digital cross-connect (DCS), and there is a yellow CGA from another DCS, and these DCSs are connected, then Bad-Card-Correlation should be asserted. Different correlation rules can lead to the assertion of one and the same correlation.

The conditional, or so-called left-hand side (LHS), part of a correlation rule uses NEs, messages, and correlations as arguments to form the rule-firing condition. The condition can contain Boolean patterns, sequences of events based on time relations, as well as event counters. The arguments for the Boolean patterns could be the following entities:

- Parameters of event messages, e.g., alarm severity level
- Event message class types, e.g., DS0 class alarm message
- Parameters of NEs, e.g., location code
- NE class types, e.g., Class 5 switch
- Connectivity and containment statements between NEs
- Temporal relations between events
- Correlations

The subsequent application of correlation rules, instantiation of correlations, and consumption of the produced correlations by the next rule describes the event propagation process.

Figure 5 illustrates how correlations and correlation rules could be described. Let's consider the following sample situation which should be detected and reported at the operator's terminal:

A carrier group alarm type "A" happened at a time ?t1 on some NE named ?ne, and during the following 1-minute interval an expected carrier group alarm type "B" did not occur at the same NE.

The events to be correlated are alarm A (?msg1) and not alarm B (?msg2). The fact that event B did not happen is formally also an event. The additional constraints are that (1) a simple network configuration constraint that both messages are coming from the same network element ?ne, and (2) a temporal constraint that the event "not alarm B" came 60 seconds later than alarm A. The first constraint is achieved by using the same reference to the network element ?ne in both messages, while the second constraint is implemented using temporal relation AFTER.

```
Rule Name: EXPECTED-EVENT-RULE
Conditions
    MSG:    ALARM-TYPE -A           ?msg1
            NE                      ?ne
    not
            MSG:ALARM-TYPE-B        ?msg2
            NE                      ?ne
    after   TIMESENT ?t             ?msg1 ?msg2 60
Actions
    Assert: EXPECTED-EVENT-CORR
                    MSG: ALARM    ?msg1

Correlation Name: EXPECTED-EVENT-CORRELATION
Lifespan 120 minutes
Requires
    MSG:    ALARM
            INSTANCE                ?ne
            TIMESENT                ?t
Parents: BASIC-CORRELATION
Children:
Template:       ?t ?ne Expected event type "B" did not happen
                during 1 minute after the alarm type "A"
Slots
    Slot: INSTANCE                  Value: ?ne
    Slot: TIMESENT                  Value: ?t
```

Figure 5 Correlation and correlation rule for an expected event situation.

If the logical condition of the rule is true for certain events in the Dynamic Event Memory, the correlation EXPECTED-EVENT-CORRELATION is asserted to the memory and a message is sent to the operator terminal. Variable ?msg1, binding all information about the ALARM-TYPE-A message, is sent from the rule condition part to the correlation asserted in the action part. The correlation has built-in slots (parameters) to store information that could be passed to the higher level correlations that use this correlation as a component. As with NEs, correlations are organized into class hierarchies. The class references are implemented through Parents and Children relationships. The EXPECTED-EVENT-CORRELATION has one parent, BASIC-CORRELATION, and no child correlations.

7 IMPACT

The event correlation model described in the previous sections is implemented in IMPACT, a general-purpose telecommunication network alarm correlation system (Jakobson and Weissman, 1993; Jakobson, Weihmayer, and Weissman, 1994). As an example of a specific implementation of the correlations discussed in Section 2.3, we will refer to the event counting correlation. There are two operators in IMPACT that are used for counting events: Timespan and Count. The operator Timespan takes as an input an event correlation pattern and a time interval and returns the count of how many times the event pattern happened during the time interval. The function of the Count operator is opposite to Timespan: It takes as an input an event correlation pattern and a given number of event counts, and returns the time interval needed to count the pattern.

IMPACT contains three major components: Application Run-Time Component, Application Development Environment, and the Network Knowledge Base. The Application Run-Time Component monitors the network events in real-time. It performs the following functions: (1) alarm message collection and parsing, (2) event correlation, and (3) execution of external procedures (test, database access, message logging, etc.). The Application Development Environment provides powerful tools for building the Network Knowledge Base. The core of the environement consists of eight editors, with a common look and feel, which are grouped into three sets of tools: Network Configuration Tools, Alarm Correlation Tools, and Network Graphics Tools.

IMPACT has been implemented using the CLIPS expert system shell (Giarratano, 1993). The graphical user interface is programmed in Tcl/TK (Ousterhout, 1990). Many time-critical functions are written in C. The system runs on various UNIX workstations, and it is integrated with two GTE network alarm management systems, SmartAlert and ISM2000. IMPACT is currently used for a land-based telecommunication network and for cellular network alarm correlation, fault diagnostics, and calling card abuse monitoring.

TRADEMARKS

UNIX is a trademark of UNIX Systems Laboratories
SmartAlert is a trademark of GTE TSI, ISM/2000 is a trademark of GTE NMO

ACKNOWLEDGEMENTS

We thank network management personnel from GTE Mobilnet, GTE NMO, and GTE TSI for valuable domain knowledge and feedback, and Dr. S. Goyal for his continuous encouragement and support. Our thanks go also to an anonynous reviewer for many useful comments and suggestions.

REFERENCES

Allen, J.F. (1983) Maintaining knowledge about temporal intervals. *Communications of the ACM*, pp. 832-853

Davis, R., Shrobe, H., and Hamscher, W. (1982) Diagnosis based on description of structure and function. Proceedings of the 1982 National Conference on Artificial Intelligence, Pittsburgh, PA, pp. 137-142

Giarratano, J. (1993) CLIPS user's guide. NASA LBJ Space Center, Software Technology Branch.

Jakobson, G., M. and Weissman (1003) Alarm Correlation. *IEEE Network*, 7 (6), pp. 52-59.

Jakobson, G., Weihmayer, R., and Weissman, M. (1994) A domain-oriented expert system shell for telecommunication network alarm correlation. In *Network Management and Control*, Volume II, (editor M. Malek), Plenum Press, New York, NY.

Ousterhout, J. (1990) Tcl: An embeddable command language, Proceedings of the Winter USENIX Conference, pp. 133-146.

BIOGRAPHIES

Gabriel Jakobson is a Principal Member of Technical Staff at GTE Laboratories, where he has been project leader of several expert systems, intelligent database, and telecommunication network management systems development projects. He received M.S.E.E. from the Tallinn Polytechnic Institute and Ph.D. in CS from Estonian Academy of Sciences in 1964 and 1971, respectively. Dr. Jakobson is the author or co-author of more than 40 technical papers in the areas of databases, man-machine interfaces, expert systems, and telecommunication network management.

Mark D. Weissman received his BS in Chemical Engineering and his BA in Computer Science from the State University of New York at Buffalo in 1983 and 1984, respectively. He is a Senior Member of Technical Staff at GTE Laboratories, where he has been a major contributor to the development of several expert systems for network management applications.

SECTION FOUR

AI Methods in Management

Intelligent filtering in network management systems

M. Möller, S. Tretter, B. Fink
Philips Research Laboratories Aachen
Weißhausstr. 2, 52066 Aachen, Germany
Tel: +49 241 6003-{510, 552}, Fax: -519
{moeller, tretter}@pfa.philips.de

Abstract

Network management systems have to handle a huge volume of notifications reporting unprompted on events in the network. Filters that reduce this information flood on a per-notification basis fail to perform adequate information preprocessing required by management application software or human operators. Our concept of intelligent filtering allows for a highly flexible correlation of several notifications: Secondary notifications can be suppressed or a number of notifications can be aggregated. An intelligent filter was implemented using a rule-based language and was applied within SDH network management. Several modules, configurable while the filter is operating, support the user considerably and with excellent runtime performance. Further development is envisaged that provides for smooth integration into management application software.

Keywords

Network Management, Event Correlation, Filtering, Synchronous Digital Hierarchy

1 INTRODUCTION

1.1 The problem

Networked systems are growing in size and complexity, which means that a vast amount of information has to be handled by their management systems. Most of this information is produced spontaneously: Notifications report on certain events within the network, e.g. a status change of a network element or an equipment malfunction. To make effective management possible - be it performed automatically by software components or carried out by the human operator - this message flood has to be preprocessed. Such preprocessing has to correlate information from different network resources and, based on these correlations, has to suppress superfluous notifications, generate lost notifications or aggregate notifications.

So far information preprocessing is mostly performed by filter modules that reduce the information flow in a context-free manner. This means that for a single notification it can be decided whether it will be suppressed or not, depending on the information it is carrying. Correlation of information from several notifications is still left to the management application or the human operator, e.g. to identify the primary message and neglect the secondary ones when a message burst is caused by a faulty component, or to condense several messages carrying superfluous details into one with more abstract information.

Intelligent filters are software components within the management system that perform this preprocessing task. They can be used to directly support the human operator as well as to separate tasks within a management application software.

1.2 Example: notifications in an SDH network

Within telecommunication networks using the new *Synchronous Digital Hierarchy* (SDH), correlation of notifications is very important. SDH has the ability to detect faults on its different capacity levels via embedded overhead information such as check sums and trail labels. In the standard information model (ITU-T G.774) the detection capabilities of the hardware (ITU-T G.783) manifest themselves as a set of termination points representing the multiplexing hierarchy and offering hooks for a management system. Within this model each termination point is able to send notifications concerning the transmission connection it is terminating.

The example in Figure 1 shows the alarm notifications sent in case of a failed transmission line with capacity STM-1 (155 Mbit/s), which is the basic transmission rate for SDH. In the example, one initial fault causes two primary alarm notifications: Two LOS (Loss Of Signal) notifications report on a loss of the carrier signal detected by the physical interfaces. But since the STM-1 carrier is able to transport up to 63 2-Mbit/s signals, up to 254 AIS notifications (Alarm Indication Signal, propagated via in-band signalling) are also sent by the termination points of the multiplexing hierarchy down to those of the affected 2-Mbit/s signals.

The example is based on the multiplexing structure for 2-Mbit/s transmission according to the ITU-T recommendation (ITU-T G.709) as it is used in Europe. The use of STM-16 (2.5 Gbit/s) transmission lines (the highest transmission capacity currently supported) would increase the number of notifications by a factor of 16.

1.3 Requirements on intelligent filters

By studying the functional capabilities of information preprocessing needed by automatic management software and human operators we identified the following main requirements:

Filtering Functionality: The filter should be able to perform notification suppression, compression and aggregation. This means it decides on the forwarding/non-forwarding of notifica-

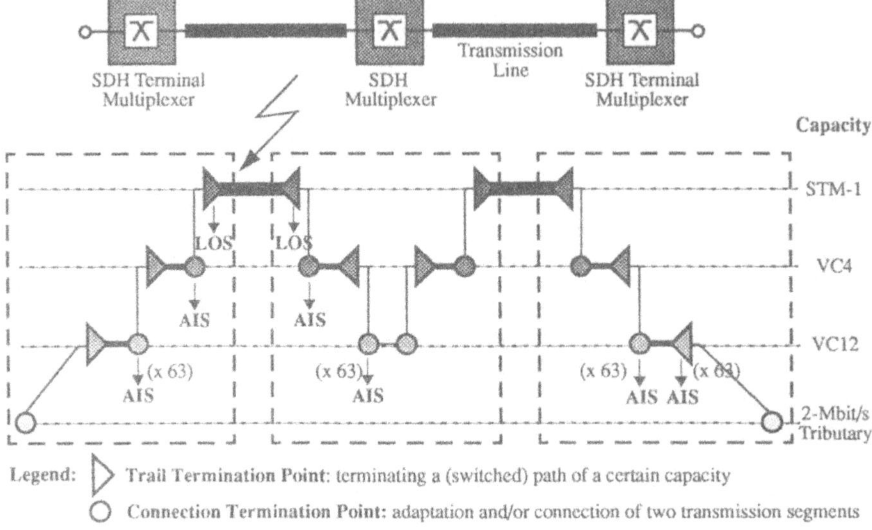

Figure 1 Example line failed on an SDH transmission line.

tions, it condenses multiple occurrence of the same notification to one and it generates notifications of higher semantical level. Suppression, aggregation and compression of notifications are based on their correlation over time and over different resources.

Ordering: The filter has to preserve the order of incoming events.
It may well be that events have overtaken each other, which can be detected by looking at their time-stamps. Nevertheless this reversed order may be of relevance for diagnosis and thus should not be corrected. The filter has to have an internal mechanism to correlate events arriving in an order deviating from the order in which they were generated.

From this and the functional requirements it follows that the filter has to have a notion of time.

Modification: A filter has to allow its modification concerning two aspects:
(1) The set of event types that shall be dealt with and the set of nodes they may come from may vary over time due to the dynamics of the networked system.
(2) Which events have to be suppressed, aggregated or generated at what time are filter parameters that may change.
The filter must allow these modifications to be made at runtime.

Performance: A filter has to guarantee a certain throughput: The time period necessary to process events (that is the decision to forward it, to suppress it or to send a generated event based on it) has to be limited.

Scalability: A filter must be applicable to networked systems of any scale and has to support the dynamic growth of the system.

So, what we call an 'intelligent filter' is a software component with three interfaces: Its *input* are notifications in a certain format, its *output* are notifications of the same format. An output notification is either generated by the filter or has been part of the input. Via the *modifi-*

cation interface the filter's behaviour can be adjusted. This definition is an extension of the discriminator object as introduced in the management standards (ITU-T X.734).

2 EXISTING APPROACHES

There are a number of significant publications on the topic of intelligent filtering (Boda, 1992, Brugnoni, 1993, Jacobson, 1993, Lewis, 1993, Pfau-Wagenbauer, 1993, Deters, 1994). As an application area all those use *fault diagnosis*. It should be noted here that of course this is the area where the most operator knowledge is available, but that in general filtering is required in other management areas as well. For example a certain network performance, deducible from several messages, is not necessarily a fault, but an item of information important for taking appropriate performance management measures. Thus we would like to apply filter components to any area where it is necessary to focus on relevant information and to discard unnecessary bits.

Most systems come as stand-alone or higher-level systems. That means that they work in addition to normal network management systems (Jacobson, 1993) or on top of them (Brugnoni, 1993). Our aim is to devise filter components that go into a management application in various quantities and at various places rather than having one big filter system. For this reason we see filters as passive components in the following sense: They work on the incoming notifications only, and do not retrieve additional information (such as attribute values) from the network. Thus a filter as such cannot perform diagnosis - the information carried by events does not in general suffice for this.

Correlation systems can have different underlying paradigms that are all taken from the area of Artificial Intelligence. Approaches that are based on Neural Networks or Case-based Reasoning rely on the fact that there is a large validated base of cases or training examples available (Lewis, 1993, Deters, 1994). However, when it comes to new technologies such as SDH, there is not enough operating experience and therefore no training set is available. Approaches that are based on models of either correct or faulty system behaviour (Pfau-Wagenbauer, 1993, Jacobson, 1993) seem easier to obtain, but have to allow for customisation at runtime in order to adapt the model to real system behaviour. We adopted the model-based, manipulative approach.

3 INTELLIGENT FILTERING

3.1 Design of the intelligent filter

The requirements for intelligent filtering led to the following design decisions:

- We chose a rule-based approach for the shallow model of notification dependences.
- Rules have to describe what shall be forwarded rather than what has to be suppressed.
- We divided the filter into modules, each of them responsible for a certain functionality and a part of the network.

- Each module is divided into three parts, describing the dependences, the topology of the network (or subnetwork) under consideration and rules for the filtering process as such. The dependences and the topology can be manipulated at runtime.
- The modules work jointly on the notification buffer, where notifications are stored for a certain time interval.

Figure 2 outlines this design. Notifications arriving from the network are preprocessed in order to obtain facts as used by the rule-based system. E.g. a message *1;27;mo.Hoern_Ac_040; mo.Hoern_Ac_040:TTP.0;comAl;191.2500;ATU.0;LossOfSignal;1* would result in the internal fact *occurred(mo.Hoern_Ac_040:ATU.0, LossOfSignal)*, stating that a loss of signal occurred at adaptation termination unit 0 of the managed object Hoern_Ac_040.

After preprocessing, the event is classified with respect to its relevance for filtering. This is done with reference to dependences and topology information contained in the filter modules or with reference to explicit discrimination constructs that are formulated on a per-event basis. This means: If a module contains rules that refer to this event, the event is stored in the notification buffer. If a discriminator construct determines that an event shall be forwarded, this event is directly passed to the postprocessing function. This case allows for context-free filtering as specified in the ITU-T standards (ITU-T X.734). If neither discriminator constructs nor modules refer to the event under consideration, the event is absorbed.

The notification buffer stores events for a predefined time-period. During this period, the different filter modules work on the buffer's content. Each module can mark events in the buffer as 'to be forwarded' or can put new events into the buffer. When the lifetime of an event in the buffer has elapsed, the event is forwarded only if it is marked accordingly. All unmarked events are absorbed.

The postprocessing function is the inverse of the preprocessing function, so that notifications leave the filter in the same format and the same order they had when they entered it.

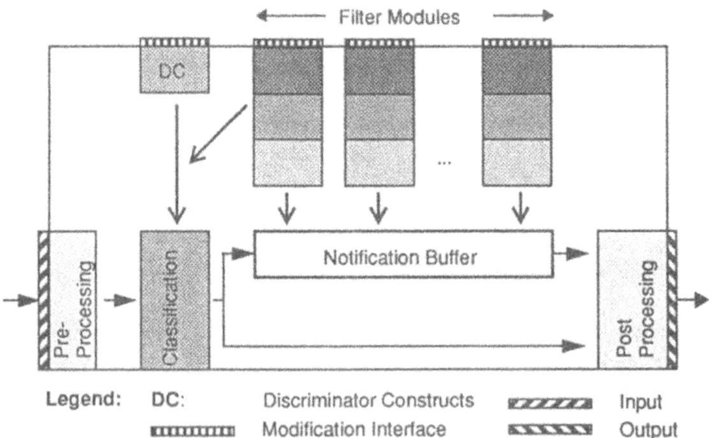

Figure 2 Intelligent filter design.

3.2 Design of a filter module

A filter module consists of 3 parts:

Topology information describe explicitly network elements and lines between them that are subject to the module's filtering functionality. Example 1 (Table 1) states that there are two adaptation termination units and that a line exists between them.

Dependences relate notifications from different nodes to each other. For instance a dependence may state that one event causes another event on the same managed object (MO, example 2 in Table 1) or that two events from different managed objects shall be aggregated to a third one (example 3).

Filtering function rules describe what the module actually performs based on topology and dependences. A causal relation as given in example 2 might lead to different actions: The *Loss of Signal* notification is forwarded and the *Alarm Indication Signal* notification is simply dropped, or the *Loss of Signal* notification is forwarded with the additional information that a certain *Alarm Indication Signal* that was a consequence of this has been dropped. This would allow the human operator to inspect the logfile later on if more detail is needed. Example 4 shows the former case, using variables for managed objects and events. Please note that all notifications that are not forwarded are absorbed by default - it is not necessary to express this.

Table 1 Examples of intelligent filter parts

Example 1: Topology Information
atu(mo.Hoern_Ac_040:ATU.0), atu(mo.City_Ac_04:ATU.3), line(mo.Hoern_Ac_040:ATU.0, mo.City_Ac_04:ATU.3)
Example 2: Causality between Events
causes(LossOfSignal,AlarmIndicationSignal)
Example 3: Aggregation of Events
aggregate(mo.Hoern_Ac_040:ATU.0, LossOfSignal, mo.City_Ac_04:ATU.3, LossOfSignal, Line_Ac_012, LineFailed)
Example 4: Filter Function Rules
causes(E_1, E_2) & occurred(MO, E_1) & occurred(MO, E_2) -> forward(MO, E_1)
Example 5: Rules with Time Annotations
(occurred(MO_1, E_1) & occurred(MO_2, E_2)/0.05)/0.01 -> forward(L, E_3)

3.3 Implementation of the intelligent filter

To implement the filter we used RTA, a rule-based language developed at Philips Research Laboratories (Graham, 1991). Within this language rules can contain time annotations that denote durations of facts or delays in firing. For example the rule given in example 5 (Table 1) states that if E_1 occurs at MO_1, and E_2 occurred at MO_2 no less than 50 ms earlier, then E_3 will be forwarded after another 10 ms.

The chosen language can be compiled. At runtime, rules can fire concurrently, and the RTA runtime system takes care of synchronisation among rules and with system time. Furthermore the language supports an interface to C so that function calls can be attached to facts. This allowed us to realize pre- and postprocessing functions as well as the classification function in C. RTA supports facts and rules being turned ON and OFF from outside. This way topology and dependences can be changed at runtime. However, 'compiled' rules mean that all possible values of variables have to be known at compilation time, so that changes at runtime can only be made within a range that is known in advance. If this range is too limited, modules have to be changed at source code level and recompiled. For this, RTA supports modules being loaded and unloaded at runtime so that a module can be exchanged while the others are still executing.

3.4 Modules for filtering in SDH networks

Each module can perform a certain filter functionality on a certain part of the network. It does so steered by dependences. In our implementation for the management of SDH networks we decided on three modules that cover the following functions:

- Simple Compression:
  All notifications of the same type from the same managed object within a given time interval are compressed into one notification.
- Causal Suppression:
  Notifications that are secondary ones are suppressed. Their identifiers are attached to the primary notification.
- Aggregation:
  Several notifications from one or more managed objects are aggregated to a new notification.

All three modules work on the entire network. However, the modules could have been configured in such a way that, for example, aggregation is only performed in one part and causal suppression in another.

3.5 The intelligent filter in use - a simple scenario

At runtime, the human operator or the management application software can configure the filter modules in various ways. A little scenario shall now demonstrate part of the overall functionality. For this, the filter is assumed to be used in a configuration as depicted in Figure 3: several managed objects are controlled by a manager (this can be a human manager or a software component). The notifications emitted by the managed objects are passed to the manager via the filter. The manager can manipulate the filter (and of course the managed objects - this is, however, not depicted here).

Let us assume that at a certain point in time no modules are loaded within the filter, and let us assume that a line fails in the network. All managed objects concerned by this will now emit notifications; since no module is loaded, the filter will not suppress anything. All notifications will arrive at the manager that has to process this information flood. The manager can now load certain modules. It starts by loading the Compression module. By default, no dependences are switched on when this module is loaded, so the manager might decide that all operational state changes to '0' and all communication alarms with the severity 'major' from the same managed object shall be compressed into one notification with the semantics 'Multiple operational state changes to '0' occurred' or 'Multiple major communication alarms occurred' at the respective node. When another line fails in the network, the manager will now receive these two notifications from each of the two managed objects that are connected by the failed line.

This type of information is not the most appropriate to diagnose the failed line. Therefore, the manager unloads the Compression module and loads the Suppression module. Some of its dependences state that when a *Loss of Signal* is reported from an MO, the same MO will send out various other notifications caused by this. In fact, when a line fails, the two adjacent MOs both send a *Loss of Signal*. With the Causal Suppression module loaded, these two messages will now be forwarded with a list of identifiers attached to each of them. The lists denote the notifications suppressed as secondary and enable the manager to look those up in the logfile should this be necessary.

In some cases only two *Loss of Signal* messages might be too little information; e.g. it might be vital to be informed about major communication alarms, even if they are secondary. For this reason the manager can switch off the dependence causes(LossOfSignal, MajorCommunicationAlarm) within the Causal Suppression module. Now the filter sends two *Loss of Signal* notifications and all major communication alarms. Of course the latter are missing in the list attached to the former.

The manager can now in addition load the Aggregation module. This effects that two *Loss of Signal* notifications are aggregated to *Line L failed* if they are emitted by two MOs that have a common line L. However, due to the semantics of the Causal Suppression module, the two *Loss of Signal* notifications are forwarded as well. To suppress these, the Causal Suppression

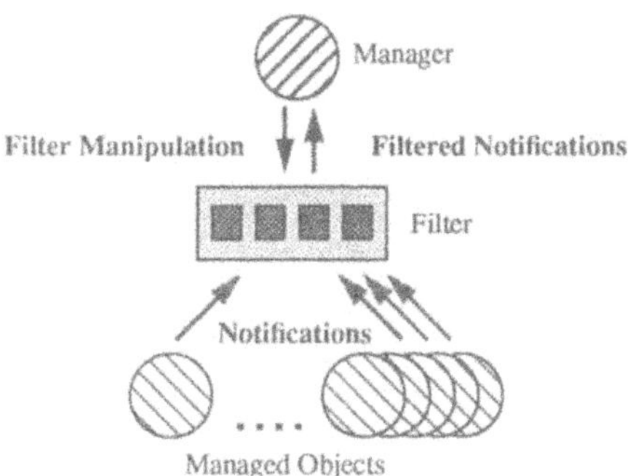

Figure 3 Filtering scenario

module has to be unloaded. The effect is that when a line fails only the message *Line L failed* is received by the manager.

This way the filter can be configured to cover various demands for information preprocessing. Should a situation occur where the filter's configuration is found to be not optimal and thus relevant information is not presented, the filter can be re-configured and re-run on the log-file off-line.

4 EVALUATION OF THE CHOSEN SOLUTION

The intelligent filter has been put to the test against an SDH network simulator and against a notification generator (Beyerlein, 1993). These experiments carried out on a SUN SPARCstation 10 under Sun OS 4.1.3 showed that the implementation is very fast: For a network with 13 network elements and 13 lines, 1000 notifications/s were sent to the filter for a period of one minute; during this minute the lag of the filter behind realtime rose linearly from 0 to 5 s. This means that during a heavy notification burst notifications left the filter not later than 5 s after they had entered it.

This excellent runtime behaviour can be attributed to the fact that the RTA language is compiled. This means the RTA compiler instantiates all rules that contain variables with possible combinations of their values. These variable-free rules can then effectively be executed at runtime. The memory space needed to do this is in the order of $(N_{MO}*T_N)^k$, where N_{MO} is the number of MOs, T_N is the number of notification types in the network under consideration and k is the maximal number of notifications that can occur in a correlation dependence. For the example mentioned above this leads to a memory consumption of 1.5 Mbyte. This means that only very few filters of this size are likely to be run at the same time.

The scenario in section 3.5 showed the high flexibility of the filter with respect to module loading and unloading as well as to setting topology facts and dependences ON or OFF at runtime. This is only possible, however, for facts and dependences that have been foreseen at compile time. For a dynamic network, though, where managed objects and relations between them are created dynamically this is not appropriate. Consider for example path objects in SDH networks: A path is a connection between two nodes that is switched via several other nodes; a path is provided to a client for a certain time period after which it does not exist any longer. Our filter would have to define beforehand as many path objects as could be present simultaneously. A more dynamic way of dealing with scalability is necessary.

Besides the fact that the filter's memory size limits the number of filters that can be integrated into management application software, the problem of how to integrate the filters at source code level has not yet been studied. So far, management application and filters are coded separately; no cross-checking can be performed before runtime. It is necessary to augment the language chosen for the development of management applications by filter constructs.

The filter is designed so as to perform multi-stage filtering (although this has not been applied in the implementation). Multi-stage filtering means that intelligent filtering is performed - again - on the filters output; e.g. aggregated messages could be aggregated another time. Two ways of implementing this can be envisaged. First, one can construct filter chains, which means directly coupling filters such that one filter's output is the next filter's input. Looking at the internal functionality (Figure 2), this would mean performing unnecessary post- and preprocessing. The second approach is to add further modules that perform higher-level filtering: These modules would work on the same notification buffer, but would only consider notifications that are deemed to be forwarded by other modules or created by them.

5 FUTURE WORK

As stated before, the implemented intelligent filter is not integrated into the management application. This is a major drawback since management applications would want to influence the intelligent filter by:

- specifying or removing filter rules according to their special needs *and*
- supplying the filter with topology information during runtime to perform event correlation.

The management applications' notification handling is currently done by the installation of *event forwarding discriminators* (EFD) in an event distributor and specialised notification handlers in the applications themselves. The EFDs allow for filtering on single notifications only. Triggered by incoming notifications, the notification handlers perform arbitrary management actions with the application's state variables and one notification's additional information as parameters. This means that for context-sensitive filtering the context would have to be coded explicitly into the application's state and that the event correlation would be mixed up with the notification handling.

The approach we envisage now is to integrate intelligent filtering into the management language used for the application creation. Briefly, this language is an extension of GDMO (ITU-T X.722) in that it also allows managing objects to be specified and makes GDMO's behaviour clauses operational (DOMAINS, 1992).

A management application that wants to correlate notifications will have to implement a filter package with application specific filter rules. These will consist of boolean expressions over facts and relations on the left-hand side. A *fact* refers to the fields of one notification only; a *relation* refers to several notifications, it can for example contain topology information. Instance variables for notifications and managed objects can be used within rules. If a rule fires, a special action for the recognized situation (stated on the right-hand side) is called with all the necessary information from the left-hand side of the rule.

Example rule for the situation *Line Failed*:

```
[(N_1.type = comAlarm) & (N_1.probableCause = LOS)]             /* fact(N_1) */
[(N_2.type = comAlarm) & (N_2.probableCause = LOS)] &           /* fact(N_2) */
line(N_1.instanceName, N_2.instanceName, L)                     /* relation(N_1, N_2, L) */
-> lineFailedHandler(N_1.instanceName, N_1.time, N_2.instanceName, N_2.time, L)
```

where N_1 and N_2 are notification variables and L is a variable for a managed object.

Topology information as referred to by relations will not be hard-coded in the filter package but will be provided by the application during runtime. Thus the application is responsible for updating the topology information according to its knowledge about the network. For the new filtering scenario see Figure 4.

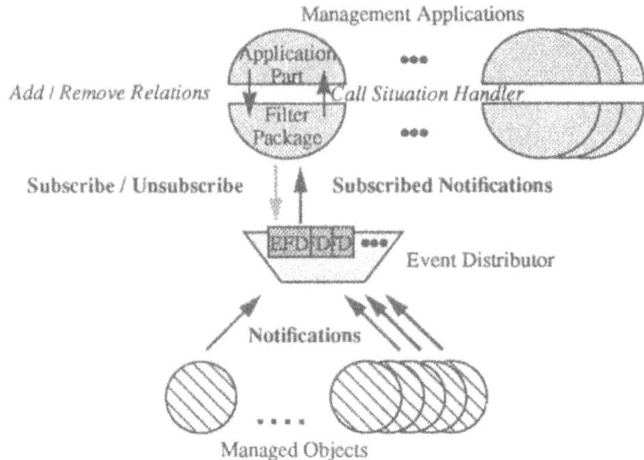

Figure 4 New filtering scenario

Consequently,

- the syntax of the application creation language will have to be extended by constructs for the specification of relations and for the filter package's rules *and*
- a runtime system has to be provided that handles a time window of incoming notifications, matches the filter rules and calls the specialised handlers.

A study should be carried out as to whether it is possible to automate the installation of EFDs in the event distributor based on the information given in the rules' facts. Then the application would be independent of the registration and only notifications that match at least one fact would be sent.

6 CONCLUSION

We have designed and implemented a powerful tool for intelligent filtering on notification streams. This has been evaluated by application to the network scenario of the Synchronous Digital Hierarchy. We have presented this application to network providers and found that there is a need for such tools and that our tool is suited for use by human operators. It can be used as a basis for professional tools enabling diagnosis and off-line logfile inspection. First concepts that allow for smooth integration of several smaller filters into our management system have been formulated. They are still to be implemented and tested.

7 REFERENCES

Beyerlein, R. (1993) Intelligent Filtering in Management Systems. Diploma Thesis Philips Research Laboratories Aachen / University of Dresden (in German).

Boda, M., Brandt, H., Gustafson, E. and Kling, L. (1992) Application of Neural Networks in Fault Diagnosis. Proc. XIV International Switching Symposium, Yokohama October 1992, pp 254-258.

Brugnoni, S., Bruno, G., Manione, R., Monatriolo, E., Paschetta, E. and Sisto, L. (1993) An Expert System for Real Time Fault Diagnosis of the Italian Telecommunications Network. Proc. IFIP 4th Int. Symp. on Integrated Network Management, San Francisco May 1993, pp 617-628.

Deters, R. (1994) Case-Based Event Correlation. Proc. 14th Int. Avignon Conference (AI 94), Paris May/June 1994.

DOMAINS (1992) DOMAINS Management Language. Deliverable D2c ESPRIT Project 5165 DOMAINS, May 1992.

Graham, M. and Wavish, P. (1991) Simulating and Implementing Agents and Multiple Agent Systems. Proc. European Simulation Multiconference, Copenhagen June 1991.

ITU-T G.709 Synchronous Multiplexing Structure. ITU-T Recommendation.

ITU-T X.722 OSI: Structure of Management Information: Guidelines for the Definition of Managed Objects. ITU-T Recommendation.

ITU-T X.734 OSI: Systems Management: Event Report Management Function. ITU-T Recommendation.

ITU-T G.774: Synchronous Digital Hierarchy (SDH) Management Information Model. ITU-T Recommendation.

ITU-T G.783: Characteristics of Synchronous Digital Hierarchy (SDH) multiplexing equipment functional blocks. ITU-T Recommendation.

Jacobson, G. and Weissman, M.D. (1993) Alarm Correlation. IEEE Network Nov. 1993, pp 52-59.

Lewis, L. (1993) A Case-Based Reasoning Approach to the Resolution of Faults in Communication Networks. Proc. IFIP 4th Int. Symp. on Integrated Network Management San Francisco, May 1993, pp 671-682.

Pfau-Wagenbauer, M. and Nejdl, W. (1993) Model/Heuristic-Based Alarm Processing for Power Systems. AI EDAM 1993 7(1), pp 65-78

The Authors

Marita Möller obtained her Diploma and Doctor's degree in Computer Science at the Technical University of Aachen, Germany. Her main areas of interest are Network Management and Artificial Intelligence.

Stefan Tretter graduated in Computer Science at the University of Kaiserslautern, Germany. He is a specialist in Telecommunications Network Management and Distributed Systems.

Barbara Fink received her Diploma in Electrical Engineering from the Technical University of Aachen, Germany, in 1967. Her key activities are architectures and computer languages.

28

NOAA - An Expert System managing the Telephone Network

R. M. Goodman and B. E. Ambrose
California Institute of Technology, Pasadena, CA91125, USA
Ph: (818)3956811 Fax: (818)5688670
email: rogo@micro.caltech.edu

H. W. Latin and C. T. Ulmer
AGL Systems

Abstract

A report is given on an expert system called NOAA, Network Operations Analyzer and Assistant, that manages the Pacific Bell Californian telephone network. Progress towards automatic implementation of expansive controls is complete. Progress towards restrictive controls is partially complete. Comments are made on current research including the use of neural networks for Time Series Prediction.

Keywords

Network Management, Telephone Network, Expert Systems, Expansive Controls, Restrictive Controls, Neural Networks

1 INTRODUCTION

Pacific Bell and Caltech have for several years been working on a real-time traffic management/expert system (Goodman, 1992, 1993). This project is called NOAA, Network Operations Analyzer and Assistant. The task of NOAA is to take information from the Pacific Bell network management computer, use it to isolate and diagnose exceptional events in the network and then recommend the same corrective advice as network management staff would in the same circumstances. A new company called AGL Systems has started up to continue the NOAA project and market it to all the Regional Bell telephone companies.

The NOAA project has several unique features:

- Provides expert system capability for complex decision making.
- Runs in real-time managing the whole of California's telephone traffic.
- Implements network controls automatically, 24 hours a day.
- Has a real-time earthquake information interface.
- Incorporates neural networks for time series prediction.

The rest of the paper gives a description of the Pacific Bell telephone network and the architecture of the Network Operations Analyzer and Assistant (NOAA) system. This is followed by sections on Expert Systems, Restrictive Controls, CUBE (Broadcast of Earthquakes), Research Aspects, and Conclusions.

2 PACIFIC BELL TELEPHONE NETWORK

In order to gain some appreciation of the network management tasks, one must have a description of the network to be managed. The Pacific Bell network in California is divided into North and South regions. Each region has a network management center associated with it. The network provides service to at least 4 million subscribers. The following network description is simplified for clarity.

The network is hierarchical. *End offices* are the exchanges that serve customers, and *tandems* are the exchanges used for traffic between end offices that are not directly connected (Bellcore, 1986). In the network as a whole, there are 15 tandems to be managed and over 400 end offices. The south is responsible for 6 of these tandems and about 200 end offices. The north is responsible for 9 tandems and about 200 end offices.

There are two types of trunk groups. *High usage* trunk groups are dimensioned to be lossy, i.e. during the busy hour they are not guaranteed to have enough capacity to carry all offered traffic. Traffic will therefore overflow onto the *Final* trunk groups which are dimensioned to provide a good Grade of Service. In general there will be a final route between each end office and its home tandem. It is these final routes that provide the backbone of the network. The final routes are therefore closely monitored by the network managers. If such a final overflows then a customer gets an 'all circuits busy - please try again later' recording. It is the goal of network managers and NOAA to eliminate such messages as much as possible.

3 NOAA ARCHITECTURE

The Architecture of NOAA is shown in Figure 1. The Pacific Bell network management system is called NTMOS (officially NetMinder/NTM OS from AT&T). NOAA is connected over an ethernet data link and appears as an ordinary operators terminal to NTMOS. NOAA then runs on a Sun workstation under UNIX. Other operations systems interfaces are planned.

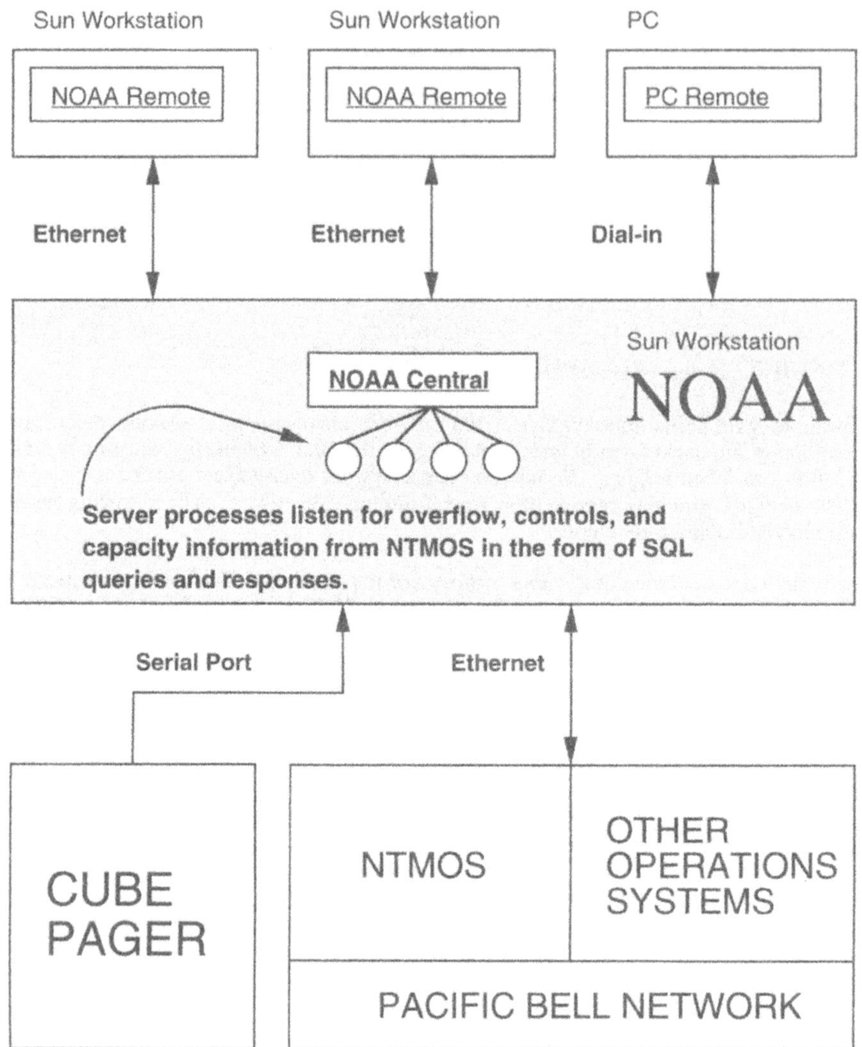

Figure 1 Architecture of NOAA.

4 EXPERT SYSTEMS

There have been other applications of expert systems to telephone traffic operations and management. For example (Sloman, 1994) lists the following among others. MAX from NYNEX and AMF from BT do fault isolation. NETTRAC from GTE and NEMESYS from AT&T do traffic management. However not all the features listed in the introduction are found in these products.

When an exception condition has been noted on a trunk route, there could be many possible explanations for it. Typically phone-ins to radio stations and TV stations may generate excess call attempts. Facilities (trunks) failures may mean that overflow shoots up on related trunk groups. Occasionally maintenance operations may interfere with the data gathering and unreliable data is returned. Random overflows can occur on individual trunk groups. Most significantly, earthquakes can cause catastrophic overflows in a metropolitan area such as Los Angeles as people instinctively try to call loved ones after a moderate quake. The demand for dial tone can exceed normal operating loads by orders of magnitude, and bring the whole network to its knees.

After diagnosing the network problem, network management staff may choose to reroute traffic elsewhere (*expansive controls*) or cut the traffic off at its source (*restrictive controls*). Currently NOAA handles expansive controls and also restrictive controls to a lesser degree.

The rules used in the program are of three separate types:

- rules that indicate which exceptions can be safely ignored. For example overflow on high usage routes is ignored;
- rules that indicate which routes can be used as candidate re-routes;
- rules that map a suggested re-route into a list of controls to effect the re-routes. E.g. certain other routes may have to be finalized first to prevent a round-robin situation. When a route is finalized, it no longer overflows onto a final route. A round robin situation is essentially a routing loop.

```
DISREGARD ANY EXCEPTIONS ON TRUNK GROUP COMMON LANGUAGES (CLLIS)
ENDING WITH "MD" (EX: LSANCA02AMD; LSANFDRCCMD).  EXCLUDE SAME WHEN
SEARCHING FOR VIA ROUTE CANDIDATES.

DISREGARD ANY EXCEPTIONS ON CLLIS INDICATING "PB" IN THE STATE
DESIGNATION (EX: OKLDPB0349T).  EXCLUDE SAME WHEN SEARCHING FOR VIA
ROUTE CANDIDATES.

DISREGARD ANY EXCEPTIONS ON THE FOLLOWING HIGH VOLUME CALL-IN CLLIS:

      HLWDCA01520 SNANCA01977 COTNCA1143A
      SIMICA11629 SNDGCA0157X
```

Table 1 Typical Network Management Rules

Some of the above rules were already written down in operators handbooks. Others were supplied by the network management staff. Examples of the rules are given in Table 1. In addition, automated rule acquisition using our ITRULE algorithm has been used to extract rules. NOAA currently contains approximately 120 rules and this number is expected to grow as interfaces to other operations systems are added.

The automatic installation of controls raises questions about how the system fares in situations that are outside the rule base. In the short term, a button is available that marks a route as a special case. Also configuration files can be tailored to prevent NOAA from dealing with certain routes. For a more permanent fix, a suggestion screen is available to the operator, and based on the operators suggestions additions are made to the rule base to allow NOAA to deal with new situations.

As with any rule based system, including a good coverage of rules in the rule base has the advantage that any rarely seen special cases are immediately recognized as special cases and appropriately dealt with. In contrast, a human operator dealing with a rarely seen special case may need to refer to handbooks and reference material before implementing a control. However for complete trust in the system, the rule base has to be extensively tested and compared with the experts analysis in a wide range of cases.

5 RESTRICTIVE CONTROLS

The work of automating expansive controls is completed to the point where NOAA is capable of automatically implementing expansive controls and indeed this feature of NOAA is taken advantage of by the network operators. The next major goal is to provide the functionality to allow restrictive controls to be automatically implemented in the same fashion.

Restrictive controls are appropriate for call-in conditions, where most of the traffic has a low probability of completion, but its presence interferes with the normal network operations. Restrictive controls are also used for earthquake situations. In an earthquake situation, 10 times the traffic that the network is dimensioned for is typically present.

Interviews have been conducted with the network management operators in an attempt to find out the action of the network management operators in response to these and other failure possibilities.

For each event, the following questions were asked:

Awareness - How does the NM operator first become aware of the problem? What NTMOS statistics might be give-aways?
Decisions - During an event, what decisions have to be made? What control options are available? Is there coordination of actions with other personnel?
Decision Support Information - What information is needed to support each of the above decisions.

The following list of failure events was considered:

- Signalling System Failure
- Transmission Cable Cut
- Switch Office Failure
- Earthquake
- Call-in Event
- Weather Event

5.1 Signalling System Failure

Signalling information is used to set up and clear calls. More recently Signalling System No. 7 (SS7) signalling has allowed more flexible routing and number translation features.

With the older Multi-Frequency (MF) signalling, the signalling information is sent on the trunk carrying the call. If the signalling runs into problems, the individual trunk group will show problems and this will be detected by NTMOS.

With the newer SS7, the signalling is carried on a separate network to a special processing node called an STP. This makes it easy to install new signalling features by changing the software at the STP. If an STP were to fail, it would be a disaster. Redundancy is therefore supplied. Each office is linked to two STPs and each STP is loaded at a maximum of 50% so that if one STP fails, the other can take over.

The exact symptoms of a signalling system problem depend on switch type. In general increased ineffective call attempts, and low holding time of calls are observed. The appropriate action is for the signalling people to fix the STP.

5.2 Transmission Cable Cut

If a cable is cut, there may be enough capacity in the network to route around the point of failure. The cable may be carrying from tens to thousands of conversations. The main indication of a cable cut would be overflow on trunk groups. However even this information may not be available if the traffic levels are low. It may be that a single cable cut can halve the capacity of the trunk group. Thus with low traffic levels, no indication of any problem may be seen.

The appropriate action is to try to reroute any overflow around the failure. If no reroute paths remain intact nothing can be done.

5.3 Switch Office Failure

A switch office failure can be caused by a number of events. There may be a fire in the location, or the power supply equipment may fail, or the switch software may perform poorly in high load situations. The tandem switches are especially important to the health of the network, because of the volume of traffic that they carry.

'Discretes' from NTMOS are a good indicator of switch problems. Discretes are updated on a 30-second interval and hence provide early warning of switch malfunctions. The machine congestion discrete and the dial tone delay discrete indicate switch problems. It may be that the problem is temporary, in which case the appropriate action is to do nothing.

With SS7, congestion limiting controls may be automatically put in place if a problem is detected in sending traffic to a particular switch. The SS7 controls need to be augmented by manual controls if there is a switch failure. The manual controls would restrict traffic entering the network if the traffic probably would not complete. The manual controls would also reroute traffic to avoid heavily congested parts of the network.

Once the situation is diagnosed and controls put in place, the next action is to call people located near to the switch to check on the state of the switch. They have the decision power for removal of the controls.

5.4 Earthquake

The magnitude of the earthquake and the closeness to populated areas make a big difference in the severity of the event from the network managers point of view. A magnitude 5.0 in Los Angeles may be more serious that a magnitude 7.0 in the Mojave desert.

For serious earthquakes, say 6.0 or more in a populated area, there are many indications of problems. The discretes will indicate machine congestion and dial tone delay from switches whose load has increased. There will be lots of trunk group overflow from all over the region as every one picks up the phone to call their in-state and out-of-state friends and relations.

The Caltech CUBE broadcast of earthquake information should provide an indication of the magnitude of the quake and the location of the epicenter.

If the network is functioning ok, the appropriate action may be to partially directionalize the trunk groups to favor outgoing calls. In this case, outgoing call attempts are favored in the battle for the available resources. Any existing reroutes are taken out. 10 times more call attempts than the network is dimensioned for are typically present.

It is the experience of the network managers that the tandem exchanges win the battle for trunk group resources more often than the end offices. If this is seen to be the case, restrictive controls are put in at the tandems to allow both tandems and end-offices equal access to the trunks. Fairness of access to limited facilities is the guiding principle.

It may be useful to implement reroutes in less affected areas.

5.5 Call-in Event

If concert tickets go on sale at 10:00am on Monday morning and there is a lot of publicity about the event, a sharp increase of network traffic may be seen. Similarly if a cable TV company suffers a cable cut and goes off the air during primetime viewing, many people will call the cable operator at the same time to complain. Most of these calls will not complete, and customers may re-dial using auto-diallers. This volume of ineffective traffic may interfere with the regular traffic by overloading the switches' and signalling systems' call processing capabilities.

This traffic is characterized by a large number of call attempts per circuit and low holding time. The tandem exchanges can provide an indication of when restrictive controls are appropriate through a hard to reach (htr) indication. This provides NOAA with information about an area code and telephone number prefix to which congestion is being experienced. NOAA can then do a table lookup to find the business that is associated with the telephone number, and place a restrictive control in all the offices in the network to cut down traffic whose destination is this number.

If a number is identified, it can do no harm to call gap the number. This won't affect calls to the number, provided the call volume is low, since its only action is to limit the number of calls accepted per 5 minute period. Even with call gaps in place, the office may be still overloaded by calls coming in from the long-distance network.

5.6 Weather Event

A weather event typically is a storm or blizzard. Weather events are characterized by a higher than usual level of traffic in the network. However the resources should still be available to handle the traffic. Thus although traffic may be 10 to 20% higher than usual, for the network managers, the only difference is a larger number of overflow exceptions to be handled. No special procedures are needed.

6 CUBE

CUBE is the Caltech / U S Geological Survey Broadcast of Earthquakes system. It provides epicenter and magnitude information of any earthquake occurring in California. In the event of a major earthquake NOAA applies a special set of rules to either scale back its recommendations or enter protective controls. Although CUBE only applies to California, the same type of system could conceivably be used to access information about other types of natural disaster, such as the National Hurricane Center's early warning system and tornado watch data.

Indications of an earthquake are first received on sensors distributed throughout California. This data is relayed to Caltech in Pasadena, where it is processed to provide epicenter location and magnitude information. Pager messages are then sent on the standard paging system to NOAA, and a data interface to the CUBE pager allows the message to be read and processed by NOAA.

Earthquake information received in real-time is displayed on NOAA's map in the form of a circle around the epicenter along with a numerical indication of the magnitude of the quake on the Richter scale. The map interface allows the operator an immediate identification of quake location and magnitude and well as identification of end-offices that may be impacted by the quake.

7 RESEARCH ASPECTS

During the course of developing NOAA, there have been opportunities for research. The involvement of the California Institute of Technology has been invaluable for investigation of these issues. Examples of the research issues that have been investigated are:

- the use of neural networks for time series prediction.
- the use of simulation to verify call saved metrics.
- the use of automated knowledge acquisition to generate rules describing correlation of exceptions.

7.1 Neural Networks

NOAA performs traffic prediction using *neural networks*. This comes under the heading of trend analysis. It can be applied to many time series found in NOAA's data structures to improve network capacity analysis and indicate potential network equipment shortages.

Neural networks have been used in applications ranging from pattern classification to associative memories. One of their main features is the ability to learn an arbitrary mapping between the network inputs and the outputs. In contrast to artificial intelligence algorithms, the learning is based on memorizing example patterns by the process of adjusting weights in the network, rather than looking up rules. Much progress has been made on the algorithms used to train neural networks (Hertz, 1991).

In this case, to aid in traffic management, the neural network was used to predict a future value of trunk occupancy on a route, based on previous readings. This provides a better indication of spare capacity for rerouting purposes and can also be used for extrapolation in the event of data not being available. The advantage of using a neural network for this application is that it can implement non-linear mappings between the inputs (in this case the previous occupancy readings) and the output (the predicted occupancy reading).

The Quickprop (Fahlman, 1988) program for network training was used as it was advertised as having faster convergence than standard backprop. The quickprop program incorporates a weight decay factor which avoids overtraining. We modified it to include linear outputs since squashing functions on the output units will not aid function fitting.

A plot of hidden unit activations gave valuable insight into the features of the data. The features that were recognized in the training set by the hidden units were traffic level and rate of change of traffic level. In particular occasional traffic spikes showed strong activation for two of the hidden units. We are researching this feature as a means of signaling unusual conditions, e.g. the start of earthquake activity. This can then be used to automatically initiate restrictive controls.

7.2 Simulation for IRR metrics

NOAA displays a running total of the number of calls saved by network controls during the course of the day.

For ORR controls, which reroute calls that overflow from a problem route, the number of calls saved during a 5 minute period is simply equal to the number of calls that overflowed from the trunk group. A correction is made for any calls that were rerouted but still failed.

For IRR controls, which reroute calls before they even attempt the problem trunk group, the number of calls saved is not so easy to derive. Instead the number depends on (i) the number of trunks in the problem route (ii) the number of trunks in high usage routes that are overflowing to the problem route (iii) the holding time of calls and (iv) the number of call attempts on the problem route. A formula was derived which gave the number of calls saved assuming a knowledge of quantities (i), (iii) and (iv). In general, quantity (ii) is difficult to obtain. Simulations showed the formula accurately estimated the calls saved over a wide range of conditions. The formula itself is based on the Erlang Blocking formula that network planners use to find the number of trunks required for a given level of traffic.

7.3 Automated Knowledge Acquisition

In the knowledge acquisition process, we have been faced with the problem of developing rules via the traditional techniques of knowledge acquisition from human experts. This is a very time consuming process in terms of human resources, particularly expert availability. We have therefore investigated various automated knowledge acquisition techniques aimed at speeding up this process. In particular we have been concerned with the automated induction of rules from network management databases. These databases include trouble ticket databases, alarms databases, and topology databases. This area of learning from examples is referred to as machine learning, and a number of statistical and neural network algorithms exist that enable rules or correlations between data to be learned. We have developed our own algorithm ITRULE (Information Theoretic RULe Engineering) (Goodman, 1992). The ITRULE algorithm possesses a number of significant advantages over other algorithms in that the rules that are generated are ranked in order of informational priority or utility. It is thus an easy matter to directly load the rules into a standard expert system shell (such as NEXPERT), utilize an inferencing scheme based on these rule priorities, and have a working expert system performing inference in a matter of minutes. We have implemented the ITRULE suite of programs on a number of platforms (Sun, Mac, PC), and linked these into a number of expert system shells (NEXPERT, KES). This approach means that the expert system developer can 'instantly' generate and run a tentative expert system with little domain expertise. This 'bootstrap' expert system can then be used to refine the rules in conjunction with the domain expert in a fraction of the time of traditional 'cold' question and answer knowledge acquisition techniques.

8 CONCLUSIONS

Over the past three years, much work has been done in interfacing NOAA to the Pacific Bell network management computer and building the infrastructure for an expert system. The rules implemented in the program have been tested by running the program on live data. The loop has been closed and NOAA now carries out controls autonomously. Clearly considerations of reliability and robustness had to be taken into account when this step was carried out. Confidence in NOAA is very high, and NOAA is regarded by network management staff as a valuable tool. In one case, where a switch had temporary problem, NOAA was able to implement 70 controls to route traffic around the switch in 15 minutes giving a much faster response than a human operator.

The ability of NOAA to diagnose problems correctly and to take the correct actions will be enhanced if the system has other information sources besides NTMOS. Two other sources being considered at present are NetMinder/NTP from AT&T which provides information about seizures of trunks, and a separate system which provides information about the SS#7 (Signalling System No. 7) signaling network.

The events of interest to the network managers are characterized by a sharp increase in traffic level or a sharp reduction in network resources. In some cases the increase in traffic level may be such that no network management controls are effective in managing the network throughput. In other cases, the scale of the event is smaller allowing re-routes or restrictive controls to bypass or reduce the problem.

There is plenty of scope for the rule-base of NOAA to be augmented to recognize these situations and take appropriate action. Some of the information to start doing this is already available from NTMOS. As interfaces to more Operations Systems become available, NOAA can begin to correlate event indications, and more effectively diagnose events.

Looking at the long term future for NOAA, the definition of a standard data format for exceptions and for statistical information about trunk group performance would help in minimizing the cost of upgrade of NOAA, as new versions of NTMOS become available. As in any network management application, standardization of data formats between applications that share the data is an important requirement. The Bellcore GR495 (Bellcore, 1993) specification of network management information transmission should go some way to filling this gap.

To summarize, network management advice is currently being generated and controls automatically implemented for the whole of the California telephone network. As the rules that generate this advice were tuned, a robust network management application was developed that relieves network management staff of most of the need to supervise the day to day running of the telephone network.

9 REFERENCES

Bellcore, Network Management Intra-LATA Network Fundamentals, *BR 780-150-122*, Issue 1, December 1986.

Bellcore, Network Management Information Transmission Requirements, *BR GR-495-CORE*, Issue 1, November 1993.

Fahlman, S. E., Faster-Learning Variations on Back-Propagation: An Empirical Study in *Proceedings of the 1988 Connectionist Models Summer School*, Morgan Kaufman, 1988.

Goodman, R. M., Smyth, P., Higgins, C. M., Miller, J. W., Rule-Based Neural Networks for Classification and Probability Estimation, *Neural Computation*, Vol. 4, No. 6, November 1992.

Goodman, R. M., Ambrose, B., Latin, H., Finnell, S., Network Operations Analyzer and

Assistant (NOAA): A real-time traffic rerouting expert system, *Globecom*, Florida, December, 1992.

Goodman, R. M., Ambrose, B., Latin, H., Finnell, S., Network Operations Analyzer and Assistant (NOAA): A hybrid Neural Network / Expert System for Traffic Management, *IFIP*, San Francisco, April, 1993.

Hertz, J., Krogh, A., Palmer, R. G., *Introduction to the Theory of Neural Computation*, Lecture Notes Vol. I, Addison Wesley, New York, 1991.

Sloman, M., *Network and Distributed Systems Management*, Addison Wesley, New York, 1994.

Dr. R. M. Goodman is a professor of Electrical Engineering in the Electrical Engineering Department of the California Institute of Technology and has been with Caltech since 1975. He holds a B. Sc from Leeds University (1968) and a Ph. D. from the University of Kent (1975).

Mr. B. E. Ambrose is currently completing a Ph. D. in Electrical Engineering at the California Institute of Technology. He holds a B. E. from University College Cork (1986) and a M. Sc. from Trinity College Dublin (1990).

Mr. H. W. Latin is a Vice President of Systems Technology with AGL Systems. Prior to co-founding AGL Systems, Mr. Latin spent 10 years with Pacific Bell in the field of network management and applications development. He holds a B. Sc. from California Polytechnic University at Pomona.

Mr. C. T. Ulmer is a Development Engineer with AGL Systems. He holds a B. Sc. (1990) and M. Sc. (1991) from the California Institute of Technology.

29

Using master tickets as a storage for problem-solving expertise

Gabi Dreo
University of Munich, Department of Computer Science
Leopoldstr. 11B, 80802 Munich, Germany
email: dreo@informatik.uni-muenchen.de

Robert Valta
Leibniz-Rechenzentrum
Barerstr. 21, 80333 Munich, Germany
email: valta@lrz-muenchen.de

Abstract

Heterogeneity and distribution of communications services and resources impose new requirements on fault management. Support staff performing fault diagnosis has to be supported with sophisticated tools, like enabling a simple and fast access to problem-solving expertise. This paper presents an approach for the storage and retrieval of problem-solving expertise by introducing the concept of a *master* ticket. The idea is to generalize information about a fault and store this information in a master ticket. Problem-solving expertise is obtained by the retrieval and the instantiation of a useful master ticket. A structure on the master ticket repository is defined by specifying relationships between master tickets, which guide the operator throughout fault diagnosis and fault recovery. The usability of the proposed concept is verified using a prototype.

Keywords

Distributed Systems, Fault Diagnosis, Trouble Ticket Systems

1 Introduction

As the heterogeneity, complexity, and distribution of communications resources, services, and applications continue to grow, the importance of being able to manage such complex environments increases correspondingly (e.g., [HeAb 94]). To cope with these requirements, new sophisticated functionalities and advanced tools to provision, manage, and maintain the network are needed. This becomes especially obvious in the area of fault management, which generally comprises fault detection, fault diagnosis, and fault recovery.

Fault management in such a heterogeneous environment has to deal with the specialization of the personnel maintaining the network, the great amount of alarms issued from a network management platform, and the ambiguous, incomplete information reported from end users in case of recognizing a trouble. Resulting potential problems are (i) difficult access to problem-solving expertise, mostly hidden in the "heads" of a few experts, (ii) the flooding of experts with events from a network management platform, and (iii) the ambiguity and incompleteness of information reported from end users.

Trouble Ticket Systems (TTSs) have been introduced to assist during all phases of fault management. Information entered and activities performed during the fault management process are documented in a trouble ticket. Basic functions of a TTS include the means for trouble ticket management and the coordination of maintenance, repair, and testing activities (e.g., [RFC 1297]). Beside the basic functions of trouble management, as described in (e.g., [ITU-T 92], [ANSI 92], [NMF 92b]), the necessity for more sophisticated functions has been recognized. For example, in [NMF 92a] the need for building knowledge databases from user experience, in [LeDr 93] the extension of TTSs to fault diagnosis, and in [VaJa 93] the deployment of group communication techniques in network management were discussed.

This paper tackles the problem of improving the general access to problem-solving expertise by introducing the concept of a *master* ticket. The idea of the master ticket concept is to generalize information about a fault and store this information in a master ticket. Problem-solving expertise for an outstanding trouble ticket is obtained by the retrieval and the instantiation of a useful master ticket. The concept of a master ticket and the relationships defined between master tickets provide a kind of a "structure" on a trouble ticket repository.

Problem-solving is a vital research topic in artificial intelligence (e.g., [Hinr 92], [Stee 90], [Aamo 91], [Koto 89]). Recently, the applicability of case-based reasoning to fault management has been investigated, for example in [Lewi 93]. The key point of this approach is to retrieve problem-solving expertise by searching for a trouble ticket which is "similar" to an outstanding ticket. The diagnostic and repair activities performed for this ticket are applied to the outstanding ticket. Difficulties of this approach are the definition of the determinators that record relevance information, and the similarity relations between trouble tickets.

The paper proceeds as follows: First, the concept and the structure of a master ticket are outlined. Subsequently, the generation and application of master tickets for the storage and retrieval of problem-solving expertise are presented. Relationships between master tickets are pointed out. In addition, we discuss the usability of the master ticket approach for the correlation of trouble tickets. A description of the prototype follows. Finally, some concluding remarks and further work are stated.

2 Master Ticket Concept

2.1 Motivation and requirements

Due to the heterogeneity of services and resources, the specialization of personnel becomes an evident problem. The motivation for the introduction of the master ticket concept results from the requirement to enable fast and simple access to problem-solving expertise. Beside this, organizational support of fault diagnosis has to be provided. When developing a concept for the access to problem-solving expertise, the integration with the network management environment has to be considered as well.

Requirements for a concept of problem-solving expertise can be structured with respect to the acquisition, storage, and retrieval of problem-solving expertise, and are as follows:

- The acquisiton of problem-solving expertise should be simple and proceed as much as possible automatically from the documented fault information in trouble tickets (cases).
- Problem-solving expertise should be stored in a *structured* library. The structure should be realized through generalized fault information. Fault-specific information is in the generalized form represented with parameters.
- The number of retrieval steps to obtain useful problem-solving expertise should be minimal.

2.2 The idea

To meet the stated requirements, we propose the concept of a *master* ticket. The idea of the master ticket is to structure the trouble ticket repository under the viewpoint of generalizing information about a fault (Fig. 1). Information, like symptoms, diagnostic activities, and repair activities is stored in a master ticket in a generalized form. Generalization means that failure-specific information, like user information, addresses of nodes or topology data contained in trouble tickets is replaced with parameters in a master ticket. An example of the information held in a master ticket would be *is_active($process, $node)*, where *$process* represents a process, *$node* the hostname or IP address of a computer system, and *is_active()* a diagnostic activity which tests whether the specified process is running on the host.

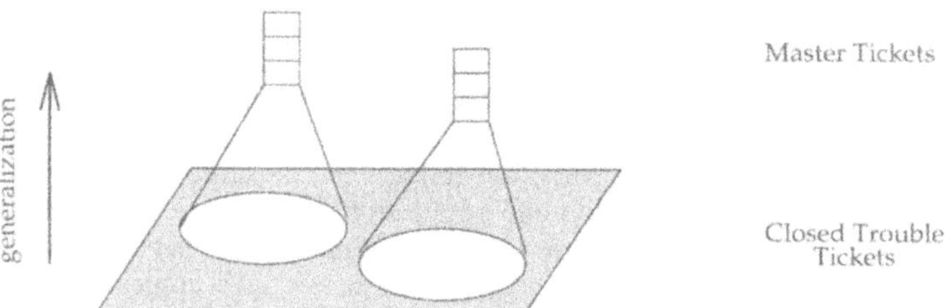

Figure 1: Master ticket concept

Retrieving problem-solving expertise is the search for an adequate master ticket. The retrieval proceeds in two steps. First, an adequate master ticket has to be determined, and second, this master ticket has to be instantiated. To instantiate a master ticket means to substitute, for example, the parameter *$node* in the previous example with an IP address and the parameter *$process* with the name of a process. Thus, the result would be to apply *is_active("named", "129.187.10.32")* as a diagnostic activity for an outstanding trouble ticket.

During fault recovery, the state of a trouble ticket switches from **open**, including only the symptom, to **closed**, including also the diagnostic activities taken, the identified fault, and the repair activities performed. If the search for a useful master ticket fails (i.e., the fault type has not yet appeared), the open trouble ticket has to be solved solely by an expert. Afterwards, the master ticket repository is updated with a new master ticket for this fault. The update of the master ticket repository proceeds also if new activities for existing faults are encountered.

To summarize, the master ticket concept consists of two steps:

1. the generation of master tickets, and the
2. application of master tickets. Subsidiary steps of the application are:
   - the retrieval of a useful master ticket, and the
   - instantiation of this master ticket.

2.3 Structure of the master ticket

Recalling that a master ticket contains generalized information about a fault, the information contained in a master ticket is as follows:

Master_ticket = [symptom(p), diagnostic_activity(p), fault(p), repair_activity(p)],

where p is an abbreviation for parameters. The first item in the master ticket is a symptom (i.e., trouble report). When considering trouble reports which are issued by end users, the symptom includes the description of the service used and whether the service (i) was not provided or (ii) not provided with the requested Quality of Service (QoS). The idea behind this classification is to decompose the symptom information into elements that allow the retrieval of a master ticket and the instantiation of a master ticket. For the retrieval of a master ticket, the service used and the classification is sufficient. However, information such as the end user who has reported the trouble and the time the trouble was recognized is of importance for the instantiation.

The parameters in the master ticket have to be substituted with concrete values. Substitution of parameters can be done in several ways:

- The operator who is diagnosing the fault retrieves the values for the parameters from the problem description provided by the end user who reported the problem.
- The operator contacts the end user to get information which cannot be retrieved from the problem description.
- The operator retrieves data from management databases, for example from an inventory system, to map a user account to the name of a user or a user location to the name of a printer.
- The operator might access the client node to retrieve client specific configuration parameters, for example the default printer.

The second item in the master ticket describes the diagnostic activity taken to diagnose the fault, which is described in the third item of the master ticket. The fourth item describes the repair activity which should be performed to recover from the diagnosed fault.

Examples of master tickets are as follows:

Master_ticket$_i$ = [
    no_printing_output
    ($client = <name of node where user starts the print job>,

$server = <name of print server used for printing>,
$printer = <name of printer>),
lpstat($client),
queuing_is_disabled($printer),
enable_queuing($printer)]
Master_ticket$_j$ = [
no_printing_output
($client = <name of node where user starts the print job>,
$server = <name of print server used for printing>,
$printer = <name of printer>),
is_reachable($client, $server),
host_is_down($server),
restart($server)]
Master_ticket$_k$ = [
telnet: Connection_timed_out
($client = <name of node where user starts telnet session>,
$server = <destination node for telnet session>),
is_reachable($client, $server),
host_is_down($server),
restart($server)]

If now a user describes a problem, like *"When I try to print a report from my workstation sun12 at our department printer, there is no output"*, such a problem would first lead to the retrieval of Master_ticket$_i$. After its instantiation, the diagnostic activity *lpstat(sun12)* is performed. If this activity reveals no problem, another master ticket — Master_ticket$_j$ — is retrieved. The diagnostic activity *is_reachable(sun12, sun-department)* is performed, which shows that node *sun-department* crashed. Thus, the problem can be solved by restarting node *sun-department*.

An important design issue regarding the contents of a master ticket is the level of specificity used to describe the diagnostic and repair activities. In the above example, a diagnostic activity is stated as *is_reachable($client, $server)*. Such a statement leaves some freedom about how reachability between a client node and a server node is really tested (e.g., using a ping or a traceroute command or checking the status information provided by a management station). However, an activity could be specified more precisely if required, which would decrease the presumed level of expertise for staff members.

2.4 Relationships between master tickets

In today's distributed environments many end user services rely on a hierarchy of underlying services. A distributed application (e.g., *remote printing*) depends on client, server and gateway processes, which themselves depend on system software and hardware. For communication between processes a transport network is required. Transit networks, networking devices and communication links must be properly configured and in operating state. Furthermore, many services rely on other distributed services, e.g., name resolution provided by a distributed name server. This hierarchical structure affects fault diagnosis because the underlying service hierarchy can be tested in a top-down or bottom-up strategy to isolate a fault whenever a problem is reported.

For our master ticket approach this has several consequences. We have to avoid a complete, exhaustive diagnosis of a service-related problem within a single master ticket for that service, because that would lead to a high redundancy (i.e., testing the transport network would be represented in all master tickets for distributed services). Instead, we not only provide master tickets for user services but also for the underlying services within our service hierarchy. As easily recognized, the service hierarchy implies a corresponding hierarchy between master tickets for the different services. For example, if a service *A* relies on a service *B*, applying master ticket *A* might lead us to the conclusion that the problem might be caused by service *B*. Thus, we can start to work on that problem by using the master ticket for service *B*.

This raises the question of how relationships between services – and thereby relationships between master tickets – should be handled within our master ticket approach:

1. Based on a framework for distributed applications we can model a service hierarchy and derive a corresponding model for our master tickets. An example of such a framework is presented in [HNG 94], which consists of application services, application-oriented services, basic distributed services, and communications services.

2. We can define relationships between master tickets in a more pragmatic way according to the procedures followed during fault diagnosis.

We decided to choose the second approach because experience shows us that it is rather difficult to define a common service architecture for an existing heterogeneous environment.

In general, the process of fault diagnosis is iterative. The availability or quality of a service is tested by testing the availability or quality of the underlying services. Testing itself is in many cases nothing else but trying to use an underlying service. In such a case the tester behaves like a normal user of the underlying service. Master tickets are therefore related by interpreting diagnostic activities as usage of a service. Relationships between master tickets are defined as follows:

- A diagnostic activity within a master ticket is interpreted as usage of a service (i.e., ping as a diagnostic activity is interpreted as usage of an IP reachability service).

- Failure of a diagnostic activity leads to a new trouble ticket, called **Internal Trouble Ticket (ITT)**, which can be further diagnosed by searching for a new master ticket.

To make sure that the diagnosis process terminates, we distinguish between

1. **Core master tickets**, which contain a fault and a repair activity.
   If the diagnostic activity of a core master ticket fails, we immediately know the fault and how to repair it (e.g., if the diode labeled *cpu* on a router's front panel is red, the cpu board is malfunctioning and has to be replaced).

2. **Relational master tickets**, which do not contain a fault and a repair activity.
   If the diagnostic activity of a relational master ticket fails (e.g., brouter *bro4cz* could not be reached), we have not yet identified the fault. We have to continue with the diagnosis process by creating a new internal trouble ticket which is further diagnosed by retrieving a new master ticket. Thus, relational master tickets are only "pointers" leading to other relational master tickets or finally to a core master ticket (Fig. 2).

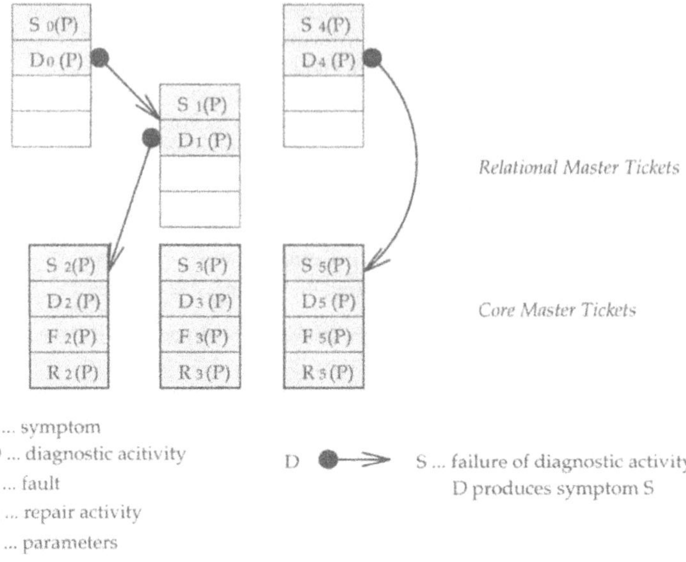

Figure 2: Relationships between master tickets

3 Generation of Master Tickets

3.1 Requirements on the trouble ticket structure

The structure of a master ticket depends to a great extent on the structure of a trouble ticket. The term "structure" of a trouble ticket means the set of fields in the trouble ticket schema, and the set of predefined selection values for each field. A trouble ticket is completely structured if information about a trouble is entered completely via predefined selection values. Requirements on the structure of a trouble ticket are stated from

1. users or help–desk staff who prefer free–form text when describing a problem and how it was solved, and

2. the procedure for the creation of new master tickets, which requires formalized and structured trouble tickets.

These requirements are almost opposite to each other. Thus, an extensive analysis of the trouble ticket structure, still acceptable by the users of a TTS, but supporting also the master ticket concept is of great importance.

Our experiences, gained in one year of usage of TTSs at the computing center, have shown that the acceptance of a TTS by the users depends to a great extent on the efficiency and speed of entering information about a problem. A desire is that the information entered should be precise, complete, and as unambiguous as possible. Unfortunately, personnel documenting the reported problems just want to enter the information as it is reported, and do not want to structure it. There are various reasons for this, like lack of time, knowledge or experience.

Realizing these problems we have provided support to the personnel by enabling a lot of information to be entered automatically by the system. For example, an assignee for an open trouble ticket is determined automatically according to the service specified and availability. We are developing a hypertext based tool, called "Intelligent Assistant", which provides very flexible and fast access to various databases, and guides the operator during the entering of information.

To fill the gap between the structure of a trouble ticket as required by the support staff and as needed by the master ticket concept, a formalization of a trouble ticket is necessary. The formalization function transforms a user trouble ticket, containing free-form descriptions, to a formalized user trouble ticket used further in the master ticket concept. Parsing the free-form description of the symptom should be performed with sophisticated lexical text analysis. If not stated explicitly otherwise, we are considering only formalized trouble tickets for the remainder of the paper.

The structure of a formalized user trouble ticket as required by the master ticket concept is shown in Fig. 3.

```
Symptom
        Service: (selection values);
        Classification: (no_service, QoS_problem);
        User: (site, location, etc.);
        Time: (time the user has recognized the trouble);
        Description: (free-text);
Diagnostic activities
        Activity(s): (selection value);
        Activity_parameters: (set of objectIds);
Fault
        Fault: (selection value);
        Fault_parameters: (set of objectIds);
Repair activities
        Activity(s): (selection value)
        Activity_parameters: (set of objectIds);
```

Figure 3: Structure of a formalized trouble ticket

3.2 Generation procedure

The generation of master tickets proceeds in two steps:

- generation of core master tickets based on product descriptions (e.g., in case a new device or application is incorporated in the network), and
- generation of core or relational master tickets based on closed trouble tickets.

The first step is performed by experts analyzing the documentation of the products and identifying the documented faults, diagnostic and repair activities.

If during the retrieval of a master ticket no useful master ticket could be obtained, an expert has to proceed with fault diagnosis without access to problem-solving expertise. During fault recovery he documents all performed diagnostic activities in the current trouble ticket. After fault recovery, the update procedure is started to generate master tickets (relational and core) for this closed trouble ticket. The updated procedure is as follows:

1. First, it is checked if a core master ticket exists for the fault diagnosed in the closed trouble ticket. If this is true, new diagnostic activities must be added to the master ticket repository by defining new relational master tickets. Note, this situation occurs if a new symptom or diagnostic activity is identified for an already documented fault.

2. In case a core master ticket could not be identified for the diagnosed fault, a new core master ticket has to be generated. Part of the information contained in the closed trouble ticket (e.g., the diagnostic activities identifying the fault, the fault itself, and the repair activities) is included in the core master ticket. The symptom, and the diagnostic activities leading to the core master ticket are included in the relational master tickets. During the generation of the relational master tickets, it is checked whether some of them already exist.

Concrete values, like IP addresses of nodes, in the closed trouble ticket are replaced with parameters in the master tickets.

4 Application of the Master Ticket Approach

For the application of the master ticket approach the key points are the efficiency of the usage and the acceptance of the concept. Accessing problem-solving expertise in the proposed concept means (i) to retrieve a useful master ticket, and (ii) to instantiate this master ticket.

The retrieval of a useful master ticket can be performed with more or less sophisticated methods. The easiest way is simple pattern matching between the symptom contained in the open trouble ticket and the symptoms contained in the master tickets. Thus, for a given trouble ticket (TT_1) with a symptom $S_1(V)$, master tickets $MT_1, \ldots, MT_n$ with the same symptom information are retrieved. Then, for each master ticket MT_i, i=1 ... n, the following steps are performed (Fig. 4):

1. All parameters of the master ticket are substituted.

2. The diagnostic activity D_i of master ticket MT_i is executed with all parameters replaced by the previously determined values.

3. If the diagnostic activity does not fail, i.e., it gives us no indication of the cause of the problem, the next master ticket is worked on.

4. If the diagnostic activity fails, we have to check whether a fault is defined for this diagnostic activity:

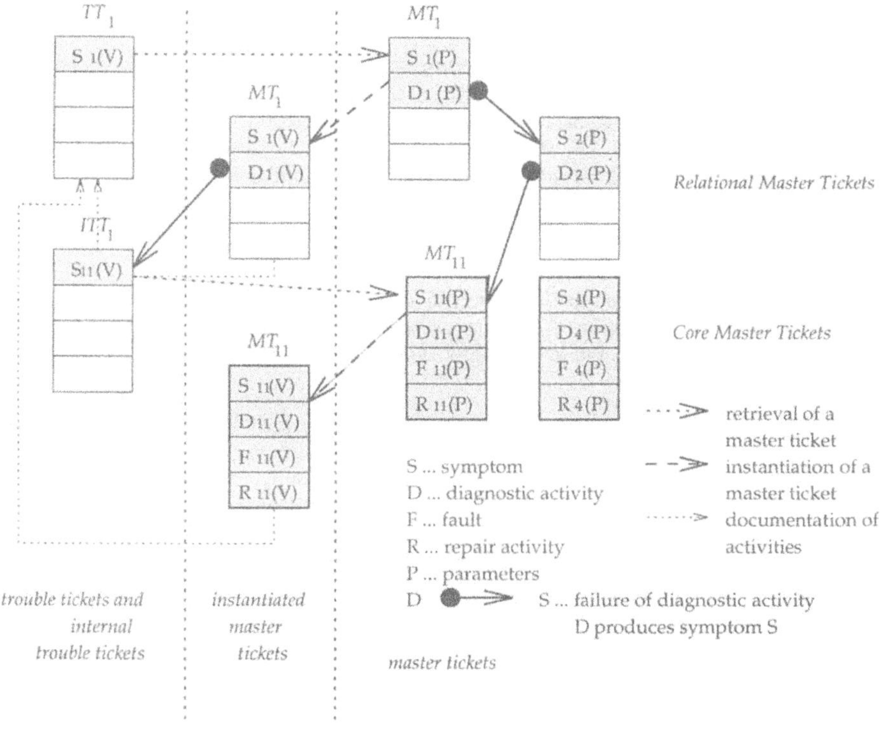

Figure 4: Application of the master ticket concept

(a) If there is no fault associated with the diagnostic activity, a new internal ticket ITT_1 which describes the negative test result as a failure of the usage of the underlying service is created.
The new internal ticket ITT_1 is then diagnosed by searching for a corresponding master ticket (e.g., MT_{11}) for the indicated service failure.

(b) If there is a fault (and a repair activity), we instantiate the fault and the repair activity. The repair activity is presented to the support staff and can be executed. The algorithm terminates.

5. If no master ticket could be retrieved, no problem-solving expertise is available for the symptom. An expert has to proceed with fault diagnosis on his own. He documents all performed diagnostic and repair activities and the identified fault in the current trouble ticket. This information is later used to create a new master ticket for the currently unknown symptom. Furthermore, the range of available diagnostic activities should be offered to him as a help.

5 Correlation in the Master Ticket Approach

In addition to recording problem-solving expertise, the master ticket approach provides a mechanism for correlating trouble tickets. Correlation is defined as the grouping of trouble tickets that are associated with the same fault. A benefit of correlation is that it prevents multiple diagnoses of the same fault.

During the application of the master ticket graph the fault diagnosis process can produce a sequence of internal trouble tickets (Fig. 4), like

$$TT_1 \rightarrow ITT_{11} \rightarrow ITT_{12} \rightarrow ITT_{13}$$

where TT_1 means an open trouble ticket, and ITT_{1i} i=1, ..., 3 are the internal tickets obtained during retrieval and instantiation. Assuming another user has reported a trouble some minutes later than the first one, then the associated sequence of internal trouble tickets would be as follows:

$$TT_2 \rightarrow ITT_{21} \rightarrow ITT_{22} \rightarrow ITT_{23}$$

The sequence of internal trouble tickets provides traces of the fault localization process. If during fault diagnosis common internal trouble tickets can be identified (e.g., $ITT_{12} = ITT_{23}$), then the originating trouble tickets TT_1 and TT_2 can be considered to be correlated. The comparison of sequences of internal trouble tickets is performed solely on a syntactical basis.

If such common internal trouble tickets could be identified, it can be decided to continue work only on one sequence of internal trouble tickets. The most promising way is to continue work with the sequence including information which have been reported from a person with high domain knowledge.

The proposed approach provides a simple but efficient method to correlate new incoming trouble reports with existing tickets. The existing tickets may or may not be already in the process of fault diagnosis.

6 Design of *MASTER*

The master ticket concept is currently implemented in a prototype, called *MASTER*, on the Application Programming Interface of the Action Request System from Remedy (version 1.2). The ARS is used by the hot line of the computing center and for research purposes at the university. The runtime environment of *MASTER* is shown in Fig. 5.

The core of *MASTER* are the programs for the text analysis, generation, instantiation, and retrieval of master tickets using the ARS API.

We use the following schemas: the *trouble ticket* schema, the *formalized trouble ticket* schema, the *internal trouble ticket* schema, and the *master ticket* schema. The *trouble ticket* schema is used by the hot line of the computing center to document trouble reports. The implementation of the formalization function is currently based on lists of negative and positive keywords. The formalized trouble ticket schema is presented to an operator as a proposal who can check the validity of the formalization. A more sophisticated text analysis could minimize the interventions of the operator. The retrieval and the instantiation of master tickets are implemented with the available ARS mechanisms, like active links or macros, and programs using the ARS API.

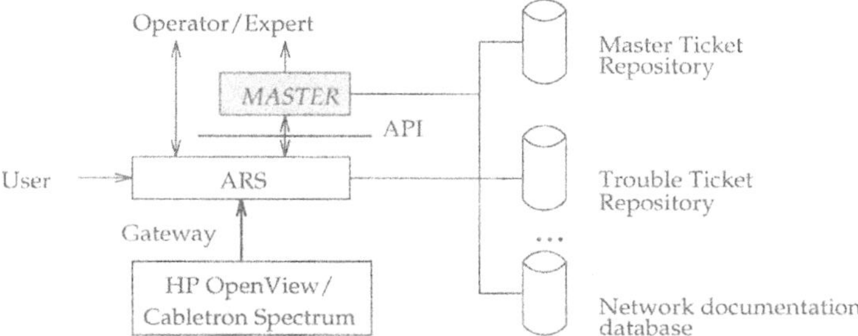

Figure 5: Environment of *MASTER*

First experimental results with the prototype have shown promising results. Of course, an extensive usage of the prototype at the computing center will answer the question whether the system will render fault management more efficient and less time-consuming.

7 Conclusions and Further Work

Heterogeneity and distribution of communications services and resources impose new requirements on fault management. Due to the specialization of the personnel maintaining the network, the access to problem-solving expertise is a vital research topic in fault management. In this paper, a solution for this problem is presented by introducing the concept of a *master* ticket. The idea of the master ticket approach is to generalize information about a fault and store it in a master ticket. Problem-solving expertise is obtained by the (i) retrieval of a useful master ticket, which is based on the procedure followed during fault diagnosis, and (ii) the instantiation of the useful master ticket.

Our further work will concentrate on the (i) development of various tools, like a tool for supporting the generation of master tickets, (ii) the feasibility of using a common service hierarchy to implement the relations between master tickets, and (iii) testing the prototype extensively at the computing center.

Acknowledgements

The authors wish to thank the members of the Munich Network Management (MNM) Team for helpful discussions and valuable comments on previous versions of the paper. The MNM Team is a group of researchers of the Munich Universities and the Bavarian Academy of Sciences. It is directed by Prof. Dr. Heinz-Gerd Hegering. We gratefully acknowledge in particular Bernhard Neumair, Victor Apostolescu, and Anja Schuhknecht, who provided valuable suggestions and advice.

References

[Aamo 91] A. Aamodt, *A knowledge-intensive approach to problem solving and sustained learning*, Ph.D. dissertation, University of Trondheim, 1991.

[ANSI 92] ANSI, *Operations, Administration, Maintanance, and Provisioning (OAM&P) - Extension to Generic Network Model for Interfaces between Operations Systems across Jurisdictional Boundaries to support Fault Management - Trouble Administration*, T1M1.5/92-01R2, 1992.

[HeAb 94] H.-G. Hegering and S. Abeck, *Integrated Network Management and System Management*, Addison-Wesley, September 1994.

[Hinr 92] T.R. Hinrichs, *Problem solving in open worlds*, Lawrence Erlbaum Associates, 1992.

[HNG 94] H.-G. Hegering, B. Neumair and M. Gutschmidt, "Cooperative Computing and Integrated System Management — A Critical Comparison of Architectural Approaches", *Journal of Network and Systems Management*, 2(3), October 1994.

[INM-III 93] H.-G. Hegering and Y. Yemini, editors, *Proceedings of the 3rd IFIP/IEEE IInternational Symposium on Integrated Network Management, San Francisco*, IFIP, North-Holland, April 1993.

[ITU-T 92] ITU-T, *Trouble Management Function - An overview*, Question 24/VII, 1992.

[Koto 89] P. Koton, *Using experience in learning and problem solving*, Ph.D. dissertation, Massachusetts Institute of Technology, 1989.

[LeDr 93] L. Lewis and G. Dreo, "Extending Trouble Ticket Systems to Fault Diagnostics", *IEEE Network Special Issue on Integrated Network Management*, 7(6):44–51, November 1993.

[Lewi 93] L. Lewis, "A Case-Based Reasoning Approach to the Resolution of Faults in Communications Networks", In [INM-III 93], pages 671–682.

[NMF 92a] "ISO/CCITT and Internet Management: Coexistence and Interworking Strategy", Issue 1.0, Network Management Forum, October 1992.

[NMF 92b] "Application Services: Trouble Management Function", Issue 1.0, Network Management Forum, August 1992.

[RFC 1297] IAB, *NOC Internal Integrated Trouble Ticket System, Functional Specification Wishlist*, RFC 1297, January 1992.

[Stee 90] L. Steels, "Components of expertise", *AI Magazine*, 11(2):29–49, 1990.

[VaJa 93] R. Valta and R. de Jager, "Deploying Group Communication Techniques in Network Management", In [INM-III 93], pages 751–763.

Biographies

GABI DREO received B.S. and M.S. degrees in computer science from the University of Maribor, Slovenia. Currently, she is a Ph.D. student at the University of Munich and a member of the Munich Network Management team, directed by Prof. Dr. Heinz-Gerd Hegering, where she does research on integrated network and system management.

ROBERT VALTA received the degree of a Diplom-Informatiker in 1984 and the degree of a Dr.rer.nat. in 1990 both from the Technische Universität in Munich. He was a research staff member at the department of Computer Science of the Technische Universität and at the Leibniz-Rechenzentrum in Munich. In 1994 he joined Softlab GmbH where he is engaged in several network and system management projects.

SECTION FIVE

Panel

30

Management of Cellular Digital Packetized Data (CDPD) Networks

Moderator: *Jock EMBRY,* Opening Technologies, U.S.A.

The Cellular Digital Packet Data (CDPD) Network extends existing data networks to mobile data devices, by using radio channels and cell sites already in place for Advanced Mobile Phone Service (AMPS). Currently being deployed throughout North America and other regions, CDPD services will enable a wide variety of applications for wireless users, such as e-mail, dispatching, mobile query, portable point-of-sale terminals, etc. The CDPD Specification calls for both existing technology, such as off the self routers, and new network elements unique to CDPD. The management part of the CDPD Specification is based on OMNIPoint 1, and adds additional ensembles and managed objects specific to CDPD.

This panel will discuss the issues and challenges associated with managing the CDPD Network, such as agent deployment, integration with existing management systems, tradeoff between proprietary and standards based solutions, and interoperability between service providers.

SECTION SIX

ATM Management

31

Object-oriented design of a VPN bandwidth management system

T. Saydam[a], J.-P. Gaspoz[b], P.-A. Etique[b], J.-P. Hubaux[b]

[a] *University of Delaware, Newark, DE. 19716, USA, tel. (1) 302 831 27 16, fax (1) 302 831 84 58, e-mail: saydam@cis.udel.edu*

[b] *Swiss Federal Institute of Technology (EPFL), Telecommunications Laboratory, 1015 Lausanne, Switzerland, tel. (41) 21 693 5258, fax (41) 21 693 2683, e-mail: gaspoz@tcom.epfl.ch*

Abstract

This paper describes the application of a general purpose object-oriented software engineering method to the design of a bandwidth management system for ATM-based virtual private networks (VPNs). Such a system allows a VPN customer to dynamically modify the bandwidth allocated to VPN connections. The design process has focused on the service management information model and interfaces required to provide that service to the customer. Object interaction graphs have been designed and class descriptions have been derived. Finally the VPN customer, value added service provider and network providers service management system interfaces have been designed and corresponding primitives are given.[1]

Keywords

VPN, ATM, TMN, object-oriented design, service management, bandwidth management

1 INTRODUCTION

One of the major trends in the evolution of current business information networking is an increasing need for high performance data communications, especially in the wide area. Provided as an alternative to dedicated leased lines networks, virtual private networks (VPNs) are gaining more and more acceptance among customers and network providers. VPNs permit to connect physically separated business sites without using dedicated resources.

The principal applications to be supported by future VPNs, that is, LAN interconnection and emerging multimedia applications, require the use of a flexible networking technology supporting a variety of services with very different quality of service requirements, in other

[1] Part of this work has been performed in the framework of the RACE project R2041 PRISM and thus has been funded by the 'Office Fédéral de l'Education et de la Science' (OFES, Switzerland)

words ATM (Asynchronous Transfer Mode). This paper will thus focus on ATM-based VPNs and more precisely on an open and very important issue in such an environment, namely bandwidth management. Indeed, multimedia applications have very different and often unpredictable bandwidth requirements which may vary over time. Moreover, ATM networks require, in general, resources to be reserved for each connection established over the network. Therefore, bandwidth management mechanisms would be very useful for the customer subscribing to the VPN service over ATM as a way to optimize resources usage and cost.

The main goal of this paper is to design a bandwidth management service, provided as an enhancement to the basic VPN service, and that allows the customer to dynamically modify the bandwidth allocated to VPN connections. A second generation object-oriented method called Fusion (Coleman, 1994) has been chosen for design purposes in order to provide a consistent approach, promoting reusability and scalability along the system design process. This design is based on the corresponding object-oriented analysis presented in (Gaspoz, 1994).

2 ATM-BASED VPN

A VPN allows to build a logical private network by using the physical public network infrastructure instead of dedicated network resources (e.g. leased lines). The service is offered as an extension and/or an alternative to a company's own network and aims at offering economic advantages as well as meeting ever changing customer needs and requirements.

ATM is a packet oriented transfer mode based on fixed length cells. It provides a non-hierarchical structure in which, cells belonging to different applications are transported commonly, independent of bit rate and burstiness. Multiplexing and switching may be performed at two levels: the virtual channel (VC) level and the virtual path (VP) level. As ATM is intrinsically a connection oriented service, communications between VPN users will be realized by Virtual Channel Connections (VCCs). This includes in general the allocation of the required resources on the user access and within the network.

The concept of virtual path allows the grouping of a set of virtual channels into a 'pipe'. VP cross-connects systems treat such bundled channels as an entity, regardless of the constituting virtual channels. In these systems virtual path connections (VPCs) are semi-permanently allocated between endpoints, thus allowing a simple and efficient management of network resources. When the cross-connected network handles connections between end nodes belonging to the same customer, it offers a virtual private network service.

The provision of VPN services over Virtual Path networks is mentioned several times in the literature (Wernik, 1992), (Verbeeck, 1992), (Gaspoz, 1994). Most of these papers refer to VPNs based on semi-permanent VPCs. In the same way, the broadband multimedia VPN considered in this paper is built by connecting each customer premise network (CPN) to every other, with the help of one or several semi-permanent virtual path connections, thus forming a logically fully meshed virtual private network, based on one or more physical networks.

3 VPN MANAGEMENT ARCHITECTURE

To support the provision of the bandwidth management service and more generally of VPN related customer network management services in an heterogeneous environment, an open and standardized management architecture based on the TMN layering principles has been considered (Figure 1). Figure 1 shows how different management systems interact with each other and with the underlying networks and network elements, to monitor and control the network resources, as well as to provide, enhance and offer network services.

According to (M.3010, 1992) the element management layer manages subsets of network elements on an individual basis and supports an abstraction of the functions provided by the network element layer. The network management layer has the responsibility for the management of all the network elements both individually and as a set. Service management is responsible for the implementation of the contractual aspects of services that are being provided to customers. Management services are provided to the customer in a client/server way. The VASP-SMS acts as a server with regards to the customer NMS (client) and as a client to the services provided by the network providers NMSs.

In the following chapters, the design efforts will focus on the management systems in the upper box, namely, the information model and the functionalities of the VASP-SMS as well as its interactions and interfaces with the CPN- and NP-NMSs in a bandwidth management perspective. To facilitate service layer information modeling, an abstract model of the VPN service under study has been established (Gaspoz, 1994). Some of its constitutive concepts are illustrated in Figure 1. For instance, a *virtual private line* is defined as a VPN end-to-end logical link connecting two CPNs and supporting the connections established between these CPNs. A *segment* is the part of a virtual private line belonging to one single management domain.

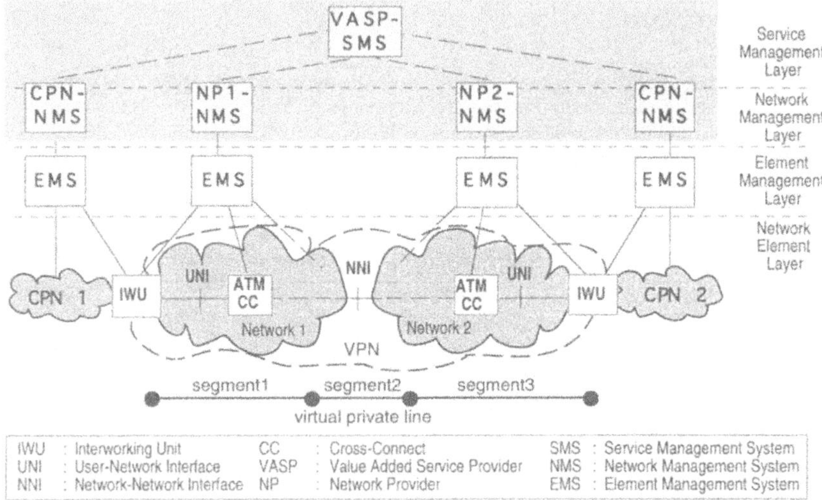

Figure 1 VPN Management Architecture.

4 VPN BANDWIDTH MANAGEMENT

4.1 Motivation

Our principal motivation in this paper is to specify and design a bandwidth management system to allow the end-users manage their bandwidth requirements. Bandwidth management plays a central role in ATM networks due to the great bandwidth access and transfer flexibility offered by this technology. From the network operator's point of view, this issue often refers to mechanisms used to protect the network against misbehaving users and avoid congestion.

Considered from the customer's point of view bandwidth management aims at optimizing bandwidth utilization. This is particularly true in an ATM context where resources have to be reserved for each connection to guarantee the required quality of service (QoS). A crucial issue in this context is to achieve the dual, yet often contradictory, goal of ensuring a high utilization of the reserved resources, while maintaining a sufficient QoS to the individual connections. The use of a bandwidth allocation scheme providing an optimal compromise between statistical multiplexing gain and loss rate is certainly of major importance in this respect. For this purpose, dynamic bandwidth management allows the user to specify the resources needed by a connection (VCC) as well as to renegotiate them during the lifetime of the connection.

4.2 Bandwidth allocation and enforcement

It results from the specification of the VPN and its related actors that bandwidth will be allocated and enforced at two different levels, the VPC level and the VCC level, in our example under the responsibility of the network providers and the service provider, respectively. Indeed, the network providers will sell VPC bandwidth to the service provider and will enforce that bandwidth to ensure the contract agreements and prevent network congestion. The service provider will in turn sell that bandwidth to the customer, but to ensure the QoS of the individual connections, bandwidth enforcement will have to be performed at the VCC level as well. Normally three traffic descriptors parameters are required for bandwidth allocation at that level, namely, peak rate, mean rate and maximum burst size.

5 OBJECT-ORIENTED BANDWIDTH MANAGEMENT DESIGN

The main focus of this study is to specify and primarily design the service management layer object classes required to provide a dynamic bandwidth management service to the customer. The interactions between the customer and the SMS (Service Management System) are only considered from a bandwidth management point of view. The object-oriented specification and design of the bandwidth management system follows the Fusion method (Coleman, 1994).

5.1 Requirements of bandwidth management

The bandwidth management system will allow the customer to monitor and dynamically control the bandwidth allocated to a VPN connection. In order for the service provider to satisfy most of the customer requests directly (i.e. without requiring from the network providers to update the virtual private line bandwidth, each time one of its connections is modified), as well as to limit the frequency of network resources reservation and release requests, some spare capacity is foreseen at the virtual private line level. Thus, when a connection is released or when the bandwidth of a connection is decreased, the amount of aggregate bandwidth that will actually be released will depend on the spare capacity available at that time.

The connection bandwidth may be increased directly if there is enough spare capacity on the virtual private line supporting that connection. If the spare capacity is smaller than the requested amount, the system transparently attempts to reserve more network resources for each segment composing the virtual private line. A request will thus be issued to each corresponding network provider to increase the bandwidth of the virtual path connection represented by each segment. The virtual private line will only be updated if all the reservation requests have been accepted.

5.2 Object model

The object model defines the static structure of the information in the system, i.e., the classes and their relationships specified to accomplish a certain task. The object model in Figure 2 is centered on a connection bandwidth update request by the customer. Each class is represented by a box with the name of the class at the top and the attributes in the lower part of the box. Relationships are represented as diamonds joined to the participating classes by arcs.

Aggregation is represented by nesting the component class into the box of the aggregate class. A number, a range, zero or more ('*'), one or more ('+') are allowed cardinality constraints.

As illustrated in Figure 2, both the *VirtualPrivateLine* and the *Connection* have a *VplBw* and a *ConnectionBw,* respectively. This 'has a' relationship is modeled as an aggregation representing a logical rather than a physical containment. For a complete treatment of object models and other specification details please refer to (Gaspoz, 1994).

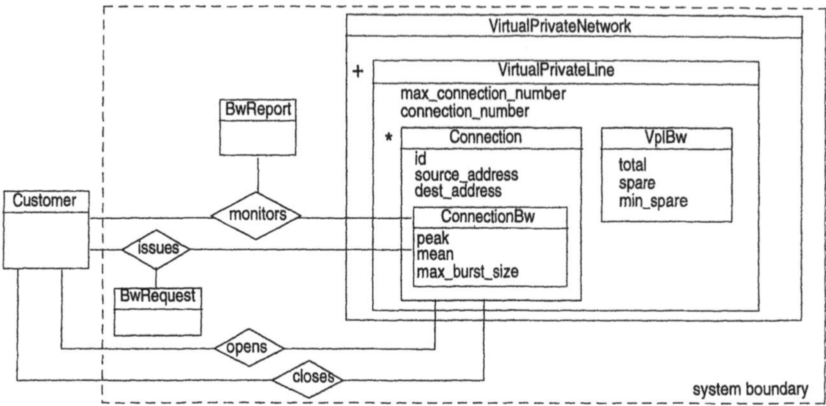

Figure 2 System Object Model for Bandwidth Update.

5.3 System interface

The object models take into account both the system under study and its environment. The next phase in the Fusion analysis process is to determine the boundary between the two, that is to say, the system interface. A useful technique for that purpose is to consider scenarios of usage.

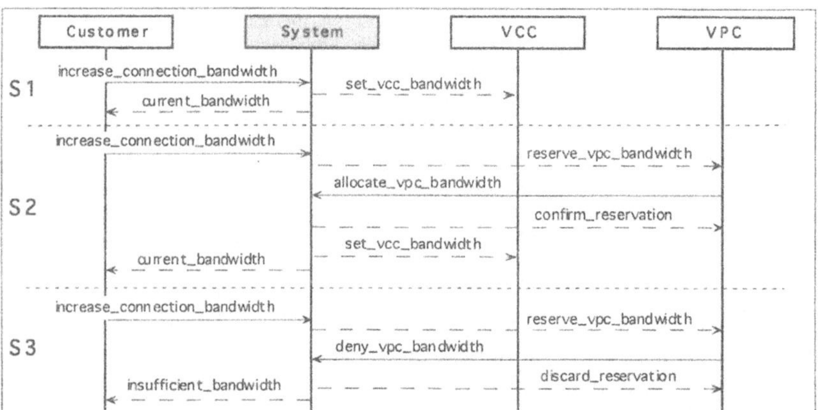

Figure 3 Scenario for Connection Bandwidth Allocation.

The Figure 3 shows a scenario for a connection bandwidth increase represented in timeline diagrams. This scenario considers three different alternatives involving three external agents. In the first one, S1, the system has enough spare capacity to satisfy the request of the customer. The other two alternatives deal with the case where the system tries to reserve additional resources from the network, either successfully (S2) or not (S3). Similar scenarios can be defined for bandwidth monitoring, bandwidth decrease, etc. One of the main benefits of these scenarios is that they allow to draw the boundary of the system, by considering the classes modeling the agents in these scenarios as external to the system (see Figure 2).

These scenarios may be generalized and formalized into life-cycle expressions, that is, regular expressions allowing to express sequences, repetition, alternatives as well as optionality and whose complete set constitutes the life-cycle model. This model specifies the allowable sequence of system operations (i.e. the input events and the effects they can have) and output events. The life-cycle model of the system under study has been developed in (Gaspoz, 1994).

Operation model

The operation model determines the system functionality as expected by the user. The behavior of each system operation is specified in a declarative way, in particular by using preconditions and postconditions. The preconditions express the conditions that must be satisfied whenever a system operation is invoked. The postconditions describe how the state of the system (i.e. the set of objects that participate in relationships as defined in the system object models) is changed by an operation and which events are sent to the agents. The operation model consists of a set of schemata, one for each system operation. The schema for the system operation 'increase_connection_bandwidth' is shown in Figure 4. The preconditions and postconditions are expressed in the 'Assumes' and 'Result' clauses, respectively.

```
Operation :    increase_connection_bandwidth
Description :  Request the connection bandwidth to be increased by a given amount

Reads :        supplied peak_amount, mean_amount : BitRate
               supplied max_burst : BurstSize, supplied conn_id : ConnectionId

Changes :      Connection with connection.id equal to conn_id, connectionbw, vplbw
               new bwrequest, new bwreport, new reservation

Sends :        virtual_channel_connection : {set_vcc_bandwidth}
               virtual_path_connection : {reserve_vpc_bandwidth}
               customer : {current_bandwidth}

Assumes :      conn_id is a valid connection identification

Result :       a bwrequest has been created and initialized with the supplied values,
                   peak_amount, mean_amount and max_burst
               If (initial vplbw.spare) is sufficient to support the requested bandwidth increase then
                   vplbw.spare has been decreased by a value computed from peak_amount,
                         mean_amount and max_burst
                   set_vcc_bandwidth has been sent to virtual_channel_connection
                   connectionbw.peak has been increased by peak_amount
                   connectionbw.mean has been increased by mean_amount
                   connectionbw.max_burst_size has been set to max_burst
                   a bwreport has been created and initialized with the final values of connectionbw
                   current_bandwidth(bwreport) has been sent to customer
               Otherwise        /* not enough bandwidth on the vpl
                   reserve_vpc_bandwidth has been sent to virtual_path_connection
                   a reservation has been created
                   reservation.status has been set to pending
                   reservation.pending_responses has been set to virtualprivateline.nb_of_segments
```

Figure 4 Operation Schema for 'increase_connection_bandwidth' system operation

The communication between the system and its environment is asynchronous, that is, the sender does not wait for the event to be received (Coleman, 1994). This assumption has a significant influence on the way system operations are specified, as, for instance, the response to an output event has to be described in a different operation schema. Moreover, behavior conditional on output events (e.g., the fact that each 'reserve_vpc_bw' should be followed by either 'allocate_bandwidth' or 'deny_bandwidth') is difficult to express in these schemata.

5.4 Designing object interaction graphs

All the models described so far are part of the Fusion analysis process. Once this step completed, the goal of the object-oriented design consists in defining how objects interact to provide the system functionality specified in the operation model. The main scope of the design phase is then to collect abstract definitions into concrete software structures, especially with respect to implementation and distribution of functionality. This distribution is captured in an object interaction graph. Each graph allows to define the sequences of messages exchanged between a set of objects to realize a given operation. The system software architecture starts then appearing as each system operation is designed. There is no unique way to design this functional distribution. Certain assumptions, design tradeoffs and choices as well as the larger system issues all influence the design process.

Object interactions are defined as procedural types of interactions. Indeed, when a message is sent to a server object, the corresponding method of its interface is invoked. This method is executed before the control is returned to the client. In other words, although the data flow may be bi-directional or unidirectional depending if a value is returned as the result of the method call, the control flow associated with such method calls is always bi-directional.

The Figure 5 shows the object interaction graph corresponding to the three system operations 'increase_connection_bandwidth', 'allocate_bandwidth' and ' deny_bandwidth'. Boxes and dashed boxes represent (design) objects and collections of objects, respectively. The arrows represent the invocation of the corresponding method on an object. A selection predicate (in square brackets) may be defined to send a message to one particular object in a collection. By default, the message is sent to all objects in the collection. Numbers define the sequencing of invocation. Method invocations labeled with the same sequence label occur in an unspecified order. Letters appended to a sequence number define alternatives.

The *vpn* has been selected as controller object, that is, the object which takes the responsibility for the given system operation. Its main role is to find out, among all the virtual private lines it contains, the *vpl* on which the given connection has been established. The central role played by the *vpl* in this design arises quite naturally from the data structures and relationships defined during the analysis. Indeed, according to the system object models specified previously (see Figure 3 and (Gaspoz, 1994)) a *vpl* object has relationships to both the active *connections* it contains and the *segments* that constitute it.

The decision as to whether the increase request may be satisfied directly or requires further resources from the network providers, is taken by the *vplbw*. For this purpose, this object has to perform a statistical computing taking into account not only the current request but also the bandwidth parameters of all the existing connections as well as the admissible loss probability. As the goal of this paper is not to elaborate on such issues, the method 'compute_bw_inc_req' is supposed to encompass this statistical computation and will not be developed further.

Careful readers have certainly noted that three operation schemata have been designed into one single object interaction graph, which is clearly not in-line with the approach recommended by Fusion. The reason why there is not a one-to-one mapping comes from the different ways objects are supposed to communicate within the system and with external agents. Indeed objects exchange messages within the system in a synchronous request/response style of interaction called sometimes interrogation (ODP, 1993). On the other hand, the system -and thus the objects that constitute it- communicates with its environment in a fire-and-forget style of interaction called announcement (ODP, 1993).

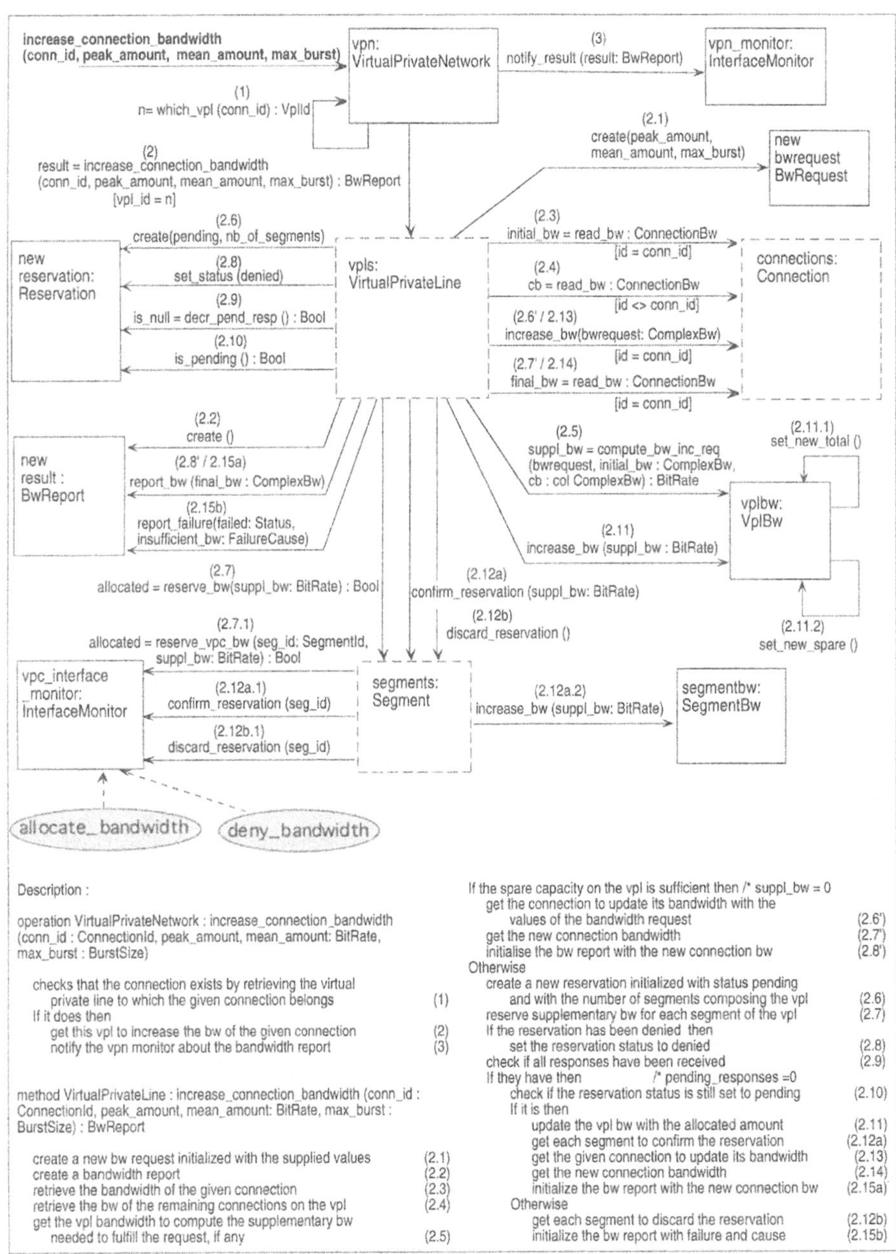

Figure 5 Object Interaction Graph for 'increase_connection_bandwidth' operation.

To keep analysis and design consistent, as well as to preserve the semantics of the object interaction graphs, this duality has been maintained. The mapping between these two types of interactions has thus to be performed at the boundary of the system by the so-called *InterfaceMonitors*. These objects get thus a more active role than initially described in the Fusion method. Concretely, they have to map each interrogation invoked on their interface into an announcement to the corresponding agent. The asynchronous response to this announcement, if any, is on its turn converted back into the result part of the initial interrogation. For instance, the two system operations 'allocate_bandwidth' and 'deny_bandwidth' are encapsulated in the boolean result of the method 'reserve_vpc_bw' invoked at the *vpc_interface_monitor.* A special notation has been introduced in Figure 5 to illustrate this situation. Thanks to these mappings, the *Interface_monitors* hide to the system objects the announcement-based style of communication of the system with its environment. Consequently, objects may communicate transparently with other objects inside or outside the system in a consistent interrogation-based way.

The previously mentioned design choices are trade-offs between simplicity and efficiency. The choice of a sequential approach which, by waiting for the network providers responses, prevents the system to process a new customer's request before the previous one is completed -according on this point to the life-cycle developed in the analysis (Gaspoz, 1994)- is certainly not the most efficient. However, it offers great advantages with respect to error handling and concurrency issues, thus leading to a much simpler design. For instance, missing responses or error messages may be considered as implementation issues of the *Interface_monitors*, i.e., dealt with by some kind of transaction processing mechanism, and need not be considered further. In the same way, two consecutive customer's requests addressing the same virtual private line will not give rise to any conflict.

On the other hand, a good improvement that is consistent with the life-cycle, would be to allow parallelism to the bandwidth requests going to the different network providers. This issue is left for further study.

5.5 Designing visibility graphs

In the previous design phase, all objects were assumed to be mutually visible. The goal of this second design step is then to determine for each class which objects the instances of the class need to reference as well as the type of reference required (Coleman, 1994).

The visibility graph for the *VirtualPrivateLine* class is shown in Figure 6. All server objects -or collections of server objects (in dashed boxes)- whose lifetime is bound to that of the *virtualprivateline* client are shown inside the class box. A dashed arrow and a double border box denote a dynamic reference (e.g. *bwrequest*) and an exclusive reference (e.g. *segments*), respectively. Constant mutability (i.e. the reference is not reassignable after initialization) is explicitly shown by prefixing the server object with the keyword 'constant'.

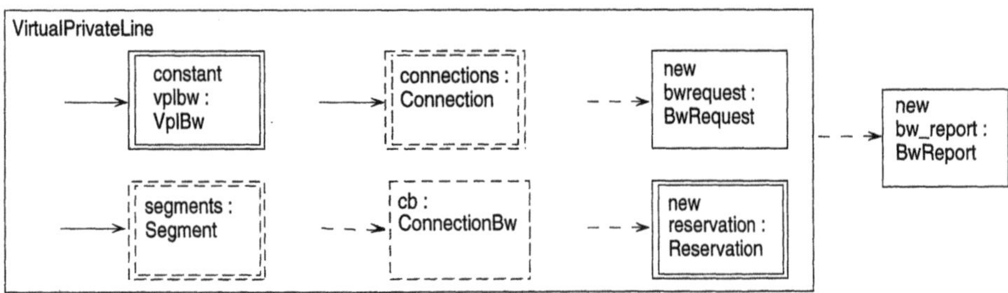

Figure 6 Visibility graph for the VirtualPrivateLine class.

5.6 Designing inheritance graphs

The inheritance considered in this document is a subtyping inheritance in the sense that objects of a subtype can only extend the properties of the supertype but not alter them. Unlike in programming languages, the focus is not on efficiency and code reuse but on simplicity of reasoning. A very useful consequence is that instances of a subclass may always be freely substituted for instances of a superclass.

A good starting point for deriving inheritance graphs is provided by the generalization and specialization relationships identified during the analysis. For instance, the classes *ConnectionBw, VplBw,* and *SegmentBw* have been identified as subclasses of an abstract class *Bandwidth.* During the design phase, it has been felt useful to partition the class *Bandwidth* into a *ComplexBw* (to deal with sustained rate allocation scheme) and a *SimpleBw* (for peak rate allocation). On the other hand, the two classes *BwReport* and *BwRequest* are used to encapsulate bandwidth related information. Using multiple inheritance, these classes may be defined as subtypes of the class *ComplexBw* and of the classes *Report* and *Request,* respectively. Thanks to this structure, it has been possible to reference objects of type *ComplexBw* and substitute them by instances of either of its subtypes (see Figure 5).

5.7 Deriving complete class descriptions

The individual design steps described in the previous sections are now all integrated into class descriptions. The complete class descriptions are final software design structures upon which implementation is based. They provide a specification of the class interface, i.e., the externally visible data attributes, object reference attributes and methods signatures, as well as of the inheritance relationship, if any. The description of the class *ConnectionBw* is presented below as an example.

```
class ConnectionBw is_a ComplexBw
attribute peak : BitRate
attribute mean : BitRate
attribute max_burst_size : BurstSize
method create ()
method delete ()
method add_bw (bw_chg : ComplexBw)
method remove_bw (bw_chg : ComplexBw) : Bool
method get_peak () : BitRate
method get_mean () : BitRate
method get_mbs () : BurstSize
endclass
```

5.8 Designing the system communication interfaces

The aim of this last step is to collect all the messages exchanged between the VASP-SMS and the NMSs that constitute its environment (see Figure 1). Mainly derived from the scenarios and the different object interaction graphs, this information allows to specify the different interfaces between these management systems in terms of the primitives exchanged (Figure 7).

On the CPN-NMS side, the interface specifies the management service offered to the customer in terms of the management functions he may invoke to perform bandwidth related management operations on his ATM-based VPN. The set of primitives that are part of the SMS - NP-NMS interface represent the service that the VASP has to request from the network providers NMS in order to provide the bandwidth management service to the customer.

A TMN conformant specification of these interfaces (X interfaces) would imply the mapping of these high level primitives into CMIP (Common Management Information Protocol) ones and would above all require a GDMO specification (X.722, 1991) of the information exchanged across the interfaces. These issues imply a level of design detail that goes beyond the scope of this paper and have not been considered further.

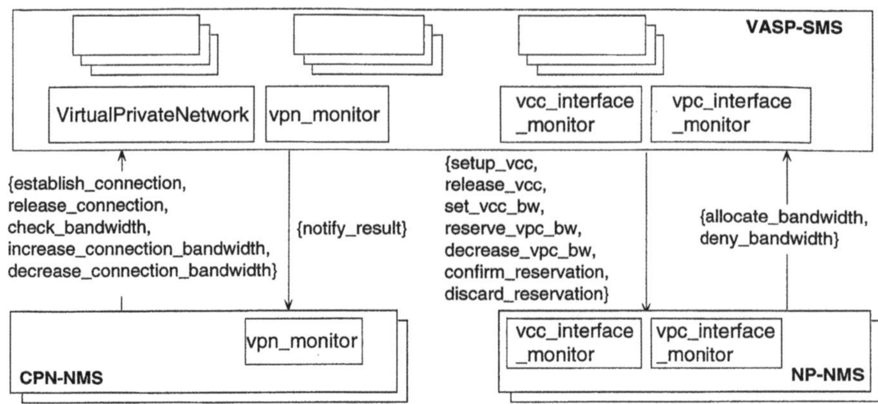

Figure 7 SMS - NMSs communication interfaces.

6 CONCLUSION

Bandwidth management is of critical importance in ATM-based networks due to the great bandwidth flexibility it offers to end-users. This paper has described the software structures that need to be implemented in the VASP-SMS to support the provision of a bandwidth management service to customers. In addition, the corresponding service required from the underlying network providers' NMSs for that purpose has also been brought into light. However, even if the work has focused on VPN bandwidth management based on cross-connected ATM networks, the model developed at the service management layer is quite abstract and general enough to be applied to other service management cases.

Although the VPN management architecture considered is based on TMN principles, the modeling approach selected in this paper provides an interesting alternative to the TMN methodology where both a functional and an object-oriented approach coexist (M.3020, 1992). Indeed, management services fulfilling the customer requirements are decomposed into management service components and management functions according to a top-down functional decomposition. Conversely, the modeling of the managed system is object-oriented, namely, all network resources are modeled as managed objects. Therefore, the mapping between the management functions and the managed objects is far from being straightforward.

On the other hand, the Fusion method retained in this paper models the entire problem domain in a consistent object-oriented way. The functionality of the system as expected by the customer is defined quite formally in the operation model, thanks to the use of pre- and post-conditions. System operations specified in this model, which are in fact similar to TMN management functions in our example, are implemented in the design phase as a set of interacting objects. The mapping of the functionality expected by the user into the object model representing the system is then realized in a very consistent and straightforward way.

The problem domain addressed in this paper involves several actors and different systems that work in parallel and interact to constitute a distributed bandwidth management system. Although this study has focused on one specific part of this distributed system, namely the VASP-SMS, the functionality needed to provide the final service is clearly distributed in the different management systems. As a software engineering method that has been developed for sequential and centralized systems, Fusion is not very well-suited to deal with the specification and design of distributed systems. Issues such as the conflict between system internal communications based on interrogation and external communications based on announcement could be dealt with in a more elegant way by using a distributed systems conformant approach all along the development process. However, the integration of some of the models advocated by Fusion into the ODP viewpoints could be a very interesting topic of further study.

7 REFERENCES

Coleman , D. et al. (1994) Object-Oriented Development: The Fusion Method, Prentice Hall.

Gaspoz, J.P., Saydam, T. and Hubaux J.P. (1994) Object-Oriented Specification of a Bandwidth Management System for ATM-based Virtual Private Networks, *proceedings of the third ICCCN conference.*

M.3010 (1992) Principles for a Telecommunications Management Network, *CCITT Recommendation M.3010.*

M.3020 (1992) TMN Interface Specification Methodology, *CCITT Recommendation M.3020.*

ODP (1993) Basic Reference Model of Open Distributed Processing (ODP), parts 1-3, *ISO/ITU-T Draft Recommendations X.901, X.902, X.903.*

Rumbaugh, J. et al., (1991) Object-Oriented Modeling and Design, Prentice Hall.

X.722 (1991) Guidelines for the Definition of Managed Objects, *ITU Recommendation X.722.*

Verbeeck, P. et al., (1992) Introduction Strategies Towards B-ISDN for Business and Residential Subscribers Based on ATM, *IEEE JSAC*, December edition.

Wernik, M. et al. (1992) Traffic Management for B-ISDN Services, *IEEE Network*, November edition.

8 BIOGRAPHY

Tuncay Saydam has been a professor of computer science at the University of Delaware since 1979. He has received his graduate degrees at Istanbul Technical University and The University of Texas at Austin. His current research interests include network management, network interconnections and object-oriented software design. Member of IEEE, Sigma XI and New-York Academy of Sciences, Dr. Saydam is author of over fifty technical articles.

Jean-Paul Gaspoz graduated in electrical engineering at the Swiss Federal Institute of Technology in Lausanne. He then worked three years at Ascom, a Swiss telecom company where he contributed to the development of an ISDN PABX. He is currently doing a Ph.D. at the Swiss Federal Institute of Technology in Lausanne and his research interests include virtual private networks, services management and distributed systems specification.

Pierre-Alain Etique graduated in computer science at the Swiss Federal Institute of Technology in Zurich. He then joined Ascom, a Swiss telecom company were he worked 3 1/2 years on the development of a PBX. He is currently with the Swiss Federal Institute of Technology in Lausanne where he is working on his Ph.D.

Jean-Pierre Hubaux graduated in computer science at the Institute of Technology of Milan. He then joined Alcatel where he worked ten years as development engineer, consultant and project manager. He has been a professor at the Swiss Federal Institute of Technology of Lausanne since 1990.

32

A TMN system for VPC and routing management in ATM networks

D. P. Griffin
Institute of Computer Science,
Foundation for Research and Technology-Hellas,
PO Box 1385, 711-10 Heraklion, Crete, Greece.
Tel: +30 81 391722, Fax: +30 81 391601
email: david@ics.forth.gr

P. Georgatsos
ALPHA Systems S.A.,
3 Xanthou Str., 177-78 Tavros, Athens, Greece.
Tel: +30 1 482 6014, 15, 16, Fax: +30 1 482 6017
email: panos@alpha.ath.forthnet.gr

Abstract

In this paper we present a VPC and Routing Management Service for multi-class ATM networks. Considering the requirements, we decompose the Management Service into a number of distinct but cooperating functional components which we map to the TMN architecture. We describe the architectural components and analyse their operational dependencies and information exchange in the context of the overall system operation.

The proposed system offers the generic functions of performance monitoring, load monitoring and configuration management in ATM networks. In addition, it provides specific functions for routing and bandwidth management in a hierarchical structure.

Keywords

ATM, TMN, performance management, routing, VPC, multi-class environment.

1 INTRODUCTION

The efficient operation of a network depends on a number of design parameters, one of them being routing. The overall objective of a routing policy is to increase the network throughput, while guaranteeing the performance of the network within specified levels. The design of an efficient routing policy is of enormous complexity, since it depends on a number of variable and sometimes uncertain parameters. This complexity is even greater, taking into account the diversity of bandwidth and performance requirements that the network must support. The routing policy should be adaptive to cater for traffic and topological changes.

Routing in Asynchronous Transfer Mode (ATM) (ITU I.150) is based on Virtual Path Connections (VPCs). A route is defined as a concatenation of VPCs, where each VPC is defined as a sequence of links being allocated a specific portion of the link capacity. It has been widely accepted that VPCs offer valuable features that enable the construction of economical and efficient ATM networks, the most important being management flexibility. Because VPCs are defined by configurable parameters, these parameters and subsequently the routes based on them can be configured on-line by a management system according to network conditions.

Since user behaviour changes dynamically there is a danger that the network may become inefficient when the bandwidth allocated to VPCs or the existing routes are not in accordance with the quantity of traffic that is required to be routed over them. To combat this, the VPC topology, the routes, and the bandwidth allocated to VPCs must be dynamically re-configured. A VPC and Routing management system is required to take advantage of the features of VPCs while ensuring that the performance of the network is as high as possible during conditions of changing traffic.

The ITU-T have distinguished between the management and control planes in the operation of communications networks (ITU I.320, I.321) and introduced the Telecommunications Management Network (TMN) (ITU M.3010) as a means of provisioning management systems with standard interoperable components according to the ISO systems management standards. The TMN should compliment and enhance the control plane functions by configuring operational parameters. The TMN should not replace the control plane and in general it has less stringent requirements on real-time response.

Although there is a significant research interest in the area of performance management on ATM particularly in routing (Sykas 1991, Gelenbe 1994), bandwidth assignment (Hui 1988, Saito 1991) and VPC management (Ohta 1992, Sato 1991), the problem of VPC and routing management remains largely open. The majority of management systems deployed today are concerned with network configuration and network monitoring and the management intelligence is provided by the human users of the management systems. There is a trend (Woodruff 1990, Wernic 1992, Geihs, 1992) to increase the intelligence of the management functions to encapsulate human management intelligence in decision making TMN components to move towards the automation of the monitoring, decision making and configuration management loop.

Within the framework of performance management this paper investigates the requirements of VPC and routing management functions for ATM based B-ISDN networks and proposes a TMN system for implementation. The ITU-T terminology (ITU M.3020) for describing Management Services is adopted. In particular the paper proposes a Management Service for VPC and routing management and decomposes it into a number of components. The design is mapped to the TMN architecture for implementation using TMN and OSI systems management principles.

Section 2 defines the VPC and Routing Management Service and section 3 discusses the environmental assumptions and constraints. Section 4 presents the decomposition into management components and outlines the rationale behind it. The mapping to the TMN architecture is also presented in this section. Section 5 details the management components and section 6 describes their interactions and their relationships. Finally section 7 presents the conclusions and identifies future work.

2 THE MANAGEMENT SERVICE

Within a multi-class ATM network environment the objective of the VPC and Routing Management Service is to guarantee network availability whilst guaranteeing that the network meets the performance requirements of the different service classes. This Management Service is beneficial to the network operator since it ensures that the network resources are used as efficiently as possible.

The VPC and Routing Management Service has both static and dynamic aspects. The static aspect is related to the design of a VPC network and a routing plan (the set of routes and selection criteria for each source-destination pair and service class) to meet predicted demand. In fact the static aspect is of quasi-static form in the sense that is invoked whenever the traffic predictions change significantly. The dynamic aspect manages the VPC network and the routing plan to cater for unpredictable user behaviour within the epoch of the traffic predictions.

This Management Service belongs to the performance and configuration management functional areas and specifically covers traffic management while its static aspects are related to the network planning functions. Figure 1 shows the relationship of VPC and Routing Management with the network, human managers (TMN users), other management functions, network customers and other network operators.

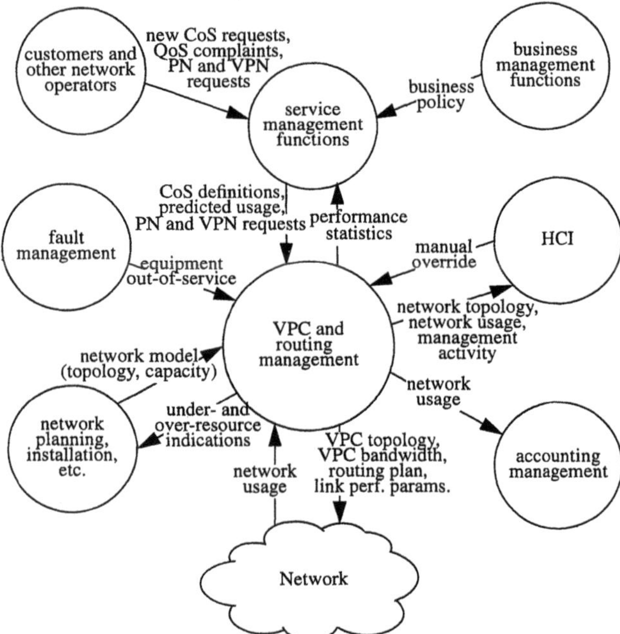

Figure 1 Enterprise view of VPC and Routing Management.

The methodology of ITU-T Recommendation M.3020 is adopted. According to this Recommendation Management Services are composed of Management Service components (MSCs) which in turn are constructed from management functional components (MFCs). MFCs are themselves constructed from management function sets (MFSs) which are groups of management functions which logically belong together. In this paper we will decompose the Management Service to identify the constituent MSCs and MFCs and show how these can be mapped to the logical TMN architecture.

3 THE ENVIRONMENT

This section describes the network environment from the perspectives of the VPC and Routing Management Service.

The managed environment is assumed to be a public ATM network offering switched, on-demand services ranging from simple telephony and file transfers to multi-media conferences.

3.1 Assumptions on the network services

Service calls are decomposed into a number of unidirectional connections. A large number of connection types are supported, for example, telephony services may be supported by a number of connection types which offer a range of qualities - different delays or call blocking probabilities. The term class of service (CoS) is used to denote a particular connection type. The CoSs are the bearer services provided by the network. The CoS definition characterises the connection type in terms of bandwidth and performance requirements.

Our work assumes that the bandwidth requirements can be characterised by mean and peak values. Alternative bandwidth parameters may be used according to the specific connection admission control (CAC) algorithms employed in the switches.

We assume the following performance parameters: cell loss probability; delay; delay jitter; and connection blocking probability (or availability). In fact these are the performance parameters the Management Service is able to influence and are of direct interest but other performance parameters may be included in the CoS definition, connection release delay for example, but these may not be influenced by this Management Service and are not considered further here.

An issue to be clarified is the relationship between the classes of the bearer services provided by the network and the four AAL classes recommended by the ITU-T (ITU I.362). The AAL provides a limited range of services, e.g. connection-oriented vs. connectionless, error recovery, re-transmissions with the assumption of a given performance of the underlying bearer service.

Our view is that there needs to be a range of bearer services of different qualities and costs to support the AAL services. This will allow decisions to be made on whether to use a comprehensive AAL with a cheap, low performance bearer service or a lightweight AAL with a higher performance bearer service (e.g. smaller cell loss ratio). This view is in accordance with the views of the ATM Forum (ATM Forum 1993) where they explicitly recommend the augmentation of the AAL service classes with a range of quality of service classes. The AAL exists in the user terminals whilst the underlying bearer service is provisioned by the network operators.

Our work concentrates on the management of the bearer services from the viewpoint of the network operator. Although AAL issues are considered from the perspective of the requirements they impose on the underlying bearer services, the end-to-end management issues of layer 4 and above are not the focus of our work.

Another important point is the role of connection oriented services with a predefined bandwidth and performance compared to that of best effort (no performance requirements) or available bandwidth (ABR) services. We recognise the requirement for all types of services but our work concentrates on the management needs of the services with predefined bandwidth and performance. Best effort and ABR services are controlled via the signalling protocols. If they are to coexist with services of defined quality on the same network, there is a necessity for the bandwidth and routing management functions to dynamically manage the partitioning of the network resources. However this is an issue for future work.

3.2 Assumptions on user behaviour

User behaviour is not constant and changes dynamically. There are two sources of variation: the type of user and the population of the users. There are potentially many different types of users characterised by the types of service they use and also by their usage patterns. The behaviour of individual users changes over time with respect to the services they use and the way that they use their services.

We assume, by virtue of the law of large numbers, that estimates of aggregate user behaviour can be made and trends can be identified in the short term (e.g. business vs. domestic traffic throughout the working day) and the medium to long term (e.g. seasonal variations, new service introduction, competition).

3.3 Assumptions on network operation

ATM networks are connection-oriented networks. Each node basically provides switching and call control (CC) functionality which includes route selection and connection admission control (CAC). Switching is done at two levels: VP cross-connects switch cells within a VPC based on the VPI; VC switches switch cells of a particular VCC between VPCs based on their VCI. VPCs are created on a semi-permanent basis by management actions whereas VCCs are created dynamically by the control plane of the network via UNI and NNI signalling. Route selection refers to the selection of a particular route upon receipt of a connection establishment request by means of a Route Selection Algorithm (RSA). CAC is required in order for the node to determine whether the connection can be accommodated on the selected route. This is done by means of a CAC algorithm that controls the VPC loading in the admissible region i.e. in the region where buffer overflow is within the bounds of a pre-defined probability (cell loss target of the CAC).

The latter two functions, route selection and CAC, are part of the control plane. However their behaviour is specified according to operational parameters which are defined and managed by the TMN.

In order to accomplish routing all possible routes towards a given destination and a particular CoS are stored local to the switches in a route selection table. For a number of reasons (increased availability, reduced vulnerability to failures, adaptivity) more than one route may exist to a destination for a CoS. The RSA in each switch searches its route selection table for entries satisfying the destination and CoS. We assume that the RSA is based on route selection parameters associated with the available routes. These parameters reflect the preference of selecting one route over another. The RSA should be as fast as possible and cause the minimum overhead to the network.

4 DECOMPOSITION

4.1 The rationale

Connection rejection is affected by two factors: the number of alternative routes and the available capacity on the VPCs. These two factors cannot be treated in isolation and the VPC and Routing management system must therefore ensure that there are sufficient numbers of routes and bandwidth on the VPCs forming the routes to guarantee network performance and availability.

As mentioned previously the Management Service should provide adaptivity to changing traffic conditions. There are two levels at which the traffic can change: cell level variations within the scope of a single connection; and connection level variations as users establish and release calls. The former is considered to be dealt with by the CAC and UPC functions of the control plane. Connections can never exceed the bandwidth parameters defined for a CoS due to the role of the UPC functions. If connections do not consume the full bandwidth the shortfall cannot be used by other connections because of the concept of pre-defined bandwidth reservation at connection set-up time which is paid for by the users. For this reason cell level variations are of no concern to this Management Service and the management of connection level variations is the main focus.

The following views of the network are useful for offering different levels of abstraction to assist the task of formulating the problem faced by the VPC and Routing Management Service.

- *The physical network* consisting of the network nodes and the transmission links.
- *The VPC network* consisting of the VC switches interconnected by VPCs.
- *The ClassRoute networks*. For each CoS, the ClassRoute network is the sub-network of the VPC network which consists only of the VPCs that belong to routes of that CoS.
- *The SDClassRoute networks*. For each CoS and a given source-destination (s-d) pair, the SDClassRoute network is the sub-network of the ClassRoute network consisting only of the VPCs that belong to the routes of the given (s-d) pair.

Having introduced the above network views the goal of the VPC and Routing Management Service can be formulated as follows:

- Given the physical network and the traffic predictions per s-d and CoS, define VPC and SDClassRoute networks so that the traffic demands are met and the performance levels specified per CoS are guaranteed.

The solution requires answers to the following questions:

- How is the VPC network constructed and how frequently will it change?
- How are the ClassRoute networks constructed and how frequently will they change?
- According to what criteria will routing be achieved in the ClassRoute networks? i.e. Given the VPC and ClassRoute networks how are the route selection parameters assigned and how frequently will they change?

The definition of the VPC and ClassRoute networks is an iterative procedure which cannot separate the two tasks involved. Routes are defined in terms of VPCs and the VPCs have been defined in order to support routing.

The VPC and the ClassRoute networks are constructed using, as input, estimates for the network traffic per s-d pair and CoS. The construction of these two networks is related to the network planning activity, whereby the topology of the physical network is defined based on longer term network traffic predictions. The design of the VPC and Routing management system should therefore cater for changes in the predictions and inaccuracies in the predictions.

Whenever the traffic predictions change, the VPC and ClassRoute networks need to be reconstructed. The level of reconstruction obviously depends on the significance of the changes. As a result, new values for VPC bandwidth may be given, or the topology of the VPC network may change (by creating and deleting VPCs) or the topology of the ClassRoute networks may change (by creating and deleting routes). Each of these reconfigurations deals with a different level of abstraction according to the network views described above. Moreover they may be performed within different time scales and they require different levels of complexity and hence computational effort. We envisage that an efficient way to deal with such reconfigurations is through a hierarchical system.

The essence of the hierarchy we propose is as follows. First the VPC bandwidth is reconfigured within the existing SDClassRoute networks. If it is not possible to accommodate the traffic predictions within the SDClassRoute networks, the SDClassRoute networks are reconfigured within the existing VPC network. If it is found that the VPC topology is insufficient for the predicted traffic then finally the VPC network is reconfigured. Ultimately it may be discovered that the physical network is unable to cope with predicted traffic and the network planning functions are informed to request that additional physical resources are deployed.

This indicates the need for having three management components: Bandwidth Allocation (for VPC bandwidth updates given SDClassRoute networks), Route Planning (for route updates given the VPC network) and VPC Topology (for VPC topology updates).

The above assumes that the traffic predictions are accurate, but as mentioned previously, this cannot be taken for granted. For this reason we introduce a lower level into the hierarchy which tries to make the initial estimates more accurate by taking into account the actual usage of the network. The lower level functionality operates within the SDClassRoute networks and redefines the VPC bandwidth and route selection parameters taking into account the actual network load. Redefinition of SDClassRoute networks and VPC topology is not done at this level since it must be as lightweight as possible. However this level will provide triggers to the higher level when it is proved that the first level estimates under or over estimate the actual situation and this cannot be resolved at this level. Even if the predictions are accurate there is still a case for lightweight lower level functions to cater for traffic fluctuations within the timeframe of the predictions.

This indicates the need for two components in the lower level: Bandwidth Distribution (for updating VPC bandwidth) and Load Balancing (for updating route selection parameters).

The proposed hierarchical system exhibits a fair management behaviour whereby initial management decisions taken with a future perspective are continuously refined in the light of current development. Apart from its fairness, such a behaviour provides a desirable level of adaptivity to network conditions.

4.2 MSCs and MFCs

The previous section indicates the following decomposition of the VPC and Routing Management Service into MSCs:

- *management of VPC topology* which is placed in a VPC Topology MFC
- *management of VPC bandwidth* which is further decomposed into:
    - a VPC Bandwidth Allocation MFC
    - a VPC Bandwidth Distribution MFC
- *management of the routing plan* which is placed in a Route Planning MFC
- *network load balancing* which is placed in a Load Balancing MFC
- *performance verification* which is placed in a Performance Verification MFC
- *traffic predictions* which are placed in a Predicted Usage Model MFC

Additionally, the following support MFCs are required:

- a Configuration Management MFC which includes the network model
- a Current Load Model MFC for providing the required network statistics
- a CAC Manager MFC for the TMN to model the CAC behaviour for dimensioning purposes
- a CoS Model MFC

4.3 Mapping to the TMN architecture

The functional architecture is based on the principles of ITU-T recommendation M.3010. Figure 3 shows the allocation of MFCs to OSFs and also places the OSFs into the architectural

VPC Bandwidth Distribution OSF can be controlled.

5.3 Load Balancing OSF

The Load Balancing OSF operates within the SDClassRoute and VPC networks defined by the Route Design OSF. The Route Design OSF reserves network resources (VPCs) and indicates their use (by defining routes); The Load Balancing OSF tries to make the best possible use (most efficient utilisation) of the reserved resources.

To achieve this, the Load Balancing OSF takes a network-wide view and tries to influence the routing decisions so that arriving connections use the routes with the highest availability. The view taken is that the routes at a node are prioritised according to their potential as being good routes. Since in our case we deal primarily with a connection-oriented network, potential refers to the availability (spare capacity) of the route to accommodate connections. This way the network load is spread as even as possible and the network availability for new connections is as even as possible (hence the name Load Balancing). Note that the above view is in accordance with the traditional view of routing according to which routing schemes are variants of shortest path algorithms

The multi-class environment that the network operates in, should be taken into account. The Load Balancing OSF should not only aim at optimising routing in the ClassRoute networks but also in the VPC network. This further justifies the need of a central component, like the Load Balancing OSF, which utilising network-wide information about every class tries to harmonise routing within each class and between classes.

5.4 Performance Verification OSF

The Performance Verification OSF is concerned with ensuring that the network meets the performance targets for the different CoSs. This is done at two levels: by monitoring the network and by accepting customer QoS complaints via the service layer's customer interface.

The connection rejection ratios per CoS and per source destination pairs are retrieved from the Current Load Model and compared to the rejection targets as specified per CoS. Customer complaints are analysed and if they are justified the Route Design OSF will be triggered. If CoSs are found to experience connection rejection ratios in excess of the target, an indication is sent to the Route Design OSF to cause the number of routes, or the bandwidth required by the routes, to be updated. The Performance Verification OSF quantifies the performance of the Route Design, the Load Balancing and the VPC Bandwidth Distribution OSFs, being the indisputable measure of their efficiency.

5.5 Predicted Usage Model OSF

This models the predicted usage of the network in terms of the numbers of connections of each CoS required between s-d pairs. The model details how the number of connections changes: hour by hour over the day; day by day over the week; and week by week over the year

Initially this is configured by the service level of the TMN but it is modified by the actual usage of the network via the Current Load Model. This is so that the predicted model becomes more accurate as experience of the usage of the network is gained.

Whenever the predicted load model indicates that the traffic will change significantly the Route Design OSF will be provided with a prediction of traffic for the next time interval. The exact definition of a significant change is a design variable to be experimented with according to the traffic changes significantly. Based on the predicted usage, the s-d predictions are mapped to VPCs within the existing SDClassRoute networks, and the minimum bandwidth required by each VPC in order to meet the predicted demand is identified. If it is impossible to allocate sufficient bandwidth for the predicted traffic within the constraints of the current SDClassRoute networks and the link capacities, the Route Planning MFC is notified.

The Route Planning MFC attempts to redesign the SDClassRoute networks on the existing VPC network, to remove bottlenecks for example. It tries to increase the number of alternative routes, using the current VPC topology. This process also identifies the new bandwidth requirements on the VPCs. In order to enhance alternative routing and to compensate for inaccuracies in the routing estimates, Route Planning may assign a set of 'back-up' routes to each CoS in addition to the primary set of routes. For a given CoS, the set of 'back-up' routes consists of the routes allocated to the higher quality CoSs. If the Route Planning MFC cannot design a new set of SDClassRoute networks to accommodate the predicted traffic due to limitations in the existing VPC network topology, the VPC Topology MFC is invoked.

The VPC Topology MFC redesigns the VPC network to meet the new requirements. New VPCs may be created to coexist with the current ones and new SDClassRoute networks will be defined so that the new VPC topology may be introduced gradually for new connections. The bandwidth requirements for the VPCs in the final VPC topology are identified and passed down to the lower MFCs. If it is not feasible to design a VPC network to satisfy the traffic demand because of limitations in the underlying physical network, e.g. not enough links, the network planning function is notified.

The Route Design OSF should cater for designing SDClassRoute networks according to the CoS requirements. CoS cell losses targets can be met by adjusting the CAC cell loss targets appropriately so as to ensure that accumulated cell losses over the links of the SDClassRoute network do not exceed those defined for that CoS. Guarantees for delay and jitter can be provided by identifying the maximum number of buffers and switches and ensuring that the SDClassRoute networks do not exceed these values. Finally CoS availability is guaranteed by being an overall optimization constraint that the iterative procedure for defining VPC and SDClassRoute networks should meet.

5.2 VPC Bandwidth Distribution OSF

Taking the current load into account, the VPC Bandwidth Distribution OSF implements the allocation of bandwidth to VPCs as requested by the Route Design OSF. The current load must be considered to avoid situations where the predicted required bandwidth is lower than the current load and hence the new bandwidth allocation would violate the assumptions made by the CAC algorithms in the network and possibly cause excessive cell losses.

In addition to implementing the policies of the Route Design OSF, the VPC Bandwidth Distribution OSF attempts to compensate for inaccuracies in the Predicted Usage Model by distributing any unallocated link bandwidth (viewed as a common pool) among the VPCs.

Unused bandwidth (allocated bandwidth minus current load) in each VPC is the criterion for redistribution to avoid situations where some VPCs are heavily utilised (and consequently there is little bandwidth available for new connections) whilst other VPCs on the same links are lightly utilised. Unused bandwidth is distributed as evenly as possible within certain constraints. For example VPCs can be assigned a class or priority attribute to indicate which VPCs should gain unused bandwidth at the expense of lower priority VPCs. VPCs used for CoSs with low blocking probabilities will be assigned higher priorities.

By varying the averaging interval for calculating the required measures, the sensitivity of the

9 REFERENCES

E.Sykas, K.Vlakos, E.Protonotarios, "Simulative Analysis of Optimal Resource Allocation and Routing in IBCNs", IEE J. Select. Areas Comm., Vol.9, No.3, April 1991.

J.Y.Hui, "Resource Allocation for Broadband Networks", IEEE J. Select. Areas Commun., Vol.6, No.9, Dec.1988.

S.Ohta, K.Sato, "Dynamic Bandwidth Control of the Virtual Path in an Asynchronous Transfer Mode Network", IEEE Trans. Commun., Vol.40, No.7, July 1992.

G.Woodruff, R.Kositpaiboon, "Multimedia Traffic Management Principles for Guaranteed ATM Network Performance", IEEE J. Select. Areas Commun., Vol.8, No.3, April 1990.

Y.Sato, K.Sato, "Virtual Path and Link Capacity Design for ATM Networks", IEEE J. Select. Areas Commun., Vol.9, No.1, Jan.1991.

M.Wernic, O.Aboul-Magd, H.Gilbert, "Traffic Management for B-ISDN Services", IEEE Network, Sept.1992.

H.Saito, K.Shiomoto, "Dynamic Call Admission Control in ATM Networks", IEEE J. Select. Areas Commun., Vol.9, No.7, Sept.1991.

E.Gelenbe, X.Mang, "Adaptive Routing for Equitable Load Balancing", ITC 14 / J. Labetoule and J.W.Roberts (Eds), 1994 Elsevier Science B.V.

ATM Forum, "ATM User-Network Interface Specification", Version 3.0, Sept. 1993.

K. Geihs, P. Francois, D. Griffin, C. Kaas-Petersen, A. Mann, "Service and traffic management for IBCN", IBM Systems Journal, Vol. 31, No. 4, 1992

ITU-T Recommendation I.320 - ISDN protocol reference model

ITU-T Recommendation I.321 - B-ISDN protocol reference model and its application

ITU-T Recommendation I.150 - B-ISDN asynchronous transfer mode functional characteristics

ITU-T Recommendation I.362 - B-ISDN ATM Adaptation Layer (AAL) functional description

ITU-T Recommendation M.3010 - Principles for a telecommunications management network

ITU-T Recommendation M.3020 - TMN interface specification methodology

David Griffin received the B.Sc degree. in Electronic, Computer and Systems Engineering from Loughborough University, UK in 1988. He joined GEC Plessey Telecommunications Ltd., UK as a Systems Design Engineer, where he worked on the CEU RACE I NEMESYS project on Traffic and Quality of Service Management for broadband networks. He was the chairperson of the project technical committee and worked on TMN architectures, ATM traffic experiments and system validation. In 1993 Mr. Griffin joined ICS-FORTH in Crete, Greece and is currently employed as a Research Associate on the CEU RACE II ICM project. He is the leader of the project group on TMN architectures, performance management case studies and TMN system design for FDDI, ATM and optical networks.

Panos Georgatsos received the B.S. degree in Mathematics from the National University of Athens, Greece, in 1985, and the Ph.D. degree in Computer Science, with specialisation in network routing and performance analysis, from Bradford University, UK, in 1989. Dr. Georgatsos is working for ALPHA Systems SA, Athens, Greece, as a network performance consultant. His research interests are in the areas of network and service management, analytical modelling, simulation and performance evaluation. He has been participating in a number of telecommunications projects within the framework of the CEU funded RACE programme.

performance of the system as a whole.

5.6 Configuration Management OSF

The configuration manager is responsible for maintaining a consistent model of the physical and logical configuration of the network. It will receive configuration actions from the other OSFs and be responsible for implementing the changes in the network. This task may involve coordination of configuration actions over a number of network elements, for example when a VPC is created. The configuration manager can provide event reports to the other OSFs whenever a configuration action has succeeded.

5.7 Current Load Model OSF

The Current Load Model monitors the network usage and calculates usage statistics according to the requirements of the other OSFs. The Current Load Model is capable of calculating peak, mean, EWMA, etc. statistics according to the specifications of the other components. It will identify the minimum number of network probes and measurements to meet the varied demands of its users.

5.8 CAC Manager OSF

The CAC Manager reproduces the CAC algorithm in the network. When supplied with a traffic mix in the form of a list of the number of connections of each CoS the CAC Manager returns the effective bandwidth of that traffic mix. The calculation has exactly the same result as the equivalent CAC algorithm in the network.

5.9 CoS Model OSF

This models the bandwidth and performance targets for each CoS (see section 3.1).

6 INTERACTIONS BETWEEN THE ARCHITECTURAL COMPONENTS

6.1 Manager-Agent relationship

Figure 3 shows the manager-agent relationships between the derived components.

The VPC Bandwidth Distribution OSF and the Load Balancing OSF are agents of the Route Design OSF. However, their operation is not totally independent, since the effect (in the network) of one of them is taken into account by the other. The VPC Bandwidth Distribution OSF looks at the current load of the VPCs, which is determined by the routing decisions, and the Load Balancing OSF looks at the availability of the VPCs which is determined by the VPC Bandwidth Distribution OSF. This indicates that some coordination needs to exist among them, to avoid possible contradictions.

The Route Design OSF and the VPC Bandwidth Distribution OSF manage the VPC network whereas the Load Balancing OSF determines how to optimise its use. It can be argued that the Load Balancing OSF complements the VPC Bandwidth Distribution OSF, in the sense that it takes advantage of the VPC bandwidth increase.

When the Load Balancing OSF is activated it assumes a stable VPC network. This implies that during the operation of the Load Balancing OSF, the VPC Bandwidth Distribution OSF and the

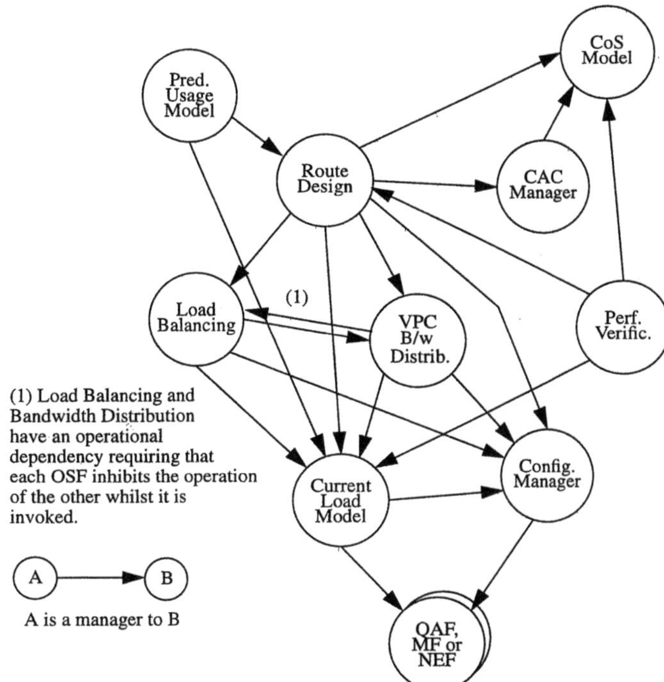

Figure 3 Manager-agent relationships between the OSFs.

Route Design OSF should be prohibited from taking actions. And conversely when the VPC Bandwidth Distribution OSF or the Route Design OSF are about to change VPC bandwidth or topology, the Load Balancing OSF should not be activated until the change has been made.

7 CONCLUSIONS AND FUTURE WORK

In this paper we dealt with a VPC and Routing Management Service for multi-class ATM networks. The proposed system offers the generic functions of performance monitoring, load monitoring and configuration management on ATM networks. In addition, it provides specific functions for routing and bandwidth management in a hierarchical structure. The components in the hierarchy differ in terms of the level of abstraction, complexity and timescale. The management functions to be invoked most frequently are close to the NEs and are as lightweight as possible to reduce management overhead. The more comprehensive functions are placed in the higher levels of the hierarchy and are only invoked when the lower levels are unable to resolve issues within the scope of their functionality and operational parameters. Such a hierarchy provides for continuous refinement of the management decisions and avoids the problems of a fully centralised approach.

The VPC and routing management system provides the following benefits to the network operator:

It allows the network to be used as efficiently as possible within the constraints of the physical resources. It will indicate when the network resources are insufficient for the traffic and hence additional resources need to be deployed. Alternatively it will show when resources are under-used and may be taken out of service or redeployed to avoid congestion elsewhere.

It implements the requirements of the service management layer to provide for users according to the business policy of the network operator. A range of service qualities and types (CBR and VBR) can be implemented for which the service management layer may charge different prices. It designs logical overlay VPC and routing networks so that the different service types can exist on the same physical network.

It distributes load as evenly as possible throughout the network to maximise the network availability and minimise disruptions in the case of failures. It can make dynamic configurations to adapt the network configuration to fluctuating traffic and make changes before they actually happen based on a Predicted Usage Model.

By building intelligence into the TMN the requirements on the NEs are simplified. The TMN functions replace the alternative of elaborate algorithms in the switches that must interact via signalling procedures to allow global network conditions to influence local algorithms. In a multi-class environment the inter-node exchange of routing information is prohibitive simply by the large number of CoSs. Therefore it increases the capacity for revenue earning traffic. By placing these functions in the TMN no additional requirements are placed on the NEs apart from the most basic of management interfaces.

The design is flexible enough to incorporate different algorithms or different levels of functionality to adapt to the specific CAC and RSAs in the network elements. Static algorithms in the elements can be transformed to quasi-static algorithms by TMN actions.

The proposed system can be used for implementing private and virtual private network services since it manages bandwidth reservation and routing within specified performance targets. Provision has been made (see Section 5.4) to provide an abstract interface to the service management functions responsible for the private services to implement their requests.

The architectural framework can be used as a testbed for testing and validating bandwidth management, routing management and load balancing algorithms.

At the time of writing algorithms for the architectural components described in this paper have been developed and the detailed design of prototypes has been completed. This work being undertaken by the RACE II ICM project. A significant portion of the system has already been implemented and demonstrated. Future work includes testing and validation of the components, the system and the architectural concepts on a real ATM testbed provided by another RACE II project (EXPLOIT) as well as in a simulated environment for scalability and extended testing purposes. The information modelling of the interfaces is based on the existing and emerging standards and where necessary, object definitions were expanded and new managed objects were defined. These extensions will be fed back into the standardisation activities.

8 ACKNOWLEDGEMENTS

This paper describes work undertaken in the context of the RACE II Integrated Communications Management project (R2059). The RACE programme is partially funded by the Commission of the European Union.

layers.

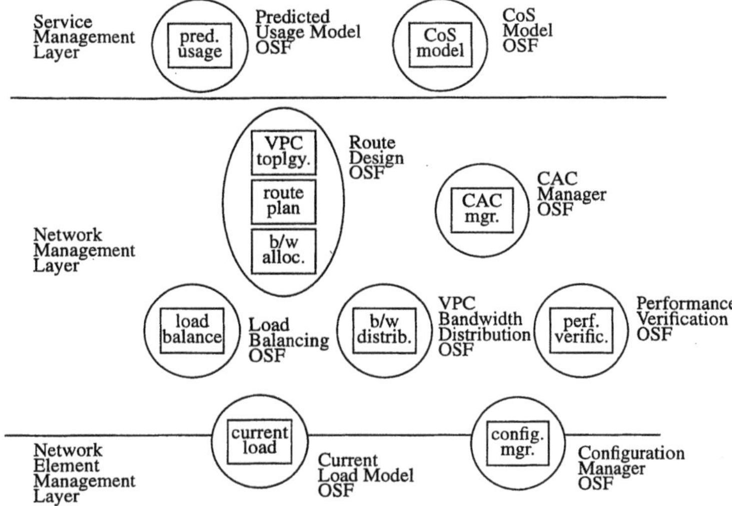

Figure 2 Mapping of MFCs to OSFs and OSFs to the TMN hierarchical layers.

By adopting a hierarchical TMN architecture we take advantage of a centralised management approach in the sense of reducing the placement of intelligence in the managed elements and so burdening their design and eventually their cost. But at the same time we use a hierarchical system to push management intelligence and frequently used management functions as close as possible to the network elements to avoid the management communications overhead inherent in centralised systems.

5 DESCRIPTION OF THE ARCHITECTURAL COMPONENTS

The functionality of the identified OSFs is briefly discussed in this section. The description is at a high level as the paper focuses on architectural rather than design issues. The OSF problems resemble the well known problems of network design, capacity sharing, bandwidth management and routing. However, these problems need to be consolidated and put into the perspective of the proposed architecture.

5.1 Route Design OSF

This OSF has both static and dynamic aspects. The static aspect is related to the network planning activity and is used to initially configure the network in terms of VPCs and routes. This part is performed at network start-up time. The dynamic aspects of its operation cater primarily for changes in the predicted network traffic and for prediction inaccuracies that could not be resolved by the lower level OSFs. As a result, the VPC and ClassRoute networks are reconfigured. The dynamic part consists of the functionality of the VPC Topology, Route Planning and Bandwidth Allocation MFCs.

The Bandwidth Allocation MFC is the first function to be invoked whenever the predicted

33

Managing Virtual Paths on Xunet III: Architecture, Experimental Platform and Performance

Nikos G. Aneroussis and Aurel A. Lazar

Department of Electrical Engineering and
Center for Telecommunications Research
Rm. 801 Schapiro Research Bldg.
Columbia University, New York, NY 10027-6699
e-mail: {nikos, aurel}@ctr.columbia.edu
Tel: (212) 854-2399

Abstract

An architecture for integrating the Virtual Path service into the network management system of future broadband networks is presented. Complete definitions and behavioral descriptions of Managed Object Classes are given. An experimental platform on top of the XUNET III ATM network provides the proof of concept. The Xunet manager is equipped with the necessary monitoring tools for evaluating the performance of the network and controls for changing the parameters of the VP connection services. Performance data from Xunet is presented to highlight the issues underlying the fundamentals of the operation of the VP management model such as the trade-off between throughput and call processing load.

Keywords

ATM, Quality of Service, Virtual Path Management, Performance Management, Gigabit Testbeds, Xunet

1. INTRODUCTION

Central to the operation of large scale ATM networks is the configuration of the Virtual Path (VP) connection services. VPs in ATM networks provide substantial speedup during the connection establishment phase at the expense of bandwidth loss due to reservation of network resources. Thus, VPs can be used to tune the fundamental trade-off between the cell throughput and the call performance of the signalling system. They can also be used to provide dedicated connection services to large customers such as Virtual Private Networks (VPNs). This important role of VPs brings forward the need for a comprehensive management architecture that allows the configuration of VP connection services and the evaluation of the resulting network performance. Furthermore, call-level performance management is essential to the operation of large ATM networks for routing decisions and for long term capacity planning.

The review of the management efforts for ATM broadband networks reveals that there has been little work regarding the management of network services. In [OHT93], an OSI-based management system for testing ATM Virtual Paths is presented. The system is used exclusively for testing the cell-level performance of Virtual Paths, and allows the control of cell generators and the retrieval of information from monitoring sensors. The system is designed for testing purposes only and does not have the capability to install Virtual Paths, regulate their networking capacity, or measure any call-level statistics.

A more complete effort for standardizing the Management Information Base for ATM LANs that meets the ATM Forum specifications is currently under way in the Internet Engineering Task Force (IETF) [IET94]. This effort focuses on a complete MIB specification based on the SNMP standard for configuration management, including VP configuration management. Performance management is also considered but at the cell level only.

The ICM RACE project [ICM93] is defining measures of performance for ATM networks both at the call and at the cell level and the requirements for Virtual Path connection management. It is expected to deliver a set of definitions of managed objects for VP management and demonstrate an implementation of the architecture.

In [ANE93] we have described a network management system for managing (mostly monitoring) low level information on XUNET III. Our focus in this paper is on managing services, in particular, services provided by the connection management architecture. In order to do so, there is a need to develop an understanding of the architecture that provides these services. The integration of the service and network management architectures can highly benefit from an overall network architecture model [LAZ93].

Within the context of a reference model for network architectures that we have previously published [LAZ92], we present an architectural model for VP connection setup under quality of service constraints. The architecture is integrated with the OSI management model. Integration here means that VPs set up by the connection management system can be instrumented for performance management purposes. The reader will quickly recognize that this instrumentation is representative for a large class of management problems such as billing (accounting), configuration management, etc.

We emphasized the following capabilities: monitoring Virtual Circuits (VCs) independently; monitoring and control of Virtual Paths; monitoring the call-level performance by computing statistics such as call arrival rates, call blocking rates, call setup times, etc.; control of the call-level performance through allocation of network resources to Virtual Paths, and control of other operating parameters of the signalling system that influence the call-level performance, such as retransmission time-outs, call setup time-outs, call-level flow control, etc.

We have tested our overall management system on the Xunet ATM broadband network that covers the continental US. Finally, we have taken measurements that reveal the fundamental trade-off between the throughput and the signalling processing load as well as other quantities of interest that characterize the behavior of broadband networks.

This paper is organized as follows. Section 2 presents the architectural framework for managing VP connection services. Section 3 describes the Xunet III experimental platform and the implementation details of the VP management system. Network experiments with the objective of evaluating the management model and the performance of the network under several configurations of the VP connection services are presented in Section 4. Finally, Section 5 summarizes our findings and presents the directions of our future work.

2. ARCHITECTURE

In this section we present an overall architectural framework for managing the performance of VP services on broadband networks. Underlying our modeling framework is the Integrated Reference Model (IRM) described in Section 2.1. The VP architecture embedded within the IRM is discussed in Section 2.2. The management architecture is outlined in section 2.3. Finally, in section 2.4 the integration of the service and network management architectures is presented.

2.1 The Integrated Reference Model

To overcome the complexity problems in emerging broadband networks - caused by the variety of communication services to be provided, the required quality of service guarantees, the large number of network nodes, etc. - there is an urgent need for integrating network management, service management and real-time control tasks into a consistent framework. To this end, we have developed an overall model for network architectures called the *Integrated Reference Model (IRM)* [LAZ92]. In this model, the key role for network integration is played by the network telebase, a distributed data repository that is shared among network mechanisms.

The IRM incorporates monitoring and real-time control, management, communication, and abstraction primitives that are organized into five planes: the network management or N-plane, the resource control or M-plane, the data abstraction and management or D-plane, the connection management or C-plane and the user information transport or U-plane (Figure 1). The subdivision of the IRM into the N-, M- and C-

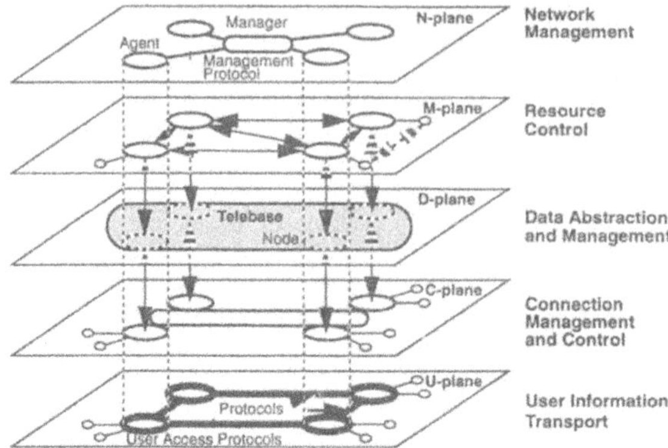

Figure 1: The Integrated Reference Model.

plane on the one hand, and the U-plane on the other, is based on the principle of separation between controls and communications. The separation between the N- and, M- and C-planes is primarily due to the different time-scales on which these planes operate.

The N- plane covers the functional areas of network management, namely, configuration, performance, fault, accounting and security management. Manager and agents, its basic functional components, interact with each other according to the client-server paradigm. The M-plane consists of the resource control and the C-plane of connection management and control. The M-plane comprises the entities and mechanisms responsible for resource control, such as cell scheduling, call admission, and call routing; the C-plane those for connection management and control. The user transport or U-plane models the protocols and entities for the transport of user information. Finally, the data abstraction and management or D-plane (the Telebase) implements the principles of data sharing for network monitoring, control and communication primitives, the functional building blocks of the N-, M-, and C- and U-plane mechanisms. (A mechanism is a functional atomic unit that performs a specific task, such as setting up a virtual circuit in the network [LAZ93]).

2.2 VP Architecture

The VP architecture closely follows the organization proposed by the IRM. It can be divided in two parts: the first part describes a model for establishing VPs, and the second presents a model for VP operation during the call setup procedure. In either case, central to the VP architecture is the D-plane of the IRM. The D-plane contains all information regarding the configuration and operational state of VPs and is used by the algorithms of the other planes both for monitoring and control operations.

The establishment of VPs is performed by the signalling system. The latter resides in the C-plane. A signalling protocol is used to establish a VP hop by hop. At every node along the route of the VP, the necessary networking capacity must be secured from the output link that the VP is traversing. The networking capacity of links is described by the Schedulable Region (SR) [HYM91], and of VPs by the Contract Region (CR) [HYM93b]. Informally, the Schedulable Region is a surface in a k dimensional space (where k is the number of traffic classes), that describes the allowable combinations of calls from each traffic class that can be accepted on the link and be guaranteed Quality of Service. The Contract Region is a region of the SR reserved for exclusive use by the VP. If the requested capacity allocation of a VP cannot be achieved, the allocated capacity at the end of the VP establishment phase is the minimum capacity available on the corresponding links (best effort resource allocation). The set of all VPs in the network, characterized by their route, Contract Region and associated configuration information, comprise the *VP distribution policy.* The VP distribution policy is stored in the D-plane.

An admission control algorithm located in the M-plane formulates the admission control policy (ACP), which is encoded as an object in the D-plane. The ACP is used by the signalling algorithm of the C-plane to make admission control decisions for incoming call requests. Thus, the VP architecture represents a connection service installed in the D-plane.

Figure 2 shows the interaction between entities in the various planes of the IRM that provide the VP con-

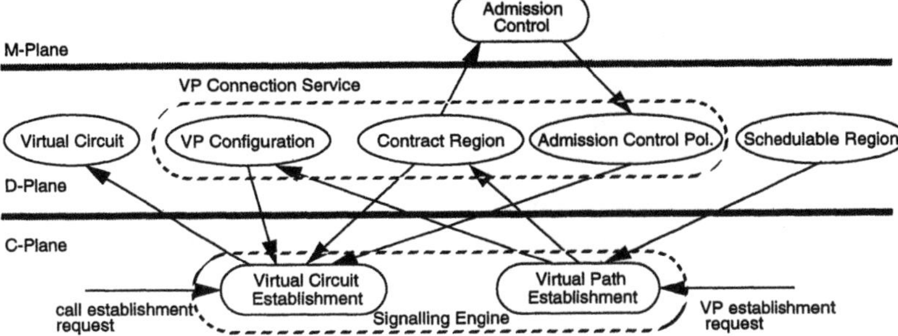

Figure 2: Flow of Information during Installation and Operation of the VP Connection Service.

nection service. During the VP establishment phase, the signalling engine creates a set of 3 objects in the D-plane: the CR, ACP and VP Configuration objects. The VP configuration object contains general VP configuration information such as the input and output port numbers, the allocation of the VCI space, the VP operational state, etc.

During the VC establishment phase, the signalling engine reads the VP configuration object to determine if the VP can be used to reach the desired destination. It also reads the CR and ACP objects to examine if the call can be admitted on the VP. When the call has been established, a Virtual Circuit object is created in the D-plane that contains all necessary information for the VC. This information includes the VP Identifier (VPI) and VC Identifier (VCI), the traffic descriptor used to allocate resources, and other parameters for performance monitoring.

VPs can be used in two ways. If the VP is terminated at the Customer Premises Equipment (CPE), the customer is controlling the VP admission controller. In this case the VP can be regarded as a dedicated virtual link (or pipe) of a rated networking capacity. A network composed of such VPs terminated at the customer premises is also known as a Virtual Private Network (VPN). The Network Manager has the capability to configure and maintain a VPN by managing the individual VP components according to the customer requirements.

Alternatively, the termination of VPs may not be visible to the network customer. In this case, VPs are used by the network operator to improve the performance of the signalling system, the availability of resources between a pair of nodes, or even improve certain call level measures of Quality of Service for the customer, such as call setup time and blocking probability.

2.3 Management Architecture

The Management Architecture builds on the OSI Management model. According to this model, network entities (either physical, like hardware modules, or logical, like virtual circuits) are mapped into "Managed Objects" for monitoring and control purposes. The managed objects are also referred to as *logical objects* and the network entities that they represent as *real objects*. A Management Agent contains the information about the managed objects in the Management Information Base (MIB). The MIB is an object-oriented database. Managed objects are characterized by a set of attributes that reflect the state of the corresponding real object and behavioral information, which defines the result of management operations on the managed object. A proprietary protocol can be used for *linking* the state of every real object to its logical counterpart in the MIB. The Manager connects to the agent(s) and performs operations on the objects in the MIB using CMIP (the Common Management Information Protocol). These operations are of synchronous nature, i.e., they are initiated by the manager who, then, waits for a reply from the agent(s). Events of asynchronous nature (notifications) such as hardware faults can be emitted from the agent(s) to the manager using the event reporting primitive of CMIP.

Management operations take place in the N-plane of the IRM (Figure 1). The MIB of every agent is located in the D-plane of the IRM. As a result, the linking of the logical objects in the MIB with real objects is done within the D-plane [TSU92]. Control operations from the manager applied to objects in the MIB are reflected in the state of the real objects of the D-plane, which in turn, affect the behavior of the algorithms in the C- and M- planes. Conversely, the change of state of the real objects in the D-plane will cause an update of the state of the logical objects in the MIB.

Therefore, in our model, monitoring and control of the VP architecture is possible by defining the appropriate managed objects in the MIB and linking them to the appropriate entities of the D-plane. What managed objects to define, how to integrate them in the D-plane and how to define their behavior will be the topic of the following section.

2.4 Integration of the VP and Management Architecture

The purpose of this section is to describe the object level model for VP management and its integration within the D-plane. Management of VPs takes place in the N-plane. The network manager decides on a VP distribution policy and implements this policy by issuing commands to the agents installed across the network. Although the capabilities to install and control VPs are essential requirements to implement a VP distribution policy, it is also essential for the manager to evaluate the performance of the network under a given VP distribution policy. For this reason, a generic performance management model (in addition to the VP management model) both at the call and the cell level becomes necessary. Note, however, that VP management operations stem from the call-level performance management model, and therefore, the VP management model can be considered as part of the latter.

The performance management model consists of a set of *quantities* that reflect network performance, and a set of *controls* that affect this performance. A set of *rules* use the performance measures to derive the necessary controls that will reach a *performance objective*. At the call level, the quantities of interest are the call arrival rate, the call blocking rate, the call setup time, the signalling load, etc.; at the cell level, the cell arrival rate, cell throughput and end-to-end delay. These quantities together with a set of controls must appear in the definition of Managed Object Classes (MOCs) for performance management.

In our model, one agent is installed at every ATM switch. The agent processes the information on call attempts from every input port. For each successful call attempt, an object of class *VcEntity* is created for the corresponding Virtual Circuit connection (VC). Each VC object contains configuration information such as the number of the input and output slot, Virtual Path Identifier (VPI) and Virtual Circuit Identifier (VCI). In ATM terminology, this implies that the VC object models the input and output Virtual Circuit Link (VCL) at every switch. Thus, the end-to-end management of a VC that spans many switches (and hence has one instance in each OSI agent at every switch) is achieved by managing the individual objects in combination. Additional attributes for each VC include the call establishment time, traffic descriptor (composed of a service class characterization and the allocated networking capacity in kilobits per second), adaptation layer information and source/destination end-user service information. The package *cellPerformancePackage* contains attributes associated with the cell-level performance related parameters, and will be described below.

The class *VirtualPath* derived from *Top* is used to describe a VP. The VP object in analogy with the VC object is comprised of an incoming and an outgoing part at every switch. At the VP source or termination point, the VP has only an outgoing / incoming part respectively. Attributes used to describe the configuration of the Virtual Path are: *vpIdentifier* (VPI), *vpSource*, *vpDestination* (VP source and termination address), *circuitCapacity* and *timeEstablished.* The VP object at the source also contains a *callPerformancePackage*, and an *admissionControllerPackage.* These will be described below.

The class *Link* is derived from *Top* and is used to model input or output network links. The mandatory attributes for this class are *linkType* (input or output), *linkModuleDescription* (describes the hardware of the link interface), *linkSource, linkDestination* and *linkSlotNumber* (the slot number the link is attached to). If it is an output link, it contains a *callPerformancePackage*, and an *admissionControllerPackage.*

The class *SourceDestination* is used to describe the call level activity between a pair of nodes, and can be used to evaluate the call level performance in an end-to-end fashion. A Source-Destination (SD) object exists in the agent if there is call-level activity between the two nodes, and the source node is either the local switch, or a directly attached User-Network Interface (UNI). The SD object contains the following attributes: *sourceNodeAddress* and *destinationNodeAddress* and a *callPerformancePackage.*

The *callPerformancePackage* is an optional package that measures the call-level performance. It is contained in all SD objects, and in some link and VP objects. For the objects of class *Link,* the package measures the activity for calls that follow the link but not a VP that uses the same link. For VP objects, the package measures the activity of call requests that use the VP. The attributes of the *callPerformancePackage* are the following: *activeCircuits*, *callArrivalRate* (average arrival rate of requests in calls/min), *callArrivedCounter* (counter of call requests), *callResourceBlockedCounter* (counter of calls blocked due to resource unavailability), *callErrorBlockedCounter* (counter of calls blocked due to protocol errors, e.g., time-outs, etc.), *callBlockingRate* (average rate of calls blocked for any reason in calls/min), *setupTime* (average time to establish the connection in milliseconds), *holdingTime* (average duration of connections in seconds), *numExchangedMessages* (average number of messages that have been exchanged to setup the connections, as an indicator of the processing required for each connection), and *measureInterval* (the time in which the above averages are computed in seconds). All quantities are measured separately for each traffic class, and then a total over all classes is computed.

The *cellPerformancePackage* measures cell-level performance. The attributes *cellTransmittedCounter, cellTransmissionRate*, *cellDroppedCounter* and *cellDroppedRate* measure the number of cells transmitted or blocked and their respective time averages. The attribute *avgCellDelay* measures the average time from the reception till the transmission of cells from the switch. The package is included in objects of class *VcEntity*, and in this case, only the cells belonging to the VC are measured. As an option, it can also be included in objects of class Link, SourceDestination or VirtualPath. In the latter case, a sum of the attributes over all VC objects that belong to the Link/SourceDestination/VirtualPath is computed, and the respective attributes of the Link/SourceDestination/VirtualPath objects are updated.

The package *admissionControllerPackage* is mandatory for output link and VP objects. It describes the state of the admission controller, which is located at the output links (for switches with output buffering) and at all VP source points. The package contains the following attributes: *networkingCapacity* (the schedulable region for link objects or the contract region for VP objects), *admissionControllerOperatingPoint* (the operating point of the admission controller given the established calls for each traffic class), *admissionControlPolicy, admissionControllerOperationalState* (enabled (call requests allowed to go through and allocate bandwidth) or disabled) and *admissionControllerAdministrativeState.*

The class *ConnectionMgmt* contains attributes that control the operation of the local signalling entity. There is only one instance of this class in every agent. Attributes of this class are the following: *signallingProcessingLoad* (an index of the call processing load observed by the signalling processor), *maxSignallingProcessingLoad* (the maximum signalling load value allowed, beyond which the signalling processor denies all call establishment requests), *signallingRetransmitTimeout* (the time-out value in milliseconds for retransmitting a message if no reply has been received), and *signallingCallSetupTimeout* (the maximum acceptable setup time in milliseconds for a call establishment. If the time to establish a circuit is more than the current value, the circuit is forced to a tear-down). The single instance of this class is also used to contain four other container objects of class *LinkMgmt*, *SourceDestinationMgmt*, *VirtualPathMgmt*, and *VirtualCircuitMgmt*. There is only one instance from each of these four classes, which is used to contain all objects of class *Link*, *SourceDestination*, *VirtualPath*, and *VirtualCircuit*, respectively.

As discussed in the previous section, the MIB of every agent resides in the D-plane. Managed Objects use the information stored in the D-plane to update their state. For example, the Managed Objects of class *VcEntity* represent the Virtual Circuit object that was created in the D-plane by the signalling system. The attributes of the managed object mirror the state of the corresponding real object. In the same manner, the MO of class *VirtualPath* contains attributes that reflect the state of the corresponding real objects (VP Configuration, Contract Region and Admissible Load Region). An MO of class *Link*, uses the object Schedulable Region (among other information), to reflect the state of the link Schedulable Region on one of its attributes. Additional processing of events (such as VC creation, etc.) inside the agent can provide the necessary call-level performance related properties (such as call arrival rates). These might not be readily available from other objects of the D-plane (see [ANE94] for more details).

The purpose of the above description was to give an overview of the managed object classes and attributes for performance management. For simplicity, we omitted the definition of associated thresholds for each performance variable that can trigger notifications in case of threshold crossing [ISO92]. Such definitions can be easily incorporated in the above model.

3. EXPERIMENTAL PLATFORM

3.1 The Xunet ATM Testbed

Xunet is one of the five Gigabit testbeds sponsored by the Corporation for National Research Initiatives. It has been deployed by AT&T in collaboration with several universities and research laboratories in the continental United States [FRA92]. The topology of the network is shown in Figure 4. The network links are currently rated at 45 Mbps and are gradually substituted with 622 Mbps links. Access at every node is provided by 200 Mbps network interfaces. A variety of standard interfaces (TAXI, HiPPI, etc.) is under development and will be available in the near future. A workstation serves as the switch Control Computer (CC) at each network node. The CC runs the switch control software that performs signalling, control and fault detection functions.

3.2 The Xunet VP Signalling and Admission Control Architecture

Xunet supports five traffic classes. Class 1 is used for high priority control messages and is given absolute priority by the link scheduler (multiplexer). Class 2 is used for Video service, Class 3 for Voice, Class 4 for priority data and Class 5 for bulk data [SAR93].

A signalling system very similar in characteristics with the CCSS#7 (Common Channel Signalling System) has been installed on Xunet. The system allows virtual circuit establishment with best effort resource allocation in a single pass. An admission controller operates on every output link. The necessary Schedulable Region and Admission Control Policy objects are downloaded from a management station. The admission control policy used is complete sharing [HYM93a].

Virtual Path establishment is also done in one pass with best effort resource allocation. When the VP has been established, an admission controller is activated at the source node of the VP that uses the allocated contract region for admission control decisions. The admission control policy is again complete sharing. A signalling channel is also established between the two VP termination points to carry call establishment requests over the VP. It operates in the same way as the signalling channel used on every physical link. As a result, from the point of view of the signalling system, VPs are considered as regular links with only minor differences.

Every Contract Region can be changed dynamically. The deallocation or allocation of additional resources is performed in the same way as in the VP establishment phase. Finally, when a VP is removed, the Contract Region is returned to the Schedulable Regions of the links along the path, all VCs using the VP are forced to termination and the VP signalling channel is destroyed.

3.3 The Xunet OSI Management System

From the five functional areas covered by the OSI management model, we have chosen to implement a configuration, fault and performance management architecture for Xunet (the remaining functional areas being security and accounting management). The configuration and fault management architecture enables us to monitor closely all the network switches for hardware faults, such as link level errors, buffer overflows, etc. The performance management architecture builds on the performance management model (managed object definitions and behavior) that was presented in the previous section.

As the basis of our OSI Management system, we have selected the OSIMIS software [KNI91]. Our implementation expanded the agent with managed objects for Xunet and the management applications to include powerful graphics that depict the operational state of the network and a control interface that facilitates the overall management task. The management applications run at Columbia University. TCP/IP sockets are used at the transport layer to connect to the agent at each site. Inside the agents, communication between logical and physical objects is achieved by using a proprietary protocol between the agent and the Xunet switch. For this purpose, we use UDP/IP packets over a local Ethernet.The structure of the system is shown in Figure 3.

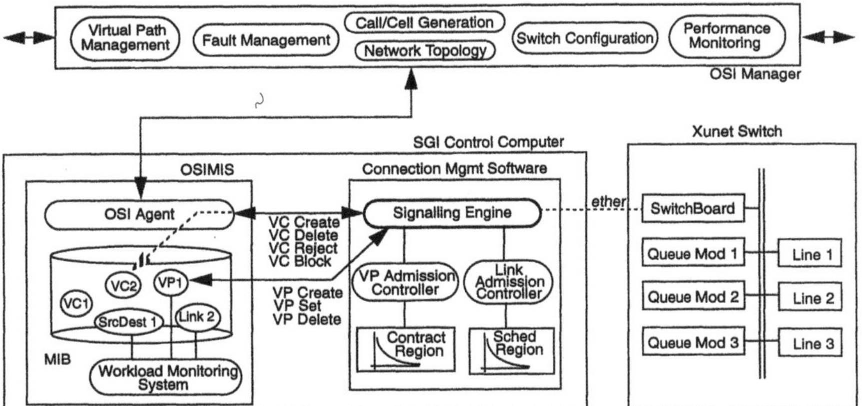

Figure 3: Structure of the Xunet Management System.

3.4 The OSI Agent

The OSI agent contains managed objects for configuration, fault and performance management. The agent consists logically of two major groups of Managed Objects.

3.4.1 Hardware Configuration and Fault Management Group (HCFMG)

For the purpose of configuration and fault management, we have implemented managed object classes for each Xunet hardware module, such as *SwitchBoard*, *QueueModule*, etc. Each module is polled at regular time intervals by the agent to detect possible faults. A hardware fault triggers a notification inside the agent, which in turn can generate a CMIS Event Report primitive if the appropriate Event Forwarding Discriminator object has been created by the manager [ISO91]. Currently, more than 300 different hardware errors can produce an equal number of event reports. This wealth of information provides the manager with extensive fault monitoring capabilities. The configuration and state of the hardware modules is obtained from the Xunet switch every 20 seconds. The information is processed internally to update the corresponding managed objects.

The set of the hardware managed objects also gives complete configuration information of every switch. The management applications can display graphically the configuration and indicate the location of every generated event report.

3.4.2 Performance Management Group (PMG)

The PMG consists of a set of managed objects that monitor closely the performance of Xunet both at the cell and at the call level. All call level information is obtained from the local signalling entity. The OSI agent receives four types of events: VC-Create, VC-Delete, VC-Blocking (no network resources) and VC-Rejection (any other cause), with the appropriate parameters. Upon a VC creation event, a Managed Object of class *VcEntity* is created inside the MIB that contains all the available information on this VC. The object is related to the appropriate *Link, SourceDestination* or *VirtualPath* objects. Every 30 seconds, the Xunet switch is scanned to compute the number of cells transmitted or dropped for each VC. At the same time we use this information to update the total number of cells transmitted or lost on a link, SD pair or VP based on the relations defined when the VC was created. The VC object is removed by a deletion event.

All 4 event types cause the agent to update some internal counters in the corresponding Link/SD/VP objects. Additional processing is performed at a regular time interval (controllable by the manager through the *measureInterval* attribute, and usually set to 30 seconds). At that time, the agent calculates the values for the performance related attributes, taking into account only the events that occurred during the past interval. For example, when a VC is created, a counter that registers the call arrivals is incremented. At the end of the 30 second period, the arrival rate is calculated, and the counter is reset to zero. All other attributes are calculated in a similar fashion.

VP management functions originate at the network manager site. When the management application issues an M-Create command with the appropriate parameters, a VP managed object inside the MIB is instantiated, and the Xunet signalling entity is informed to initiate a VP setup procedure. VPs can be subsequently modified by the M-Set command operating on the appropriate object, and deleted with an M-Delete command.

Parameters of the signalling entity are controlled through M-Set operations on attributes of the *ConnectionMgmt* object. Each Set operation causes a control message to be sent from the agent to the signalling entity.

3.5 The OSI Manager

Xunet is currently monitored and controlled through a Motif/X-Toolkit-based application. The same application is used to run controlled call and cell generation experiments on Xunet. It consists of six tightly interconnected subsystems (Figure 3). Every subsystem contains the appropriate display tools and management interface functions for a specific management task:

1. Switch Configuration: Displays the hardware configuration of the switch using information from the objects of the HCFMG.
2. Fault Management: Receives OSI Event reports from the agents, that are related to hardware problems, and uses the Switch Configuration subsystem's display functions to inform the manager about the nature and location of the problem.
3. Network Topology: Displays a map of the network, with all switches, links and attached user-network interfaces. The displayed objects can be selected and observed independently. Support is also provided for displaying the route and configuration information of VPs.

4. Virtual Path Management: The manager is able to create and subsequently control VPs with M-Create and M-Set operations. The VP control task is guided by the observations obtained from the Performance Monitoring system.
5. Performance Monitoring: Collects the information that is provided by the PMG objects in each node and displays it using the functions of the Network Topology subsystem. The information can be either displayed in textual form, or graphically. In the latter case, we use time series plots that are updated in real-time. The plots allow us to observe the performance "history" of the network and the effects of VP management controls.
6. Call and Cell Generation: The Xunet signalling entities contain a call generation facility. A managed object inside the local agent makes it possible to control the call generation parameters in terms of destination nodes, call arrival rate and call holding time on a per traffic class basis. The call generation system can also be linked to the Xunet cell generator for real-time cell generation.

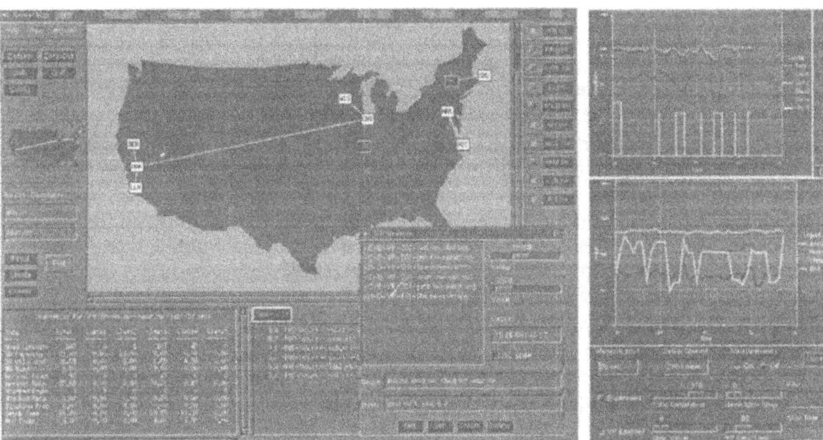

Figure 4: The Xunet Management Console displaying the call level performance.

4. PERFORMANCE

We are currently using the management system to run controlled experiments on Xunet to study the call level performance of the network, such as the performance of the signalling system and the network throughput under various VP distribution policies. Call level experiments consist of loading the signalling system with an artificial call load. A Call Generator on every switch produces call requests with exponentially distributed interarrival and holding times. In the remaining of this section we will focus on the objective of performance management at the call level and will demonstrate results from various call-level experiments conducted on Xunet.

4.1 Semantics of Performance Management

The objective of performance management at the call level can be summarized in the following:

- Minimize call blocking due to unavailability of network resources. This unavailability can be caused by several factors including a faulty link, a poor VP distribution policy, a routing malfunction, an overloaded signalling processor, etc.

- Minimize the call setup time. The call setup time is perceived by the user as a measure of the quality of service offered by the network. High call set up times may prompt the user to hang-up leading to loss of revenue for the service provider.

Increasing the bandwidth of a VP results in reducing the signalling load on the network, but also in a possibly reduced network throughput. Our main goal is to evaluate this fundamental trade-off between network throughput and signalling load and choose a VP distribution policy that results in the best overall performance.

The manager collects measurements in regular time intervals, and evaluates network performance, either in a per SD-pair basis or by looking at individual nodes, links or VPs. If the performance is not satisfactory (high blocking, high call setup times and high signalling load), the manager can apply the following controls:

1. Create a VP between two nodes and allocate resources to it. This action alleviates the intermediate nodes from processing call requests and decreases the call setup time.
2. Delete a VP responsible for the non-satisfactory performance. This course of action may be taken because the maximum number of VP terminations has been reached and new VPs cannot be created in the system, or because there is no offered load to the VP, or because a new VP distribution policy has been decided and the VP topology must change.
3. Change the allocated networking capacity of a VP either by releasing a part of or increasing the allocated resources. This control is performed when the load offered to the VP has been reduced or increased.
4. Change signalling parameters, such as the time-out for call setups, the time-out for message retransmissions and the maximum allowed signalling load (which is a window-type control on the number of requests handled by the signalling processor). These parameters affect the call blocking rates, but also the average call setup time.

With the above in mind, the call-level experiments have been separated in two major phases. In the first phase, we measure the performance of the signalling system without using VPs. This experiment allows us to find suitable values for the parameters of the signalling entities that give the highest call throughput. The second phase builds upon the first phase, and attempts to determine the call throughput by measuring the performance of the network with VPs in place.

4.2 Performance of the Signalling System for Virtual Circuit Set-Up

In this experiment, the network was loaded with an artificial call pattern. Our goal was to measure the performance of the signalling system under heavy call arrivals. For each call, the signalling system sets up a VC by traversing all the nodes in the path and patching the appropriate connections in each node. Call generation is controllable for each destination in terms of the call arrival rate and call holding time for each of the five traffic classes supported by Xunet. The network was homogeneously loaded from five sites (Murray Hill, Rutgers U., U. of Illinois, U.C. Berkeley and Livermore) with Poisson call arrivals and an exponential holding time with a mean of 3 minutes. We used only video calls (assumed to consume a peak of 6 Mbps/call) and voice calls (64 Kbps/call) in a ratio of 1:10 (i.e., the arrival rate of voice calls is 10 times greater). All the links in the experiment described here are of 45 Mbps capacity. The schedulable region (SR) of each link assumed to be given by a two dimensional hyperplane. We used peak rate allocation, and according to this scheme, the SR can accommodate a maximum of 7 video calls or 703 voice calls. The admission control policy used was complete sharing [HYM93a].

Figure 5 shows the measurements obtained by gradually increasing the call generation rate. Each measurement (throughput, call setup time, call blocking etc.) is computed by adding (or averaging) the respective measurements for the video and voice traffic. Both the call throughput and call blocking due to resource unavailability (the "Throughput" and "Blocked Percentage" curves) rise almost linearly with the call arrival rate. The sudden drop in the total call throughput is due to the overloading of the signalling system with call setup requests. At that point, the call setup time and the percentage of calls blocked due

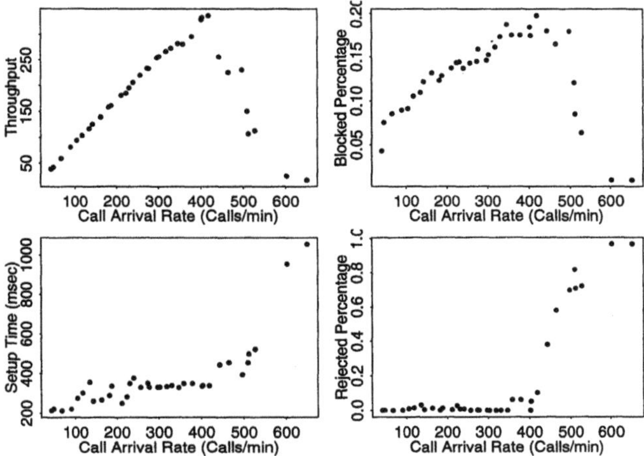

Figure 5: Performance of the Signalling System.

to congestion of the signalling system (the "Rejected Percentage" plot) start to rise sharply. The "BlockedPercentage" curve drops because the strain has now been moved from network transport to call setup, and thus, calls contend for signalling resources rather than networking resources. During overload, only a small percentage of the total call attempts is actually established, and therefore, the probability that these calls will find no networking capacity available is diminished. In the extreme situation, all calls are blocked while the networking capacity of all links is unused.

The congestion situations seem to appear first at the Newark and Oakland switches, that are the first to become overloaded with call request messages. It is therefore essential for the network manager to regulate the call arrival rate at the entry points in the network. This can be done by setting an appropriate value for the *maxSignallingProcessingLoad* attribute of the ConnectionMgmt object. The signalling load is computed from the number of signalling messages received and transmitted from the switch in the unit of time. If the load reaches the *maxSignallingProcessingLoad* value, a fraction of the incoming call requests are discarded. We have found experimentally that by restricting the signalling load to about 450 messages per minute at the nodes connected to the call generators, the network operates within the capacity of the signalling processors.

4.3 Performance Trade-off due to VP Capacity Allocation

This experiment had the objective of studying the trade-off between the network throughput and the allocation of networking capacity to VPs. It was performed on the east coast segment of the network. This four node segment consists of two nodes in MurrayHill (called MHEX and MH), one node in Newark (NWRK), and one at Rutgers University (RUTG) connected in tandem.

The generation ratio between Class 2 and Class 3 calls was 1:100. The call arrival rate from each call generator was kept fixed throughout the experiment. The generator at MHEX produces traffic to NWRK (180 calls/min) and RUTG (210 calls/min). The generator at MH produces only traffic to RUTG at 180 calls/min. The generator at RUTG produces traffic to NWRK at 180 calls/min. One VP is established from MHEX to RUTG (Figure 6).

Only the traffic from MHEX to RUTG is allowed to follow the VP. Calls that find the VP full and the calls from other traffic generators follow a hop by hop call setup procedure, that is, trying to secure bandwidth on each hop. By contrast, calls that follow the VP contest for bandwidth only at the VP source node. The

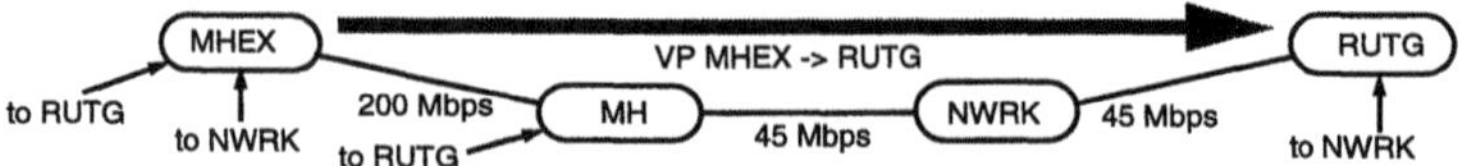

Figure 6: Network Topology for the VP experiment.

capacity of the VP varies from 0 to 100 percent of the capacity of the smallest link on the VP (the MH-NWRK link which is rated at 45 Mbps). When the VP capacity is at 100% only calls from MHEX to RUTG are allowed, since all other calls find no available resources to proceed to their destination. When the VP capacity is reduced to 0, all calls are attempting the regular VC setup procedure. Figure 7 shows the obtained measurements. The throughput curve reveals that the maximum throughput is attained when

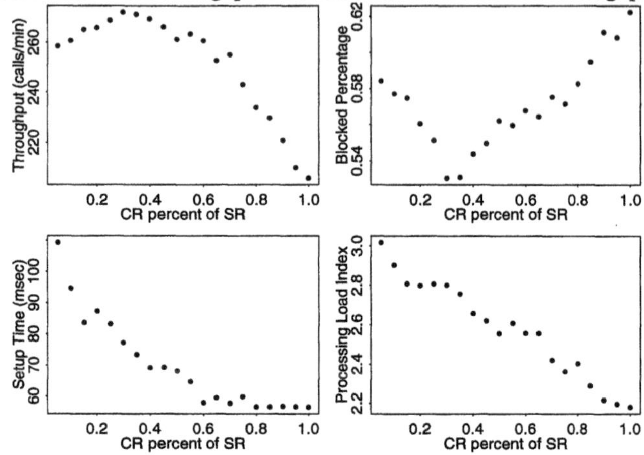

Figure 7: Virtual Path performance vs. allocated networking capacity.

the VP contract region is approximately 30 percent of the link schedulable region. This happens because below that value, an increasing amount of call requests from MHEX to RUTG find the VP full and use the regular VC setup procedure, thereby forcing the signalling entity at MH and NWRK into an overload state, that causes high call setup times and higher blocking. When the VP contract region increases above 30 percent, the throughput drops slowly as the extra capacity allocated for the VP is partially unused, and as a result a larger percentage of the interfering traffic (that does not follow the VP) is blocked. The fourth plot depicts the average number of signalling messages needed to establish (or reject) an incoming call. The numbers drop as the VP increases in capacity, as calls from MHEX to RUTG follow the VP and use less hops to reach the destination.

5. SUMMARY AND FUTURE WORK

A basic model for the performance management of VP connection services for ATM broadband networks was presented. A set of managed object classes following the OSI standard for network management with complete attribute structure was proposed. The call-level model enables the network manager to retrieve information from agents installed in the network, make decisions based on the configuration and performance observations, and apply a set of controls if the observed performance can be improved. These controls include setting the operating parameters of the signalling code and changing entirely or in part the distribution of the VPs in the system.

Our model was fully implemented on the Xunet ATM testbed. The manager is able to observe the call level performance of Xunet from a dedicated management tool. We have presented some aspects on the call level performance of Xunet and demonstrated the behavior of the network when VPs are in use.

We are currently working on an algorithm for an automated tool that observes the offered call load and the call-level performance related properties and makes decisions regarding the VP distribution policy and the operating parameters of the signalling software. Such a system will significantly facilitate the performance management task for a network with a large number of nodes and VPs.

This work was funded in part by NSF Grant CDA-90-24735, and in part by a Grant from the AT&T Foundation.

REFERENCES

[ANE94] Nikos G. Aneroussis and Aurel A. Lazar, "Managing Virtual Paths on Xunet III: Architecture, Experimental Platform and Performance", CTR Technical Report #369-94-16, Center for Telecommunications Research, Columbia University, 1994, URL: "ftp://ftp.ctr.columbia.edu/CTR-Research/comet/public/papers/94/ANE94.ps.gz".

[ANE93] Nikos G. Aneroussis, Charles R. Kalmanek and Van E. Kelly, "Implementing OSI Management Facilities on the Xunet ATM Platform," *Proceedings of the Fourth IFIP/IEEE International Workshop on Distributed Systems: Operations and Management*, Long Branch, New Jersey, October 1993.

[FRA92] A.G. Fraser, C.R. Kalmanek, A.E. Kaplan, W.T. Marshall and R.C. Restrick, "Xunet 2: A Nationwide Testbed in High-Speed Networking," *Proceedings of the IEEE INFOCOM'92*, Florence, Italy, May 1992.

[HYM91] Jay M. Hyman, Aurel A. Lazar, and Giovanni Pacifici, "Real-time scheduling with quality of service constraints,"' *IEEE Journal on Selected Areas in Communications*, vol. 9, pp. 1052-1063, September 1991.

[HYM93a] Jay M. Hyman, Aurel A. Lazar, and Giovanni Pacifici, "A separation principle between scheduling and admission control for broadband switching," *IEEE Journal on Selected Areas in Communications*, vol.11, pp. 605-616, May 1993.

[HYM93b] Jay M. Hyman, Aurel A. Lazar, and Giovanni Pacifici, "Modelling VC, VP and VN Bandwidth Assignment Strategies in Broadband Networks", *Proceedings of the Workshop on Network and Operating Systems Support for Digital Audio and Video*, Lancaster, United Kingdom, November 3-5, 1993, pp. 99-110.

[ICM93] ICM Consortium, "Revised TMN Architecture, Functions and Case Studies", ICM Deliverable 5, 30 September 1993.

[IET94] Internet Engineering Task Force, "Definition of Managed Objects for ATM Management", Internet Draft Version 7.0, March 9, 1994.

[ISO91] Information Processing Systems - Open Systems Interconnection, "Systems Management - Fault Management - Part 5: Event Report Management Function," July 1991. International Standard 10164-5.

[ISO92] Information Processing Systems - Open Systems Interconnection, "Systems Management - Performance Management - Part 11: Workload Monitoring Function", April 1992. International Standard 10164-11.

[LAZ92] Lazar, A.A., "A Real-Time Management, Control and Information Transport Architecture for Broadband Networks", in *Proceedings of the 1992 International Zurich Seminar on Digital Communications*, Zurich, Switzerland, March 1992.

[LAZ93] Lazar, A.A. and Stadler, R.,"On Reducing the Complexity of Management and Control in Future Broadband Networks", *Proceedings of the Fourth IFIP/IEEE International Workshop on Distributed Systems: Operations and Management*, Long Branch, New Jersey, October 1993.

[KNI91] George Pavlou, Graham Knight and Simon Walton, "Experience of Implementing OSI Management Facilities," *Integrated Network Management, II* (I. Krishnan and W. Zimmer, editors), pp. 259–270, North Holland, 1991.

[OHT93] Ohta, S., and Fujii, N., "Applying OSI System Management Standards to Virtual Path Testing in ATM Networks", *Proceedings of the IFIP TC6/WG6.6 Third International Symposium on Integrated Network Management*, San Francisco, California, 18-23 April, 1993.

[SAR93] H. Saran, S. Keshav, C.R. Kalmanek and S.P. Morgan, "A Scheduling Discipline and Admission Control Policy for Xunet 2", *Proceedings of the Workshop on Network and Operating Systems Support for Digital Audio and Video*, Lancaster, United Kingdom, November 3-5, 1993.

[TSU92] Tsuchida, M., Lazar, A.A., Aneroussis, N.G., "Structural Representation of Management and Control Information in Broadband Networks", *Proceedings of the 1992 IEEE International Conference on Communications*, Chicago IL., June 1992.

Nikos G. Aneroussis was born in Athens, Greece in 1967. He received the Diploma in Electrical Engineering from the National Technical University of Athens, Greece, in May 1990, and the M.S. and M.Phil. degrees in Electrical Engineering from Columbia University, New York, NY in 1991 and 1994. Since 1990, he is a graduate research assistant in the department of Electrical Engineering and the Center for Telecommunications Research at Columbia University, where he is currently pursuing the Ph.D. degree. His main research interests are in the field of computer and communication networks with emphasis on management architectures for broadband networks and network performance optimization. He is a student member of the IEEE and a member of the Technical Chamber of Greece.

Aurel A. Lazar is a Professor of Electrical Engineering and the Director of the Multimedia Networking Laboratory of the Center for Telecommunications Research, at Columbia University in New York.

Along with his longstanding interest in network control and management, he is leading investigations into multimedia networking architectures that support interoperable exchange mechanisms for interactive and on demand multimedia applications with quality of service requirements.

A Fellow of IEEE, Professor Lazar is an editor of the ACM Multimedia Systems, past area editor for Network Management and Control of the IEEE Transactions on Communications, member of the editorial board of Telecommunication Systems and editor of the Springer monograph series on Telecommunication Networks and Computer Systems. His home page address is http://www.ctr.columbia.edu/~aurel.

SECTION SEVEN

Telecommunications Management Network

34

Modeling IN-based service control capabilities as part of TMN-based service management

T. Magedanz
Technical University of Berlin
Open Communication Systems, Hardenbergplatz 2, 10623 Berlin, Germany
Phone: +49-30-25499229, Fax: +49-30-25499202

Abstract

IN and TMN standards represent the key constituents of future telecommunication environments. Since both concepts have been developed independently, functional and architectural overlaps exist. The harmonization and integration of IN and TMN is therefore currently in the focus of several international research activities. This paper investigates the thesis as to whether IN service features may be substituted by corresponding TMN management service capabilities. This means that service control of telecommunications services could be regarded as being part of the functional scope of TMN service management. Therefore this paper analyses the relationship between IN service control and TMN service management and examines, if and how TMN concepts with respect to functional and architectural aspects could be used as a basis for the provision of IN-like service capabilities for a variety of communication applications in a unified way.

Keywords

Customer Profile Management, IN, Service Control, Service Management, TMN

1 INTRODUCTION

In the light of a broad spectrum of different bearer network technologies (i.e. PSTN, ISDN, B-ISDN) the service-oriented network architecture of the *Intelligent Network (IN)* concept is intended to unify the creation, provision and control of advanced telecommunication services above these heterogeneous networks in a highly service-independent manner. Hence, it can be considered as the basic "network" architecture for the realization of sophisticated telecommunication services in the coming age. The *Telecommunications Management Network (TMN)* provides the world-wide accepted ultimate framework for the unified management of all types of telecommunication services and the underlying networks in the future. It provides the basis for the modeling of management services, management information and related management interfaces. Both concepts were standardized at the beginning of the 1990's within the international standards bodies [Q.12xx], [M.3010].

IN and TMN are closely related in the future telecommunications environment, since they cover complementary aspects, i.e. service creation, provisioning and management [Maged-93a]. Nevertheless, both concepts are not harmonized with respect to functionality, architecture and methodologies. Consequently a harmonization and integration of both concepts is strongly required for the target telecommunication environment and therefore subject of several international research activities and the standards bodies. Generally two evolutionary steps can be identified for that integration:

1. The application of TMN concepts for the management of IN services and networks in the medium term time frame, since the first set of IN standards has not addressed this issue.
2. The long term integration of IN and TMN within a common platform allowing the integrated creation, provision and management of future telecommunication services, comprising both communication and related management capabilities represents the ultimate target scenario.

This paper is related to the long term IN/TMN integration and proposes a new integration approach of IN and TMN concepts, taking into account the findings of research related to the medium term TMN-based management of INs [Maged-93c]. Comparing the increasing scope of emerging TMN (service) management services with the capabilities offered by IN services it could be recognized that there is an overlap of functionality since IN service features focus on the control and management of bearer transmission services (e.g. telephony). The reason for this functional overlap between IN and TMN stems from the fact that most of the IN service features were designed many years ago, when standardized (service) management concepts were not available, while facing market needs for enhanced "bearer" service capabilities and emerging customer control requirements. Consequently the IN could be regarded as a short term realization of a *"service management network"*.

In contrast to existing approaches for the long term integration of IN and TMN [NA-43308], this paper proposes a different evolution scenario from current IN environments towards a long term telecommunications environment taking into account the increasing significance of Open Distributed Processing (ODP) standards and emerging results of the TINA Consortium. Therefore this paper studies the relationship between IN service control and TMN service management in more detail and investigates if and how TMN concepts could be used for the provision of IN-like service control capabilities.

The basic idea for this approach is to model (IN) service data, i.e. the *"customer profile"* located in the *Specialized Data Function (SDF)*, as management information in a service related *Management Information Base (MIB)* and to provide access to this information via standardized management protocols. This means that IN *service logic programs* will be substituted by TMN *management services*, which requires a replacement of the IN *Service Control Function (SCF)* by a TMN *Operations System Function (OSF)*. The advantage of this idea is that no distinction has to be made between service control and service management, since future TMN systems could provide also IN-like service control ("call management") capabilities in a uniform way to a variety of future telecommunication services.

Therefore the following section examines the relationship between IN service control and TMN service management in more detail. Section 3 provides a brief comparison of IN and TMN functional capabilities. Section 4 provides a possible mapping of IN and TMN architectures, indicating how TMN functional elements could be used to provide IN-like management capabilities for arbitrary bearer services. An example for a TMN-based realization of the *Time Dependent Routing (TDR)* service features will illustrate the adopted approach in section 5. Section 6 outlines the future perspectives. A short summary concludes this paper.

2 IN SERVICE CONTROL VERSUS TMN SERVICE MANAGEMENT

The historical separation of *service control (SC)* and *service management (SM)* has to be reviewed in the light of enhanced customer control capabilities offered by advanced telecommunication services and the enlarging scope of TMN management systems. Unfortunately there exists no unique definition of the relationship between service control and service management in the literature and a fuzzy borderline exists. Nevertheless, we try to illustrate the historical difference.

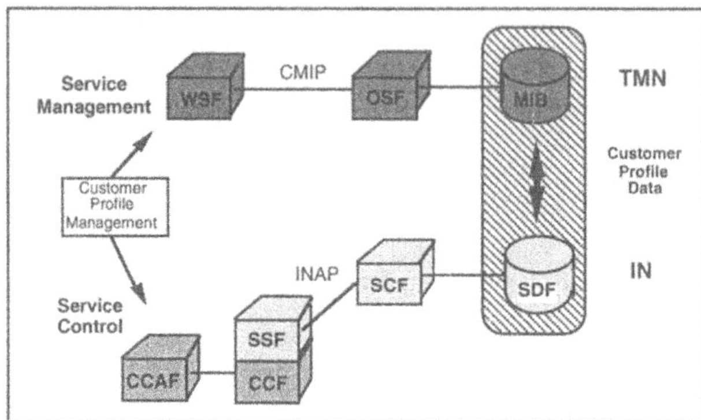

Figure 1 Customer Profile Management at the borderline between SC and SM.

Typically the term **service control** will be used for the real-time interpretation of (customer-specific) service data during service execution and for the manipulation of that service data for a specific customer. IN is regarded as a typical concept for the provision of service control capabilities in a network-independent way. A centralized *Service Control Function (SCF)*, hosting the IN service logic program, interacts with a *Specialized Data Function (SDF)*, which hosts in accordance with the subscribed service features of an IN service the customer-specific data in a *"Customer Profile"*. Interactions between the switches, i.e. a *Service Switching Function (SSF)*, and the SCF, which are required for IN call processing, are realized via the signalling network, i.e. the *IN Application Protocol (INAP)* [INAP-93]. A particular service feature, *Customer Profile Management (CPM)*, allows customers to perform limited modifications on their service parameters in the customer profile, e.g. for rearrangement of call forwarding numbers. This access will be realized like a normal IN call via the signalling network, as indicated in Figure 1 (lower part).

On the other hand **service management** means administering a service on a global base, i.e. for all customers by a service provider or service subscriber mostly without any real-time constraints. TMN represents the world-wide accepted concept for the provision of service management capabilities. The TMN management services are hosted by an *Operations System Function (OSF)* which accesses a corresponding *Management Information Base (MIB)* hosting the required management information, modeled as Managed Objects. Access to the OSF from a *Workstation Function (WSF)* is realized via OSI´s *Common Management Information Protocol (CMIP)* [ISO-9596-1]. Typical areas of service management are Accounting Management (e.g. billing) and Performance Management (e.g. QoS), but also Configuration Management, comprising service installation and reconfiguration as well as customer administration.

In addition RACE TMN projects [H400] have defined a new management functional area, referred to as *Customer Query & Control (CQ&C)*, which allows customers to read and/or modify specific (service) management information via a WSF. This means that besides the installation and modification of IN service triggers and IN service logic programs, etc. access to the customer profile data is also subject to TMN service management. This is also depicted in Figure 1 (upper part). But this requires modeling the customer data also as management information in a *"Customer Management Profile"*.

For the following considerations it has to be stressed that there are two access types to the customer profile data:

- *service management access*, i.e. the initialization, customization and manipulation of the customer profile data by the customer or service provider. This access has only limited real-time constraints although some modifications, e.g. an user registration update, should go into effect immediately. This is one major attribute of IN services.
- *service control access*, i.e. the interpretation of the customer profile data during the service execution of an IN service for "controlling" a (bearer) service. This access is required by the SCF for service (feature) execution and is subject to real-time constraints.

Taking recent TMN-based IN management approaches into account, there is the general trend towards duplication of the customer service data in two separate profiles; one *"customer management profile"* within the TMN system supporting service management access and a corresponding IN *"customer profile"* in the SDF for service control access. This approach necessitates a mapping of data modifications initiated in each of both profiles. Based on the assumption that IN services can be regarded as specific (bearer) connection management services (see next section for more details), it seems sensible to use the customer management profile also for the service control access! In addition, it has to be studied, if the SCF could be modelled as a specific (real time) OSF, where IN services will be modelled as specific TMN management services. This will be addressed in the following two sections.

3 MAPPING OF IN SERVICES TO TMN MANAGEMENT SERVICES

Based on the previous considerations, in particular the usage of the customer management profile data for service execution access, it is necessary to investigate the functional relationships between IN services and TMN service management services in more detail. A mapping of IN services and service features [Q.1211] to corresponding TMN management services [M.3020], [M.3200] requires a careful analysis of IN services and service features with regard to their management functionality. Such an analysis has been performed within the BERKOM II project "IN/TMN Integration" [BERKOM-94a].

This analysis was primarily concerned with the assignment of the IN services and service features to the *Telecommunication Management Functional Areas (TMFAs)* [CFS-H400]. The TMFAs have been defined for structuring the management requirements in telecommunications networks. Thus, they provide a means for defining appropriate TMN management services. The management functionality has been determined by the criterion whether an IN service or service feature has management related tasks that can be assigned to one or more of the ten identified TMFAs as depicted in Figure 2.

The motivation for not mapping the IN services and service features directly to the TMN management services is to gain a management functionality analysis that is universal and independent of the management functionality of specific TMN management services. In addition, it has to be stressed that today only a few management services have been defined for specific application areas. The analysis leaves the common management requirements of IN services and service features out of consideration. For example, service accounting requires functions for collecting and processing service accounting information to generate bills for service usage.

The analysis has revealed that a lot of the IN services (features) contain complex management functionalities. For most of the service features an assignment to the TMFAs could be made. However, a one to one mapping, i.e. the assignment of the management functionality of an IN service or service feature onto one TMFA, could seldom be found. Mostly, the management functionality is assigned to more than one TMFA leading to the assumption that an IN service or service feature can only be replaced by more than one TMN management service. Examining the IN services and service features closely it is striking that a lack of exact description and specification makes the analysis of the management functionality very difficult. Nevertheless, through the analysis of the individual IN service and service feature, the inherent management functionality could be clarified and determined.

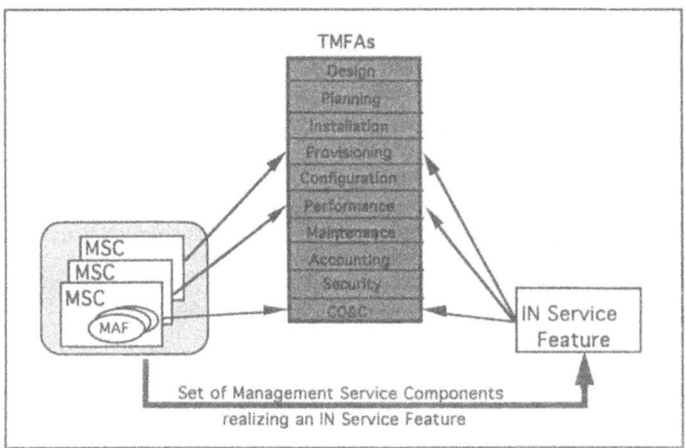

Figure 2 Mapping of IN services to TMN management services based on TMFAs.

In general, a set of management service components is needed to realize the functionality of an IN service feature. For most of the examined IN services and service features TMN management services and management service components or functions, respectively, could be found providing the management functionality of the original. After comparing the IN service features with the enlarging scope of TMN management services (MSs) the following mapping of service features (SFs) to corresponding management services can be derived:

- *Customer Q&C management MSs* could be used for general modifications of subscriber profiles, replacing the Customer Profile Management SF.
- *Configuration management MSs* could be used for realizing flexible network access and routing procedures, replacing Private Numbering Plan, Origin/Time-Dependent Routing, One Number, Call Distribution SFs, etc.
- *Accounting management MSs* could be used for flexible accounting procedures, replacing Premium Charging, Split Charging and Reverse Charging SFs.
- *Security management MSs* could be used for flexible screening options, replacing Closed User Group, O-/T- Call Screening, Authentication and Authorization Code SFs.
- *Performance management MSs* could be used for the provision of customer specific service statistics replacing the Call Logging and Statistics SFs.

4 MAPPING OF IN AND TMN ARCHITECTURES

Based on the previous considerations it becomes clear that IN services features provide many (service) management related capabilities and could be replaced by TMN management services from a functional perspective. This section addresses the architectural issues for a TMN-based provision of IN-like service features. Taking into account the considerations of the introduction, basically two major steps can be distinguished.

4.1 TMN-Based Service Management of INs

The first step represents the application of TMN concepts for IN service management. Within this approach the IN functional entities *Service Management Agent Function (SMAF)* and *Service Management Function (SMF)* will be replaced by the TMN functional entities *Workstation Function (WSF)* and *Operations System Function (OSF)* for Service Management, respectively. The service execution related IN functional entities, namely SSF, SCF, SDF, and SRF will have to be modeled as TMN *Network Element Functions (NEFs)* in order to manage the service related data on these elements, e.g. service triggers, service logic programs, service data, etc. This is illustrated in Figure 3.

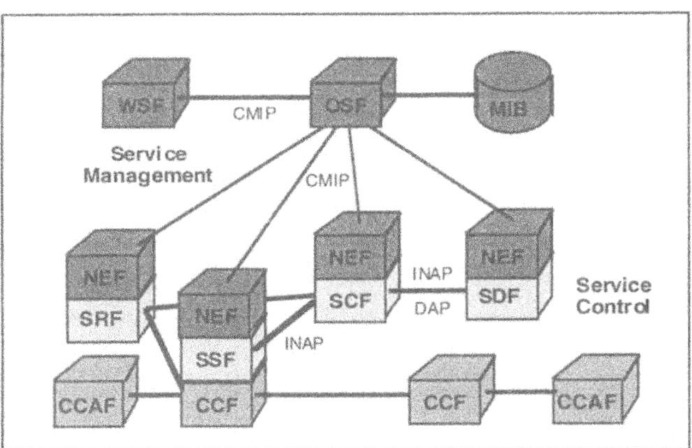

Figure 3 TMN-based service management of IN.

In addition, IN *Customer Profile Management (CPM)* capabilities, originally modeled as IN service feature, will have to be realized by TMN *Customer Query & Control (CQ&C)* management services (CMIS versus INAP). This is due to the fact that most IN service features require complex data manipulation capabilities, e.g. *Time Dependent Routing (TDR)* table initialization, which could not be realized in an efficient way via the IN CPM service feature from a simple telephone set. This means that there will be a *"Customer Management Profile"* in a corresponding *Management Information Base (MIB)* either co-located with the SDF or within the corresponding OSF-S. The development of specifications for the TMN management of IN is currently within the focus of several research projects within EURESCOM and RACE. In addition, the IN standardization related to IN CS-2 is also investigating this important topic. Interested readers are referred to [Maged-93b] for more information on this subject.

4.2 TMN-based (IN) Service Control

Based on the considerations of section 3 and the fact, that most of the data required by the IN service features will be modeled in addition as management information in order to be managed by TMN management service, e.g. CQ&C services, it seems to be straight forward to make use of this management data for service control support. This means that there should be an integration of service control and service management approaches and concepts. It has to be stressed that there exists no one-to-one mapping of IN service execution related functional elements and TMN functional elements, due to the conceptual differences (IN function orientation versus TMN object orientation). Nevertheless, specific IN functional elements could be replaced step by step through TMN elements.

The first IN functional element to be replaced is the SDF. Based on the considerations of section 2 it seems likely, that the **IN SDF will become a TMN MIB**, containing the customer management profile. This means that there will be only one customer profile which could be used for both service management and service control access, as depicted in Figure 4. The profile data will by accessed via an appropriate management protocol, i.e. the *Common Management Information Protocol (CMIP)* [ISO-9596-1], for both service management and service control access. Hence the SCF will have to access that data via CMIP instead of the *IN Application Protocol (INAP)* [INAP-93] or the *Directory Access Protocol (DAP)* [X.500]. The prerequisite for this approach is the availability of fast CMIP and MIB implementations. One possible solution may be to implement CMIP on top of the signalling network (i.e. on top of TCAP).

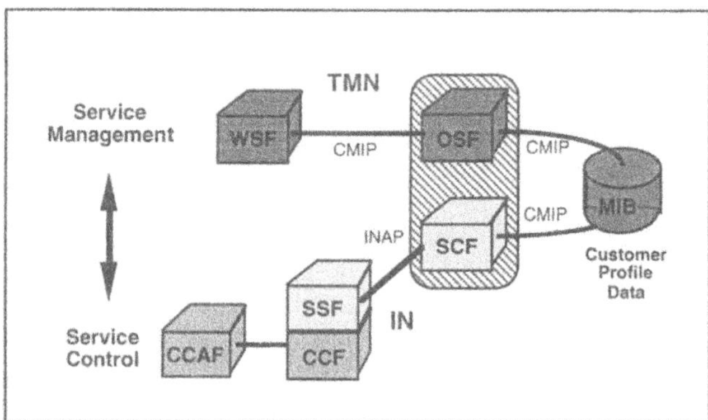

Figure 4 Common customer profile for service management and service control.

In order to realize IN services (i.e. service logic programs) by means of TMN service management services, the traditional **IN SCF has to be replaced by a TMN OSF**, which will run the corresponding TMN-based service control applications, as illustrated in Figure 5. This means that IN service control capabilities will be realized by appropriate TMN management services (including MSCs, MAFs, FEs) and corresponding MOs. This step represents the ultimate evolution step from the function-oriented IN environment towards a long term, object-oriented telecommunications world, such as postulationed by the emerging TINA-C initiative. It has to be stressed, that the notion of the term "OSF" in this context is a little provocative, but it is used to stress the basic idea of this evolutionary approach, namely to use the same (service management) concepts for both service management and service control. This approach is totally in line with the TINA-C approach of using management concepts for both management applications (such as TMN services) and telecommunications applications (such as IN services) [Pavón-94].

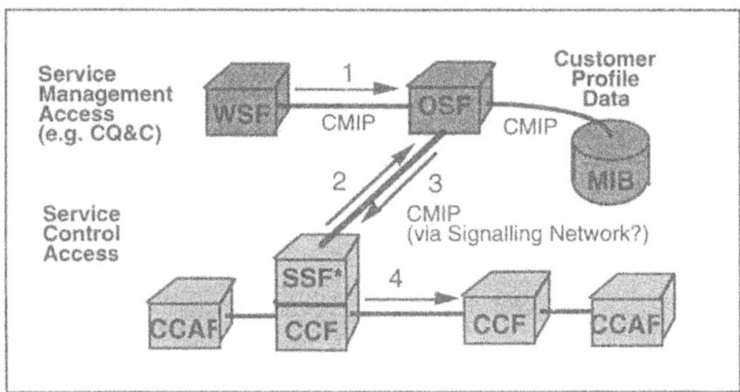

Figure 5 Target scenario for TMN-based service control.

However, there will probably no single OSF-S for running both service management and service control applications in reality. When realizing this approach it seems most likely, that there will be separate "Managers" or "Agents" for service management and service control in order to cope with the real time constraints of service control, as depicted in Figure 6. Therefore the author proposes a dedicated **Service Control Agent (SCA)** that will run the appropriate TMN-based service control applications, whereas a **Service Management Agent (SMA)** will run the corresponding service management services. Both agents will use the common customer profile located in the MIB. A similar approach for B-ISDN service control can be found in [Fukada-94].

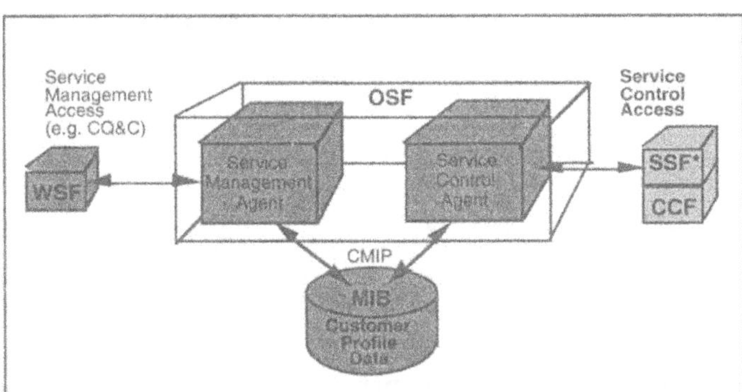

Figure 6 TMN-based service control agent and service management agent.

It has to be stressed that both "agents" will act in both "manager" and "agent" roles according to the OSI "manager-agent" paradigm; the term "agent" has only been selected in analogy to system components within an OSI environment, such as a "Directory System Agent". In TINA-C these components would probably referred to as "managers", such as "Session Manager" or "Service Manager" [Gatti-84].

The most challenging aspect of this scenario is the communication between traditional (IN-based) switch architectures (i.e. the SSF) and the new (TMN-based) SCA. In order to make use of CMIP instead of INAP between SSF and SCA it is necessary to introduce a new component in the SSF. Therefore the SSF is now called SSF*. This component is called **Basic Call Agent (BCA)**, which has to recognize (based on an adapted "call model") that additional service control is required by the SCA. This is indicated in Figure 7.

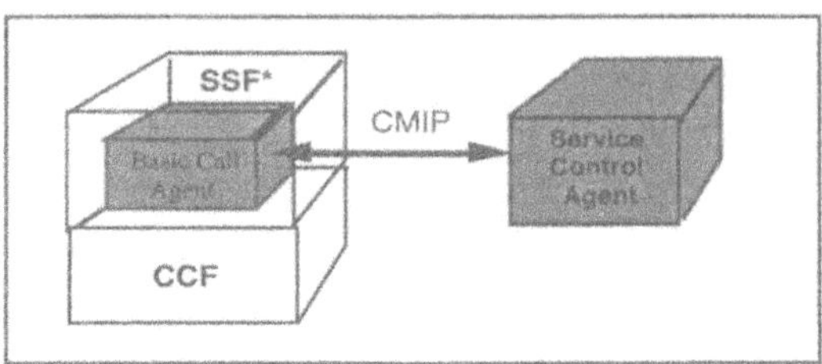

Figure 7 TMN-based basic call agent required in SSF*.

In order to support the real time requirements for service control CMIP may be implemented on top of CCS7. It has to be noted that in advanced service scenarios, such as multimedia conferencing services, much more complex call/connection models are required. The development of the SSF* and the related BCA, as well as the definition of appropriate Interworking Functions (IWFs) for supporting access to the SCA from standard switch architectures are areas for further studies.

Note that this issue represents also the basic evolution problem from current IN architectures towards TINA, since the rudimentary IN call model is not applicable to complex multimedia service sessions, and hence requires major enhancements in the switches (see [Brown-94], [Gatti-94]). Therefore, TINA has defined a corresponding *"Communication Session Manager"* which makes use of a *"Connection Manager"* in order to cope with the emerging separation of connection and call (in multimedia context replaced by the more appropriate term "session") control.

5 A REALIZATION SCENARIO

It is the purpose of this section to demonstrate how IN service features and thus IN services could be realized by TMN concepts. Therefore the **Time-dependent Routing (TDR)** service feature has been chosen as an example.

5.1 TMN-based TDR Service Management

The TDR service feature is representative of all service features that act on *"table"* Managed Objects (MOs). The operations that are performed on table MOs are almost identical. The *"TDR"* MO, representing a subclass of the table MO which is contained in a *"Service"* MO within the customer profile, has first to be created and initiated by the *"Create/Delete Table"* and *"Set Table"* Functional Elements (FEs). Before the customer is allowed to access the TDR MO (e.g. for adding a table entry), security functions check his/her identification and authorization. Then the customer can access the TDR table MO for modifying it (see arrow 1 in Figure 5 in section 4.2).

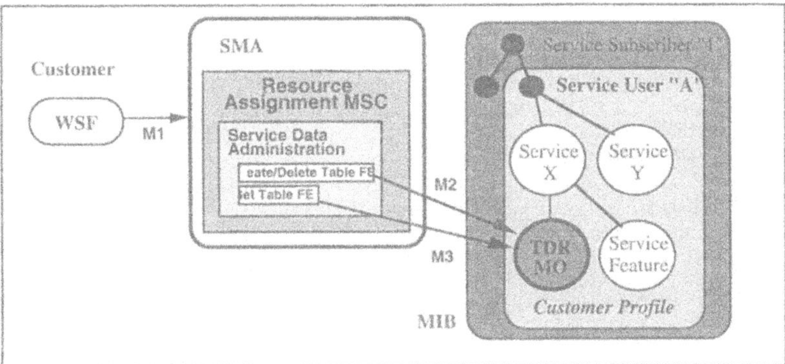

Figure 8 TMN-based service management (CQ&C) for the TDR service feature.

Since the TDR service feature is only one component of an IN service (e.g. UPT), the provisioning of TDR will mostly take place during service provisioning unless the customer decides to add this feature to a service he/she is already provided with. The customer requests by means of appropriate CQ&C management services via a WSF the service provider's OSF, namely the Service Management Agent (SMA), to provide the TDR service feature (M1). The procedure for TDR provisioning is depicted in Figure 8. A *"Resource Assignment"* Management Service Component (MSC) is addressed for the provisioning of service resources. The *"Create/Delete Table"* FE, as component of the *"Service Data Administration"* Management Application Function (MAF), initiates an instantiation of the TDR (Table) MO within the MIB (M2). If this operation is successful the TDR MO is initialized by the "Set Table" FE. In addition, the customer could modify and (de)activate specific entries of the TDR MO by corresponding MSCs, which reuses the *"Set Table"* FE.

5.2 TMN-based TDR Service Control

The TMN-based service control access for TDR looks similar to this scenario, with the "WSF" entity in Figure 8 replaced by a SSF*, the "SMA" replaced by a Service Control Agent (SCA). In addition, different MSCs, MAFs and FEs will be used. The following information flows could be identified in this scenario. The SSF*, namely the Basic Call Agent (BCA), will recognize during call set-up (based on the dialled number) the need for external service control support and hence will request by means of an appropriate MSC (e.g. "Find Call Destination") via CMIP support from the SCA. This means that the BCA has to act in a "manager" role in order to contact the SCA.

The SCA determines by corresponding service control MSCs, which service for which user is requested and identifies the corresponding User MO and the corresponding Service MO (by interpreting the dialled number). The Service MO itself (by an appropriate MO action) or a corresponding MAF will then check which service features, such as TDR, are activated by the customer within the customer profile, and finally determines the appropriate destination number by requesting the TDR MO for appropriate routing information. The result will be passed back to the BCA. The information flows (2 + 3) in Figure 5 indicate how the SSF* will obtain the required information, where the OSF embodies both SMA and SCA.

It has to be stressed that the relationship between management services and their components, i.e. MSCs and MAFs, and managed objects is a subject of ongoing research. The consequent application of a fully object-oriented approach for the MOs will probably eliminate the MSCs and MAFs to a large extend, since most of these functionalities will be embodied in future MOs by corresponding MO operations, i.e. actions. This means that the service control applications will be moved to the MOs themselves.

6 FUTURE PERSPECTIVES

The approach presented has been adopted within the BERKOM-II Project "IN/TMN Integration" undertaken by the Technical University of Berlin for the Deutsche Telekom Berkom (De•Te•Berkom). The objective of this project is the development of a TMN-based *Personal Communication Support System (PCSS)*. This PCSS is based on an enhanced TMN platform, being part of the "Y" Platform [Zeletin-93] and offers IN-like service control capabilities, such as user registration and call handling procedures, supporting personal mobility and service personalization for an open set of (multi-media) communication services in a distributed office environment. All customer related data (e.g. user location, call handling data, etc.) will be stored in generic *user service profiles*, modeled as management data in an integrated X.500/X.700 system. This flexible profile integrates the data required for personal communications for all the services a user has subscribed to. Access to that profile for both customer control (i.e. profile manipulation) and communication services control (i.e. during service execution) will be realized by management services (components) via a common PCSS application programming interface. More information on the PCSS can be found in as described in [Eckardt-95] and [Berkom-94b].

7 SUMMARY

The IN can be regarded as the right concept for solving today's service provision requirements. But current IN concepts are limited in functionality and methodology, e.g. the function oriented nature of IN, since the trend in telecommunications is towards openness, reusability and in particular object orientation. Although the IN capability set approach allows for stepwise enhancements of IN functionalities it seems douptful, whether IN could keep the pace of evolution, in particular in the light of rapid progress in ATM deployment and multimedia service provisioning. Hence a major paradigm shift is required for IN evolution in the future.

Obviously there exist severe overlaps between IN and TMN. TMN is already based on object orientation, although the areas of management service design, creation and realization are still under development. Due to the ongoing integration of the telecommunications environment and the increasing availability of powerful management concepts, systems and services in the near future, it seems likely that IN concepts could be replaced in the long term by TMN concepts for telecommunication services provision. The basic advantage of this approach is that no separation of service control, i.e. core service capabilities, and service management has to be made, which is in line with TINA-C objectives. This has been illustrated in this paper.

8 ACKNOWLEDGEMENTS

The ideas presented in this paper have been developed within the BERKOM II project "IN/TMN Integration" performed at the Department for Open Communications Systems at the Technical University of Berlin for Deutsche Telekom Berkom (De•Te•Berkom). In addition the author thanks Jaqueline Aronsheim-Grotsch, who has investigated the management aspects within the IN services and service features.

9 REFERENCES

[Berkom-94a] Berkom II Project "IN/TMN Integration", Deliverable 4: "Study on the TMN-based Realization of IN Capabilities", De•Te•Berkom, Germany, Berlin, June 1994

[Berkom-94b] Berkom II Project "IN/TMN Integration", Deliverable 5: "State of the Art in Personal Communications and Overview of the PCSS", De•Te•Berkom, Germany, Berlin, November 1994

[Brown-94] D.K. Brown: "Practical Issues Involved in Architectural Evolution from IN to TINA", International Conference on Intelligent Networks (ICIN), Bordeaux, France, October 1994

[CFS-H400] RACE Common Functional Specification (CFS) H400: "Telecommunications Management Functional Specification Conceptual Models: Scopes and Templates", November 1992

[Eckardt-95] T. Eckardt, T. Magedanz: "The Role of Personal Communications in Distributed Computing Environments", 2nd International Symposium on Autonomous Decentralized Systems (ISADS), Phoenix, Arizona, USA, April 25-26, 1995

[Fukada-94] K. Fukada et.al.: "Dual Agent System using Service and Management Subagents to Integrate IN and TMN", International Conference on Intelligent Networks (ICIN), Bordeaux, France, October 1994

[Gatti-94] N. Gatti: "IN and TINA-C Architecture: a Service Scenario Analysis", International Conference on Intelligent Networks (ICIN), Bordeaux, France, October 1994

[INAP-93] ETSI DE/SPS-3015: "Signalling Protocols and Switching - IN CS-1 Core Intelligent Network Apllication Protocol (INAP)", Version 08, Mai 93

[ISO-9596-1] ISO/IEC/IS 9596-1 I ITU-T Recommendation X.711: Information Processing Systems - Open System Interconnection - Common Management Information Protocol Definition (CMIP), 1991

[M.3010] ITU-T Recommendation M.3010: "Principles for a Telecommunications Management Network", Geneva, November 1991

[M.3020] ITU-T Recommendation M.3020: "TMN Interface Specification Methodology", Geneva, November 1991

[M.3200] ITU-T Recommendation M.3200: "TMN Management Services: Overview", Geneva, November 1991

[Maged-93a] T. Magedanz: "IN and TMN providing the basis for future information networking architectures", in Computer Communications, Vol.16, No. 5, May 1993

[Maged-93b] T. Magedanz et.al.: "Managing Intelligent Networks the TMN Way: IN Service versus Network Management", RACE International Conference on Intelligence in Broadband Service and Networks (IS&N), Paris, France, November 1993

[Maged-93c] T. Magedanz: "Towards a Common Platform for Future Telecommunication and Management Services - Some Thoughts on the Relation between IN and TMN", Invited Paper at Korea Telecom International Symposium (KTIS´93), Seoul, Korea, November 1993

[NA-43308] ETSI DTR/NA-43308: "Baseline Document on the Integration of IN and TMN", Version 3, September 1992

[Pavón-94] J. Pavon et.al.: "Building New Services on TINA-C Management Architecture", International Conference on Intelligent Networks (ICIN), Bordeaux, France, October 1994

[Q.12xx] ITU-T Recommendations Q.12xx Series on Intelligent Networks, Geneva, March 1992

[Q.1211] ITU-T Recommendation Q.1211: "Introduction to Intelligent Network Capability Set 1", Geneva, March 1992

[X.500] ITU Recommendation X.500 / ISO/IEC/IS 9594: Information Processing - Open Systems Interconnection - The Directory, Geneva, 1993

[Zeletin-93] R. Popescu-Zeletin et.al.: "The "Y" platform for the provision and management of telecommunication services", 4th TINA Workshop, L´Aquila, Italy, September 1993

35

Handling the Distribution of Information in the TMN

Costas Stathopoulos, David Griffin, Stelios Sartzetakis
Institute of Computer Science,
Foundation for Research and Technology - Hellas (ICS-FORTH),
PO Box 1385, Heraklion, GR 711-10, Crete, Greece.
tel.: +30 (81) 39 16 00, fax: +30 (81) 39 16 01.
e-mail: stathop@ics.forth.gr, david@ics.forth.gr, stelios@ics.forth.gr

Abstract

This paper proposes a solution for mapping managed resources (network elements, networks) to the managed objects representing them. It supports an off-line, dynamic negotiation of Shared Management Knowledge in the TMN. Given a method for globally naming managed resources, managers identify the resource they want to manage as well as the management information they require. The manager's requirements are then mapped to the agents which contain the managed objects. From the global name of the agent, and knowledge about the management information that the agent supports, the manager can construct the global distinguished name of managed objects.

The approach uses the OSI Directory where information about managed resources as well as agents and managers is stored. An architecture is described which provides a means of identifying in a global context which agent contains the required management information. Additionally, the architecture provides the abstraction of a global CMIS and the function of location transparency to communicating management processes to hide their exact physical location in the TMN.

Keywords

TMN, systems management, manager/agent model, shared management knowledge, global naming, directory objects, managed objects, location transparency.

1 INTRODUCTION

The M.3010 Telecommunications Management Network (TMN) recommendation (ITU M.3010) describes a distributed management environment: management information for physically distributed network resources and services provided over a large geographical area is maintained on a large number of distributed agents. These agents interact with a variety of management applications over the TMN. The collection of managers and agents (or, with a single name, management processes) in the TMN interact according to the OSI manager/agent model (ISO/IEC 10040). The management information is kept in the agents and consists of managed objects structured hierarchically (ISO/IEC 10165-1) forming the Management Information Tree (MIT). Network resources (network elements or networks) and services being managed, are represented by the managed objects.

A typical TMN implementation may have hundreds of agents. There are proposals (ISO/IEC 10164-16, Sylor 1993, Tschichholz 1993) for the global naming of managed objects. These proposals assume a priori knowledge of which specific agent contains the managed objects in question. This mapping is straightforward in the case where the agent is running on the same system as the managed resources, but in the TMN the mapping may not be as obvious. The general case in the TMN is a "hierarchical proxy" paradigm where Q Adaptors (QAs), Mediation Devices (MDs), and Operations Systems (OSs) are located in separate systems from the Network Elements (NEs). Additionally, the TMN is involved in managing more abstract resources than simple NEs, for example a management process may be interested in networks, services and lower level management processes.

This paper deals with the functionality needed by the TMN in order to efficiently answer the following basic questions: Given a particular managed resource or service that we want to manage (i.e. perform a particular management operation) which is the agent that contains the managed object(s) needed in our management operation? Given that agent, what is the management information base (MIB) it supports? What is the address where the agent is awaiting for requests?

Actually, each of the above questions corresponds to some Shared Management Knowledge (SMK) interrogation (ISO/IEC 10040, NM/Forum 015). Our approach is to provide a global way for referring to elements of the SMK. In order to do so, we use the OSI Directory to register elements of the SMK (such as the mapping from resources to agents, the presentation addresses of management processes and their supported MIBs). Thus, we can achieve an off-line, dynamic SMK negotiation between the management processes.

This paper describes appropriate Directory schemata for storing information about network resources, agents (including the MIBs they support) and managers in the Directory. As a major part of this work, we propose an architecture based on the OSI manager/agent model and the OSI Directory Service. We show how a global Common Management Information Service (CMIS) can be realized and implemented by using this architecture.

We propose a mechanism for supporting the basic function of location transparency. This is one of the distribution transparencies (ITU X.900) necessary in a distributed environment and refers to a location-independent means of communication between management processes, hence hiding their exact location in the TMN.

The OSI Directory Service standard (ITU X.500) describes a specialized database system which is distributed across a network. The Directory contains information about a large number of objects (e.g. services and processes, network resources, organizations, people). The overall information is distributed over physically separated entities called *Directory Service Agents (DSAs)* and consists of directory objects structured hierarchically forming the Directory Information Tree (DIT). The distribution is transparent to the user through the use of *Directory Service Protocol (DSP)* operations between the DSAs. Each directory-user is represented by a *Directory User Agent (DUA)* which is responsible for retrieving searching and modifying the information in the Directory through the use of *Directory Access Protocol (DAP)* operations. The basic reasons for choosing the Directory as the global SMK repository are:

- It provides a global schema for naming and storing information about objects that are highly distributed. For example, every management process in the world can be registered with a unique name (i.e. its Distinguished Name (DN)).
- It provides powerful mechanisms (e.g. searching within some scope in the DIT using some filter) for transparently (through the use of DSP operations between DSAs) accessing this global information.
- One of the major objectives of the OSI Directory, since it was recommended, was to provide an information repository for OSI application processes. For example, by keeping the locations (i.e. OSI presentation addresses) of the various application entities representing the application processes within the OSI environment.

In the following section we describe a way for globally naming managed objects based on registering the management processes in the DIT while in the third section we propose the enhanced manager/agent model that interfaces with the OSI Directory. Putting it all together, section 4 describes the mapping from resources to managed objects and how our enhanced manager/agent model supports the SMK negotiation between two management processes. Next, we present the abstraction of a global CMIS and a location transparency mechanism. Finally, section 6 gives an overview of the implementation of the mechanisms described in this paper.

2 GLOBAL DN FOR MANAGED OBJECTS

The OSI Directory can be used for globally naming application processes in a distributed environment. Any kind of application process can be represented by a directory object that contains information about the process (provided that this information is relatively static). Thus, any application process acting either in the manager or agent role can be globally named. Bearing

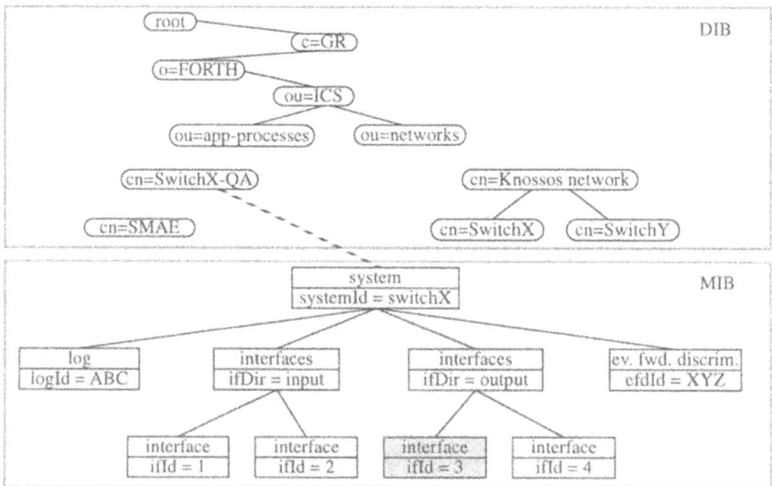

Figure 1 Global DNs for managed objects

in mind that the managed objects use a similar hierarchical naming structure as the directory objects, a common global name space can be realised for both the managed objects and directory objects (Sylor 1993, Tschichholz 1993, and recently ISO/IEC 10164-16).

Figure 1 depicts an example of managed objects named in the global context. Consider the management process that is registered in the Directory Information Base (DIB) with DN:

{C=GR, O=FORTH, OU=ICS, OU=app-processes, CN=SwitchX-QA}

maintaining an MIB containing managed objects that represent some network element (e.g. an ATM switch). Consider a managed object, containing information about interface 3 of the network element, with Local Distinguished Name (LDN) that is, a DN within the scope of the local MIB:

{systemId = SwitchX, ifDir = output, ifId = 3}

This managed object can now be named globally with DN:

{C=GR, O=FORTH, OU=ICS, OU=app-processes, CN=SwitchX-QA, systemId = SwitchX, ifDir = output, ifId = 3}

3 ENHANCING THE MANAGER/AGENT MODEL

In the previous section we described how we can globally name managed objects by exploiting the OSI Directory. In this section we enhance the basic OSI manager/agent model (ISO/IEC 10040) so that a management process can make use of the Directory Service in order to perform systems management functions on the managed objects in the global context.

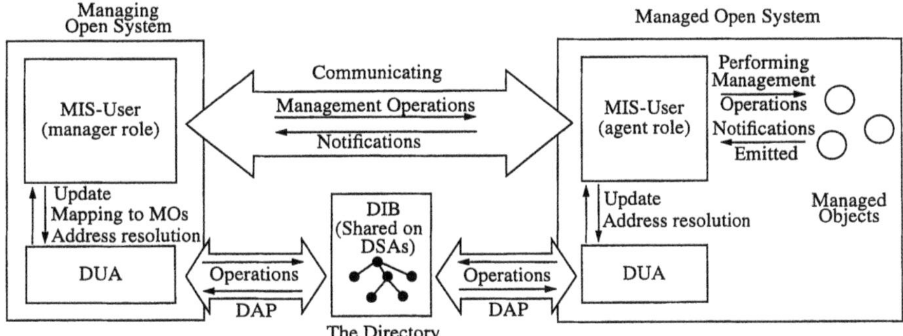

Figure 2 Enhanced manager/agent model

Figure 2 depicts the enhanced manager/agent model. Every open system includes a special purpose DUA. This DUA is responsible for retrieving and updating the information kept in the Directory by issuing DAP operations to the DSAs. In general, the management process uses the DUA for the following:

- Updating the Directory: Management processes should have the capability of updating the Directory by creating, changing or deleting directory objects that represent themselves or other management processes as well as their associated application entities. Although every management process will be able to perform directory updates for its own entry (e.g. on startup an attribute that marks the process as "running" might be set), it is likely that only special management processes that are responsible for the management of the TMN will fully support this function. These management processes are also responsible for updating the directory objects for the resources with information such as the DN of the management process(es) (acting in the agent role) that represent these resources.

- Mapping to Managed Objects: Every management process acting in the manager role, eventually needs to perform some mapping from the resources it wants to manage to the managed objects (representing the resource) that contain the needed information. This procedure is described in the next section.
- Address Resolution. Every management process, that wishes to make an association with a peer management process, needs a mechanism for finding the presentation address (PSAP) of an application entity representing the latter. Since this address is not always the same for a specific management process, a location transparency mechanism is needed for association establishment. Such a mechanism is described in section 5.

4 MAPPING FROM RESOURCES TO MANAGED OBJECTS

Systems Management deals with management information for physically distributed network resources provided over a large geographical area (divided to many management domains). In general, the relationship between resources and managed objects that represent them is many to many. This means that not only a resource is represented by many managed objects (each one providing a different view of the resource) but also a managed object may represent a collection of resources. Hence, there is no straightforward way for mapping between resources and managed objects that represent them. The knowledge of such a mapping in the TMN is very critical and is actually part of the shared management knowledge because it contains information that must be shared among management processes.

For example, consider the network management case where some decision has to be made about a network reconfiguration due to some network failure. Certain information about the network resources (e.g. network topology information) has to be known in order to discover an optimum reconfiguration solution. This means that having identified the resources that have to be reconfigured, the managed systems that contain the managed objects representing these resources have to be contacted and the appropriate management operations need to be performed. Thus, there must be a way to map from an a priori known resource to some managed object that represent some view of this resource.

In this section we assume a TMN where the management processes communicate based on our enhanced manager/agent model. We describe the information that we have to keep in the Directory for the resources and the management processes and how the latter can use it for performing the above mentioned mapping. Bearing in mind the global naming of managed objects described in section 2, we are going to provide that mapping in the global context.

4.1 Negotiation of Shared Management Knowledge

The mechanisms described in this section provide support for an off-line dynamic negotiation of a part of the shared management knowledge (SMK). In general, the SMK refers to the common knowledge between the application processes performing systems management. This includes (but is not limited to):

- protocol knowledge (e.g. supported application context)
- knowledge of the supported functions (e.g. which management service is provided and to what extent)
- managed object knowledge (e.g. classes and instances)
- knowledge about the relationships between the functions and the managed objects
- knowledge about the mapping of resources to managed objects

For example, to enable management communication between two management processes, prior knowledge such as the MIBs they support and the management activities they can perform is needed. This information can be obtained from the Directory. We use the term "off-line" negotiation of the SMK because it happens prior to the association establishment. It is also dynamic because it happens at run-time. Every management process can update the Directory with management knowledge information and, thus, dynamically modify the SMK that is available to every process.

4.2 Registering network resources in the Directory

Throughout this paper, we use the term "network resources", or, simply, resources, to denote either network elements (e.g. switches) or groups of interconnected network resources (i.e. networks). Given the above definition, network resources can be thought as a containment hierarchy where we have networks containing other (simpler) networks as well as network elements which are always the leaf nodes of a conceptual containment tree.

The Directory can be used for storing information for network resources by registering a directory object for each resource. The containment hierarchy described above together with the existing Directory structure can provide a naming schema for unambiguously identifying network resources. Currently, there is no standard Directory schema for registering network resources in the Directory, although there is ongoing work to that direction (Mansfield 1993) and it is expected that appropriate schemata will exist in the future.

Figure 1 depicts an example of registering a simple network (with two network elements) in the Directory with DN *{C=GR, O=FORTH, OU=ICS, OU=networks, CN=Knossos Network}*.

4.3 Registering the TMN in the Directory

The Telecommunications Management Network is a, possibly separate, network that interfaces a telecommunications network at several different points in order to exchange information with it for management purposes. The TMN is intended to provide a wide variety of management services through the support of a number of management functions.

The TMN physical architecture is composed of a variety of building blocks: Operations Systems (OSs), Mediation Devices (MDs), Q Adaptors (QAs), Data Communication Networks (DCNs), Network Elements (NEs) and Workstations (WSs). Each one of the above building blocks contains a number of TMN functions. For a detailed description of the TMN building blocks and their functions refer to (ITU M.3010).

According to the (ITU M.3020) the overall network management activity is decomposed in areas called TMN management services. The constituent parts of a TMN management service are called TMN management service components. The smallest parts of a TMN management service are the TMN management functions (e.g. performance monitoring).

Additionally, the management functionality may be considered to be partitioned into layers with each layer concerned with some subset of the total management activity. A four-layer management functionality has been identified consisted of the following layers:

- network element management layer, which is concerned with the management of network elements, and supports an abstraction of the functions provided by the network elements,
- network management layer, which is concerned with the management of all the network elements, as presented by the previous layer, both individually and as a set,
- service management layer, which is concerned with how the network level information is utilized to provide a network service, the requirements of a network service and how these requirements are met through the use of the network, and
- business management layer which has responsibility of the total enterprise and is the layer where agreements between operators are made.

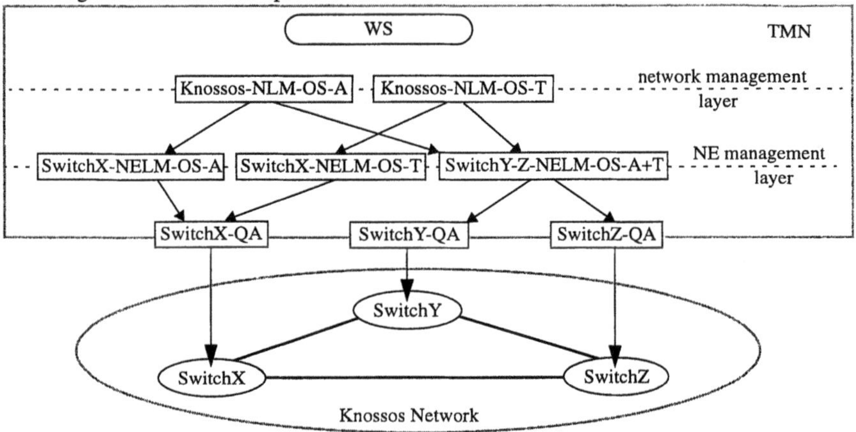

Figure 3 An example of a TMN for a simple network

In the OSI environment we can think of the TMN as a collection of systems management application processes (SMAPs) each one containing one or more Systems Management Application Entities (SMAEs) as defined in (ISO/IEC 10040) in order to accomplish communication between them.

Consider a management domain administered by the organisational unit registered in the Directory with the DN:

{c=GR, O=FORTH, OU=ICS},

a simple network, named "Knossos Network", within the above organizational unit consisted of three switches (NEs) registered in the Directory under the subtree with DN:

{C=GR, O=FORTH, OU=ICS, OU=networks, CN=Knossos Network}

and a TMN in this organizational unit consisted of the following SMAPs (i.e. management processes) (See Figure 3, "An example of a TMN for a simple network"):

- three QAs containing managed objects for the three network elements. (Although a QA may

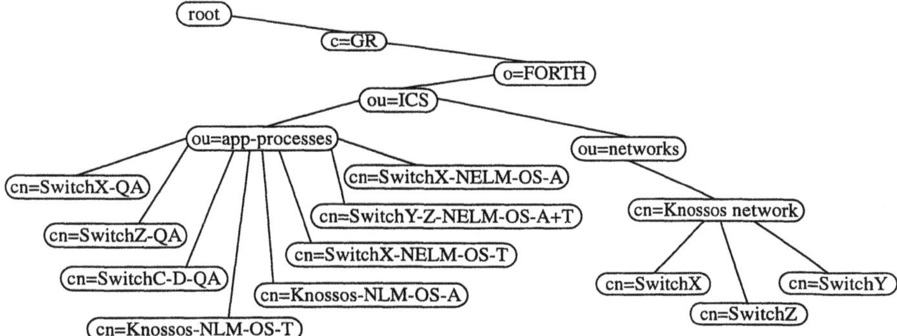

Figure 4 Registering the TMN in the DIB

contain managed objects for more than one network element we only show the simple one-to-one case in this example.),

- three network element level management OSs (NELM-OSs) each one managing a number of network elements in respect to one or two of the *accounting management* and *traffic management* management services. One NELM-OS (namely, the **Switch-Y-Z-NELM-OS-A+T**) manages the two QAs for SwitchY and SwitchZ for both accounting and traffic management. The other two NELM-OSs manage SwitchX, each one for a different management service (namely, **Switch-X-NELM-OS-A** for accounting management and **Switch-X-NELM-OS-T** for traffic management),
- two network level management OSs (NLM-OSs), each one managing the network in respect to one of the above two management services by connecting to the appropriate NELM-OSs (thus, the **Knossos-NLM-OS-A** is for accounting management while the **Knossos-NLM-OS-T** is for traffic management), and
- a WS that is able to manage the resources by connecting to the appropriate OSs or QAs.

In order to register these SMAPs in the Directory we create entries with each entry containing information for a single SMAP. This assigns a global name to every SMAP. Figure 4 depicts the DIT after registering the entries for our TMN example (the network resources are also shown registered in a hierarchy). SMAPs are organized as children of the "*cn=app-processes*" entry under the ICS' entry. Note that we do not register the processes in a hierarchy since this information is going to be obtained from the management services they provide (which includes the management layer on which the SMAPs operate). Also, note that for every SMAP we have to register the entries that contain information about the SMAEs representing the SMAP. Although not depicted in figure 4, these will be registered below the SMAP they represent.

4.4 The Approach to the Mapping Problem

In the TMN, SMAPs (WSs, OSs, QAs, MDs) could be located in separate systems from the resources they represent. This means that even though we know the resource that we want to manage, this does not give any information about the agent that keeps the managed objects that represent the resource. Additionally, there may be more than one agent providing different management services for a resource. In the mapping problem introduced at the beginning of this section we assume that we initially know a global name (namely, the DN of a directory object) for the resource that we want to manage.

Our basic requirement is to provide to every SMAP, acting in the manager role, a mechanism for identifying in the global context the managed objects representing a given resource. Our approach involves the following two-step procedure:

1. Given the DN of a resource and a description of the requested management information that includes:
   - the management service that we want to perform (this will normally be a TMN management service e.g. traffic management),
   - an MIB-independent description of the managed object(s) (this can be based on some abstract description of the object class and the semantics of every managed object. Mechanisms for describing and discovering management information are currently under standardization (ISO/IEC 10164-16)),

   find out the DN of the SMAP that maintains the requested managed object(s) based on the needed management service and by performing a DAP read operation on the resource's directory entry.
2. Perform a DAP read operation on the SMAP you found in the previous step (in case of more than one match, a choice is made based on the MIB that the matching SMAPs support) and identify the LDN(s) of the requested managed object(s) based on
   - the MIB supported by the SMAP and
   - the MIB-independent description of the managed object(s) we have.

   Form the global DN(s) of the managed object(s) you are interested in by concatenating the LDN(s) with the DN of the SMAP.

In order to perform the above procedure, every directory object that represents a resource (either a network or network element) must have a multi-valued attribute that provides the DN of a SMAP that provides some management service for the resource and also identifies which management service this is. That is, a pair of the form: *(DN of agent, Management Service)*. The name of this attribute is "*responsibleSMAP*" and is multi-valued (i.e. many SMAPs can keep managed objects for a single resource in respect to some management service).

Our approach also requires that the following information is kept in every directory object that represents a SMAP:

- an attribute that provides the MIB that the SMAP supports. The name of this attribute is "*supportedMIB*" and is multi-valued (i.e. many MIBs can be supported on a single SMAP). This attribute is present only on SMAPs that are acting on the agent role.
- an attribute that denotes the TMN building block that the SMAP implements. The name of this attribute is "*TMNBuildingBlock*" and is single-valued.
- an attribute for the management service provided by the SMAP. The name of the attribute is "tMNMS" and is multi-valued (i.e. many management services can be provided from a single SMAP).

The value for the *supportedMIB* attribute is a DN. This is the ideal case where the management information is registered under some well-known part of the DIT. The reader can refer to (Dittrich 1993) which describes an approach for registering management schema information in the

Directory. Also, (ISO/IEC 10164-16) recommends the appropriate directory objects for registering the above information in the Directory.

Every directory object that belongs to the standard *applicationEntity* object class should also have attributes with information about the characteristics of the Common Management Information Service Element (CMISE) and the Systems Management Application Service Element (SMASE) of the SMAE. These attributes are discussed in section 5 and are fully described in (ISO 10164-16).

An appendix at the end of this paper contains the ASN.1 definitions for the new attributes. Note that the list for the TMN management services is definitely not complete but rather a small subset of the existing management services (ITU M.3200). Also, since a management service is composed of management service components which, in turn, perform a number of management functions, a Directory schema can be used for registering the hierarchy of the existing TMN management services in the DIT. Finally, every SMAP belongs to the *managementProcess* object class, a subclass of the standard *applicationProcess* class.

5 OSI SYSTEMS MANAGEMENT IN THE GLOBAL CONTEXT

In the previous section we described how the Directory can be used to identify the agent containing specific management information about specific managed resources and how the information about the MIB that the agent supports can be used in the construction of globally unique DNs of the required managed objects. We now show how an OSI SMAP can use DNs in order to issue management operations and notifications in the global context. Additionally, we describe a mechanism for providing location transparency in the proposed manager/agent model (see Figure 2, "Enhanced manager/agent model") for communicating SMAPs.

5.1 The Global CMIS

The Common Management Information Service (CMIS) definition (ISO/IEC 9595) states that, following association establishment between a manager and an agent, the manager issues management operations (while the agent can issue notifications) within the scope of a specific association using LDNs to identify the required managed objects. We can now provide an interface of a global CMIS where the users of the service simply issue CMIS requests using the DN of the managed objects without dealing with the association establishment procedure. For example, (using a simplified semantic notation) a managing open system can issue:

M-GET(DN, attribute_list [,other parameters])

rather than:

A-ASSOCIATE(PSAP_of_agent, &ASSOCIATION_ID)

M-GET(ASSOCIATION_ID, LDN, attribute_list [,other parameters])

which requires that the presentation address (PSAP) of the agent is already known.

On the other hand, a managed open system (i.e. an agent) that reports some notification to a process acting in the manager's role can send the global DN of the managed object that emits the report rather than the LDN. Figure 5 depicts how two management applications communicate using the interface of the Global CMIS.

The Global CMIS uses the Directory to provide a location transparency function. This not only

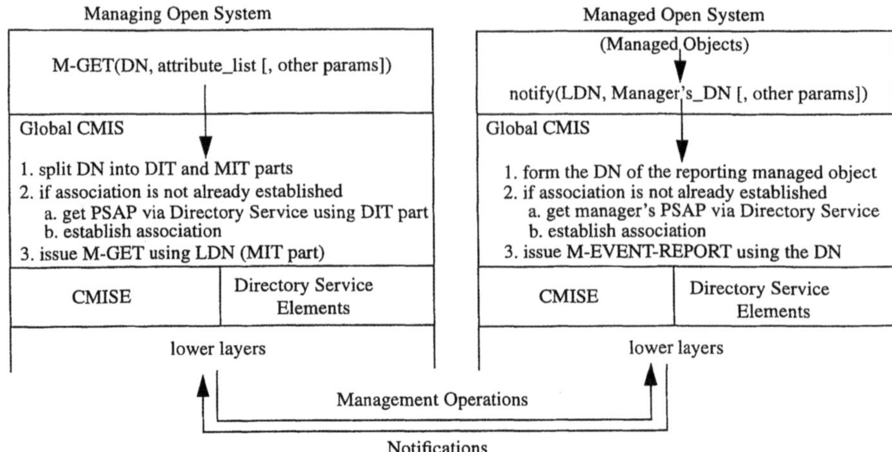

Figure 5 The global CMIS

relieves the management application from the concern of establishing associations with the correct agent but also hides the physical location (PSAP) of the required agents. The management application can assume that managed objects are part of a global and seamless MIB and are identified by their DNs.

5.2 Providing Location Transparency

Location transparency is a basic mechanism in a distributed environment (ITU X.900). In the TMN, it provides a means for finding the address of SMAPs in a location independent way. Bearing in mind that the location of a SMAP may change over time (e.g. a QA for some ATM-switch that is running on machine X might migrate to some other machine if X crashes), we conclude that location transparency should be supported in a TMN. Since the location of a SMAP does not change very frequently, the OSI Directory is appropriate for storing, retrieving and modifying location information for SMAPs.

5.2.1 The Location Transparency Mechanism

The basic requirement for a location transparency mechanism is that, given a SMAP's name, it should provide a means of identifying the location (i.e. OSI presentation address) where the systems management application entity (SMAE) representing that SMAP is awaiting either for management operations or notifications. In the TMN though, there is the possibility that a SMAP is represented from more than one SMAEs. For example, consider the case of a NELM-OS (like the ones depicted in Figure 3) that can act as a manager (by issuing management operations to a QA) and an agent (by serving management requests issued by a NLM-OS) at the same time. Furthermore, there is the possibility that a SMAP supports more than one interoperable interface meaning that a different SMAE might be present for every interface. Additionally, a SMAP that provides some management service can implement a number of management functions. These management functions will be provided by a number of SMAEs representing the SMAP. Bearing these in mind, a location transparency mechanism involves choosing among a number of SMAEs representing the SMAP we wish to communicate with.

In order to provide this functionality, the following information should be kept in every directory object that represents an SMAE:

- the application context supported from the communicating entity. The standard attribute *supportedApplicationContext* will be used for this purpose.
- the presentation address (PSAP) where this SMAE is located. The standard attribute *presentationAddress* will be used for this purpose.

Additionally, every SMAE directory object should contain information regarding the systems management application service element (SMASE) and the common management information service element (CMISE) in the SMAE. The Directory auxiliary object classes *sMASE* and *cMISE* are defined in (ISO 10164-16) for this purpose. They contain attributes that provide information about the supported systems management application service (SMAS) functional units (FUs), the supported management profiles, the supported CMIP version and the supported CMIS FUs on every SMAE.

In our current implementation, every SMAP has the ability to update (either by issuing a DAP modify or DAP add or DAP remove operation) the directory objects that represents itself and its corresponding SMAEs. These update operations take place on start-up or on shut-down of a SMAP. Having the above information about SMAEs registered in the Directory, each SMAP (either in the manager or agent role) can establish an association with a named SMAP after identifying the PSAP of the appropriate SMAE by performing the following (step 2a in figure 5):

1. Given the DN of the SMAP it wishes to associate with, it performs a DAP search under the following conditions:
   - the DN of the SMAP is used as the base object for the search
   - search for objects with the standard application context name "*systems-management*" (defined in ISO 10040)
   - search for objects that support the interoperable interface through which it wishes to communicate (by checking the supported CMIP version and the supported CMIS FUs)
   - search for objects that perform a specific management function in the opposite role (by checking the supported SMAS FUs and the supported management profiles)

   which should return the value of the *presentationAddress* attribute of the matching SMAE.

6 IMPLEMENTATION

The network management platform that is used in the implementation is the OSIMIS platform (Pavlou 1993), developed by the University College of London, which conforms to the CMIP/CMIS standards (ISO/IEC 9595, ISO/IEC 9596). The Directory Service implementation is based on the ISODE Directory System QUIPU (Kille 1991) version 8.0. A first implementation of the location transparency mechanism has been incorporated into the latest OSIMIS distribution. A full implementation of the mechanisms described in the previous sections is on progress. The performance of the overall system depends heavily on the performance of the QUIPU system which has been analysed and proved satisfactory for our purposes (see also Hong 1993).

7 ACKNOWLEDGMENTS

This work is supported by the CEU RACE project R2059 ICM (Integrated Communications Management). The authors would like to thank all the ICM members for their feedback and support.

8 REFERENCES

Dittrich, A. et al. (1993) Representation of Management Schema Information, TR 6331 GMD-FOKUS.

Hong, W. J. et al. (1993) Integration of the Directory Service in the Network Management Framework, Proceedings of the IFIP TC6/WG6.6 III.

Kille, E. S. (1991) Implementing X.400 and X.500: The PP and QUIPU Systems, Artech House, Boston MA.

Mansfield, G. et al. (1993) Internet working draft: Charting Networks in the Directory.

NM/Forum 015 (1992) Shared Management Knowledge Requirements, OMNIPoint 1.

Pavlou, G. et al. (1993) The OSI Management Information Service User's Manual, Version 1.0.

Sylor, M. (1993) Junction Objects, Proceedings of the IFIP TC6/WG6.6 III.

Tschichholz, M. and Donnelly, W. (1993) The PREPARE Management Information Service, Proceedings of the 5th RACE IS&N Conference.

ISO/IEC CD 10164-16 (1994) Information Technology - Open Systems Interconnection - Systems Management: Management Knowledge Management Function, Denmark.

ISO/IEC 10040 (1992) Information Technology - Open Systems Interconnection - Systems Management Overview, Geneva.

ISO/IEC 10165-1 (1992) Information Technology - Open Systems Interconnection - Structure of Management Information.

ISO/IEC 9595 (1991) Information Technology - Open Systems Interconnection - Common Management Information Service Definition.

ISO/IEC 9596 (1991) Information Technology - Open Systems Interconnection - Common Management Information Protocol Specification.

ITU Recommendation M.3010 (1992) Principles for a Telecommunications Management Network.

ITU Recommendation M.3020 (1992) TMN Interface Specification Methodology.

ITU Recommendation M.3200 (1992) TMN Management Services: Overview.

ITU Recommendation X.900-series (1992) Open Distributed Processing.

ITU Recommendation X.500-series (1988) The Directory.

9 APPENDIX

```
responsibleSMAP ATTRIBUTE
WITH ATTRIBUTE-SYNTAX responsibleSMAPSyntax
MULTI VALUE
```

```
responsibleSMAPSyntax ::= SEQUENCE {
        DistinguishedName, -- DistinguishedName is defined in the standard
        tMNManagementService }
tMNManagementService ::= ENUMERATED {
        Customer Administration (0),
        Management of the security of the TMN (1),
        Traffic Management (2),
        Switching Management (3),
        Accounting Management (4),
        Restoration and Recovery (5) }
managedResource OBJECT-CLASS
        SUBCLASS OF Device -- Device is defined in the standard
        MAY CONTAIN {responsibleSMAP}
supportedMIB ATTRIBUTE
WITH ATTRIBUTE-SYNTAX DistinguishedNameSyntax
MULTI VALUE
tMNMS ATTRIBUTE
WITH ATTRIBUTE-SYNTAX tMNManagementService
MULTI VALUE
tMNBuildingBlock ATTRIBUTE
WITH ATTRIBUTE-SYNTAX tMNBlockSyntax
SINGLE VALUE
TMNBlockSyntax ::= ENUMERATED {
        NE (0), QA (1), MD (2), SL-OS (3), NL-OS (4), NE-OS (5), WS (6) }
managementProcess OBJECT-CLASS
        SUBCLASS OF applicationProcess -- applicationProcess is defined in the standard
        MUST CONTAIN {TMNBuildingBlock}
        MAY CONTAIN {supportedMIB, tMNMS}
```

10 BIOGRAPHY

Costas Stathopoulos received the B.Sc. degree in Computer Science from the University of Crete, Greece in 1992. In 1993 he began the M.Sc. degree at the same university in collaboration with the Advanced Networks, Services and Management Group of the ICS-FORTH, Greece where he also works as a Research Assistant on the CEU RACE II ICM project from 1993. He is involved in the project group for TMN platform extensions, and specifically in providing distribution transparencies and metamanagement support. His main research interests are internetworking, network management, directory services and distributed systems.

David Griffin received the B.Sc. in Electronic Engineering from Loughborough University, UK in 1988. He joined GEC Plessey Telecommunications Ltd., UK as a Systems Design Engineer, where he worked on the CEU RACE I NEMESYS project on Traffic and Quality of Service Management for broadband networks. He was the chairperson of the project technical committee and worked on TMN architectures, ATM traffic experiments and system validation. In 1993 Mr. Griffin joined ICS-FORTH in Crete and is currently employed as a Research Associate on the CEU RACE II ICM project. He is the leader of the project group on TMN architectures, performance management case studies and TMN system design for FDDI, ATM and optical networks.

Stelios Sartzetakis received his B.Sc. degree in Physics and Mathematics from Aristotelian University of Thessaloniki in 1983, and his M.Eng. in Systems and Computer Engineering from Carleton University of Ottawa, Canada in 1986. He worked doing research in communication protocols in Canada. He joined ICS-FORTH in 1988. Today he is research scientist in the networks group responsible for CEU RACE projects in ATM broadband telecommunications networks and services management. Mr. Sartzetakis is responsible for FORTH's telecommunications infrastructure at large. He was principal in the creation of FORTHnet, a multiprotocol, multiservice network, the first Internet access provider in Greece. He served as an independent consultant to private companies and public organizations.

36

Testing Management Applications with the Q_3 Emulator

Kari Rossi, Sanna Lahdenpohja
Nokia Telecommunications Oy
P.O. BOX 33 02601 Espoo
tel: +358–0–5060 3857
fax: +358–0–5060 3876
email: kari.rossi@ntc.nokia.com

Abstract

Testing Q_3 based management applications is often a laborious and complex task. The Q_3 emulator agent (Q3E) is a tool for improving the effectiveness of testing the semantic functionality of management applications. An emulator agent is able to participate in OSI network management communication as the agent part: an emulator agent is an OSI agent in every sense, but it emulates to be running in a network element. For testing purposes, the operation of emulator agents can be controlled using the Q_3 emulator language (QEL) designed to decrease the test case design and implementation effort of management applications. In QEL, managed objects can be created or deleted, their action behaviours can be defined, and the sending of spontaneous events can be caused. Based on QEL definitions, the Q3E is able to respond automatically to requests from management applications. For the management application there is no difference: the agent, whether in network element or an emulator, responds similarly and handles the same managed objects.

Keywords

Testing Q_3 applications, testing CMIS/CMIP applications, Q_3, CMIS, CMIP, GDMO

1 INTRODUCTION

Testing Q_3 (ITU–T, 1992) based management applications is a demanding task and often requires significant development effort. One of the main reasons for this is the inherent complexity of the Q_3 interfaces and the specification formalisms Guidelines for the Definition of Managed Objects (GDMO) (ISO, 1992[2]) and Abstract Syntax Notation One (ASN.1) (ISO, 1990). Testing requires also deep knowledge and skill of both the management application and testing practices. In addition, it may be impractical or even impossible to maintain a realistic testing environment for the testers due to the high costs. Therefore, in order to decrease the development and testing effort and costs tools that support high level abstractions are needed. Unfortunately, the abstraction level of most currently available tools, such as XOM/XMP (X/Open, 1991) (X/Open, 1992) are low.

The Q_3 emulator agent (Q3E) (Rossi and Toivonen, 1994) is a high level tool for testing the semantic functionality of management applications. An emulator agent can be used to test management applications in an operation environment close to the real environment: the CMIS (ISO, 1991[2]) messages sent and received correspond to the real messages, and Q3E can emulate a network of managed objects. Q3E is not targeted at the OSI protocol or interoperability testing (ISO, 1991).

In this paper we first summarize the background of the Q_3 interface and the objectives of the Q3E. Section 4 describes the functionality of Q3E and section 5 explains how Q3E is used for testing management applications. Section 6 presents the conclusions.

2 BACKGROUND

The management concept of the Q_3 emulator agent is based on the Telecommunications Management Network (TMN) information architecture (ITU–T, 1992). The principles of the architecture are object oriented and are based on the OSI systems management concepts (ISO, 1992[1]), and the fundamental concepts are managed objects, manager and agent roles.

In the model the managed network and devices are structured into managed objects which have attributes, operations and notifications. Network management applications are distributed: an agent provides an object oriented view in the terms of managed objects of the resources it manages, and the manager issues management requests to the managed objects of the agent, and receives notifications from these managed objects.

The standardized interface between the manager and agent is Q_3. The managed objects are specified in GDMO and the attributes of managed objects in ASN.1. Each device type managed by Q_3 needs its own GDMO object model characterizing the special properties of the device. The communication protocol used for exchanging operation requests of managed objects is CMIS and CMIP (ISO, 1991[1]).

3 OBJECTIVES

The objectives of the Q_3 emulator agent are the following:

- Provide automation for the semantic testing of Q_3 management applications. OSI protocol and interoperability testing are out of the scope of this application, they are tested using other tools.
- Support testing of a network: Q3E has to support the emulation of a network consisting of many network elements.
- Q3E has to be programmable by an interpreted script language.
- Communication has to be based on OSI protocols, and CMIS, CMIP, ASN.1 and GDMO have to be fully supported.
- The system architecture has to be based on automatic code generation from GDMO and ASN.1 templates.

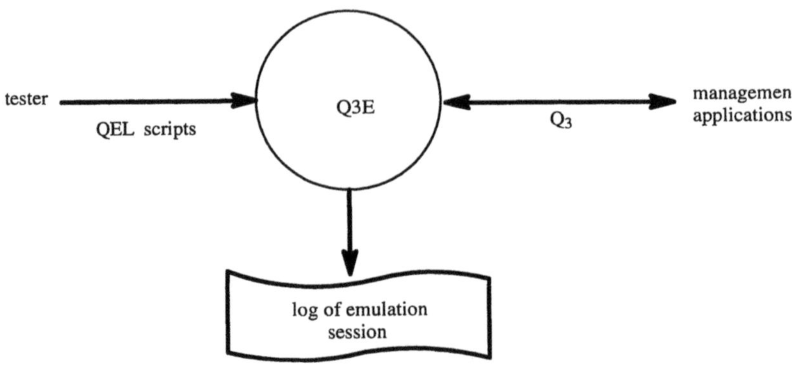

Figure 1 Controlling the execution of Q_3 emulator.

4 Q_3 EMULATOR AGENT FUNCTIONALITY

4.1 User Interface

The use of the Q3E is based on operation requests: the tester operates the emulator by writing a Q_3 emulator language (QEL) script, and when the script is ready, he submits the script to the Q3E for execution. For examining and monitoring the results of the execution of scripts and CMIS management operations the tester uses the Q3E log file. QEL scripts can be executed both interactively at run time or as batch scripts. See Figure 1.

From unix shell, the tester submits a QEL script to the Q3E using the **qrc** program. For instance, the unix command (1) executes the QEL script 'event.qel':

qrc event.qel (1)

4.2 QEL Language

Managed Objects

In QEL, managed object classes are referred to with the names given in GDMO templates. Managed object instances are referred to by distinguished names that are relative to the global root, as shown in the example (2):

[/, networkId = 1, managedElementId = 53] (2)

Distinguished names can also be constructed by specifying the path relative to another object such as a QEL variable. Managed object instances are stored in the Management Information Base (MIB) in the unix file system as ASCII files.

Attributes of object instances are referred to with the dot notation. For instance, the attribute State of the managed object (2) is referred to by:

[/, networkId = 1, managedElementId = 53].State (3)

Creating and Deleting Managed Objects

The tester can create managed objects using the **create** command. As a result of the command, the object is updated to the MIB. The parameters are similar to those of the CMIS create request. The tester can delete objects with the **delete** command.

Overriding Operations of Managed Objects

CMIS indications are served automatically by the Q3E. For changing this default behavior, QEL scripts can be assigned to CMIS indications to be called by Q3E when serving indications. Operations can be overridden based on object instance or class, or globally.

In the script (4) the **set** operation of the class 'trailTerminationPoint' is changed to run the script 'disabledEvent.qel' (5). The purpose of the script (5) is to send an event if the 'operationalState' attribute of the managed object is disabled:

```
change–operation trailTerminationPoint {
      set = "disabledEvent.qel"
};                                                                    (4)

-- script 'disabledEvent.qel'

if (%mo–instance.operationalState = "disabled") then
      event–req send {
            mode = non–confirmed,
            mo–class = %mo–class,
            mo–instance = %mo–instance,
            event–type = "communicationsAlarm"
      };
else
      emulate;          -- executes the default emulation operation of CMIS set
end–if;                                                               (5)
```

When changing the way to serve indications, the tester can call the automatic emulation using the **emulate** command. This is useful when the tester wants to extend the default emulation behavior as in scripts (4) and (5).

QEL Variables and Expressions

Two basic types of variables are supported: integer and string. In addition, variables of any ASN.1 type defined in the ASN.1 specification files can be created. Variables are declared by the **declare** command.

Integer expressions can be constructed with the arithmetic operators (+, –, *, /, %) from other integer expressions. String operators are + (concatenation) and – (removes the first occurrence of the second string from the first). Expressions can be grouped with parenthesis.

Assignment statements are begun with the **let** keyword. Variables may be assigned values of compatible types: type cast to integer is achieved by **integer()**, and to string with **string()**. For instance, strings 'prefix' and 'nodeId' and an integer 'i' are declared and assigned values by the script (6):

declare integer: i;
declare string: prefix, nodeId;
let i = 100;
let prefix = "node_";
let nodeId = prefix + *string*(i); (6)

As a result of the script (6) the value of the variable 'nodeId' is "node_100".

QEL language provides two sets of predefined variables: global variables, beginning with '$', and references to CMIS indication parameters, beginning with '%'. The advantage of QEL variables is that they are more general and easier to use than absolute values since they contain emulator context specific information. References to the CMIS parameters of indications can be used when sending responses. This makes it possible to set appropriate context sensitive default values for the response parameters.

The **let** command is also used for assignment of attribute values of managed objects, e. g.

let $mo.systemTitle = {pString "node_5"}; (7)

In the script (7) the value of the attribute 'systemTitle' is an ASN.1 string, but its type is an ASN.1 choice. **$mo** is the CMIS indication parameter referring to the managed object of the latest CMIS indication.

CMIS Commands

QEL provides commands for direct CMIS control: **create–rsp**, **delete–rsp**, **get–rsp**, **set–rsp**, **action–rsp** for sending CMIS responses and **event–req** for sending event report requests. For instance, script

get–rsp send {
  mo–class = *$mo–class*,
  mo–instance = *$mo–instance*,
  current–time = *$current–time*,
  attr–list = {delay = 10, bufferSize = 21}
}; (8)

sends a CMIS get response in which the managed object class and instance are the same as in the get indication, and attribute list contains two attributes 'delay' and 'bufferSize'. **$current–time** is a predefined QEL variable.

QEL supports also the sending of linked responses and CMIS error messages.

Delays and Timing

In order to closer emulate the response times of network elements, the tester can define a delay for specified managed object instances. Delay specifies the time in seconds to wait before executing the response. In the example below, response delay will be 10 seconds:

```
set-delay {
        [/, networkId=1, managedElementId=53] = 10
};                                                              (9)
```

In order to time the scripts to be executed by the emulator the tester can use unix scripts.

Control Structures

Conditionality can be represented with the **if** structure, and repetition in turn with the **loop** and **exit–loop** commands. The script (10) demonstrates one way to implement a 'for' loop from 1 to 10:

```
declare integer i;
let i = 1;
loop
        -- do the job here...
        if (i = 10) then
                exit-loop;
        end-if;
        let i = i + 1;
end-loop;                                                       (10)
```

A script can be run from another script with the **run** command, and a script can be exited with the **return** statement. The only way to 'pass parameters' is to use global variables.

4.3 System Architecture

The Q3E system architecture is based on C++–code generation tools (see Figure 2). The Q3E emulators are generated from the same GDMO and ASN.1 definitions used to specify the management interface between the management applications and the actual network elements. Two tools are used in the generation: the GDMO compiler of the Q3++ framework (Pohja et al., 1993) generates code from the GDMO definitions, and the Q3++ ASN.1 compiler produces the code necessary for handling ASN.1 types. The generated code is compiled and linked with the static Q3E code and with the Q3++ CMIS/CMIP communication library.

At run time Q3E consists of two unix processes, Q3E Run Script (QRC) client and the Q3E Emulator Server (QES) (see Figure 3). These two processes communicate through unix datagram sockets. QRC client program is the run time user interface of the emulator, and the QES server emulates the managed objects. The QES server listens to the requests of the QRC client process and CMIS indications of the management applications: QES executes each request or indication completely before starting the execution of the next request or indication.

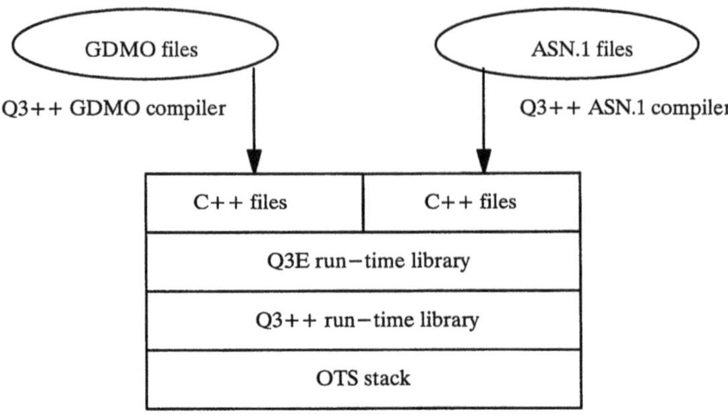

Figure 2 Q3E system architecture: generation principle.

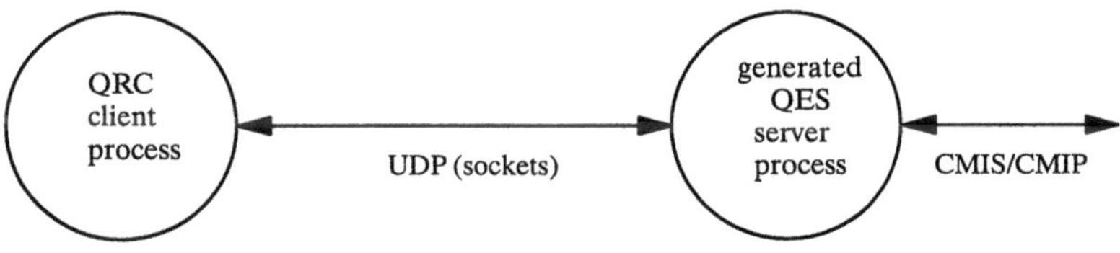

Figure 3 Q3E system architecture: run−time processes.

As the application programming interface towards the OSI stack two interfaces are supported, Nokia OSI stack, and HP OpenView XOM/XMP. In both cases, HP OTS/9000 stack provides the transport services. The current hardware platform is HP 9000 System 700*.

5 TESTING MANAGEMENT APPLICATIONS

5.1 Setting up the Testing Environment

The testing environment at the OSI agent side consists of the OSI stack, QRC client process, QES server process and Q3E MIB storing the managed object instances as ASCII files. MIB is most conveniently created with a QEL script, but it can be also created manually or by a unix shell script.

When setting up the testing environment, the first step is to generate the QES server process binary code from the GDMO templates and ASN.1 definitions specifying the management interface, and to configure the OSI stack used by the emulator.

* HP and OpenView are trademarks of Hewlett−Packard Co.

5.2 Principles of Writing Test Scripts

The test cases have to be planned carefully. Once they are written they can be repeated and reused. If no scripts are given to the Q3E, it receives indications and responds to them according to the standards and it's MIB. Other behavior of Q3E has to be defined using QEL scripts.

The Q3E is able to keep log file on various aspects of its operation. Since the log file is the only output from the Q3E, it is also the main source to be checked for the test results. Consequently, test planning should include also logging targets.

The QEL scripts implement the test cases and depend thus on the management application and the testing objectives. In the following paragraphs examples of different kinds of test cases are given.

Startup Scripts

If a Q3E were invoked without a startup script it would in most cases not be usable due to lack of information of the managed object instances. The purpose of a Q3E startup script is to define:

- the MIB containing the managed object instances of the emulated network elements;
- managed object class action behavior;
- the default usage of CMIS parameters;
- emulator specific defaults, e.g., logging parameters.

Testing CMIS Semantics

The most elementary test scripts respond to simple CMIS requests of management applications. The MIB is sufficient for many test cases testing the correct behavior of management applications, e. g., creating object instances, setting attribute values and combinations of CMIS parameters. For testing CMIS defined errors, it is convenient to override the default behavior of some object instances or classes to respond with CMIS errors.

Failure Scripts

A QEL failure script should be written for each kind of failure of a network element. A failure script executes all the emulator commands modelling a fault, such as modifying the managed object instances of the emulator to represent the new faulty state or sending an event or a set of events for the management application to inform of the failure. For example, the script 'communicationsFailure.qel' (11) changes the 'operationalState' and 'probableCause' attributes and sends an event report:

```
-- script 'communicationsFailure.qel'

[/, networkId = 10, equipmentId = 2].operationalState = "disabled";
[/, networkId = 10, equipmentId = 2].probableCause = "ProbableCause:{ localValue : 8}";
-- {localValue: 8} means loss of signal

event-req send {
   mode = confirmed,
   mo-class = "equipmentX",
   mo-instance = [/, networkId = 10, equipmentId = 2],
   event-type = "communicationsAlarm",
   event-info = asn1[AlarmInfo : {
            probableCause localValue : 8,
            perceivedSeverity major,
            notificationIdentifier 20}]
};                                                                    (11)
```

Combining Unix Scripts with QEL Scripts

QEL scripts can be combined with unix scripts to achieve even more complex tests. It is possible for example to create dynamically new QEL scripts, execute QEL scripts periodically, or automate a test session. For instance, the C shell script (12) calls the communications failure script (11) once in a minute:

```
while (1)
      qrc communicationsFailure.qel
      sleep 60
end                                                                   (12)
```

Unix shell must be used in this timing test case. If only QEL were used, the QEL script would block the execution of other QEL scripts and CMIS indications in the QES emulator server process because the QES executes one script (and CMIS indication) at a time.

Emulating a Network in Failure Scripts

The **mo–class** parameter of the event report request command defines the distinguished name of the object instance. Since the attribute values in the distinguished name can be arbitrary QEL expressions, it is possible to program a script sending events from multiple managed objects. If event report requests are sent in a loop, the loop variable can be used to construct attribute value assertions for the distinguished name. As a result, the management application receives event report indications from seemingly different managed objects.

In the script (13) a 'for' loop is implemented using the generic **loop** command and the variable i as a loop counter. An event is sent in every iteration. The distinguished name varies according to the value of the loop counter 'i': the last attribute–value–assertions in the distinguished names will be equipmentId = "node_1", equipmentId = "node_2", etc.:

```
declare integer: i;
let i = 0;

loop
      if (i = 100) then
            exit–loop;
      else
            let i = i + 1;
      end–if;
      event–req send {
            mode = non–confirmed,
            event–type = "communicationsAlarm",
            mo–class = "equipmentY",
            mo–instance = [/, networkId = 1, equipmentId = "node_" + i],
            event–time = $current–time,
            event–info = asn1[AlarmInfo : {
                  probableCause localValue : 2,
                  perceivedSeverity minor,
                  notificationIdentifier 20,
                  additionalText "Equipment Y specific fault text!"}]
      }
end–loop;                                                                    (13)
```

Generating Side Effects

With QEL scripts it is possible to extend emulation of network elements from the simple receiving of indications and sending of responses and event requests. When receiving an indication, a special QEL script can be executed which changes other attributes of the managed object or modifies other managed objects in the MIB than requested by the CMIS indication.

For example, the scripts (14) and (15) extend the behavior of the create indication of 'managedElement' class to create a 'log' object instance for the current managed object:

```
change–operation managedElement {
    create = "createManagedElement.qel"
};                                                                           (14)

– – script 'createManagedElement.qel'

let nextLogId = nextLogId + 1;      – – global variable, initialized to 0 in the startup script
create {
      mo–class = log,
      mo–instance = [%mo–instance, logId = nextLogId],
      attr–list = {operationalState = "enabled"}
};
emulate;                                                                     (15)
```

5.3 Performance Testing

The performance and stability of management applications can be tested using scripts that cause heavy communication loads. This could for instance be achieved with event generation scripts that send events at fast rate. Another alternative would be using several Q3Es.

6 CONCLUSIONS

This paper has discussed the testing of Q_3 based management applications. The testing of management applications is a demanding task, because, among other things, the Q_3 interfaces and the specification formalisms GDMO and ASN.1 are complex. Therefore testing tools are needed to decrease the effort and costs involved. The Q_3 emulator agent tool covers the semantic testing part of Q_3 management applications.

The main advantage of using Q3E lies in the reduction of testing costs. The testing costs affected are those for test equipment, man power, training and testing time. This is achieved because Q3E provides a high abstraction level for the testing personnel, and new emulators can be generated at short notice with minor effort.

The first version of Q3E that supports event sending over XMP has been in use since February –94. Initial experiences have been encouraging. This first version has been generated for three network element types, and their generation required about one day's effort from one person. The first two emulators are used by development teams in module testing and the third by a system testing group. The complete version of the Q3E will be released during first half of 1995.

The system architecture has been proven to be sound. The generation mechanism makes Q3E suitable for testing a very wide range of management applications. A considerable engineering effort was however required to achieve this kind of generality.

7 REFERENCES

ISO (1990) Specification of Abstract Syntax Notation One (ASN.1). ISO/IEC 8824, ITU–T Recommendation X.208.

ISO (1991^1) Common Management Information Protocol. ISO/IEC 9596–1, ITU–T Recommendation X.711.

ISO (1991^2) Common Management Information Service Definition. ISO/IEC 9595, ITU–T Recommendation X.710.

ISO (1991^3) Conformance Testing Methodology and Framework. ISO/IEC 9646–1.

ISO (1992^1) Systems Management Overview. ISO/IEC 10040, ITU–T Recommendation X.701.

ISO (1992^2) Structure of Management Information Part 4: Guidelines for the Definition of Managed Objects. ISO/IEC 10165–4, ITU–T Recommendation X.722.

ITU–T (1992) Principles for a Telecommunications Management Network. ITU–T Recommendation M.3010.

Pohja, S., Kaski, J. and Nurmi, E. (1993) Application Programming Interface for Managed–Object Communications, in *IEEE First International Workshop on Systems Management*, Los Angeles.

Rossi, K. and Toivonen, H. (1994) Q3E: Q_3 Emulator Agent, in *1994 IEEE Network Operations and Management Symposium*, Orlando.

X/Open Company Ltd (1991) OSI–Abstract–Data–Manipulation API (XO*M)*. X/Open CAE Specification.

X/Open Company Ltd (1992) Management Protocols API (XMP). X/Open Preliminary Specification.

8 BIOGRAPHY

Kari Rossi received his M. S. and Licentiate of Technology degrees in Computer Science at Helsinki University of Technology in 1986 and 1991. Mr. Rossi was the R&D project manager of the Q3E and Q3++ GDMO++ compiler projects. He is currently the R&D project manager of Nokia OMC for Fixed Network project which is developing a management system for Nokia DX 200 switches.

Sanna Lahdenpohja received her M. S. in Computer Science at Turku University in 1992. She was a senior engineer in the Q3E project. Currently she is a senior engineer in the Nokia OMC project.

9 ACKNOWLEDGEMENTS

Hannu Toivonen, Timo Posio, Lasse Seppänen, Saku Rahkila, Marko Setälä, Markku Rehberger and Susanne Stenberg have been working in the project team and have contributed essentially to Q3E.

The Q3E project has been partially funded by the Technology Development Centre of Finland (TEKES).

37

Application of the TINA-C Management Architecture

L. A. de la Fuente (TELEFONICA I+D, TINA-C Core Team Member)
M. Kawanishi (OKI, TINA-C Core Team Member)
M. Wakano (NTT, TINA-C Core Team Member)
T. Walles (BT, TINA-C Core Team Member)
C. Aurrecoechea (Columbia University, Bellcore Summer Student)

c/o Bellcore, 331 Newman Springs Rd.,Red Bank, NJ 07701,USA;
Phone: +1 908 758 5653; Fax: +1 908 758 2865; E-mail: alberto@tinac.com

Abstract

This paper presents the characteristics of the TINA Architecture and the TINA Management Architecture, the main information concepts that appear in the Network Resource Information Model, and how the Management Architecture is applied in the definition of management services for the Free-Phone telecommunication service.

Keywords

Management architecture, network resource information model, connection management, resource management, computational viewpoint, management service, free-phone service

1 INTRODUCTION

TINA-C (*Telecommunication Information Networking Architecture Consortium*) is a consortium formed by network operators, telecommunication equipment suppliers and computer suppliers, working on the definition of a software architecture to support the *rapid* and *flexible* introduction of new telecommunication services, as well as the ability to manage these services and the networks that support them in an *integrated* way. This architecture aims to be *independently* evolvable from the underlying switching and transport infrastructure that allows for the construction and deployment of applications independently from specific network technologies. The application *interoperability* in the TINA architecture is supported by a distributed processing environment which enables software components to interact across different network domains in a distribution transparent way.

In TINA-C, service is understood in a broad sense that includes the traditional concepts of telecommunication service -any service provided by a network operator, a service provider, etc., to customers, end-users or subscribers- and management service -any service needed for control, operation, administration and maintenance of telecommunication services and of networks used to provide these telecommunication services-. The management services in the TINA context refer to operations on network resources and also on telecommunication services. Moreover, in TINA-C the basis on which telecommunication and management services are specified, designed or provided, is the same. In this sense, TINA integrates both concepts and, as a result, approaches focusing in both areas such as IN and TMN, are integrated together with ODP concepts in the TINA Architecture (Chapman et al., 1994).

2 THE TINA ARCHITECTURE

The TINA Architecture is a consistent set of concepts and principles that can be used to design and implement any telecommunication software application, which may be contained within a single computing node or distributed among several heterogeneous computing nodes. They are classified in the TINA Architecture in four technical areas that, by extension, are also called architectures: Service, Network, Computing and Management Architecture (Figure 1).

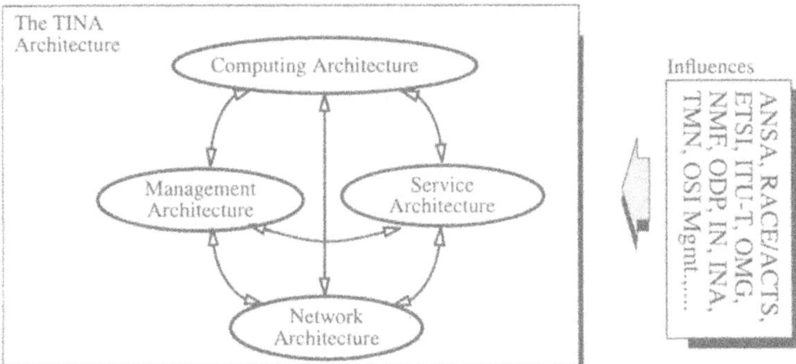

Figure 1 The TINA Architecture.

The Computing Architecture provides the basis for interoperability and reuse of telecommunication software through a set of modelling concepts that facilitate the specification, design and deployment of distributed telecommunication software components in a technology-independent way. It also defines a Distributed Processing Environment (DPE) that provides the support for the distributed execution of such software components and offers distribution transparency to them. The modelling concepts are defined for the Information, Computational and Engineering viewpoints of the ODP standards (Rec. X.901, 1993). The information modelling concepts focus on the definition of information-bearing entities (information objects), their relationships and the rules and constraints that govern their behaviour. The computational modelling concepts focus on the description of a system as a set of software components (computational objects) which are candidates for distribution, and in their interaction. The engineering modelling concepts focus on the infrastructure required to support distribution transparent interworking of software components, how software components are bundled in placement and activation units, how these units communicate, and how computing resources are allocated to each unit. These modelling concepts will be used in the specification of each of the architectures.

The Service Architecture defines a set of concepts and principles for specifying, analysing, reusing, designing, and operating service-related telecommunications software components.

The Network Architecture provides a set of generic concepts that describe transmission networks in a technology independent way. At one end, it provides a high level view of network connections that can be used by the services to satisfy their connectivity needs. At the other end it provides a generic (i.e., technology independent) description of (network) elements that can be specialised to particular technologies and characteristics.

The Management Architecture provides the general principles and concepts for management in TINA. It follows the TINA information and computational modelling concepts and takes results from several standards and recommendations as inputs. For instance, the OSI Management functional areas separation (Rec. X.700, 1992) and manager-agent relationships (Rec. X.701, 1992), the ITU-T TMN functional layering (Rec. M.3010, 1993), and the transport layering and partitioning (Rec. G.803, 1992). Results from other relevant fora and consortia are also taken into account like, for instance, the OMNI*Point* 1 results (NMF, 1992) for the TINA trouble ticketing functionality.

The management principles and concepts are applied to the Service, Network and Computing Architectures to obtain the desired management functionality. In other words, in TINA each of these architectures is responsible for the management of the resources, elements and/or components that are under their scope. It is outside the scope of the Management Architecture the definition of the concrete management activities in the Service, Network and Computing Architectures.

3 TYPES OF MANAGEMENT ACTIVITIES IN TINA

The Management Architecture is applicable to all types of management activities within TINA. These activities are classified as telecommunication management and computing management.

In a TINA consistent environment, the applications that can be found running on that environment are applications of telecommunication services, and applications of management of the telecommunications services (service management applications) and of the underlying telecommunication networks (network and network element management applications). Telecommunication management is the management of telecommunication services and the underlying telecommunication networks.

Computing management involves the management of the computers, platform and transport facilities that form the distributed environment (infrastructure) in which the TINA applications may run. The management of the software (applications, in general terms) that runs on that distributed environment is also inside its scope. Therefore, computing management can be further divided in:

- Software management (e.g., deployment, configuration, instantiation, activation, deactivation and withdrawal of software), including management of the TINA applications from the software point of view (i.e., applications seen as a set of software components). Management here does not concern with what the applications are doing nor application specific management.
- Infrastructure management, including DPE management, management of the infrastructure transport facilities (kernel transport network), and computer environment management.

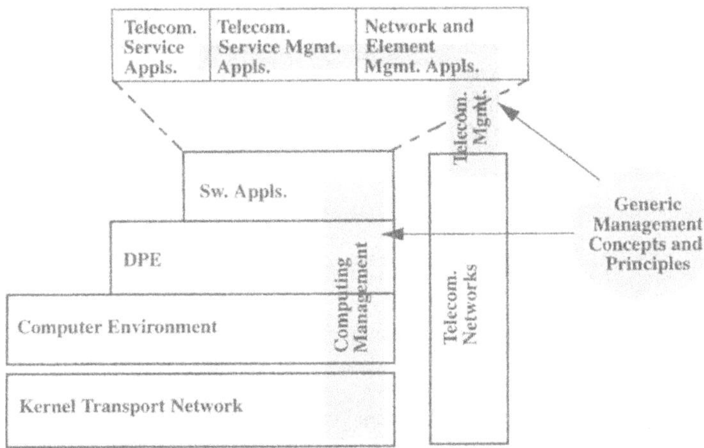

Figure 2 Types of management activities in TINA.

Therefore, computing management is under the scope of the Computing Architecture, and telecommunication management is under the scope of both Service and Network Architectures in the following way: the Service Architecture is responsible for the management of the services, and the Network Architecture is responsible for the management of the network elements and networks. Computing, Service and Network Architectures perform the management activities applying and extending and/or refining the generic principles and concepts of the Management Architecture.

This paper focus on telecommunication management activities and will describe, in the following sections, how the Management Architecture concepts are applied for the management of the Network Architecture, focusing on the connection management functionality. Then, a service scenario will exemplify the usage of that management functionality by a telecommunication service, the Free-Phone Service.

4 MANAGEMENT APPLIED TO THE NETWORK ARCHITECTURE

This section describes the application of the management functional areas and the TMN layers to the Network Architecture. It also describes the results of the application of the TINA information and computational modelling concepts in the NRIM and the definition of the connection management functionality, respectively.

Management Functional Areas in the Network Architecture

As stated previously, the Management Architecture follows the functional area organization defined in the OSI Management, namely fault, security, accounting, performance and configuration management. Considering the special relevance of the latter in all the management activities and that management of connections is a fundamental activity in all networks, two new functional areas replacing and specializing it have been identified: connection management and resource configuration management (Figure 3). This refinement is valid only for the management of the Network Architecture (for the management of the Service Architecture the five "classical" functional areas are used). Although TINA embraces all the areas, the work done so far has been focused in the following areas: fault management, connection management, resource configuration and accounting management. A brief description of each of these functional areas follows.

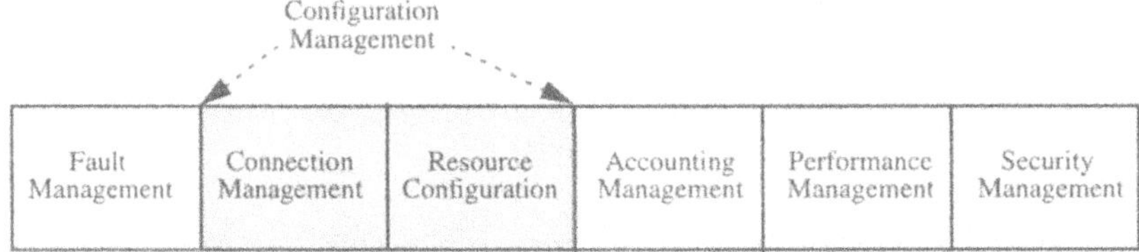

Figure 3 TINA functional areas for the management of the Network Architecture.

Fault Management is responsible for the following activities: Alarm surveillance (that collects and logs alarm notifications from the network resources), fault localization (that analyses the collected alarm information, detects the alarm root cause, and notifies to the alarm surveillance service clients), fault correction (that deals with the resources in which a root alarm is detected in order to restore or to recover them from the fault condition), testing (that invokes a test capability of a resource upon request and it may also support the test of series of resource), and trouble administration (that reports the troubles due to fault conditions and tracks their status).

Connection Management is responsible for providing the functionality required to deal with the setup, maintenance and release of connections, including the specification of a connection model, the signalling and routing methods, the management of the resources needed for the connections, and the methods for handling resource failures and overloads. The Connection Management functionality will be described with more detail in this paper.

Resource Configuration is responsible for the identification and location of resources and the associations among them. Its functionality includes: installation support (installation and removal of network resources, including the establishment of relationships between network resources), provisioning (assignment/release and activation/deactivation of network resources), and status and control (configuration information, including topological and inventorial view of network resources as well as the maintenance of those information).

Concerning Accounting Management, a model for accounting management has been proposed in TINA. This model covers metering (identification and recording of information relevant to the usage of resources in a meaningful way), charging (establishment of charges for the use of the resources from the metered information, including the usage of tariffs in order to calculate the charges) and billing aspects. Note that billing is an user-related activity and, thus, it is under the scope of the management activities in the Service Architecture, although this functional area must provided the network accounting information to the Service Architecture accounting management functional area in order to be allow the latter to generate the billing for the use of the network resources.

TMN Functional Layers

Network Architecture management comprises the Network Management Layer (NML) and the Network Element Management Layer (EML), since both networks and (network) elements are the resources being considered in the Network Architecture. Relationships with the Service Management Layer (SML) have been identified (as the previously mentioned about the transfer of accounting information from the NML to the SML). Although the Business Management Layer is outside the scope of the Network Architecture, the policies and agreements made at this level have a strong influence in the management functionality.

The Network Resource Information Model

The information model defined in TINA for the Network Architecture is the Network Resource Information Model (NRIM). It contains the object classes needed for the representation of network resources and is transmission and switch technology independent (the information specification is independent of the technology of the network resources, e.g. SDH, SONET or ATM). It supports different types of services (e.g., VPN, PSTN, multimedia, multiparty). The NRIM is concerned with how individual network element entities are related, topographically interconnected, and configured to provide and maintain end-to-end connectivity. This model is used by telecommunication and management applications.

The main sources of input for this model are the ITU-T Rec. G.803 for the concepts of layering and partitioning (although this recommendation is focused on SDH, these concepts are extended to cope with other network technologies), and the Generic Network Information Model (Rec. M.3100, 1992) object classes for the management aspects, that has been extended with new object classes describing aspects not covered by M.3100, which is mainly oriented to network element management. As M.3100 is switching and transmission technology independent, the resulting information model is generic enough to be applicable to existing models describing network element aspects (e.g., G.774 SDH information model). Currently, the NRIM contains the common managed object classes relevant for the Connection Management, Resource Configuration and Fault Management functional areas. The information model is presented in several fragments, as in the M.3100 Recommendation. They are defined Quasi-GDMO, that is a GDMO (Rec. X722, 1991) based notation tailored for its use in TINA. Figure 4 shows the relationship between the NRIM and the TMN layers. As it is depicted, the NRIM doesn't define a model for network elements, only relationships to existing standards in this area (some of these standards also appear in the Figure 4).

The nine NRIM fragments are: Connection Graph (gives a high-level abstraction of the network as seen by service applications; using this fragment, the applications can express their needs of network resources in a simple manner), Network (shows the overall structure of a network and basic concepts such as layering and partitioning), Connectivity (shows the different types of connections that can be established across the network), Termination Point (the connectivity relates two or more end-points to each other; the end-points are called termination points and they can also be viewed as access points for the user to the network),

Resource Configuration (shows the support objects used by resource configuration management), Fault Management (shows the support objects used by fault management), Adapter (describes the adapter functionality between layer networks), Domain (shows the relationships between different domain concepts (administrative domain, connection management domain, etc.), and Reuse (shows how the existing standards from which NRIM inherits have been reused).

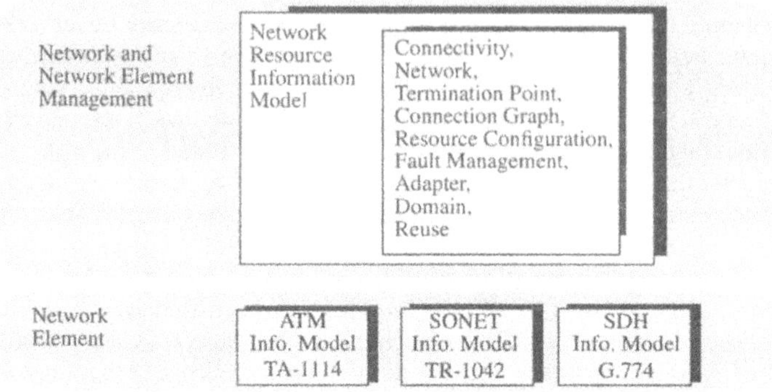

Figure 4 NRIM and TMN Layers.

In order to better understand the Connection Management functionality and the service example described in the next sections, the first three fragments will be briefly explained here. The Connection Graph (Figure 5) is an object which uniquely describes the connectivity between ports, independent of how it is achieved and independent of the underlaying technology. The connection graph is also a container for the other objects. The line represents a unidirectional connectivity between one source port and one or more sink ports. A branch object is associated with the sink ports. Line 1 between port 1 and port 3 in Figure 6 is an example of a point-to-point connection. Line 2 between the source port 2 and the sink ports 4 and 5 is an example of a point-to-multipoint connection. The vertex object represents a grouping of ports and provides a general mechanism for describing resources with capabilities to process information. A vertex may represent a network resource, a third party owned (or controlled) resource, a software resource or an end-user resource.

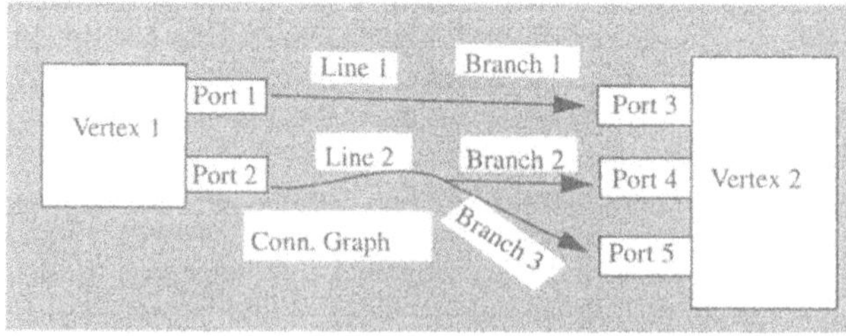

Figure 5 Connection Graph.

A network can be described as a set of layer networks. Each layer network represents a set of compatible inputs and outputs that may be interconnected and characterized by the information that is transported. A layer network (Figure 6) contains topological links and subnetworks. The Connectivity in it consists of trails, connections and subnetwork connections. A trail transfers validated information between end points of the layer network. A subnetwork connection describes the connectivity between termination points of a subnetwork. A connection describes the connectivity between two subnetworks. A number of connections may be bundled to form a topological link. Each subnetwork may be further broken down into more subnetworks and connections interconnecting them.

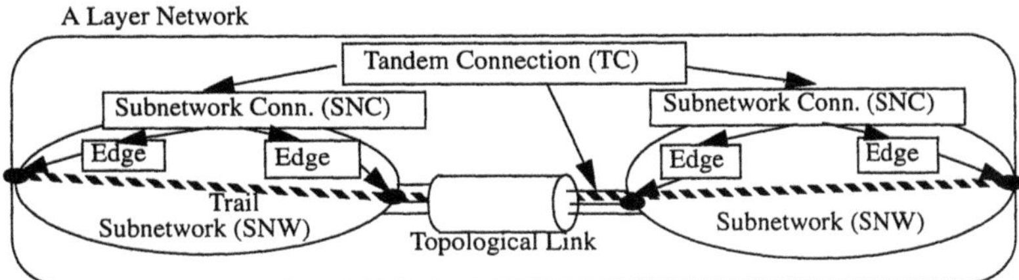

Figure 6 Connectivity across a Layer Network.

Connection Management

The TINA Connection Management (CM) functionality (Bloem et al., 1994) provides to the telecommunication services the necessary connectivity between terminals or processing nodes, and/or connectivity between computational objects. To the management services it provides the connectivity needed to access specific network elements (to be tested, for instance) and also the connectivity needed to support the desired management policies (re-routing policies in case of failures, etc.). To the DPE, as client of this functionality, CM will provided the necessary connectivity when DPE instances in TINA nodes need a connection to exchange information.

Its activities can be classified in the following three main types: Connection Manipulation (creation, modification, and destruction of network connections including locating connection end points and control of network resources), Connection Resource Management (identification of resources used to implement connections and management of the information needed to select resources and routes through the network), and Administrative Control (control and monitoring of connection management procedures for both network operator and customer use -not defined yet in TINA-).

CM defines a set of computational objects which support connectivity needs of both telecommunication and management services at several levels of abstraction. CM functions only reside in the Element Management Layer and the Network Management Layer. Functions above and below these layers are outside the scope of CM. Figure 7 shows an example of the CM functionality modelled as computational objects.

The shaded computational objects in it are inside the scope of the CM functionality. The SSM is one of the possible clients of this functionality and is out of the scope of the Network Architecture. The computational objects in the NEL model the physical transmission and switching equipment. CSM and CC offer an interface oriented to the service components in terms of operations on connection graphs. LNC and CP offer an interface in terms of trails, tandem-connections, subnetwork connections and termination points:

- Communication Session Manager (CSM). Defined at the top level, is the object which provides the service for setting up, maintaining and releasing logical connections. The term logical stresses the fact that its specification refers to computational object interfaces instead of addressable points in the network. Connectivity requirements are specified in terms of a Logical Connection Graph, which is a subclass of the Connection Graph (CG) described previously, supporting distribution and network structure and technology abstractions.
- Connection Coordinator (CC) provides interconnection of addressable termination points of networks. Connectivity requirements are specified in terms of a Physical Connection Graph, a subclass of the CG described previously. The specification of the connection comprises the termination point addresses and the characteristics of the connection, e.g., quality of service parameters, but it is independent of information concerning the underlying transmission and switching technology and the structure of the underlying networks.
- Layer Network Coordinator (LNC) provides interconnection of termination points of a layer network. There is a LNC for each domain in a layer network. A LNC receives request for trails in its layer network and has federation capabilities with LNCs of other domains in the layer network.
- Connection Performer (CP) provides interconnection of termination points of a subnetwork, that is, subnetwork connections. There are two classes of connection performers depending on the management layer at which it is used, e.g., network and network element.

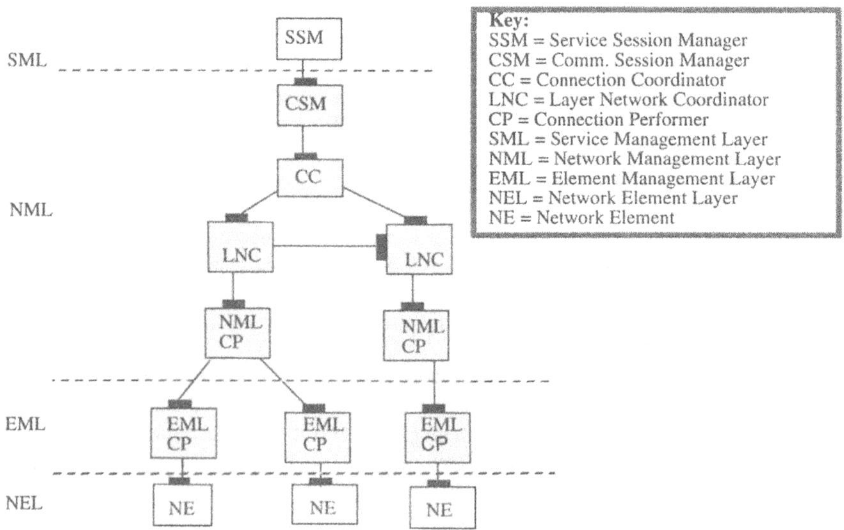

Figure 7 Connection Management Computational Model.

5 USE OF MANAGEMENT FUNCTIONALITY IN THE FPH SERVICE

The FPH service is an example of an IN service that illustrates how a telecommunication service can be provided as a TINA service. It is characterized by two main IN service features, one number (ONE) -that allows a subscriber with two or more terminating lines in several locations to have a single telephone number- and reverse charging (REVC). Figure 8 shows the use of management components (CSM, Accounting Manager and Subscription Manager).

One possible scenario of a FPH service session is the following in a time order:

- service_req(): The user with telephone number A sends a FPH request to User Agent (UA) A. UA is a representation of an End User in the network and mainly supports the End User to access services and to interface telecommunication services. The UA also manages the mobility, customization and security of End Users, and interworking with the End User System.
- check(): User Agent A checks the availability of the FPH service of subscriber S with the Subscription Manager which is considered a specialization of Configuration Management. The Subscription Manager is made up of several computational objects. However, they are shown here as only one. Subscription Agent, Subscription Register, and Service Description Handler play specific service dependent/independent role in a distributed manner for subscription management. The Subscription Manager maintains the relationships between services, service providers, and subscribers. A subscription is one relationship between them and is defined by the subscriber. End Users are listed in a subscription list.
- create(): After the UA receives approval from the Subscription Manager (based on the subscription to which the End User A belongs), it asks the FPH Service Factory (SF) to create the FPH service. The FPH SF creates the User Session Manager (USM) for End User A and a FPH SSM. A USM manages the resources which are locally used by a specific End User. SSM manages the resources which are commonly used by End Users. The SSM supports service execution, joining of End Users, and negotiation among End Users. The USM supports the SSM in the execution of local functions (e.g., the control of sinks and sources of information associated to the End User) and it specialized the service (control) interface offered to the End User System on the basis of usage context of that specific End User.
- resolve(): The FPH SSM gives the identity of Subscriber S to the Subscription Manager and obtains the interface reference of UA B based on the subscription of Subscriber S. This subscription will contain the ONE and REVC information. Derivation of UA B is based on the ONE info. and charging on the subscriber of End User B is based on the REVC information.
- join-in-session(): The FPH SSM requests User Agent B to accept the start of FPH Service.
- session-invitation: User Agent B confirms with End User System B that it will join the FPH Session. A User Agent needs to know which End User System is appropriate for End User may obtain more than on End User Systems. This mechanism is not detailed in this scenario.
- create(): User Agent B requests FPH SF to create USM B. USM B is dedicated to End User B, executes local management for End User B, and supports FPH SSM.
- charge-configure(): The FPH SSM asks the Accounting Manager to establish a charging configuration for this service (the subscriber of the called user (S) is going to pay 100% of the charge). The Accounting Manager keeps the charging configuration identification, receives all the information about resource usage and calculates the charges and bill.

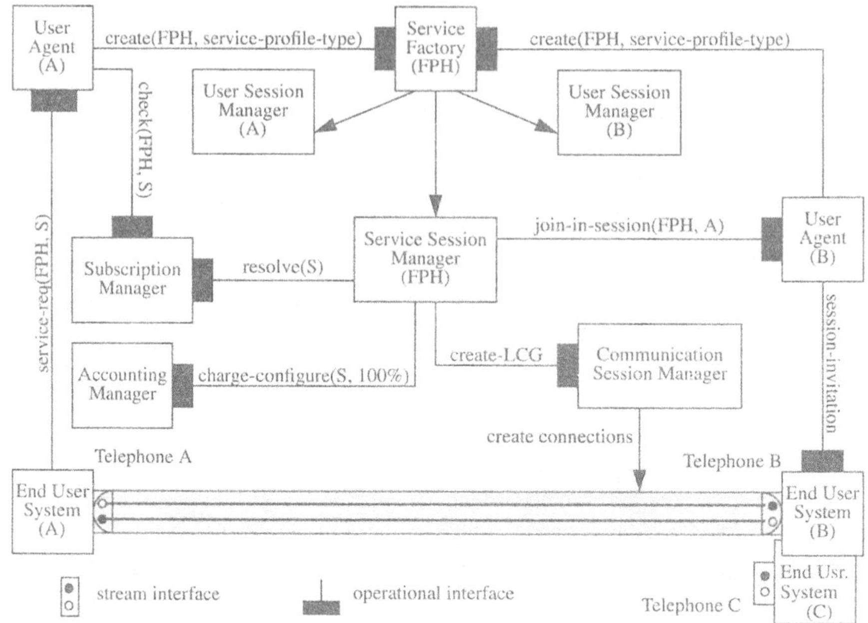

Figure 8 Computational Model of the FPH Service.

- create-LCG: The FPH SSM requests the CSM to create a Logical Connection Graph and the connection is set up between the stream interfaces of Users A and B, as it has been described in the previous section.

Deletion of these objects is not shown in this scenario. Life cycle management of these objects relies on DPE services. The identification of these objects are based dependency of several aspects: service, user, and subscriber. Management of heterogeneity is cover by, for instance, a USM for a End User.

6 SUMMARY

TINA-C is an international consortium working on the definition of a software architecture for a rapid and flexible introduction of both telecommunication and management services, independently of the underlying switching and transport infrastructure. This paper has presented the TINA-C Management Architecture (that combines concepts from the TMN and OSI Management standards) and some of the results of its application to the Network Architecture: the Network Resource Information Model and the Connection Management functionality. To illustrate the principles and the ideas behind the TINA-C Architecture, the usage of these results by a TINA service (the Free-Phone Service) has also been presented.

7 REFERENCES

Bloem, J., et al. (1994) The TINA-C Connection Management Architecture, TINA'95, Melbourne, Australia, Feb. 13-16, 1995.

Chapman, M., Dupuy, F. and Nilsson, G. (1994) An Overview of the Telecommunications Information Networking Architecture, TINA '95, Melbourne, Australia, Feb. 13-16, 1995.

ITU-T Rec. G.803 (1992) Architectures of Transport Networks Based on the Synchronous Digital Hierarchy, Geneva.

ITU-T Rec. M.3010 (1993) Principles for a Telecommunication Mgmt. Network, Geneva.

ITU-T Rec. M.3100 (1992) Generic Network Information Model, Geneva.

ITU-T Rec. X.700 (1992) Management Framework for Open Systems Interconnection (OSI) for CCITT Applications, Geneva.

ITU-T Rec. X.701 (1992) OSI - Systems Management Overview, Geneva.

ITU-T Rec. X.722 (1991), Guidelines for the Definition of Managed Objects, Geneva.

ITU-T Rec. X.901 (1993) Basic Reference Model of Open Distributed Processing - Part 1: Overview and Guide to Use, Geneva.

Network Management Forum - NMF (1992), OMNIPoint 1, Morristown, New Jersey.

8 BIOGRAPHIES OF THE AUTHORS

Cristina Aurrecoechea received her Master Degree in Industrial Engineering at the Basque Country University (Bilbao, Spain). From 1987 until 1991 she worked in Telefónica (the Spanish PTT) as software engineer in the management of a SNA/X.25 wide area network. She obtained her Master Degree in Electrical Engineering in 1992 at Columbia University, where she is currently a PhD student at the Center of Telecommunication Research (CTR).

Luis A. de la Fuente received his Master Degree in Telecommunication Engineering in 1987 and his Specialist Degree in Communication Software Design in 1989, both from the Politechnical University of Madrid (Spain). He joined Telefónica I+D in 1988, where he has been working on specification and design of new network and service management systems for the Spanish PTT. He have been participating in several EURESCOM projects, and he is also representative of his company in the NMF. He joined the Core Team on February 1994.

Motoharu Kawanishi received a B.E. from Meiji University, Japan, in 1983, and a M.E. in Computer Information from Stevens Institute of Technology, USA, in 1994. In 1983, he joined OKI Electric Industry Co., Ltd., Japan, where he has been working on software development for ISDN switching systems. He is TINA-C Core Team member since April 1993.

Masaki Wakano entered NTT in 1989 after he finished his Master Degree in Electronical Engineering at Kobe University, Japan. He has been working in developing OSI management system for NTT's business networks and in the application of CMIP to the next generation transport network. Since 1993, the first year of the TINA Consortium, he is Core Team member. He is now investigating service operations of multi-media services at Network Operation Systems Laboratory in NTT, Japan.

Tony Walles has been working in BT for over 30 years. His last activities have been the development of System X and SS#7 signalling in BT Research Labs (Ipswich). He has been also tutor at BT Vocational Training facility on digital switching and signalling systems. He also participated in the CASSIOPEIA. project. He is Core Team member since January 1994.

PART THREE

Practice and Experience

SECTION ONE

Agent Experience

38

Exploiting the power of OSI Management for the control of SNMP-capable resources using generic application level gateways

Kevin McCarthy, George Pavlou, Saleem Bhatti
Department of Computer Science, University College London,
Gower Street, London, WC1E 6BT, UK.
tel.: +44 71-380-7215, fax: +44 71 387 1397
e-mail: k.mccarthy, g.pavlou, s.bhatti @cs.ucl.ac.uk
José Neuman De Souza
Department of Computer Science, Federal University of Ceara,
Pici Campus, 60000 - Fortaleza - Ceara, BRAZIL.
tel.: +55 85 226-4419, fax: +55 85 223-1333
e-mail: neuman@lia.ufc.br

Abstract

A major aspect of Open Systems' network management is the inter-working between distinct Management architectures. This paper details the development of a generic object oriented application level gateway that achieves seamless coexistence between OSI and SNMPv1 management systems. The work builds upon the Network Management Forum's 'ISO/CCITT and Internet Management Coexistence' activities. The power of the OSI Systems Management Functions is made available for the management of SNMPv1 based resources, bringing fully event driven management to the SNMP domain.

Keywords

OSI, SNMP, Q-Adapter, Gateway.

1 INTRODUCTION

Whether driven by technological merit, simplicity of development or government profiles, considerable investments have been made and will continue to be made into the provision of network management solutions based on the two dominant management architectures, namely SNMPv1 [RFC1155, RFC1157, RFC1212] and OSI [X701,X720]. They exist together so they must be made to coexist, so as to achieve global inter-working across heterogeneous platforms in the management domain.

It is the authors' contention that coexistence can most readily be achieved by selecting a semantically rich reference model as the basis for this inter-working. Such an approach can then be readily extended to encompass both up and coming technologies such as CORBA [OMG91], together with architectures that have not yet bridged the synaptic gap in the collective minds of standards bodies and manufacturers' consortia.

The collaborative work of the Network Management Forum's (NMF) ISO/CCITT and Internet Management Coexistence (IIMC) activities has provided a sound basis to our efforts in achieving coexistence through automated application level gateways. Through out this paper we shall use the terms 'proxy', 'application level gateway' and 'Q-Adapter' [M3010] synonymously, to indicate the automated translation of information and protocol models, so as to achieve the representation of management objects defined under one proprietary paradigm under that of an alternative model, namely OSI.

The development of the gateway has been undertaken by the RACE Integrated Communications Management (ICM) project, to achieve Network Element management of non-OSI resources. Partners from VTT (Finland), Prism (France), CET (Portugal) and UCL (UK) have been principally involved with this effort. ICM has a mandate to demonstrate the feasibility of integrating Advanced Information Processing technologies for Telecommunication Management Networks. The gateway has been developed using the Object Oriented power of UCL's OSI Management Information Service development platform [Pavlou, 1993].

2 COMPARISON OF THE OSI AND SNMP MODELS

The approaches are a result of two distinct (some would say diametrically opposed) underlying tenets. The Internet-standard Network Management Framework is based on the notion of universal resource deployment. This may be alternatively stated by the fundamental axiom:

> "*The impact of adding network management to managed nodes must be minimal, reflecting a lowest common denominator*" [Rose, 1991]

In contrast the OSI standardization process attempts to achieve an all encompassing framework, to meet any future management requirements. Since OSI standardization is a self-perpetuating process a great deal of thought was initially placed into the underlying object oriented model so as to allow for the planned continual expansion.

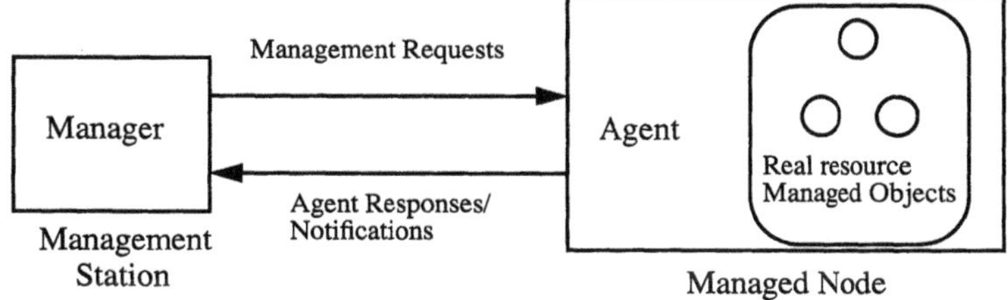

Figure 1 The Manager/Agent model.

If we consider the manager/agent model shown in Figure 1, then under SNMP the burden of management would be placed firmly on the management station, with only minimal impact on the more numerous managed nodes. Under OSI a more significant load is placed on the agents due to a greater expectation of the capabilities of managed nodes.

Both camps set out with the same overall aim of achieving the effective management of heterogeneous resources. One took a pragmatic approach and achieved exceptional market acceptance, the other attempts to provide a complete solution at the expense of its complexity.

2.1 Management information

Each agent provides a management view of their underlying logical and physical resources, such as transport connections and power supplies, to the managing applications. Managed Objects provide an abstract view of these real resources. The Managed Object data is held in a management database called the Management Information Base (MIB). Both SNMP and OSI define schemata for the description of Managed Object MIB data, namely the Structure of Management Information (SMI)[RFC1155, RFC1212] and the Guidelines for the Definition of Managed Objects (GDMO) [X722].

The OSI information model is object-oriented and permits the refinement of existing Managed Object templates via inheritance, see Figure 2. Refinement may occur due to an increase in the capabilities of a given Managed Object, perhaps due to the evolution in the technology of the underlying resource. The OSI model supports allomorphism, which facilitates the management of a given object as if it was an instance of any of the object classes in its inheritance hierarchy, thus permitting managing applications that have been coded to an earlier version of the information model, to continue to exercise control.

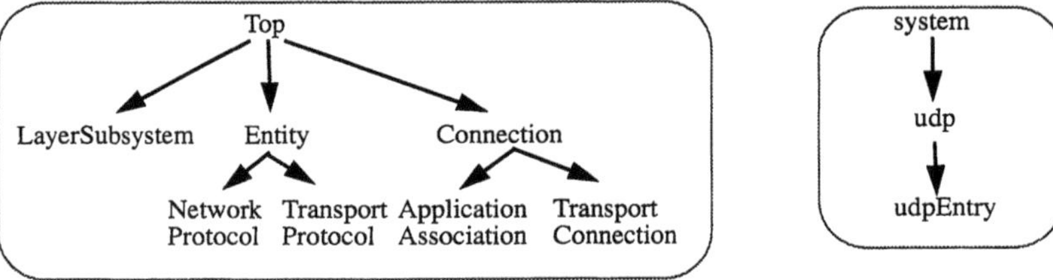

Figure 2 An example of inheritance and containment hierarchies.

The aggregation relationships between managed objects, such as "kind-of" and "part-of", are described by Name Binding containment descriptions. These containment descriptions yield a Managed Object instance hierarchy which is termed the Management Information Tree (MIT), see Figure 2. The MIT facilitates globally unique instance naming via Distinguished Names.

SNMP's object-based information model is simpler than its OSI counterpart so as to reduce the complexity of the agent implementations. SNMP objects represent single, atomic data elements that may be read or written to in order to effect the operation of the associated resource. The SNMP SMI permits the variables to be aggregated into lists and tables but there is no mechanism provided by SNMP to enable the manager to operate on them as a whole. Object identifiers are used to achieve object instance naming, see Figure 3. The syntaxes that each variable may hold are a very much reduced subset of the unlimited syntaxes that are permitted by the OSI model.

iso(1) org(3) dod(6) internet(2) mgmt(1)
mib-2(1)
tcp(6) udp(7)
tcpConnTable(13) udpTable(5)
tcpConnEntry(1) udpEntry(1)
tcpConnLocalPort(3) udpLocalPort(2)

Figure 3 An Internet management MIB object identifier instance naming tree.

2.2 Protocol operations

OSI makes a distinction between the service offered by a layer and the underlying protocol that achieves those services, whilst SNMP makes no such distinction. OSI management's services and protocol are defined by the Common Management Information Service [X710] and the Common Management Information Protocol [X711] respectively.

In placing the emphasis for Manager/Agent communications between asynchronous interrupt-driven and polling based approaches, SNMP selected 'trap-directed polling', whilst OSI adopted an event driven approach. Upon an extraordinary event the SNMP agent emits a simple Trap notification to its manager, which must then initiate MIB polling to ascertain the full nature of the problem. Since Traps are both simple and unacknowledged their generation places only a small burden on the managed node. The manager may still need to poll important attributes periodically if the set of supported Traps are not sufficient to indicate the occurrence of all important error conditions. CMIS supports extremely expressive and optionally acknowledged event reports, to the managing application, via the M-Event-Report operation, thus removing the need for any additional polling. The onus is placed on the OSI agent to inform the manager of significant events.

The requirement for simplicity extends to the number and complexity of SNMPv1 protocol operations, compared with their OSI counterparts. CMIS operations may include specifications for 'scope' and 'filter' so that the operation may be applied to some subset of the agent's managed objects. Scoping selects a sub-tree from the agent's MIT and filtering specifies a criteria, such as 'those routing entries with a valid status', to select from the scoped objects.

M-Get and M-Set are provided to retrieve and update attribute values. Since the usage of scoping and filtering means that the number of responses to an M-Get (which are received in linked replies) will not necessarily be known when the request is sent, an M-Cancel-Get operation is provided to prevent the possibility of the manager being over loaded. M-Create and M-Delete cause the creation and deletion of managed objects. M-Action facilitates the execution of any supported imperative command such as 'start diagnostic tests'.

SNMPv1 supports the retrieval of management information via Get and Get-Next primitives, the latter facilitating MIB traversal. Retrieval responses are limited to a single packet, which ensures that the manager will not be overloaded with response data, at the expense of requiring multiple retrieval requests to traverse an entire table. The Set primitive is used to update MIB objects which, via side-effects, achieves the control of imperative actions and the creation or deletion of table entries.

2.3 Transport mappings

Although SNMP is transport protocol independent, the connectionless User Datagram Protocol is the principal transport for SNMP. The "end-to-end" argument [Saltzer et al, 1984] makes a very strong case for leaving the selection of aspects such as the transport protocol to the application level, since only the application (in this case management) has a complete appreciation of its transport requirements.

By selecting a connectionless protocol such as UDP the management implementation is free to produce its own timeout and retransmission mechanisms. At times of network congestion the SNMP implementation can then configure an appropriate level of retransmissions to increase the chances of successful management when the network itself is failing. The application can more readily determine when some form of out-of-bands communication is essential. This approach requires that each SNMP implementation must attempt to produce its own transport mechanism that will not end up accentuating any problems of network congestion.

OSI management is association based, requiring association establishment and removal phases, in addition to the transfer of management requests. It should be born in mind that manager/agent associations are intended to be held open for a period of time, thus spreading the cost of the association over a number of management requests. The Transport level implementation, whether TPx or TCP (if RFC1006 is followed), is entrusted with achieving the efficient delivery of management messages whatever the underlying network conditions.

2.4 Generic functionality

OSI management standardization has greatly surpassed the more *ad hoc* efforts of the SNMP community in defining functionality through an ever growing series of Systems Management Functions (SMFs). The Event-Reporting SMF [X734] permits the managing application to create Event-forwarding-discriminators at the agent, which control the selection and destination of all event reports that the agent generates. A related SMF is the Log-Control function [X735], which permits event logging according to manager configurable criteria.

To reduce requirements for remote polling and data retrieval the Metric Monitor [X738] and Summarization SMFs [X739] have been developed. Together they permit manager applications to configure agents to undertake localised polling, threshold checking, data summarization and statistical analysis.

The X.500 Directory [X500] provides a global, hierarchically structured data repository. By incorporating the Directory into the OSI management model, the distributed transparencies, such as faults, replication and location transparency, can be achieved.

3 MANAGEMENT COEXISTENCE

At an early stage in the design of the gateway the decision was made to build upon the work that has been undertaken by the Network Management Forum's ISO/CCITT and Internet Management Coexistence (IIMC) activities. The IIMC package currently consists of five documents [IIMCIMIBTRANS, IIMCOMIBTRANS, IIMCMIB-II, IIMCPROXY and IIMCSEC]. Two of these documents are of the greatest significance to our work, namely 'Translation of Internet MIBs to ISO/CCITT GDMO MIBs' and 'ISO/CCITT to Internet Management Proxy'.

As intimated above, although it was ourrintent to follow the IIMC specifications in full, a number of instances arose where we selected options that either differed with or continued on from the IIMC work. For example the IIMC define a 'stateless' proxy, whilst our gateway is 'stateful' and can thus take advantage of caching. Other issues such as achieving maximum efficiency in the protocol translation, improved automation and the inter-working with non-conformant SNMP agent implementations, have been given greater consideration by our research.

3.1 Mapping the Management Information Model

The procedures used in converting a MIB defined under the SNMP SMI into one using ISO GDMO templates are those defined by the IIMC [IIMCIMIBTRANS]. We shall first consider an example of mapping the OBJECT-TYPE templates for the 'udpTable' and 'udpEntry' objects into the corresponding GDMO Managed Object Class (MOC) and Name Binding templates.

```
udpTable OBJECT-TYPE
            SYNTAX SEQUENCE OF UdpEntry
            ACCESS    not-accessible
            STATUS    mandatory
            DESCRIPTION "A table containing UDP listener information."
::= { udp 5 }
udpEntry OBJECT-TYPE
            SYNTAX UdpEntry
            ACCESS not-accessible
            STATUS mandatory
            DESCRIPTION "Information about a particular current UDP listener."
            INDEX { udpLocalAddress, udpLocalPort }
::= { udpTable 1 }
UdpEntry ::=
            SEQUENCE {
                  udpLocalAddress IpAddress,
                  udpLocalPort INTEGER (0..65536)
            }
```

The semi-automatic IIMC Internet MIB translation procedures then produce:-

```
udpEntry MANAGED OBJECT CLASS
      DERIVED FROM "Rec. X721|ISO/IEC 10165-2:1992":top ;
      CHARACTERISED BY udpEntryPkg PACKAGE
      BEHAVIOUR udpEntryPkgBehaviour
      BEHAVIOUR DEFINED AS
            !BEGINPARSE
                        REFERENCE !!
                        This managed object class maps to the "udpEntry" object
                        with object identifier {udpTable 1} in module
                        RFC1213-MIB!!
                        DESCRIPTION !!
                        Information about a particular current UDP listener. !!
                        INDEX       RFC1213-MIB.udlLocalAddress,
                                    RFC1213-MIB.udpLocalPort
```

```
            ENDPARSE!;;
        ATTRIBUTES
            udpEntryId          GET        -- IIMC naming attribute --,
            udpLocalAddress     GET,
            udpLocalPort        GET ;;;
        REGISTERED AS { iimcAutoTrans 1 3 6 1 2 1 7 5 1 };
udpEntry-udpNB NAME BINDING -- RFC 1213-MIB --
            SUBORDINATE OBJECT CLASS udpEntry AND SUBCLASSES;
            NAMED BY SUPERIOR OBJECT CLASS udp AND SUBCLASSES;
            WITH ATTRIBUTE udpEntryId;
            BEHAVIOUR udpEntry-udpNBBehaviour
            BEHAVIOUR DEFINED AS
            !BEGINPARSE
                    INDEX       RFC1213-MIB.udpLocalAddress,
                                RFC1213-MIB.udlLocalPort;
            ENDPARSE!;;
            REGISTERED AS { iimcManagementNB 1 3 6 1 2 1 7 5 1 }
```

It is worth emphasising certain aspects of the above translation. Firstly, information that is contained within the SNMP SMI, but can not be directly represented by the corresponding GDMO, is held in 'BEHAVIOUR' clause 'PARSE' statements, e.g. the objects used for entry indexing. Secondly, conceptual table objects (i.e. those that do not contain any MIB variables, such as the MIB-II 'udpTable' object), are not mapped to GDMO MOCs. This means that the 'udpEntry' MOC is bound directly below 'udp'.

A fundamental requirement when mapping between management models is the ability to translate between a CMIS Distinguished Name (DN) and their equivalent SNMPv1 MIB Object Identifier (OID). The Relative Distinguished Name components of DNs consist of either an ASN.1 NULL, for single instanced managed object classes, or an ASN.1 SEQUENCE of the INDEX variables contained in the corresponding SMI OBJECT-TYPE template.

The following is an example of a full DN:

```
{{ systemId              = "uk.ac.ucl.cs.synapse" }
   ipId                  = NULL }
   ipNetToMediaEntryId   = SEQUENCE {
                              ipNetToMediaIfIndex {2},
                              ipNetToMediaNetAddress {128.16.8.170}
                           }}
```

Should we need to refresh the 'ipNetToMediaType' attribute for the MOC defined by this DN, then we first obtain the IIMC defined OID for this OSI attribute, namely { iimcAutoObjAndAttr.1.3.6.1.2.1.4.22.1.2 }. The leading 'iimcAutoObjAndAttr' sub-identifiers are removed, before appending on the SMI instance sub-identifiers, which for this case are '2.128.16.8.170', yielding the correct SNMPv1 OID. Producing the OID for a single instanced MOC would have required the appending of the '.0' sub-identifier instead.

The reverse mapping from SMI OID to CMIS DN must be undertaken when translating Traps to Event-Reports. The correct system object is determined by checking the Trap source address and community strings that have been registered for a given remote system. The hierarchical MIB information for the MIBs supported by this remote system is then traversed for all bar the instance sub-identifiers. The instance sub-identifiers are then converted to either a NULL or SEQUENCE syntax as in the example DN above.

In terms of the TMN standards [M3010] the information model produced by the IIMC translation rules is Q_x rather than Q_3. For example the GDMO produced for an ATM switch MIB would be semantically similar to, but not exactly the same, as that produced by the ITU, leading to a requirement for a Mediation Function to achieve a full Q_3 interface.

3.2 Protocol mapping

When translating CMIS operations to SNMP requests it is immediately apparent that a one to many mapping exists, due to the presence of scoping and filter parameters. An efficient mapping requires the minimisation of the number of generated SNMP requests, especially since MIB traversal using the Get-Next primitive necessitates a wait state until the response to the current retrieval request is received before a further request can be emitted. The number of object instances listed in an SNMP request must be maximised. Attempts to achieve this may cause a 'Too-Big' error response, leading to the generation of smaller requests. The managed objects that are present in the scoped MIT sub-tree must be refreshed in a top-down manner. The filter can then be applied to their state to permit selection of those instances that will have the current CMIS operation applied.

Since usage of the SNMPv1 Get-Next does not cause an error response if the specified object(s) are not instantiated at the SNMPv1 agent (unless the MIB has been fully traversed), it is utilised in preference to the Get primitive, unless a refresh is required for a single table entry object. When retrieving the variables of a single instanced object the corresponding SNMP Object Identifiers for each OSI attribute instance are determined without including a trailing '.0' sub-identifier, so that a Get-Next request can be utilised.

Retrieving all the entries within a table requires the generation of an initial request that specifies the OID for each table entry attribute, but excludes any trailing instance sub-identifiers, which yields the first table entry instance. The OIDs from this response are used as the parameters to the second Get-Next request, so as to retrieve the second table entry, and so on until the table has been fully traversed. An important optimization can be achieved when refreshing existing tables since multiple Get-Nexts can be fired off in parallel, each starting from a different existing entry, e.g. from every fifth entry.

Providing that the GDMO Managed Object Class and Attribute templates indicate that the operation may be applied, CMIS M-Create, M-Delete and M-Set operations are all mapped to SNMP Set requests. To ensure that the semantics of the original CMIS requests are not infringed, M-Set requests that would cause the creation or deletion of a multi-instanced SNMP object are prevented.

SNMP Traps are mapped to an 'internetAlarm' [IIMCIMIBTRANS] CMIS Event-Report. This notification contains the list of name/value pairs that are provided by the Trap's list of variable-bindings. The proxy is also required to determine the Distinguished Name of the object instance that is associated with each variable-binding. The completed event report must then be forwarded to any manager that has previously requested such reports, and may also be logged locally.

4 THE OSIMIS PLATFORM SUPPORT

The OSI Management Information Service [Pavlou, 1993] provides a generic and extensible management platform. The support provided for the development of management agents is known as the Generic Managed System (GMS). The GMS is a fundamental aspect of OSIMIS as it provides the facilities to construct agents as it supports the rich facilities of the OSI management model e.g. scoping, filtering and multiple replies. A primary advantage in selecting the OSIMIS platform is the provision of a large number of Systems Management Functions (SMFs). These include the Event-Report-Management [X734], Log-Control[X735], Metric-Monitor [X739] and Summarization [X738] SMFs.

5 THE ICM OSI/SNMPV1 GATEWAY

A fundamental design requirement for the gateway is to achieve seamless inter-operability between TMN management Operations Systems (OS) or Mediation Functions (MF) and SNMPv1 managed resources. An efficient mapping is essential given the fact that the gateway introduces an intermediate hop in the manager/agent communication path, see Figure 4.

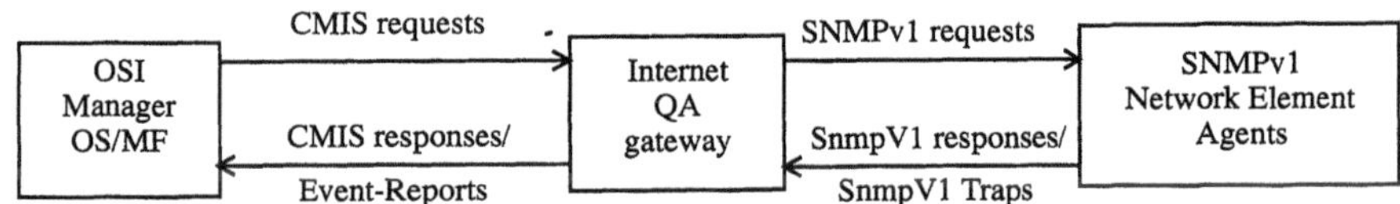

Figure 4 Manager/Agent communication paths.

5.1 The Internet Q-Adapter (IQA) gateway in operation

An underlying aim of our research is to maximise the level of automation in generating a Q-Adapter that proxies for the desired remote SNMPv1 agents. Three stages are required, namely *translate*, *convert* and *run*, see Figure 5.

• Translation involves the usage of VTT's SMI to GDMO converter ('imibtool') to produce an OSI representation of the MIB that is to be managed.

• Conversion yields a simplified MIB description using the GDMO compiler.

• Run - the gateway reads in the simplified input file MIB description(s) and is ready to provide an 'OSI view' of the SNMPv1 managed real resources.

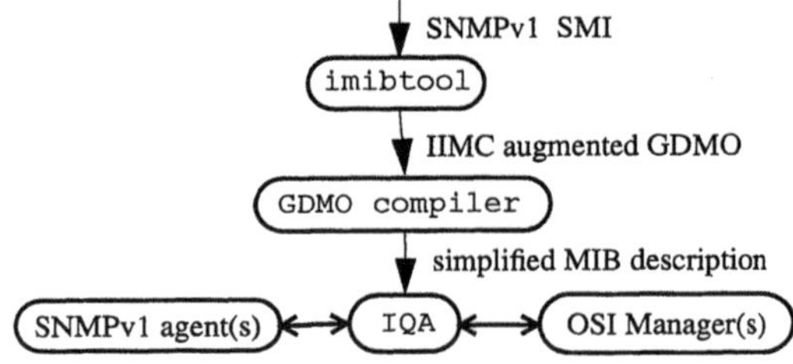

Figure 5 The Internet Q-Adapter's execution cycle.

5.2 Implementation aspects.

The structural decomposition of the Internet Q-Adapter (IQA) gateway is shown in Figure 6. We shall now endeavour to describe these components in some detail. At start-up an instance of each of the IQA `system`, `proxySystem`, `cmipsnmpProxy` and `remoteSystem` classes is instantiated. The `proxySystem` object represents the gateway's local resources, whilst the `remoteSystem` object(s) represents the remote SNMPv1 systems.

The `cmipsnmpProxy` object reads the initial configuration requirements and creates a `cmipsnmpProxyAgent` object, a `remoteSystem` object and an `SnmpRMIBAgent` object for each remote SNMPv1 system. The `remoteSystem` objects can only be created successfully if a poll of the remote SNMPv1 agent receives a response. The `SnmpRMIBAgent` objects encapsulate an SNMPv1 protocol interface.

A tree of `SnmpImageMO` MOs, corresponding to objects held at the remote SNMPv1 agent, will be built up below the respective `remoteSystem` objects in response to incoming Cmis requests. MO class descriptions are held within the `SnmpImageMOClassInfo` meta-class instances, which are themselves constructed into an MIT during the initialisation phase.

The `SnmpImageMOs` utilise the meta-class information to determine whether the corresponding SNMP SMI objects are single or multiply instanced. If multiply instanced then the INDEX attributes are indicated so that retrieved object values can be converted into Relative Distinguished Names in Cmis responses. Meta class information is also kept on which attributes are supported by the remote SNMPv1 agents so that the IQA does not request attributes that have been determined to be non-existent already.

Asynchronous Trap to Event-Report translation utilises the `proxyKS` object. This is an instance of an OSIMIS Knowledge Source and listens on the incoming SNMPv1 Trap UDP ports (e.g. 162), that have been configured for each remote SNMPv1 agent.

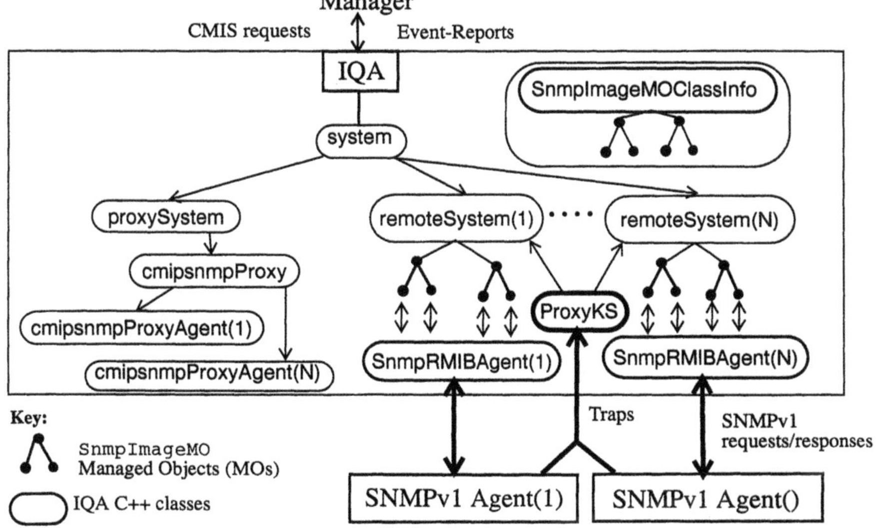

Figure 6 The IQA's structural decomposition.

5.3 Initial performance trials

Since the supply of management information is time critical, we have carried out a number of performance trials to confirm the validity of utilising the IQA for management purposes. We shall consider two comparative cases in the retrieval of the 'ip' and 'tcp' groups [RFC1213] from a remote ISODE SNMPv1 agent. The comparisons are against direct SNMPv1 retrieval using the 'dump' command of ISODE's 'snmpi' manager application, see Figure 7.

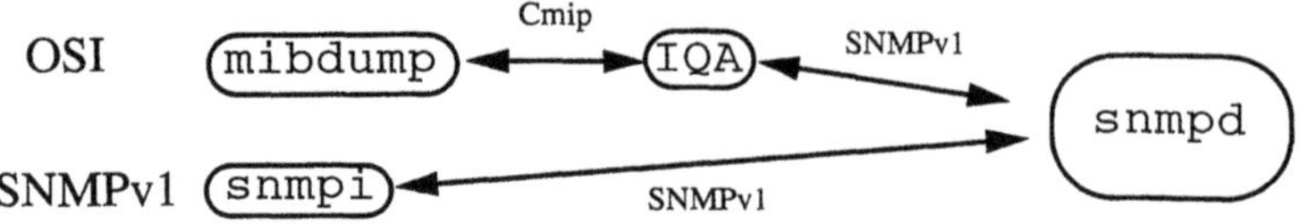

Figure 7 The test components for the SNMPv1 vs OSI trials.

Table 1: SNMPv1 versus OSI data retrieval timings

Test case	Manager	Test runs	Minimum (s)	Maximum (s)	Mean (s)
IP	SNMPv1	21	2.540	2.767	2.602
IP	OSI	25	1.976	4.233	2.284
TCP	SNMPv1	35	1.699	2.275	1.831
TCP	OSI	46	1.545	3.653	1.754

Notes : The OSI timings do not include the association setup and tear down components, which are around 0.2s and 0.02s respectively. Clearly these components can be amortized over far larger data transfers than have been considered in these trials. Any SNMPv1 test runs where no response was received have been excluded.

6 MANAGEMENT SCENARIO EVALUATION

A centralised manager application is required to poll outlying agents to determine whether the values of certain MIB objects have exceeded some threshold. Let us now consider how this requirement can be achieved using both SNMPv1 and OSI managers. Usage of an Internet Q-Adapter means that the management protocol utilised at the real resource is not relevant.

6.1 Using an SNMPv1 manager

If the manager polls too rapidly then it is in danger of taking a significant share of the transmission path's capacity. Whilst if the manager does not poll frequently enough then there is every chance that the event that it was monitoring for, so as to permit it time to take evasive action, will be missed.

Even if the remote agents have a hard-wired enterprise specific Trap generation capability for certain thresholds then the unconfirmed UDP Trap may not even reach the manager. Also the manager can not remotely configure the agent to monitor a threshold that has not been hard-wired in.

The manager might be utilising a remote monitoring agent [RMON94] to achieve its goals, but this is limited to transmission paths that offer a promiscuous mode of operation.

6.2 Using an OSI manager

Localised polling can be remotely configured by the creation of metric monitor objects at the OSI agent or Internet Q-Adapter. Should the value or some weighted average of the values of a monitored attribute cross a defined threshold then an event report will automatically be emitted without further management intervention. This idea is taken significantly further by the Summarization Function, which facilitates the summarization and statistical analysis of the data contained within the agent's MIB, without the need to upload considerable amounts of data so that analysis can be undertaken at the managing application.

Since OSI supports both confirmed Event Reports and the designation of a backup event sink should the primary location fail, the OSI agent can be informed when its report has reached an appropriate manager, or can re-direct the report elsewhere if the first manager is off-line. Even if we take the worst case scenario, when the transmission path itself goes down, then the generic OSI logging facilities still permit the management application to ascertain the agent state up to and beyond the failure, just as soon as the path is reinstated.

7 CONCLUDING REMARKS.

Until the day arrives when a single Network Management architecture reaches 100% market penetration, there will always be a necessity to achieve meaningful inter-working between diverse management paradigms. The authors' research has attempted to meet this goal for the OSI and SNMPv1 models in a highly automated manner.

We have found that the OSI's powerful management functionality can be utilised successfully in enriching the SNMPv1 information model, by providing generic functions such as localised polling, remotely configurable event generation criteria and logging. The SNMP community wishes to retain the simplicity of their agents and by utilising generic OSI Q-Adapters *the agents can remain simple*, whilst *the managers can be presented with a very powerful management architecture* - the best of both worlds ?

Acknowledgements

The research work detailed in this paper was produced under the auspices of the Integrated Communication Management (ICM) project, which is funded by the European Commission's Research into Advanced Communications in Europe (RACE) research program. The authors would like to acknowledge the work of Jim Reilly of VTT (Finland) who achieved a significant level of automation with his SMI to GDMO MIB converter. James Cowan of UCL must be congratulated for developing the innovative GDMO compiler. It would be remiss of us to sign off without re-emphasising our appreciation to the NMF and in particular Lee LaBarre, Lisa Phifer and April Chang, for the excellence of the IIMC document package.

8 REFERENCES

[IIMCMIBTRANS] Lee LaBarre (Editor), Forum 026 - Translation of Internet MIBs to ISO/CCITT GDMO MIBs, Issue 1.0, October 1993.

[IIMCSEC] Lee LaBarre (Editor), Forum 027 - ISO/CCITT to Internet Management Security, Issue 1.0, October 1993.

[IIMCPROXY] April Chang (Editor), Forum 028 - ISO/CCITT to Internet Management Proxy, Issue 1.0, October 1993.

[IIMCMIB-II] Lee LaBarre (Editor), Forum 029 - Translation of Internet MIB-II (RFC1213) to ISO/ CCITT GDMO MIB, Issue 1.0, October 1993.

[IIMCOMIBTRANS] Owen Newman (Editor), Forum 030 - Translation of ISO/CCITT MIBs to Internet MIBs, Issue 1.0, October 1993.

[M3010] ITU M.3010, Principles for a Telecommunications Management Network, Working Party IV, Report 28, 12/91.

[OMG91] The Common Object Request Broker: Architecture and Specification, OMG Draft 10 December 1991.

[Pavlou, 1993] Pavlou G., The OSIMIS TMN Platform: Support for Multiple Technology Integrated Management Systems, Proceedings of the 1st RACE IS&N Conference, Paris, 11/93

[RFC1006] M.Rose, D.Cass, Request for Comments: 1005, ISO Transport Services on top of the TCP, Version3, May 1987.

[RFC1155] M.Rose, K.McCloghrie, Request for Comments: 1155, Structure and Identification of Management Information for TCP/IP-based Internets, May 1990.

[RFC1157] J.Case, M.Fedor, M.Schoffstall, J.Davin, Request for Comments:1157, A Simple Management Protocol (SNMP), May 1990.

[RFC1212] M.Rose, K.McCloghrie (editors), Request for Comments:1212, Concise MIB Definitions, March 1991.

[RFC1213] K.McCloghrie, M.Rose (editors), Request for Comments:1213, Management Information Base for Network Management of TCP/IP-based internets: MIB-II, March 1991.

[RMON94] S.Waldbusser, Internet Draft, Remote Network Monitoring MIB, June 1994.

[Rose, 1991] Rose M., The Simple Book, An introduction to Management of TCP/IP-Based Internets, Prentice-Hall, 1991.

[Saltzer et al, 1984] J.H.Saltzer, D.P.Reed and D.D.Clark, End-To-End Arguments in System Design, ACM Transactions on Computer Systems, Vol.2, No. 4, November 1984.

[X500] ITU X.500, Information Processing, Open Systems Interconnection - The Directory: Overview of Concepts, Models and Service, 1988.

[X701] ITU X.701, Information Technology - Open Systems Interconnection - Systems Management Overview, 7/91

[X710] ITU X.710, Information Technology - Open Systems Interconnection - Common Management Information Service Definition, Version 2, 7/91

[X711] ITU X.711, Information Technology - Open Systems Interconnection - Common Management Information Protocol Definition, Version 2, 7/91

[X720] ITU X.720, Information Technology - Structure of Management Information - Part 1: Management Information Model, 8/91.

[X722] ITU X.722,Information Technology - Structure of Management Information: Guidelines For The Definition of Managed Objects, January 1992.

[X734] CCITT Recommendation X.734 (ISO 10164-5) Information Technology - Open Systems Interconnection - Systems Management - Part 5: Event Report Management Function, 8/91.

[X735] CCITT Recommendation X.735 (ISO 10164-6) Information Technology - Open Systems Interconnection Systems Management - Part 6: Log Control Function, 6/91

[X738] Revised Text of DIS 10164-13, Information Technology - Open Systems Interconnection - Systems Management - Part 13: Summarization Function, March 1993.

[X739] ITU Draft Recommendation X.739, Information Technology - Open Systems Interconnection - Systems Management - Metric Objects And Attributes, September 1993.

9 BIOGRAPHIES

Kevin McCarthy received his B.Sc. in Mathematics and Computer Science from the University of Kent at Canterbury in 1986 and his M.Sc. in Data Communications, Networks and Distributed Systems from University College London in 1992. Since October 1992 he has been a member of the Research Staff in the Department of Computer Science, involved in research projects in the area of Directory Services and Broadband Network/Service Management.

George Pavlou received his Diploma in Electrical, Mechanical and Production Engineering from the National Technical University of Athens in 1982 and his MSc in Computer Science from University College London in 1986. He has since worked in the Computer Science department at UCL mainly as a researcher but also as a teacher. He is now a Senior Research Fellow and has been leading research efforts in the area of management for broadband networks, services and applications.

Saleem N. Bhatti received his B.Eng.(Hons) in Electronic and Electrical Engineering in 1990 and his M.Sc. in Data Communication Networks and Distributed Systems in 1991, both from University College London. Since October 1991 he has been a member of the Research Staff in the Department of Computer Science, involved in various communications related projects. He has worked particularly on Network and Distributed Systems management.

José Neuman de Souza holds a PhD degree at the Pierre and Marie Curie University (Paris VI). He worked on the european projects PEMMON (ESPRIT programme), ADVANCE (RACE I programme) and ICM (RACE II programme), as a technical member and his contribution is related to the heterogeneous network management environment with emphasis on the TMN systems. He participated closely with the UCL group in developing the Internet Q-Adapter. He is currently a researcher at the Federal University of Ceara-Brazil and his research interests are in distributed systems, network management and intelligent networks.

39

MIB View Language (MVL) for SNMP

Kazushige Arai
2nd Development Department
Data Communications Division
NEC Corporation,
1131 Hinode, Abiko, Chiba, 270-11, Japan
Phone: +81-471-85-7650
arai@dcd.trd.tmg.nec.co.jp

Yachiam Yemini*
Computer Science Department
Columbia University,
New York, NY, 10027, USA
Phone: +1-212-939-7123
yemini@cs.columbia.edu

Abstract

This paper introduces "MIB view language (MVL)" for network management systems to provide capability of restructuring management information models based on SNMP architecture. Views concept of database management systems is used for this purpose. Our MVL can provide "atomic operation" feature as well as "select" and "join" features to management applications without changing SNMP protocol itself.

Keywords: Network Modeling; Views; SNMP

1 Introduction

Network management agents provide a data model of element instrumentation to the network management system (NMS). For SNMP agents[1], this data model is captured by respective MIBs, defined in terms of the structure of managed information (SMI) language [RFC1155]. From a perspective of traditional database technology [EN89] a MIB can be viewed as a database of element instrumentation data. The protocol provides a data manipulation language (DML) to query MIBs and the SMI provides a data definition language (DDL) to define the MIB schema structures. Management applications executing at the NMS can access and manipulate MIB data using the protocol query mechanisms.

A central difficulty in developing management applications is the need to bridge the gap between the data models rigidly defined in MIB structures and the data model required by an application. As a simple example consider a fault management application which requires data on health measures [GY93] associated with a network element. These health measures may be computed by sampling MIB variables sufficiently fast. For example, the error-rate associated with an interface can be computed by sampling the respective error counter and computing its derivative. Ideally, the agent should export a data model of health parameters that can be accessed and manipulated by the fault management application. However, the specific data model required can vary from element to element and among different installations, and over time. The MIB designers can not possibly capture the large variety of possible health functions in a rigid MIB.

Of course, it is possible for the application to retrieve raw MIB data and compute the health data model at the NMS. This solution can be highly inefficient and unscalable as it would force

*Supported by ARPA contract F19628-93-C-0170.

[1] The techniques and concepts introduced by this paper are cast within the framework of SNMP. They could be mapped to the GDMO framework of CMIP where they would play an equally important role. This mapping will be described in future work.

excessive polling of MIB data. Furthermore, it does not allow various applications that execute at multiple NMS to share the computations of the health data model effectively. In a multi-manager environment such sharing of data models is of great significance.

An alternative approach, developed in this paper, is to support effective computations of user-defined data models — *views* — at the agent side. The ability to define computed views of data has found a broad range of applications in traditional databases. View definition and manipulation capabilities are integral components of virtually all database systems. This paper proposes to extend the SMI and agent environment to support similar view computations to meet the need of management applications to transform raw MIB data into useful information.

The health data model, for example, could be defined in terms of the proposed MIB view language (MVL). The MVL computations could be delegated [YGY91] to the agent's environment, or to a local manager. Views can be organized in and accessed through agent's MIB. Applications could use standard SNMP queries to access and retrieve these view definitions. One can thus consider view MIB as a programmable layer of transformations of raw MIB data into information required by remote management applications.

There is an approach to transform MIB data, especially for OSI SMI architecture [SB93]. In this paper, we concentrate on SNMP architecture and are introducing actual MVL.

In following sections, we describe what can be done with views (Section 2) and then provides actual MVL specifications (Section 3).

2 Views of Managed Data

A database system includes an intrinsic data model defined by its schema. This intrinsic model is used to provide effective access to stored data, anticipating certain patterns of use by applications. Often, however, applications require a different data model than the one that is stored. A view provides a mapping of the intrinsic stored-data model to the data model needed by the application.

The data model created by a view can be considered as a *virtual MIB*. A virtual MIB is computed by an agent and may be accessed by SNMP managers like any other MIB. The examples illustrate various applications of virtual MIBs.

An application may wish to correlate data in multiple related tables. In a relational databases such correlation is accomplished by computations of a join. Consider for example the MIBs used in a terminal server device. The physical layer of the terminal server, described by the RS232 MIB [RFC1317], includes a table (`rs232SyncPortTable`) describing the physical ports. The logical layer of the terminal server, described by the point-to-point protocol (PPP) MIB [RFC1471], includes a table (`pppLqrTable`) describing logical link objects. These tables are depicted in Figure 1(a) and (b). Suppose now that a management application wishes to correlate the status of logical links and of the physical ports which they use. This is important in isolating faults that are manifested by managed objects associated with both layers. To accomplish such correlation one would want to compute a join of the respective tables (Figure 1(c)).

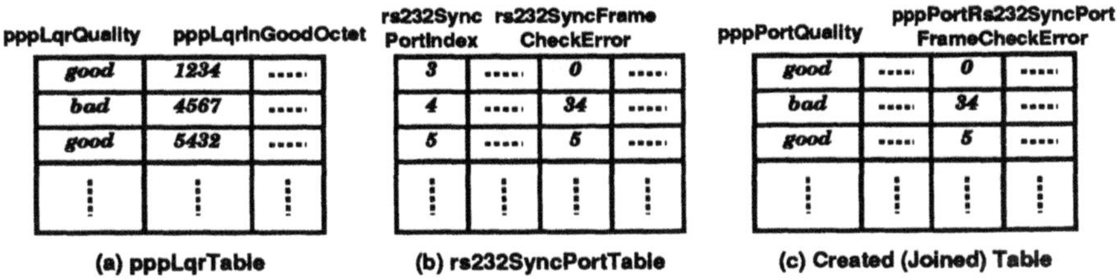

pppLqrQuality	pppLqrInGoodOctet	
good	*1234*	.....
bad	*4567*	.....
good	*5432*	.....
⋮	⋮	⋮

(a) pppLqrTable

rs232Sync PortIndex		rs232SyncFrame CheckError	
3	.....	*0*	.....
4	.....	*34*	.....
5	.....	*5*	.....
⋮	⋮	⋮	⋮

(b) rs232SyncPortTable

pppPortQuality		pppPortRs232SyncPort FrameCheckError	
good	.....	*0*	.....
bad	.....	*34*	.....
good	.....	*5*	.....
⋮	⋮	⋮	⋮

(c) Created (Joined) Table

Figure 1: Example of "join" Tables.

In contrast with databases, neither SNMP nor CMIP provide a mechanism to correlate data by computing joins. In the example, the fault analysis application will have to retrieve both tables from the terminal server agent and compute the join. This computation of a join is very inefficient as much more data than needed will be retrieved and processed by the application. Moreover, it can lead to serious errors. Retrievals of tables by SNMP is not an atomic operation. Each GET-NEXT access will retrieve the current data in the respective tables. If attributes stored in the table change during retrieval the table images at the application side will reflect multiple versions of the respective MIB tables. The fault analysis routine may be mislead by the data to identify the wrong faults. Problem management could exacerbate the problems rather than resolve them. The problem of computing a join of table as an atomic action commonly occur in other network management scenarios. For example, resolution of routing problems typically involves correlation of routing, address translation and other configuration tables. It would be thus very useful to support effective computations of atomic joins.

Views can be used to perform such computations efficiently. A view computation could obtain an atomic (or a very good approximation of it) snapshot of the respective tables and then join them at the agent side. The joined table is a part of a virtual view MIB. It could be accessed by applications for retrievals via GET-NEXT (or GET-BULK) as any other MIB table. Atomic retrievals, of course, can be important even when tables are not joined. A view could be used to generate an atomic snapshot of a MIB table in the virtual MIB which could then be retrieved by managers.

Views may be used, similarly, to select objects that meet certain filtering criteria of interest. Selective retrievals are provided by CMIP via filters passed to agents as part of queries. In contrast, SNMP does not permit filtering of data at the source. Consider the terminal server example. Suppose one wishes to retrieve logical link data for all troubled links (defined by some filtering conditions on link status). At present, it is necessary to retrieve the entire logical links table and perform the filtering at the manager. This is inefficient and presents great difficulty in searching large tables (e.g., of thousands of virtual circuit objects in AtomMIB [ATOMMIB]). A view could be defined over the logical link table to perform the filtering required by the manager. A GET-NEXT access to this view will retrieve the next logical link that meets the filtering criteria. This can be used to augment SNMP with selective retrievals without any changes to the protocol. Furthermore, this method of filtering could be more efficient than the one pursued by CMIP since the filters are delegated ahead of access and require no parsing and interpretation during access time.

Views may be used to support participatory management of complex multi-domain networks. Consider for example a collection of private virtual networks (PVN) sharing a common underlying physical network. Such PVN are commonly used by telecommunication service providers as means to partition bandwidth among multiple organizations. NMS responsible to manage the various PVN must share access to agents of the underlying network elements. At the same time, their access should be limited to monitor and control the resources in their respective PVN. It is thus necessary to provide each PVN with a view of the actual MIBs. SNMPv2 [RFC1442] provides a "context" mechanism to support projection view of a MIB. A party may be authorized to access a subset of the MIB. Views significantly extend this mechanism to support not only projections but also computed data. The virtual MIBs accessed by PVN may hide some of the underlying network features to prevent PVN from compromising sensitive resource data.

Views may be used to support atomic actions in a multi-manager environment. In the multi-manger environment, it is difficult to ensure atomicity of actions invoked from several managers. With SNMP architecture, a side-effect of a SET operation is used to invoke an action. This operation may take one or more parameters which control a behavior of the action. When an action invoked by setting an value to an object (trigger object), agent may treat one or more other objects as parameters related to the action (parameter objects). But a parameter object set by one NMS may be modified by other managers before the previous manager invokes the action by setting trigger object. This can lead to incorrect behaviors. A view can define the action trigger and its parameters as an atomic group. This will associate with the group a queue of action requests. Each SET invoked by a manager to any object in the group will be queued. When all object SET requests by a given manager have been received in the queue the action is invoked atomically. Should two managers access the action concurrently, their actions are serialized by the queuing mechanism.

Views could also provide a beneficial mechanism to protect access to data. A view can be used to define the data model and access rights available to certain applications. This is routinely used in databases to secure data access. SNMPv2 has this capability but view could provide it even with SNMPv1. However, a full discussion of view applications to secure management is beyond the scope of this preliminary paper.

Finally, views may be used to simulate abstraction/inheritance relations among SNMP objects, similarly to the object model provided by CMIP. For example, a view could define a port object and its properties as a common abstraction of various port objects in different MIBs. The abstract port properties could be mapped by the view (simulating inheritance) to properties of the specific port objects in the MIB. Similarly, one can use views to model containment relations among objects. These features, however, is beyond the scope of this paper.

In summary, a view could be used to support extensive computations over MIBs (correlations and filtering of data), atomicity of data access and actions, access control and object abstractions/inheritance and containment. These capabilities are summarized in Figure 2.

FEATURE	DESCRIPTION
CORRELATION	Join tables to create new table which contains correlated data.
ATOMIC RETIEVE	Generate atomic snapshot of a MIB table which can be retieved atomicly.
FILTERING	Select data which meet a filtering condition at agent side.
SELECT PARTIAL MIB	Provide partial access to each manager in multi-manager environment.
ATOMIC ACTION	Garantee atomic invocation of actions in multi-manager environment.
SECURE ACCESS	Define access rights to each management application.
OBJECT-ORIENTED MODEL	Simulate data abstructions and representation of containment relationships.

Figure 2: Summary of View Features.

3 MIB View Language

This section introduces the *MIB View Language (MVL)*. The goal of MVL is to provide a minimal extension of SNMP's SMI [RFC1155, RFC1442] that supports:

- definitions of the structure of view objects
- conversion of data from real MIB objects to compute view objects.

Traditional database systems use SQL, the data manipulation language, to define views. For example,

```
CREATE VIEW View1
AS SELECT     T.Attr1, T.Attr2
   FROM       Real_Table T
   WHERE      T.Attr3=0 AND T.Attr4=1
```

These SQL expressions accomplish definitions of the structure of view objects and their computations from real objects simultaneously, using a SELECT-FROM-WHERE construct. MVL develops a similar approach to view definitions, adapted to the SMI.

View definitions in MVL are compiled by a MVL compiler into appropriate agent computations and MIB structures for the view MIB. Access to a view MIB by a manager is indistinguishable from access to any other MIB.

An important consideration in implementing views is the organization of a view MIB and access within a complex multi-MIB agent environment. There are a few issues that an implementation architecture must address.

1. how does a manager query of a view MIB get processed
2. how are computations of a view MIB executed
3. how do view computations access real MIB objects
4. how are views delegated to an agent environment

A comprehensive discussion of the architectural options to address these questions is beyond the scope of this paper. We provide here a brief summary of one possible solution. View computations are encapsulated in a view agent. A view agent can function as a subagent within a multi-agent environment. An SNMP query of a view will be communicated by the master agent to the subagent (e.g., using one of a number of mechanisms currently available such as SMUX, WINSNMP, or other extensible agent mechanisms). The view agent is entirely responsible to compute the views. Views can be delegated to the view agent using the management by delegation mechanisms [YGY91]. Figure 3 depicts the overall organization of the different components of a typical SNMP management environment extended with view mechanisms.

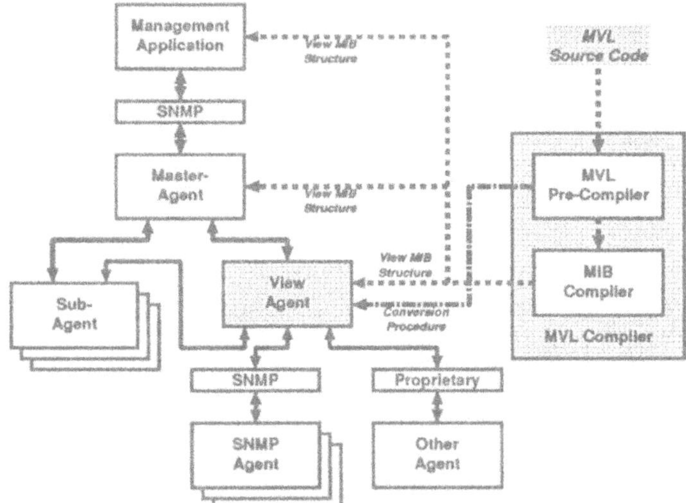

Figure 3: View Agent and MVL compiler.

Notice that a view agent may act as a manager and use proprietary or standard protocols to access remote agents and retrieve data needed in computing views. This may be accomplished by functions, invoked through view computations, to access and retrieve remote data.

3.1 The VIEW-TYPE Macro and View Function

The data structure of view object can be defined by a modified **OBJECT-TYPE** macro of SNMP. We called it a **VIEW-TYPE** macro. The only difference between view objects and real objects is that the value of the view object is computed from existing managed objects. Therefore we introduce **VIEW-FUNCTION** block to specify how to compute the value of a view object. **COMPUTED-BY** clause is used to bind view object and view function. Figure 4 illustrates a view definition using a **VIEW-TYPE** macro and a **VIEW-FUNCTION** block.[2]

```
viewObject1 VIEW-TYPE                      function1 VIEW-FUNCTION
    SYNTAX       INTEGER                       SELECT realObject1
    MAX-ACCESS   read-write                    WHERE  realObject1 <= 100
    STATUS       current
    DESCRIPTION  "Example of view
                  definitions"
    COMPUTED-BY  function1
    ::= {view 1}
```

Figure 4: VIEW-TYPE and VIEW-FUNCTION.

3.1.1 COMPUTED-BY Clause

The **COMPUTED-BY** clause declares the function name which computes the value of the object by accessing existing objects. If no **COMPUTED-BY** clause appeared in **VIEW-TYPE** macro, value of the view object is persistently stored into the view MIB instead of being computed from existing real objects. In this case the value of the object should be initialized by a manager application or through a **DEFVAL** clause.

3.1.2 VIEW-FUNCTION Block

Each function declared by a **COMPUTED-BY** clause has to be defined using a **VIEW-FUNCTION** block. This defines conversion procedures of data. A **VIEW-FUNCTION** block is based on SQL. It consists of two clauses, "**SELECT**" and "**WHERE**".

SELECT Clause The SELECT clause defines how to access values of existing objects in computing a view object. Note that, the existing objects specified here may be other view objects. The following operators are available for computing selection:

[ ] operator Indexing column object in table structure (See Section 3.2)

– > operator To access a column object from an object identifier of conceptual row object.

+, –, *, / operator Arithmetic operators for calculation

WHERE Clause The **WHERE** clause specifies a condition that filters the instances of objects accessed by the **SELECT** clause. The following conditional operators are available:

AND operator Logical AND operator

OR operator Logical OR operator

[2] In this paper, we use syntax of SNMPv2 SMI as an example. But MVL is also applicable to SNMPv1 with few modification.

NOT operator Logical NOT operator

IN operator Compare two object identifiers (OIDs) whether right hand OID is included in the left hand OID's subtree

=, <>, <, >, <=, >= operator Compare magnitude of two expressions

[ ], − >, +, −, *, / operator All these operators described with `SELECT` clause can be used in `WHERE` clause

The key word "`SELF-INDEX`" is used as an index value of `[ ]` operator to specify index value of the view object its self. (See Section 3.2)

3.2 Computing Join and Selection View Tables

Specifying the computations of view tables from real tables is particularly challenging. One must identify how conceptual rows in the view table relate to respective rows in the real tables. Of particular significance is the computation of the index of a view table. The simplest case is when a view table uses a column from a real table as its index. This is illustrated by the following example.

```
viewIfIndex VIEW-TYPE
    SYNTAX INTEGER
      :
    COMPUTED-BY func_ifIndex
      :

func_ifIndex VIEW-FUNCTION
    SELECT ifIndex[SELF-INDEX]
```

In here `viewIfIndex` is the index column of the view table and `ifIndex` is a column of a real MIB table. The notation `[SELF-INDEX]` is used to specify the index of the real MIB table containing `ifIndex`. Of course, one must ensure that the values in `ifIndex` can be suitably used as an index (i.e., they are key for the view table).

Consider now the case where the view table is created by selecting a subset of conceptual rows from the real table. This may be used to filter row entities using appropriate filtering condition. For example, `ifOperStatus` represents the operational status of interface objects and a value 1 represents that an interface is operational. [RFC1213, RFC1573] The following example creates a view table that includes index values for all operational interfaces. A manager accessing this view table via GET-NEXT could retrieve index values for operational interfaces only.

```
func_column1 VIEW-FUNCTION
    SELECT ifIndex[SELF-INDEX]
    WHERE  ifOperStatus[SELF-INDEX] = 1
```

We now illustrate how to specify join views using MVL expressions. Consider two tables, `ifTable` [RFC1213, RFC1573] and `atmInterfaceConfTable` [ATOMMIB] whose index column is `ifIndex`. We wish to create a view table that join the two tables using their common index values and containing the common index column, followed by `ifSpeed` of `ifTable` and then the `atmInterfaceMaxVpcs` and `atmInterfaceMaxVccs` of `atmInterfaceConnTable`. This is depicted in Figure 5 and is accomplished by the MVL specification in Figure6.

3.3 Computing Atomic Operations in MVL

Supporting invocation by managers of actions at agents is of great significance in management. Remote actions can be used to control configuration (e.g., partition hub ports, establish permanent virtual circuits through a switch or configure collection of statistics by a remote monitor) or invoke

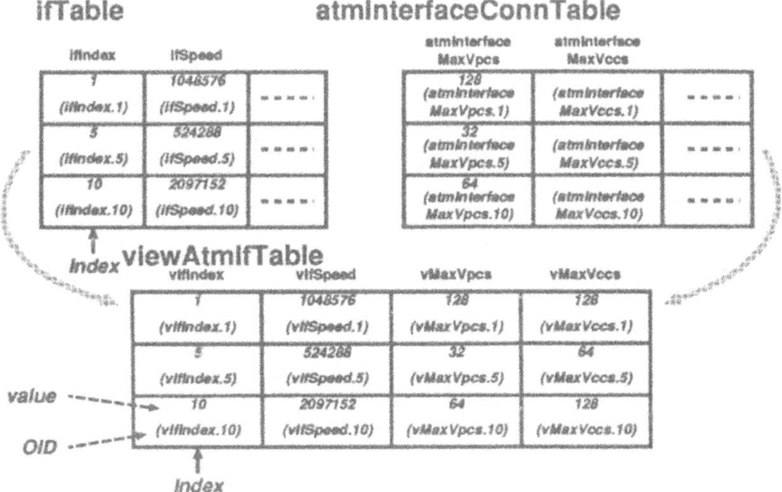

Figure 5: Join Tables.

diagnostic procedures. CMIP, therefore, supports explicit invocation of remote procedure calls. In contrast, SNMP utilizes side-effects of SET to invoke agent procedures. This implicit invocation of remote procedures is seriously limited in passing parameters to actions. A manager would have to ensure that parameters are set prior to triggering the execution of an action that uses them. In a multi-manager environment interference among managers trying to invoke a parameterized actions could lead to erroneous actions. One manager could reset the values of parameters just set by another manager who issues an action triggering request.

The parameterized action model of SNMP may be best viewed as a form of supporting transactions among managers and agents. The problem is, accordingly, that of supporting concurrency control of such transactions to assure their serializability. Interference among managers can lead to non-serial execution schedules.

Currently, there are several approaches to realize the parameterized atomic action. For example, using "lock" variable to control write access to parameters of action is the most popular method for the concurrency control.

Managers must check the lock variable before modifying the parameters and if it is not set, the manager set the variable to lockout access from other managers. The agent keeps an ID of the manager which set the variable, and does not accept SET access from other managers. This method still have a chance of conflict to access lock variable itself.

Another example to realize the parameterized atomic action is introduced by using row creation of SNMPv2. With this method, each manager which invokes an action creates a new row which contains parameters of the action. Since the other managers does not know the identifier of the new row, this manager can set the parameters without conflict from other managers. This method is fit for an action like create a new virtual circuit. But it may not be appropriate for changing parameters of some services. And this method cannot be applied to current SNMP.

MVL provides a simple and generic mechanism to support concurrency control of SET transactions by multiple managers. MVL uses the **ATOMIC-GROUP** construct to accomplish this. We use the example in Figure 7 to illustrate the atomic execution mechanisms of MVL.[3]

[3]This example is based on an action of virtual path (VP) cross-connect establishment described in [ATOMMIB],

```
viewAtmIfTable VIEW-TYPE
    SYNTAX SEQUENCE OF
            vAtmIfTableEntry
    MAX-ACCESS not-accessible
    STATUS current
    DESCRIPTION "ATM Interface Table"
    INDEX {vIfIndex}
    ::= {view 1}

vAtmIfTableEntry VIEW-TYPE
    SYNTAX VAtmIfTableEntry
    MAX-ACCESS not-accessible
    STATUS current
    DESCRIPTION "Conceptual row"
    ::= {viewAtmIfTable 1}

VAtmIfTableEntry ::= SEQUENCE {
    vIfIndex INTEGER,
    vIfSpeed Gauge,
    vMaxVpcs INTEGER,
    vMaxVccs INTEGER
    }

vIfIndex VIEW-TYPE
    SYNTAX INTEGER
    MAX-ACCESS read-only
    STATUS current
    DESCRIPTION
      "ifIndex from ifTable"
    COMPUTED-BY func_vIfIndex
    ::= {vAtmIfTableEntry 1}

func_vIfIndex VIEW-FUNCTION
    SELECT ifIndex[SELF-INDEX]
    WHERE  ifType[SELF-INDEX] = 37

vIfSpeed VIEW-TYPE
    SYNTAX Gauge
    MAX-ACCESS read-only
    STATUS current
    DESCRIPTION
      "Interface speed"
    COMPUTED-BY func_vIfSpeed
    ::= {vAtmIfTableEntry 2}

func_vIfSpeed VIEW-FUNCTION
    SELECT ifSpeed[SELF-INDEX]
    WHERE  ifType[SELF-INDEX] = 37

vMaxVpcs VIEW-TYPE
    SYNTAX INTEGER
    MAX-ACCESS read-write
    STATUS current
    DESCRIPTION
      "Max. number of VPCs."
    COMPUTED-BY func_vMaxVpcs
    ::= {vAtmIfTableEntry 3}

func_vMaxVpcs VIEW-FUNCTION
    SELECT atmInterfaceMaxVpcs
                        [SELF-INDEX]
    WHERE  ifType[SELF-INDEX] = 37

vMaxVccs VIEW-TYPE
    SYNTAX INTEGER
    MAX-ACCESS read-write
    STATUS current
    DESCRIPTION
      "Max. number of VCCs"
    COMPUTED-BY func_vMaxVccs
    ::={vAtmIfTableEntry 4}

func_vMaxVccs VIEW-FUNCTION
    SELECT atmInterfaceMaxVccs
                        [SELF-INDEX]
    WHERE  ifType[SELF-INDEX] = 37
```

Figure 6: Example of MVL definition to join tables.

```
vVpConnConf VIEW-TYPE
    SYNTAX TimeTicks
    MAX-ACCESS read-write
      :
    ATOMIC-GROUP { vVpConnLowIfIndex,
                   vVpConnLowVpi,
                   vVpConnHighIfIndex,
                   vVpConnHighVpi
                   vVpConnAdminStatus,
                   vVpL2HOperStatus,
                   vVpL2LOperStatus }

vVpConnLowIfIndex VIEW-TYPE
    SYNTAX INTEGER
    MAX-ACCESS read-write
      :
    COMPUTED-BY func_LowIfIndex

func_LowIfIndex VIEW-FUNCTION
    SELECT
      atmVpCrossConnectIndex[SELF-INDEX]

      :

vVpConnAdminStatus VIEW-TYPE
    SYNTAX INTEGER
    MAX-ACCESS read-write
      :
    COMPUTED-BY func_AdminStatus

func_AdminStatus VIEW-FUNCTION
    SELECT
      atmVpCrossConnectAdminStatus
                        [SELF-INDEX]
      :

vVpL2HOperStatus VIEW-TYPE
    SYNTAX INTEGER
    MAX-ACCESS read-only
      :
    COMPUTED-BY func_L2HOperStatus

func_L2HOperStatus VIEW-FUNCTION
    SELECT
      atmVpCrossConnectL2HOperStatus
                        [SELF-INDEX]
      :
```

Figure 7: Example of Atomic Group.

The view object **vVpConnConf** is defined as an *atomic object* with a value of **TimeTicks**. The **ATOMIC-GROUP** declaration binds a group of view objects to an action (transaction) associated with **vVpConnConf**. The group of view objects is called *atomic group*. When a manager starts to invoke the atomic action, it would first SET the **vVpConnConf** with a value of time-out by which time the atomic action is canceled. Once the atomic object is SET, all subsequent SET accesses to any objects in the atomic group by the same manager are queued until the view agent has obtained a SET for all objects in the atomic group whose access is **read-write** (In this example, **vVpConnLowIfIndex**, **vVpConnLowVpi**, **vVpConnHighIfIndex**, **vVpConnHighVpi** and **vVpConnAdminStatus**). At that time, the view agent executes all these SETs in the order defined by their request-id. When the last SET request is executed, the action is invoked at real agent (In this example, virtual path is connected). The other SET requests are used to set parameters of the action.

After finishing all SET request executions, the view agent will execute GET requests to all read-only objects in the atomic group (In this example, **vVpL2HOperStatus** and **vVpL2LOperStatus**). These read-only objects are used to return results of the atomic action. The view agent takes an atomic snapshot of these values for subsequent GET and GET-NEXT accesses by the manager. The snapshots will be deleted either through another SET request by the same manager or through a timer (**vVpConnConf**) expiration on the time-out. If the timer expires before all objects are SET, the atomic action will be canceled and the agent dose not execute any SET request issued by the manager.

If all the variables in the **ATOMIC-GROUP** are **read-only**, then the agent interprets the SET request to the atomic object as an atomic retrieval initiation. It takes a snapshot of all these objects in the **ATOMIC-GROUP** and stores them. All subsequent GET or GET-NEXT access to these variables by

but it is slightly modified from actual MIB definitions. Because, original definitions have their own way to provide atomic action with row creation techniques described above.

the manager that issued the SET retrieves this atomic snapshot.

MVL also provides another capability of atomic retrieval which is called ***asynchronous update***. Suppose that there are two or more counters defined in a MIB (a real MIB). And they are being updated concurrently. There is no guarantee that two counter values which are retrieved by manager are consistent. Because, the values may be updated by agent after the manager retrieves one counter value and before retrieves another one. Retrieving two values on different cycles of updating may make inconsistency between these values.

MVL uses `UPDATE-GROUP` construct to prevent this inconsistency. Example in Figure 8 is used to illustrate the asynchronous update mechanism.

```
updateErrorCounts VIEW-TYPE
   SYNTAX TimeTicks
   MAX-ACCESS read-write
      :
   UPDATE-GROUP { vIfInErrors[SELF-INDEX],
                  vIfOutErrors[SELF-INDEX] }
   UPDATE-CONDITION
     IF ifOperStatus[SELF-INDEX] = 1
      :
```

```
vIfInErrors VIEW-TYPE
   SYNTAX Counter
   MAX-ACCESS read-only
    :
   COMPUTED-BY func_inErrors

func_inErrors VIEW-FUNCTION
   SELECT ifInErrors
                [SELF-INDEX]
```

Figure 8: Example of Update Group.

An `updateErrorCounts` is declared as an object controls asynchronous update with a value of `TimeTicks`. The `UPDATE-GROUP` declares a group of view objects which are updated simultaneously. The view agent updates a snapshot of the view objects in the group by getting values through view functions of each view objects. By default, interval time of update is given as a value of object defined with `UPDATE-GROUP` declaration (In this example, value of `updateErrorCounts`).

Since all member objects of `UPDATE-GROUP` are updated simultaneously, managers always retrieve values with in a same update cycle by GET access. If agent receives GET request during it is updating values, the request is queued and responded after finishing update process. On the other hand, if update timing is reached during process of responding GET access, the update process is delayed after completion of the response.

`UPDATE-CONDITION` clause specifies conditions of updating members of the `UPDATE-GROUP`. In an example of Figure 8, the agent updates the member objects only if a value of `ifOperStatus` is equal to 1 (UP). Any other condition identical to `WHERE` clause can be specified with `IF` clause. `TIMING` clause can also be specified to determine more complex update schedule. For example: `TIMING { 10:00, 12:00, 14:00, 16:00 }` declares that the member objects are updated at 10:00, 12:00 and so on. With this declaration, the value of `updateErrorCounts` is ignored.

4 Conclusion

Introducing views concept of database management systems into managed object definitions of network management systems provides capability of restructuring network models. This capability makes it is possible that network models can be modified to be best for each management application. Especially, views provides "select" and "join" features of database management systems and they make development of network management application easier and they can also reduce traffics between manager and agent nodes.

We introduced "MIB View Language (MVL)" for SNMP architecture which can be used without changing any protocols between manager and agent. With MVL compiler, we can produce MIB structure for view agent and view functions which convert existing data models to view models.

MVL and view agent also provide atomic operation features. With these features atomic invocation of actions and asynchronous update of view objects, which is not available with current SNMP architecture, can be achieved without changing SNMP protocol itself.

References

[ATOMMIB] M.Ahmed, K.Tesink, Editor, "Definitions of Managed Objects for ATM Management, Version 7.0", *Internet Draft*, 1994

[EN89] R.Elmasri, S.Navathe "Fundamentals of Database Systems", The Benjamin/Cummings Publishing Company, Inc., 1989

[GY93] G.Goldszmidt, Y.Yemini, "Evaluating Management Decisions via Delegation", *Integrated Network Management, III*, Elsevier Science Publishers B.V. (North-Holland), 1993

[RFC1155] M.Rose, K.McCloghrie, "Structure and Identification of Management Information for TCP/IP-based Internets", *RFC-1155*, 1990

[RFC1213] K.McCloghrie, M.Rose, "Management Information Base for Network Management of TCP/IP-based Internets: MIB-II", *RFC-1213*, 1991

[RFC1317] B.Stewart, Editor, "Definitions of Management Objects for RS-232-like Hardware Devices", *RFC-1317*, 1992

[RFC1351] J.Davin, J.Galvin, and K.McCloghrie, "SNMP Administrative Model", *RFC-1351*, 1992

[RFC1442] J.Case, K.McCloghrie, M.Rose, S.Waldbusser, "Structure of Management Information for version 2 of the Simple Network Management Protocol (SNMPv2)", *RFC-1442*, 1993

[RFC1471] F.Kastenholz, "The Definitions of Managed Objects for the Link Control Protocol of the Point-to-Point Protocol", *RFC-1471*, 1993

[RFC1573] K.McCloghrie, F.Kastenholz, "Evolution of the Interfaces Group of MIB-II", *RFC-1573*, 1994

[SB93] S.Bapat, "Richer Modeling Semantics for Management Information", *Integrated Network Management, III*, Elsevier Science Publishers B.V. (North-Holland), IFIP 1993.

[YGY91] Y.Yemini, G.Goldszmidt, S.Yemini, "Network Management by Delegation", *Integrated Network Management, II*, Elsevier Science Publishers B.V. (North-Holland), IFIP 1991

40
The Abstraction and Modelling of Management Agents *

Graeme S. Perrow, James W. Hong, Hanan L. Lutfiyya,
and Michael A. Bauer
Department of Computer Science
University of Western Ontario
{graeme,jwkhong,hanan,bauer}@csd.uwo.ca

Abstract

Management agents play an important role in distributed systems and network management. Agents are used to gather information, create, delete, and change the state of managed objects, and forward notifications of events from managed objects to managers. All management agents perform the same basic operations, yet there is no precise specification of the capabilities and architecture of *generic* management agents. As a result, developing management agents at present is difficult and time-consuming. This paper presents the design of a generic management agent and describes the architecture and service interface of such an agent. We also present an implementation of a management agent creation tool for automating the creation of management agents (CMIP, SNMP and other) which all bear the generic agent architecture. The use of this tool greatly reduces the time needed, and therefore the cost of developing management agents.

[**Keywords:** management agents, general agent architecture, CMIP agent, SNMP agent, extensible agent, automated agent development tool]

1 Introduction

Management systems contain three main types of components that work together: *managers*, which make decisions based on collected management information, *management agents*, which collect management information, and *managed objects*, which represent actual system or network resources being managed. Management agents perform operations requested by managers and notify managers of pre-determined events of interest to the manager. Agents are said to operate "on behalf of" managers, so that the manager's workload is greatly reduced, the load is distributed around the system or network, and efficiency is increased.

Agents play an important role in any management system. Agents are used to gather information, create, delete, and change the state of managed objects, and forward notifications of events

*This research work is supported by the IBM Center for Advanced Studies and the Natural Sciences and Engineering Research Council of Canada.

from managed objects to managers (see Figure 1). A management agent is defined as an entity that provides a mechanism that performs management operations on managed objects and emits notifications on behalf of managed objects.

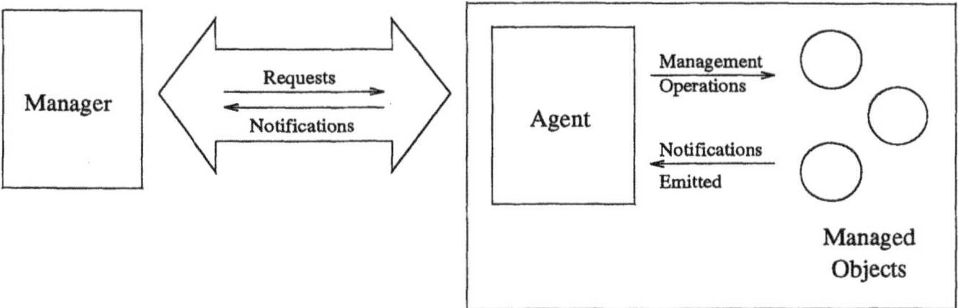

Figure 1: Manager-Agent-Managed Object interaction

To date, most of the research on systems and network management has concentrated on management protocols such as CMIP [6, 14] and SNMP [1, 2, 15]. There has also been a lot of work done on managed object definition; both the OSI [7] and the Internet [13] have created a managed object specification language for their respective management frameworks. These languages can be used to describe and define managed objects. There are even compilers available to parse these definitions and generate code that implements the managed object [10]. These compilers greatly facilitate the development of managed objects. However, relatively little work has focussed on facilitating the development of management agents.

Presently, the development of management agents is difficult, time-consuming and *ad hoc*. There are many decisions that must be made in the development of agents, such as, what services to offer, what relationships the agent should have with the environment (*i.e.*, hardware or software resources, user interface, *etc.*). Part of the reason for the difficulty in developing management agents is that these design issues have not been separated out from implementation details. For example, there is a set of services that is required from all or most agents; these include accepting monitoring and control requests from managers, executing these requests, returning results, notifying the manager of pre-determined events of interest and communicating with other entities. These services are independent of the underlying management protocols provided by the environment.

Some work has been done in the area of developing management agents. Both Bull [8] and DEC [9, 16] provide frameworks to create basic agents that handle management requests built into the standard management protocols (SNMP and CMIP). However, some of the operations (such as self-description, logging, user-defined) that we believe are key requirements for management agents are missing, and adding these operations to the agents would be a non-trivial task. Management by Delegation (MbD) [3, 4] has the opposite problem: creating large, powerful agents is easy, but if a small, simple agent is required, it would be far too large and intrusive to be useful. What is needed is a way to create agents so that the creation of a basic, simple agent is just as easy as creating a larger, more powerful agent with dynamically extensible functionality.

To facilitate the development of agents we have identified a generic architecture for agents. This architecture describes the services that the agents should or could provide, the components comprising an agent and how the components satisfy the services. The result is a specification of the capabilities of a *generic* management agent. This specification aids in the development of new management agents, since it saves the developer from designing the components, services, and interface of the agent to be created, and allows the developer to concentrate on customization of the agent. We then show that based on this architecture that a good deal of the creation of an agent can be automated based on a few pieces of user-supplied information, such as the management protocol, the basic management operations, and the resources to be managed.

The rest of this paper is organized as follows. Section 2 discusses the functional requirements of management agents. Section 3 presents our model of the architecture and service interfaces of a generic management agent. Section 4 describes a prototype management agent creation tool that can generate CMIP and SNMP agents automatically with the user's inputs from its graphical user interface. Section 5 concludes the paper with a summary and some future work.

2 Functional Requirements of Management Agents

In this section, we identify the key functional requirements for a management agent. An agent must be able to respond to requests from managers, other agents, and managed objects. The following classes of operations must be supported:

1. **Self-Description Operations:** An agent should be able to describe itself to other entities (*e.g.*, managers or other agents), so that they can discover what operations a particular agent is capable of, and what resources it can manage.

2. **Common Management Operations:** These represent operations common to the standard management protocols. They include operations to get information from managed resources, set information within managed resources, send commands to managed resources, and create and delete managed objects. Note that creating and deleting managed objects refers to creating and deleting abstract entities representing the managed resources, not the real resources themselves.

3. **User-Defined Operations:** Agents will require specific user-defined operations, such as performing different types of analysis on collected information. Hence, there must be some way to incorporate user-defined operations into an agent. Agents should also be able to execute services periodically, *i.e.*, a manager could request that this agent collect a certain piece of information every five minutes, without the need for the manager to send a request every five minutes.

4. **Logging Operations:** Agents should be able to log requests, notifications, and other management information for analysis, security, or statistical purposes.

3 The Generic Agent

This section presents an architecture for a generic management agent. It describes the components of the agents and the services that the agent provides. Using this architecture, the need for the developer to spend weeks or months designing the agent is greatly reduced. It allows agent developers to create any type of management agent quickly and easily. The user specifies the type of agent desired and its capabilities, and the code for the agent is generated.

3.1 Architectural Components

In this section, we describe each of the components of the architecture, as well as the information necessary to automate creation of that component. The components of a generic management agent are shown in Figure 2. When a request arrives, it is first given to the Coordinator, where it is parsed. It is then passed onto the Request Verification component, where it is verified. The request is then sent onto the appropriate component where it is executed, and any results are returned to the requester.

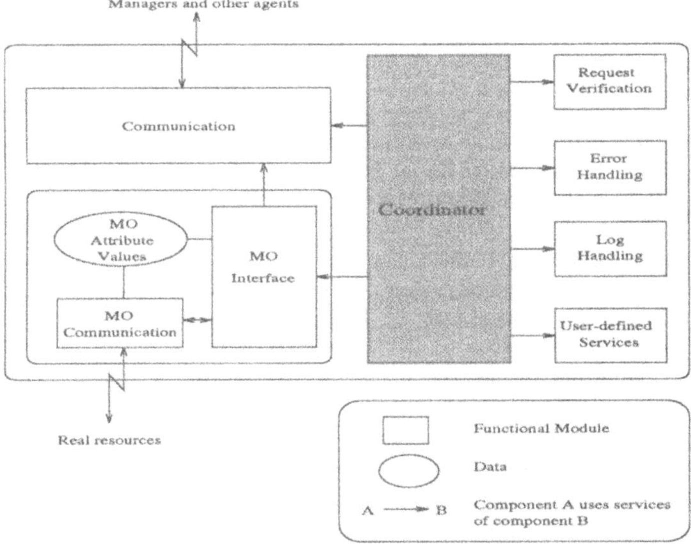

Figure 2: The Generic Agent Architecture

- **Coordinator:** The Coordinator is the central component of the agent. It parses requests and passes required information to the appropriate component. To be able to parse the requests, knowledge of the management protocol that is being used is required. The Coordinator must also know which of the services listed in Section 3.2 are provided by this agent, in order to be able to describe the agent to other management entities.

- **Communication:** The communication component provides communication services to other components, such as services to allow the agent to receive requests, return information to the requester, and forward notifications from managed objects to other entities. The management protocol to be used is all that this component needs to know.

- **Request Verification:** This component verifies each incoming request to make sure that the following conditions hold:
  - this agent supports the request,
  - any managed objects that are referenced are valid,
  - the requester is authorized to make a request, and
  - the requester has permission to perform the given request with the given managed objects.

  To check these conditions, this component must have knowledge of which agent services are supported, as well as which managed object classes can be managed. It does not, however, need to know which management protocol is being used.

- **Managed Object Interface:** This component contains the managed objects, which are abstractions of the managed resources. Each managed object "represents" a single resource for the purposes of managing it. Note that a managed resource may or may not be a physical resource. For example, a management system could have a managed object representing an application, which itself contains several managed objects each representing a different process which is part of that application. This component provides the interface for the agent to interact with the managed objects which, in turn, interact with the resources they represent.

  This component does not need to know the management protocol that is being used, since it does not have to "communicate" per se with managed objects. Managed objects may need to communicate with external resources, but the method or protocol that they use to do this communication is a decision made by the managed object developer, and is not relevant to the design of the agent. The only information that the agent developer must supply in order to create this component automatically is the list of management operations which should be supported.

- **Log Handling:** This component contains services that allow the agent to log information, such as management requests (get information, set information, creation of a managed object, *etc.*), the execution of dynamic services (see Section 3.2.3), or notifications received from managed objects. The Log Handling component can be entirely automated with no information from the user.

- **User-Defined Services:** This component stores the services that have been added to the agent, and provides a way for the agent to execute these requests. The execution could involve simply running the service and returning the result, starting the service as a background process, or even providing a multi-tasking environment in which the services run [4].

- **Error Handling:** This component hides some of the details about the management protocol from the rest of the agent. It translates internal error codes into a format recognizable by a manager, which can then diagnose and possibly correct the problem. Knowledge of the management protocol to be used by this agent is necessary for this component's creation.

An obvious advantage to having modular components is that each component is not dependent on the implementation details of the other components. For example, to automate the Error Handling component, we only need knowledge of the management protocol. Whether or not the agent supports, for example, user-defined services or logging has no effect on the code for this component. This code reuse implies another advantage: increased reliability. The same Error Handling code is used in *all* agents with the same protocol regardless of other services offered and managed object classes supported. If this code is tested and found to be stable in a particular implementation, it can be assumed that the code will work in other implementations, since the implementation details of the other components do not affect this one. Once the code is stable, it can also be optimized, resulting in faster and more efficient agents.

3.2 Agent Service Interface

This section describes the operations that comprise the service interface. These operations represent the vast majority of the operations that management agents provide to other management entities, such as managers. There are, of course, exceptions: agents which provide specialized operations that are not listed here. However, using the User-defined operations described below, even these agents can fit into this architecture, with the specialized operations added as user-defined services.

There are four classes of operations offered: *self-description* operations that allow the agent to describe itself and its capabilities to other management entities; *common management* operations that are used to manipulate managed objects; *user-defined* operations that allow an agent's capabilities to be extended; and *logging* operations that allow the agent to log requests, notifications, and other management information. The detailed specification of the interfaces described below can be found in [11].

3.2.1 Self-Description Operations

Self-description operations can be used by managers, managed objects, or other agents to ask an agent for information about itself.

- **DescribeMyself:** Allows the agent to describe itself and its (built-in) capabilities.
- **GetMOList:** Lists the managed objects that are being managed. This operation basically traverses the tree of managed objects, and returns the list of objects. A parameter can be given that allows the requester to limit or control the traversal of the managed object tree.
- **ListPeriodicServices:** Allows a management entity to query an agent to find out which services have been scheduled to run as periodic services. A periodic service is a service which has been dynamically added and has been scheduled to be executed periodically (*e.g.*, every five minutes).
- **ListServices:** Allows a manager to query an agent to find out what services have been added.

3.2.2 Common Management Operations

Management operations can be used by managers and other agents to create and delete managed objects, perform actions on managed objects, and get and set attribute values within managed objects, and can also be used by managed objects to send notifications of events to managers.

- **Action:** Send an action command to one or more managed objects. Actions are defined within managed objects, and are operations that managed objects perform on themselves. For example, a file managed object may have actions called `rename`, `touch`, `remove`, or `copy`.
- **Create:** Creates a new managed object of the specified class, with the specified name, and with the specified attribute values (if present).
- **Delete:** Deletes the specified managed object(s).
- **ForwardNotification:** Allows a managed object to send a notification to a manager.
- **Get:** Returns the value of the specified attribute(s) from the specified managed object(s).
- **Set:** Sets the value of each of the specified attribute(s) in each of the specified managed object(s) to the specified value.

3.2.3 User-Defined Operations

User-defined operations allow the agent's functionality to be extended. "Services" can be statically or dynamically added to the agent, and can be executed at any time thereafter. These services can also be scheduled to be periodically executed.

- **AddNewService:** Dynamically adds new functionality to the agent. Each agent has three types of operations that it can perform: static operations (this list), static user-defined operations (code which is written by the user and is linked in with the agent at build-time), and dynamic user-defined operations (called "services" – user-written code that is added to the agent dynamically, at run-time). This operation allows a management entity to send a program to an agent, telling the agent to add the program to its list of services. That service is then accessible by any management entity, by using the ServiceHandle assigned to it by this operation.
- **ExecuteService:** Executes the specified (previously added) service once and returns the results.
- **StartPeriodicService:** Schedules the specified service to be executed at the beginning of each successive period of specified length. Any service that has been added using AddNewService can be used in a periodic service. If the service specified returns a value, the return value after each execution will be ignored. Note that the service is started and runs to completion each time it is executed; is it not "woken up" periodically.
- **StopPeriodicService:** De-schedules the specified periodic service.

3.2.4 Log Operations

Logging operations allow the agent to log information about events that have occurred or about operations that have been performed on managed objects [5].

- **DescribeLog:** This operations allows a management entity to query a log that has been started to find out its current state, size, etc.
- **LockLog:** Locks or unlocks the specified log. Locking a log allows records currently in the log to be read, but no new records can be added.
- **StartLog:** Create a new log and begin logging. When started, the log is both enabled and unlocked.
- **StopLog:** Stops the specified log. Once a particular log has been stopped, it cannot be restarted.

4 Prototype Implementation

As a proof of concept, we have developed a prototype tool that can automatically generate management agents (CMIP and SNMP) which possess the general agent architecture introduced in Section 3. In this section, we describe this prototype called the Management Agent Creation Tool (MACT) and how it is used to create management agents.

4.1 The Management Agent Creation Tool

MACT is a Motif-based tool that allows the user to specify the type of agent desired using a simple graphical user interface [11, 12]. The tool allows the user to enter the required information, ensures that the information entered is valid, and then generates the code for the specified agent.

As described in Section 3.1, the following information must be supplied by the user (an agent developer) in order to be able to generate the agent code:

- which managed object classes should be supported,
- which agent operations should be supported, and
- which management protocol will be used.

MACT makes it easy for the user to specify this information. When a user starts MACT, a main window requesting the user to select the management protocol (either CMIP or SNMP) is displayed. When the user selects CMIP, the user interface for generating CMIP agents, as shown in Figure 3, is displayed. The MACT user interface for generating SNMP agents is shown in Figure 4. When the user completes the input of the desired agent including the target platform (*e.g.*, Sun4 or AIX) and the create operation is requested, MACT generates all the source code including a Makefile into a directory. All the user has to do is to exit the tool and type `make` to generate an executable of the desired agent.

Currently, the OSIMIS implementation of the CMIP protocol and the ISODE implementation of the SNMP protocol are supported. We describe these in more detail below.

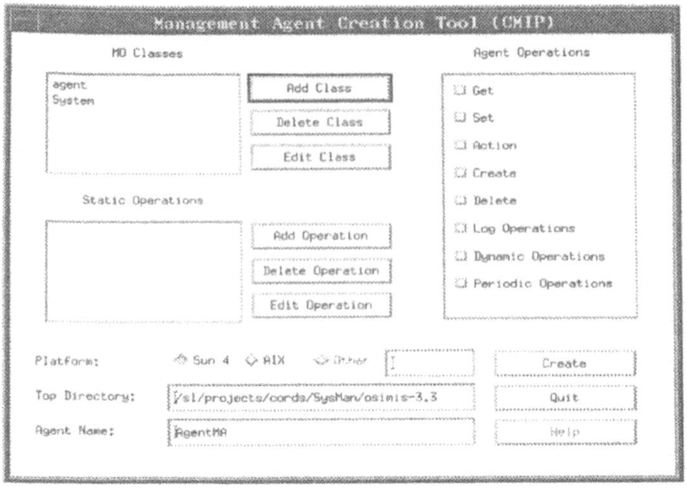

Figure 3: MACT User Interface for Creating CMIP Agents

4.2 OSI Management Agents

The OSI Management agent that MACT generates is based on OSIMIS (OSI Management Information Service) Version 3.3, which is an implementation of the OSI management framework, developed at University College London [10]. It was developed using the ISO Development Environment (ISODE), and provides a GDMO compiler, which generates C++ code for managed objects from specifications made in the GDMO language.

The architecture of the OSIMIS agent is somewhat less structured than the architecture we presented in Section 3.1. There is a class called Coordinator, but it provides the functionality of our Communication component. There is also a class called CMISAgent, which provides the functionality of our Coordinator. One method within the CMISAgent class acts as a (limited) request validation component, while another method is the error handling component. There are no user-defined service nor log handling components. The GDMO compiler provided with OSIMIS ensures that all managed objects have an identical interface, which provides the functionality of the Managed Object Interface component. Unfortunately, each component in the OSIMIS agent is highly dependent on the implementation details of the other components.

OSIMIS agents provide the same management operations as listed in Section 3.2.2. However, none of the other services listed in Section 3.2 are offered by OSIMIS agents. Logs are supported, but only for notifications, and they are driven by the managed objects, not the agent.

To validate our architecture, the OSIMIS agent was enhanced to include the other agent operations outlined in Section 3.2 that were not already present. A Log Handling component and operations were added, making the agent capable of logging requests without generating a notification. The log information is now generated by the agent, not by the managed object. User-defined service and self-description operations have also been added to the OSIMIS agent. Note that, as stated in Section 3.1, the code to provide the log, user-defined service, and periodic service operations is protocol-independent; in fact, the same code is used for these operations in both the CMIP and SNMP agents.

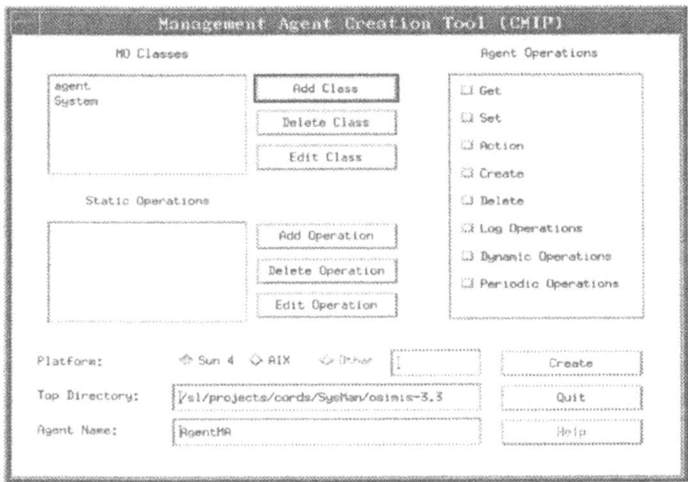

Figure 4: MACT User Interface for Creating SNMP Agents

4.3 SNMP Management Agents

The SNMP agent that MACT generates is based on the ISODE (ISO Development Environment) version 8.0. The SNMP agent was not designed or developed using an object-oriented methodology. As such, there is no clear distinction between components, and both error handling and request validation are done *ad hoc*. As in the OSIMIS agent, there are no user-defined service or log handling components. The Managed Object Interface consists of a list of C functions, each of which can get or set the values of specific groups of variables. It seems that every part of this agent is highly dependent on the implementation details of many other parts.

Nevertheless, we can identify functions and sections of code that act as the Coordinator and the communication component. It would not be difficult to write a single request validation routine and a single error handling routine, which would reduce the large amount of code repetition present within the agent.

The only operations that SNMP agents provide are Get, Set, Trap (for sending notifications), and Get-Next, which is used in the GetMOList operation. None of the other services listed in Section 3.2 are offered by SNMP agents. The SNMP agent has been enhanced to include other agent operations outlined in this paper that were not already present. Log Handling, User-Defined Service, and Self-Description operations have all been added to the SNMP agent. The Action, Create, and Delete operations have not been implemented since they are not supported in SNMP agents. The SNMP standard specifies that variables are created only on agent initialization, and cannot be created afterward, nor can they be destroyed. SNMP also does not support actions on variables, and so the Action operation is also unnecessary.

5 Concluding Remarks

We have described the role and importance of management agents within management systems. We then outlined the requirements for generic management agents, and presented a general architecture for these agents. We have identified four kinds of basic services which are common to all agents, and the interfaces to each of those services. We have also outlined the information that is required from the agent developer in order to be able to create the desired agent.

We have developed a prototype tool called MACT that automates much of the development process of management agents. Using MACT will greatly reduce the time needed, and therefore the cost of creating management agents, and will eliminate the need for the agent developer to "re-invent the wheel". Because the code is reused in different agents, it is more robust than an *ad hoc* solution. One of the most important benefits of using MACT is that most agents will not require much (if any) code written by the agent developer. The only code that needs to be written is the code for the managed objects to access real resources being managed and the code for any user-defined routines, which can easily be added to the agent.

MACT has been sparingly used by our group members for developing various CMIP and SNMP management agents. We have used MACT to generate a number of management agents including UNIX system management agent, a generic distributed application management agent as well as a number of specific application management agents. Our generic management agent combined with MACT, in our opinion, provides an excellent framework for providing "extensible" agents.

We are in the process of enhancing the functionality of MACT. We hope to develop and add to it a managed object class library browsal and definition tool, which would allow the user to browse through the existing managed object classes, modify existing or define new managed object classes on the fly. We plan to develop and experiment with both the static and dynamic operations to extend the capabilities of agents for various purposes. We also plan to develop more management agents using MACT for network, system and application management.

References

[1] J. Case, M. Fedor, M. Schoffstall, and C. Davin. A Simple Network Management Protocol (SNMP). *Internet Request for Comments 1157*, May 1990.

[2] J. Case, K. McCloghrie, M. Rose, and S. Waldbusser. Introduction to version 2 of the Internet-standard Network Management Framework. *Internet Request for Comments 1441*, April 1993.

[3] Germán Goldszmidt. On Distributed System Management. In *Proceedings of the 1993 CAS Conference*, pages 637–647, Toronto, Canada, October 1993.

[4] Germán Goldszmidt, Shaula Yemini, and Yechiam Yemini. Network Management by Delegation - the MAD Approach. In *Proceedings of the 1991 CAS Conference*, pages 347–359, Toronto, Canada, October 1991.

[5] ISO. Information Technology – Open Systems Interconnection – System Management – Part 5: Event Report Management Function. *International Organization for Standardization, International Standard X.736*, November 1990.

[6] ISO. Information Technology – Open Systems Interconnection – Systems Management Overview. *International Organization for Standardization, International Standard X.701*, June 1991.

[7] ISO. Information Technology – Structure of Management Information – Part 4: Guidelines for the definition of managed objects. *International Organization for Standardization, International Standard X.722*, July 1991.

[8] Paul Miller. Bull's CMIP Agent Development Kit – A Platform for the Rapid Development of CMIP Agents & Objects. In *Proceedings of NOMS94*, Orlando, FL, February 1994.

[9] Oscar Newkerk, Miriam Amos Nihart, and Steven K. Wong. The Common Agent – A Multiprotocol Management Agent. *IEEE Journal on Selected Areass in Communications*, 11(9):1346–1352, December 1993.

[10] G. Pavlou, S.N. Bhatti, and G. Knight. The OSI Management Information Service User's Manual. Version 1.0, February 1993.

[11] G. S. Perrow. The Abstraction and Modelling of Management Agents. *MSc. Thesis*, Dept. of Computer Science, University of Western Ontario, London, Ontario, Canada, September 1994.

[12] G. S. Perrow, J. W. Hong, M. A. Bauer, and H. Lutfiyya. MACT User's Guide Version 1.0. *Technical Report 434*, Dept. of Computer Science, University of Western Ontario, London, Ontario, Canada, September 1994.

[13] M. Rose and K. McCloghrie. Structure and Identification of Management Information for TCP/IP-based Internets. *Internet Request for Comments 1155*, May 1990.

[14] Marshall T. Rose. *The Open Book: A Practical Perspective on OSI.* Prentice Hall, Englewood Cliffs, NJ, 1990.

[15] Marshall T. Rose. *The Simple Book: An Introduction to Internet Management, Second Edition.* Prentice Hall, Englewood Cliffs, NJ, 1994.

[16] M. Sylor and O. Tallman. Applying Network Management Standards to System Management; the Case for the Common Agent. In *Proceedings of the IEEE First International Workshop on Systems Management*, Los Angeles, CA, April 1993.

About the Authors

Graeme S. Perrow is currently working as a software engineer at Comnetix Computer Systems, Mississauga, Ontario. He received his BMath in Computer Science from the University of Waterloo in 1992 and his MSc in Computer Science from the University of Western Ontario in 1994. His research interests include network management, software engineering and information systems. He can be reached via electronic mail at gperrow@comnetix.com.

James W. Hong is a research associate and adjunct professor in the Department of Computer Science at the University of Western Ontario. He received his BSc and MSc from the University of Western Ontario in 1983 and 1985 respectively and his doctorate from the University of Waterloo in 1991. He is a member of the ACM and IEEE. His research interests include distributed computing, software engineering, systems and network management. He can be reached via electronic mail at jwkhong@csd.uwo.ca.

Hanan L. Lutfiyya is an assistant professor of Computer Science at the University of Western Ontario. She received her B.S. in computer science from Yarmouk University, Irbid, Jordan in 1985, her M.S. from the University of Iowa in 1987, and her doctorate from the University of Missouri-Rolla in 1992. She is a member of the ACM and IEEE. Her research interests include distributed computing, formal methods in software engineering and fault tolerance. She can be reached via electronic mail at hanan@csd.uwo.ca.

Michael A. Bauer is Chairman of the Department of Computer Science at the University of Western Ontario. He received his doctorate from the University of Toronto in 1978. He has been active in the Canadian and International groups working on the X.500 Standard. He is a member of the ACM and IEEE and is a member of the ACM Special Interest Group Board. His research interests include distributed computing, software engineering and computer system performance. He can be reached via electronic mail at bauer@csd.uwo.ca.

SECTION TWO

Platform Experiences

41

The OSIMIS Platform: Making OSI Management Simple

George Pavlou, Kevin McCarthy, Saleem Bhatti, Graham Knight
Department of Computer Science, University College London, Gower Street, London, WC1E 6BT, UK
tel: +44 71 380 7215 fax: +44 71 387 1397
e-mail: g.pavlou k.mccarthy s.bhatti g.knight @cs.ucl.ac.uk

Abstract

The OSIMIS (OSI Management Information Service) platform provides the foundation for the quick, efficient and easy construction of complex management systems. It is an object-oriented development environment in C++ [Strou] based on the OSI Management Model [X701] that hides the underlying protocol complexity (CMIS/P) and harnesses the power and expressiveness of the associated information model [X722] through simple to use Application Program Interfaces (APIs). OSIMIS combines the thoroughness of the OSI models and protocols with advanced distributed systems concepts pioneered by ODP to provide a highly dynamic distributed information store. It also combines seamlessly the OSI management power with the large installed base of Internet SNMP [SNMP] capable network elements. OSIMIS supports particularly well a hierarchical management organisation through hybrid manager-agent applications and may embrace a number of diverse technologies through proxy systems. This paper explains the OSIMIS components, architecture, philosophy and direction.

Keywords

Network, Systems, Application Management, Distributed Systems, Platform, API

1 INTRODUCTION AND OVERVIEW

OSIMIS is an object-oriented management platform based on the OSI model [X701] and implemented mainly in C++ [Strou]. It provides an environment for the development of management applications which hides the details of the underlying management service through object-oriented Application Program Interfaces (APIs) and allows designers and implementors to concentrate on the intelligence to be built into management applications rather than the mechanics of management service/protocol access. The manager-agent model and the notion of managed objects as abstractions of real resources are used but the separation between managing and managed systems is not strong in engineering terms: a management application can be in both roles and this is particularly true in situations where a management system is decomposed according to a hierarchical logical layered approach.

In fact, OSIMIS was designed from the beginning with the intent to support the integration of existing systems with either proprietary management facilities or different management models. Different methods for the interaction with real managed resources are supported, encompassing loosely coupled resources as it is the case with subordinate agents and management hierarchies. The fact that the OSI model was chosen as the basic management model facilitates the integration of other models, the latter usually being less powerful, as is the case with the Internet SNMP [SNMP]. OSIMIS provides already a generic application gateway between CMIS and SNMP [Pav93a] while a similar approach for integrating OSI management and the OMG CORBA framework [OMG] may be pursued in the future.

OSIMIS uses the ISODE (ISO Development Environment) [ISODE] as the underlying OSI communications mechanism but it may also be decoupled from it through the XOM/XMP [XOpen] management API. The advantage of the ISODE environment though is the provision of services like FTAM and a full implementation of the OSI Directory Service (X.500) which are essential in complex management environments. Also a number of underlying network technologies are supported, namely X.25, CLNP and also TCP/IP through the RFC1006 method. These constitute the majority of currently deployed networks while interoperation of applications across any of these is possible through Transport Service Bridging.

OSIMIS has been and is still being developed in a number of European research projects, namely the ESPRIT INCA, PROOF and MIDAS and the RACE NEMESYS and ICM. It has been used extensively in both research and commercial environments and has served as the management platform for a number of other ESPRIT and RACE projects in the TMN and distributed systems and service management areas. OSIMIS has been fully in the public domain until version 3.0 to show the potential of OSI management and serve as a benchmark implementation; later versions are still freely available to academic and research institutions for non-commercial use.

Components and Architecture

OSIMIS as platform comprises the following types of support:

- high level object-oriented APIs realised as libraries
- tools as separate programs supporting the above APIs (compilers/translators)
- generic applications such as browsers, gateways, directory servers etc.

Some of these services are supported by ISODE and these are:

- the OSI Transport (class 0), Session and Presentation protocols, including a lightweight version of the latter that may operate directly over the Internet TCP/IP
- the Association Control and Remote Operations Service Elements (ACSE and ROSE) as building blocks for higher level services
- the File Transfer Access and Management (FTAM) and Directory Access Service Element (DASE)
- a ASN.1 compiler with C language bindings (the pepsy tool)
- a Remote Operations stub generator (the rosy tool)
- a FTAM service for the UNIX operating system
- a full Directory Service implementation including an extensible Directory Service Agent (DSA) and a set of Directory User Agents (DUAs)
- a transport service bridge allowing interoperability of applications over different types of networks

OSIMIS is built as an environment using ISODE and is mostly implemented in the C++ programming language. The services it offers are:

• an implementation of CMIS/P using the ISODE ACSE, ROSE and ASN.1 tools

• an implementation of the Internet SNMP over the UNIX UDP implementation using the ISODE ASN.1 tools

• high-level ASN.1 support that encapsulates ASN.1 syntaxes in C++ objects

• an ASN.1 object-oriented meta-compiler which uses the ISODE pepsy compiler to automate to a large extent the generation of syntax C++ objects

• a Coordination mechanism that allows to structure an application in a fully event-driven fashion and can be extended to interwork with similar mechanisms

• a Presentation Support service which is an extension of the coordination mechanism to interwork with X-Windows based mechanisms

• the Generic Managed System (GMS) which is an object-oriented OSI agent engine offering a high level API to implement new managed object classes, a library of generic attributes, notifications and objects and systems management functions

• a compiler for the OSI Guidelines for the Definition of Managed Objects (GDMO) [X722] language which complements the GMS by producing C++ stub managed objects covering every syntactic aspect and leaving only behaviour to be implemented

• the Remote and Shadow MIB high level object-oriented manager APIs

• a Directory Support service offering application addressing and location transparency services

• a generic CMIS to SNMP application gateway driven by a translator between SNMP and OSI GDMO MIBs

• a set of generic manager applications (MIB browser and other)

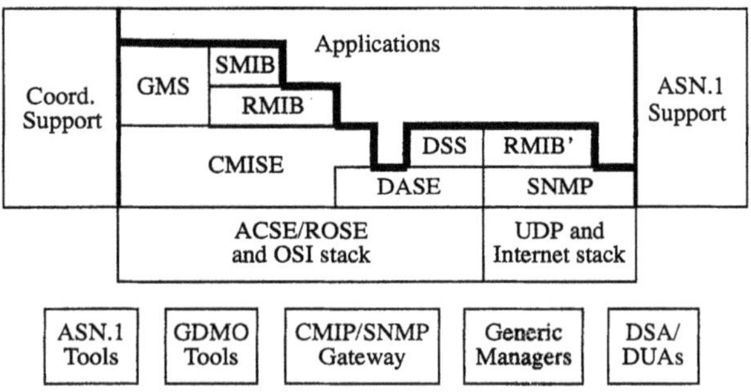

Figure 1 OSIMIS Layered Architecture and Generic Applications.

The OSIMIS services and architecture are shown in Figure 1. In the layered part, applications are programs while the rest are building blocks realised as libraries. The lower part shows the generic applications provided; from those the ASN.1 and GDMO tools are essential in providing off-line support for the realisation of new MIBs. The thick line indicates all the APIs an application may use. In practice though most applications use only the Generic Managed System (GMS) and the Remote MIB (RMIB) APIs when acting in agent and manager roles respectively, in addition to the Coordination and high-level ASN.1 support ones. The latter are used by other components in this layered architecture and are orthogonal to them, as such they are shown aside. Directory access for address resolution and the provision of location transparency may or may not be used, while the Directory Support Service (DSS) API provides more sophisticated searching, discovery and trading facilities.

2 THE ISO DEVELOPMENT ENVIRONMENT

The ISO Development Environment [ISODE] is a platform for the development of OSI services and distributed systems. It provides an upper layer OSI stack that conforms fully to the relevant ISO/CCITT recommendations and includes tools for ASN.1 manipulation and remote operations stub generation. Two fundamental OSI applications also come with it, an extensible full Directory Service (X.500) and File Transfer (FTAM) implementations. ISODE is implemented in the C programming language and runs on most versions of the UNIX operating system. ISODE does not provide any network and lower layer protocols e.g. X.25, CLNP, but relies on implementations for UNIX-based workstations which are accessible through the kernel interface. The upper layer protocols realised are the transport, session and presentation protocols of the OSI 7-layer model. Application layer Service Elements (ASEs) are also provided as building blocks for higher level services, these being the Association Control, Remote Operations and Reliable Transfer Service Elements (ACSE, ROSE and RTSE). These, in conjunction with the ASN.1 support, are used to implement higher level services. In engineering terms, the ISODE stack is a set of libraries linked with applications using it.

ASN.1 manipulation is very important to OSI distributed applications. The ISODE approach for a programmatic interface (API) relies in a fundamental abstraction known as Presentation Element (PE). This is a generic C structure capable of describing in a recursive manner any ASN.1 data type. An ASN.1 compiler known as pepsy is provided with C language bindings, which produces concrete representations i.e. C structures corresponding to the ASN.1 types and also encode/decode routines that convert those to PEs and back. The presentation layer converts PEs to a data stream according to the encoding rules (e.g. BER) and the opposite. It should be noted that X/Open has defined an API for ASN.1 manipulation known as XOM [XOpen] which, though similar in principle to that of ISODE, is syntactically very different. Translations between the two are possible and such an approach is used to put OSIMIS applications over XOM/XMP.

3 MANAGEMENT PROTOCOL AND O-O ABSTRACT SYNTAX SUPPORT

OSIMIS is based on the OSI management model as the means for end-to-end management and as such it implements the OSI Common Management Information Service/Protocol (CMIS/P). This is implemented as a C library and uses the ISODE ACSE and ROSE and its ASN.1 support. Every request and response CMIS primitive is realised through a procedure call. Indications and confirmations are realised through a single "wait" call. Associations are represented as communication endpoints (file descriptors) and may be multiplexed to realise event-driven policies.

The OSIMIS CMIS API is known as MSAP (Management Service Access Point). It was conceived much before standard APIs such as the X/Open XMP were specified and as such it does not conform to the latter. Having been designed specifically for CMIS and not for both CMIS and SNMP as the XMP one, it hides more information and may result in more efficient implementations. Higher-level object-oriented abstractions that encapsulate this functionality and add much more can be designed and built as explained in section 6. OSIMIS offers as well an implementation of the Internet SNMPv1 and v2 which is used by the generic application gateway between the two. This uses the socket API for Internet UDP access and the ISODE ASN.1 support.

Applications using CMIS need to manipulate ASN.1 types for the CMIS managed object attribute values, action, error parameters and notifications. The API for ASN.1 manipulation in ISODE is different to the X/Open XOM. Migration to XOM/XMP is possible through thin conversion layers so that the upper layer OSIMIS services are not affected. Regarding ASN.1 manipulation, it is up to an application to encode and decode values as this adds to its dynamic nature by allowing late bindings of types to values and graceful handling of error conditions. From a distributed programming point of view this is unacceptable and OSIMIS provides a mechanism to support high-level object-oriented ASN.1 manipulation, shielding the programmer from details and enabling distributed programming using simply C++ objects as data types.

4 APPLICATION COORDINATION SUPPORT

Management and, more generally, distributed applications have complex needs in terms of handling external input. Management applications have additional needs of internal alarm mechanisms for arranging periodic tasks in real time (polling etc.) Furthermore, some applications may need to be integrated with Graphical User Interface (GUI) technologies which have their own mechanisms for handling data from the keyboard and mouse. In this context, the term application assumes one process in operating systems terms.

There are in general different techniques to organise an application for handling both external and internal events. The organisation needs to be event driven though so that no resources are used when the system is idle. The two major techniques are:

a. use a single-threaded execution paradigm

b. use a multi-threaded one

In the first, external communications should follow an asynchronous model as waiting for a result of a remote operation in a synchronous fashion will block the whole system. Of course, a common mechanism is needed for all the external listening and demultiplexing of the incoming data and this is a part of what the OSIMIS Application Coordination Support provides. In the second, many threads of control can be executing simultaneously (in a pseudo-parallel fashion) within the same process, which means that blocking on an external result is allowed. This is the style of organisation used by distributed systems platforms as they are based on RPC which is inherently synchronous with respect to client objects performing remote operations to server objects.

An additional problem in organising a complex application concerns the handling of internal timer alarms: most operating systems do not "stack" them i.e. there can only be one alarm pending for each process. This means that a common mechanism is needed to ensure the correct usage of the underlying mechanism.

OSIMIS provides an object-oriented infrastructure in C++ [Pav93b] which allows to organise an application in a fully event-driven fashion and a single- threaded execution paradigm, where every external or internal event is serialised and taken to completion on a "first-come-first-served" basis. This mechanism allows the easy integration of additional external sources of input or timer alarms and it is realised by two C++ classes: the Coordinator and the Knowledge Source

(KS). There should always be one instance of the Coordinator or any derived class in the application while the Knowledge Source is an abstract class that allows to use the coordinator services and integrate external sources of input or timer alarms. All external events and timer alarms are controlled by the coordinator whose presence is transparent to implementors of specific KSs through the abstract KS interface. This model is depicted in Figure 2.

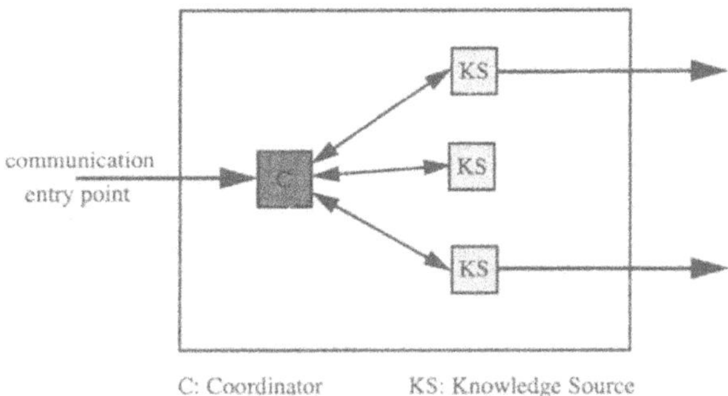

Figure 2 The OSIMIS Process Coordination Support Model.

This coordination mechanism is designed in such a way as to allow integration with other systems' ones. This is achieved through special coordinator derived classes which will interwork with a particular mechanism. This is achieved by still controlling the sources of input and timer alarms of the OSIMIS KSs but instead of performing the central listening, these are passed to the other system's coordination mechanism which becomes the central one. Such an approach is needed for Graphical User Interface technologies which have their own coordination mechanisms. In this case, simply a new special coordinator class is needed for each of them. At present, the X-Windows Motif and the TCL/TK interpreted scripting language are integrated.

5 THE GENERIC MANAGED SYSTEM

The Generic Managed System (GMS) [Pav93b] [Kni91] provides support for building agents that offer the full functionality of the OSI management model, including scoping, filtering, access control, linked replies and cancel-get. OSIMIS supports fully the Object Management, Event Reporting and Log Control Systems Management Functions (SMFs), the qualityofServiceAlarm notification of the Alarm Reporting one and partly the Access Control, Metric and Summarization objects. In conjunction with the GDMO compiler it offers a very high level API for the integration of new managed object classes where only semantic aspects (behaviour) need to be implemented. It also offers different methods of access to the associated real resources, including proxy mechanisms, based on the Coordination mechanism.

The Generic Managed System is built using the coordination and high level ASN.1 support infrastructure and most of its facilities are provided by three C++ classes whose instances interact with each other:

- the CMISAgent, which provides OSI agent facilities
- the MO which is the abstract class providing generic managed object support
- the MOClassInfo which is a meta-class for a managed object class

The GMS library contains also generic attribute types such as counter, gauge, counterThreshold, gaugeThreshold and tideMark and specific attributes and objects as in the Definition of Management Information (DMI), which relate to the SMFs. The object-oriented internal structure of a managed system built using the GMS in terms of interacting object instances is shown in Fig. 3.

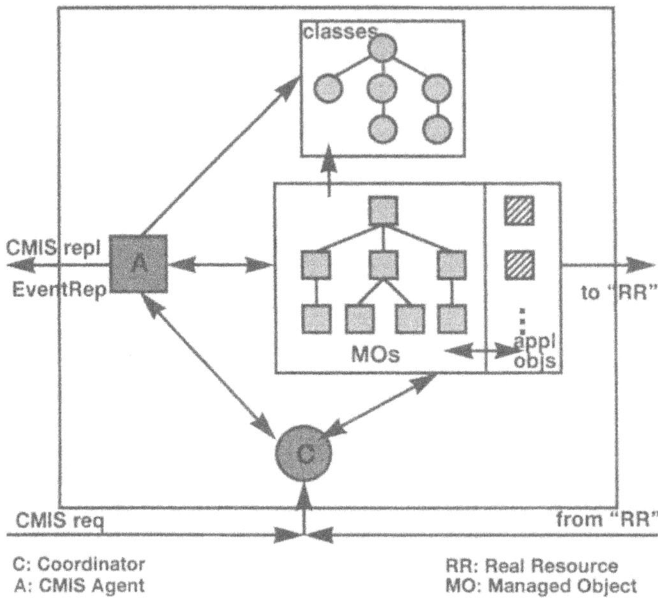

Figure 3 THE GMS Object-Oriented Architecture.

5.1 The CMIS Agent

The CMISAgent is a specialised knowledge source as it has to accept management associations. There is always one only instance of this class for every application in agent role. Its functions are to accept or not associations according to authentication information, check the validity of operation parameters, find the base object for the operation, apply scoping and filtering, check if atomic synchronisation can be enforced, check access control rights and then apply the operation on the target managed object(s) and return the result(s)/error(s).

There is a very well defined interface between this class and the generic MO one which is at present synchronous only: a method call should always return with a result e.g. attribute values or error. This means that managed objects which mirror loosely coupled real resources and exercise an "access-upon-external-request" regime will have to access the real resource in a synchronous fashion which will result in the application blocking until the result is received. This is only a problem if another request is waiting to be served or if many objects are accessed in one request through scoping. Threads would be a solution but the first approach will be a GMS internal asynchronous API which is currently being designed. It is noted that the CMISAgent to MO interface is bidirectional as managed objects emit notifications which may be converted to event reports and passed to the agent.

5.2 Managed Object Instances and Meta-Classes

Every specific managed object class needs access to information common to the class which is independent of all instances and common to all of them. This information concerns attributes, actions and notifications for the class, initial and default attribute values, "template" ASN.1 objects for manipulating action and notification values, integer tags associated to the object identifiers etc. This leads to the introduction of a common meta-class for all the managed object classes, the MOClassInfo. The inheritance tree is internally represented by instances of this class linked in a tree fashion as shown in the "classes" part of Figure 3.

Specific managed object classes are simply realised by equivalent C++ classes produced by the GDMO compiler and augmented manually with behaviour. Through access to meta-class information requests are first checked for correctness and authorisation before the behaviour code that interacts with the real resource is invoked. Behaviour is implemented through a set of polymorphic methods which may be redefined to model the associated real resource. Managed object instances are linked internally in a tree mirroring the containment relationships - see "MOs" part of Figure 3. Scoping becomes simply a tree search while special care is taken to make sure the tree reflects the state of the associated resources before scoping, filtering and other access operations. Filtering is provided through compare methods of the attributes which are simply the C++ syntax objects or derived classes when behaviour is coded at the attribute level.

5.3 Real Resource Access

There are three possible types of interaction between the managed object and the associated resource with respect to CMIS Get requests:

1. access upon external request
2. "cache-ahead" through periodic polling
3. update through asynchronous reports

The first one means that no activity is incurred when no manager accesses the agent but cannot support notifications. In the second requests are responded to quickly, especially with respect to loosely coupled resources, but timeliness of information may be slightly affected. Finally the third one is good but only if it can be tailored so that there is no unnecessary overhead when the agent is idle.

The GMS offers support for all methods through the coordination mechanism. When asynchronous reports from a resource are expected or asynchronous results to requests, it is likely that a separate object will be needed to demultiplex the incoming information and deliver it to the appropriate managed object. It should noted here that an asynchronous interface to real resources driven by external CMIS requests is not currently supported as this requires an internal asynchronous interface between the agent and the managed objects. These objects are usually referred to an Internal Communications Controllers (ICCs) and are essentially specialised knowledge sources.

5.4 Systems Management Functions

As already stated, OSIMIS supports the most important of the systems management functions. As far as the GMS is concerned, these functions are realised as special managed objects and generic attribute and notification types which can be simply instantiated or invoked. This is the case for example with the alarm reporting, metric and summarization objects. In other cases, the GMS knows the semantics of these classes and uses them accordingly e.g. in access control and event and log control. Notifications can be emitted through a special method call and all the subsequent notification processing is carried out by the GMS in a fashion transparent to application code. In the case of object management, the code generated by the GDMO compiler together with the GMS hide completely the emission of object creation and deletion notifications and the attribute change one when something is changed through CMIS.

Log control is realised simply through managed object persistency which is a general property of all OSIMIS managed objects. This is implemented using the GNU version of the UNIX DBM database management system and relies on object instance encoding using ASN.1 and the OSI Basic Encoding Rules to serialise the attribute values. Any object can be persistent so that its values are retained between different incarnations of an agent application. At start-up time, an agent looks for any logs or other persistent objects and simply arranges its management information tree accordingly.

5.5 Security

General standards in the area of security for OSI applications are only now being developed while the Objects and Attributes for Access Control Systems Management Function is not yet an International Standard. Nevertheless, systems based on OSI management have security needs and as such OSIMIS provides the following security services:

- peer entity authentication
- data origin authentication and stream integrity
- access control

These were developed in the ESPRIT MIDAS project to cater for the security of management of a large X.400 mail system [Kni94] and will also be used in the RACE ICM project for inter-TMN security requirements on virtual private network applications. Peer entity authentication relies on public key encryption through RSA as in X.509. Data origin authentication is based on cryptographic checksums of CMIP PDUs calculated through the MD5 algorithm. Stream integrity is provided in a novel way that is based on a "well-known" invokeID sequence in ROSE PDUs. It should be noted that as CMIP does not make any provision for the carrying of integrity checksums, these are carried in the ROSE invokeID field. Finally access control is provided through the implementation of the relevant SMF.

6 GENERIC HIGH-LEVEL MANAGER SUPPORT

Programming manager applications using the CMIS API can be tedious. Higher object-oriented abstractions can be built on top of the CMIS services and such approaches were initially investigated in the RACE-I NEMESYS project while work in this area was taken much further in the RACE-II ICM project [Pav94].

The Remote MIB (RMIB) support service offers a higher level API which provides the abstraction of an association object. This handles association establishment and release, hides object identifiers through friendly names, hides ASN.1 manipulation using the high-level ASN.1 support, hides the complexity of CMIS distinguished names and filters through a string-based notation, assembles linked replies, provides a high level interface to event reporting which hides the manipulation of event discriminators and finally provides error handling at different levels. There is also a low level interface for applications that do not want this friendliness and the performance cost it entails but they still need the high-level mechanisms for event reporting and linked replies.

In the RMIB API there are two basic C++ classes involved: the RMIBAgent which is essentially the association object (a specialised KS in OSIMIS terms) and the RMIBManager abstract class which provides call-backs for asynchronous services offered by the RMIBAgent. While event reports are inherently asynchronous, manager to agent requests can be both: synchronous, in an RPC like fashion, or asynchronous. In the latter case linked replies could be all assembled first or passed to the specialised RMIBManager one by one. It should be noted that in the case of the synchronous API the whole application blocks until the results and/or errors are received while this is not the case with the asynchronous API. The introduction of threads or coroutines will obviate the use of the asynchronous API for reasons other than event reporting or a one-by-one delivery mechanism for linked replies.

While the RMIB infrastructure offers a much higher level facility than a raw CMIS API such as the OSIMIS MSAP one or X/Open's XOM/XMP, its nature is closely linked to that of CMIS apart from the fact that it hides the manipulation of event forwarding discriminators to effect event reporting. Though this facility is perfectly adequate for even complex managing applications as it offers the full CMIS power (scoping, filtering etc.), simpler higher-level approaches could be very useful for rapid prototyping.

One such facility is provided by the Shadow MIB SMIB) support service, which offers the abstraction of objects in local address space, "shadowing" the real managed objects handled by remote agents. The real advantages of such an approach are twofold: first, the API could be less CMIS-like for accessing the local objects since parameters such as distinguished names, scoping etc. can be just replaced by pointers in local address space. Second, the existence of images of MOs as local shadow objects can be used to cache information and optimise access to the remote agents. The caching mechanism could be controlled by local application objects, tailoring it according to the nature of the application in hand in conjunction with shared management knowledge regarding the nature of the remote MIBs. Issues related to the nature of such an API are currently investigated in the ICM project. The model and supporting C++ classes are very similar to the RMIB ones. The two models are illustrated in Figure 4.

Both the RMIB and SMIB support services are based on a compiled model while interpreted models are more suitable for quick prototyping, especially when similar mechanisms for Graphical User Interfaces are available. Such mechanisms currently exist e.g. the TCL/TK language/ widget set or the SPOKE object-oriented environment and these are used in the RACE ICM project as technologies to support GUI construction.

Combining them to a CMIS-like interpreted scripting language can lead to a very versatile infrastructure for the rapid prototyping of applications with graphical user interfaces. Such languages are currently being investigated in the ICM and other projects.

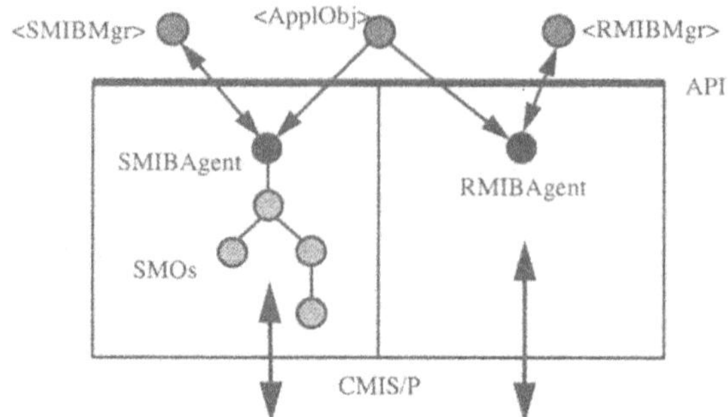

Figure 4 The Remote and Shadow MIB Manager Access Models.

7 DIRECTORY SUPPORT SERVICES AND DISTRIBUTION

Management applications need to address each other in a distributed environment. The OSI Directory Service [X.500] provides the means for storing information to make this possible. Its model structures information in an object-oriented hierarchical fashion similar to that of OSI management. This object-oriented information store can be highly distributed over physically separate entities known as Directory Service Agents (DSAs). These communicate with each other through a special protocol and requests for information a DSA does not hold can be "chained" to all the other DSAs until the information is found.

This information can be accessed through Directory User Agents (DUAs) which talk to the local domain DSA through the Directory Access Protocol (DAP) while chaining guarantees the search of the global information store.This model is very powerful and closely resembles that of OSI management. From an information modelling perspective, the latter is a superset of the X.500 one and could be used to much the same effect. It is the chaining facility though that distinguishes the two and makes X.500 more suitable as a global information store.

Directory Services can be used for application addressing in two different styles: the first resolving Application Entity Titles (AETs) to Presentation Addresses (PSAPs) in a static fashion with the second introducing dynamic "location transparency" services as in distributed systems platforms. In the first level of X.500 usage, the static information residing normally in a local database is converted into directory objects and stored in the directory. This information becomes then globally accessible while central administration and consistency maintenance become fairly simple. This approach is adequate for fairly static environments where changes to the location of applications are infrequent. For more dynamic environments where distributed applications may often be moved for convenience, resilience, replication etc., a highly flexible solution is needed. This is provided in the form of location transparency services, wherever these are appropriate. It should be noted that these services may not be appropriate for the lowest management layer (Network Element), as the same application may exist at multiple sites.

Location transparency is implemented through special directory objects holding location, state and capability information of management applications. The latter register with it at start-up time and provide information of their location and capabilities while they deregister when they exit.

Applications that wish to contact another one for which the know its logical name (AET), they contact the directory through a generic "broker" module they contain and may obtain one or more locations where this application runs. Further criteria e.g. location may be used to contact the right one. Another level of indirection can be used when it is not the name of an application known in advance but the name of a resource. A special directory information model has been devised that allows this mapping by following "pointers" i.e. Distinguished Names that provide this mapping. Complex assertions using the directory access filtering mechanism can implemented to allow the specification of a set of criteria for the service or object sought.

8 APPLICATIONS

OSIMIS is a development environment; as such it encompasses libraries providing APIs that can be used to realise applications. Some of these are supported by stand-alone programs such as the ASN.1 and GDMO compilers. Generic management applications are also provided and there are two types of those: semantic-free manager ones that may operate on any MIB without changes and gateways to other management models. OSIMIS provides a set of generic managers, graphical or command-line based, which provide the full power of CMIS and a generic application gateway between CMIS/P and the Internet SNMP.

8.1 Generic Managers

There is a class of applications which are semantic-free and these are usually referred to as MIB browsers as they allow one to move around in a management information tree, retrieve and alter attribute values, perform actions and create and delete managed objects. OSIMIS provides a MIB browser with a Graphical User Interface based on the InterViews X-Windows C++ graphical object library. This allows to perform management operations and provides also a monitoring facility. It is going to be extended with the capability of receiving event reports and of monitoring objects through event reporting. This has recently being re-engineered in TCL/TK.

OSIMIS provides also a set of programs that operate from the command line and realise the full set of CMIS operations. These may be combined together in a "management shell". There is also an event sink application that can be used to receive event reports according to specified criteria. Both the MIB browser and these command line programs owe their genericity to the generic CMIS facilities (empty local distinguished name {} for the top MIB object, the localClass facility and scoping) and the manipulation of the ANY DEFINED BY ASN.1 syntax through the table driven approach described in section 3.

8.2 The Generic CMIS/SNMP Application Gateway

The current industry standard for network element management is the Internet SNMP, which is a simplified version of the OSI CMIP. The same holds for the relevant information models; the OSI is fully object-oriented while the SNMP supports a simple remote debugging paradigm. Generic application gateways between them are possible without any semantic loss for conversion from CMIS to SNMP as the latter's operations and information model are a subset of the OSI ones. Work for standards in this area has been driven by the Network Management Forum (NMF) while the RACE ICM project contributed actively to them and also built a generic application gateway based on OSIMIS.

This work involves a translator between Internet MIBs to equivalent GDMO ones and a special back-end for the GDMO compiler which will produces run-time support for the generic gateway. That way the handling of any current or future MIBs will be possible without the need to change a single line of code. It should be added that the generic gateway works with SNMP version 1 but it will be extended to cover SNMP version 2. The current approach for the gateway is stateless but the design is such that it allows the easy introduction of stateful optimisations.

9 EPILOGUE

OSIMIS has proved the feasibility of OSI management and especially the suitability of its object-oriented concepts as the basis for higher-level abstractions which harness its power and hide its complexity. It has also shown that a management platform can be much more than a raw management protocol API together with sophisticated GUI support which is provided by most commercial offerings. In complex hierarchical management environments, object-oriented agent support similar to that of the GMS and the associated tools and functions is fundamental together with the ability to support the easy construction of proxy systems. Higher level manager support is also important to hide the complexity of CMIS services and allow the rapid but efficient systems realisation. OSIMIS has also shown that object-oriented distributed systems concepts and the protocol-based management world can coexist by combining the OSI Directory (X.500) and Management (X.700) models.

OSIMIS projects a management architecture in which OSI management is used as the unifying technology which integrates other technologies through application level gateways. The OSI management richness and expressive power guarantees no semantic loss, at least with respect to SNMP or other proprietary technologies. The emerging of the OMG CORBA distributed object-oriented framework is expected to challenge OSI management in general and platforms such as OSIMIS but there is potential for harmonious coexistence. Research work is envisaged in supporting gateways to CORBA systems and vice-versa, OSI management-based systems over CORBA, lightweight approaches to avoid the burden and size of OSI stacks through service relays, interpreted policy languages, management domains, sophisticated discovery facilities etc.

Acknowledgements

Many people have contributed to OSIMIS to be mentioned in this short space here. James Cowan of UCL though should be mentioned for the innovative design and implementation of the platform independent GDMO compiler, Thurain Tin, also of UCL, for the excellent RMIB infrastructure and Jim Reilly of VTT, Finland for the SNMP to GDMO information model translator that was produced over a week-end(!) and the first version of the metric objects. This work was carried out under the RACE ICM and NEMESYS and the ESPRIT MIDAS and PROOF projects.

10 REFERENCES

[Strou] Stroustrup B., The C++ Programming Language, Addison-Wesley, Reading, MA, 1986

[X701] ITU X.701, Information Technology - Open Systems Interconnection - Systems Management Overview, 7/91

[X722] ITU X.722, Information Technology - Structure of Management Information - Part 4: Guidelines for the Definition of Managed Objects, 8/91

[SNMP] Case J., M. Fedor, M. Schoffstall, J. Davin, A Simple Network Management Protocol (SNMP), RFC1157, 5/90

[Pav93a] Pavlou G., S. Bhatti and G. Knight, Automating the OSI to Internet Management Conversion Using an Object-Oriented Platform, IFIP Conference on LAN/MAN Management, Paris, 04/93

[OMG] Object Management Group, The Common Object Request Broker: Architecture and Specification, Document Number 91.12.1, Revision 1.1, 12/91

[ISODE] Rose M.T., J.P. Onions, C.J.Robbins, The ISO Development Environment User's Manual version 7.0, PSI Inc. / X-Tel Services Ltd., 7/91

[XOpen] X/Open, OSI-Abstract-Data Manipulation and Management Protocols Specification, 1/92

[Pav93b] Pavlou G., Implementing OSI Management, Tutorial Presented at the 3rd IFIP/IEEE ISINM, San Francisco, 4/93, UCL Research Note 94/74

[Kni91] Knight G., G. Pavlou, S. Walton, Experience of Implementing OSI Management Facilities, Integrated Network Management II, ed. I. Krishnan / W. Zimmer, pp. 259-270, North Holland, 1991

[Kni94] Knight G., S. Bhatti, L. Deri, Secure Remote Management in the ESPRIT MIDAS Project, IFIP Upper Layer Protocols, Architectures and Applications conference, Barcelona, 5/94

[Pav94] Pavlou G., T. Tin, A. Carr, High-Level Access APIs in the OSIMIS TMN Platform: Harnessing and Hiding, Towards a Pan-European Telecommunication Service Infrastructure, ed. H.J. Kugler, A. Mullery, N. Niebert, pp. 181-191, Springer Verlag, 1994

[X500] ITU X.722, Information Processing - Open Systems Interconnection - The Directory: Overview of Concepts, Models and Service, 1988

11 BIOGRAPHIES

George Pavlou received his Diploma in Electrical, Mechanical and Production Engineering from the National Technical University of Athens in 1982 and his MSc in Computer Science from University College London in 1986. He has since worked in the Computer Science department at UCL mainly as a researcher but also as a teacher. He is now a Senior Research Fellow and has been leading research efforts in the area of management for broadband networks, services and applications.

Kevin McCarthy received his B.Sc. in Mathematics and Computer Science from the University of Kent at Canterbury in 1986 and his M.Sc. in Data Communications, Networks and Distributed Systems from University College London in 1992. Since October 1992 he has been a member of the Research Staff in the Department of Computer Science, involved in research projects in the area of Directory Services and Broadband Network/Service Management.

Saleem N. Bhatti received his B.Eng.(Hons) in Electronic and Electrical Engineering in 1990 and his M.Sc. in Data Communication Networks and Distributed Systems in 1991, both from University College London. Since October 1991 he has been a member of the Research Staff in the Department of Computer Science, involved in various communications related projects. He has worked particularly on Network and Distributed Systems management.

Graham Knight graduated in Mathematics from the University of Southampton in 1969 and received his MSc in Computer Science from University College London in 1980. He has since worked in the Computer Science department at UCL as a researcher and teacher. He is now a Senior Lecturer and has led a number of research efforts in the department. These have been concerned mainly with two areas; network management and ISDN.

42

Experiences in Multi-domain Management System Development

D Lewis
Computer Science Department, University College London
Gower St., London, WC1E 6BT, U.K., tel: +44 1713911327, fax: +44 1713877050, e-mail: d.lewis@cs.ucl.ac.uk

S O'Connell and W Donnelly
Broadcom Eireann Research Ltd
Kestrel House, Clanwilliam Place, Dublin 2, Ireland, tel: +35 316761531, fax: +35 316761532, e-mail: soc@broadcom.ie, wd@broadcom.ie

L Bjerring
TeleDanmark KTAS
Teglholmsgade 1, DK-1790 Copenhagen V, Denmark, tel: +45 33993279, fax: +45 33261610, e-mail: lhb@ktas.dk

Abstract

The deregulation of the global telecommunications market is expected to lead to a large increase in the number of market players. The increasing number of value added data services available will, at the same time, produce a wide diversification of the roles of these players. Subsequently the need for open network and service management interfaces will become increasingly important. Though this subject has been addressed in some standards (e.g., ITU-T M3010) the body of implementation experience is still relatively small. The PREPARE[1] project has, since 1992, been investigating multi-party network and service management issues focusing on a multi-platform implementation over a broadband testbed. This paper reviews the problems encountered and the methodologies followed through the design and implementation cycle of the project.

Keywords

Multi-domain management, TMN, implementation methodologies, management platforms

---

[1] This work is partially sponsored by the Commission of the European Union under the project PREPARE, contract number R2004, in the RACE II programme. The views presented here do not necessarily represent those of the PREPARE consortium.

1. INTRODUCTION

The RACE II project PREPARE has investigated the development of a Virtual Private Networks (VPN) services using heterogeneous, multi-domain, multi-technology, broadband network management systems. This culminated, in December 1994, with the public demonstration of an implementation of such a system working over a broadband testbed network. The complexity of such a combined service and network management system and the large number of key players involved in the VPN service (i.e. network providers, third party service providers, customers and end-users) made it clear from the outset that a development methodology to support the full design and implementation cycles of the service was required. It is the aim of the authors to present an overview of the approach taken by PREPARE in realising this prototype VPN service, in order to provide some insight into how to address such problems of inter-domain management system development in future Integrated Broadband Communications networks.

2. PROJECT AIMS

The PREPARE project was proposed with the aim of investigating network and service management issues in the multiple bearer and value added service provider context of a future deregulated European telecommunications market. The specific example selected for implementation in PREPARE was of a Value Added Service Provider (VASP) co-operating with multiple bearer service providers to deliver a VPN service to a geographically distributed corporate customer. In order that these investigations had a realistic focus a broadband testbed network was assembled over which the VPN service would be demonstrated. This testbed consisted of several different but inter-working network technologies. Each of these sub-networks possessed its own network management system that was developed according to the principles laid down in the ITU-T Telecommunications Management Network (TMN) recommendations (ITU-T, M.3010) and using platforms supporting the OSI CMIP mechanism (ITU-T, X.700). The investigations into such multi-domain management involved the development of an architecture that allowed these separate network management systems to co-operate in providing end-to-end management services. This architecture was also developed to be conformant with the TMN reference model.

The make-up of the project consortium added a further important and realistic aspect to these investigations in that many project partners play roles that will be relevant to the realisation of future multi-domain management. The project partners and their relevant roles are:

- a network operator (KTAS), interested in integrating wide area network management with multi-domain service management based on TMN principles,
- a network equipment vendor (NKT Electronik), interested in the management of Metropolitan Area Networks (MANs) and the management of heterogeneous network inter-working,
- a customer premises network and management platform vendor (IBM: Token Ring and Netview/6000), who are interested in using their products in a multi-domain environment,

- a vendor of network management platforms (L.M. Ericsson A/S in co-operation with Broadcom Eireann Research), interested in the application of the TMOS Development Platform to value added service provision,
- researchers into advanced network management techniques (University College London, Marben and GMD-FOKUS), interested in applying their platforms to the multi-domain environment,
- researchers into multimedia applications (University College London), interested in the interactions of these applications with service and network management.

Each project partner, therefore, brought to the project their own specific interests, sometimes overlapping but often different or even contradictory. Therefore, though we were not operating in a true commercial environment, the view points of the customer, the value added service provider, the bearer service provider, the end user and management platform vendor were all genuinely represented. We can therefore assert that the methods we chose in arriving at our implementation were not purely influenced by the needs of a collaborative research project but reflect an environment in which future broadband management systems will be defined.

3. MULTI-DOMAIN MANAGEMENT SYSTEM DEVELOPMENT

The process of defining management services and information models in an environment that contains several different types of player and corresponding administrative domain has received some theoretical attention but the body of actual experience with large scale developments is still very limited. This section reviews the standardised methodologies available for management system design and their relevance to the PREPARE work. It then describes the process actually followed in PREPARE to develop a multi-domain management system.

3.1 Standardised Methodologies

The need for a methodology to support the identification and specification of the management requirements and capabilities related to the management of telecommunications networks, equipment and services is well understood by the standards and other related bodies. The main methodologies proposed to date include the ITU-T's M.3020 (ITU-T, M.3020), the Network Management Forum's Ensemble concept (Network Management Forum, 1992) and ISO's ODP framework (ITU-T, X901).

The TMN interface methodology, as defined in M.3020, forms part of the wider TMN management framework as defined in the M.3000 series of recommendations. The methodology is primarily designed to aid the specification and modelling of management functionality at any well-defined TMN interface.

Though in general the standards concentrate on the specification of generic solutions for general management problems, there is a need to tailor these solutions to solve specific management issues. The Network Management Forum group proposes the use of the Ensemble concept as a solution. The Ensemble approach is to select from the pool of standards outputs a solution appropriate the management problem and to enhance these with other support items (management information libraries and profiles) to produce maximum effectiveness. An ensemble template is provided in OMNIPoint 1 recommendation.

The ODP framework provides five key viewpoints and corresponding languages to support the specification of the problem domain. These are the enterprise, information, computation, engineering and technology viewpoints.

The major difference between the Ensemble and TMN methodology process is the scope of the two methods. The scope of the Ensemble is more focused in that ensembles are defined for specific management problems whereas M.3020 aims more at generic solutions, being intended more for use by standardisers rather than customer implementors. The Ensemble concept also defines conformance and testing requirements. The ODP framework is complementary to both methodologies in that the five viewpoints may be applied in both cases to enhance their approaches.

The major limitations of all these approaches in the case of PREPARE are that they either do not have sufficient scope or, in the case of ODP, are too general and the mapping onto TMN is not well defined. Furthermore the PREPARE project required a methodology that covered the service specification, design and implementation phases of the demonstrator work, whereas the scope of these methodologies only covers part of the specification and design process. Finally, and significantly for PREPARE, the three approaches are designed implicitly more to support single system design. None of the methodologies provide sufficient specific support for designing and implementing co-operative, multi-domain management systems. These facts resulted in no standard methodology being adopted for PREPARE. This was compounded by the fact that the pressure to provide an implemented result over-rode the desire to follow methodologies that were at that time immature and therefore not well understood by the project members. The project required instead that a mixture of the three approaches be taken. In effect it was realised that a pragmatic approach was necessary that would be primarily driven by the experience accumulated by the project members as a result of their involvement in similar work in other projects (e.g., RACE 1 Research Program). This approach is detailed in the following section.

3.2 The PREPARE Methodology

From the outset, the project followed a plan consisting of the following stages:

1. The definition of the management scenarios we wished to demonstrate, together with the supporting TMN architecture, management service definition and information models. This was conducted through 1992.
2. The implementation of the intra-domain systems required to manage the individual sub-networks making up the demonstrator testbed and the implementation and integration planning for the inter-domain management components, conducted through 1993.
3. The testing of the inter-domain components and their integration with the intra-domain management components and the actual testbed network. This work culminated in a public demonstration event in December 1994.

The broadband testbed used for the VPN management service consisted of an ATM WAN, ATM multiplexers, a DQDB MAN, a Token Ring LAN, and multimedia workstations. The enterprise context in which the VPN service was assumed to operate dictated that the WAN and MAN were separate public networks while the ATM multiplexers and Token LANs were Customer Premises Networks (CPNs). Both the public networks and the CPNs had their own separate management Operation Systems (OSs). To provide the VPN management service a separate third party Value Added Service Provider OS was introduced. This coordinated VPN resources management via X-interfaces to the public network OSs and provided customer access and control to the VPN service via X-interfaces to the CPN OSs (see figure 1).

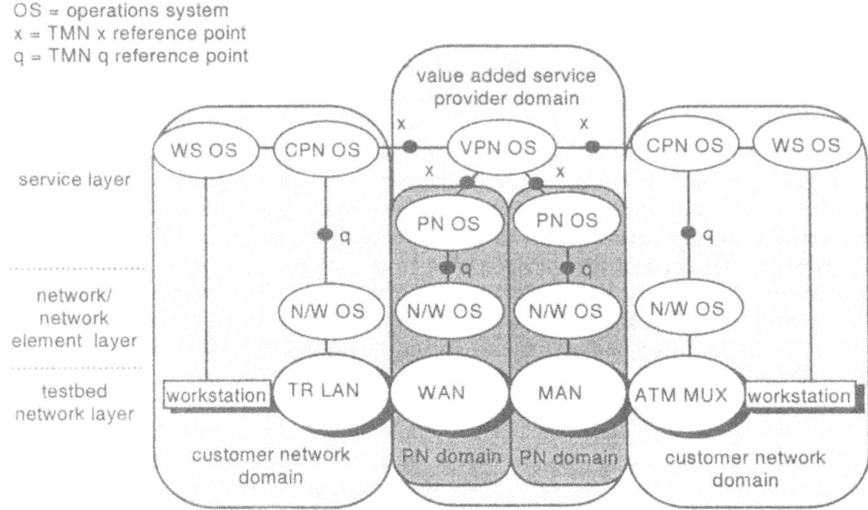

Figure 1.: PREPARE TMN Architecture

The fact that a different project partner was to implement the management systems for each of the different public networks and CPNs emphasised from the beginning of the project the administrative and human communication problems encountered in attempting to develop multi-domain management systems. This led to an emphasis on the X-interface where the different organisation's management systems had to interact.

Against this background the first stage of the work proceeded with four different groups being formed to generate; management scenario definitions, a TMN based management architecture, management service definitions and management information model definitions. The objectives of these groups were respectively as follows:

- The aim of the scenarios group was to produce a set of scenarios that would detail what would be demonstrated over the testbed network. Due to the large number of participants, components and requirements involved, these scenarios were essential in order to focus the work onto a manageable subset of demonstrable operations while at the same time presenting a coherent and realistic description of what was to be demonstrated.
- The architecture group had the task of interpreting the TMN recommendations in order to produce an implementable framework that specified how the components in the different domains should be interfaced to each other in order to provide end-to-end services.

- The management services group had to define a set of services that operated between the different management domains in accordance to the Abstract Service Definition Convention recommendation (ITU-T, X.407).
- The work required from the information modelling group consisted of defining the information models required by the various OSs that were involved in inter-domain relationships, according to the Guidelines for the Definition Managed Objects recommendation (ITU-T, X.722).

Due to restrictions of time and man-power these group's activities were in general conducted in parallel. At the beginning of 1993 a review was conducted of the work performed in the first stage and its suitability for supporting the implementation work. The output from the scenarios group described the roles of the human users and organisations involved in the VPN service as well as the motivations for the operations performed. This was supplemented by documentation of the commercial service that the VPN provider should provide to its customers. The architecture group identified all the management components required for the intended end-to-end VPN services and the different interfaces required within a TMN framework. It soon became apparent that the scenarios contributed greatly to everyone's understanding of the problem while the architecture was generally agreed upon as being suitable for the implementation of the VPN service. However it was also recognised that the outputs from the management services and information modelling groups suffered in many respects. Firstly these two sets of output were not mutually consistent, nor were they totally aligned with the output of the scenarios and architecture groups. Co-ordinating this work while running the groups in parallel had proved too complex a task given the man-power available. Secondly it was felt that, given the goal of demonstrating the scenarios; the service and information model specification were not complete and did not contain the level of detail required by the implementors. For example, although the detailed GDMO specification of all the agents in the architecture was essential, the managed object (MO) behaviour descriptions could not accurately convey the functionality of the operation systems which needed to be supported. Furthermore it was felt that a complete ASDC description of the management services would still require much additional integration with the information model to satisfy the implementors.

A path was therefore chosen which involved abandoning the further definition of management services and concentrating on refining the scenarios. The existing scenarios were therefore refined from a level where they described the player's roles and their relationships, to a state where the same scenarios were described in terms of OSs with detailed descriptions of the management information flowing between them. Adopting this technique, a full GDMO specification for the whole inter-domain information model was quickly arrived at. This approach also had the intrinsic advantages of ensuring that all information modelling was directly focused on the desired implementation areas and provided an informal but relatively brief description of the functionality associated with the information model.

The entire information model for all inter-domain components was maintained in a single document referred to as the Implementor's Hand Book (IHB). It was apparent that although the aim at this stage of the design work was to arrive at a stable version of the information model, there would inevitably be changes required to the IHB as our understanding of the problem grew. For this reason the IHB was maintained as a living document. This task was made considerably easier with the help of Damocles a GDMO parsing and checking tool developed by GMD-FOKUS. This was used to check the IHB for GDMO syntax errors, open references but more importantly it checked for consistency and completeness throughout the information model. This was especially useful considering the number of partners involved in

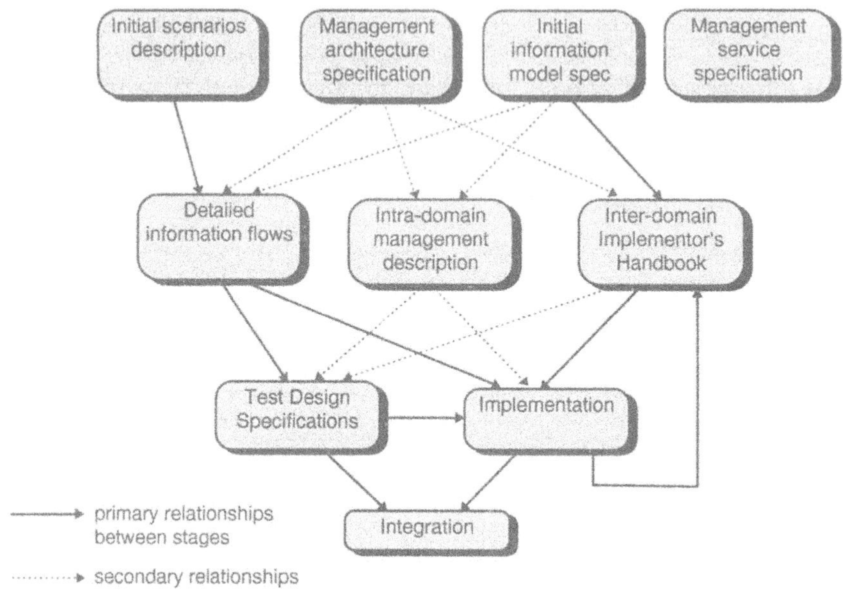

Figure 2.: Overview of Inter-domain Management System Development Methodology adopted in PREPARE

the writing of this document. A mechanism for requesting updates or modifications to the information model was also adopted since changes inevitably effected more than one partner's implementation work.

The IHB did not address intra-domain issues. However since each of the partners involved in intra-domain component implementation was represented during the scenario refinement and inter-domain information modelling, this work could be performed separately. The more difficult inter-domain modelling therefore became the principle group activity in the project, while the intra-domain definitions and implementations were the responsibility of individual partners.

As the IHB became stable and the inter-domain implementation began, the planning for integrating of the various hardware and software components commenced. This was conducted broadly following the IEEE standard 829-1983 (IEEE, 1983) which involved the generation of Test Design Specifications (TDSs) for all tests that would involve components from more than one partner. When this was performed for inter-domain management software components some interesting effects were observed. Firstly the refined scenarios proved to be ideal templates for defining the interactions that should be tested, ensuring once again that the work performed directly supported the final aims of the project. Secondly, the TDSs were written to a level of detail that defined the actual CMIS primitives that should be exchanged between the OSs and the syntactical information required. This process of writing the TDS to such a level of detail provided much valuable insight for the implementors, in that it raised many issues that had not yet been recognised and allowed these problems to be resolved before the implementation work had progressed too far.

To summarise therefore, the method followed in PREPARE was focused on achieving a demonstrable result in a limited time frame. It was heavily influenced by its multi-domain context and the requirement to co-ordinate the different partners involved in the work. Figure 2 summarises the approach adopted.

4. IMPLEMENTATION PLATFORMS

In addition to the development methodology, another key factor in management system design is the choice of platform. Due to a combination of individual partner's interests in this area and the large monetary investment often required in network management platforms, no single platform was adopted by the project. Instead each partner was free to select one, provided the platform was able to support (PREPARE, 1992): a Q3 and X TMN interface, the development of manager and agent management applications and the implementation of custom managed object classes.

The following platforms were used in the PREPARE testbed:

OSI Management Information Service (OSIMIS): This was developed by the University College London (UCL, 1993) as a result of participation in a number of EU funded projects from the RACE and ESPRIT research programs. An object oriented API is provided for implementing management applications working in either the agent or manager roles. Within PREPARE, OSIMIS has been used to implement the Inter-Domain Management Information Service (IDMIS), (RACE, 1993- H430), Q-adapters for nodes of the ATM WAN and ATM multiplexer and the OS that provided network management facilities and a service level X-interface for the DQDB MAN.

Netview/6000: The management information associated with the Token Ring is made available to other OSs via IBM's NetView/6000 management system.

OpenView: Hewlett-Packard's OpenView CMIP development environment was used to develop the OS that managed the ATM multiplexer based CPNs at the VPN service level.

Telecommunication Management and Operations Support (TMOS): This platform developed by L.M. Ericsson was used by L.M. Ericsson and Broadcom Eireann Research to develop the VASP OS and its operator's user interface.

In order to test and adjust the various platforms so that they could interchange management data using CMIP, a test MO (based on the Network Management Forum test object) was initially used. This MO contained the basic GDMO structure of a generic managed object (i.e., packages, notifications, attributes, etc.) so that when implemented over the various platforms the interchange of its management data could be tested and any problems identified.

A number of different platform related problems were identified while implementing this test managed object and during the subsequent development of the different OSs. These included the variation in the use of name bindings varied with each platform. For example, the information model within the TMOS platform starts with the network object being at the top of the containment tree whereas in the OSIMIS platform the standardised system MO is at the top of the containment tree. To overcome this, a translation function was necessary.

5. OPEN ISSUES

The experience of the PREPARE project in designing and implementing its VPN services reinforces the fact that realising inter-domain services is an extremely complex issue and requires the support of a methodology to integrate the service specification, design and implementation processes. The PREPARE approach provides a window into the type of issues that need to be addressed in inter-domain management system development and some of these are outlined below.

5.1 Inter-domain Management and TMN

Where practical the project has attempted to base its approach on the work of the standards bodies. In particular the project's approach to defining an implementation architecture to support its design and implementation work is mainly based on the TMN architectural framework. The main conclusion of the project was that the TMN framework could support the design of inter-domain service management systems. However, having a view on the future IBC environment which emphasises dynamicity and openness it is clear that the framework requires extension to provide support for a number of issues. This includes support for a globally available information service for storing, accessing and maintaining globally relevant information. A typical example is information about service providers, their offered capabilities, contact names and addresses, and "operational" information, e.g., communications addresses of OSs, information models and other information related to shared management knowledge. The OSI Directory provides a standardised approach to implementing the required technologies (RACE, 1994- D370). An approach to using the Directory in this way is demonstrated in PREPARE with the IDMIS system. This however has implications for the TMN Architecture. A proposal to; add a Directory System Function and corresponding d-reference point to the functional architecture; add Directory Objects to the information architecture and add Directory components like Directory System Agents (DSA) and Directory Access Protocol (DAP) to the physical architecture, has been presented to ITU SG IV, (Q.23/Q.5 meeting, May 1994) and subsequent meetings. We expect it to be reflected in future versions of M.3010 (Bjerring, 1994).

5.2 Security

Security within the PREPARE VPN management framework and particularly TMN is an important issue that has not been addressed so far within the project. Generally security refers to the application of an appropriate set of logical and physical measures in order to ensure the availability, accountability, confidentiality and correctness of the management data accessible to other TMN-like systems (RACE, 1994- H211). Open Network Provisioning (ONP) is expected to be introduced by the European Public Network Operators (PNOs) by the late 90's. In technical terms, the ONP concept emphasises the need to define and adopt open non-discriminatory standardised interfaces to the underlying public network infrastructure for the provision of new value added services (Plagemann, 1993). To address this new trend in the public telecommunications industry a high degree of security is necessary to reduce the possibility of large monetary losses being suffered by customer commercial organisations, the various PNOs and service providers as a result of allowing the use of services like VPN, etc. For example, US telecommunications fraud is currently estimated to be in excess of $2.5 billion per annum (Wallish, 1994).

5.3 Use of Open Platforms

As discussed above, the realisation of inter-domain services requires that the various service developers need to support the concepts of shared management knowledge and inter-operability over open interfaces. However if a customer already possesses a management platform they will be very reluctant to implement additional applications in order to get management access to a value added service which they are buying. Instead they will require the value added service provider to provide the service management application in a format compatible with their existing platform, in much the same way that LAN and router equipment manufactures are starting to do now. This would only be viable for the value added service provider if an open API of some form was available across all platforms. This has already been addressed to an extent by X/Open with the XMP/XOM API (X/Open, 1992), however in a multi-domain environment, issues of management application interaction to provide end-to-end services and support for inter-domain security and location transparency still needs to be addressed.

6. FURTHER WORK

In 1993 the PREPARE project received additional resources to sponsor an extension of its work in 1994 and 1995. This new work has two main aims; first to extend the physical testbed from Denmark, were it is currently situated, to include ATM sites in London and Berlin (Lewis, 1994), and secondly to extend its multi-domain TMN investigation to more complex multi-player situations, including the addition of multimedia teleservices and their management requirements. As part of the latter aim the project must go through another cycle of specification of demonstrator goals, architecture definition, information modelling, implementation and integration. This has to be performed in about half the time of the previous cycle and may prove more problematic since there are potentially more inter-domain relationships in the anticipated architecture. However the experience gained by project members in the work described in this paper should greatly mitigate these problems and has already led to a work-plan that follows the same scenarios centred development path. This work will give us an opportunity to investigate the integration of both the existing management systems into the ones being developed. This will be done both through he reuse of the VPN management system already developed, and also through the inclusion of more of the standardised information models that are now available.

7. CONCLUSION

The experience of the PREPARE project is that the development of multi-domain management systems is a very complex task made mainly so by the presence in the development process of more than one party. It was found that though some standardised methodologies exist, none at this time address the complexity of multi-domain systems, nor do they address all the stages of the development cycle. PREPARE has therefore developed its own pragmatic approach to the development of such systems. This approach is centred around the establishment of a set of scenarios that embody the core aims of the system being developed and therefore ensure that all work remains explicitly focused on those aims. By documenting scenarios at a high level initially, any conflicts between the requirements of different parties may be identified and resolved early on in the development process. These scenarios are then refined into detailed information flows as part of the information modelling process and finally they provide the basis for integration and test documents. PREPARE has found this method well suited to developing, with limited resources, multi-domain management systems that satisfy core requirements. The project will reuse this method in a new cycle of multi-domain management system development it is currently embarked upon.

REFERENCES

ITU-T Recommendation X.407, Abstract Service Definition Convention.

Bjerring, L.H., Tschichholz, M. (1994), *Requirements of Inter-Domain Management and their Implications for TMN Architecture and Implementation*, Proc. of 2nd RACE IS&N Conference, Aachen.

RACE Common Functional Specification D370 (1994), *X.500 Directory Support for IBC Environment* (Draft).

PREPARE (1992), *D2.2A Open Architecture and Interface Specification*, CEC Deliverable No. 2004/IBM/WP2/DS/B/002/b1.

ITU-T Recommendation X.722, *Guidelines for the Definition of Managed Objects*

RACE Common Functional Specification H221 (1994), *Security of Service Management Specification.*

RACE Common Functional Specification H430 (1993), *The Inter-Domain Management Information Service (IDMIS).*

IEEE (1983), *Standard for Software Test Documentation*, IEEE Std. 829.

Lewis, D., Kirstein, P. (1994), *A Testbed for the Investigation of Multimedia Services and Teleservice Management*, Proceedings of the 3rd International Conference on Broadband Islands.

ITU-T Recommendation M.3010 (1992), *Principles for a TMN.*

ITU-T Recommendation M.3020, *TMN Interface Specification Methodology.*

Network Management, Forum (1992), *OMNIPoint1 Specifications and Technical Reports*, Book 1 & 2.

ITU-T, Draft Recommendation X.901 (1993), ISO/IEC JTC 1/ SC 21/ N 7053, *Basic Reference Model of Open Distributed Processing* - Part 1 Overview and Guide to Use, December.

Plagemann, S. (1993), *Impact of Open Network Provisioning ONP on TMN*, Proceedings of the RACE IS&N Conference, Paris.

UCL (1993), *The OSI Management Information Service*, Version 1.0, for system version 3.0, University College London.

Wallish, P. (1994), *Wire Pirates,* Scientific American.

ITU-T, X.700, *OSI Systems Management*, X.700- Series Recommendations, OSI Systems Management.

X/Open (1992), OSI-Abstract-Data Manipulation and Management Protocols Specification,

BIOGRAPHY

David Lewis graduated in electronic engineering from the University of Southampton in 1987 and worked as an electronic design engineer for two years. In 1990 he gained a Masters in computer science from University College London where he subsequently stayed as a research fellow in the Computer Science Department. Here he has worked on primary rate ISDN hardware development and Internet usage analysis before joining the PREPARE project in which he has worked both in B-ISDN testbed definition, integration of multimedia applications and development and implementation of inter-domain management systems. He is currently conducting a part-time Ph.D. on the management of services in an open service market environment.

Sean O'Connell qualified in 1991, with an honours degree in Computer Science from the University College Dublin (UCD) following the completion of his scholarship funded final year project in secure E-Mail. He took up a research position with Teltech Ireland at UCD where he spent two years working on various security related projects including secure FTAM, the Security Management Centre, the AIM Project SEISMED and his masters degree. He left UCD in September '93 to join Broadcom Eireann Research where he is currently working on PREPARE and related security projects. His main areas of interest include cryptography, open systems security, OSI management, TMN and ATM technology.

Willie Donnelly graduated in 1984 from Dublin Institute of Technology with an honours degree in Applied Sciences (Physics and Mathematics). In 1988 received a Ph.D. in Particle Physics from University College Dublin. From 1988 to 1990 he worked with the design and implementation of Industrial control and Monitoring systems. In 1990 he joined Broadcom Eireann Research is currently the group leader in the Network Management group and the project manager for the Broadcom team in PREPARE. He also active in the management aspects of a number of Eurescom projects (European PNO organisation). His main area of interest is the application of TMN to support ATM network management.

Lennart H. Bjerring graduated in 1987 as electronics engineer in Denmark. Since then he has been working for TeleDanmark KTAS partly in Systems Technology, partly in R&D. His main work area has been network management systems specification, implementation and operations in the Danish PSPDN, and, in recent years, participation in pan-European telecommunications management related projects. He joined the PREPARE project in 1992, working mainly on TMN-based inter-domain management architecture definition, information modeling, and definition of IBC-based Virtual Private Network (VPN) services.

43

Designing a distributed management framework– An implementer's perspective

M.FLAUW - P.JARDIN
CEM Technical Office
DIGITAL EQUIPMENT CORPORATION
SOPHIA ANTIPOLIS - 06901 - FRANCE
Tel: +33 92 95 54 26 Fax: +33 92 95 58 48
flauw@vbo.mts.dec.com - jardin@vbo.mts.dec.com

Abstract

The distributed organisation and topology of telecommunications networks impose management solutions which are themselves distributed. The direction of such solutions is clearly indicated by the ITU-T TMN architectural framework which is fundamentally based on an Object Oriented paradigm.

The development of distributed solutions poses real technical challenges to vendors. This paper addresses the issues that an implementer of management solutions must consider. It discusses the perceived requirements and trade-offs that have to be faced in the design of a distributed framework.

The essence of DIGITAL's distributed Telecommunications Management Information Platform (TeMIP) is presented.

Keywords

Distributed management, Object-oriented framework, TMN, implementation

1. INTRODUCTION

The size and complexity of telecommunications networks and their continuing evolution have created interesting challenges for network managers and network management solution developers. The business and political environment creates tremendous pressure on network operators and telecommunications service providers towards delivering maximum quality of service at minimum cost. This has generated requirements for integrated management environments in which the various Operations Systems involved will exchange faster and safer critical, quasi real time, information.

In such a context a number of alternatives are offered. Pragmatism forces us to recognise that no single overarching approach can be adopted and even in standards based approaches (e.g. OSF DME), a number of different APIs, modelling languages and messaging systems are proposed. The interworking of these different technologies translates into a number of gateways.

In this complex and evolutionary situation, DIGITAL with its distributed Telecommunications Management Information Platform (TeMIP), has taken both a well architected and pragmatic approach. TeMIP is an evolution of the DECmcc framework that is specifically designed for managing telecommunications networks.

This paper offers an implementer's viewpoint which shows the constraints and often conflicting requirements that a management framework implementer must face.

2. AN OBJECT-ORIENTED FRAMEWORK

2.1 Object orientation for network management

Significant research and development are being directed to the area of computing enterprise management. Its importance manifests itself in a number of conferences and publications. The pure Object Orientation initially defined and used for programming languages has been (sometimes loosely) adapted for defining management solutions. It has been researched by consortia such as RACE, TINA-C [1], formalised through standardisation activities (ISO [2], ITU-T [3], X/Open [4], ETSI [5], T1 [6]) and realised through implementations such as the DIGITAL TeMIP framework presented in this paper.

The object-oriented analysis methodologies proposed for problem analysis and design [7] have inspired the development of management solutions. In particular, a specific methodology has been defined for these contexts (ITU-T M.3020 [8]). The approach was retained for TMN as it contains essential characteristics such as:

- the ability to define generic specifications that can be adapted to local situations with the concepts of inheritance and polymorphism,
- the ability to hide implementation details by decoupling specification from implementation aspects and focusing on object interfaces (a concept of great interest for the integration of legacy systems).
- the abililty to present different levels of abstraction. This methodology provides a 'zooming effect' allowing to gradually focus on more and more detailed aspects

Defining a management solution in an object-oriented fashion imposes the recognition of the natural dichotomy that prevails in this context:

- The managed resources are generally physically dissociated from the managing systems. The OSI management [2] and the SNMP [9] models have formalised this by introducing the concepts of *Manager* and *Agent*. An object- oriented approach will consist of modelling the managed resources as objects and making them visible via agents.
- On the Manager side, the managing application(s) may themselves be modelled and implemented as objects. They may be distributed as suggested by the ODP approach [10] as a set of interactive objects. These application objects may be very different in nature, e.g. computing components, database servers, user interfaces, communication servers, etc.

Consequently , as depicted in Figure 1-b, a management solution can be designed as the interaction between a number of fairly different objects. TeMIP is a globally object-oriented framework in which all these object classes are modelled under the single modelling approach defined by EMA (DIGITAL's Enterprise Management Architecture [11],[12]). Each class is implemented via one or several '*Management Modules*' (MMs).

The implementation of these classes as generic re-entrant MMs gives a set of building blocks that may gradually be loaded ('*enrolled*') into the framework. An idle system is simply a juxtaposition of classes/ modules that may potentially communicate with each other. An application actually becomes alive at run time when the relevant classes begin to interwork by invoking each others' services via an *Object Request Broker* (ORB) as depicted in Figure 1 -a.

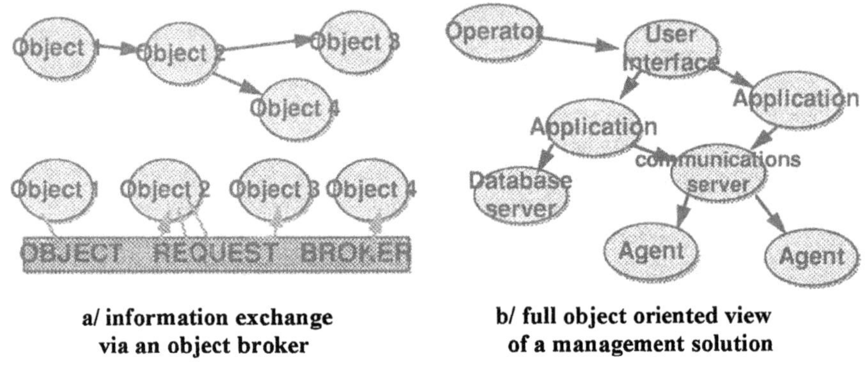

a/ information exchange via an object broker

b/ full object oriented view of a management solution

Figure 1: Co-operating Objects.

A TeMIP based management solution is a collaborative organisation of class instances. In a monolithic implementation each class is instantiated only once, while in a distributed implementation certain classes may be replicated on the various nodes. Implementation details are further discussed in section 5.

2.2 Implementing objects as 'management modules'

One of the fundamental principles of the TeMIP architecture is that each Object (implemented by a Management Module) supports three types of interface: a '*service*' interface which groups the directives used to access to the methods of each object, a '*client*' interface which the object may invoke to access the services of other objects and a '*management*' interface which groups the directives used to access to specific methods dedicated to the management of the object itself (i.e. the Management Module. This approach, which is depicted in Figure 2, is under consideration by TINA-C ([1], [13]).

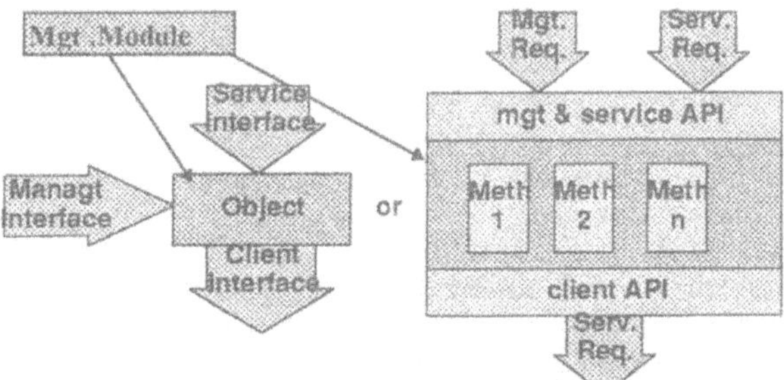

Figure 2: Objects Implemented as management modules.

The TeMIP architecture supports a common object specification language to specify the object interfaces and a common Application Programming Interface (API) through which any object gives access to the methods it supports (service or management methods) or accesses the services of any other known objects. This API, as depicted in Figure 3, is actually the Dynamic Invocation Interface of the ORB that finds the location of the invoked service based on the supplied parameters and dispatches the request accordingly. This Object Request Broker is called the *Information Manager* in TeMIP terminology.

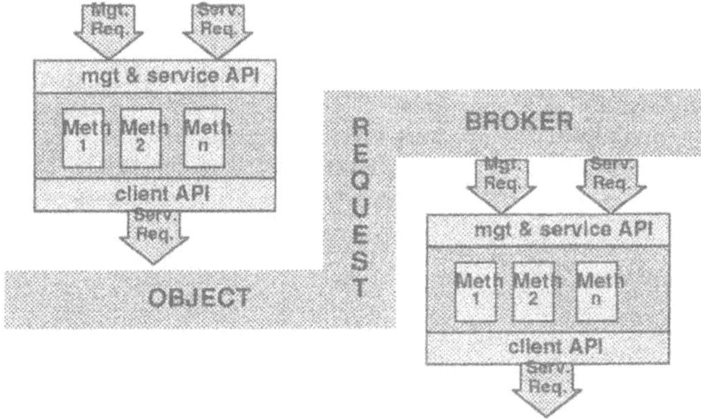

Figure 3: Inter Objects communication.

3. REQUIREMENTS FOR DISTRIBUTED MANAGEMENT

The development of an object oriented framework is conceptually satisfying but is only useful for the management of telecommunications networks if it fulfils the constraining requirements of such an environment. The essential difficulty in developing integrated management solutions lies in the accumulation of stringent functional and non functional requirements which will influence the design and implementation strategies such as:

- Geographical span: The size and geographical span of these networks impose the development of solutions that allow the functions to be partitioned and located as close as possible to the systems they manage. The management of telecommunications networks is generally partitioned and hierarchical. The management framework must be flexible enough to cope with various topologies.

- Magnitude: The management solution allows a large number of users (several tens) to monitor and control a very large number of resources (hundreds of thousands of object instances).
- Scalability: The network size, configuration and technologies keep changing. The management systems must be able to track and support these evolutions. The introduction of newer technologies must be possible in a stepped approach and without disrupting the service.
- Reliability: The deployment of distributed solutions with some replicated components should provide a form of network fault-tolerance that hardware fault-tolerant systems alone cannot satisfy.
- Openness and ability to integrate legacy systems: The continuing evolution of technologies is accompanied by their long duration, resulting in very heterogeneous environments. Openness can therefore be interpreted as the ability to adapt to an heterogeneous environment with a strong preference for standard-based solutions when applicable.
- Access control/Security: The retained trust model implies that the system is protected at the boundary i.e. all security checks are done at User Interface or Gateway level. The distributed solution must be protected against malicious and erroneous users, while the intra network protection may be reduced.
- Performance: it is expected that a distributed solution will bring obvious benefits in terms of load distribution and throughput. It is realised that, by their nature, distributed topologies may entail slight degradations in terms of response times (due to hopping). This impact must be minimised.
- Manageability: Distributing a system implies adding an additional degree of complexity. This must be taken care of at the management level.

4. THE APPROACH TO DISTRIBUTION

Distributing the management solution is essential. A number of techniques may be used each with some advantages and some inconveniences. Ideally, one unified approach should be adopted when the objective is software reusability and performance. In practice, the existence of legacy systems, the multivendor environment and the lack of consensus on any particular solution have led to a number of often overlapping proposals. Since one of the prime requirements is the ability to integrate solution components, this situation has created an unfortunate potential for the proliferation of gateways.

4.1 The ideal situation: The universal 'interoperable interface'

Achieving maximum integration and reaching a good level of performance, may be obtained by adopting an universal approach based on a common modelling technique and a minimum set of 'reference points' which translate into well-defined interfaces.

The issue has been identified in the TMN architectural model (M.3010 [3] and the NMF architecture [14]) as one of defining '*interoperable*' interfaces between co-operating components.

The ideal interoperable interface is object oriented. It should be topology independant (WAN or LAN based), compact (support of wildcarded operations), flexible (support solicited/sync. and unsolicited/async messages), efficient (support of atomic requests), secure, etc.

4.2 The actual situation: A versatile integration framework

Despite its obvious merits, the use of one single unifying global architecture can no longer be realistically considered in the TMN context. For historical reasons and diversified requirements, the ideal interface was never really agreed at the standard level. Instead, several variants emerged both at modelling level (OSI GDMO [15], SNMP SMI [9] or CORBA IDL [17]) and at stack level (CMIP, SNMP, RPC over OSI or IP). The support of multiple legacy systems additionally imposed a range of proprietary protocols, and thus the logical conclusion was to abandon the idea of a universal interoperable interface.

Some consortia such as the NMF [16] are proposing a series of options that leave the solution designers to make their choice based on environmental constraints and operational objectives. It endorses the OSF DME model ([18],[19]) which decouples the intra/inter application aspects (DME framework) from the manager-agent interface (Network Management Option) based on:

- The CORBA [17] or RPC models [10] which have been designed for handling synchronous type requests. They neither fully support complex interactions e.g. with atomic semantics nor, for the time being, provide satisfactory support for unsolicited information (event notifications).
- The manager-agent models ([2],[9]) which reflect the fact that management operations are fundamentally asymmetrical. This presents some drawbacks when two systems need to interwork as peers [14].

The solution designer will actually tend to organise his solution as the co-operation of '*technology or integration islands*', each of which offers a high level of internal homogeneity and consistency. The technology provider will have to offer a well architected integration framework that allows the interworking of these islands via a series of gateway mechanisms.

4.3 Gateways issues

Frameworks must implement multiple gateways and proxy type mechanisms in order to support the various approaches actually used in the marketplace. In some cases, the retained approaches are functionally overlapping and present the unfortunate characteristic of having adopted different modelling languages and underlying protocol stacks.

Integrating the various approaches requires defining non trivial mapping mechanisms such as those defined to integrate CMIP and SNMP ([20], [21]), or CMIP and CORBA [22]. In a similar vein, the integration of legacy systems, most of which are currently controlled and monitored via formatted ASCII message sets, imposes the nontrivial exercise of developing mapping functions such as the TMN 'Q adaptor' ([3], [23]).

These mechanisms typically imply stack interoperation, syntactic and semantic mappings and specification language translations. This proliferation of gateways leads to a more complex support and management of the solution and slower information transfers.

5. DISTRIBUTING THE TEMIP FRAMEWORK

In the context where a management solution becomes a patchwork of integrated islands coupled by various gateways, an integration paradigm must be retained for each island. Various design centers may be retained.

The visual integrator approach allows supporting various technologies in parallel but provides minimum to no application interworking capabilities. The tightly coupled mode of integration based on an unifying but constraining achitecture maximises applications synergy and reusability [24]. Both approaches have been retained in TeMIP.

The tightly coupled integration based on the EMA architecture (see Section 2), defines a Management Modules (MMs) hierarchy with *Access Modules* (AMs), which provide the connectivity with the agents/managed resources, *Function Modules* (FMs) which provide value added services and *Presentation Modules* (PMs) which interface with the users (human beings or applications).

Whatever approach is retained, the magnitude of such networks imposes the partitioning and maintenance of different contexts within the integration islands. This section discusses how the EMA based islands can be distributed in response to the magnitude and scalability requirements (see also [25]).

A monolithic TeMIP application is called a *director*. In a distributed topology, each node becomes a director as depicted in Figure 5.

5.1 Remote Call request interface

Distributing the framework entails a remote interobject interface ([26],[27]). This interface is referred to as the 'call request' interface. It actually offers the services of an ORB (the Information Manager) as described in section 2. It is a dynamic interface in the sense that the call is formally request- independent (the same call procedure is always invoked) and fully qualified by the call parameters which specify the operation (V/verb), the class instance (E/entity) and the operation parameters (P). The key of identifying a given service is the tuple [V,E,P].

The Information Manager processes the request arguments and acts as a client to establish a RPC binding with the appropriate server. Full location transparency is obtained by identifying the director associated with the target instance (and its supporting MM) via the Distributed Name Services of the framework. This is depicted in Figure 4. The reader who is familiar with the OSF DME architecture will realise that this approach is conceptually the same as that defined for the DME.

The dispatching mechanism is based on dispatch tables that are common in all directors. These dispatch tables are automatically updated in all directors when the solution is augmented with new objects i.e. when MMs are extended to offer new services or when new MMs are '*enrolled*' in the framework. The information in the dispatch tables is used to efficiently compute the management module entry point that provides the requested service.

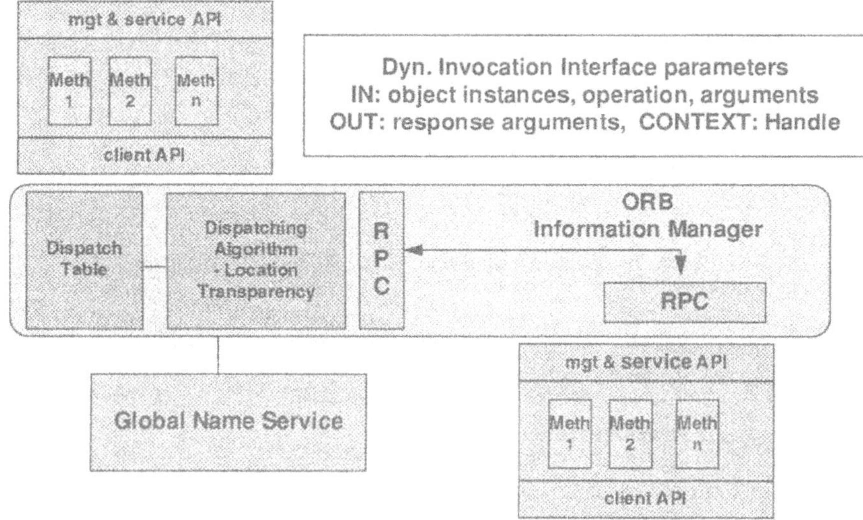

Figure 4: Inside the Object Request Broker.

5.2 Domain and Entity Access Distribution

In the TeMIP architecture, the 'call request' interface is offered in two variants: a *'call function'* interface used to access value-added services such as those provided by the FMs and a *'call access'* interface used to access the managed object services through the AMs. These two interfaces may both be remotely located for different reasons and will allow the system/solution designer to implement a distributed topology that best serves his operational objectives and environmental constraints.

The use of a remote 'call access' allows the AMs to be located as close as possible to the managed objects/agents. This may be imposed by some technical constraints such as the colocation of the AM software with a non distributed data store or by the type of protocol used between the AM and the Agent. *Entity access distribution* is supported by associating a target object instance with the director that supports its access mechanism in the global name space.

The remote 'call function' is used for load sharing purposes. It allows partitioning work and data, using dynamic grouping criteria realised as *'domains'*. The concept of domain has been designed as a dynamic user-defined grouping of object instances. It can be used to reflect a user's sphere of interest or management policies.

This powerful feature is largely used within TeMIP for historical data collection, alarms monitoring and displaying of information. Basing distribution on domains means that client/server type configurations can be built with optimum and flexible (dynamic) partitioning of the workload. For example, a given FM may be duplicated on different directors and be assigned responsibility for the work that pertains to a certain domain. *Domain based distribution* is supported by associating in the global name space a given director with the domain(s) it is in charge of.

The two forms of distribution are illustrated by Figure 5.

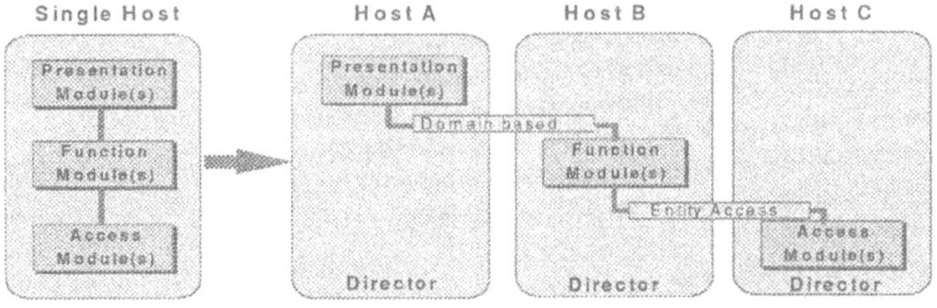

Figure 5: Domain based and Entity Access Distribution.

The Object Request Broker determines in real time where is the target module located by identifying the remote director associated with the target object instance or the domain for which the call request is issued.

5.3 Data aspects considerations

Two systems can only communicate when they share a common interpretation of the entities they are communicating about. As depicted in Figure 6 this knowledge is generally represented as data which can be subdivided into:

- Metadata representing the classes static information: TeMIP uses a common dictionary model to represent all its metadata. A copy of the dictionary is replicated in each director.
- Configuration data: A distributed system solution must ensure that all managers can reference and access a given object instance. A common network-wide instance name repository is necessary for the ubiquitous and persistent registration of the instantiated classes. TeMIP employs a Distributed Directory Service which provides a global name space.
- Private data: Each object may locally store private data. This data may be made public by the MM via its service interface. TeMIP provides an object-oriented data storage, known as the *Management Information Repository* (MIR). The use of this data storage mechanism is an implementation choice: a designer may decide to maintain the data in non object oriented public files. For example, TeMIP trouble tickets are stored in a relational database and accessible via SQLnet. Note that policy related data (domains, alarm rules, operation contexts etc..) which are modelled as objects fall into this category.
- System configuration data: This data represents the topology of the distributed manager. The MM instance data is maintained as private data by each director and the dispatch tables are replicated in each director.

The automatic replication of global information, the possible replication of functions and the use of fault-tolerant hardware for the support of critical private data are the components of high availability solutions

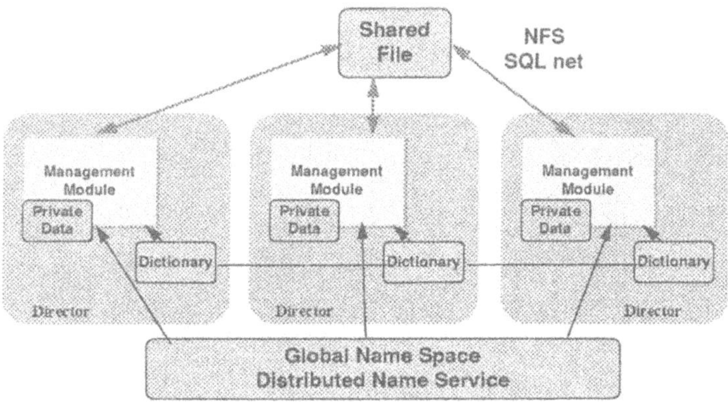

Figure 6: Handling data in a distributed topology.

5.4 Managing the manager

The management of very large networks imposes complex distributed solutions which themselves need to be managed. A distributed TMN solution, on which the stability of the target telecommunications network relies, becomes itself a network that needs to be configured and managed. This issue is well known and identified in M.3010 as the '*self management*' of the TMN.

A self- management function is quite straight-forward to implement using the TeMIP distributed architecture because of the following essential characteristics:

1. As described in section 2 and Figure 2, the design and implementation of TeMIP as an object-oriented distributed framework implies that each module is itself an object that can be managed via its management interface (see Section 2.2).
2. The concept of domains can be adequately used by the system manager (see section 5.2). The system management activity can be isolated (physically and from a security standpoint) by grouping the managed objects representing the managed directors into a dedicated domain.
3. The distributed TeMIP architecture relies on the services of a particular object, the *Framework MM,* which is in charge of the consistency and stability of each director. This MM is designed to survive system crashes, reactivate long lasting processes and re-establish inter process bindings.
4. The Framework MM also maintains a view of the connectivity with other directors that it interworks with. This is depicted in Figure 7 where the Framework MMs are labelled 'Fn'.

As depicted in Figure 8, the combination of the above features allows the easy management of the TeMIP framework by its own applications. For example, the basic TeMIP Alarm Handling function may be applied to a particular domain composed of the directors and their associated MMs to extract and collect the relevant information from the MMs themselves (considered as managed objects) and build a view of the system behaviour.

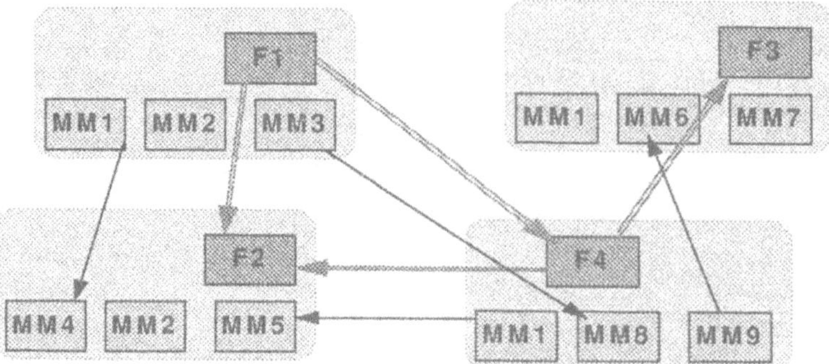

Figure 7: Inter Directors communication.

The flexibility of the framework leaves to the system manager the choice of deploying the management application on a separate director or exercising it within an existing director.

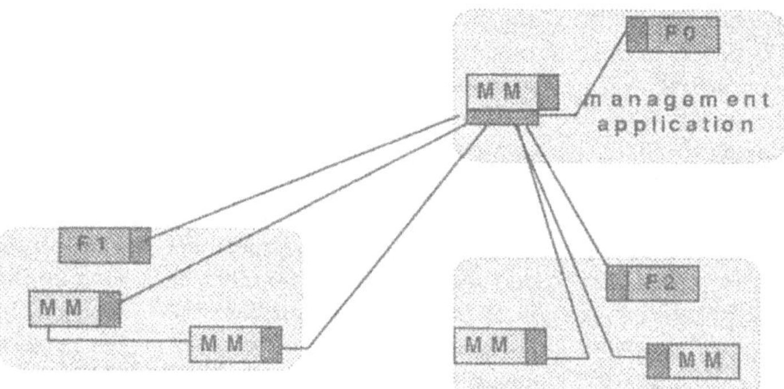

Figure 8: Managing the management solution.

5.5 Resolving trade-offs

In an ideal world an unified approach should bring significant simplification. In practice the implementation of a distributed framework that fulfils all previously identified requirements can only be complex.

The solution must be available now and nearly as cheap as commodity software. It must support all the latest standards including those still in development and perform well. It must also be user friendly and fully reliable. It must be easy to configure and deploy, scalable and flexible to support network evolution. It should be transparent for application developers, and so on. These are essentially conflicting requirements.

We decided to retain system performance as the prime objective and the driving principle for resolving a number of trade-offs. These include, but are not limited to:

- Grouping activities (domains) to allow load sharing ,
- Supporting a high-speed event communication subsystem,
- Choking off the incoming traffic as early as possible by means of a set of distributed filters,
- Efficient dispatching mechanisms for direct connection between source and target modules. This minimises the round trip delay of the call by avoiding going through intermediate routers or end-point mappers,
- Minimising access time for critical information such as dictionary data or real time data storage,
- Grouping information with concepts such as attribute groups, attribute partitions, event partitions,and support of wild carded operations,
- Strong authorisation and audit trail mechanisms at the periphery (PM level) where time is less critical and reduced internal controls (authorisation and access control on RPC bindings only).

In order to fulfil future requirements when the enabling technologies become available, all essential APIs have been frozen so as to protect the existing applications while allowing rapid swapping of the underlying technology (e.g. dictionary, object data base, communications or name server technology).

6. A VERY FLEXIBLE APPROACH - SOME SCENARII

In summary, TeMIP fully exploits the benefits of object-orientation. The use of a common dictionary allows the development of fully data-driven modules (i.e. generic functions which do not need recoding when new classes are added). The use of a distributed name service allows to locate and access to the object instances on or via any director.

Usage and policy independance have been adopted as driving principles for the development of generic modules. No rule or algorithm that normally depend on operational objectives based on local policies are hardcoded which implies immediate code availability and reusability in various environements.

These essential characteristics allow the use of TeMIP in various scenarii:

- Remote User Interfaces (PMs) acting as client running on separate machines can access functions loctaed on a number of 'heavy weighted' servers. This will allow off hour work reorganisation that transfers responsibility to a remote system (critical situation, week ends, etc.). A variant of this scenario can be achieved by means of X-display mechanisms e.g. to support PC-based user interfaces.
- Instrumention of distributed topologies with multiple servers that allow work partitioning can be achieved via domain based distribution. It may be based on:
  - → Policies, operational objectives and skills. A given user has restricted access to the only services that correspond to his skills and job.
  - → Geographical constraints. If the network is split into several regions with a management center for each region, the domains containing the objects related to a given region can be associated with the management center of that region.
  - → Architectural choices such as those retained fro the TMN ([3], [28]).

→ Resource off-loading of functions that are CPU bounds or I/O bounds (database servers) on dedicated systems/directors.

- Developing a front-end approach in which some access modules and communication servers are used to concentrate agent traffic, using entity access distribution to group all entities of a given type on dedicated communications server(s) (OSI/CMIP, SNMP, ASCII etc.).

7. CONCLUSIONS

The requirements of large, complex telecommunications networks motivate the research and development of integrated management and distributed solutions. The context is essentially heterogeneous with a slow evolution towards open interfaces.

The design and implementation of distributed frameworks must consequently integrate a number of legacy components as well as emerging de jure and de facto standards which, in many cases, are incompatible.

It is most probable that the ultimate implementation of a fully integrated TMN will actually be a patchwork of internally consistent technology islands interconnected via multiple gateways. The idealistic goal of global integration based on an overarching model will probably never be reached.

Today a few technologies are capable of fulfilling the long list of stringent and sometimes conflicting requirements. DIGITAL's TeMIP is one of these. It was designed from the beginning as an integration framework: It is architected to support multiple protocols and its distributed implementation has been designed to take into account additional essential functional and non functional requirements such as manageability, security and performance.

8. REFERENCES

[1] TINA-C, 'Definition of a service Architecture - draft document', October 1993.

[2] ISO 10040/ITU-T X.701: Information Technology - Open Systems Interconnection - 'Systems Management Overview '- 1992.

[3] ITU-T Recommendation M.3010: 'Principles for a Telecommunications Management Network (TMN)' - 1992.

[4] X/Open, 'Systems Management - Managed Object Guide (XOMG)' - X/Open ref. G302 - 1993.

[5] ETSI GSM Recommendation 12.00, 'Objectives and structure of the PLMN management' - 1993.

[6] ANSI T1.210, 'Operations, Administration, Maintenance and Provisioning-Principles of Functions, Architectures and protocols for TMN Interfaces', 1992.

[7] G.Booch ,'Object oriented design with applications', Benjamin Cummings 1993.

[8] ITU-T Recommendation M.3020: 'TMN Methodology', 1992.

[9] Marshall T.Rose, 'The simple Book - An introduction to Management of TCP/IP based internets', Prentice Hall.

[10] ISO JTC1/SC21 DIS 10746-1.2.3, ITU-T Draft Recommendation X.901/2/3: 'Basic Reference Model for Open Distributed Processing - Parts 1 to 3', 1994.

[11] Digital Equipment Corp., 'Enterprise Management Architecture - General Description', Order No EK-DEMAR-GD-001, 1989.

[12] C.Strutt and M.Sylor, 'DEC's Enterprise Management Architecture', Network and distributed systems Management by M.Sloman', Addison Wesley 1994.

[13] P.Jardin, 'The TINA service component approach', TINA-C newsletter, May 1994.

[14] NMF, Forum 004 - 'Forum Architecture' - 1990.

[15] ISO 10165-4/ ITU-T X.722, OSI Management Info. Services - SMI part 4: Guidelines for the Definition of Managed Objects.

[16] NMF, Forum 026 - 'Omnipoint Integration Architecture', Issue 1, July 1994.

[17] OMG, 'The Common Object Request Broker: Architecture and specification' - OMG document No. 91.12.1, Revision V1.1, Dec. 1991.

[18] OSF, 'OSF Distributed Management Environment (DME) Architecture', Open Software Foundation, May 1992.

[19] M.Autrata and C.Strutt, 'DME Framework and Design', Network and distributed systems Management by M.Sloman', Addison Wesley 1994.

[20] P.Kalyanasundaram and A.S.Sethi, 'An application Gateway Design for OSI-Internet Management', Proceedings of the 3rd IEEE/IFIP Integrated Network Management Symposium, 1993.

[21] NMF, Forum TR107 - 'ISO/CCITT and Internet management: coexistence and interworking strategies', 1992.

[22] X/Open, 'GDMO to OMG-IDL translation algorithm', review draft, 1994.

[23] L.Aubertin and T.Bonnefoy, 'Q-adaptor function for customer administration in a switch', IEEE Network Operations and Management Symposium Proceedings February 1994.

[24] P.Jardin, 'Benefits of applying the TMN methodology to Management platforms development', DIGITAL internal paper pending publication.

[25] C.Strutt, 'Dealing with scale In an Enterprise Management Director', Proceedings of the 2nd IEEE/IFIP Integrated Network Management Symposium, 1991.

[26] C.Strutt, 'Distribution in an Enterprise Management Director', Proceedings of 3rd IEEE/IFIP Integrated Network Management symposium, 1993.

[27] Digital Equipment Corp., 'TeMIP Framework System Reference Manual' - order No AA-PD5LE-TE and AA-Q9HGA-E, November 1994.

[28] S.Aidarous, Carey Anderso et all, 'The role of the EML in Network Management', IEEE Network Operations and Management Symposium Proceedings February 1994.

9. THE AUTHORS

- Marc FLAUW is a member of the TeMIP technical office. He has driven a number of network management projects. He is one of the key architects of the TeMIP platform.

- Pierre JARDIN is a member of the TeMIP technical office. As one of the architects of the TeMIP platform, he is in charge of AD activities and participates to a number of standardisation bodies such as ITU-T SG4.3, ETSI NA4 and ETSI SMG6.

SECTION THREE

Panel

44

Can Simple Management (SNMP) Patrol the Information Highway ?

Moderator: *Edward PRING,* Advantis, U.S.A.

Panelists: *Fred BAKER,* Cisco Systems, U.S.A.
Doug BOBKO, AT&T Paradyne, U.S.A.
Bob NATALE, American Computer and Electronics, U.S.A.

The Internet is the Information Superhighway. The Internet's native language for management is the Simple Network Management Protocol. Is SNMP up to the job of managing it?

The Internet is evolving in many dimensions simultaneously, and the need for effective management is ever more critical. Traditionally a loosely managed network inter-connecting educational and scientific institutions on a "best effort" basis with no guarantees, the Internet is rapidly morphing into a mission-critical resource for businesses that offer commercial services to customers around the world. At the same time, the technological foundation underlying the Internet is expanding to accommodate unprecedented growth and to support new applications with demanding communications requirements.

SNMP and the products based upon it are evolving, too. How will they deal with the conflicting needs of security and management as private networks partitioned by firewalls become increasingly dependent upon services available only in the public Internet ? How will they scale beyond management of communications infrastructures to management of online services as distinctions between networks and systems blur ? How will they integrate with other protocols and products to enable the automation needed to handle growth and complexity and diversity as management domains increasingly overlap?

As leading members of the SNMP standardization process and developers of products based on those standards, the panelists are highly qualified to address these issues. They will offer unique insights from their professional perspectives, share their personal experiences, and field questions from the audience.

SECTION FOUR

Management Databases

45

An Active Temporal Model for Network Management Databases

Masum Z. Hasan
zmhasan@db.toronto.edu

Computer Systems Research Institute
University of Toronto
Toronto, Canada M5S 1A1

Abstract

The purpose of a network management system is to provide smooth functioning of a large heterogeneous network through monitoring and controlling of network behavior. ISO/OSI has defined six management functionalities that aid in overall management of a network: configuration, fault, performance, security, directory and accounting management. These management functionalities provide tools for overall graceful functioning of the network on both day-to-day and long-term basis. All of the functionalities entail dealing with huge volumes of data. So network management in a sense is management of data, like a DBMS is used to manage data. This is precisely our purpose in this paper to show that by viewing the network as a conceptual global database the six management functionalities can be performed in a declarative fashion through specification of management functionalities as data manipulation statements.

But to be able to do so we need a model that incorporates the unique properties of network management related data and functions. We propose a model of a database that combines and extends the features of *active* and *temporal* databases as a model for a network management database. This model of a network management database allows us to specify network management functions as *Event-Condition-Action* rules. The *event* in the rule is specified using our proposed *event specification language*.

1 Introduction

A network management (NM) system supporting all the six functionalities of configuration, fault, performance, accounting, security and directory management has to deal with huge volumes of data that are resident on the *management station(s)* and on the *managed entities* distributed over the network.

The system generally has to deal with two types of data: *static* and *dynamic*. Static data either never change or change very infrequently. The topology of the network, hardware and software network configurations, customers information etc. and the stored *history traces* of both dynamic and static data constitute the static portion of the NM-related data. The rapidly changing dynamic data embodies the current behavior of the network. A *Management Information Base (MIB)* defines the schema of the dynamic data to be collected for a particular network entity. The dynamic data distributed over the network is not visible to the network management station until they are collected. The past and present static and dynamic data form a conceptual global database which allows a management station to see the global picture of the network.

The management of a network is generally performed through two activities: *monitoring* and *controlling*. Monitoring is performed for two purposes: collection of data *traces* for current and future analysis and watching for interesting *events*. An occurrence of an event or a set of interrelated events may cause further monitoring or controlling action.

An event can be a "happening" (for example, *link down*) in the network or a pattern of data appearing in the network. The later being called a *data-pattern event* in [WSY91]. An example of a data pattern event is the crossing of a *threshold* value of a MIB variable. A data pattern event may also be defined as a more complex pattern involving more than one variables and managed entities. A set of interrelated events is called a *composite event* or *event pattern*. The interrelationship of network management events are generally *temporal*. For example, a *composite* (alert) event may be defined which occurs when the interval during which three successive server *overload* events occur is overlapped with the interval of three successive observation of large packets on the local net from unauthorized destination or the *first* crossing (up) of a rising threshold *since* the crossing (up) of a falling threshold.

Monitoring action can be performed either by asynchronous event notification (*trap*) or through periodic *polling*. Polling can be considered as an event whose occurrence at regular intervals triggers retrieval.

Both data traces and events may be stored selectively for future analysis. A *temporal database* is required for this purpose.

From the discussion above we conclude that the nature of NM data and functionalities require a model of a database that incorporates novel features of both *active* and *temporal* databases, since active databases allow one to specify events whose occurrence trigger actions and temporal databases allow one to manipulate temporal data. We propose such a model where the NM functions are specified as declarative *Event-Condition-Action* (ECA) statements. In this system, data pattern events and any other NM functions can be specified as declarative data manipulation statements. We have developed an *event specification language* (ESL) for defining composite events used in the E part of ECA. Our ESL incorporated with a temporal data manipulation language (used in the C and A part of ECA) provides us with a sophisticated declarative language for use with a database that requires active and temporal features, such as, a network management database.

The rest of the paper is organized as follows. In Section 2 we describe the features of active and temporal databases and our proposed model of a network management database. The ESL language with examples of ESL expressions and an example of an implementation of an ESL operator is discussed in Section 3. In Section 4 we provide a number of example specifications of NM functions using ECA rules. We compare our work with others in the literatures in Section 5 and conclude in Section 6.

2 Model of a Network Management Database

Before discussing our proposed model of a network management database we first discuss the features of active and temporal databases.

2.1 Active Databases

Conventional DBMSs are passive in that they manipulate data only when requests from applications are made. On the other hand, an *Active* DBMS (ADBMS) provides facilities for specifying actions or database operations to be performed automatically in response to certain events and conditions. Active behavior in an ADBMS is achieved through *Event-Condition-Action* (ECA) [MD89] rules. The rules state that when the specified event(s) occurs and the condition holds, perform the action. A condition is defined over the state of the database and its environment (for example, transaction causing the event). An action can be an arbitrary program or a database operation.

The following *primitive* events are generally supported in an ADBMS: 1) events relating to database manipulation operations, such as, retrieve, insert, delete, update; 2) transaction events; 3) absolute and relative time events; 4) in object-oriented databases method or function execution events; and 5) explicit or abstract events that are raised explicitly by the application (programmer). We also add in the list of primitive events the *data-pattern* events. A data-pattern event is specified using a database query language, for example, SQL. An event may have typed formal arguments which are bound to actual values when the event is detected. For example, the insert event may have as arguments the name of the relation and the inserted tuple.

An event is an occurrence in the database, it's environment and application's environment and can be considered as a point in time where time is modeled as a discrete sequence of points. It is desirable for many applications to react not only to current events but also to a composition or selection of events occurring at different time points. An *event algebra* allows one to specify *composite* events consisting of other primitive and composite events by means of algebra operators. A composite events expression operates on a history of events. So a composite event expression formed using algebra operators allows one to express relationship between events in the temporal dimension. The composite event happens when the specified relationship as defined by the algebra operators is detected in the event history. Petri net [GD94] or finite state machines [GJS92] can be used to model the language operators and detect composite events expressed as event expressions.

2.2 Temporal Databases

A *temporal* database in [ea93] is defined as a database that supports some aspect of time, not counting user-defined time. In other words, a TDBMS "understands" the notion of time and provides temporal operators that allow one to specify temporal queries. A temporal database contains the history of the modeled world as opposed to the traditional *snapshot* database where the past states of the database are discarded.

A temporal database contains two types of entities: *events* and *intervals*. An *event* is an instantaneous occurrence with an implicit time attribute indicating when that event occurred. Since time is generally considered as discrete, the notion of "instantaneous" requires definition. A term called *chronon* which is the shortest duration of time supported by a TDBMS, that is, a nondecomposable unit of time, is defined in [ea93]. An event occurs at any time during the chronon interval. In the network management domain we need the support for multiple choronons associated with each event entity or relation. The need for the support of multiple chronons is mentioned in [ea94]. An *interval* is the time between two events. It may be represented by a set of contiguous chronons [ea93].

2.3 Network Management Databases

Network management consists of monitoring and controlling the behavior of a network, which require the presence of sophisticated mechanism for the specification of events and correlated events occurring at different time points and specification of rules for dealing with these events.

Both primitive and composite events may need to be saved in the database as events or intervals for current or future manipulation. Timestamped *trace* data which may or may not be considered as events may also need to be stored in the database. The later is called a *trace collection* in [WSY91]. The underlying datastore is thus a *temporal* database capturing the history of snapshots of network behavior. So a model of a database that combines the features of both active and temporal databases is well suited for network management databases.

The question then arises, how to specify polling, data pattern events, composite events and trace collection in a declarative way.

By considering the network as a database, the data pattern events can be specified as data manipulation statements in any declarative database language, for example, SQL. In [CH93] we specified data pattern events as GraphLog queries.

Management action is performed by monitoring on the network database. Polling or sampling is one form of monitoring. Monitoring action then consists of the following: 1) fetch the attributes specified in the *select* statement of the DML at each poll interval, 2) as data arrive, evaluate the query. If the evaluation succeeds, the data pattern event is generated. In case of trace collection, the DML statement will insert the arrived tuples in the database. The system may delegate the above functions to managed entities, if it knows that the entities can perform the functions themselves. The entities then report back the events to the manager.

This is how monitoring for a data pattern event or trace collection will be specified in our system:

E: poll at regular intervals
C: TRUE
A: Evaluate DML statement

Polling and composite events will be specified using our proposed ESL which is the subject of the next section. We specify polling in the E part as a composite event, because it is a time event occurring at regular intervals. By specifying it as a composite event using ESL we control how polling will be performed. A graphical view of the ECA mechanism is shown in Figure 1.

2.3.1 Special events

1) *poll(X)*, where X is an unique id of an ECA rule. This event may be used to start a polling action or execution of any action at regular intervals. 2) *deactivate (X)*, where X is an unique id of an ECA rule. This event may be used to *deactivate* a perpetually running instance of an event expression. Note that both *poll* and *deactivate* are events, not procedures. These events can be generated through a special function called $generate(e)$.

3 Event Specification Language

In this section we describe a language for specifying *composite* events. We define a number of *operators* which are used for composing primitive, other composite events and intervals into higher level composite events and intervals. The operators are chosen so that they are useful for specifying events and intervals selections, compositions and correlations in a number of advanced application domains.

In our intended applications domain events happen in parallel in the distributed entities. It is possible to order the events totally at the central site where they are collected for processing. But this does not allow us to detect arbitrary temporal ordering, for example, overlap of intervals

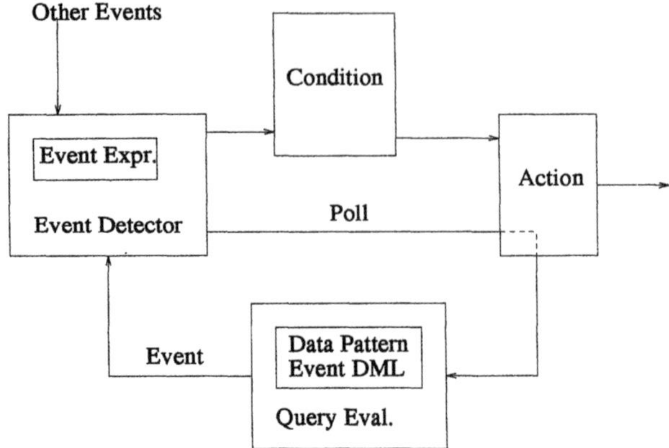

Figure 1: Graphical View of the ECA Mechanism

during which events happen. A total ordering in the event history is assumed in [GJS92]. We use Petri Net as implementation model of ESL expressions. Petri net allows reasoning about partial order of computation.

3.1 ESL Operators and Expressions

We define a number of basic operators, we think are useful for a number of applications requiring active database support. Details about the language and its implementation can be found in [Has94].

- $E = e_1 \ominus e_2$, Operator $\ominus$ defines the event that occurs when either of e_1 or e_2 occurs.
- $E = e_1 \oplus e_2$, E occurs when both of the events occur in any order.
- $E = e_1$ **fb** e_2, Event E happens when e_2 occurs any time after the occurrence of e_1.
- $E = e_1$ **se** e_2, Event E happens when e_1 occurs strictly after e_2 in the successive chronon points associated with the events.
- $E = e_1$ **in** I, E is signalled when e_1 happens in the interval I which is open at the right.
- $E = I_1$ **ol** I_2, E happens when the two intervals I_1 and I_2 overlap.
- $E = e_1$ **ne** I, E happens if e_1 does not happen in the interval I which is open at the right, E is signalled at the end point I_e of I.
- n **nth** e, E happens when n number of e events happen.
- $E =$ **first**(e), This operator selects the first e event from a series of *consecutive* or *concurrent* e events in the event history.
- $E =$ **last**(e), If an interval is not specified, then **last**$(e) \equiv e$.

- An interval between two events e_1 and e_2 is specified as $[e_1, e_2]$. The interval is open on the right.

We will now provide a number of useful additional operators.

- e_3 **fs** e_1 = **first(last(**e_3**) fb** e_1), specifies first e_1 event since (after) the recent e_3. Since this event may fire at each e_1 after the recent e_3, the *first* qualifier is necessary.

In the network management domain *persistence* of an event in an interval may be of interest. Since the model of time is discrete, rather than continuous, persistence has to be defined in terms of the discrete model of time. If an event happens at *all* chronon points associated with the event in the specified interval, then that event is said to persist for that interval.

- e_1 **pe** I = (...((e_1 **se** e_1) **se** e_1)...) **se** e_1) **in** I, defines the *persistence* of an event, which happens when e_1 events happen in strict sequence at each chronon point in the interval I.

3.2 Implementation Model of ESL Operators

In this section we will provide an implementation model of the ESL operators using *colored Petri net* (CPN).

A CPN is a directed graph with two kinds of nodes, *places* P and *transitions* T, interconnected by *arcs* A. Arcs may be inscribed with *arc expressions* and transitions with *guard expressions.* A *colored token* of a CPN, as opposed to simple Petri net, can carry complex information. Places are depicted as circles and transitions as vertical line segments.

The behavior of a CPN is described as follows: A transition fires, when it is *enabled.* A transition is enabled when the variables of *input* arc expressions can be bound with appropriate tokens or colors present on the input places and evaluated, and the guard (if present) evaluates to true. When a transition fires, tokens are removed from the input places and placed on the output places. The number of removed/added tokens and the colors of these tokens are determined by the value/type of the corresponding input and output arc expressions evaluated with respect to bindings in question.

Figure 2 shows the CPN implementation of e_2 **fs** e_1. The upper portion of the figure corresponds to **last(e**$_2$**)** before the first e_1 appears. Since the last e_2 token is removed from p_1 when t_3 fires, all e_1s appearing after the firing and until the occurrence of next e_2, will be removed. A is an *auxiliary* place which is *marked* initially. Any e_1s appearing before e_2 will be removed. If both t_1 and t_2 are enabled concurrently, then we resolve the firing sequence in favor of the *terminator* event e_1, that is, t_2 will fire first, thus removing the e_1 event.

3.3 Example ESL expressions for NM

We will now give a number of examples showing how the above operators can be used for declaratively specifying interesting events of interest in the network management domain.

- A *server_underutilized (su)* event follows a router *congestion (co)* event within 2 minutes.

|| (*co* **fb** *su*) **in** [*co*, (2 **nth** *minute*)] ||

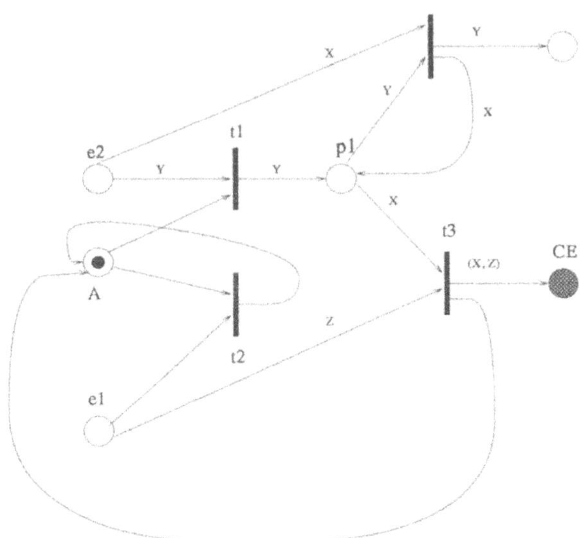

Figure 2: Petri Net Model of e_2 **fs** e_1

- *Polling* or *Sampling* is an important function in network management.

  An event of polling every 2 minutes for 1 hour can be specified as follows:

  $$CE_3 = (2 \textbf{ nth } \textit{minute}) \textbf{ in } [\textbf{last}(\textit{poll}(X)), 60 \textit{ minute})]$$

  The timer is started when the (recent) *poll* event is detected. The expression is then used to control the duration of the timer that emits (time) events every 2 minutes.

  In some cases, polling may be stopped when requested explicitly. Following expression CE_4 polls every two minutes in an interval delimited by the *poll* and *deactivate* events.

  $$CE_4 = (2 \textbf{ nth } \textit{minute}) \textbf{ in } [(\textit{poll}(X), \textit{deactivate}(X))]$$

- If the expression "$value \geq threshold$" is contained in the definition of an event, then the event will be generated at each sampling interval as long the value remains high. An ECA rule using this event will fire the action repeatedly which may be undesirable. What we need is some filtering mechanism to prevent this. For example, *first event since some other event* or the *hysteresis mechanism* as defined in the RMON specification [Wal]. The mechanism by which small fluctuations are prevented from causing alarm is referred to in the RMON specification as *hysteresis* mechanism.

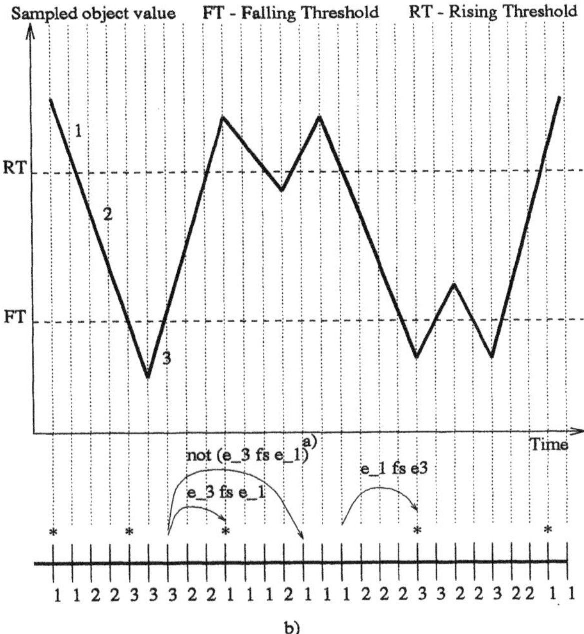

Figure 3: Specification of Hysteresis Mechanism

Hysteresis mechanism is best explained through the Figure 3.a (similar to the figure in [Sta93], we modify it to suit our purpose). As the rules for the hysteresis mechanism stipulates only the events marked as stars (*) will be reported. We assume that the events are reported at each sampling interval. Then the hysteresis mechanism can be specified as follows.

$$CE_6 = (e_1 \textbf{ fs } e_3) \ominus (e_3 \textbf{ fs } e_1)$$

A large number of interesting event patterns can be specified using ESL as opposed to programming or hardcoding limited set of rules in the system (like the hysteresis mechanism only in RMON). For example, if we consider Figure 3, events (such as, *server_overload*) in the region 1 may persist for long time. But that persistence event will not be generated by the hysteresis mechanism, thus leaving no room for taking action to alleviate the problem.

4 Example ECA Specifications

We now provide a number of example specifications of NM functions employing ESL, active and temporal databases concepts in an unified framework.

The SQL query Q1 in the rule RL1 below defines a *server_underutilized* (S_U) data pattern event.

```
RL1:
        E: CE4
        C: TRUE
        A: Q1
Q1:
GENERATE S_U (HOST, TCPINSEGS) AS
        SELECT HOST, TCPINSEGS
        FROM MIB_TCP
        WHERE HOST_TYPE = 'server'
                AND (TCPINSEGS - PREVIOUS(TCPINSEGS))
                    ≤ falling_threshold
```

Note that, Q1 refers to both static configuration data (topology information) and dynamic MIB data of managed entities. The implementation will evaluate the query over the configuration database once and filter out the servers. The servers will then be polled for *tcpInSegs* MIB variable values and as data arrive the crossing of threshold value will be checked. We assume that the underlying temporal database supports a temporal operator called *previous* which returns the last reported tuple (fetched in the previous poll). ECA rule RL1 specifies that the MIB_TCP tables are polled every two minutes until a *deactivate* event happens. Event expression CE_4 discussed in the previous section will serve the purpose. We assume that *poll(RL1)* event is generated initially.

Q1 can be specified as a *trace collection* which collects the traces in a table. Rule RL2 defines this trace collection.

```
RL2:
        E: CE4
        C: TRUE
        A: Q2
Q2:
INSERT INTO SERV_TCP_TRACE (HOST, TCPINSEGS)
        SELECT HOST, TCPINSEGS
        FROM MIB_TCP
        WHERE HOST_TYPE = 'server'
```

The following rule RL3 then specifies the generation of the S_U events. The *insert* is a database manipulation event.

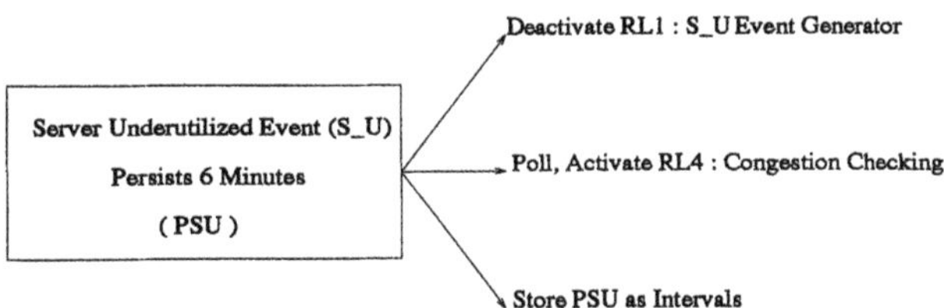

Figure 4: Diagramatic View of RL5

```
RL3:
        E: insert (SERV_TCP_TRACE, HOST, TCPINSEGS)
        C: (TCPINSEGS - PREVIOUS(TCPINSEGS))
               ≤ falling_threshold
        A: generate (S_U (HOST, TCPINSEGS))
```

We will now write an ECA rule (RL5) for the specification of the following. Watch for the *persistence* of S_U events for, say, 6 minutes. If it persists, then check for *congestion* on the routers that are on the way between the server and its clients. To detect congestion start evaluating for 1 hour every 2 minutes the corresponding data pattern event query (the corresponding rule RL4 is not shown for brevity). Deactivate the generation of S_U events and store the persistence of S_U events (PSU) as intervals in the database. A diagramatic view of RL5 is shown in Figure 4.

```
RL5:
        E: PSU (int (Self), H, V) = persist (S_U(H, V), 6 minute)
        C: TRUE
        A: Q5 AND
           generate (poll (RL4)) AND
           generate (deactivate (RL1)) AND
           INSERT INTO SERV_UNDUTIL_PERSIST PSU
```

Query Q5 filters out the routers between the server and its clients. We do not show query Q5 here. Similar query can be found in [CH93]. The routers found are passed to the query portion of RL4. PSU is defined as an interval. The interval is calculated using the *int* operator on the persistent composite event PSU. Operator *int* returns the timestamps of the end points of an interval.

5 Related Work

The database issues for network management similar to the ones discussed in this paper have also been considered in [WSY91]. We provide a more uniform and consistent framework for specifying *data pattern events* and *trace collections*, that is, as ECA rules. They provide a separate mechanism for specifying trace collections. The main difference with our work is in our proposed composite event specification language, ESL. Their work lacks such an event specification language. As a result, polling and other composite events can not be specified in their system, that could control uniformly the collection of data pattern events, traces and other actions, as is done in our system. We also provide a consistent mechanism to collect events and traces in a *temporal database*. The notion of *persistence* is mentioned in their work, but no formal definition of it is provided. The MANDATE MIB project [HBNRD93] also addresses similar network management database issues. But the proposal for a unified framework for incorporating active and temporal databases concepts in a network management database similar to ours is lacking in their work. The work in [Shv93] discusses only the issues of a *static* (historical) temporal database for network management data.

6 Conclusion

We have proposed a model for network management database where the network management functions are specified as *Event-Condition- Action* rules. In proposing the model we have considered unique properties of NM data and functionalities. We have designed a temporal event and interval specification language that allows us to specify composite or (temporally) interrelated events.

Work is in progress to implement efficiently the ESL operators. Visual specification of ESL expressions and visualization of event detection process will be helpful in many application domain, including network management. We are working towards that goal. As a future work we plan to incorporate real-time or hard-deadline issues in the language.

Acknowledgments

I would like to thank Prof. Alberto Mendelzon of University of Toronto for his fruitful suggestions and support. I also thank Prof. William Cowan of University of Waterloo for his support. I specially thank Michael Sam Chee of Bell Northern Research, Ottawa, Canada for his many suggestions.

The work was supported by The Natural Sciences and Engineering Research Council of Canada and the Information Technology Research Centre of Ontario.

References

[CH93] Mariano Consens and Masum Hasan. Supporting network management through declaratively specified data visualizations. In H.G. Hegering and Y. Yemini, editors, *Proceedings of the IEEE/IFIP Third International Symposium on Integrated Network Management, III*, pages 725–738. Elsevier North Holland, April 1993.

[ea93] C. Jensen et. al. Proposed temporal database concepts - may 1993. In *Proceedings of the International Workshop On an Infrastructure for Temporal Databases*, pages A–1–A–29, June 1993.

[ea94] N. Pissinou et. al. Towards an infrastructure for temporal databases, report of an invitational ARPA/NSF workshop. Technical Report TR 94-01, Department of Computer Science, University of Arizona, March 1994.

[GD94] S. Gatziu and K. Dittrich. Detecting composite events in active database systems using petri nets. In *Proceedings of the Fourth International Workshop on Research Issues in Data Engineering*, pages 2–9, February 1994.

[GJS92] N. Gehani, H. Jagadish, and O. Shmueli. Composite event specification in active databases: Model and implementation. In *Proceedings of the 18th International Conference on Very Large Data Bases*, 1992.

[Has94] Masum Z. Hasan. Active and temporal issues in dynamic databases. PhD Thesis Proposal, Department of Computer Science, University of Waterloo, Waterloo, Ontario, Canada, 1994.

[HBNRD93] J. Haritsa, M. Ball, J. Baras N. Roussopoulas, and A. Datta. Design of the MANDATE MIB. In H.G. Hegering and Y. Yemini, editors, *Proceedings of the IEEE/IFIP Third International Symposium on Integrated Network Management, III*, pages 85–96. Elsevier North Holland, April 1993.

[MD89] D. McCarthy and U. Dayal. The architecture of an active data base management system. In *Proceedings of the ACM-SIGMOD 1989 International Conference on Management of Data*, pages 215–224, 1989.

[Shv93] A. A. Shvartsman. An historical object base in an enterprise management director. In H.G. Hegering and Y. Yemini, editors, *Proceedings of the IEEE/IFIP Third International Symposium on Integrated Network Management, III*, pages 123–134. Elsevier North Holland, April 1993.

[Sta93] W. Stallings. *SNMP, SNMPv2, and CMIP, The Practical Guide to Network Management Standards*. Addison-Wesley Publishing Company, Inc., 1993.

[Wal] S. Waldbusser. Remote network monitoring management information base. RFC 1271, Carnegie Mellon University.

[WSY91] O. Wolfson, S. Sengupta, and Y. Yemini. Managing communication networks by monitoring databases. *IEEE Transactions on Software Engineering*, 17(9):944–953, September 1991.

About the Author

Masum Z. Hasan is a Research Associate at the Computer Systems Research Institute, University of Toronto and a Ph.D candidate in the Department of Computer Science, University of Waterloo. He obtained BEng, MEng in Computer Engineering from former USSR and MMath in Computer Science from University of Waterloo. His research interests are in active temporal databases, network management, networked document browsing/searching, distributed and parallel programming environment, visualization. Mr. Hasan has worked for industry both in Bangladesh and Canada.

46

ICON: A System for Implementing Constraints in Object-Based Networks

Shravan K. Goli
Dept. of Comp. Sc., and ISR, Univ. of Maryland, College Park. Currently at Microsoft Corporation.
Jayant Haritsa
IISC, Bangalore, India, and ISR, University of Maryland, College Park.
Nick Roussopoulos
Dept. of Computer Science, ISR, and UMIACS, University of Maryland, College Park.

Abstract

A vitally important step in network configuration management is to check the validity of updates made to data elements in the Management Information Base (MIB). For example, if an operator mistakenly configures a ninth port on an eight port card, the MIB should both detect and prevent this error. In this paper, we focus on the problem of checking MIB update validity and introduce the design of ICON (Implementing Constraints in Object-Based Networks), a proposed network constraint management system. In ICON, constraints are expressed through rules, which are based on the Event–Condition–Action paradigm. Rules and events are integrated cleanly into the object model by treating them also as objects.

1 Introduction

In enterprise communication networks, the network operator's interface to the network is through a Management Information Base (MIB). The MIB stores all management-related data such as network and system configurations, accounting information, and trouble logs. A vitally important step in network *configuration* management is to check the validity of updates made to MIB data elements. For example, if an operator mistakenly configures a ninth port on an eight port card, the MIB should both detect and prevent this error. In this paper, we focus on the problem of checking MIB update validity, which can be viewed as a specific instance of the general problem of *constraint management* in database systems. In particular, we introduce the design of ICON (Implementing Constraints in Object-Based Networks), a proposed network constraint management system intended for use in the *object-based* MIB of the PES (Personal Earth Station) network, a proprietary product of

Hughes Network Systems, Inc., Germantown, Maryland, U.S.A. We also discuss here the integration of ICON with the PES data model. A simplified ICON system prototype has been developed and integrated with a graphical user interface.

2 Examples of Constraints

A sample set of typical network management constraints is shown in Figure 1.

The constraint that attempting to configure more than 8 ports on an 8-port card exceeds the physical limitations of the card is expressed in Figure 1(a). Another type of constraint is shown in Figure 1(b) – here, the LAN type between communicating HUB and REMOTE LANs should be the same, that is, they should both be ethernet or both be token ring. In Figure1(c), it is mandated that the only legal values for a modem's baud rate attribute are 2400, 4800 and 9600. Finally, Figure1(d) states that only certain operators are allowed to make updates to parameters of network switches.

From the above examples, we observe that network management constraints have a variety of dimensions:

1. Constraints may be *physical* as in Figure 1(a), or *logical* as in Figure 1b.
2. Constraints may refer to a *single* object as in Figure 1(a) or span *multiple* objects as in Figure 1(b).
3. Constraints may be checked *immediately*, that is, as soon as the update is made, as in Figure 1(c) or *deferred* to a later time (e.g. completion of a related set of updates), as in Figure 1(b).
4. Constraints may apply *universally* to all applications accessing an object, as in Figure 1(c), or be *selectively* enforced based on the application accessing the object, as in Figure 1(d).

3 Incorporating Constraints in OO DBMSs

Constraint maintenance in object oriented databases differs from that of relational databases in many aspects, as discussed in [1, 2, 9, 11] . A detailed discussion of the differences between relational and object-oriented database management systems with respect to constraint management is also given in [15].

Virtually all research into constraint management in object-oriented database systems has assumed that the constraints are expressed in the form of *rules*. Several OO systems that support such rules are described in the literature. For example, in [9, 11], rule management for Ode, a product of AT&T, has been described. In Ode, constraints are associated with class definitions. Ode has extended C++ [17] to O++ which provides

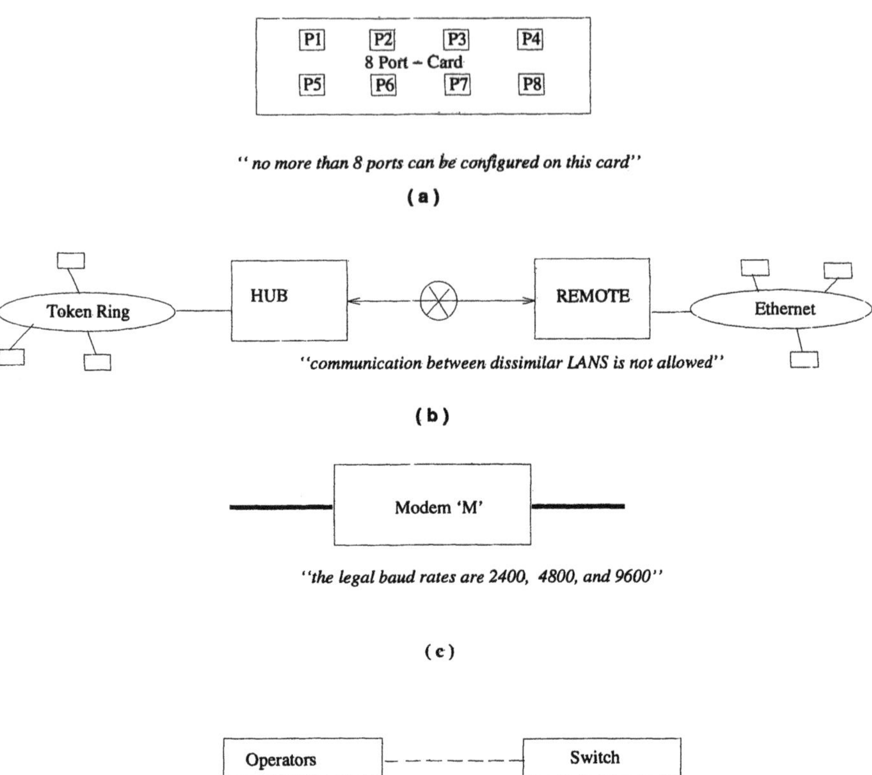

Figure 1: Constraint Examples

facilities for associating constraints with an object. The specified constraints are checked every time an instance of that class is updated, a new instance is created, or an old instance is removed. Each constraint is expressed as a two-tuple <condition, action>. Whenever the constraint condition is violated, the action code is executed and the constraint is again tested. Detailed examples of how to express network management constraints in Ode are given in [15].

A different approach to constraint management, called ADAM, is described in [7]. Unlike Ode, where constraints are specified as a part of the object definition, the rules (or constraints) in ADAM are also treated as objects similar to the other objects in the system. Relationships can be established between the monitored objects and their associated rules. Each rule object maintains a list of monitored objects, and at the same time, the monitored objects also maintain a list of rules on them thus forming a two–way–relationship. In [15], a few examples of how to express network management constraints in ADAM are provided.

Yet another approach to constraint management, called Sentinel, is described in [1, 2]. The Sentinel approach captures the advantages of both Ode and ADAM, and extends them to provide significantly new features. It supports both constraints specified along with class definitions (as in Ode) as well as constraints specified as separate objects (as in ADAM). This has features to build rules spanning multiple objects which is difficult to do in both Ode and ADAM. In the following section, we discuss how several features of Sentinel were used in building the ICON system.

4 The Design of ICON

In this section, we present the design of the ICON network constraint management system, designed for use in the object-based MIB of the PES network (a satellite based network), a proprietary product of Hughes Network Systems, Germantown, Maryland, USA. The ICON design has taken most of its features from Sentinel (discussed in the previous section) and adapted them for the special requirements of the network management domain. The integration of ICON with the PES data model is described in Section 5, while the implementation details of the ICON prototype are presented in Section 7.

4.1 PES

In this section we give a brief description of the PES network. This network is composed of a hub, the systems control center and multiple remotes as shown in Figure 2.

- The hub provides centralized communication management for the remotes. All traffic between the remotes must pass through the hub; traffic cannot be passed from one remote to another directly over the satellite link.

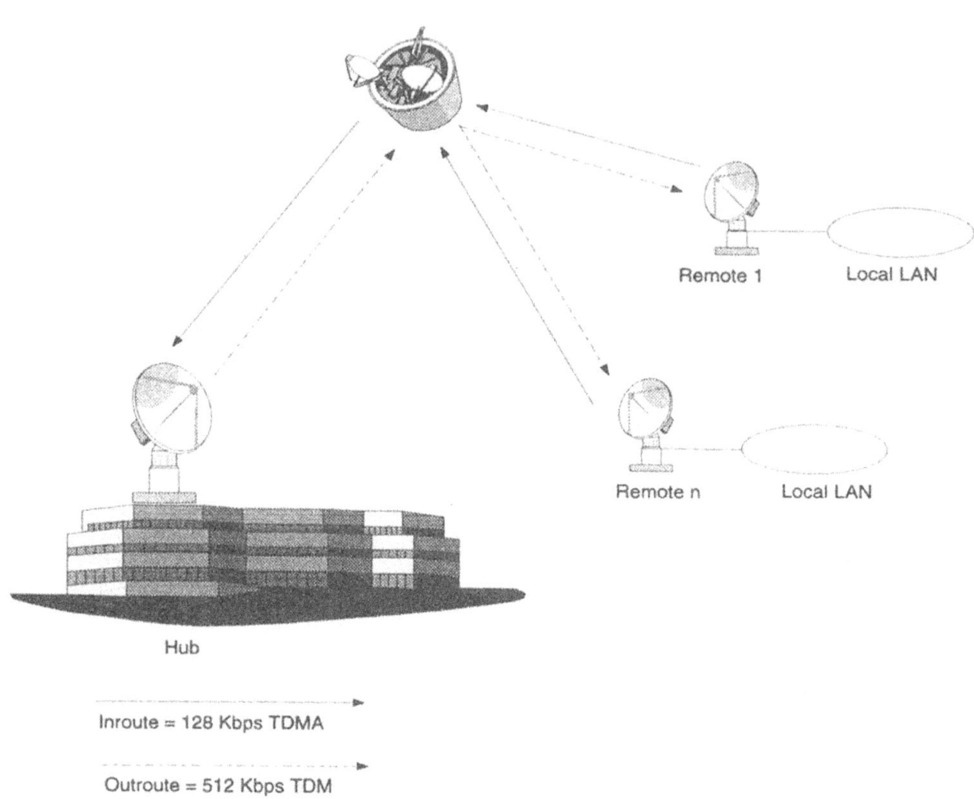

Figure 2: PES network

- The systems control center (SCC) controls the network, i.e. all management of this network occurs from the SCC which is thus conceptually centralized but may be distributed in practice. The SCC and Hub is usually co-located. Management is done through operator consoles through which operators configure, monitor and control the network.
- The remotes are geographically dispersed sites that contain *remote node equipment*. The remote equipment is typically attached to customer equipment such as LANs, computers and workstations. Customer equipment is connected to the network via remote ports.

Information is exchanged between the remote sites over a satellite link through the hub. Remote to hub transmissions travel over *inroutes*, while hub to remote transmissions travel over *outroutes* as shown in Figure 2. Thus, remote to remote travels from the remote to hub over an inroute, then from the hub to the other remote on an outroute. Of course, all transmission must be relayed through the satellite.

4.2 Design Details

We consider first the problems of how constraints are specified, how they are stored, and how they are evaluated and enforced. In ICON, constraints are expressed through *rules*. Each rule is composed of an *event*, a *condition*, and an *action* (this is also known as the E-C-A paradigm [4]). Occurrence of the event triggers the rule, the condition is a boolean check, and the action is executed if the condition is satisfied. This rule definition is illustrated in the following example, which is used during configuration of the HUB module of the PES network to check for uniqueness of data port card (DPC) names:

```
class DPC{

  private:  char dpcname[20];
  public:   Set_Name(char *name);

  rule dpcname_uniq;
  when Set_Name(name)                  /* event */
     if not_unique(dpcname)            /* condition */
        then highlight(dpcname_field); /* action */
```

In the above example, a class `DPC` is defined for data port cards. The rule `dpcname_uniq` monitors the configuration of `DPC` objects, and is triggered whenever a DPC object invokes the method `Set_Name`. A check is then made as to whether or not the 'dpcname' is unique. If the name is not unique then the action routine 'highlight()' is called to indicate the error to the operator.

It is important to note that the term event used here does not refer to network events, but refers to *database events*. In our object-oriented framework, database events consist primarily of *object method invocations*. With respect to configuration management, database events would mainly be initiated by operator actions. More generally, however, we expect that network event messages received by the MIB during network operation could lead to the generation of one or more database events.

4.3 Rule Specification

In ICON, rules are implemented as *first-class objects*, following the Sentinel approach. With this approach, rules can be created, modified, and deleted in the same manner as other objects, thus providing a uniform view of rules in an OO context. Second, rules are now separate entities that exist independently of other objects in the system. This allows for rule definitions to exist even when the object classes on which the constraints operate do not exist. Third, each rule has an object identity, thereby allowing rules to be associated with other objects. Finally, an extensible system is provided due to the ease of introducing new rule attributes or operations on rules.

4.4 Event Specification

In ICON, events are implemented as *first-class objects*, as in Sentinel. This implementation is chosen because events exhibit the properties of objects in terms of having state, structure and behavior. The state information associated with each event includes the occurrence of the event and the parameters computed when the event is raised. The structure of an event consists of the events it represents, while the behavior consists of specifying when to signal the event. By making events to be objects, events can be created, deleted, modified, and designated as persistent similar to other types of objects. The introduction of new event types and attributes is easily incorporated. Also, events spanning distinct classes can be expressed in a clean fashion. Complex events can be constructed using a hierarchy of event operators such as conjunction, disjunction, etc.

4.5 Event Generation

In ICON, each object is allowed to declare some subset of its public method interface to be *reactive*. This means that an event message is generated whenever a method in this reactive subset is invoked by the object. These event messages are propagated to other objects by a mechanism described in the following subsection. Event messages have the following structure:

Event Message = Oid + Class + Parameters

Here, *Oid* denotes the object identifier of the object generating the message, *Class* denotes the class of this object, and *Parameters* denotes the set of parameters with which the method is invoked.

4.6 Rule – Event Association

Rules are associated with events through a *subscription* mechanism. This mechanism allows rules to dynamically subscribe to the events generated by reactive objects. After the subscription takes place, the rule is informed whenever an object it subscribes to issues an event. Each reactive object maintains a list of its subscribed rules.

The above subscription mechanism results in some advantages: First, the runtime rule checking overhead is reduced since only those rules which have subscribed to an event are checked when that event is generated, that is, rule checking is localized (or distributed). Second, a rule is defined only once and it can be associated with any number of reactive objects. This is more efficient than defining the same rule multiple times and applying each rule to one type of object. Finally, rules triggered by events spanning distinct classes can be expressed. This is accomplished by a rule subscribing to the events generated by instances of different classes.

4.7 Object Classification

In ICON, objects are classified into the three categories described in Sentinel: passive, reactive, and notifiable. These categories and their relationship to events are described below.

Passive objects: These are regular C++ objects. They can invoke methods but do not generate events. Objects which do not need to be monitored fall into this category.

Reactive objects: Objects on which rules may be defined are made reactive objects. Once a method is declared as an event generator, its invocation will be propagated to other objects. Thus, reactive objects communicate with other objects via event generators.

Notifiable objects: Notifiable objects are those objects capable of being informed of the events generated by reactive objects. Therefore, notifiable objects become aware of a reactive object's state changes and can perform operations as a result of these changes. All rules are notifiable objects. There is an m:n relationship between reactive objects and notifiable objects, that is, a reactive object instance can propagate events to any number of notifiable object instances and a notifiable object instance can receive events from several reactive object instances.

4.8 ICON Example

The basic paradigm in ICON is that events are produced by reactive objects (producers) and they are consumed by notifiable objects (consumers). An example of Producer/Consumer paradigm is shown in Figure 3, taken from [1, 2]. Here, object P generates (produces) a primitive event event1 and sends it to a rule R1. The rule passes

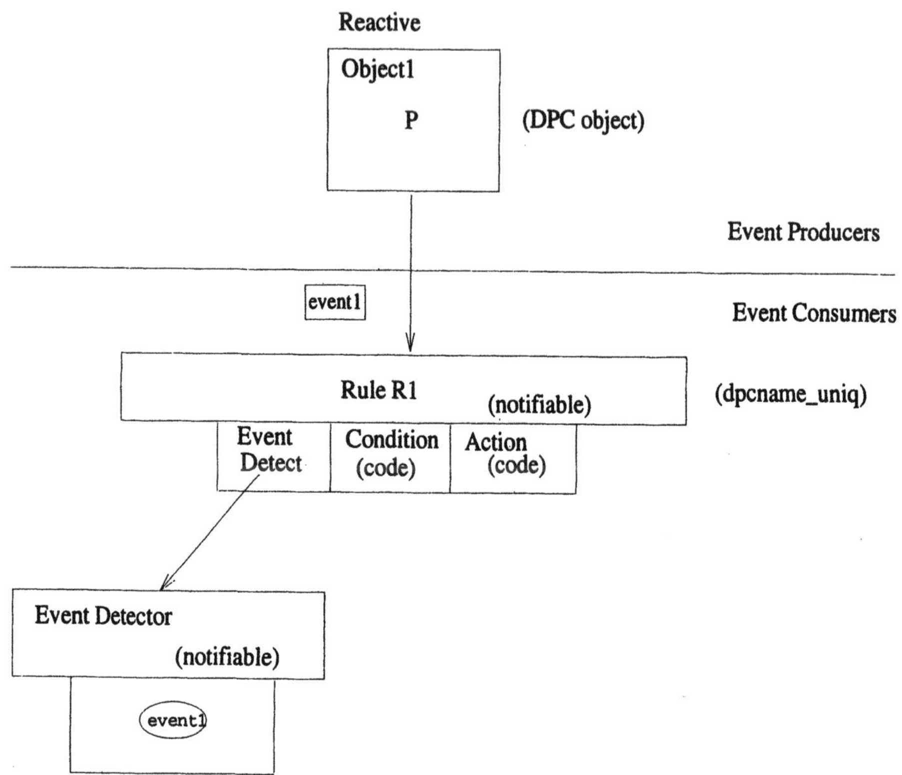

Figure 3: Event Producer/Consumer Analogy

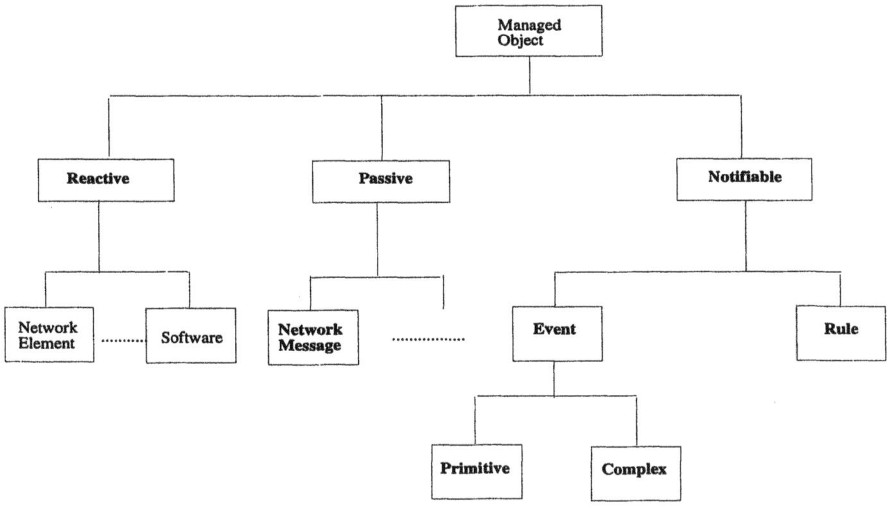

Figure 4: Integration with PES data model

(consumes) the event to the event detector for storage and event detection, and if the event is detected, the rule checks the condition and takes appropriate actions. In this example, P is a reactive object, event1 is a primitive event, and R1 is a notifiable object. With reference to the earlier example in section 4.2, P is of type DPC and R1 is 'dpcname_uniq'.

4.9 Summary

The above design of ICON provides for: (i) rule definitions to be independent from the objects which they monitor, (ii) rules to be triggered by events spanning sets of objects, possibly from different classes, and (iii) objects to dynamically determine which object state changes they should react to and associate a rule object for reacting to those changes. Essentially, this separates the object and rule definitions from the event specification and detection process. This aids in building a modular and extensible system.

5 Integration with Data Model

We now describe how the constraint management in ICON could be integrated with the PES data model. The details of the PES data model are given in [6]. For the purpose of integration, we have subdivided the top-level managed object class of the PES data model into a reactive class, a passive class, and a a notifiable class, as shown in Figure 4. All the elements of the PES data model which need to generate database events fall into the reactive class. Similarly, all the constraint management objects are in the notifiable subclass. The remaining objects, which do not generate any databse events or which do not have any rules imposed on them, fall into the category of passive objects.

6 Implementation details

As mentioned earlier, a prototype of a simplified version of ICON has been developed. The prototype implementation is discussed in detail in [16]. The [16] describes in detail the implementation of Reactive, Notifiable, Event and Rule classes. It also describes a simple algorithm for ICON which we discuss in the section below. In addition, it also describes MOTIF/Galaxy [13] versions of graphical user interface developed for ICON. Some impelmentation examples are also described. The prototype was developed on the object-oriented database platform provided by ObjectStore, a commercial OO-DBMS [18, 14, 5].

6.1 ICON Algorithm

We show here, in pseudo-code, a simple algorithm which describes the mechanism of event generation and rule checking in ICON:

```
algorithm_ICON()
{
       Whenever a Reactive method is accessed, at some point in its
       processing, a method called Notify() is used to send a message
       to all Rules subscribed to that reactive object;

       For each rule subscribed {
         if (the rule is enabled) {
           pass the event to the rule's event detector;
             if (event detected)    {
               check the condition of the rule;
                 if (condition is satisfied)
                 perform the action; } }
       }
       return the appropriate value;
}
```

6.2 Graphical User Interface

As described in earlier sections, rules exist as independent objects and they subscribe to reactive objects. Each reactive object maintains a list of currently subscribed rules, and a list of rules which are valid but currently unsubscribed. It is possible to dynamically change these lists and, thereby, the behaviour of the monitored objects themselves. In our implementation, we have developed a Motif user interface to help the user of the system decide the behaviour of the monitored objects. The main advantage of this is that it helps in customization of the product to match different users requirements. Note that if the rules were implemented in the traditional way of writing code wherever required, this kind of dynamic behaviour would not have been possible. For any change, the code would have to be not only re–written but also recompiled. This would create several problems in network MIBs since recompilation requires database access

References

[1] Anwar, E., Maugis, L., and Chakravarthy, S. (May 1993) A New Perspective on Rule Support for Object-Oriented Databases. *ACM SIGMOD*, 99-108.

[2] Anwar, E. (1992) Supporting Complex Events and Rules in an OODBMS: A Seamless Approach. *Master's Thesis, Univ. of Florida* .

[3] Bauzer Medeiros, C. and Pfeffer, P. (1990) A mechanism for Managing Rules in an Object-oriented Database. Altair Technical Report.

[4] Chakravarthy, S. et al. (July 1989) HiPAC: A Research Project in Active, Time Constrained Database Management. *TR XAIT-89-02*, Xerox Advanced Information Technology, Cambridge, MA.

[5] Class Notes, ObjectStore , Object Design, Inc.

[6] Datta, A. and Ball, M. (1993) MOON :- A Data Model for Object Oriented Network management. To be published, ISR, University of Maryland at College Park.

[7] Diaz, O., Paton, N. and Gray, P. (Sept. 1991) Rule Management in Object-Oriented Databases : A Unified Approach. *Proc. of VLDB, Barcelona*, 317-26.

[8] Dupuy, A. et al (March 1991) NETMATE: A Network Management Environment. *IEEE Network Magazine.*

[9] Gehani, N. and Jagadish, H. (Sept. 1991) Ode as an Active Database: Constraints and Triggers. *Proc. of VLDB, Barcelona*, 327-36.

[10] Haritsa, J. et al (1993) Design of the MANDATE MIB. *Integrated Network Management, III*, Elsevier Science Publishers, 85-96.

[11] Jagadish, H. and Qian, X. (Aug. 1992) Integrity Maintenance in an Object-Oriented Database. *Proc. of VLDB, Vanouver*, 469-80.

[12] Klerer, S. (March 1988) The OSI Management Architecture: an Overview. *IEEE Network*, 2(2).

[13] Plaisant, C., Kumar, H., Teittinen, M. and Shneiderman, B. (1994) Visual Information Management for Network Configuration. *TR 94-48*, ISR, University of Maryland at College Park.

[14] Reference Manual, ObjectStore Release 2.0, Object Design, Inc.

[15] Shravan, G., Jayant, H. and Nick, R. (1994) Integrity Constraints in Configuration Management. *TR 94-62*, ISR, University of Maryland at College Park.

[16] Shravan, G., Jayant, H. and Nick, R. (1994) A System for Implementing Constraints in Object-based Networks. *TR-xxx under preparation*, ISR, University of Maryland at College Park.

[17] Stroustrup, B. (1986) The C++ Programming Language. Addison-Wesley.

[18] User Guide, ObjectStore Release 2.0, Object Design, Inc.

[19] Yemini, Y. (May 1993) The OSI Network Managment Model. *IEEE Communications Magazine*, 20-29

[20] Zdonik, S. and Maier, D. (1990) Object Oriented Fundamentals. *Readings in Object-Oriented Database Systems*, 1-32.

Shravan K. Goli received the B.E. degree in Computer Science and Engineering from the Osmania University, Hyderabad, India, in 1992, and the M.S. degree in Computer Science from the University of Maryland, College Park in 1994. During 1992-1994, he was a Graduate Fellow at the Institute for Systems Research, University of Maryland, College Park. He is currently working in Microsoft Corporation, Redmond, WA. He previously worked with Hughes Network Systems, Germantown, MD. His research interests include distributed systems, network protocols, network management and object oriented database systems.

Jayant R. Haritsa received the B.S. degree in Electronics and Communications Engineering from the Indian Institute of Technology, Madras, India, in 1985, and the M.S. and Ph.D. degrees in Computer Science from the University of Wisconsin, Madison in 1987 and 1991, respectively. During 1991-1993, he was a Post Doctoral Fellow at the Institute for Systems Research, University of Maryland, College Park. He is currently an Assistant

Professor in the Supercomputer Education and Research Centre and in the Department of Computer Science and Automation at the Indian Institute of Science, Bangalore, India. During 1988 and 1990, he spent summers at the Microelectronics and Computer Technology Consortium and at the IBM T.J. Watson Research Center, respectively. Dr. Haritsa's research interests include database systems, real-time systems, network management and performance modeling. He is a member of IEEE and ACM.

Nick Roussopoulos received the B.A. degree from the National University of Athens, Greece, and the M.S. and Ph.D. degrees from the University of Toronto. He has worked as a Research Scientist at IBM Research at San Jose, and as faculty with the Department of Computer Science at the University of Texas at Austin. Since 1981 he has been with the University of Maryland, where he is a Professor of the Computer Science Department and the Institute of Advanced Computer Studies. His research area is in database systems, multi-databases and interoperability, engineering information systems, geographic information systems, expert database systems, and software engineering.

47

Implementing and Deploying MIB in ATM Transport Network Operations Systems

Tomoaki Shimizu, Ikuo Yoda and Nobuo Fujii
NTT Optical Network Systems Laboratories
1-2356 Take, Yokosuka-shi, 238-03 JAPAN
E-mail : shimizu@ntttsd.ntt.jp

Abstract

TNMSKernel, a network operations system development platform, can be used to produce a Management Information Base (MIB) in conjunction with a database management system. A previous study used a RDBMS (Relational Database Management System) and an OODBMS (Object-Oriented Database Management System), to implement two functionally equivalent MIB functions. However, these MIB implementations are not suitable for network elements such as digital cross-connect systems and subscriber line terminals. Because processing capabilities including processing power, memories and disk I/O speeds for TMN operations interface attached to them are limited. These problems are solved by implementing a new MIB on the main memory technique. The proposed method offers sufficient performance comparing with methods using RDBMS and OODBMS. Furthermore, this paper describes a strategy of selecting the best MIB implementation for each sub-system in an ATM transport network operations system. The effectiveness of the strategy is confirmed through an experiment on a prototype ATM transport network operations system.

Keywords

MIB, TMN, OSI, Main Memory Resident Database, ATM, Network Element

1 INTRODUCTION

The authors have developed an operations system development environment called "TNMSKernel" to efficiently realize transport network operations systems based on TMN (Telecommunications Management Network) standards. The TMN standards provide the operation interface specifications by which multiple carriers can realize telecommunications network interoperability through their operations systems(CCITT M.3010, 1992). The interface specifications utilize the OSI systems management standards, which include the management information model and the common management information services/protocol (CMIS/CMIP) (CCITT X.701, 1992)(CCITT X.711, 1992). TNMSKernel is now being used to develop an ATM transport network operations system (Yata, 1994)(Yoda, 1994). While

TNMSKernel provides several functions for the rapid development of operations systems, the purpose of this paper is to describe its implementation of MIB (Management Information Base).

As an operations system consists of several sub-systems with different roles, the operational performance of each sub-system determines the total system performance. The performance of each sub-system strongly relies on MIB performance that depends on the MIB implementation and the processing capacity of the sub-system. Two MIB implementations have already been developed and tested on TNMSKernel (Yoda, Sakae, 1992)(Yoda, 1993). Other MIB implementation results can be found in (Dossogne, 1993)(Huslende, 1993). All these studies focused on using RDBMS (Relational Database Management System) or OODBMS (Object-Oriented Database Management System) to implement the MIB function. These approaches seem reasonable only for fairly large sub-systems that we can expect will have the large processing power needed to run the DBMSs. The DBMSs, while they do offer some advantages, impose quite high penalties in terms of computing capacity. Thus, it is difficult to apply the previous implementations to sub-systems with less computing power such as network elements.

What is needed is, therefore, a MIB implementation that is more efficient than the previous approaches. It should produce a MIB that runs quickly, handles all regular management functions, and is suitable for low processing capacity sub-systems.

This paper proposes a MIB implementation that uses the main memory of the managed open system. This realizes a high performance MIB within a limited processing capability. Next, three MIB implementations are evaluated using experimental CMIS operations. Then, the optimal MIB deployment strategy is assessed for the ATM transport network operations system. To do this, the technical requirements of the sub-systems are analyzed. Finally, the effectiveness of the proposed MIB implementation and the deployment strategy is confirmed through an experiment on a prototype system.

2 REQUIREMENTS FOR MIB IMPLEMENTATION

2.1 MIB functions

MIB consist of managed object instances and their definitions written in the GDMO format (CCITT X.701, 1992)(CCITT X.722, 1992). There are four main MIB functions from the viewpoint of the OSI systems management.

- Management of object instances and attributes: The MIB shall effectively store object instances and attribute values including relationships. It shall also provide a sophisticated information retrieval mechanism.
- Management of containment relationships: Managed object instances named by the containment relationship constitute a tree structure called MIT (Management Information Tree). Since CMIS operations pinpoint the managed object instance to be operated based on this naming rule, the MIB shall map MIT onto MIB.
- Scope and filter: In order for the manager to point to a managed object instance in the managed open system, the base managed object instance, scope and filter parameters are used. The MIB shall have mechanisms to scope managed object instances and filter them by the attribute values specified by the manager.
- Management of transaction: Multiple managing open systems can independently and simultaneously access the same set of management information. Thus, transaction control becomes a critical requirement. The MIB shall ensure the consistency of the management information control functions includes atomic operations for exclusive and transaction controls.

In addition to the above mentioned functional requirements, the MIB shall have a performance sufficient to handle millions of managed object instances that are spatially distributed. Maintainability and reliability are also required to realize stable network operation.

2.2 MIB on the platform

TNMSKernel is the software platform proposed by the authors to develop TMN based operations systems efficiently. TNMSKernel consists of MIB, Human Machine Interfaces, and communication functions including CMIP. These capabilities are supplied as C++ object class libraries (Yoda, 1994). Figure 1 depicts the original TNMSKernel configuration. MIB was originally realized either on a relational database management system(RDBMS) or on an object oriented database management system(OODBMS). Since TNMSKernel provides the MIB Application Program Interface(API), which conceals the database management system's characteristics, programmers can write managed object behavior programs without knowledge of the internal MIB implementation scheme. Mediation programs (Data Base Access API) are used to hide the differences between database management systems. Furthermore, a managed object instance caching mechanism is provided between APIs to improve MIB performance.

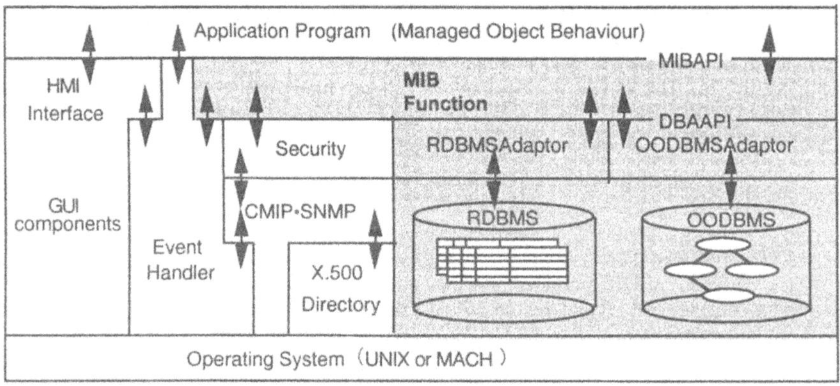

Figure 1 TNMSKernel configuration.

3 MIB IMPLEMENTATIONS

TNMSKernel makes use of database management systems to implement MIB functions. The previous research considered only commercially available RDBMS and OODBMS. These are discussed briefly in the following section. Section 3.2 introduces the new idea of Main Memory Resident MIB (MMR-MIB).

3.1 MIB implementations based on ready-made DBMSs

RDBMS

RDBMS manages data as tables using mathematical relationships (Ullman, 1988). MIB implementation rules based on an RDBMS are described below (Yoda, 1993).

- Use an internal object identifier (AOI: Agent Object Identifier) to identify a managed object instance within the MIB.
- Define an attribute table to handle managed object attribute values and define a table key with using an AOI.
- Use a table of AOI pairs to form relationships between containing managed objects and contained managed objects.
- Generate appropriate SQL code from the base object instance, scope and filter parameters specified in the operation request, and perform the management operation.

By mapping managed object instances and attribute values onto RDBMS tables, MIB uses RDBMS as the management information storage tool. To handle managed object instances with the attributes stored in the various relation tables, the table JOIN operation (Ullman, 1988) is needed to perform each management operation. This operation heavily loads the managed system and degrades transaction performance (Yoda, 1993). On the other hand, RDBMS offers an advantage in that the software program is relatively simple but offers powerful scope and filtering procedures. This is because the SQL powerfully supports relationship operations.

OODBMS

OODBMS permanently stores object instances of complex data structures(Chorafas,1993). This database management system is adequate for the store complex data structures usually found in TMN managed object definitions. The following are the MIB implementation rules for OODBMS implementation.

- Form permanent pointers to indicate managed objects and attributes
- Use pointers to form managed object instance - attribute relationships.
- Realize the containment relationship by having containing and contained object instance pointers in each managed object. The collection management function supported in the OODBMS is used to ensure that the containing managed object has a 1 to N relationship with contained object instances.
- Find managed object instances using pointer navigation from the base object instances using the conditions specified by scope and filter parameters and perform management operations on them.

The use of OODBMS minimizes the amount of program code needed to realize the MIB function, especially in generating the data schema since managed object instances of complex data structures can be directly stored in the database. Furthermore, pointer processing yields good performance in the management operation on a single managed object instance. By contrast, simultaneous operations on a large number of managed object instances are negatively affected by the clustering effect of object instances on the storage media. Thus, the MIB performance can vary greatly depending on the characteristic of the operation. In addition, the lack of standard query mechanisms such as SQL on RDBMS, increases the amount of program code needed for condition handling.

3.2 Main Memory Resident MIB

The OSs (Operating Systems) implemented in most network elements offer only limited capabilities and do not support ready-made DBMS packages. Even if the DBMS is supported, the processors of network elements have relatively low CPU power and low disk I/O speeds. Thus, it is difficult to obtain sufficient processing performance. This means that basing MIB on a ready-made database management system, RDBMS or OODBMS, is suitable only when the systems have processing power sufficient to handle large numbers of instances.

What is needed is to replace the ready-made DBMS with another technique; the best candidate is memory resident data handling (Ammann, 1985)(Molina, 1992). It can be used to achieve portable, high performance MIBs that can run on a system with limited CPU power. With this approach, high performance data processing is obtainable by placing and processing managed object instances in the main memory. Figure 2 depicts the concept of the proposed Main Memory Resident MIB (MMR-MIB). This method is configured with the following seven implementation rules.

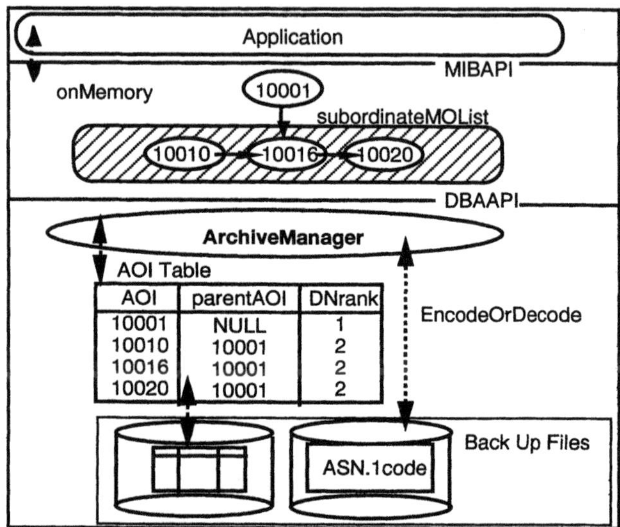

Figure 2 MMR-MIB implementation concept.

- MIB schema: In order to handle managed object instances on the main memory, class definitions themselves are used as the schema information. The class definitions used by application programs are generated from GDMO definitions with using the GDMO translator (Yoda, Minato, 1992).
- Management of managed objects and attributes: The managed object instance is instantiated from the class definition in the MIB schema. Access to the managed object instance and attributes is achieved through the containing managed object pointer and the distinguished attribute.
- Management of containment relationships: The managed object instance has the containing managed object pointer and manages the contained managed object instance pointer group as a unidirectional list. Furthermore, the AOI table is introduced to specify the managed object instance depth on the MIT. This table includes the managed object instance AOI, the containing managed object instance AOI, and the rank of the instance in the MIT.
- Scope and filter: In order to point to the managed object instance, the managed object instance pointing mechanism is furnished. This mechanism processes the logical operators in the scope condition, the filter condition, and AVA(attribute value assertion).

- Management of transaction: Managed object instance entries including "commit", "abort", "prepare", etc. are employed to realize the atomic operation capability based on two-phase commitment. This capability maintains the data integrity of managed object instances.
- Backup operation: Managed object information is backed-up on nonvolatile memory after the completion of each transaction to avoid the loss of management information. In particular, ArchiveManager object is used to manage AOI table and attribute data encoded in ASN.1 during back-up operations.
- Indexing: The performance of managed object instance access is improved by creating Hash or AVLTree indexes.

3.3 MIB implementation evaluation

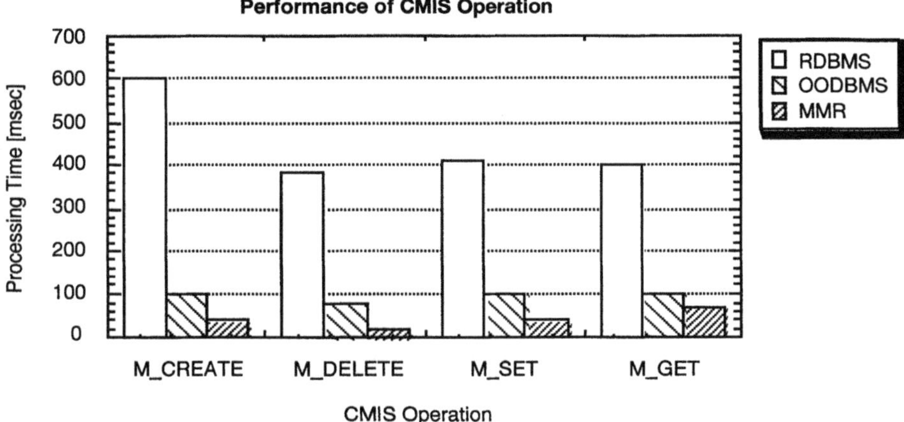

Figure 3 CMIS operations performance.

The above mentioned three implementations were used to implement the same basic MIB, and transaction processing time was measured for each implementation. The MIB, which runs on a UNIX server with RISC processor, stored managed object instances with the attributes of all possible ASN.1 basic data types. Figure 3 shows the average operation times spent to perform CMIS M_CREATE, M_DELETE, M_GET and M_SET operations on a managed object instance as invoked. This result confirms that MMR-MIB achieves the best performance for every examined operation. Regarding M_CREATE operation, MMR-MIB was ten and two times faster that the RDBMS and OODBMS version. In other words, this result shows that MMR-MIB will provide a similar performance to the RDBMS and OODBMS version on control systems that have one tenth and half processing power of RISC processor. While the MMR-MIB implementation supports fewer managed object instances than the other methods, this is not a significant problem for network element applications because network elements manage predictable number of instances and some of them do not require to be persistent in the storage. The performance improvement obtained here is caused by redundant DBMS functions for MIB application elimination such as data schema conversion and ASN.1 data encoding and decoding.

4 MIB DEPLOYMENT FOR AN ATM TRANSPORT NETWORK OPERATIONS SYSTEM

As described in the previous section, each MIB implementation has its advantages and disadvantages. Thus, which is best for each sub-system in a network operation system depends on the technical requirements of the sub-system. This section clarifies sub-system requirements and introduces the strategy of MB assignment. As an example of a network operation system, let us consider an ATM transport network operations system.

4.1 System architecture

Figure 4 depicts the ATM transport network operations system architecture considered in this paper. The hierarchical operations system architecture is adopted (Yoshida, 1992) to increase operation performance and to conform to the TMN standards (CCITT M.3010, 1992). This architecture consists of four layers: the resource layer, the resource control layer, the resource management layer, and the operation scenario management layer. Each layer has sub-systems with MIBs, which store management information to be exchanged through CMIP. The management layers are detailed below.

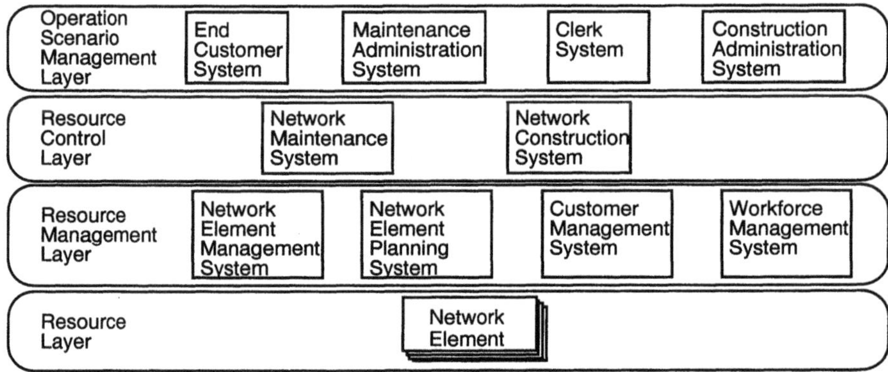

Figure 4 ATM transport network operations system architecture.

1. Resource layer: The sub-systems in the resource layer provide the upper layer sub-systems with a management view of the resources concerned. For example, defects detected by the network element are transformed into alarm notifications. The network element is a potential sub-system in this layer.
2. Resource management layer: The sub-systems in the resource management layer control the management information provided by the resource layer sub-system and generate the management view of logical resources. The network element management system, the network element planning system, the customer management system and the work force management system are located in this layer.
3. Resource control layer: The sub-systems in the resource control layer control the management information of physical and logical resources to provide management views to sub-systems in the operation scenario management layer. Each management view considers

one component of the management scenario. This layer includes the network maintenance and operations system as well as the network construction system.

4. Operation scenario management layer: The sub-systems in the operation scenario management layer perform management scenarios by controlling the sub-systems of the resource management layer. The end customer control system, the maintenance administration system, the clerk system, and the construction administration system are located in this layer.

4.2 Managed object model

Since each sub-system uses CMIP to exchange management information, the management information in each sub-system MIB is modeled as managed objects specified according to GDMO. The managed objects appearing in each management layer are discussed below.

1. Resource layer: As abstractions of network element resources, termination points of SDH Trail and virtual path(VP) Trail, connection information, and packages of equipment and software are modeled as managed objects.
2. Resource management layer: As abstractions of the network, SDH Trails and VP Trails are modeled as managed objects. To handle customer information, the customer's name, the contact phone number, etc. are also modeled.
3. Resource control layer: As components of the management scenario, the procedures used to manage trouble restoration and network construction are modeled as managed objects.
4. Operation scenario management layer: Since sub-system processes of this layer do not play agent roles, there is no managed object except the identifier of agent systems to initiate management operation.

4.3 Requirements and limitations of systems

The technical requirements of each layer are described below.

1. Resource layer: A sub-system in this layer is always a managed open system and stores only the management information that represents the components of the sub-system itself. For this reason, it is easy to predict the number of managed objects when designing the system. Moreover, a large storage resource (memory or disk) is not necessary to store these objects. On the other hand, rapid processing is necessary for the sub-systems, particularly for the case of transmitting alarm notification. Another requirement is that the cost of the sub-system should be low. This is because a large number of sub-systems exist in this layer. To satisfy this requirement, the capabilities of the processing machine used in the sub-system must be limited in terms of processing performance attributes such as CPU power or disk I/O speed. For example, existing network elements are using micro-processor based processor machine such as Motorola 68030, Intel 80386 with limited memories less than 128M bytes.
2. Resource management layer: Sub-systems in this layer must frequently communicate with other sub-systems in other layers and manage a large amount of management information. Therefore, sub-systems must be implemented with server class multipurpose machines.
3. Resource control layer: The requirements are similar to those of the resource management layer; server class multipurpose machines are needed.
4. Operation scenario management layer: Sub-systems in this layer mainly handle the human-machine interface that allow human operators to access operation functions. This does not require large scale database handling. The multipurpose workstation is probably the best candidate.

4.4 MIB deployment strategy

Table 1 compares the three MIB implementations using the following parameters.
• Processing speed: The processing time needed for CMIP operation.
• Transportability: Program transportability among different machines.
• Software installation cost.
• Storage capacity: The available capacity to store management information.
• System maintainability.

We note that the optimal MIB implementation depends on system-specific requirements. According to the requirements described in Section 4.3, the MIB for sub-systems in the resource management and control layers should be based on ready-made DBMSs (RDBMS or OODBMS) because it needs to handle a lot of management information and facilitate system integration and conversion functions. Meanwhile, sub-systems in the resource layer should adopt MMR-MIB to reduce the implementation cost and to increase the operation performance.

Table 1 Comparison of MIB versions

Parameters	*RDBMS-MIB*	*OODBMS-MIB*	*MMR-MIB*
processing speed	low	medium	high
transportability	no	no	yes
software cost	high	high	low
storage capacity	large	large	medium
maintainability	excellent	good	fair

5 MIB EVALUATION

5.1 Evaluation method

In order to clarify the effectiveness of MMR-MIB in the ATM transport network operations system, a prototype system was developed using the TNMSKernel and evaluated in terms of management processing time. Figure 5 illustrates the target ATM transport network and its management system. The network consists of the ATM cross-connect system(ATM-XC), the ATM subscriber line terminal(ATM-SLT), and the digital subscriber unit(DSU) located in the customer premises. This design has the ATM-SLT manage the DSU while the ATM-SLT and the ATM-XC manage physical resources of the network such as packages and termination points. The network management system (NMS) controls the ATM virtual path (ATM-VP) Trails and the SDH Trails established between network elements. In addition to these components, a debug manager was deployed to initiate the NMS.

In this prototype, the MIB in the NMS was implemented on an OODBMS while the network elements used MMR-MIB. RISC-based UNIX workstations were used as the processing machines of each sub-system in this experiment. The communication protocol between components was CMIP over TCP/IP. The Directory access function was used to realize location transparency of managed objects (Minato, 1993) .We examined the following two operation scenarios to evaluate the processing time.

1. SDH Trail Creation: This creates SDH Trail managed objects between ATM-XC and ATM-SLT as well as ATM-SLT and DSU. This also creates termination point managed objects such as VP Adaptors, VP connection termination points (VPCTP).

2. VP Trail Creation: This creates VP Trail from SDH Trails by obtaining the bandwidth of each SDH Trail and establishing the appropriately sized cross-connection in ATM-XC.

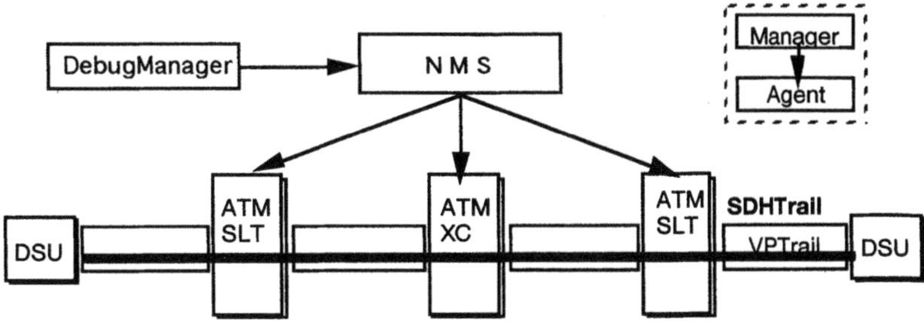

Figure 5 ATM transport network operations system architecture.

5.2 Evaluation

A number of CMIP operations made in this experiment are indicated in Table 2 M_GET operations were used to check the availability of network resources. M_SET operations were used to unlock managed object administrative state. M_ACTION operations were used to create multiple VPCTP and cross-connections in SLT and XC while M_CREATE operations were used to create multiple VPCTP of DSU.

Table 2 CMIP operations in the experiment

Operations .	*agents*	*M_GET*	*M_SET*	*M_CREATE*	*M_ACTION*
SDH Trail Creation	NMS			1	1
(XC/SLT)	XC	10	4	8	1
	SLT	13	3	8	1
SDH Trail Creation	NMS			1	
(SLT/DSU)	SLT(DSU)	15	3	260	
VP Trail Creation	NMS			5	1
	XC	3			1
	SLT(DSU)	14	2	2	2

Table 3 indicates the average operation processing time for each operation scenario. Managed object creation time for SLT/DSU SDH Trail is large because the SLT manages both termination points on SLT and DSU and it requires 260 VPCTP creations in DSU. The VP Trail Creation time was smaller than SDH Trail Creation time. This is because number of termination point creations are smaller than SDH Trail Creation case. This result verifies that operation performance is sufficient. This operation performance can also be improved by reducing the availability check sequences in each Trail Creation. It also confirms the validity of the proposed MIB deployment strategy. Since a previous experiment on a RDBMS-MIB yield processing times of 10 to 20 seconds(Yata, 1994), the proposed method offers improved operation process time.

Regarding the required MIB size in network elements, it was clarified that an XC needs more than a half million managed object instances to represent its operation function. Thousand managed object instances among them need to be persistent on the storage and less than thirty thousand managed object instances need to be visible at the same time. In order to reduce required memory size on XC, we introduced a virtual managed object representation technique to make managed object visible on the MIB memory space when they are needed. Those object instances are reloaded on to the MIB memory by programs. By using this method, it was confirmed that XC requires less than 70M bytes memory to realize its operation function. This is within the range of network element processing capability.

Table 3 Operation processing time

Operations	*Average processing time(sec.)*
SDH Trail Creation (XC/SLT)	4.3
SDH Trail Creation (SLT/DSU)	7.5
VP Trail Creation	3.6

6 CONCLUSION

This paper has considered implementing the MIB function on three types of database management systems: RDBMS, OODBMS, and the newly proposed main memory resident technique. The performances of each implementation were evaluated by realizing the same MIB function on "TNMSKernel". The MIB based on the main memory resident technique offers significantly improved performance, which makes it suitable for relatively small systems such as network elements. An MIB deployment strategy was proposed for a hierarchical ATM transport network operations system architecture. Experimental results confirmed that excellent performance is achieved by adopting the appropriate MIB method in each sub-system.

7 ACKNOWLEDGEMENT

Authors wish to thank Dr. Ikuo Tokizawa for their support and Dr. Tetsuya Miki for his encouragement. Authors also thank Mr.Kouji Yata, of Telecommunications Software Headquarters, NTT, for his great help in implementing the experimental systems.

8 REFERENCES

Ammann, A.C., Hanrahan, M.B. and Krishnamurthy, R. (1985) Design of memory resident DBMS. IEEE COMPCOM.

CCITT Recommendation M.3010 (1992) Principles for a Telecommunications Management Network (TMN).

CCITT Recommendation X.701 (1992) | ISO/IEC 10040 (1992), Information Technology - Open Systems Interconnection - Systems management overview.

CCITT Recommendation X.711 (1992) | ISO/IEC 9596-1 (1991 (E)), Information Technology - Open Systems Interconnection - Common Management Information Protocol Specification - Part 1: Specification, Edition 2.

CCITT Recommendation X.722 (1992) | ISO/IEC 10165-4 (1992), Information Technology - Open Systems Interconnection - Structure of Management Information: Guidelines for the Definition of managed objects.

Chorafas, D.N. and Steinmann, H. (1993) Object-Oriented Databases, PTR Prentice Hall, Englewood Cliffs,New Jersey.

Dossogne, F. and Dupont, M.P. (1993) A software architecture for Management Information Model definition, implementation and validation. Integrated Network Management, III(C-12), San Francisco.

Huslende, R. and Voldnes, I. (1993) A Q3 Interface for Managing a National Telecommunication Network: Motivation and Implementation. ICC'93, Geneva.

Minato, K., Yoda, I. and Fujii, N. (1993) Distributed Operation System Model using Directory Service in Telecommunication Management Network. GLOBECOM'93, Houston.

Molina, H.G. and Salem, K. (1992) Main Memory Database Systems: An Overview. IEEE Trans. on Knowledge & Data Engineering 4(6), 509-516.

Ullman, J.D. (1988) Principles of Database and Knowledge-Base systems. Co. Computer Science Press.

Yata, K., Yoda, I., Minato, K. and Fujii,N. (1994) ATM Transport Operation System Based on Object Oriented Technologies. GLOBECOM'94, San Francisco.

Yoda, I., Minato, K. and Fujii, N. (1992) Development of transmission networks operation systems programs by GDMO Translator. Techinical Report of IEICE CS92-54,Japan.

Yoda, I., Sakae, K and Fujii, N. (1992) Configuration of a Local Fiber Optical Network Management System based on Multiple Manager Systems Environment. NOMS'92, Nashville.

Yoda, I. and Fujii, N. (1993) Method for Constructing a Management Information Base (MIB) in Transmission Network Operations: Electronics and Communications in Japan.76, 21-33.

Yoda,I., Yata, K. and Fujii, N. (1994) Object Oriented TMN Based Operations Systems Development Platform. SUPERCOMM/ICC'94, New Orleans.

Yoshida, T., Fujii, N. and Maki, K. (1992) An Object-oriented Operation System Configuration for ATM Networks. ICC'92, Chicago.

9 BIOGRAPHY

Tomoaki Shimizu was born in Kanagawa, Japan, in April 1965. In 1988, after receiving his B.S. degree in electronics engineering from Musashi Institute of Technology, Tokyo, Japan, he joined Nippon Telegraph and Telephone Corporation. He has been engaged in the development of private network management systems and currently in the research on transmission network management systems, TMN based operation systems and modeling and implementation of MIB.

Ikuo Yoda was born in Tokyo, Japan, in 1963. He received the B.S. and M.S. degrees in electronics engineering from Waseda University in Tokyo, Japan, 1986 and 1988. In 1988, he joined Nippon Telegraph and Telephone Corporation(NTT's) Transmission systems Laboratories. Since then, he has been engaged in the research on transmission network management systems, TMN based operation systems and modeling and implementation of MIB.

Nobuo Fujii received the B.E. and M.E. degrees in applied physics from Osaka University in 1977 and 1979, respectively. In 1979, he joined NTT. Since then, he has been engaged in the research and development of control system for digital cross-connect systems, the high speed digital leased line system, and the telecommunications network operations system. He is currently running a research group in NTT Optical Network Systems Laboratories. He is a member of the IEEE.

SECTION FIVE

Managed Objects Relationships

48
Towards Relationship-Based Navigation

Subodh Bapat
BlacTie Systems Consulting
16441 Blatt Blvd., Suite 203, Ft. Lauderdale, Florida 33326, USA
Phone: +1 305 389 8347 *Email:* `bapat@gate.net`

Abstract
This paper builds upon the OSI General Relationship Model and presents mechanisms to perform relationship-based navigation among the managed object classes of the OSI Management Information Model. Examples demonstrate how such relationship-based navigation through the semantic network can permit extended reasoning and inferencing during network management.

1 INTRODUCTION

In the OSI General Relationship Model (GRM) [X.725], a relationship among two or more managed object classes is specified using a *managed relationship class*. A managed relationship class describes the characteristics of the managed relationship independent of the actual classes that may participate in that relationship. Such characteristics include the roles which its participant managed objects play, the cardinalities with which they participate in the relationship, the behavior of the relationship, and any additional constraints and dependencies that may govern the participation of managed objects in that relationship.

The participation of a specific set of managed object classes in a managed relationship is described in a *role binding*. A role binding asserts that a particular managed relationship holds between particular managed object classes, and also indicates the roles played by each participant managed object class in the relationship. A role binding also specifies additional behavior, constraints on roles, and the conditions under which participant managed objects can enter and exit the managed relationship.

The same relationship class may be used in different role bindings to bind different groups of managed object classes in relationships. For example, the roles of a backup relationship — such as `backs-up` and `is-backed-up-by` — may be defined in the `backup` managed relationship class, independent of the managed object classes that participate in such relationships. Once this relationship class is established, one role binding may bind the `dialUpCircuit` managed object class in the `backs-up` role with the `dedicatedCircuit` managed object class in the `is-backed-up-by` role (indicating that a dial-up circuit may back up a dedicated circuit, as is typical in private data networks), while another role binding may bind the `serviceControlPoint` managed object class in the `backs-up` role with the `adjunctProcessor` managed object class in the `is-backed-up-by` role (indicating that an SCP may back up an Adjunct Processor, as is typical in Intelligent Network architectures).

Although relationships are formally specified using the template notation of the General Relationship Model, it is often helpful for comprehension to depict them graphically as well. We depict them using *extended Kilov diagrams*. In a Kilov diagram [Kilo94a, Kilo94b], the relationship construct is indicated in a rectangle, as are the participating object classes; the triangle construct "`Rel`" between them indicates that the classes participate in the indicated relationship. In this paper, we represent each role binding with a Kilov diagram, extending it to also depict the roles which the participant classes play with each other.

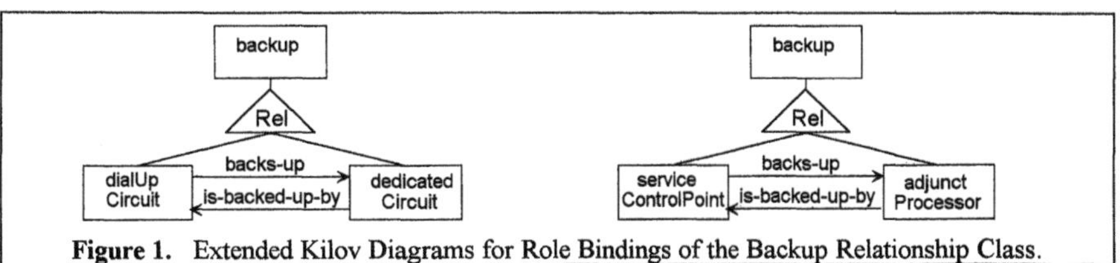

Figure 1. Extended Kilov Diagrams for Role Bindings of the Backup Relationship Class.

The formal specification of managed relationship classes and role bindings is performed using templates defined for that purpose. For example, the `backup` managed relationship class (simplified for the purposes of this paper) may be defined as

```
backup RELATIONSHIP CLASS
  BEHAVIOUR backupBehaviour BEHAVIOUR DEFINED AS
    "Backup object assumes failover operation for backed-up object"
  ROLE backs-up REGISTERED AS {...}
  ROLE is-backed-up-by REGISTERED AS {...}
REGISTERED AS {...};
```

Different role bindings may now be established for this relationship class. As one example, assuming that `dialUpCircuit` and `dedicatedCircuit` are both managed object classes defined and registered elsewhere in their own `MANAGED OBJECT CLASS` templates, the following role binding establishes the required backup relationship:

```
dialUpCkt-backsUp-dedicatedCkt ROLE BINDING
  RELATIONSHIP CLASS backup
  BEHAVIOUR circuitBackupBehaviour BEHAVIOUR DEFINED AS
    "dialUpCircuit assumes failover operation for dedicatedCircuit"
  ROLE backs-up RELATED CLASSES dialUpCircuit AND SUBCLASSES
  ROLE is-backed-up-by RELATED CLASSES dedicatedCircuit AND SUBCLASSES
REGISTERED AS {...};
```

The same relationship may be established between other managed object classes using other role bindings. The templates above have been intentionally simplified to keep the focus on roles played participant object classes; role cardinality constraints, for example, have not been specified. The complete `RELATIONSHIP CLASS` template and `ROLE BINDING` template define many additional characteristics of a relationship. The `RELATIONSHIP CLASS` template specifies the *constraints* which must be satisfied by managed object instances in order to be participants in the relationship. It also specifies various *dependencies* which describe how the participation of a managed object instance in the relationship is influenced by its participation in other relationships. It specifies *relationship management operations* that may be performed, e.g. relationship establishment, binding, querying, notification, unbinding, and termination. The managed relationship class template also specifies the conditions governing the *dynamic entry* and *dynamic departure* of a participant managed instance in an established relationship.

Aside from binding managed managed object classes is a relationship, the `ROLE BINDING` template also specifies various ways of representing the relationship. For example, a relationship may be represented by a separate *relationship object* (an instance of the relationship class) whose attributes indicate the names of the participating managed objects. Such an implementation is typical for relationships having a many-to-many role cardinality. A relationship may alternatively be represented by "pointer attributes" within each participant managed instance, whose value indicates the other managed object instance(s) to which that object is currently bound. An implementation using such conjugate pointers is typical for many relationships having a one-to-one role cardinality. A `ROLE BINDING` template also specifies an *operations mapping*, which indicates how relationship management operations map to ordinary systems management operations on managed object classes. For example, in a conjugate pointer implementation, the relationship management operation to unbind the relationship may simply map to the systems management operation of setting the values of the conjugate pointer attributes in the participant managed objects to null.

Although all these aspects are important for the complete specification of a relationship, we will not concentrate on them in this paper, as our focus is on the semantics of *relationship-based navigation*. For simplicity, our examples will omit those clauses of the `RELATIONSHIP CLASS` and `ROLE BINDING` templates which are not relevant to semantic extensions we propose; it ought to be borne in mind, however, that a complete (compilable) specification of a relationship should include all the required template clauses.

In this paper, we introduce new concepts in the modeling of relationships by exploiting special properties of the *roles* of a relationship class. By defining operations on roles, we can enhance the GRM to include several semantically useful concepts. These concepts allow us to express extended relationships within our model precisely and succinctly.

2 CONCEPTUAL BACKGROUND

A *virtual relationship* is a relationship whose existence can be inferred from other relationships [Bapa94]. A virtual relationship is not created by relationship establishment; it is dynamically computed and resolved within the management information repository from existing established relationships. The supporting relationships which give rise to a virtual relationship are termed *base relationships*. A virtual relationship implicitly arises when the roles of its base relationships have certain special *properties*.

We define an *actual relationship* as a relationship which cannot be inferred from the properties of roles of other relationships, and therefore must be explicitly created by the architect using a `ROLE BINDING` template. The base relationships which give rise to a virtual relationship may be actual relationships, or may themselves be virtual [Bapa93b].

A *virtual relationship instance* is formed by the set of object instances which participate in the virtual relationship. A virtual relationship does not make existing objects participants in a new relationship. Rather, objects which are already participants in *actual relationship instances* become automatic participants in virtual relationship instances, because of the special properties of the roles they play in their actual relationships. Thus, although a virtual relationship may have instances, it can never be *imperatively established*; only actual relationships can.

Therefore, operations such as `BIND`, `UNBIND`, `ESTABLISH` and `TERMINATE` are illegal on a virtual relationship. A virtual relationship is *automatically* established and terminated as and when its supporting base relationships are established and terminated. Objects are *automatically* bound and unbound in a virtual relationship as and when they are bound and unbound in its supporting base relationships. Any change made to its supporting actual base relationships will be automatically reflected in the virtual relationship, since the virtual relationship instances are, in effect, dynamically resolved from actual base relationship instances every time they are queried. As far as the user is concerned, the `QUERY` operation works exactly the same way on a virtual relationship as it does on an actual relationship.

3 PROPERTIES OF ROLES

A virtual relationship arises as a consequence of special *properties* possessed by the roles of its supporting base relationships. A property of a role is a shorthand mechanism for specification reuse, which allows us to define many extended relationships from a single construct. By indicating the properties a role possesses, we create a mechanism which captures within a single relationship class *more* semantics than just the usual association between participant object classes, their cardinalities, participation constraints, and roles. By specifying our knowledge of the special properties of roles in extension clauses of the `RELATIONSHIP CLASS` template, we can compile into our management information repository the ability to perform extended navigation through relationship semantics.

There are five important properties which the roles of a relationship class may possess:

- The *Commutativity* property;
- The *Transitivity* property;
- The *Distribution* property;
- The *Convolution* property; and
- The *Implication* property.

It is important to emphasize that these properties belong to the roles of relationship classes, and not to role bindings. Thus, if the roles of a relationship class possess these properties, they will be operative in all role bindings in which that relationship class is used.

4 COMMUTATIVE VIRTUAL RELATIONSHIPS

A *commutative virtual relationship* is a relationship which arises from the *commutativity* property of the roles in its base relationship class. We define commutative roles as follows:

> *A pair of roles* `{r1,r2}` *of a relationship class is said to be* ***commutative*** *if, given two managed object classes* `A` *and* `B`*, and a role binding in which* `A` *plays role* `r1` *with* `B` *and* `B` *plays role* `r2` *with* `A`*, then it can be inferred that* `B` *plays role* `r1` *with* `A` *and* `A` *plays role* `r2` *with* `B`.

This states that any role binding of a base relationship class with commutative roles automatically implies the existence of another role binding in which the roles are "flipped around".

Many examples of commutative virtual relationships exist in network modeling. The most common examples of these are *connectivity* and *interconnectivity* relationships between instances of network devices. These relationships can be traversed from either direction in the semantic network: if an object a_0 is interconnected (perhaps via many intermediate nodes) to object b_0, then the repository can infer (e.g. in a topology display application) that object b_0 is interconnected to object a_0.

Another common example of commutative relationships are *mutual backup relationships*. It is important to understand that *all* backup relationships are not commutative. For example, the `backup` relationship class we defined earlier with the roles `backs-up` and `is-backed-up-by` did not have the commutativity property. Being non-commutative, this relationship is not mutual; it is only a one-way relationship. When we use this relationship to inform the repository that managed object class `A` plays the role `backs-up` with managed object class `B` (or that `B is-backed-up-by A`), the repository *cannot* infer that `B backs-up A` (or its reciprocal, `A is-backed-up-by B`).

We might not wish to invest the roles `backs-up` and `is-backed-up-by` with the commutativity property, because in some contexts (as in the `dialupCkt-backsUp-dedicatedCkt ROLE BINDING` above) it may be used as a non-commutative (one-way) backup relationship. We specify a new relationship class — say, the `mutualBackup` relationship class. (Since relationship classes may derive from each other due to inheritance, it is possible for the `mutualBackup` relationship class to derive from the `backup` relationship class. This is omitted here for simplicity.) Assume that the `mutualBackup` relationship class has the roles `mbacks-up` and `is-mbacked-up-by`, standing for "mutually backs up" and "is mutually backed up by". We invest *these* roles with the commutativity property. Thus, this relationship class carries more semantics than the one-way `backup` relationship class. (In a later section, we will see how the semantics of a single `mutualBackup` relationship can be made to imply the semantics of the one-way `backup` relationship in both directions.)

```
mutualBackup RELATIONSHIP CLASS
  ROLE mbacks-up COMMUTATIVE REGISTERED AS {...}
  ROLE is-mbacked-up-by COMMUTATIVE REGISTERED AS {...}
REGISTERED AS {...};
```

Consider an example of a `signalTransferPoint` object class, instances of which are generally deployed in "mated pairs" with each other in an Intelligent Network. By cross-linking pairs of `signalTransferPoint` objects, we provide redundancy in the signalling network. This may be specified as a relationship with `mutualBackup` roles. If so, it implies that for each relationship instance between pairs of `signalTransferPoint` objects, the repository can also infer a commutatively derived virtual relationship instance.

```
stp-mbacksUp-stp ROLE BINDING
  RELATIONSHIP CLASS backup
  ROLE mbacks-up RELATED CLASSES signalTransferPoint AND SUBCLASSES
  ROLE is-mbacked-up-by
                  RELATED CLASSES signalTransferPoint AND SUBCLASSES
REGISTERED AS {...};
```

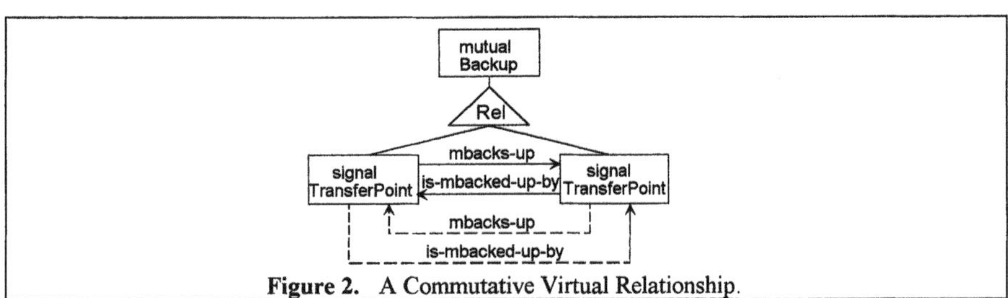

Figure 2. A Commutative Virtual Relationship.

5 TRANSITIVE VIRTUAL RELATIONSHIPS

A *transitive virtual relationship* is a relationship which arises from the *transitivity* property of the roles in its base relationship class. We define transitive roles as follows:

A pair of roles `{r1,r2}` *of a relationship class is said to be* ***transitive*** *if, given the managed object classes* `A`, `B`, *and* `C`, *a first role binding in which* `A` *plays role* `r1` *with* `B` *and* `B` *plays role* `r2` *with* `A`, *and a second role binding in which* `B` *plays role* `r1` *with* `C` *and* `C` *plays role* `r2` *with* `B`, *then it can be inferred that* `A` *plays role* `r1` *with* `C` *and* `C` *plays role* `r2` *with* `A`.

This definition implies that relationship roles are transitive if, given a common "linking" participant, they can be "chained together". Well-known examples of transitive relationships are `interconnectivity` relationships which are important for fault diagnostics and topology display purposes. For example, if we model the `interconnectivity` relationship class with the roles `interconnects-to` and `is-interconnected-to`, then we know that that if `A` `interconnects-to` `B` and `B` `interconnects-to` `C`, it follows that `A` `interconnects-to` `C`.

There are other examples of transitive relationships in network modeling. Consider a network of electronic mail application processes, which exchange electronic mail messages among themselves over local and wide-area computer networks. These application processes may all have different implementations and protocols: some could be Message Transfer Agents, some could be mail handling demon processes, and so on. Although these mail handlers may use different standards, through the use of .programs like `sendmail` or other electronic mail gateways with address translation mechanisms, they may all have the ability to forward mail to each other.

The `mailForwarding` relationship class may be modeled with the roles `{forwards-mail-to, receives-mail-from}`. Clearly, this relationship is commutative. We could further provide additional information about this relationship to the repository by specifying this relationship as being *transitive*. This implies that once we create any two different `mailForwarding` role bindings between any three different managed object classes (such as `x400mta`, `smtpDemon` and `uucpDemon`), we also automatically create a transitively derived virtual relationship instance between the third pair.

```
mailForwarding RELATIONSHIP CLASS
  ROLE forwards-mail-to COMMUTATIVE TRANSITIVE REGISTERED AS {...}
  ROLE receives-mail-from COMMUTATIVE TRANSITIVE REGISTERED AS {...}
REGISTERED AS {...};

x400-forwardsto-smtp ROLE BINDING
  RELATIONSHIP CLASS mailForwarding
  ROLE forwards-mail-to RELATED CLASSES x400mta AND SUBCLASSES
  ROLE receives-mail-from RELATED CLASSES smtpDemon AND SUBCLASSES
REGISTERED AS {...};

smtp-forwardsto-uucp ROLE BINDING
  RELATIONSHIP CLASS mailForwarding
  ROLE forwards-mail-to RELATED CLASSES smtpDemon AND SUBCLASSES
  ROLE receives-mail-from RELATED CLASSES uucpDemon AND SUBCLASSES
REGISTERED AS {...};
```

Because the `mailForwarding` relationship class has transitive roles, given the role bindings above the repository can automatically infer a role binding for the `mailForwarding` relationship between `x400mta` and `uucpDemon`, even though such a role binding has not been explicitly specified in the information model.

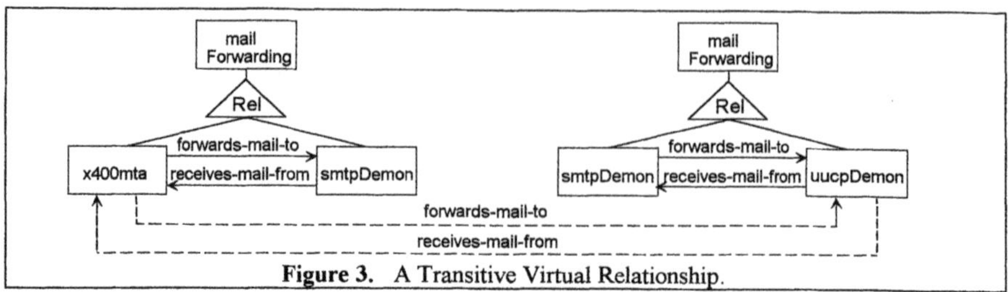

Figure 3. A Transitive Virtual Relationship.

6 DISTRIBUTIVE VIRTUAL RELATIONSHIPS

A *distributive virtual relationship* is a relationship which arises from the *distribution* property of the roles in its base relationship class. We define distributive roles as follows:

> *A pair of roles* `{r1,r2}` *of a relationship class is said to* ***distribute over*** *another pair of roles* `{r3,r4}` *of another relationship class if, given the managed object classes* `A`, `B`, *and* `C`, *a first role binding in which* `A` *plays role* `r1` *with* `B` *and* `B` *plays role* `r2` *with* `A`, *and a second role binding in which* `B` *plays role* `r3` *with* `C` *and* `C` *plays role* `r4` *with* `B`, *then it can be inferred that* `A` *plays role* `r1` *with* `C` *and* `C` *plays role* `r2` *with* `A`. *The role* `r1` *is said to* ***distribute ahead*** *of role* `r3`, *and the role* `r2` *is said to* ***distribute behind*** *role* `r4`.

This definition states that, given a common "linking" participant object class, the *distributing* roles `{r1,r2}` *distribute over* the *distributand* roles `{r3,r4}`. That is, the roles `{r1,r2}` can be virtually established between `A` and `C`.

As an example, suppose we wish to model geographic information about `equipment` objects. We model this as a `housing` relationship between the `location` object class and the `equipment` object class, in which `location` plays the role `houses` with `equipment`, which reciprocates by playing `is-housed-at` with `location`. (The class `location` is modeled as a separate class because it could have its own attributes, such as `streetAddress`, `telephoneNumber`, and so on.)

There may also be a `termination` relationship between the `equipment` object class and the `circuit` object class, in which `equipment` plays the role `terminates` with `circuit`, which plays the role `is-terminated-at` with `equipment`. Occasionally, rather than knowing which `equipment terminates` a given `circuit`, it may be useful for certain outside-plant engineers to know the physical address or location where the circuit `is-terminated-at`. Ordinarily, we would have to perform two queries for this information: one to determine which instance of `equipment` the `circuit is-terminated-at`, and another to determine which instance of `location` that `equipment is-housed-at`.

To make this more concise, we may simply say that the roles of the `termination` relationship class *distribute over* the roles of the `housing` relationship class. More specifically, this means that the `is-terminated-at` role distributes *ahead of* the `is-housed-at` role. This creates a *distributive virtual relationship* between `circuit` and `location`: thus, if instance c_0 of `circuit is-terminated-at` instance d_0 of `equipment`, and instance d_0 of `equipment is-housed-at` instance l_0 of `location`, we may infer (e.g., for topology display purposes) that `circuit` c_0 `is-terminated-at location` l_0.

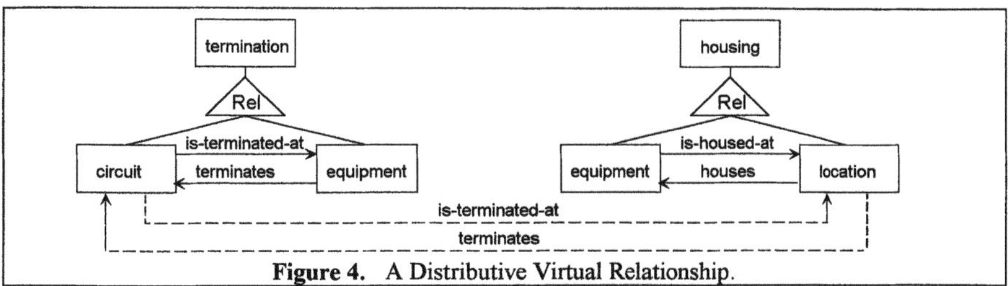

Figure 4. A Distributive Virtual Relationship.

The same semantics could be equivalently stated in terms of reciprocal roles. We could say that the `terminates` role *distributes behind* the `houses` role. This means that if some `equipment terminates` a `circuit`, and some `location houses` that `equipment`, then that `location` also `terminates` the `circuit`.

```
housing RELATIONSHIP CLASS
  ROLE is-housed-at REGISTERED AS {...}
  ROLE houses REGISTERED AS {...}
REGISTERED AS {...};

termination RELATIONSHIP CLASS
  ROLE is-terminated-at DISTRIBUTES AHEAD OF is-housed-at
      REGISTERED AS {...}
  ROLE terminates DISTRIBUTES BEHIND houses REGISTERED AS {...}
REGISTERED AS {...};

equipment-terminates-circuit ROLE BINDING
  RELATIONSHIP CLASS termination
  ROLE is-terminated-at RELATED CLASSES circuit AND SUBCLASSES
  ROLE terminates RELATED CLASSES equipment AND SUBCLASSES
REGISTERED AS {...};

location-houses-equipment ROLE BINDING
  RELATIONSHIP CLASS housing
  ROLE is-housed-at RELATED CLASSES equipment AND SUBCLASSES
  ROLE houses RELATED CLASSES location AND SUBCLASSES
REGISTERED AS {...};
```

Given the definitions above, the repository can automatically infer the existence of a role binding of the `termination` relationship class between `circuit` and `location`, even though such a role binding is not explicitly created by the architect. In general, relationships may distribute over base relationships regardless of whether the base relationships are actual or virtual.

Since a virtual relationship instance may be queried exactly like an actual relationship instance, this implies that if we queried an instance of a `circuit` for the `location` where it terminated (that is, we tracked the `location` object to which it is bound via its `is-terminated-at` role) we would directly get the correct instance of `location`, without having to compose any relational joins in our query to include the intermediate `equipment` object class. Under conventional modeling, some form of a join condition between entities would be required in the query in order to elicit the desired response — even if the implementation platform for the management information repository is not relational.

A little reflection indicates that a transitive virtual relationship is a special case of a distributive virtual relationship in which both the distributing and distributand roles are the same.

7 CONVOLUTE VIRTUAL RELATIONSHIPS

A *convolute virtual relationship* is a relationship which arises from the *convolution* property of the roles in its base relationship class. We define convolute roles as follows:

> *A pair of roles* {r1,r2} *is said to* ***convolute from*** *the roles* {r3,r4} *and* {r5,r6} *if, given three participant managed object classes* A, B, *and* C, *a first role binding in which* A *plays role* r3 *with* B *and* B *plays role* r4 *with* A, *and a second role binding in which* B *plays role* r5 *with* C *and* C *plays role* r6 *with* B, *then it can be inferred that* A *plays role* r1 *with* C *and* C *plays role* r2 *with* A. *The role* r1 *is said to* ***convolute above*** *the roles* r3 *and* r5, *and the role* r2 *is said to* ***convolute below*** *the roles* r4 *and* r6.

This definition states that, under certain circumstances, two role bindings of separate relationship classes may give rise to a virtual relationship with an entirely different pair of roles.

Consider an example in which an information service provider operates a data network which supplies multiple information services, such as stock and bond price quotations, to multiple clients. Clients such as brokerage firms subscribe to these information services. The terms of a typical subscription require the information service to automatically provide the information of interest to the client. The quotation service downloads this information to the trading computers of all employees of the corporate client. Because the information of interest is dynamic and useful only within a small life span, each trading computer downloads this information directly from the service provider, rather than from any internal central redistribution database owned and operated by the client company itself. If the client adds more trading computers, they too will receive information directly from the service provider.

We model the `subscription` relationship class with the roles `subscribes-to` and `is-subscribed-to-by`. Thus, we might model a role binding in which a `quotationService` object class `is-subscribed-to-by` a `brokerageFirm` object class. Further, we might also model a `tradingComputer` object class which participates in an `ownership` relationship with the `brokerageFirm` class. The `brokerageFirm` class plays the role `owns` with the `tradingComputer` class, which reciprocates with the role `is-owned-by`. We also know that the `quotationService` class must play a `downloads-to` role with respect to the `tradingComputer` class. This would be a `download` relationship, in which the `tradingComputer` reciprocates with the role `downloads-from`.

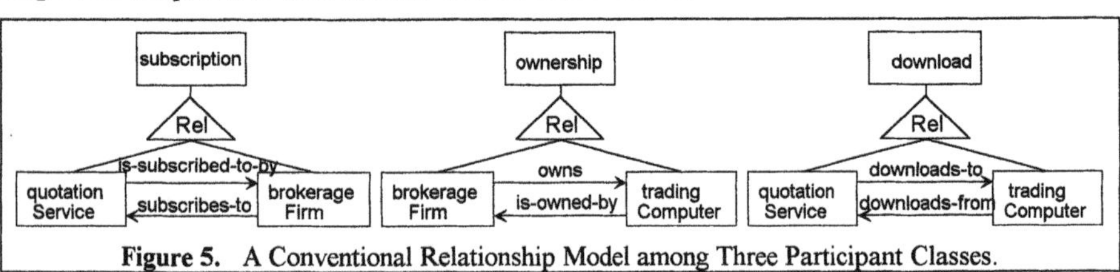

Figure 5. A Conventional Relationship Model among Three Participant Classes.

In this situation, each time the `brokerageFirm` adds a new `tradingComputer`, we must create a new instance of the `ownership` relationship between these two classes. Ordinarily, we must also establish a separate instance of the `download` relationship between the `quotationService` object class and the same instance of `tradingComputer`. We will always have to ensure that all such parallel relationships are consistently maintained. Every time the `brokerageFirm` decommissions or scraps a `tradingComputer`, we must delete the ownership relationship instance with that tradingComputer. We must then ensure that we also delete the corresponding instance of the download relationship between the quotationService object and the same tradingComputer. We would have to maintain this consistency using some mechanism external to the relationships, since we have no mechanism *within* the relationships to *automatically shadow* the changes of one set of relationship instances in another.

We can eliminate this problem entirely by defining the download relationship to be a *convolute virtual relationship* which *convolutes from* the subscription and ownership relationships. The role downloads-to *convolutes above* the first role is-subscribed-to-by played by quotationService with brokerageFirm and the second role owns played by brokerageFirm with tradingComputer. The reciprocal role downloads-from *convolutes below* the two roles subscribes-to and is-owned-by. With this specification, a virtual relationship instance of the download relationship is automatically created or destroyed every time an actual relationship instance of the ownership relationship is created or destroyed. (Or, several instances of the download relationship are automatically created or destroyed each time a subscription relationship is created or destroyed.)

```
subscription RELATIONSHIP CLASS
  ROLE subscribes-to REGISTERED AS {...}
  ROLE is-subscribed-to-by REGISTERED AS {...}
REGISTERED AS {...};

ownership RELATIONSHIP CLASS
  ROLE owns REGISTERED AS {...}
  ROLE is-owned-by REGISTERED AS {...}
REGISTERED AS {...};

download RELATIONSHIP CLASS
  ROLE downloads-to CONVOLUTES ABOVE is-subscribed-to-by AND owns
      REGISTERED AS {...}
  ROLE downloads-from CONVOLUTES BELOW subscribes AND is-owned-by
      REGISTERED AS {...}
REGISTERED AS {...};

firm-subscribes-to-infoService ROLE BINDING
  RELATIONSHIP CLASS subscription
  ROLE is-subscribed-to-by
    RELATED CLASSES quotationService AND SUBCLASSES
  ROLE subscribes-to RELATED CLASSES brokerageFirm AND SUBCLASSES
REGISTERED AS {...};

firm-owns-computer ROLE BINDING
  RELATIONSHIP CLASS ownership
  ROLE owns RELATED CLASSES brokerageFirm AND SUBCLASSES
  ROLE is-owned-by RELATED CLASSES tradingComputer AND SUBCLASSES
REGISTERED AS {...};
```

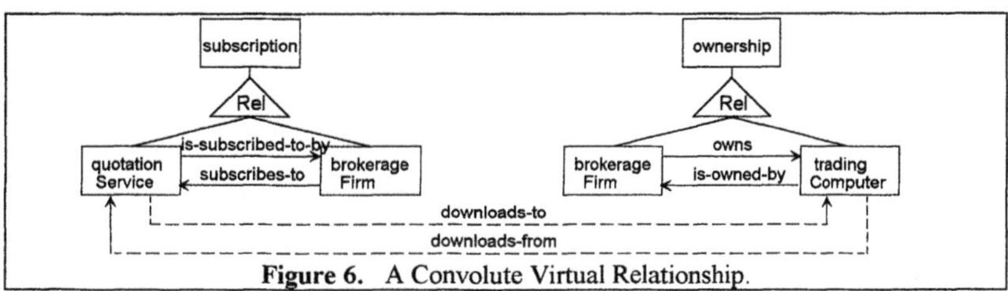

Figure 6. A Convolute Virtual Relationship.

Given the definitions above, the repository can automatically infer the existence of a role binding of the download relationship class between quotationService and tradingComputer, even though such a role binding is not explicitly created by the architect.

A little reflection indicates that a distributive virtual relationship is a special case of a convolute virtual relationship in which the convolute virtual roles are the same as the base distributing roles.

8 IMPLICATE VIRTUAL RELATIONSHIPS

An *implicate virtual relationship* is a relationship which arises from the *implication* property of the roles in its base relationship. We define implicate roles as follows:

> *A pair of roles* {r1,r2} *of a relationship class is said to* **implicate** *another pair of roles* {r3,r4} *if, given two participant managed object classes* A *and* B, *and a role binding in which* A *plays role* r1 *with* B *and* B *plays role* r2 *with* A, *it can be inferred that* A *plays* r3 *with* B *and* B *plays role* r4 *with* A.

This definition states that a role binding with the base roles automatically *implicates* (implies) the existence of a virtual relationship between the same participant managed objects with different roles.

Implication is actually one of the most general forms of virtual relationships. It is important to note the details of its definition: the definition does not require that the implicate roles have the same *properties* as the base roles. It is possible that the base roles have properties such as commutativity, transitivity, and so on, which do not necessarily carry over to the implicate roles. It is also possible that the implicate roles possess properties which the base roles do not. Generally, the implicate virtual relationship conveys different semantics than its base relationship. A few examples will clarify this point.

An implicate virtual relationship arises when the base relationship acts as a "stronger" expression or shorthand for a collection of several "looser" relationships. In general, implicate virtual relationships arise because they indicate consequential roles which exist because the base roles exist. As a general guideline, if consequential roles arise from a single base relationship, they are specified as an implicate virtual relationship. If consequential roles arise from two base relationships with a common "linking" participant, they are specified as a transitive, distributive, or convolute virtual relationship.

Consider the backup relationship class whose roles are backs-up and is-backed-up-by. These roles are not *transitive*: if in one role binding object class A backs-up object class B and in another role binding object class B backs-up object class C, it is not true that object class A backs-up object class C. This is because if an instance of C fails, it is not true that an instance of A will take over.

However, these roles gives rise to the implicate roles of the pointOfFailure relationship class, i.e. {is-a-point-of-failure-for, has-point-of-failure}. Thus if A backs-up B and B backs-up C, then A is-a-point-of-failure-for B and B is-a-point-of-failure-for C. These roles convey different semantics than the roles of the backup relationship class, and also have different properties: they *are* transitive. While it is not the case that A backs-up C, due to the transitivity of A is-a-point-of-failure-for B and B is-a-point-of-failure for C, it is true that A is-a-point-of-failure-for C. Further, the implicate roles also have a *distribution* property which its base roles do not have: the role has-point-of-failure distributes ahead of the contains role which component objects play with composite objects. If C has-point-of-failure A and A contains D, then it is true that C has-point-of-failure D. Thus, it is possible to have implicate virtual relationships with completely different properties than their supporting base relationships.

```
pointOfFailure RELATIONSHIP CLASS
  ROLE has-point-of-failure TRANSITIVE
           DISTRIBUTES AHEAD OF contains REGISTERED AS {...}
  ROLE is-a-point-of-failure-for TRANSITIVE
           DISTRIBUTES BEHIND is-contained-in REGISTERED AS {...}
REGISTERED AS {...};
```

```
backup RELATIONSHIP CLASS
  ROLE backs-up IMPLICATES is-a-point-of-failure-for REGISTERED AS {..}
  ROLE is-backed-up-by IMPLICATES has-point-of-failure REGISTERED AS {..}
REGISTERED AS {...};
```

We established earlier a role binding for the backup relationship class between the managed object classes dialUpCircuit and dedicatedCircuit. The repository can now automatically infer the existence of an implicate pointOfFailure role binding between the same two classes.

Implicate virtual relationships are a powerful mechanism to capture extended semantics in a concise manner. In the example above, we create only the three role bindings for the actual relationships A backs-up B, A contains D, and B backs-up C. Because the backup relationship gives rise to implicate virtual relationships which have a distribution property, the three actual relationships generate four virtual relationships, as shown below.

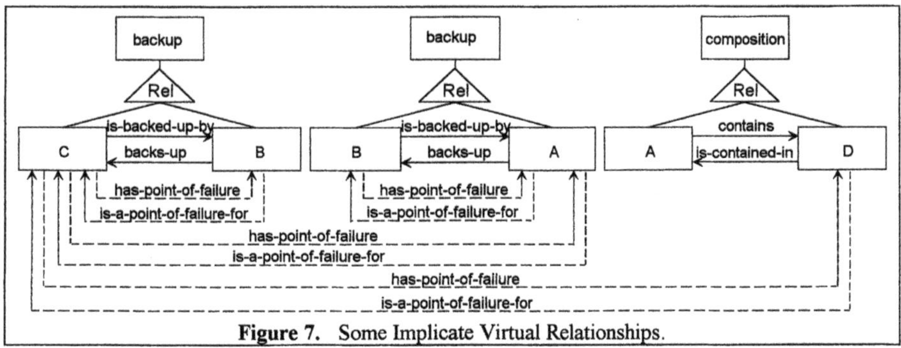

Figure 7. Some Implicate Virtual Relationships.

If we query an instance of C for all its points of failure (that is, all its related objects via the has-point-of-failure role) the response will include the instances of B, A, and D. In fact, due to the transitivity of containment, the transitivity of pointOfFailure, and the distribution of pointOfFailure over containment, the response will include *all* component objects of A, *all* component objects of B, the transitive closure of A's has-point-of-failure role (that is, all objects which may back-up A, their back-ups, and so on) and all *their* components as well. By simply specifying the correct properties for relationship roles, we can equip network management applications with the power to navigate through an extensive semantic network in our management information repository.

Implicate virtual relationships are sometimes used to "break down" commutative relationships into two one-way relationships where necessary. For example, the roles of the `mutualBackup` relationship class can be broken down into the roles of two one-way `backup` role bindings. This can be accomplished by specifying that the roles `{mbacks-up, is-mbacked-up-by}` implicate *both* the one-way role pairs `{backs-up, is-backed-up-by}` and `{is-backed-up-by, backs-up}`:

```
mutualBackup RELATIONSHIP CLASS
  ROLE mbacks-up COMMUTATIVE
          IMPLICATES backs-up, is-backed-up-by REGISTERED AS {...}
  ROLE is-mbacked-up-by COMMUTATIVE
          IMPLICATES is-backed-up-by, backs-up REGISTERED AS {...}
REGISTERED AS {...};
```

9 CONCLUSION

Virtual relationships provide an effective mechanism for extending relationship semantics. Because of their ability to automatically shadow the changes of one set of relationship instances in another, they reduce the potential for inconsistency. If an object is virtually bound to another via a chain of supporting actual relationships, we can query the object for its virtually bound object exactly as we query it for an actually bound object. The run-time environment in the repository internally and transparently resolves the virtual relationship in terms of its chain of supporting actual relationships. This eliminates the need for us to compose any relational joins in our query, which otherwise can be quite complex. Consequently, virtual relationships considerably enhance the semantic richness of our model [Bapa93a].

It is important to remember that virtual relationships arise as properties of the roles of a relationship class, and not in role bindings. All the properties of the roles of a relationship class continue to hold in in *every* role binding of that relationship class. A role binding cannot choose to "drop" certain properties of roles of its relationship class, nor can it invest those roles with new properties which hold only in that particular role binding.

We present below a concise summary of the types of virtual relationship we have defined using informal logical expressions. In these expressions, `A`, `B`, and `C` are managed object classes, and `r`, `s`, and `t` are roles of relationship classes. The construct `r(A,B)` is interpreted as a role binding, and is read as "r is the role played by `A` with `B`". If " ∧ " is read as "*and*" and "→" is read as "*gives rise to*", then

- *Commutative Virtual Relationship*: `r(A,B)` → `r(B,A)`
- *Transitive Virtual Relationship*: `r(A,B)` ∧ `r(B,C)` → `r(A,C)`
- *Distributive Virtual Relationship*: `r(A,B)` ∧ `s(B,C)` → `r(A,C)`
- *Convolute Virtual Relationship*: `r(A,B)` ∧ `s(B,C)` → `t(A,C)`
- *Implicate Virtual Relationship*: `r(A,B)` → `s(A,B)`

Virtual relationships provide us with a robust mechanism to enforce consistency between a chain of links in a semantic network of objects. The presence of virtual relationships enables us to drop certain constraints which would otherwise be imposed across the semantic network.

For example, a requirement which traverses many links in the semantic network, such as: "*The operator responsible for addressing an alarm generated by a network device must be an employee of the outsourcing vendor who administers the location which houses that network device*" is normally specified in most systems of knowledge as a sequence of multiple *constraints*.

These constraints are explicit: the user must specify how to enforce them by equating values of identifier attributes of *pairs of objects* across binary links in the semantic network.

A role binding asserts a relationship between instances of object classes as a statement of a fact, just as an attribute value assertion is a statement of a fact. By a logical conjunction of such facts, we can *infer* the existence of other facts across *multiple* links in the semantic network. By specifying virtual relationships between objects such as `operator`, `alarm`, `equipment`, `outsourcingVendor`, and `location` extending over actual roles such as `is-responsi-ble-for`, `is-generated-by`, `is-employed-by`, `administers`, and `houses`, the semantic constraint above falls out automatically and does not have to be explicitly specified. Because virtual relationships automatically reflect the changes of one relationship in another, they provide us with the ability to extend the "reach" of nodes in the semantic network to nodes other than their immediate neighbors. As such, they are powerful mechanism to facilitate extended navigation, reasoning, and inferencing within the management information repository.

10 REFERENCES

[Bapa93a] Bapat, Subodh, "Richer Modeling Semantics for Management Information", *Integrated Network Management III: Proceedings of the 1993 IFIP International Symposium on Integrated Network Management*, pp. 15-28.

[Bapa93b] Bapat, Subodh, "Towards Richer Relationship Modeling Semantics", *IEEE Journal on Selected Areas in Communications*, 11(9), Dec. 1993, pp. 1373 - 1384.

[Bapa94] Bapat, Subodh, *Object-Oriented Networks: Models for Architecture, Operations, and Management*, Prentice-Hall, 1994.

[Kilo94a] Kilov, Haim, and James Ross, *Information Modeling: An Object-Oriented Approach*, Prentice-Hall, 1994.

[Kilo94b] Kilov, Haim, "Generic Concepts for Modeling Relationships", *Proceedings.of the IEEE Network Operations and Management Symposium (NOMS) 1994.*

[X.725] "Information Technology - Open Systems Interconnection - Structure of Management Information - Part 7: General Relationship Model", ITU-T Rec. X.725, 1994.

11 BIOGRAPHY

Subodh Bapat is Principal of BlacTie Systems Consulting, and has worked with several network equipment vendors and telecommunications carriers in the areas of applying object-oriented modeling techniques to network architecture, and to the development of network management software. As a lead architect and implementer of standards-based network management systems, he made leading contributions in the area of applying object-oriented techniques to the architecture of networking equipment and to information modeling for databases used in network management and operations support. His involvement extended over the complete product life-cycle, including the architecture, design, development, testing, and maintenance phases. Subodh is the author of "***Object-Oriented Networks: Models for Network Architecture, Operations and Management***," (Prentice Hall, 1994, 757 pp.), a state-of-the-art book which demonstrates how the application of second-generation object-oriented modeling techniques can lead to sophisticated, intelligent, and highly automated network systems. He has published several articles in leading technical journals and has presented papers at major industry conferences. He has been awarded a number of patents in the area of implementing network management software.

49

Testing of Relationships in an OSI Management Information Base

Brigitte Baer*
University of Frankfurt
Department of Computer Science
P.O.Box 111932
D-60054 Frankfurt/Main
Germany
baer@informatik.uni-frankfurt.de

Alexander Clemm* †
Munich Network Management Team
University of Munich
Department of Computer Science
Leopoldstr. 11b
D-80802 Munich
Germany
clemm@informatik.uni-muenchen.de

Abstract

In open distributed environments such as in OSI network management, a procedure of conformance testing is essential for increasing the level of confidence that component implementations from different sources actually meet their specifications as a prerequisite for their ability to interact as intended. This applies not only to OSI communication protocols but also to open management information. In particular, this includes relationships between managed objects, an aspect which has been largely ignored so far but which deserves particular attention and which we therefore focus on in this paper. Using the OSI General Relationship Model as a basis, we discuss how respective conformance requirements can be identified which serve as a starting point for the development of test cases.

1 Introduction

Conformance testing addresses the problem of how to determine whether the behavior that an implementation exhibits conforms to the behavior defined in its specification. The issue of conformance testing is of particular importance in open environments where components from different sources and manufacturers have to interwork. Here, a procedure of conformance testing can be substantial in increasing the level of confidence that an implementation acts according to its specification and that it will be able to interact in an open environment with other components as expected.

The problem of conformance testing also applies to the OSI network management arena for which openness of implementations of many different vendors and their ability to interwork is required. Besides conformance of management protocol implementations (such as CMIP [14]), for which ordinary protocol conformance testing methodologies [15] apply conformance of management information to its specification is a key issue. This involves the testing of the *Management Information Base* (MIB) with its *Managed Objects* (MOs) that represent the underlying network resources to be managed. Conformance of a MIB is an assumption for the proper functioning of management applications which operate on MOs and directly depend on the correct implementation of these MOs.

First approaches for testing the conformance of MOs can be found in [7,9,12]. Those approaches all have in common that they look at MOs in isolation; they do not cover aspects that involve

*The authors' work has also been supported by IBM European Networking Center, Heidelberg, Germany.
†A.Clemm is now with Raynet GmbH, Munich, Germany.

combinations of MOs or the context of the MIB as a whole. However, MOs are not isolated from each other but maintain relationships reflecting the interworking and dependencies among the underlying network resources. The importance of relationships has been acknowledged by work on the ISO *General Relationship Model* (GRM) [18] and other activities [5,3,19]. The GRM is essentially an 'attachment' to the basic information model. It allows for an additional specification of those aspects of MOs that relate them to other MOs in order to document those aspects in a more formal manner and to add structure to models of management information as a whole. Although the GRM has some shortcomings [5], it provides an important supplement to the OSI information model and will be referred to in the further discussion.

Independent of the existence of the GRM, relationship aspects must be considered in conformance testing as they are in any case present in a MIB. This has already been recognized in [1] where a 'relationship view' has been introduced as an integral part of a conformance testing methodology for MOs. Formal specification of relationship aspects using the GRM makes the task of determining their conformance requirements and deriving according test cases easier than basing the task on informal MO behavior specifications only. The purpose of this paper is to investigate the subject of relationship conformance testing with respect to the GRM. This includes to examine the conformance requirements that can be derived from the aspects specified in the GRM and to address the problems associated with the development of test cases for relationships.

To set the stage, we will first summarize the basic concepts of the GRM in section 2. A general knowledge of OSI management and the OSI information model with its *Guidelines for the Definition of Managed Objects* (GDMO) [16,17] is assumed. Section 3 gives an overview over conformance testing concepts. In section 4, we use a classification scheme to systematically identify relationship conformance requirements that result from those relationship aspects that are formally specified in the GRM. These requirements form the basis for the derivation of abstract test cases for relationships. The according process is explained in section 5 by a relationship example dealing with an ATM cross connection. Some conclusions are offered in section 6.

2 The general relationship model

The aim of the GRM is to provide additional specification means for the definition of relationships in a formal manner. This concerns for instance MO attributes referring to other MOs or constraints concerning the joint behavior of MOs [19] in behavior specifications. The representation and management of relationships per se as part of a MIB are like before based on the well known basic OSI management concepts. Thus, the GRM is an attempt to eliminate shortcomings associated with the specification of relationships between MOs in the conventional plain OSI information model while leaving it in itself unaffected.

According to the GRM, relationships between MOs are modeled independently of MOs in terms of *Managed Relationships.* A MO bound in a relationship is known as a *participant.* Common characteristics of relationships are summarized in *Managed Relationship Classes* (MRCs) for which new templates are provided. MRCs can but do not have to be derived from one or more other MRCs.

MRCs allow to specify certain constraints among participants. For this purpose, *roles* are used to model the properties of various related participants in a relationship. To play a given role, a MO may be required to possess a certain set of characteristics, specified in terms of a MO class (MOC) that any participant in that role will have to be compatible with. A *role cardinality* is used to specify how many MOs may participate in a given role in any one relationship. Also, roles can be specified to be 'dynamic' if MOs are allowed to enter and/or leave a relationship without affecting its existence, as opposed to static roles where MOs remain participants in a relationship for its entire life span. In addition, in a behavior part any other aspects can be defined in natural language text for which no formal specification means are provided.

MRCs are defined independently of the representation of the relationship in a MIB. A so-called *role binding* template is provided which can be used to specify how a certain relationship is represented as part of management information. For this purpose, for each role the class(es) of MOs that can participate in the relationship in that role are specified and whether that includes subclasses. Relationship instances can be represented as part of management information in the following ways:

- **Name bindings:** A relationship is represented by naming, i.e., in a given relationship the participants in one role (subordinates) are contained in a participant (superior) of another role. The role binding identifies one or more name bindings that represent the relationship.
- **Attributes:** A relationship is represented by relationship attributes which participating MOs in a given role have to support. Their values identify related participants in other roles.
- **MOs:** The relationship is represented by dedicated MOs of a certain class. As a result, a relationship is explicitly represented in a MIB in terms of an instance of a relationship MOC called *relationship object*. All relationship MOCs have to be derived from the standardized MOC *relationshipObjectSuperClass*.
- **MO operations:** A relationship is implicitly represented by means of systems management operations. The behavior description in the role binding has to define the meaning of these operations when applied to participants of the relationship.

Role bindings also specify the effects of abstract *relationship operations* and their mapping to systems management operations. Relationship operations include e.g. operations to establish and terminate relationships, to bind and unbind MOs to/from a relationship, and to retrieve information about relationships. One or more mappings are allowed for the same operation. A behavior clause is used to define the semantics of each operation. The abstract relationship operations are not to be confused with relationship services in the sense of a 'relationship management function'; all they do is state in which way certain management operations that operate on MO aspects are to be interpreted from a relationship perspective.

In addition, a role binding allows to specify the effects associated with the dynamic departure of a participant in a relationship: whether it may not depart unless other roles have no participants, whether related MOs in other roles are to be deleted as a consequence, or whether the related MOs are released from the relationship. Access to certain attributes or actions can be prohibited. A behavior part describes any other impacts imposed as a consequence of the role binding. Several role bindings can be defined for a single MRC, reflecting different ways that the same kind of relationship is represented for different MO classes.

3 Conformance testing concepts

The purpose of conformance testing is to increase the probability that different OSI (protocol) implementations are able to interwork. In the *Conformance Testing Methodology and Framework* ([15]), conformance testing is defined to be the assessment process for determining whether the externally visible behavior of an OSI implementation conforms to the behavior required by its specification. A real system is said to exhibit conformance if it complies with the *conformance requirements*, e.g. certain capabilities or allowable behaviors, defined in the corresponding OSI standard in its communication with other real systems.

In order to harmonize the process of testing and certification for OSI implementations, the framework provides a methodology for specifying conformance test suites and defines procedures to be followed by implementation providers and test houses. A standardized test notation, called ***Tree and Tabular Combined Notation*** (TTCN), is proposed for the development of abstract test suites. TTCN aims at providing a common language in which test cases for various implementations can be expressed on an abstract level. Abstract test cases specify a series of actions (***test events***) that are needed to test a specific conformance requirement. The entirety of all test cases for a certain protocol specification forms the test suite. The use of standardized test suites and common procedures for testing the conformance of OSI implementations leads to comparability and acceptance of test results.

Although devoted to OSI protocols, the test case development and conformance assessment process described in the framework can also be applied to other OSI implementations, especially to MOs. A MO is said to exhibit conformance if it complies with the conformance requirements of its corresponding specification. Testing a MO for conformance requires the externally visible behavior of MOs to be observed by applying operations and analyzing their effects.

In [2], an architecture suitable for MO conformance testing is described. A test system in the role of a manager is responsible for executing test cases based on sending and receiving CMIS [13] requests to an agent in which the MOs to be tested are embedded (see Figure 1). If possible, resource specific test requests may be used to drive the resources in order to observe the reactions of MOs to real effects. A positive test verdict is only assigned if the responses received comply with the expected responses defined in the test cases. The test results are summarized in a test report. Conformance of agents and CMIS is presupposed because these components can be dealt with separately from MO testing [1]. Basing test events on standardized CMIS service primitives allows for the use of TTCN for the definition as well as the standardization of abstract test cases for MOs.

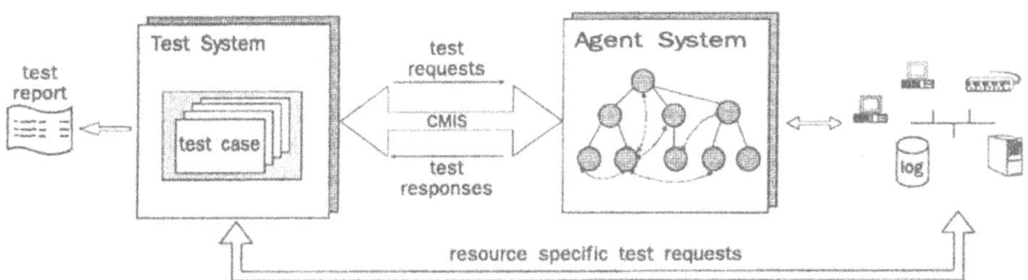

Figure 1: MO test architecture.

In order to structure the test case development process for MOs, a distinction is made between three different views. This concept requires to focus on MOs in isolation, to address the interactions between related MOs, and to take into account the consistency of a MO with its underlying resource. The MO conformance testing concepts can not be introduced in length within this paper. For further details it is referred to [2].

4 Analysis of relationship aspects

4.1 Specification requirements for relationships

In the context of the OSI information model, specification and conformance testing are related in the following sense (see Figure 2):

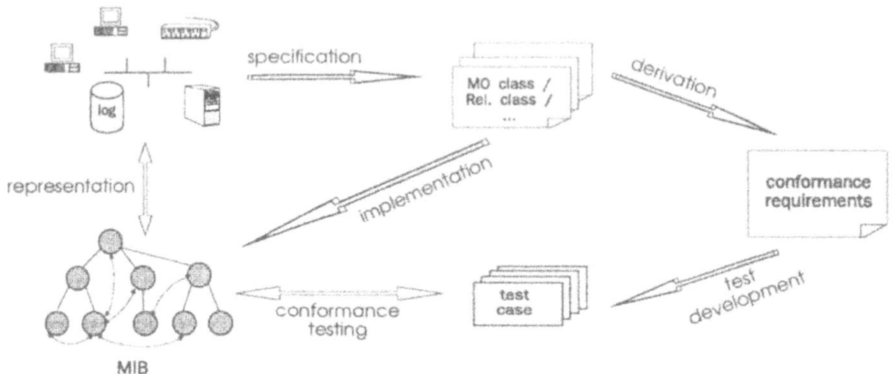

Figure 2: Relation between specification and conformance testing.

- Specification looks at aspects of the managed resources and represents them by means of the information model using dedicated specification tools.
- Conformance testing looks at specified aspects and checks whether the behavior exhibited by the management information conforms to the behavior defined in the specification.

Accordingly, the very same aspects that are relevant for specification are also relevant for conformance testing. A classification of the various aspects being involved in MO relationships has been presented in [5] as a basis for the evaluation and derivation of MO relationship specification means. This same classification can serve as the basis for the derivation of conformance requirements. Aspects of relationships can be grouped along the following perspectives:

- **Structure:** This perspective covers aspects of relationships that are concerned with describing them as a part of management information, i.e., the way they provide associations between the MOs they relate and the rules according to which they add structure to the MIB as a whole. This includes e.g. aspects such as properties of relationship participants (i.e. roles), for instance prerequisites that a MO has to fulfill in order to be allowed to participate in a relationship in that role.
  With respect to the GRM, this perspective covers also aspects concerning the instantiation of relationships. This is because the modeler is not only responsible for the specification of abstract relationship properties but also for the representation of those relationships as part of the MIB. Aspects such as role cardinalities stating how many MOs may participate in a role in any one relationship instance or constraints imposed on the leaving and joining of relationship instances by MOs have to be considered as well. (A relationship approach with a different philosophy [4] keeps instantiation aspects transparent to modeler and application and instead hides them in an information layer in order to provide better 'data independence' – here such aspects do no apply.)
- **Effects:** This perspective is concerned with effects of relationships on participating MOs, as relationships often imply that operation of one MO affects the other. For instance, if a

MO participates in a relationship, it may no longer be deleted because of that relationship. Another example is that a MO is deleted as a side effect of the deletion of another MO it is related to. It also includes possible dependencies of MO attributes on other participants of a relationship, e.g., of an operational state attribute of a MO that is functionally dependent on another MO.

- **Management:** This perspective covers aspects of relationships that relate to their need to be managed and accessed as part of management information; for instance, whether a relationship is subject to manipulation by management operation.
- **Object Orientation:** Those aspects deal with the embedding of relationships into the (object oriented) OSI information model; for instance aspects related to inheritance.

[5] also mentions a fifth perspective, 'network management context' that deals with particular management application requirements for dealing with relationships. This, however, is of no importance with respect to the GRM as it applies less to the OSI information than to the OSI functional model.

4.2 Generic conformance requirements

Test objectives for abstract test cases are aligned with conformance requirements of a certain specification. Conformance requirements have to be determined before starting to develop test cases. As proposed in [15], conformance requirements should be part of the conformance clause of a standard. Looking at OSI information modeling standards, explicit conformance statements are still missing today. Therefore, these have to be added as extensions to the standard documents. In the meantime, efforts have been started to define so called *Managed Object Conformance Statement* (MOCS) proformas as extensions to standardized MOCs and *Managed Relationship Conformance Statements* (MRCS) proformas for MRCs. Such proformas focus on static MO/relationship capabilities, such as the support of packages or relationship operations in an implementation. However, these proformas do not cover the complete set of conformance requirements of a MO or a relationship. For instance, requirements resulting from the behavior part of a specification are outside the scope of these documents.

The specification requirements introduced in the previous chapter are used as a starting point for the derivation of conformance requirements. This is because aspects relevant for specification also lead to aspects that are subject to testing. Correct specification is presupposed in this discussion as ensuring the consistency of a specification is not subject to conformance testing. In the following, we investigate which generic conformance requirements result from the various relationship perspectives with respect to the specification means of the GRM, independent of the particular representation of a relationship in the MIB:

Structure:

- Requirements concerning relationship participants:
  In order for a MO to participate in a given relationship role, its characteristics must be compatible with the characteristics for that role, i.e., the MO class referenced in the MRC.
- Requirements concerning relationship and relationship instance:
  - ◇ The required role cardinality must not be violated.
  - ◇ If roles are static, participants are not allowed to enter or leave an established relationship instance.
  - ◇ MOs must not be related with each other if there is no role binding that would allow instances of their classes to be related in the respective roles in that particular class of relationship.

  As a consequence, any operation that would violate these constraints must be rejected.

There are other common requirements resulting from relationship aspects that are not part of the formal specification but can be expressed in relationship behavior clauses. We want to name a few to give an impression of what further requirements relationships can imply:

- ◇ A MO may only be allowed to enter or leave a certain relationship role if the state of the MO (i.e. certain attribute values) corresponds to the state required in the specification.
- ◇ In order to fulfill a certain role in a given relationship, a MO can be required to fulfill some role in another relationship. A MO can also be prohibited from participating in instances of different MRCs simultaneously.
- ◇ A MO may be allowed to enter or leave a given relationship only if other MOs enter or leave the relationship simultaneously.

Effects: (on participants)

- An attribute of a relationship participant must not be altered if specified in the respective role binding as 'restricted'. Operations attempting to manipulate such attributes must be rejected.
- Actions of relationship participants must not be performed (and accordingly have to be rejected) if specified in the respective role binding as 'restricted'.
- A participant of a relationship must not be deleted if the respective role binding specifies for the respective role 'only-if-none-in-roles' and other MOs are in the specified roles.
- When deleting a relationship participant, related MOs in other roles must be deleted if specified in a 'deletes-all-in-roles' clause in the respective role binding.
- When deleting a relationship participant, related MOs in other roles must no longer participate in the according relationship instance if specified in a 'releases-all-in-roles' clause in the respective role binding.

Again, further requirements can result from relationship aspects expressed in relationship behavior clauses, e.g., any dependencies between attribute values of related MOs.

Management: Relationship management solely occurs as an indirect effect of management of MOs. The role binding defines the mapping of abstract relationship operations to systems management operations. The conformance requirements associated with this perspective refer to the correctness of systems management operations when applied to relationship instances. In particular, this concerns preconditions and postconditions associated with a relationship operation as specified in the behavior clause of the corresponding operations mapping.

Object Orientation: A MRC derived from other MRCs inherits their characteristics. With the kind of strict inheritance defined for the GRM, conformance requirements of relationship superclasses apply to relationship subclasses. Conformance requirements resulting from inherited features are grouped along and added to the perspectives explained previously.

The representation of a relationship determines to which extent relationship information is explicitly available in a MIB and how it can be monitored/controlled by management applications or a test system, respectively. Therefore, the representation independent conformance requirements explained above translate into representation dependent conformance requirements for the respective relationship representations. For instance, a conformance requirement related to a bounded role cardinality by a number n can translate to the conformance requirement that e.g. the set-valued attribute representing that relationship must not contain more than n members.

It should be noted that there is a different kind of relationship information available in a MIB when using different representations for the same relationship. The representation by means of a relationship object is the most powerful alternative. It provides information about the relationship class, its name, and the role binding in use while other representations do not.

Furthermore, the representation by MOs and attributes do have in common that it is possible to directly identify participants in roles. This information is only implicitly available when using name bindings and can hardly be obtained when representing relationships by management operations. Management operations therefore represent the weakest alternative for expressing relationship information in a MIB. An important consequence is that the conformance requirements can differ for the same kind of relationship for different representations of the relationship.

5 Test case development for relationships

5.1 The ATM cross connection relationship

Exemplary, we have extracted relationship information from an object catalogue for the management of an *Asynchronous Transfer Mode* (ATM) cross connection [8]. For the relationship information expressed in the MOCs of the catalogue, explicit MRCs and role bindings have been defined using the specification tools of the GRM. These relationship specifications are used as a starting point for the development of abstract test cases. The first step in this process is to determine the conformance requirements which have to be derived from the MRC and the role binding specifications. This task is guided by the relationship perspectives explained in the previous chapter. The conformance requirements then provide the basis for the second step, the development of abstract test cases for relationships. This proceeding will be explained for a specific example.

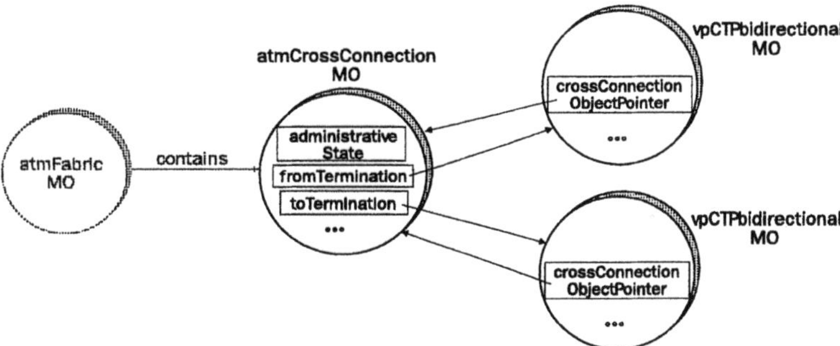

Figure 3: MOs involved in the establishment of an ATM cross connection relationship.

In [8], a **vpCTPbidirectional** MOC is defined to model a virtual path termination point where a virtual path link connection is originated and terminated. An **atmCrossConnection** MOC is specified to represent a relationship between two instances of **vpCTPbidirectional**. On instantiation of an **atmCrossConnection** MO, a virtual path link connection is established between two **vpCTPbidirectional** MOs. The values of two attributes (**toTermination** and **fromTermination**) of the **atmCrossConnection** MO refer to the cross connected MOs. In addition, the cross connected **vpCTPbidirectional** MOs provide an attribute (**crossConnectionObjectPointer**) pointing back to the **atmCrossConnection** MO. The deletion of the **atmCrossConnection** MO terminates the cross connection and the pointers to the **atmCrossConnection** MO have to be deleted in both participants. An instance of the MOC **atmFabric** is responsible for managing the establishment and release of all ATM cross connections for an ATM cross connected network element. For instance, if the establishment of a new ATM cross connection is requested the **atmFabric** MO creates a new **atmCrossConnection**

```
crossConnection RELATIONSHIP CLASS
   BEHAVIOR ...;
   ROLE toTerminationPoint ROLE CARDINALITY (1..1) REGISTERED AS ...;
   ROLE fromTerminationPoint ROLE CARDINALITY (1..1) REGISTERED AS ...;
REGISTERED AS ...;

crossConnectionRepresentation ROLE BINDING
   RELATIONSHIP CLASS crossConnection;
   BEHAVIOR ...;
   RELATIONSHIP OBJECT atmCrossConnection
   ROLE toTerminationPoint
      RELATED CLASSES connectionTerminationPointBidirectional AND SUBCLASSES
         RELATED BY RELATIONSHIP OBJECT USING ATTRIBUTE toTermination;
   ROLE fromTerminationPoint
      RELATED CLASSES connectionTerminationPointBidirectional AND SUBCLASSES
         RELATED BY RELATIONSHIP OBJECT USING ATTRIBUTE fromTermination;
   OPERATIONS MAPPING
      ESTABLISH MAPS TO OPERATION ACTION atmConnect OF atmFabric
         WITH BEHAVIOR ...;
      TERMINATE MAPS TO OPERATION ACTION disconnect OF atmFabric
         WITH BEHAVIOR ...;
      ...
REGISTERED AS ...;
```

Figure 4: Cross connection relationship specification.

MO which is contained in the `atmFabric` MO. Figure 3 shows the MOs that are involved in the establishment of an ATM cross connection relationship. For further details of the MOCs introduced it is referred to [8].

The MOCs explained above lead to the specification of a `crossConnection` relationship class depicted in Figure 4. There, two roles for the `crossConnection` relationship class are defined, `toTerminationPoint` and `fromTerminationPoint`. In both roles only one participant is allowed to take part in a `crossConnection` relationship. Although not using the specification tools of the GRM, the specifier(s) of the object catalogue have decided to represent an ATM cross connection by an explicit relationship object. This results in the representation by relationship object `atmCrossConnection` in the role binding for the `crossConnection` relationship class (see Figure 4). The 'related classes' constructs for both roles prescribe that instances of the MOC `connectionTerminationPointBidirectional` or any subclasses may participate in the relationship. As `vpCTPbidirectional` is an indirect subclass of `connectionTerminationPointBidirectional`, instances of `vpCTPbidirectional` are allowed to participate in both roles in the relationship.

5.2 Derivation of conformance requirements

The conformance requirements for the `crossConnection` relationship are derived from the specification depicted in Figure 4 and are grouped along the identified relationship perspectives. To our experience, it is easier to derive conformance requirements from formal relationship specifications than from informal relationship specifications only. As the resulting conformance requirements for the `crossConnection` relationship can not be introduced in length within this paper, only excerpts are listed below:

Structure:

- The role cardinality (1..1) must not be violated for either the **toTerminationPoint** role nor the **fromTerminationPoint** role. I.e., the value of the **toTermination** attribute and the value of the **fromTermination** attribute in an **atmCrossConnection** MO have to refer to a single participant.
- Participants can not enter or leave an established **crossConnection** (because it is a static relationship).
- In order for a MO to participate in a **crossConnection** relationship in the **toTerminationPoint** or **fromTerminationPoint** role, the MOC of the potential participant must be **connectionTerminationPointBidirectional** or a specialization of this MOC.

Effects: (on participants)

- The value of the **crossConnectionObjectPointer** attribute of a participant in the **crossConnection** relationship has to be the name of the corresponding **atmCrossConnection** MO.
- On deletion of a MO participating in the **crossConnection** relationship, the corresponding **atmCrossConnection** MO has to be deleted (behavior requirement). As a result, the related MO in the other role is released from the relationship.
- If the value of the administrative state of the **atmCrossConnection** MO is 'locked' no traffic can pass through cross connected MOs participating in this relationship (behavior requirement).

Management:

- On establishment of a new **crossConnection** relationship, i.e. requesting the action **atmConnect**, an instance of the MOC **atmCrossConnection** has to be created and a participant in each role has to be bound. The value of the **toTermination** attribute has to be the name of the participant in the **toTerminationPoint** role and the value of the **fromTermination** attribute has to be the name of the participant in the **fromTerminationPoint** role.
- On termination of a **crossConnection** relationship, i.e. requesting the action **disconnect**, the corresponding **atmCrossConnection** MO has to be deleted.

Apart from deriving conformance requirements for each relationship separately (***intra relationship requirements***), there may be effects specified for the participant in one relationship that will also cause effects on participants in other relationships. This kind of conformance requirements, which we call ***inter relationship requirements***, can only occur if relationship specifications allow that MOs can participate in different relationships simultaneously. Suppose a **dependency** relationship which requires that a participant in a **parent** role can only be deleted if all MOs in the **dependent** role are deleted as well. One of the dependent MOs however also participates in a **crossConnection** relationship in role **toTerminationPoint** for which the condition 'releases-all-in-roles fromTerminationPoint' has been specified. As a result of requesting the deletion of the MO in the **parent** role, not only the MOs in the **dependent** role have to be deleted but the MO participating in the **crossConnection** relationship in role **fromTerminationPoint** has to leave its relationship. The results presented in [21] remind us that having tested each relationship in isolation does not necessarily imply that this is also sufficient for testing the composition of relationships if these relationships are interdependent. Therefore, inter relationship requirements have to be taken into account in conformance testing as well. When comparing the testing of related MOs with testing clusters of objects in object oriented programs, this conclusion is also acknowledged by work on object oriented program testing approaches. In [20], it is stated that special attention has to be paid to classes of which instances can be bound to more than one cluster.

5.3 Development of abstract test cases

Testing of related MOs is based on the observation and manipulation of MOs making use of systems management operations only. This requires access to all MOs involved in the relationship to be tested. Each conformance requirement identified has to be addressed in one or more test cases. Abstract test cases for relationships heavily depend on the mapping information contained in role bindings. In particular, this applies to test events for requesting relationship operations and test events for observing the reactions in related MOs that have to be mapped to corresponding systems management operations. Figure 5 shows a simplified example test case defined in TTCN focusing on the requirement that a MO can only participate in a `crossConnection` relationship if the MOC of the potential participant corresponds to `connectionTerminationPointBidirectional` or a specialization of this class.

Test Case Dynamic Behavior				
Test Case Name : crossConnection_establish_with_invalid_participant				
Group :				
Purpose : verify that it is not possible to bind a participant in a cross connection relationship if the class of the participant does not correspond to connectionTerminationPointBidirectional or any subclass				
Default :				
Comments :				
Nr	**Label**	**Behavior Description**	**Constraints Ref**	**Verdict**
1		+preamble		
2		!MActionRequest START Timer	atmConnectReq	
3	L1	?MEventReportIndication		
4		GOTO L1		
5		?MActionConfirm CANCEL Timer	atmConnectCnf	(PASS)
6		+postamble		
7		?OTHERWISE CANCEL Timer		(FAIL)
8		+postamble		
9		?TIMEOUT		(INCONC)
10		+postamble		

Figure 5: Example TTCN test case for the `crossConnection` relationship.

The TTCN test case consists of a header containing overview information like a test case name, the test purpose etc. and a body for the test case behavior. The body is partitioned into different columns. In a *Behavior Description* column, test events to be sent to the system under test and its possible responses are defined. Send events are indicated by a *!*. A *?* is used to denote receive events. A so-called *preamble* describes a sequence of test events needed to drive the system under test into a state from which the test body will start. The so-called *postamble* sets the system back to a stable end state after the test body has been executed. An entry in the *Constraints Ref* column refers to a specification of the data values (parameters) to be transmitted in a send event or expected as part of a received event. In the *Verdict* column, a verdict for the received test event is given.

In our example test case in Figure 5, a `MActionRequest` is sent to an agent which is responsible for invoking an action on an instance specified in the corresponding constraint `atmConnectReq` (see behavior line 2). According to this constraint, the action **atmConnect** has to be called on an instance of the MOC **atmFabric** requesting a new cross connection to be established between two MOs. Due to space restrictions, the actual constraints can not be depicted. In this example, we assume that one of the participants specified in the constraint **atmConnectReq** does not match the required class for its role. Different receive events have to be distinguished as a result of the **MActionRequest**. As MOs can issue notifications asynchronously, event reports can be received. As the purpose of the test case does not focus on notifications, these are ignored in a loop until any other event is received (see behavior line 3 and 4). If a **MActionConfirm** event occurs and the data received complies with the data specified in the constraint **atmConnectCnf**, the test case verdict **PASS** is assigned. In this example, the error message 'mismatchingInstance' is expected stating that an incorrect participant given in the request has lead to the rejection of the action. In the case that a **MActionConfirm** with invalid data values or any other event is received (see behavior line 7), the test case verdict is **Fail**. In order to take into account that no response is sent from the agent, a timer is started whenever sending a new test event (see behavior line 2). A **TIMEOUT** event is generated by the test system indicating that no events have been received within the timer interval. According to [15], timeout events lead to the test case verdict **INCONCLUSIVE** (see behavior line 9).

When defining relationships between resources, the correctness of the resulting conformance requirements have to be verified during the relationship testing process. However, under certain circumstances there can be conformance requirements which do not necessarily have to be addressed in the testing process. This is the case if a relationship conformance requirement only focuses on physical relationships between resources, or in the terms of [6] on *descriptive* aspects of relationships. Suppose the following example: A dependency relationship between two MOs has been modeled that represents a functional dependency of their underlying resources. A requirement for this relationship could be that if the operational state of one resource changes to 'disabled' this has also to be the case for the dependent resource. Assuming the proper functioning of the resources, the state values of the corresponding MOs will have to change to 'disabled' as well. If the MOs participating in the dependency relationship behave really as images of their underlying resources (this should be the case having tested the MOs in isolation), there is no need to test such kind of conformance requirements.

The overall goal is to develop abstract test cases which 'cover' the intra and inter relationship requirements identified for each relationship in an object catalogue. The abstract test cases developed for the conformance requirements are used for testing the relationships in a whole MIB. Clearly, a test case can only address aspects that have explicitly been defined in a specification. If there exists a relationship between resources that is not specified in the model, the influences of this relationship can not be included in the testing process. The test suite for an object catalogue (including MOC, name binding, MRC, and role binding definitions) comprises the set of all abstract test cases developed for testing MOs in isolation combined with the abstract test cases developed for relationships. The difficulties of dealing with resources in the testing process have already been discussed in [2].

6 Conclusion

In this paper, we have discussed the subject of conformance testing in OSI network management with respect to relationships occurring between MOs in a MIB. Despite its high relevance, relationship conformance testing has been ignored so far, possibly because dedicated concepts for the treatment of relationships have for a long time been missing in OSI management. We have classified generic conformance requirements according to the perspectives put forward in

[5] for the specification of relationships, which refer to the same aspects that have to be checked during a procedure of conformance testing. We have explained how from a formal relationship specification appropriate conformance requirements can be derived. The resulting conformance requirements form the starting point for the development of abstract test cases for relationships. This process has been carried out for an example relationship derived from the object catalogue for an ATM cross connection.

The test case development process for the relationship specifications defined for the ATM cross connection MOCs is supported by a prototype test system for MIBs allowing for the definition of abstract test cases in TTCN and its automatic execution. The test system is based on an existing protocol conformance test tool (*Automated Protocol Test System/2* [11]) for which an extension has been implemented providing for the exchange of CMIS service primitives between test system and a management system [10]. The test system provides the platform for the practical application of our concepts with respect to management information testing. In particular, the test cases developed for the ATM cross connection MOCs will be applied to a prototype MIB which is being implemented as part of an European research project (RACE II PREPARE) dealing with the cooperative end to end service management across heterogeneous Integrated Broadband Communication Networks. Finally, it should be noted that the procedure of testing relationships introduced in this paper is not only of interest for conformance testing but can also aid in an integrated development/testing life cycle of MIB implementations.

Acknowledgements

We wish to thank our colleagues, the research staff directed by Prof. Geihs at the University of Frankfurt, the Munich Network Management Team of the Munich Universities directed by Prof. Hegering, and IBM ENC's system and network management department.

References

[1] B.Baer, *A Conformance Testing Approach for Managed Objects,* 4th IFIP/IEEE Int. Workshop on Distributed Systems: Operations & Management, Long Branch, New Jersey, USA, October 1993.

[2] B.Baer, A.Mann, *A Methodology for Conformance Testing of Managed Objects,* 14th Int. IFIP Symposium on Protocol Specification, Testing, and Verification, Vancouver, BC, Canada, June 1994.

[3] S.Bapat, *Towards Richer Relationship Modeling Semantics*, IEEE Journal on Selected Areas in Communication Vol.11 No.9, December 1993.

[4] A.Clemm, *Incorporating Relationships into OSI Management Information*, 2nd IEEE Network Management and Control Workshop, Tarrytown, NY, September 1993.

[5] A.Clemm, *Modellierung und Handhabung von Beziehungen zwischen Managementobjekten im OSI-Netzmanagement*, Dissertation, University of Munich, June 1994.

[6] A.Clemm, O.Festor, *Behaviour, Documentation, and Knowledge: An Approach for the Treatment of OSI-Behaviour*, 4th IFIP/IEEE Int. Workshop on Distributed Systems: Operations & Management, Long Branch, New Jersey, USA, October 1993.

[7] CTS3-NM, *Methodology Report on Object Testing*, The Establishment of a European Community Testing Service for Network Management, Deliverable 3, Brussels, Directorate-General XIII-E4, April 1992.

[8] ETSI, *B-ISDN Management Architecture and Management Information Model for the ATM crossconnect*, ETSI/NA5 WP BMA, April 1994.

[9] EWOS PT-16, *Framework for conformance and testing of network management profiles*, Report 1 of EWOS/EG NM/PT-16, June 1992.

[10] W.Herrnkind, *Design und Implementierung einer Erweiterung eines Konformitätstestwerkzeugs für den Einsatz in OSI-Netzmanagementsystemen*, Diploma Thesis (in German), University of Frankfurt, Department of Computer Science, January 1995.

[11] IBM, Automated Protocol Test System/2 User's Guide, SV40-0373-00, June 1993.

[12] ISO, *Final Answer to Q1/63.1 (Meaning of Conformance to managed objects)*, ISO/IEC JTC 1/SC 21 N 6194, May 1991.

[13] ISO, *Information Processing Systems - Open Systems Interconnection - Common Management Information Service Definition*, ISO Int. Standard 9595, second edition, 1991.

[14] ISO, *Information Processing Systems - Open Systems Interconnection - Common Management Information Protocol - Part1: Specification*, ISO Int. Standard 9596-1, second edition, 1991.

[15] ISO, *Information Processing Systems - Open Systems Interconnection - Conformance Testing Methodology and Framework*, ISO Int. Standard 9646, 1991/92.

[16] ISO, *Information Technology - Open Systems Interconnection - Management Information Services - Structure of Management Information - Part 1: Management Information Model*, ISO Int. Standard 10165-1, January 1992.

[17] ISO, *Information Technology - Open Systems Interconnection - Management Information Services - Structure of Management Information - Part 4: Guidelines for the Definition of Managed Objects*, ISO Int. Standard 10165-4, January 1992.

[18] ISO, *Information Technology - Open Systems Interconnection - Management Information Services - Structure of Management Information - Part 7: General Relationship Model*, ISO Draft Int. Standard 10165-7, March 1994.

[19] H.Kilov, J.Ross, *Generic Concepts for Specifying Relationships*, IEEE/IFIP 1994 Network Operations and Management Symposium, Orlando, Florida, February 1994.

[20] J.D.McGregor, T.D.Korson, *Integrated Object-Oriented Testing and Development Processes*, Communications of the ACM, Vol. 37 No. 9, September 1994.

[21] E.J.Weyuker, *The Evaluation of Program-Based Software Test Data Adequacy Criteria*, Communications of the ACM, June 1988.

50

DUALQUEST: An Implementation of the Real-time Bifocal Visualization for Network Management

Shoichiro Nakai, Hiroko Fuji, and Hiroshi Matoba
C&C Research Laboratories, NEC Corporation
1-1, Miyazaki 4, Miyamaeku, Kawasaki 216 JAPAN

Tel:+81-44-856-2314, Fax:+81-44-856-2229

E-Mail: nakai@nwk.cl.nec.co.jp, fuji@nwk.cl.nec.co.jp, matoba@mmp.cl.nec.co.jp

Abstract

Most of the current network management systems employ graphic-user-interfaces for the network visualization purposes. These are well suited for both small- and medium-size networks. For a large-size network, hierarchical multi-window-based network visualizations are usually used; however, tracing a long path (i.e., composed of a huge number of nodes) may meet some difficulties because it must be at first divided into several segments displayed segment-by-segment in several windows. In addition, window manipulations, such as opening and closing operations, are quite complex. To overcome the disadvantages of the multi-window network visualization, we proposed a real-time bifocal network visualization that is capable of displaying both the context and all details of a network within a single window (Fuji, 1994). This paper enhances that approach and describes an implementation, called DUALQUEST, that was installed in a workstation equipped with a frame buffer memory proposed in (Matoba, 1990) for real-time bifocal image processing.

Keywords

Graphic-user-interface, Network visualization, Bifocal display, Fish-eye view

1 INTRODUCTION

At present, graphic-user-interfaces are widely used to facilitate realization of network management functions. For example, NMS (Network Management System) reports a change of network status to the operators by alternating visual attributes of the graphic symbols displayed on a monitor screen (Cunningham, 1992). If the size of a managed network increases, more symbols must be displayed on the same screen. With regard to this, the hierarchical multi-window presentation (see Figure 1) was proposed. (Hewlett Packard, 1992) Although it is well suited for tracing paths composed of several nodes, paths comprising a huge number of nodes can only be traced on a segment-by-segment basis. This requires passing through several (separate) windows to trace such a path from its origin node to its destination node. Furthermore, the effect of *overlapping windows* may cause missing some important information. Thus, operators must perform many complex (opening, moving, and closing) window operations to obtain the desired information.

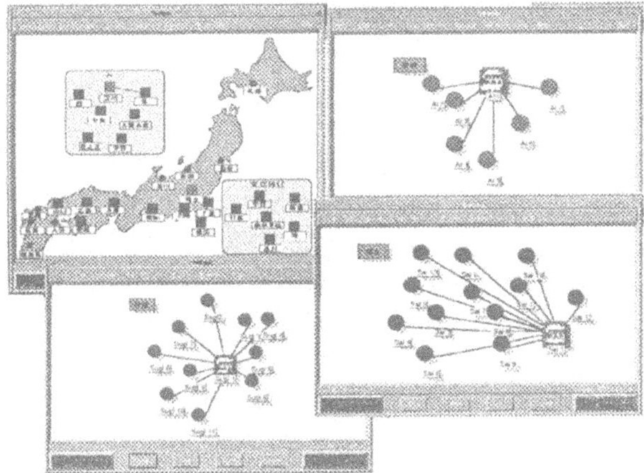

Figure 1 Multi-window graphic-user-interface style.

To overcome those difficulties, we proposed (Fuji, 1994), an approach that uses a bifocal display for providing both the network's context and details within a single window. This paper describes an implementation of it. The implementation, called DUALQUEST, was installed in a workstation equipped with a frame buffer memory for real-time bifocal image generation. For the performance evaluation and comparison purposes, we tested (with the aid of an event simulation program) both DUALQUEST and the hierarchical multi-window presentation in the presence of network alarms caused by, for instance, network element failures.

The paper is organized as follows. At first, we present the bifocal network visualization and compare it with the hierarchical multi-window visualization (Section 2). Next, DUALQUEST is introduced (Section 3). Then, we describe an experiment that was done to examine performance of those two methods (Section 4). Finally, we discuss some results obtained in the experiment.

2 BIFOCAL PRESENTATION VS. MULTI-WINDOW PRESENTATION

Hierarchical multi-window presentations are often used to handle networks which are too large to be meaningfully displayed within a single window. In the approach proposed in (Hewlett Packard, 1992), the complete topology of a managed network is displayed within a single window, while details of the network can be displayed within other windows. This may result in some difficulties for the operator; two of them are now briefly discussed.

Since multiple windows overlap each other, an amount of information can be lost. If a significant information is lost, a network operator must perform complex maneuvers to recover it. Another problem appears when the operator is going to trace a network path that comprises a large number of nodes because a single window displays only one segment of the path. Thus, the operator must monitor several windows to recognize such a path.

To display a large amount of data within a limited area, the bifocal display approach was proposed and analyzed in (Leung, 1989., Sarkar, 1992., or Brown, 1993). For instance, according to (Leung, 1989), a single window covers nine distinct regions as was shown in Figure 2; at any time, one of those regions can be enlarged while the others must be compressed to accommodate the enlargement. This is illustrated in Figure 2, the area 'a' is enlarged to the area 'A,' while 'b,' 'c' and 'd' are compressed to 'B,' 'C' and 'D,' respectively. As it is shown in Figure 2, a bifocal image can be generated by combining the data obtained from four different types of images (Misue, 1989).

Advantages of the bifocal approach can be summarized as follows.

- Since views are generated through expanding one area and compressing the others, no objects are missed at any time.
- Since all objects are viewed continuously, the whole nine regions can be easily traversed.

These advantages make bifocal display attractive for a network management user interface. Since, at any time, every object is displayed in a single window, the operator can continuously monitor the status of all network elements. In addition, the operator can traverse network connections displayed in several regions. This plays a key role especially for node-to-node connection management.

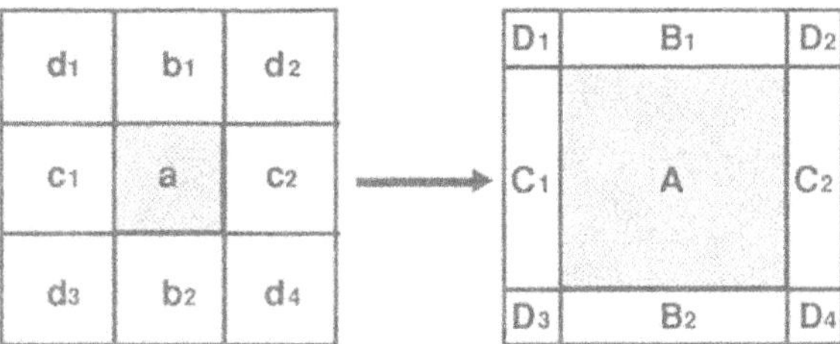

Figure 2 Illustration of the bifocal image generation.

3 DUALQUEST

In the bifocal display applications for network management purposes, such as fulfilling the alarm surveillance task, any area should be simply enlarged by clicking the mouse at an appropriate point on the screen. Since real-time responding to network notifications and operator's actions is required, we proposed the real-time bifocal network visualization using a frame buffer memory (Fuji, 1994). Then, the idea has been enhanced and resulted in an implementation, here called DUALQUEST.

3.1 Rearrangement of network nodes

Displays of a major city network usually contain many overlapping nodes and links; see, for instance, Figure 3a. To eliminate the overlapping effect and to use a screen more efficiently, a rearrangement of network nodes is required (see Figure 3b).

3.2 Presentation guideline

To determine network topology information that should be provided by the bifocal display, a presentation guideline is needed. Generally, two types of network views can be provided by DUALQUEST: the initial view and the enlarged view. To simplify an information display, both node names and link symbols corresponding to the local communication lines are not included to the initial view but they appear within the enlarged view that is, a view generated by the bifocal display using a frame buffer memory. As a result,

- every node name, and

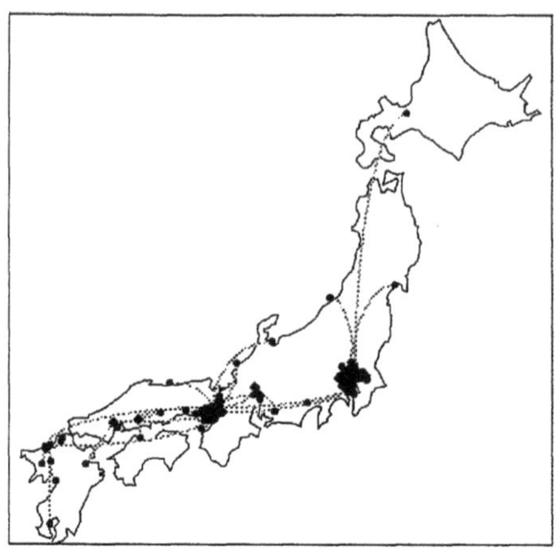

(a) Network before rearrangement

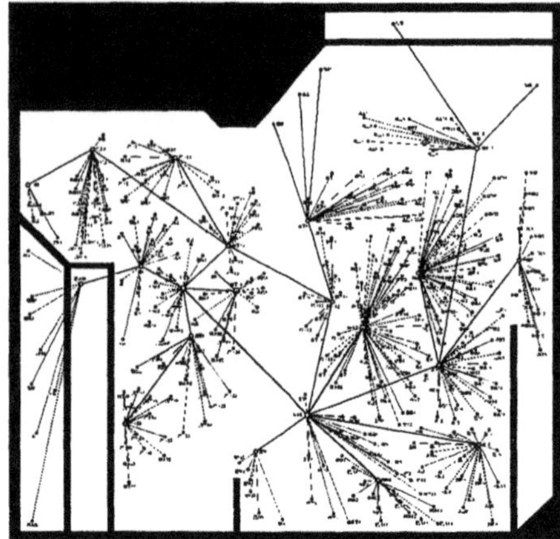

(b) Network after rearrangement

Figure 3 Rearranging network nodes.

- all network connections, including both backbone and local lines,

are given in the detailed section of the enlarged view.

Figure 4 shows an example in which the names of nodes and node connections are displayed in detail. In the bifocal network visualization, operators can continuously monitor the status of all network elements in a single window only. In addition, since picture continuity is maintained, network connections are displayed in full detail and the operator can easily trace them. Currently, the full network display is achieved with 900 × 900 pixels, while any individual section, displayed with 300 × 300 pixels in the initial view, may be enlarged to 600 × 600 pixels in the enlarged view.

3.3 Real-time bifocal image generation

DUALQUEST is equipped with a frame buffer memory that enables generating bifocal images in real-time. The frame buffer memory is provided with five planes: four image planes, for storing image data, and one plane, for the buffer control (Matoba, 1990). Every pixel-space of the buffer control plane contains address of the image plane whose data should be represented by an appropriate pixel of the bifocal image generated. The bifocal image consists of nine distinct regions; each of them is demarcated in the buffer control plane. Since regions of the same character are characterized by the same magnification (see Figure 2), it is possible to generate a

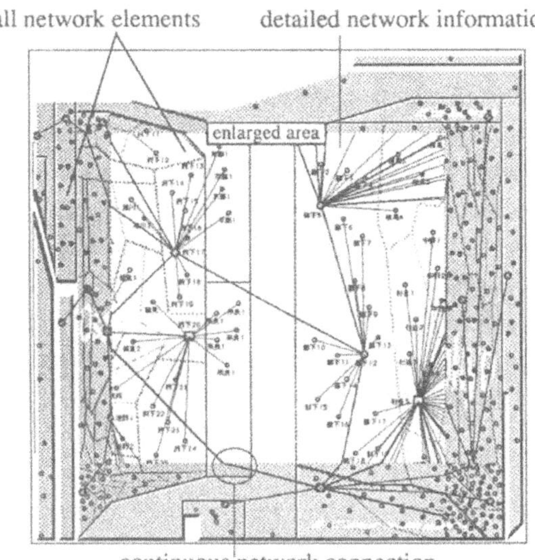

Figure 4 Bifocal network visualization.

bifocal image with only four types of image. Thus, as depicted in Figure 5, a complete bifocal image can be constructed by combining the data of the enlarged image 'A' with those of the images 'B,' 'C,' 'D,' ..., and 'I' of the three compressed peripheral images. Every pixel-space of the buffer control plane is given by the address of an appropriate image plane. According to the previously described presentation guideline, the enlarged image includes complete information of a network topology, while the compressed peripheral images exclude node names and local lines.

Because all the above operations for bifocal image generation are done in hardware, they can be accomplished instantly at each mouse click. As compared to a software operation, here no computation time is required. Due to this, users can easily traverse network topology as well as they can continuously trace paths of any length. In addition, a larger number of events can be processed within the same period of time since the saved computational time can be spent for fulfilling another task(s).

The current version of DUALQUEST supports fulfilling the alarm surveillance task in a similar way as that described in (Cunningham, 1992); the steps are as follows.

- If an alarm occurs, some symbols corresponding to nodes or links start blinking.
- By making click on the point of interest, the surrounding area appears within the enlarged view.
- The operator can observe the status of all events in a detailed area and follow any

change, such as back-up, or recovery, of it. This is simply indicated by changing symbol's color.

In fulfilling the alarm surveillance task, the operator can perform such operations as back and forth changes between both the initial and the enlarged views or a change in location of the detailed section. Since all areas remain still visible on the original display, the operator is able to detect other alarms that occur in the compressed network areas and then observe them in more detail. This is illustrated in Figure 6, where at first the surrounding area 'A' is enlarged (see Figure 6a), and then the surrounding area 'B' is enlarged (see Figure 6b). Even then when the detailed section is being focused on the area 'A,' alarms within the area 'B' can be noticed at the time of their occurrence. Thus, all alarms can be seen in those enlarged views; however, some of them may be shrunk in the compressed view.

4 EXPERIMENT

To compare the real-time bifocal network presentation (DUALQUEST) with the hierarchical multi-window presentation, we conducted an experiment similar to that proposed in (Mayhe, 1992) for evaluating window style. We selected a sample network, comprising 400 nodes, and an event simulation program that controls the time interval (5, 10, or 15 seconds) between two consecutive events. Then we invited ten users, including 5 people having no experience with network management systems, to take part in the experiment. Their goal was to fulfill the alarm surveillance tasks by using both the multi-window presentation and DUALQUEST. Operations performed by those users were simultaneously recorded by (i) video cameras, (ii) an *eye-mark-recorder* tracing any movement of the human eye-sight, and (iii) a device sampling mouse operations. All those participants were asked to fill out survey forms twice; before the experiment was started and after it was completed.

4.1 Multi-window presentation system

To implement the multi-window presentation system, we used HP OpenView (Hewlett Packard, 1992). In this implementation, we categorized the sample network into 28 groups, where each group aggregated from 10 to 18 nodes. As a result, a two-layer presentation system was built; that is, the network was presented by using 28 group symbols in one window (corresponding to the first layer), while the details of any group were displayed in the second window (corresponding to the second layer). Each of these windows could be opened by clicking the appropriate group symbol on the first layer window. An example window layout in this multi-window presentation system is illustrated in Figure 1.

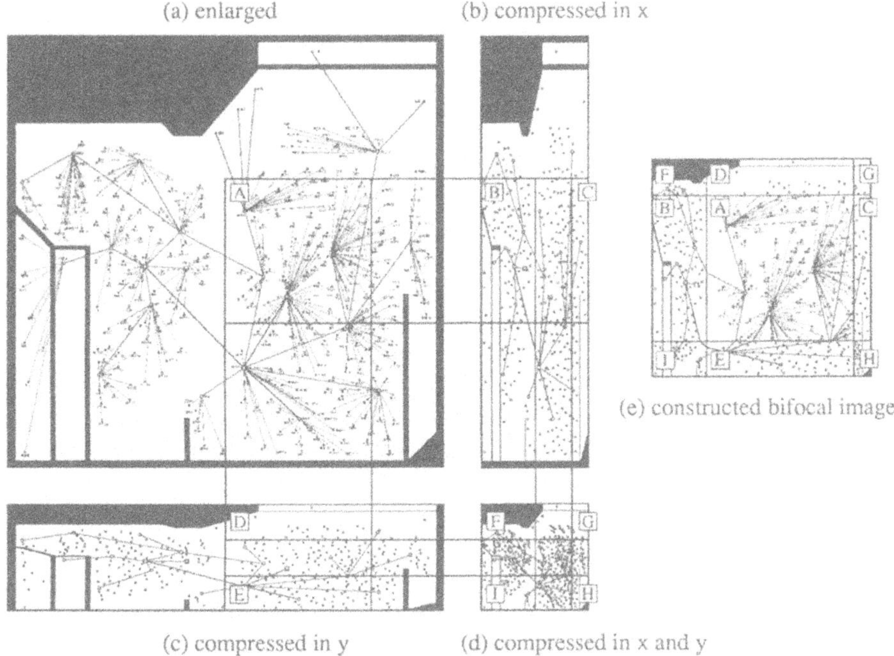

Figure 5 Bifocal image generation using frame buffer memory

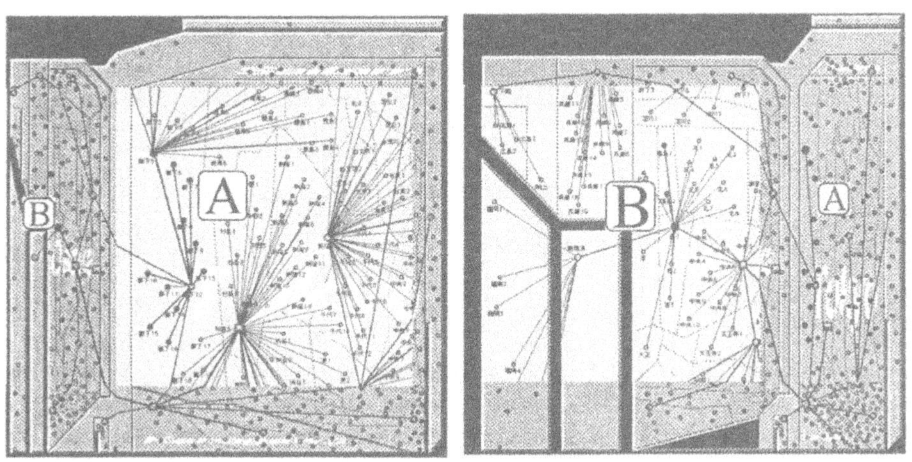

(a) Area 'A' enlarged, area 'B' unenlarged (b) Area 'A' unenlarged, area 'B' enlarged

Figure 6 Screen examples of alarm surveillance.

4.2 Results

As a result of the experiment, DUALQUEST was slightly better than the multi-window presentation system in terms of the time necessary to detect an alarm and the number of alarms not detected within the assumed period of time. For instance Table 1 gives results obtained for the 5 second slots. We think the lack of a performance significant difference between the tested systems was mainly caused by using only two layers of the hierarchical multi-window presentation system.

The two other major results from the experiment can be summarized as follows.

- Nine from the ten users of the ten, among them *all inexperienced users*, reported finding it easier to discover alarms on DUALQUEST because they were able to perform their tasks in a single window only without complex window operations. (However their first impression on DUALQUEST was not a positive one, since they are used to multi-window style GUI.)
- Smaller windows seem to be more suitable than the whole screen for detecting alarms at the first stage of the alarm surveillance.

The former confirms that even an inexperienced user can operate DUALQUEST, the latter suggests us introducing some user opinions in the further work on DUALQUEST.

Table 1 Experimental results.

Metric	DUALQUEST	Multi-window System
Mean time to identify all failed nodes	4.56 seconds 1.12 seconds*	4.81 seconds 2.03 seconds*
Mean rate of oversight**	12%	17%

* standard deviation
** ratio of undetected alarms to all displayed alarms

5 CONCLUSION

An implementation (DUALQUEST) of the bifocal display concept in network management systems has been presented and discussed. Since the display is generated by a hardware system using a frame buffer memory, DUALQUEST provides the real-time image. The bifocal display allows the operator to follow all status changes by monitoring a single window on the screen. This advantage of DUALQUEST was confirmed by participants of an experiment that was done to compare the new system with the conventional (the hierarchical multi-window presentation) one.

Acknowledgment

The authors wish to thank Y. Hara of NEC Corp. for his technical support and discussion, and would like to give our special thanks to M. Yamamoto, S. Hasegawa, and H. Okazaki, all of NEC Corp. for the encouragement.

REFERENCES

Brown, H.M., Meehan, R.J. and Sarkar, M. (1993) Browsing Graphs using a Fish-eye View. In proceedings of ACM INTERCHI'93.

Cunningham, P.J., Rotella, J.P., Asplund, L.C., Kawano, H., Okazaki, T., and Mase, K. (1994) Screen Symbols for Network Operations and Management. In proceedings of the Third of Network Operation and Management Symposium.

Fuji, H., Nakai, S., Matoba, H., and Takano, T. (1994) Real-time Bifocal Network Visualization. In proceedings of the Forth of Network Operation and Management Symposium.

Hewlett Packard. (1992) HP OpenView Windows User's Guide. Manual Part Number: J2136-90000.

Leung, K.Y. (1989) Human-computer Interface Techniques for Map Based Diagrams. In proceedings of the Third International Conference on Human-Computer Interaction.

Matoba, H., Hara, Y. and Kasahara, Y. (1990) Regional Information Guidance System based on Hypermedia Concept. SPIE Vol. 1258 Image Communications and Workstations.

Misue, K. and Sugiyama, K. (1989) A method to display the whole and detail in one figure. 5th Symposium on Human Interface.

Sarkar, M. and Brown, H.M. (1992) Graphical Fish-eye Views of Graphs. In proceedings of ACM SIGCHI'92 Conference on Human Factors In Computing Systems.

Mayhe, D.J. (1992) Principles and Guidelines in Software User Interface Design.

Shoichiro Nakai received his B.E. and M.E. degrees from Keio University in 1981 and 1983, respectively. He joined NEC Corporation in 1983, and has been engaged in the research and development of local area networks, distributed systems, and network management systems. He is currently Research Specialist in the C&C Research Laboratories.

Hiroko FUJI received her B.E. degree in mathematics from Kyusyu University in 1990. She joined NEC Corporation in 1990, and has been engaged in research on network management. She is currently working in the C&C Research Laboratories.

Hiroshi Matoba received his B.E. degree in Mathematical Engineering and Instrumentation Physics from Tokyo University in 1985, respectively. He joined NEC Corporation in 1985, and has been engaged in research and development of graphic acceralators for workstations. He is currently an assistant manager in the C&C Research Laboratories.

51

A framework for systems and network management ensembles

E. D. Zeisler
The MITRE Corporation
7525 Colshire Drive; MS W549; McLean, VA 22102; USA
Phone: (703) 883-5768; FAX: (703) 883-5241;
ezeisler@mitre.org

H. C. Folts
Defense Information Systems Agency
10701 Parkridge Boulevard; Reston, Virginia 22091-4398; USA
Phone: (703) 487-3332; FAX: (703) 487-3351;
foltsh@cc.ims.disa.mil

Abstract

A richness of systems and network management technology has been defined by standards. The ensembles method developed by the Network Management Forum (NMF) joins the standards with operational functions used by the enterprise resource manager. In order to ensure that the total enterprise is considered, a framework is required that will tie NMF ensembles to a wider (scalable) management mission. This paper sets out a framework for selection of management ensembles.

Keywords

Domain, ensemble, managed objects, scenario, Telecommunications Management Network (TMN)

1 BACKGROUND AND OVERVIEW

1.1 Problem statement

An Ensemble is an OMNI*Point* NMF reusable, implementation specification. This specification is made up of requirements, scenarios, and managed objects, plus references to standard information models, and conformance test descriptions.

Ensembles are written and approved through the NMF Ensembles Working Group (EWG). There are a number of implemented OMNI*Point* ensembles, which rely on the Telecommunications Management Network (TMN) ITU-T standards. For example see document Forum 017, 1992, Reconfigurable Circuit Service - Configuration Management.

In fact, to provide a detailed representation for both communications and managed data, an ensemble includes actual CMIS/CMIP commands, which access one or more managed objects. The method has successfully provided specifications to the subcontractor(s) who build from them. However, current ensembles provide a solution to a specific (limited), network management problem. The method cannot enable one ensemble to be related to another in a cohesive manner; nor does it provide an enterprise context for the selection or build of multiple ensembles. Enterprise in the functional sense, encompasses: (a) realtime ITU-T X.700 series of specific management functions (for performance, security, accounting, configuration or fault management); as well as (b) non realtime planning, engineering and service provisioning, among others.

1.2 Concept and background

As shown in Figure 1, an ensemble, by itself, can provide a window into the real operational functions like alarm surveillance, and into the managed resources, for a class of equipment. The ensemble matrix shows types of managed resources on one axis, while the management functions are shown on the other axis. One or more of these resources and functions can be used in an ensemble to support a specific business objective.

Concept

The proposed new method (dotted lines in Figure 1) will represent ensemble 'sets' as domains or partitions for the delegation of management responsibility according to enterprise policy. Earlier work has demonstrated how very large-scale distributed systems can be managed using domain and policy concepts [ESPRIT, 1993]. To this end, controlling interfaces is predominantly a matter of structuring different types of organizations, geographic areas, groups of users, or managed technology *into domains.* Figure 1 shows how ensembles can be further characterized: by policy, domains, services, and features. In general, a service or feature could be provided in some domains, but not others. The point is that, with hundreds of networks, equipment types, services, features and management functions, an ensemble specification/build must be *tailored to a domain of interest*.

Background

The genesis of the framework comes from implemented OMNI*Point* 0 RCS ensembles. These ensembles led to the British Telcom CONCERT system, which operates with an end-to-end view of a complete network, and further, enables interworking with a range of network management systems, as exemplified in Newbridge Network's ConnectExec system for managing a T1 MUX [Gamble, 1993; Newbridge, 1994]. To develop our domain-based framework, we examined the shared management knowledge (SMK) utilized by the just-mentioned implementation systems; we selected and added objects (or object subclasses) for what we call the 'core managed objects'. Experience has shown that it is better to define a core Management Information Base (MIB) first, containing only essential objects; later, if experience demands, other objects can be added [Rose, 1991].

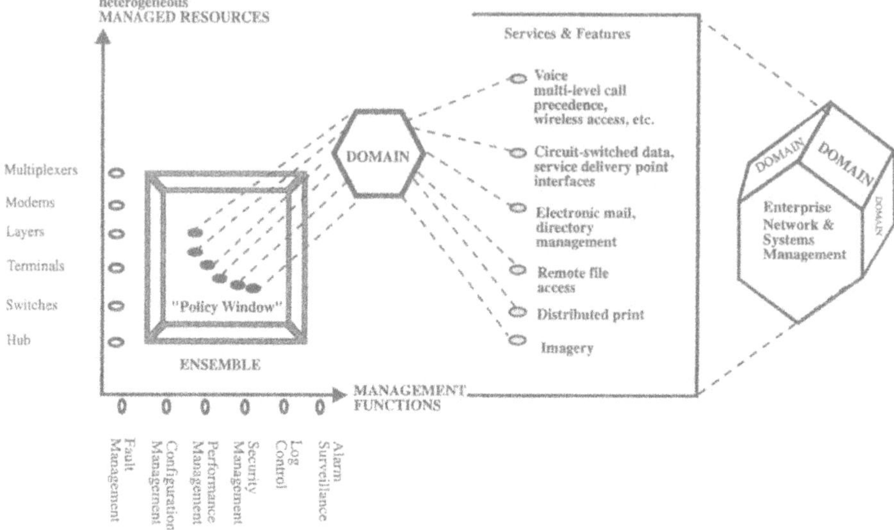

Figure 1 Domain Concept for Ensemble Selection.

Table 1 shows how the designer can associate policy, core managed objects, and a functional area, in this case ***performance management,*** for a given domain [Newbridge, 1994; Forum 022, 1992]. The chart shows:

- A 'service description' legend at the upper left, providing a domain association,
- Columns - 'core' Managed Objects (MOs) that can be used in managing the voice service;
- Rows - the performance policy; each cell relates a policy to the objects; the objects will contain the behavior and attributes required for policy institution; and,
- The U (update) and R (retrieval) notations, which set the stage for more detailed specification (i.e., specifying the low-level protocol operations to access objects).

To summarize, the 'core' objects, packaged along with the standards that guide or restrict the use of those objects, not only can express the union of objects across domains, but also can be coupled with a strategy for intra-domain coordination.

Criteria

The standard core objects are expressed in OMNI*Point* 1 syntax. Given a base set for all domains, any ensemble must provide **one or more** of the 'core' objects. Other criteria follow:

- **Monitoring and control** - some core objects will be used by other objects, i.e., notifications pass through the sieve (discriminator object) to create events (event records object) that can be logged (log object) or reported. It's possible that only one object (the system object) would be used, as the top of a hierarchy that defines additional subclasses, for layer management.

Table 1 Sample of Core Managed Objects Related to Corporate Policy

DOMAIN - Subnetwork SERVICE Description Voice service and router switch to support interconnections **Performance Management Policy** / **Core Managed Objects**	*Network Object*	*Op Equipment Object*	*Facility Object*	*Discriminator Object (Event Reporting Sieve)*	*Log Object (Event Log Object)*	*Event Records Object {alarm, state change, etc. }*	*Location Object*	*Agent CME Object*	*Circuit Object [TRM Groomed Circuit}*	*Forum Test Object*	*TRM Terminating Equipment Object*	*Function Object {Terminating Point Object}*	*System Object*
	1	2	3	4	5	6	7	8	9	10	11	12	13
Report overall transmission availability - see Note	*			✓	✓	U		✓				R	*
Report backbone circuits & nodes for T1 failing to meet management thresholds; i.e., time out/in, total outage time, number of outages	*	R	R	✓	✓	U		✓	R		R	R	*
Detect long-term trends in degraded transmission performance	*	R	R	✓	✓	U		✓	R		R	R	*
Correct long-term degraded transmission i.e., signal quality per bit-error-rate, frame slips, errored seconds, cyclic redundancy check, bipolar violations	*	U	U	✓	✓	U		✓	U		U		*

Table Legend

Note:
Includes objects for connection_oriented_transport_protocol_layer_entity, connectionless_network_protocol_layer_entity

* = use for naming hierarchy

R= retrieval, U= update

✓ = use without attribute retrieval or update

( )= OMNIPoint 0 reference

{ }= use with subclass

- **Extensions** - other MOs could be added to an ensemble like the function object, service object, policy object and security object. For instance, the service object in the draft 'Customer Administration Configuration Management' ensemble is used for voice service types including bearer, supplementary and teleservice. A **rule of thumb** is that the lower-level managers in a domain hierarchy may have a limited number of core object extensions (data attributes/values).

Of relevance, our framework does **NOT** attempt to provide a step-by-step procedure for developing an ensemble.

Goal

As mentioned, the NMF ensembles being developed describe managed objects using a specific small (limited) management task area. The domain framework intends to enable development of open systems in a wider management task area by defining a 'domain (partition)' as a small task area and correlating multiple 'domains'. Here, relationships of domains are hierarchical, and some domain management facilities are needed for negotiations between domains.

1.3 Terms and concepts

Overall, a taxonomy of domains is vital to our framework. **Domains** are sets of objects/entities/things that may be logical or physical. For the purposes of this paper, we are primarily interested in **management domains**, which restricts this generic definition to the extent that the objects/items contained within the set are all subject to the same management agent(s). Another way of putting this is that the domain defines the <u>span of control</u> or <u>sphere of influence</u> of management. Additional requirements (e.g., the managed objects must be finite, named/ catalogued, etc.) have been set forth elsewhere.

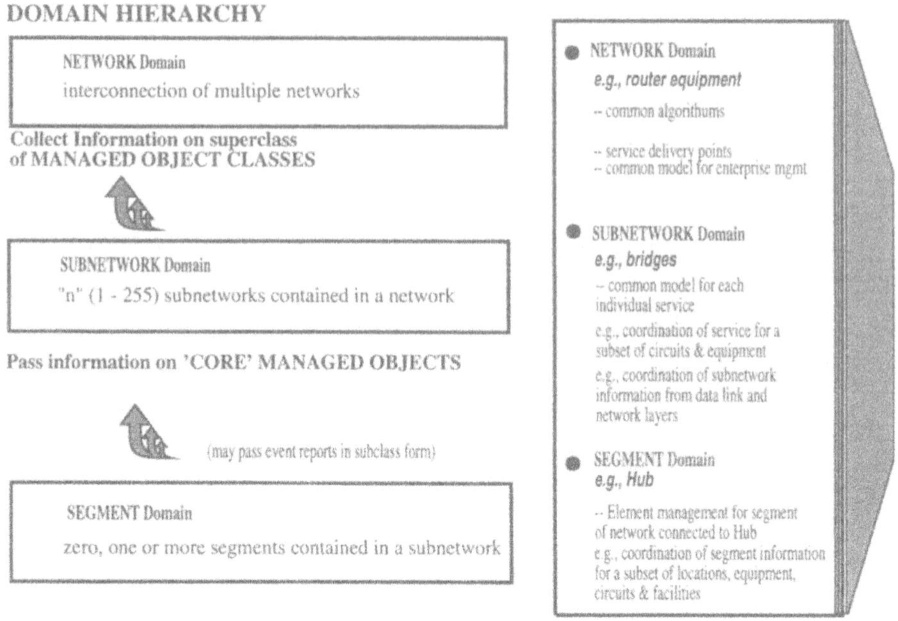

Figure 2 Domains for Connection-oriented Networks.

Physical domains

There are acknowledged differences on how to standardize and implement domains in the marketplace (ISO CD 10164-19.2) [Moffet, 993]. As used herein, a **physical domain** consists of a set of real-world objects within a boundary. An example of such a physical domain is all of the computers and peripheral devices contained within one building to comprise a computer system. Another physical partitioning concept could use a domain 'triple' set (networks, subnetworks, segments) for managing different types of switches: one domain manages a public branch exchange (PBX); another domain manages a router; the domains are interconnected to support an end-to-end network management service. Section 3 will apply this triple in a scenario. Figure 2 illustrates the triple domain hierarchy. The relationship of domains in a hierarchy reflects a tailored, open path within a wider networked environment.

Logical domains

Logical domains may be broken down in as many ways as one may logically group either real-world objects (or representations of real-world objects) or logical resources. To this purpose, a domain hierarchy could be logically partitioned as follows:

- **Disjoint or independent** - a set of domains that do not interact in any way with each other;
- **Overlapping** - at least two management domains exist–each containing its set of managed objects, but some or all of the objects in each set (where the two domains intersect) are subject to the management of each of the domains; and,
- **Nested** - an outer domain exists which contains its set of managed objects and some of its objects may be within an inner domain; the objects in the inner domain are subject to the

management of that domain, but they are also still subject to the control and are owned by the outer domain.

Further, there must be rules for setting up domains: for instance, a rule might state there can be only one domain manager (coordinator) for each partition in a nested domain.

2 FRAMEWORK MODELS

Besides domain design, different models may provide new perspectives for ensembles selection:

- ***Service model*** - in this model, requirements for managing components are broken out by services, such as mail, software distribution, printing, remote file access, etc.;
- ***Work flow model*** - this model shows organizational requirements based on management controls, input/outputs, processes, subprocesses, or systems; and,
- ***OMNIPoint business model*** - this is an administrative view of the managed components for the customer, supplier and service provider; associated with ITU-T TMN interface points [Q821, 1993]

The two models that are described herein exploit multiple perspectives for the framework: the first is a generic network model, which describes Connectivity at a high level; the second, is a model for the ITU-T TMN interface points (see Section 3.3).

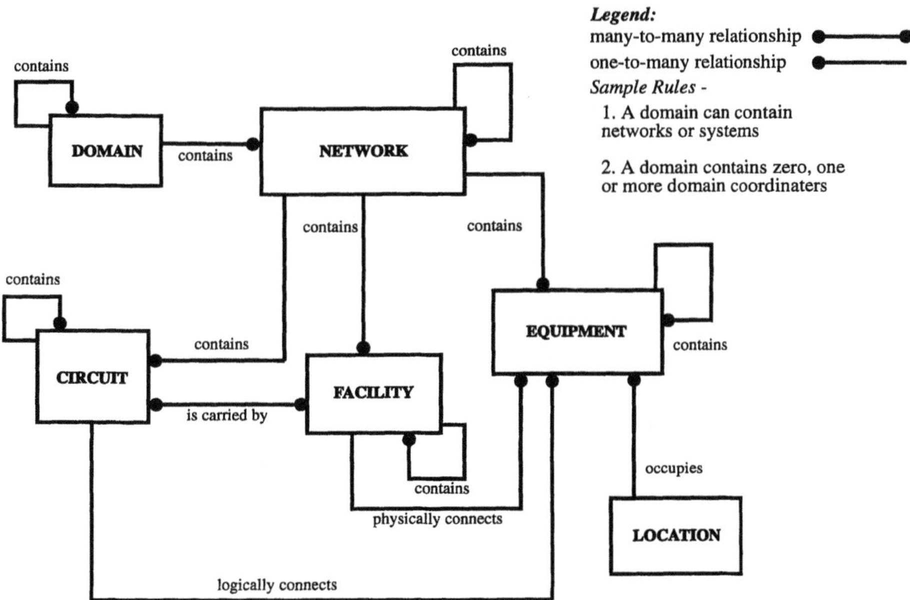

Figure 3 Generic Model of the Network.

2.1 Generic network model

Figure 3 uses a subset of the 'core' objects for describing connectivity. This kind of generic model of connectivity is essential to the generation of uniform operations, planning, engineering, service provisioning and maintenance. The specific resource properties and attributes for each Figure 3 entity are specified in the OMNI*Point* NMF Libraries.

First, the model can be used to describe connectivity for any number of domain types. That is, the entity-relationship diagram relates each domain to its component networks/subnetworks/ segments, circuits, facilities, equipment, and locations.

Second, besides populating the model for network connectivity, it could be expanded for systems information: the domain entity could contain a system entity, which is separate from the network entity, and relates to other entities that are used for system management.

Third, the model identifies the building blocks for a component (resource) hierarchy for an ensemble. Ultimately, service-users may control and observe the values of the attributes of these resources. An example is given next to show the distinction between circuit and facility, as used in the model [Kennedy, 1991].

The **circuit** managed object refers to a *connection that is independent of the means of carrying the signal.* Instances of this managed object carry information from point to point and preserve its content. The bandwidth (information/data rate) of a circuit may be known, fixed or variable.

The **facility** managed object refers to the physical means of carrying a signal; e.g., a connection composed of different media (coaxial or fiber cable) in an ordered sequence with intervening equipment (for optical-electrical conversion, reshaping, or regeneration). The signal rate and medium of a facility may be known and fixed and is implied by the facility type. Thus, the bandwidth of a facility is not dynamically alterable. Facilities are used to carry circuits:

- one facility object instance can carry many circuits;
- one circuit object instance may be carried by many facilities;
- a facility *physically* connects to one or more equipment's; and,
- a circuit *logically* connects to one or more equipment's.

2.2 Modeling tools

The Integrated Computer Aided Manufacturing (ICAM) Definition (IDEF) graphic representations of the network model employed an IDEF-based modeling tool, since the IDEF family of standards is nonproprietary and public domain:

- the generic network model plus an expanded model (template) with the business rules for domain activities and entities, were created using IDEF1X data modeling syntax;
- high-level management functions were decomposed using an IDEF0 function modeling tool [DISA, 1993].

2.3 Outcomes

Requirements were modeled to develop a generic template for all domains. The full template includes a characterization of domain activities and an operational model (i.e., domain manager, shield, domain agent, and target resources). A shield is an intelligent control between the manager and agent, which hides nonrelevant details from the manager. In general, the shield supports intelligence for visibility control, computation, aggregation and/or transformation [ESPRIT, 1993].

Modeling supported a kind of 'intermediate analysis', or translation, for the sampled requirements. Further, the IDEF rigor imposed should ensure consistency for our next phase, which will produce a prototype for domain ensemble(s).

3 DOMAIN SCENARIOS

Now that we have described our framework and goals for the OMNI*Point* Network Management Forum EWG, we will look at how we actually used the concept. A typical domain requirement for a large packet-switched data system would be '<u>the service shall be capable of supporting a hierarchy of logical subnetworks with independent addressing and management domains. At least three levels of hierarchical networks, with at least 256 logical subnetworks at each level, shall be supported</u>'.

3.1 Voice service for organization independent views

In this particular scenario, the Figure 2 domain breakout (networks, subnetworks, segments) is applied to meta-management of voice communications management systems. The domain triple, is compatible with other models, in particular for TMN administrative roles (service provider, subscriber). Figure 4 has three rings: the outer ring represents the End-User subnetwork domain, the middle ring shows the Local Exchange Carrier (LEC) subnetwork domain, and the inner ring represents the Interexchange Carrier (IC) backbone subnetwork domain. The views (labeled one, two, and three) illustrate activity for the different players, respectively:

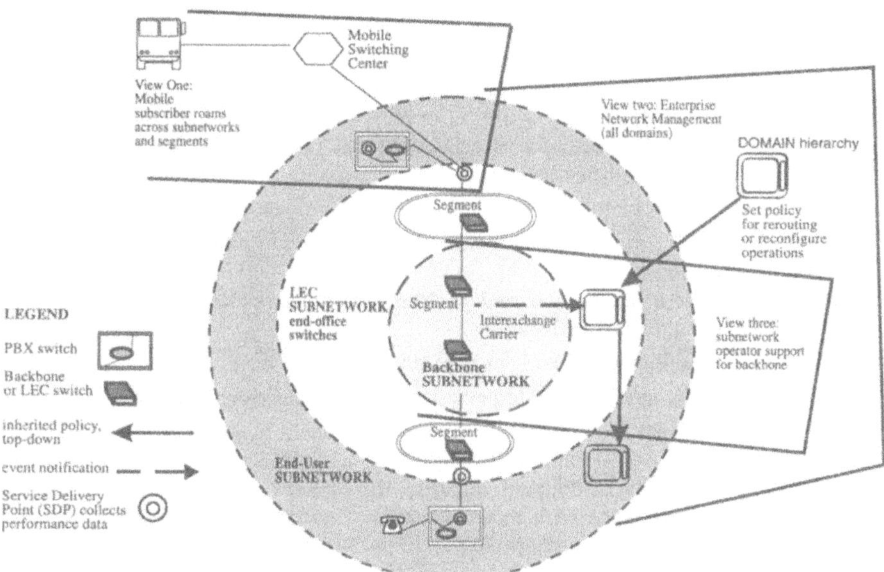

Figure 4 Voice Service Scenario for Performance Management.

- view one - a customer's (subscriber) view from the end-user subnetwork;
- view two - a service provider's enterprise view of the switching network and all domains;
- view three - an internal operator's view, in this case providing support in a service provider role, for the backbone subnetwork domain.

The Figure 4 partitions (rings) and organizationally-independent views were used, to apply the triple (network, subnetwork, segment) domains design. Of benefit, overlapping organizationally-independent domains enable scalable management:

- ability to institute rights over another manager's domain (constraints);
- flexibility for inherited behavior for a sphere of interest; and,
- ability to harmonize management policy for heterogeneous, distributed resources.

That is, a user could be supported by:

- whichever domain operator controls the switches/trunks at the time a problem occurs
- whichever domain triggered a network fault, performance or accounting error.

In general, any of the Figure 4 switches could be involved in management activity, since they constitute a component hierarchy for the end-to-end switched voice service: (a) either of the end-user PBX switching nodes; (b) either of the LEC subnetwork switching nodes; and (c) either of the backbone switches for the IC switching node.

3.2 Sample scenario: voice service and performance management

Again, Figure 4 illustrates just one possible domain scenario. The performance problem (all trunks are busy) is due to a switch overload in the backbone subnetwork domain. Consequently, problem management impacts the mobile subscriber and must be resolved by the backbone subnetwork operator. Note that because all circuits are busy, the subscriber can't call in the problem, and the trouble ***must*** be detected by the network management system.

Domain interaction

First, in the innermost ring, a state notification or fault is generated concerning service to the mobile subscriber. Simultaneously, a user error message to a domain management station in the outer ring (end-user subnetwork) associates the call problem with one of the backbone switches, and further identifies a cause code.

Second, to resolve the problem, the view analysis will coordinate among:

- **the End-User, LEC, and backbone subnetwork switch domains** for performance and state information; and,
- **the enterprise network manager and backbone subnetwork domain**, for backup plans for bandwidth/signal/channel allocations.

Third, something must happen in the backbone ring to resolve the problem.

- the backbone domain manager could perform rerouting operations in the backbone subnetwork domain (i.e., using view three); and,
- a dynamic rerouting solution could require coordination between views two and three.

Last, through the hierarchy of domain managers, the subnetwork and segment domains affected will receive knowledge of a reconfiguration. Further, the coordinated information could be used *not only* to resolve the performance problem, but also to adjust customer usage billing.

Ensemble requirements

Intelligence in the domain agent is needed, to correlate performance and state information. Also, there can be policy related to thresholds for blocked subscriber calls or related to who needs to be notified (e.g., the stations at the access switch).

Ordinarily, in realtime when a trunk group becomes unavailable due to congestion and there is limited available resources, a mobile user's calls will be dropped based on user priorities (multilevel precedence and preemption activity). Thus, a series of dropped calls or error messages (call incomplete) could occur until the threshold for critical congestion is exceeded.

Domain coordination functions

As shown in Figure 4 (with workstation screens), each subnetwork domain 'manager' has capability to institute *end-to-end service* policy and policy for reporting performance, as directed by the enterprise network domain manager. For instance, the domains will regularly:

- pass performance logs between domains with overlapping managed resources;
- use alternative routes only as short-term solutions, according to policy; or,
- send performance event reports for long-term trend analysis; these could be collected at the different Service Delivery Points (SDPs) in the network, i.e., at the interface between the customer and service provider.

Service Delivery Point requirements

- Filter collected statistics on degraded transmission performance, and,
- Collect network statistical data on traffic flow.

Benefits

In summary, the domain concept can enable the operator directly responsible for resolving a problem to become involved, rather than going through intervening 'organizational domains'. Note that getting a window into service provider subnetworks or internal organizations will require setting up service level agreements [Moffet, 1993].

3.3 Model for Telecommunications Management Network interface points

Figure 5 maps our scenario to the TMN architecture (M.3000 series of recommendations from ISO/ITU-T). The maturation process of the TMN standards will continue to accumulate importance for service providers and network equipment vendors for the remainder of this decade. Therefore, a benefit of the domain framework for Systems and Network Management Ensembles is that it could be used to set goals for integrating TMN into the enterprise requirement.

Essentially, a TMN is a network to provide surveillance and control of another network. The management network may be separate from, or share elements with the network it controls. Figure 5 references the scenario domains, to identify standardized TMN interfaces.

Function blocks for interfaces are [Shrewsberry, 1994]:

WSF - workstation function to interpret TMN information to the human user

OSF - operations system function to process information related to management

QAF - Q adapter function for protocol conversion to a standard TMN interface

NEF - telecommunications functions for the network element (managed device), including the MIB and associated management applications

MF - mediation function to store, adapt, and filter detailed information between an OSF and a NEF or QAF.

A **Reference Point** (RP) classified for the message communication between any two function blocks is [Shrewsberry, 1994]:

f - attachment for a workstation function; used here as an X Windows interface to a human user

m - class between a QAF and its non-TMN managed resources; typically an older telecom-type interface like Bellcore's Translation Language 1 (TL-1) or the Telemetry Byte-Oriented Serial (TBOS) protocol

q - a class between an OSF, QAF, MF, or NEF for standardized interoperability (e.g., between the network management, element management ,and managed element) or between pairs of each; a q3 is a fully standards-compliant interface that uses a CMIP manager/agent pair on OSI protocol stacks. A qx is a 'not quite q3', where conformance problems arise in the network element layer (between switches in the subnetworks and the domain manager) or where the embedded system is small or uses SNMP

x - a class for providing interoperability between an OSF and a similar function in another management network; an interface between the administrative domains (service providers and customers) uses core objects to pass event reports or invoke maintenance.

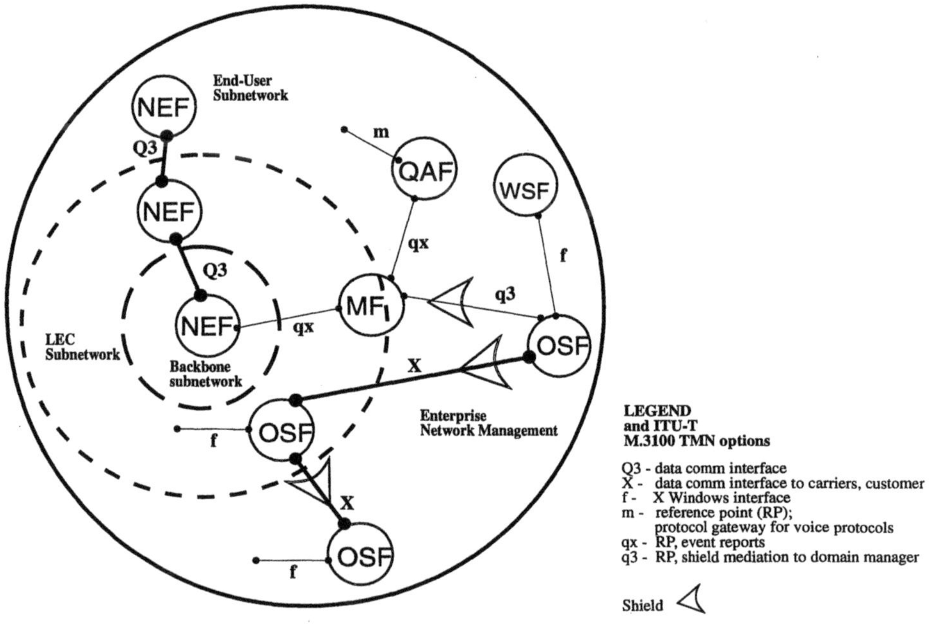

Figure 5 A Global Set of Requirements.

In Figure 5, upper case and emboldened lines, distinguish an interface from a RP. A RP becomes an interface when it occurs at a location requiring data communications between hardware elements. Significantly, all the managing software in a layer of the TMN architecture, resides at the application layer of the OSI reference model; the agents and services are not tied specifically to the underlying layers of the OSI protocol stack.

The Figure 5 shields separate managing and managed components (e.g., hardware devices) across domain boundaries; as a result, the manager works through the shield, to invoke/forward SMK to another domain or target element management system. For instance, 'shield' computations (located in the agent) could centralize bit-error-rate (BER) calculations.

4 CONCLUSIONS

As network connectivity grows, there is a need to extend the scope of management from a few nodes to a global environment, where many networks are interconnected. We have designed a framework to make open systems development in a wider task area easier and cheaper. Service providers invest immense amounts of money with suppliers for technology that will support their service strategies; therefore, we predict significant cost savings if a domain hierarchy is adaptable to managing services for heterogenous, multi-vendor devices.

odeling for ensemble negotiations between domains focuses on how to effectively share management knowledge across provider and customer networks. Further, the framework enabled us to identify three potential candidates for future prototyping and ensemble specification: (a) Customer interfacing TMN domains for intelligent correlation of switch faults; (b) Performance Management Ensemble for correlating BER across domains; and, (c) Customer interfacing TMN domains for intelligent correlation of trouble tickets.

In the next phase, we plan to prototype parts of the triple domain set in a physical design for the optimal number and type of domains. The prototype will use the generic network model to specify logical connectivity; a shield and domain MOCs will be identified, for policy institution. In prototyping TT and/or fault correlation, we intend to investigate the impact of the management protocol, as regards *how many managed nodes the domain coordinator (station manager) can manage*. It is assumed that coexistence between SNMP and CMIP is required, and both COTS and ensemble (new) development code will be used.

REFERENCES

Defense Information Systems Agency (DISA) (1993) Communications System Network Management: Network Flow Diagrams. AT&T/Seta Team, Reston, VA.

ESPRIT Project No. 5165 (1993) Distributed Open Management Architecture in Networked Systems (Domains). Deliverables D2a, V1.0 and D3, V1.0.

Gamble, R. (1993) Generic Agent CME. British Telecom CONCERT OMNI*Point* 0. Belfast Engineering Center, Ireland.

Kennedy, T.W., Riegner, S.E.M. (1991) An Object Oriented Model for the Operations, Administration and Maintenance of FAA Telecommunications. MITRE Corp., McLean,VA.

Moffet, J.D., Sloman, M., Twidle, K.P., Varley, B.J. (1993) Domain Management and Accounting in an International Cellular Network. IFIP Transactions III.

Newbridge Networks (1994) MainStreet Connect Exec, Technical Reference. Rel 5, Generic SBL115.

NMF OMNI*Point* Forum 017 (1992) Reconfigurable Circuit Service: Configuration Management Ensemble V1.0., Morristown, NJ.

NMF OMNI*Point* Forum 022 (1992) NM Forum Mapping from Release 1 to OMNI*Point* 1. Morristown, NJ.

Q.821 CCITT Rec and NMF OMNI*Point* (1993) Strategic Framework. Morristown, NJ.

Rose, M.T. (1991) Network Management is Simple: you just need the "right" framework! IFIP Transactions II.

Shrewsberry, J.K. (1994) TMN in a Nutshell, V1.01. WilTel, Tulsa, OK.

BIOGRAPHY

Elizabeth D. Zeisler

Liz Zeisler, a lead scientist with the MITRE Corporation, has over 20 years of lifecycle development experience in information systems, database and network management. She has degrees from Cornell University (B.F.A.), Computer Processing Institute (A.A.), State University of New York at Buffalo (M.E.) and American University (M.S.).

Harold C. Folts

With 35 years experience in telecommunications, Hal Folts currently serves as a senior systems engineer for network and systems management applications in the Defense Information Systems Agency. He has been involved over the past years with the development of many international standards for data communications and open systems. He has a BSEE from Tri-State University and an M.S. in Systems Management from the University of Southern California.

SECTION SIX

Managed Objects Behaviour

52

MODE: a Development Environment for Managed Objects based on Formal Methods

Olivier Festor[1]
Centre de Recherche en Informatique de Nancy
Bâtiment LORIA, BP-239, 54506 Vandoeuvre-lès-Nancy, FRANCE, Tel (+33).83.59.20.48, Fax (+33), E-mail: festor@loria.fr

Abstract

The need for mechanisms and techniques to describe formally managed objects behaviour in the Open Systems Interconnection (OSI) Management Framework has been recognized in various places. Building a formal specification of managed objects forces the designer to be more rigourous and allows a better understanding of what has been done. But the development of such specifications is a difficult and time consuming task which must be supported by a powerful set of tools. Moreover the effort invested in the development of the formal specification should pay off in some way during the Management Information Base development process.

In this paper, we present a development environment based on the formal mechanisms we include with the Guidelines for the Definition of Managed Objects (GDMO) notation to allow Managed Object behaviour to be formally described. This environment is intended to improve the process of building a formal description of OSI based Management Information Bases and provide several tools to exploit this formal description during the whole development process.

Keywords:

Behaviour, Development Environment, Formal Description Techniques, GDMO, Network Management.

[1]The authors work is also supported by IBM European Networking Center, Heidelberg, Germany

1 INTRODUCTION

One of the most important and complex task of OSI based management application builders is the design and modelling of the network components they want to manage. To facilitate the specification of such network components, the GDMO notation has been standardised within ISO (ISO-10165.4 1992) and is today widely accepted and used as the description technique for Managed Object (MO) design and specification.

The need to formally describe MO behaviour and provide guidelines for using the various specification templates of GDMO in a more systematic fashion, leading to clearly structured, coherent models has been expressed in (Kilov 1992). A first attempt for a design method based on a detailed study of behaviour classification has been proposed in (Clemm & Festor 1993).

The effort invested in formalizing Managed Object behaviour can be used to derive a better product and to automate certain steps of the development process. Accordingly, this formalization effort can only be accepted and used if it is part of a well defined development process and supported by an integrated development environment which provides tools for exploiting these new functionalities in the development process.

In this paper we present the MODE (Managed Object Development Environment) development environment. This set of tools is based on the development process used in Formal Description Techniques (FDT) based approaches and supports both standard GDMO and the formal extensions we proposed to the behaviour part of the notation.

The remainder of this paper is organised as follows: the next section describes the purpose and main goals of the development environment. Section 3 provides some features of the formal mechanisms we have adopted to extend the GDMO notation. Section 4 contains the description of the Management Information Base (MIB) design tool. Section 5 presents the MIB design application. Section 6 is concerned with a validation tool which allow formally described MOs to be interactively simulated. Section 6 provides information on the status of the environment and some future directions are discussed. Finally, a summary of the presented work is given.

2 PURPOSE OF THE ENVIRONMENT

In the last years, several tools and software environments based on the standardised GDMO notation have been proposed ((Dossogne & Dupont 1993, Wittig & Pfeiler 1993)) and today several products are available on the market. Most of them have nice features, several advantages and probably some limitations. However, none supports formally described behaviour for MOs and MIBs and thus tool support for this aspect of the development process is missing.

When the decision was taken to start the development of the MODE integrated tool-set, our goal was not to produce "yet another development environment" but to implement tools in order to validate our concepts on how behaviour should be formally described and how this formal part could improve the whole development process of MIBs. We concentrated our work on trying to provide both validation and test generation tools in an early stage of the development process which have not been considered, due to the lack of formalism in the standard, in most other toolkits. Thus we can say that these tools can be considered as extensions to other development environments rather than concurrent ones.

3 THE FORMAL BACKGROUND

The MODE development environment currently supports the extensions we have proposed to GDMO in a language called LOBSTERS (Festor 1994). Most concepts we developed for the integration of LOBSTERS into GDMO can be easily applied to other formalisms which are or about to be standardised in the OSI framework. After a summary of the LOBSTERS concepts, the link to other FDTs is discussed. Then a short overview of the selected development process is presented. The definition of this process is done to identify which support tools are expected in our environment.

3.1 LOBSTERS

LOBSTERS is the acronym of "Language for Object Behaviour Specification based on Templates and Extended Rule Systems".The notation is a compatible extension of the standard GDMO notation. In LOBSTERS, the static parts of objects (attributes, operation and action signatures, packages, ...) are exactly the same templates are those defined in GDMO. The formal behaviour part in LOBSTERS is based on an extended version of the Communicating Rule Systems FDT (Mackert & Neumeier-Mackert 1987). The notation based on a set of rules for describing behaviour has been extended with object-oriented features such as inheritance (Festor & Zoerntlein 1993). As the CRS Formal Description Technique supports the standardised ASN.1 notation and provides several operators to access and manipulate ASN.1 typed variables (Schneider 1992), the link with the static part of GDMO was trivial. A first approach to the integration of the rule mechanisms into the behaviour templates of the GDMO notation has been proposed and the result of this integration is the LOBSTERS FDT. It is fully compatible with the standard GDMO notation, i.e. it can be parsed with standard parser as well as extended ones, and provides facilities to specify formally the MO behaviour.

One of the main problems encountered during the integration was how behaviour specifications should be distributed over the various templates of a Managed Object, i.e. packages, conditional packages, attributes, actions, other MOs,inheritance, etc... This problem was resolved by defining a methodology for the development of behaviour specification and more generally by defining a rigourous approach to the specification of Managed Objects. This approach is based on specialization concepts and scope limitation in each kind of behaviour template present in a MO definition (Clemm & Festor 1993). Based on this approach, an algorithm for collecting different behaviour parts within a Managed Object definition was designed. This algorithm takes all distributed behaviour parts and builds one rule set by connecting the different rules through basic predicate logic operators such as AND/OR. Through the use of this algorithm it is possible to determine the behaviour specification for any given MO and thus open some issues to validate, test and verify extended GDMO specifications which is not possible with the standard notation alone.

In addition to formal behaviour specification, LOBSTERS also provides a simple mechanism to specify formally the presence requirements of conditional packages. Based on basic first order predicates these formalized conditions are very helpful in increasing the automation in the development tools. Especially all generation tools can through these expressions detect automatically which conditions are to be met to generate the code associated with a given package and generate, for each MO, the validity mechanism which will check on a create-request if the given package-requirements conform to the standard specification of the MO.

3.2 Dealing with other formalisms

As mentioned above, the behaviour in LOBSTERS is described using rules where the condition and effects parts are specified with first order logic predicates. This mechanism is similar to the pre/post conditions approach of VDM (Jones 1990) or Z (Spivey 1989). Thus, most concepts developed for the integration of the rule-based approach into the GDMO notation can be also mapped onto the integration of another formalism based on predicates. Especially the distribution method and its associated collection algorithm can be applied to these Formal Methods as well.

The impact of a new formalism on the environment concerns at this time only small parts of the code. The development tools could easily integrate a formal behaviour description based on another FDT if the mechanisms are integrated into GDMO in a way that is compatible with the notation. This is important to consider at a time where new FDTs are in the standardization phase within ISO.

3.3 The formal development process

When an environment for the development of a particular type of system is designed it is always based on a well defined development process. In our approach we have based the MODE environment on the process depicted in Figure 1. This process is the one used in most Formal Description Techniques based approaches.

The first stage of the MIB design consists of the creation of a high level specification of the desired MIB (*formalization* stage). To achieve this task, the MIB designer uses both the requirements defined for its task and a set of existing MO definitions in order to reuse some of them in his model. Thus, the first tools that are required to help the MIB designer in the process are tools which facilitate the selection of MO classes from a set which can be very large and specification support tools which allow new specification to be built and integrated into the available set. These tools are mainly parsers and graphical user interfaces to access in a user friendly way all information necessary to the performance of this first task resulting in a first model of the MIB.

When a first formal model of the desired MIB is designed, the work of the MIB designer consists in a systematic refinement of this specification (*specialization*). This

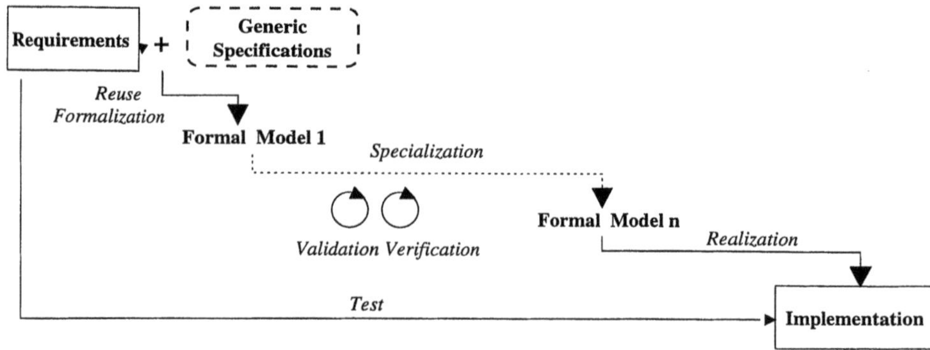

Figure 1: The FDT-based development process.

refinement can be iterated several times until the specification is precise enough to be implemented. Tool support for these stages concerns mainly validation and verification tools which guarantee that all constraints are met. These tools can be either provers or specification simulators.

When the specification is precise enough to be implemented, so-called *realization* tools are used. These tools are in most cases, code generators and compilers. Finally, when the test of the implementation towards the requirements has to be performed (*test* phase), both test execution tools and, in previous stages of the design, test generation methods and tools are required.

All these tools facilitate the work of the MIB designer, ease the whole development process and thus, justify the use of formal methods for the MIB specification.

4 THE MODE ENVIRONMENT

The MODE environment consists of two main tool-sets. The first one, called the front-end part, allows MO designers to create, parse and load managed object specifications. As the behavioural extensions proposed in LOBSTERS are fully compatible with the standard notation, the specifications which have to be parsed by the front-end can be either standard or contain LOBSTERS conditional predicates and formal behaviour parts. These tools are used intensively in the first phases of the development and are also helpful in all refinement steps where syntax parsing is necessary. Based on this front-end part, several other applications can be built and integrated in the environment. These are either validation or code generation tools.

4.1 Architecture and front-end

As depicted in figure 2, the front-end part of the environment contains, three basic parsing tools which are a GDMO parser, a second pass behaviour parser which extracts formal

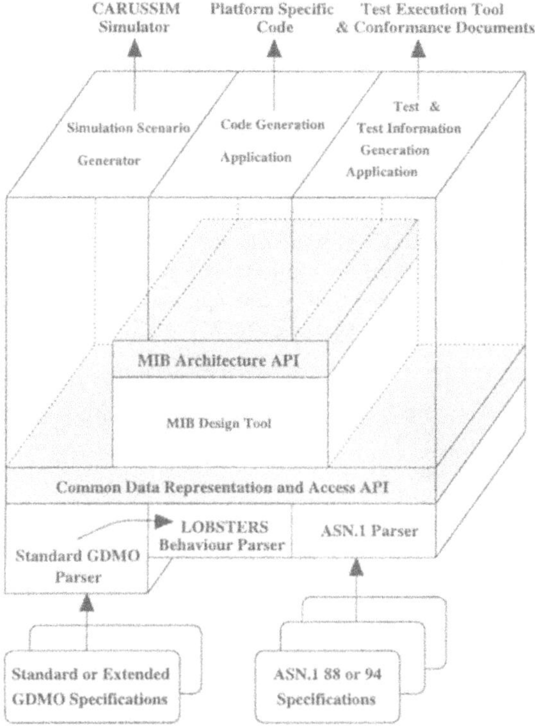

Figure 2: Architecture of the MODE environment.

expressions from the LOBSTERS specifications, and an ASN.1 parsing tool. The ASN.1 compiler is an extension of the SNACC compiler developed at the University of British Columbia (Sample 1993). It was extended with a back-end coding ASN.1 specifications in the common intermediate representation allowing these specifications to be exploited by the simulator. Some work has also been done in supporting new features in the notation according to the new ASN.1 draft. The integration of the behaviour parser was facilitated by the encapsulation of formal specifications into the basic behaviour description template of the standardized GDMO notation. This could be resolved in a more elegant way by adding specific formal behaviour templates into the GDMO notation.

These tools are used to parse and load specifications. All the information extracted from these parsing steps are stored in an internal C++ representation and accessible by tools through a well defined Application Programming Interface (API).

Based on this API several tools have been designed and implemented. These applications help the MIB designer in specifying, editing and validating his model. These tools are a MIB design application which is presented in the next section, a simulation scenario generator and a stepwise simulation tool. Several other applications can be built over the MIB design application, e.g. code generators or test generation tools. As we have focussed our attention on the early stages of the development, the code generation was not considered yet. However some work is going on in our group on this area.

4.2 The MIB design tool

The MIB design tool is the first application built over the Common representation API. This tool provides several facilities to MIB designers for template edition, modification, syntax checking through the previously mentioned parsers, and interactive MIB building. This application provides facilities to support both reuse and formalization steps. This application serves as a front-end for all other application tools which are the scenario generator, code generators and a test design tool.

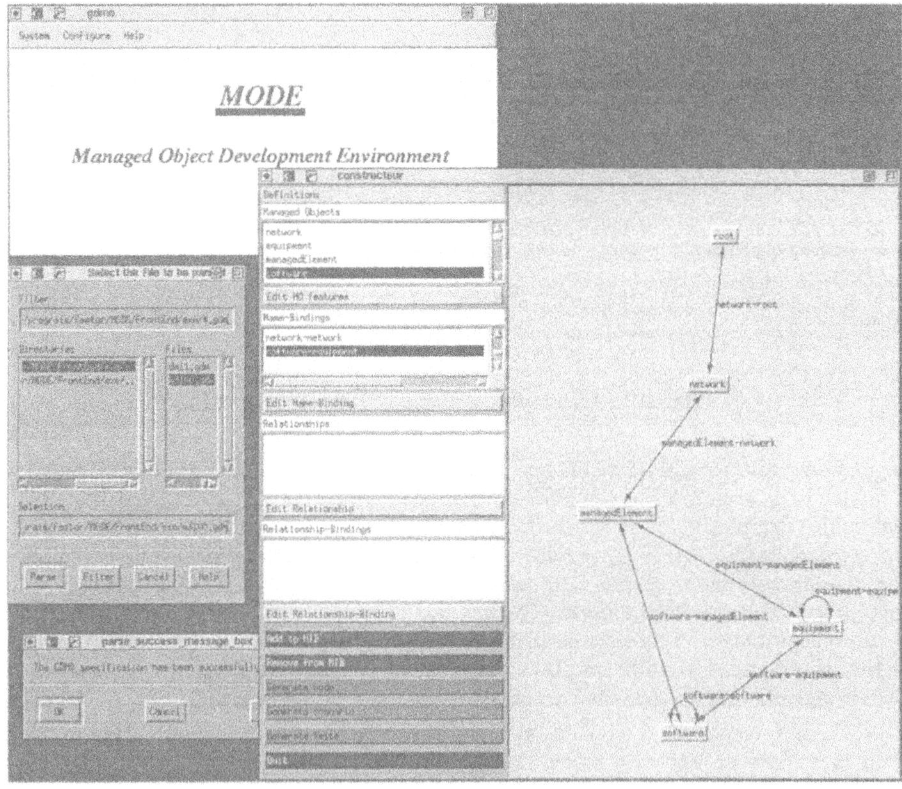

Figure 3: Interface of the MIB design tool

Figure 3 contains a screen shot of the main window from the MIB design tool. Following features are supported by the application:

- edit: several definitions can be edited in a user friendly way. These definitions can be MO classes, name-binding definitions, relationships or relationship bindings. Internal features such as attributes, actions, can also be edited but not at this level.
- add a managed object class to the MIB: if there is only one possible name-binding and if the container object is still in the expected MIB schema, the object is added to the MIB (e.g. the `managedElement` MO can only be inserted into the MIB if the `network` MO is present). If several Name-Bindings are candidate the user selects which ones are supported and all selected ones are added to the MIB. Note that the static semantics check of the definitions is performed at this level whereas the syntax check is performed at the parser level. Thus, an MO can only be added if all definitions (packages, attributes,..) are fully defined. This allows the working MIB to be always consistent.
- add an additional name-binding: if both container object classes and contained ones still exist within the MIB additional name-binding can be added to the MIB (e.g. the `equipmentequipment` name-binding can be added after the `equipment` MO was inserted).
- remove objects and/or name-bindings from the MIB: several MOs or name binding can be removed from the MIB architecture. When a MO which contains several other ones is removed, then all MO which are associated through a Name-binding to the one removed are also removed from the MIB as wee as the concerned Name-Bindings. This is done for MIB consistency requirements. For example, if the `managedElement` is removed from the MIB depicted in figure 3, then both `software` and `equipment` MOs as well as the related name-bindings (`software-managedElement`, `equipment -managedElement`, `software-software` and `equipment-equipment`) are removed too.

The MIB design application also provides facilities to define in an interactive way all parts of new MOs as well as their formal behaviour.

Concerning the relationships between MOs, the tool supports the relationship model defined in (Clemm 1993). Here also relationships can be added/removed from the MIB. However in the area of relationships, the integration of additional support such as code generation and test generation are not yet supported.

At each step during the design process, the containment tree of the current MIB architecture is visible in a user friendly way and all objects, name-bindings and relationships can be accessed.

The application has been implemented in C++. The graphical user interface was developed using OSF/MOTIF. The current MIB architecture is accessible through an API. This API is used by the application tools to collect the information they need to perform their task. We will now present one of the application tools which exploits both the presented MIB architecture and the formalized behaviour part from LOBSTERS. This application is the simulation environment.

5 THE SIMULATION TOOL

The first application we realized was a tool to simulate a MIB based on its behaviour specification. This application can be used for validation and verification of the specification of designed MIBs, to test whether they really exhibit the desired behaviour. This application can be used in several steps of the development process and is directly based on the formal description of the behaviour.

5.1 The purpose of simulation

Simulation is a way for MIB designers to analyse a model of their system in an early stage of the development. It allows the focus on different abstraction levels from basic interactions with the environment to a detailed study of internal states and interaction parameter values.

Applied to MIB scenarios, this can be very helpful to validate single Managed Objects instances (testing the interfaces and states of an object isolated from its environment), validate MIB hierarchies scenarios (testing a whole or a part of a MIB in order to check consistency of the model) and validate the management interface (adequacy of the access service to the MIB model). These validations steps have been defined and the simulator has been used for this purpose in the design of MIBs for Virtual Private Networks (see (Preuss 1993, Schneider, Preuss & Nielsen 1993)).

5.2 The CARUSSIM tool

The CARUSSIM (Communicating Automated Rule Systems SIMulator) was originally developed to be used in the area of protocol validation (Eschebach 1991). In order to adapt the simulator to the new features necessary for MOs, several enhancements have been made to the kernel (Frot, Lecorguillé, Lefranc & Orain 1993) and to the data (ASN.1) representation and manipulation part (Orain 1993). Especially full support of ASN.1 constructs such as sets, choices, sub-typing, object identifiers and value notations frequently used in MO specifications have been implemented in the tool.

In Figure 4, the graphical user interface of the simulator is depicted. It shows a simulation of a specialized Managed Element derived from the M.3100 TMN managed Object catalog. All attributes of the MO called ComputerMO are visible (operationalState, administrativeState, ...). The MO is accessible through three interfaces which are the management interface on which operations can be invoked on the MO, the notification interface through which the MO issues notifications, and an internal interface through which the MO is influenced from the real resource it models (here modelled as an environment rule system).

Through the interactive simulation, all operations such as activation, locking, ... can be performed on the MO and the MIB designer can check if the observed behaviour is the one expected. Naturally the tool can be used to simulate a more complex MIB which contains several MOs built around a containment tree, elaborated with the MIB design tool.

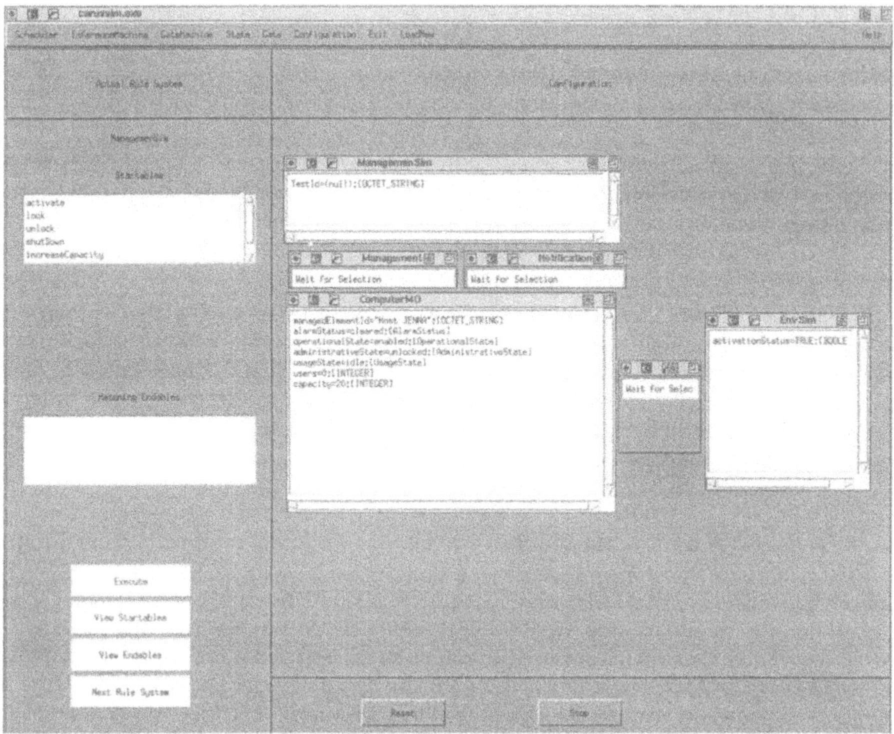

Figure 4: The simulator graphical user interface.

6 STATUS AND FUTURE WORK

The main features of the MODE environment have by now been implemented. The available features are the GDMO parser, the LOBSTERS behaviour parser, the MIB design tool, the simulation scenario generator, and the scenario interactive simulator. Provision of the test generation and code generation tools is planned for mid 95. Some work has also started on a more extensive support for relationships and their influence on code and test generation.

As LOBSTERS is not a standardised notation for describing formally behaviour, we are now starting on applying the methods used for the integration of the rule based approach into GDMO, on other formal methods. In this area we are planning to integrate some features of either VDM or Z into GDMO and add these new methods to the environment. It seems that the more powerful tools for these formal methods are the ones developped around VDM. In order to provide a complete development environment based on formal methods like we started with MODE, some investigations are going on in the area of mapping the LOBSTERS concepts onto the use of VDM-SL in GDMO.

7 CONCLUSION

In this paper, we have presented a development environment for Managed Objects which is based on GDMO and the extensions we proposed in LOBSTERS for the formal specification of the behaviour part.

We have shown that the formal behaviour description with LOBSTERS can be exploited during various stages in the development process: for Management Information Base design, validation, code generation not only for static but also behaviour parts. As a result the development environment is to that respect more powefull than approaches that ignore the very important aspect of behaviour.

The use of formal methods in the development of OSI-based MIBs is heavily dependent on the availability of development tools which provide additional facilities to support and exploit the formal development process. The MODE environment is a first step toward this goal. However a lot of additional work has still to be done to apply the concepts of LOBSTERS and MODE to formal methods that are currently subject to standardization. This task is yet going on in our group.

8 ACKNOWLEDGMENTS

The author wishes to thank Alexander Clemm for his careful reading of this paper; Wilko Eschebach who developed the CARUSSIM simulator for his precious help during the extension; David Orain who spend many months on improving the whole environment and especially the ASN.1 part of the tool; Dr. Juergen Schneider and Thomas Preuss for having encouraged my work and for their feedback with respect to their application of the formalism and for testing the tools in their project; Dr. Georg Zoerntlein for its advices during the design of LOBSTERS.

9 REFERENCES

Clemm, A. (1993), "Incorporating Relationships into OSI Management Information". 2nd IEEE Network Management and Control Workshop, Tarrytown, NY, 21-23 September 1993.

Clemm, A. & Festor, O. (1993), "Behavior, Documentation and Konwledge: An Approach for the Treatment of OSI-Behavior", *in* 'Fourth International Workshop on Distributed Systems: Operations and Management'. October 5-6, 1993, Long Branch, New-Jersey, USA.

Dossogne, F. & Dupont, M. (1993), "A software architecture for Management Information Model definition, implementation and validation", *in* H. Hegering & Y. Yemini, eds, 'Integrated Network Management, III (C-12)', Elsiever Science Publishers B.V. (North-Holland), pp. 593–604. Proc. of the IFIP TC6/WG6.6 3rd. Int. Symp. on Integrated Network Management, San francisco, Ca., 18-23 April, 1993.

Eschebach, W. (1991), "Interpretative Ausfuehrung kommunizierender Regelsysteme". Master Thesis, University of Kaiserslautern, Germany, 1991.

Festor, O. (1994), "OSI Managed Objects Development with LOBSTERS". Fifth International Workshop on Distributed Systems: Operations and Management, 12-16 Septembre 1994, Toulouse, France.

Festor, O. & Zoerntlein, G. (1993), "Formal Description of Managed Object Behavior - A Rule Based Approach", *in* H. HEGERING & Y. YEMINI, eds, 'Integrated Network Management, III (C-12)', Elsiever Science Publishers B.V. (North-Holland), pp. 45–58. Proceedings of the IFIP TC6/WG6.6 Third International Symposium on Integrated Network Management, San francisco, California, USA, 18-23 April, 1993.

Frot, J., Lecorguillé, H., Lefranc, J. & Orain, D. (1993), "CRUSADE: un environnement de développement de protocoles". Industrial Project Report, Ecole Supérieure d'Informatique et Applications de Lorraine, 1993.

ISO-10165.4 (1992), '"Structure of Management Information - Part 4: Guidelines for the Definition of Managed Objects"'.

Jones, C. (1990), *"Systematic Software Development Using VDM (second edition)"*, Prentice Hall.

Kilov, H. (1992), "Understand→Specify→Reuse: Precise Specification of Behaviour and Relationships.". IFIP/IEEE Int. Workshop on Distributed Systems: Operations and Management, Munich, October 1992.

Mackert, L. & Neumeier-Mackert, I. (1987), "Communicating Rule Systems", pp. 77–88. Proc. 7th. Int. Symp. on Protocol Specification, Testing and Verification, H. Rudin, C.H. West (editors), North-Holland 1987.

Orain, D. (1993), "A New ASN.1 Compiler for the CRUSADE Environment". Master Thesis, Ecole Supérieure d'Informatique et Applications de Lorraine, 1993.

Preuss, T. (1993), "Management von virtuellen privaten Netzen". Master Thesis, University of Magdeburg,1993.

Sample, M. (1993), *"SNACC 1.1: A High Performance ASN.1 to C/C++ Compiler"*. Master Thesis, University of British Columbia.

Schneider, J. (1992), *"Protocol Engineering: A Rule-based Approach"*, Vieweg.

Schneider, J., Preuss, T. & Nielsen, P. (1993), "Management of Virtual Private Networks for Integrated Broadband Communication". Proc. ACM SIGCOMM '93, San Francisco, CA, september 1993.

Spivey, J. (1989), *"The Z Notation"*, Prentice Hall.

Wittig, M. & Pfeiler, M. (1993), "A Tool Supporting the Management Information Modeling Process", *in* H. Hegering & Y. Yemini, eds, 'Integrated Network Management, III (C-12)', Elsiever Science Publishers B.V. (North-Holland), pp. 739–750. Proc. IFIP TC6/WG6.6 3rd. Int. Symp. on Integrated Network Management, San Francisco, Ca., 18-23 April, 1993.

10 BIOGRAPHY

Olivier Festor received the Master Thesis degree in Computer Science from the University of Nancy I, Nancy, France, in 1990 and the Ph.D. degree in Computer Science from the University of Nancy I, Nancy, France in 1994.

During his Ph.D., he spent three years at the IBM European Networking Center in Heidelberg, Germany, researching application of formal methods in the area of OSI-based Network Management. He is now working as a researcher at the Centre de Recherche en Informatique de Nancy, Nancy France. His current interests are in the fields of Network Management, Formal Description Techniques, MIB specification notations and development environments.

53
Management Application Creation with DML

Barbara Fink, Helmut Dercks, Peter Besting
Philips GmbH - Research Labs
P.O.Box 1980, D-52021 Aachen, Germany
Tel.:+49.241.6003-509 Fax:+49.241.6003-519
E-mail:fink@pfa.philips.de

Abstract

This paper presents the current state of the DOMAINS Management Language (DML) which was in its first version developed in the ESPRIT project DOMAINS and enhanced thereafter. DML is an extension of the ISO standard GDMO offering a formal and executable behaviour. The language features, the corresponding compiler and the embedding management architecture are explained. In addition, experiences gained with employing DML for non-trivial applications is reported on. Although DML has not yet reached full maturity, it is a very useful tool that successfully assists application developers. The approach of combining a specification language with an implementation language proved to be very helpful: It allowed to use already standardised GDMO specifications and to convert them into executable programs with relatively little programming effort.

Keywords

Network management, management language, managed object, management application creation, DOMAINS, DML, GDMO, GDMO compiler.

1. INTRODUCTION

The market for network systems is rapidly growing, and the increasing complexity of network systems calls for a well structured management system consisting of a generic management platform and individual applications. In order to facilitate the efficient development of applications independent of the underlying platform, a management language is needed that

- provides appropriate high-level expression means to the management application programmer for efficient and reliable application development,
- hides application irrelevant concepts and the implementation of the underlying software and hardware components, and that
- can be translated automatically into an executable program.

A first step meeting the first two requirements was made with the ISO/IEC standard "Guidelines for the Definition of Managed Objects - GDMO"[1]. However, GDMO focuses on specification in contrast to implementation. The current standard is restricted to module interface and structuring descriptions, whilst the managed object's semantic, i.e. the behaviour description, is postponed. Current GDMO applications typically wrap the behaviour as plain English text in comments. Recent standardization efforts discuss to use Formal Description Techniques - e.g. SDL [2], Z, VDM, or LOTOS [3] - for the behaviour. The GDMO extension

LOBSTER [4] attempts, as well, to integrate formal behaviour parts into the GDMO. It is based on extended CRS (Communicating Rule Systems). Here, the behaviour of a MO (Managed Object) is defined as the sequence of all observable interactions with its environment. All these approaches focus primarily on rigorous specification without concern of the final implementation. In contrast the tool DAMOCLES [5] is more technique oriented. It contains a MO Browser which gives a structured overview of all existing MO Classes and a GDMO Template Editor which guides the programmer in writing syntactically correct and semantically consistent GDMO specifications. However, none of these approaches achieves automatically generated executable programs.

It is commonly agreed that there is a strong and increasing demand for the formalization of GDMO behaviour. In addition, the authors believe that the method to be used should allow automatic, unambiguous translation into executable code which can run on different target platforms. This latter requirement is considered extremely important as there are already various standardized specifications in GDMO (as e.g. the Generic Network Information Model [6] or the SDH NE Information Model [7]), the implementations of which should result in identical effects when being used and controlled by different management systems.

Motivated by the reasons stated above and last but not least by the need for efficient management application creation, the high level management language DML (DOMAINS Management Language) was developed. It was in its first version developed in the scope of the ESPRIT Project 5165 DOMAINS (cp. [8], [9] and [10]), enhanced continuously thereafter and extensively used for various applications.

The following chapter gives an overview of the management architecture containing DML. We then introduce the language and its compiler followed by experiences gained when using the language for non-trivial applications. Finally we discuss future enhancements and still open issues.

2. THE EMBEDDING MANAGEMENT ARCHITECTURE

2.1 The Management Model

The embedding management architecture goes back to the DOMAINS project mentioned above. One of its basic principles enhances the OSI Manager-Agent model by the concept of *domains*: Domains are used to recursively decompose the overall management task into subtasks. A domain comprises a manager and the set of resources to be managed. Depending on the complexity of the management task, a managed resource can be a simple real resource or again an entire lower level domain. This way domains are used to build up a management hierarchy. The manager at the top of the system plays the manager role in accordance to the OSI manager. Managers at the bottom controlling real resources can be seen as agents in the sense of OSI. The managers on the intermediate levels control managers on a lower level while at the same time being managed by those on a higher level. Taking the recursiveness into account, both managed and managing components must be treated uniformly: DOMAINS introduced for their representation the concept of the *Kernel.*

Due to a possible overlap of domains, resources may be controlled by several managers. This leads to an m:n relationship between managers and resources where, in general, different managers have different views of one and the same resource. This idea is supported by the *Shield* concept. Whereas the Kernel represents the complete behaviour of a manager or resource, a Shield represents an interface only and that precisely tailored to the needs of the superior manager.

2.2 The Management Platform

This section describes the overall management platform, which DML is a part of. As depicted in Figure 1 the application independent stack consists of the hardware, an operating system, a distributed processing system, the DOMAINS machine and finally the DML language with its compiler. In our implementation, ANSAware of APM[1], itself residing on UNIX, is used as basis for the DOMAINS machine. Whereas ANSA provides distribution transparency and basic communication facilities the DOMAINS machine adds specific functionality such as services for event handling or notification registration. In addition, the DOMAINS machine supports dynamic object class and instance creation.

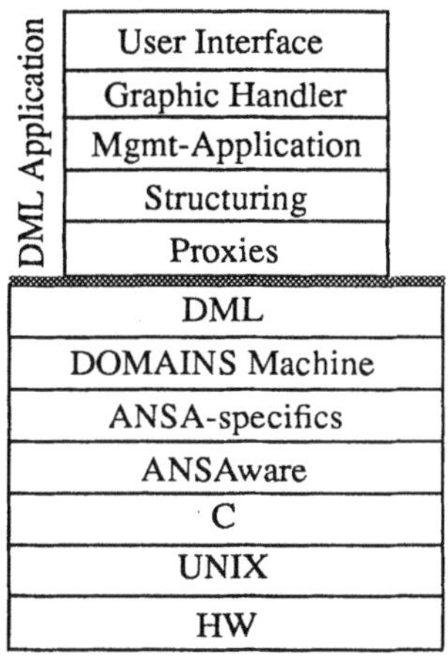

Figure 1: Management Platform

The DML compiler translates DML programs in the ANSA programming language IDL/DPL enriched with function calls for specific DOMAINS services.

A typical DML application is structured itself in several layers: The proxies are a collection of predefined specifications normally agreed upon by standardization committees. To guide the developers in designing management systems, structuring guidelines [11] have been developed. These define different views of the Management Information using the management model outlined in section 2.1. The management application - a complete specification of a management problem in DML - has to rely on these design guidelines and structuring principle. Finally, the graphical user interface can be seen as a specialized manager, who has communication paths to (possibly) all other managers in the system.

3. LANGUAGE FEATURES

3.1 Principles

DML's primary goal is to provide upward compatibility to the ISO standard GDMO to the greatest possible degree. Minor deviations were accepted in order to achieve a first running version within a given time schedule.

We start with a brief review of the basic GDMO features. Managed objects are specified by

- *Attributes* determining the object's state,
- *Actions* that can be coerced by managers through invocations, and
- *Notifications* that are issued by the managed objects to indicate, for example, attribute value changes.

From these features *Packages* can be built which in turn can be used as the building blocks of *Managed Object Classes*. A set of *templates* give proformas for specifying these features according to their external view.

1. ANSAware is a trademark of APM Architecture Projects Management Limited, Cambridge

In GDMO the formal specification is restricted to syntax aspects. DML realizes extensions with respect to the application scope and the semantics.

Managed and managing objects

The standard considers only management targets, i.e. managed objects, whereas the management activities exercised by managers are not treated. In contrast, the recursive DOMAINS management model - according to which a managed object may itself exercise management control on lower level managed objects - requires a common model for both managed and managing objects. Thus DML supports the description not only of managed resources but of managers as well. This puts extended requirements on the expression power of the behaviour clauses.

Different kinds of object classes

DML supports the DOMAINS Management Architecture by introducing different kinds of object classes, i.e. *Kernel* -, *Shield* - and *Support Object Classes* (cp. Section 2.1).

Operational and declarative behaviour language

As mentioned earlier a formal GDMO behaviour description is currently still missing. Therefore the new language was enhanced by an operational behaviour language yielding a general-purpose object-oriented programming language. For special purposes - cp. notifications in Section 3.5 - also declarative behaviour description is supported.

Integrating ASN.1

GDMO employs the ISO standard ASN.1 [12] for the description of data types, however, as semantics is not treated at all, the notation for accessing variables and/or their substructures is omitted. DML, now, has integrated a subset of ASN.1 into the behaviour language supporting convenient and type-save access to ASN.1 data.

3.2 Data Types

In accordance to GDMO, data types are distinguished from object classes. They conform to the ASN.1 standard. Conceptually DML comprises the full set of ASN.1, though the current language version is restricted to a subset only, comprising the entire set of simple types - e.g. BOOLEAN, INTEGER -, structures (SEQUENCE) and lists (SEQUENCE OF). With the intention to increase program reliability, untyped pointers or ANY are not supported in DML.

3.3 The DML Object Classes

In order to sufficiently support the management model outlined in section 2.1, DML distinguishes between three kinds of object classes:

- *Kernel Object Classes*
- *Shield Object Classes*
- *Support Object Classes.*

Kernel objects are used to represent managers and managed resources. Shield objects may represent Shields only that are inserted between a manager and a managed resource. From the language's point of view the Shield object has restricted functionality as compared to the Kernel object. Most of the Shield object's functionality is transparent to the application programmer. Its essential function is forwarding invocations and notifications. In the case of external resources residing in foreign systems, protocol transformations may be involved, hidden to the application programmer. However, in the current implementation protocol transformations are not realized. Support objects are foreseen for auxiliary tasks, such as mathematical functions, data base handling.

3.4 Object structuring

GDMO and thus DML offers various concepts and techniques for structuring objects.

Templates

The technique of templates serving as building blocks and supporting code re-usability is unrestrictedly adopted from GDMO. In addition we also allow inline-coding of templates. This method supports the traditional inline block-structured programming style. It is preferably used if otherwise control over a great number of small separate templates would be lost.

Inheritance

DML supports multiple inheritance as does GDMO. However, our current implementation does not inhibit repeated inheritance, i.e. there is no check if one and the same template is inherited several times.

Object References

Object structuring may also be achieved according to client/server modelling. In DML, objects can be accessed location transparently by their user-given name or by a typed variable that contains an object reference.

Polymorphism

The strong typing concept with static type checking supports dynamic binding that copes with polymorphism. The DML polymorphism concept is based on inheritance analogous to Eiffel [13], i.e. any inheriting class can be taken as its base class.

Action Templates

There are three ways to define actions: by direct, *deferred* or *external* specification. Deferred actions are adopted from Eiffel. The behaviour specification of these actions has to be specified in the inheriting classes. External actions are provided to link foreign programming languages to DML. Currently C is being supported.

3.5 Inter Object Communication

Objects interact with each other by means of *Invocations* and *Notifications*. The basic difference between these two types is the addressing concept: Invocations are explicitly addressed to their final destination, where they implicitly activate the corresponding *action*. In contrast, spontaneously emitted notifications do not know their final destination. One or several interested objects may register for certain notifications. Thus the destination object must take initiative for receiving a notification.

DML has introduced the concept of *Notification Handlers* specifying the reactions upon received notifications.

Actions are executed due to invocations and notification handlers due to notifications.

Invocations are sent from objects in the *manager* role to objects in the *resource* role for the purpose of *controlling* resources. The notification flow is in opposite direction, it is used for *monitoring* resources.

Invocation Types

DML distinguishes between

- synchronous, blocking invocations, called *CALL,*
- synchronous, non-blocking invocations, called *FORK*, and
- asynchronous, non-blocking invocations, called *CAST*.

All three types can pass arguments to their destination. The first and second one support reply arguments as well. In the case of a CALL the invoking program thread is suspended until the reply is received, whilst after a FORK and CAST the program thread is immediately continued, resulting in concurrently running actions. Any time after having issued a FORK invocation, the invoker can request the reply.

Notification Types

Notifications can pass arguments to the receiver(s). Unlike GDMO, DML does not support confirmed notifications. Reply parameters cannot be returned. In this case DML's restriction with respect to GDMO was deliberately undertaken. Notification confirmation is not considered necessary in the employed management model.

Notification emission specification can be

- imperative by the *NOTIFY* command or
- declarative by a logical expression over attributes.

As soon as the logical expression becomes true, the corresponding notification is emitted. This way attribute value change notifications can be naturally specified. The current implementation does not support declarative notification specifications.

Notification Registration

As stated above, objects playing a manager role must register for notifications in order to receive them. Selection criteria are the notification type, the emitting object class or object instance. In this way a manager may register for a certain notification type regardless of its source, or for a certain notification type sent by all instances of a certain class, or for a certain notification type sent by a certain object instance. The registration is dynamic, it can be cancelled again.

The registration command denotes also the notification handler, i.e. the program piece that is to be executed upon reception of the notification.

3.6 Attributes

Attributes are part of the external interface. They are accessible from other objects according to specified operations as e.g. GET or REPLACE. This aspect corresponds to the GDMO standard.

Additionally, attributes must be related to the object's own behaviour. From the object-internal view attributes are common data with full visibility according to their type. Whilst object-external access is restricted to the attribute as a whole, the object itself may access also individual data components and perform operations on them - e.g. multiplications - as defined for the specific type.

For denoting individual data components the familiar dot-notation and/or bracketed indices are applied.

3.7 Behaviour Description

DML comprises a general-purpose behaviour language. With the requirement "easy to learn and easy to use" it contains only very few and safe constructs. Eiffel was taken as a model for the notation of expressions, assignment-, conditional- and loop-statements. Transactions and special statements for object interaction as mentioned in section 3.5 are added.

Object-common data were already mentioned in the previous section. We introduced also local data for individual behaviour templates to be used as temporary local working variables.

3.8 Assertions

DML supports runtime semantics checks. There are built-in default checks - e.g. on list bounds - as well as user-defined assertions. For the latter the Eiffel concept is adopted: Action-behaviours can be enhanced by asserted pre- and post-conditions. User-defined *exception handlers* are executed if the assertions are violated.

3.9 Example

This section presents extracts from a DML program listing. The Fabric object selected represents the switching unit in a transmission network node. Its basic task is to control the set-up and release of cross-connections between pairs of termination points. Most of the program is self-explaining, some extra comments (beginning with a double hyphen) were added for convenience.

```
--              *** Fabric.dml ***
-- These are instructions for the pre-processor to include certain files.
USE "DML_Standard"  -- This file contains DML standard definitions etc.
USE "TypeDefs"      -- TypeDefs contains general ASN.1 type declarations.
USE "ProxyMO"       -- This one is used for inheritance.
USE "Adapter"       -- The Adapter object is the link to the managed network.
...

--              *** Fabric KERNEL Template ***
Fabric KERNEL OBJECT CLASS
    DERIVED FROM ProxyMO;
    MANAGING PART CHARACTERIZED BY fabricManagingPackage;
;
fabricManagingPackage PACKAGE
    ATTRIBUTES
        tpPool              GET,
        crossConnections    GET,
        adapterName         GET,
        ...
    ;
    ACTIONS
        connect,
        disconnect,
        ...
    ;
;
```

```
--              *** ATTRIBUTE Templates ***
tpPool                  ATTRIBUTE WITH ATTRIBUTE SYNTAX MOIds;;
crossConnections        ATTRIBUTE WITH ATTRIBUTE SYNTAX XConnections;;
adapterName             ATTRIBUTE WITH ATTRIBUTE SYNTAX OCTET STRING;;
...;;

--              *** ACTION Template ***
connect ACTION
   BEHAVIOUR connectBeh BEHAVIOUR DEFINED AS
   @           -- Identifies beginning of our formalised behaviour extending GDMO.
      VARIABLES
         iloop              INTEGER,
         loopFlag           BOOLEAN,
         connectRequest     Request,          -- Request is declared in TypeDefs.
         connectReply       Reply,            -- Reply is declared in TypeDefs.
         stdReply           StandardReply,    -- StandardReply is declared in TypeDefs.
         adapterRef         Adapter,          -- Object references are declared in this way.
         ...
      ;
      DO
         ...
--              *** fill message structure and send request to adapter ***
         connectRequest.modsimMsg[0] := "connect";    -- Assigning a value to a data structure.
         connectRequest.modsimMsg[1] := xConnection.from.instance;
         connectRequest.modsimMsg[2] := xConnection.to.instance;
         adapterRef := adapterName;                   -- Assigning a value to an object reference.
         CALL adapterRef.sendRequest(connectRequest ->connectReply);       -- Action invocation.
         ...
--              *** remove the "to" tp from the tpPool ***
         loopFlag := TRUE
         FROM iloop:=0;
         UNTIL (iloop >= LENGTH(tpPool)) OR (loopFlag = FALSE)
         LOOP
            IF tpPool[iloop].instance = xConnection.to.instance
            THEN
               REMOVE(tpPool[iloop]);                 -- Predefined access method REMOVE.
               loopFlag := FALSE;
            ENDIF;
            iloop := iloop + 1;
         ENDLOOP;
         ...
         RETURN stdReply;
      END      -- End of DO range.
   @;          -- Identifies end of our formalised behaviour extending GDMO.
   ;           -- End of BEHAVIOUR
   WITH INFORMATION SYNTAX xConnection : XConnection;-- ACTION input parameter
   WITH REPLY SYNTAX StandardReply;                  -- ACTION reply parameter
;              -- End of ACTION
...
```

The implementation shown here follows the specification of the fabric object according to the ITU standard M.3100 [6].

4. LANGUAGE ARCHITECTURE AND COMPILER

The DML language incorporates and integrates the GDMO specification language, the ASN.1 notation and declarative and procedural statements to express the behaviour. It allows the complete specification of management applications, that - once compiled - are executable.

The template-oriented language supports piece-wise compilation. The compilation unit is a DML file and the programmer is free to collect several templates into one DML file.

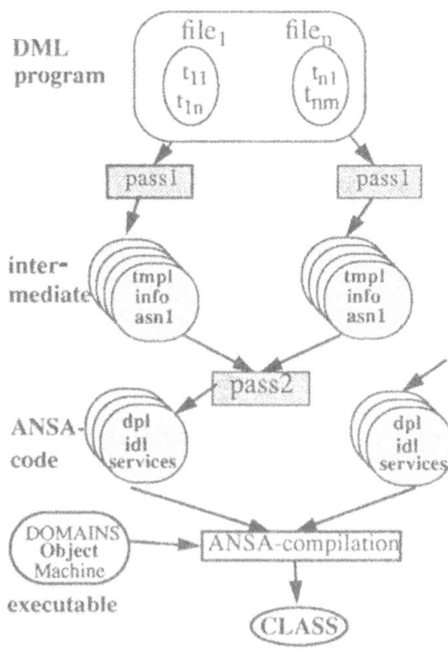

Figure 2: Compiler Structure

The compiler analyses the template descriptions and stores them in an internal repository. Data structures, described in ASN.1 notation, are mapped to C structures. Access, manipulation and assign functions are automatically generated for them.

The templates are bound together to object classes, which can be instantiated during run-time. The necessary anchors to compose the templates are stored in so called *info files* (one per object class).

The compiler is composed of two passes (cp. Figure 2). The first one is responsible for syntax checks. It builds the specific template files, ASN.1 mappings and the info files. The second pass is dedicated to semantic checks of templates and packages and their inter-relation.

The backend part generates code in the ANSA interface- and programming language IDL and DPL. It also produces support files for memory allocation/de-allocation and ASN.1 data handling. The ANSA compiler takes care of processing DPL and IDL files and linking the output with earlier generated service routines to a complete class description that can be started and instantiated by the DOMAINS machine (cf. section 2.2).

5. EXPERIENCES WITH DML

This section presents first-hand experiences with DML that were gained during the development and test of several management applications in different scenarios. DML significantly facilitates management application creation during the specification and implementation phase. The following subsections provide detailed evidence for this statement, but they also point out the main handicaps that have to be overcome in future versions of the language.

5.1 Application Scenarios

The three major DML application scenarios we refer to are related to the management of the freephone services in an intelligent network, to a fault, configuration, and service management system for an SDH network, and to the management of an ATM switching system. The SDH application resulted in a system that was presented to the public on the last CeBIT fair in Hanover in March '94. It is structured in 50 object-classes described in about 20.000 lines of DML code.

5.2 User Friendliness

DML is user friendly from various points of view:

- Short learning period of only few and simple but powerful basic constructs for data representation and control statements. Complex data structures can be accessed via a familiar point and index notation.
- Uniform programming style enforced through predefined template structures.
- Self-documentation and good readability of the program code.

5.3 Safer Code Production

The features that guarantee a safer code production than is achieved with other programming languages like C or C++ can be grouped along four main aspects:

- Raising the application programming abstraction level and freeing the application programmer from routine tasks like memory allocation and de-allocation.
- Reduction in number of code lines by an order of magnitude due to the high abstraction level.
- Restriction to safe language constructs and strong typing.
- Runtime semantics checks e.g. to prevent array overflow or use of null references.

5.4 Integration of Standard Specifications

Industrial organizations (like the ATM Forum) and standardization bodies (like ITU and ISO) put much effort into the design of open interfaces and standardized information models. These models are specified along the guidelines of GDMO. Due to DML's GDMO compatibility, these specifications can serve directly as a first code version. What is left over is the behaviour which is just given as comments in plain English and which has to be replaced by corresponding DML code. This extended GDMO code is then fed into the compiler to produce the executables. Compared to other approaches where the GDMO specifications are first translated into e.g. C or C++ code which then has to be extended with C or C++ behaviour parts, our correspondence of specification language and implementation language guarantees a much smoother program development process.

5.5 DML shortcomings

Using DML in practical applications also revealed some of its limitations and disadvantages. First to mention is that not all of the originally designed language features are supported by the current compiler.

A nuisance is the excessive use of semi-colons which terminate declarations, statements, packages, etc. which, however, is prescribed by GDMO.

A more serious shortcoming is the only very basic input and output functionality that is currently provided. And finally, testing and debugging is not yet sufficiently supported.

6. FUTURE ENHANCEMENTS

Desired enhancements can be grouped according to activities concerning the language definition and compiler and to the tools supporting the application programmer.

Language definition

New and/or enhanced concepts to be developed comprise:

- object persistency,
- combination of enhanced declarative and imperative description methods,
- intelligent notification filters,
- notion of time.

Tools

A window-oriented template editor should guide and assist the programmer in writing syntactically correct applications. A still open issue is an adequate debugging tool suited for a distributed environment.

7. CONCLUSION

DML is a high level management language that extends GDMO with a formal and executable behaviour. Experiences gained with several applications showed that DML significantly simplifies management application creation during the specification and implementation phases.

The main conclusions from these experiences are:

- DML has evolved into a useful tool
- DML is extremely user-friendly
- DML supports safer code production
- DML offers the right level of abstraction to the application programmer
- DML is capable of integrating standard specifications.

Desired enhancements towards more sophisticated tools for editing and debugging could even more increase the productivity of developers.

8. REFERENCES

[1] ISO/IEC 10165-4 - ITU-T X.722
Information Technology - Structure of Management Information
Part 4: Guidelines for the Definition of Managed Objects
1993

[2] ITU-T Recommendation Z.100
Specification and Description Language (SDL)
Geneva, 1992

[3] ISO 8807
LOTOS: A Formal Description Technique based on the Temporal Ordering of Observable Behaviour
1987

[4] O. Festor
OSI Managed-Object Development with LOBSTER
Proceedings of
5th IFIP/IEEE International Workshop on Distributed Systems: Operation and Management (DSOM'94)
1994

[5] M. Wittich, M. Pfeiler
A Tool supporting the Management Information Modelling process
IFIP Transactions C-12
Integrated Network Management, III
Elsevier Science Publisher B.V. (North-Holland)
1993

[6] ITU Draft Recommendation M.3100
Generic Network Information Model
1992

[7] ITU G.774
Synchronous Digital Hierarchy (SDH) Management Information Model for the Network Element View
1992

[8] DOMAINS Management Language
Final Deliverable of the ESPRIT Project 5165 DOMAINS
Distributed Open Management Architecture in Networked Systems
April 1993

[9] DOMAINS Management Architecture
Final Deliverable of the ESPRIT Project 5165 DOMAINS
Distributed Open Management Architecture in Networked Systems
April 1993

[10] A. Fischer, M. Herpers, D. Holden, S. Sievert
The DOMAINS Management Language
Integrated Network Management, III
Proceedings of ISIMN Symposium in San Francisco, USA, April 1993
IFIP Transactions, North-Holland
1993

[11] Eike Gegenmantel
Generic Information Structure for SDH Management
International Journal of Network Management, Vol 4, Number 1
March 1994

[12] ISO 8824
Information processing systems - Open Systems Interconnection - Specification of Abstract Syntax Notation One (ASN.1)
1987

[13] Bertrand Meyer
Object-oriented Software Construction
Prentice Hall International,
1988

9. BIOGRAPHY

The authors work in the Architectures and Systems department at Philips Research Laboratories in Aachen, Germany. Their main focus is directed on network management.

B. Fink received in 1967 a Diploma in Electrical Engineering from Technische Hochschule Aachen, Germany. Her key activities are architectures and computer languages.

H. Dercks graduated in computer science from the Technische Hochschule Aachen, Germany in 1978. He is specialist in systems engineering and compiler development.

P. Besting holds a master's degree and a PhD in Physics from University Bonn. His main interest is application creation and transmission and switching technologies.

54
Formal Description Techniques for Object Management

J. Derrick, P. F. Linington and S. J. Thompson
Computing Laboratory, University of Kent, Canterbury, CT2 7NF, UK. (Phone: + 44 227 764000, Email: {jd1,pfl,sjt}@ukc.ac.uk.)

Abstract

Open network management is assisted by representing the system resources to be managed as objects, and providing standard services and protocols for interrogating and manipulating these objects.

Application of formal techniques can make the specifications more precise, reducing the ambiguity inherent in natural language, and can automate some or all of the process of implementation and testing. This paper examines the use of formal description techniques to the specification of managed objects. In particular we examine the relative merits of two formal languages, Object-Z and RAISE, which have been proposed as suitable for use in object management.

Keywords: Managed objects, Formal methods, Open Distributed Processing.

1 Introduction

Large scale open systems require open management to integrate their components, which may have been obtained from a number of sources; the cost of system administration will depend to a large extent on how easy it is to perform this management integration. The creation of open network management depends upon there being a common representation for the resources being managed. This can be achieved by the creation of a suitable family of managed object definitions.

Different implementations of these managed objects, the agents that give access to them and the managers that control them need to interwork. Confidence in these implementations can be increased by testing. However, this testing is expensive and time consuming, because it is labour intensive.

At present the nature of the resources to be managed and the behaviour they are expected to exhibit are expressed in natural language, structured and organized using a simple specification technique set out in the Guidelines for the Definition of Managed Objects (GDMO) [GDMO]. The informal nature of this technique makes the implementation and testing of managed objects expensive, because much skilled effort is needed to interpret the specifications and construct suitable tests.

Formal description techniques offer the prospect of improved quality and cost reduction by removing errors and ambiguities from the specification and automating aspects of both implementation and testing. There are potentially large benefits to be gained from this. The number of managed objects already specified is large and can be expected to grow during the next few years until there are several thousand. These will range from objects whose behaviour is standardized internationally, through various levels of industry agreement to a wide range of vendor specific objects. Interworking will depend on specification and testing and product cost will depend on the efficiency of these processes.

However, the techniques and languages for formal description are not widely understood by the majority of implementors, and the choice of a suitable language for the application concerned is an important factor in their introduction and acceptance.

Two languages have recently been proposed for the specification of managed objects; they are Object-Z, based on the well-established Z language, and RAISE. They both have the necessary expressive power for such specifications, although they differ in the approaches taken in a number of areas. This paper examines their key features. It also reviews the tools available to support the languages, particularly with reference to the writing of managed object specifications and to the construction of tests from them.

However, the ultimate test of the acceptability of the techniques is the extent to which potential users are prepared to apply them. It is clear from consultation undertaken with the network management community that familiarity, perceived stability and relation to current practice are amongst the keys to success. Given that both languages have the necessary technical capabilities, selection should be based on the likely ease of uptake.

Action to promote the application of formal techniques in this area is timely; thousands of managed object specifications will be written and processed over the next few years, and the benefits of the formal techniques must be demonstrated before the bulk of the specification work takes place if they are to have the maximum impact.

The paper is structured as follows. The background and the requirements for the specification of managed objects are summarized in section 2. A review of RAISE and Object-Z is presented in sections 3 and 4. Tool provision is discussed in section 5. Language standardization and managed object requirements are discussed in sections 6 and 7. Section 8 discusses the testing process, and we present some conclusions in section 9.

2 Overview of network management

The OSI management framework and its associated standards have developed over a number of years [FormMan,GDMO]. The approach taken is object-oriented, involving the encapsulation of the resources to be managed as *managed objects*. These objects are manipulated in a unified way by using the common management information services and protocols and are defined in an informal way using the guidelines for the definition of managed objects and associated common object and attribute definitions.

The management framework can now conveniently be seen as a special case of the more general problem of distributed processing, and it is reviewed here in terms of the ODP terminology (from ISO/IEC 10746 parts 2 and 3), where appropriate.

2.1 The CMIS model

The Common Management Information Service (CMIS) describes the communication mechanism which links a manager to an agent, which in turn gives access to a set of managed objects. The CMIS model (derived, in part, from the Systems Management Overview) is essentially a computational model which describes the interaction between the manager and agent, and necessary aspects of the interaction between the agent and the managed objects to which it gives access. The agent provides for a degree of coordination of the management information within a system, allowing control of scoping, filtering and discrimination mechanisms. This interface includes operations invoked by the manager which allow it to: create new managed objects (subject to resource constraints); delete existing managed objects; get attribute values; set attribute values; perform actions; cancel an incomplete get operation; and for the agent to report events to the manager.

The emphasis on the CMIS as the basis for management standardization reflects the OSI emphasis on interconnection, rather than system structure. A distributed systems view of management would now de-emphasize this particular interaction in favour of a model giving equal visibility to the communication between agent and managed object (and between managed objects under a single agent).

2.2 Notations for specifying managed objects

The family of management standards includes an informal notation for specifying the behaviour of managed objects - the "Guidelines for the Definition of Managed Objects" (GDMO) [GDMO]. This notation provides a framework for the declaration of properties (based on options from the management object model) and structuring of informal statements of behaviour. The notation allows definition of one or more object classes which express the common properties of their members. The GDMO notation supports:

- identification of the class being defined;
- multiple inheritance from a set of superclasses, or structured specification by association of the definition with some packages of specification (which do not, of themselves, necessarily constitute complete objects);
- attributes or attribute groups, including the role of the attribute in the CMIS structure, the default values to be applied on creation and any restrictions on the way the attribute values may be modified;
- actions, enumerating all those possible and defining their parameterization;
- notifications, enumerating all those possible and defining their parameterization;
- behaviour, which indicates when actions or notifications are appropriate and the way in which their sequence is constrained; behaviour also defines the way that attribute values are updated by the occurrence of actions or notifications, and the way that parameter values carried by actions or notifications are determined by attribute values.

The specification may draw upon common object or attribute definitions, which also include informal descriptions of the corresponding types.

The management object model includes support for two hierarchies: an inheritance hierarchy supporting reuse and refinement of specifications, and a containment hierarchy associated with the interpretation of object creation and deletion actions. It also supports 'fairly arbitrary' assertions of compatibility called allomorphisms.

The ODP viewpoints [ISO10746] can be used to group together different concerns in managed object definitions. In the longer term this approach may simplify the relation of managed object definitions for OSI profiles.

2.3 The requirements for the formal description of managed objects

From the above discussion it can be seen that a formal description technique which is to be used for the specification of managed objects and of their manipulation will need to support:

1. the naming and name binding mechanisms for the managed objects;
2. the inheritance and containment relationships between objects and the ability to create templates representing these relations;
3. the definition of sets of actions and notifications, with their parameterization;
4. the definition of attribute types, including initial or default values and range restriction, matching rules and links to supporting abstract syntax definitions;
5. behaviour, in terms of rules for the occurrence of actions or notifications and the relation of their parameters to object attributes;
6. rules for the creation and deletion of objects;
7. rules for the concurrency constraints implicit in the use of CMIS.

In addition, future development may require statements of the interaction between managed objects which are independently defined. Capturing the full meaning of the linkage between the resources in a complex system will increasingly imply the statement of the effect that the changes applied to one managed object will have on others.

3 The Languages Z and Object-Z

The Z specification language [Spivey] has been developed over the past ten years, and is based upon set theory and first-order predicate calculus. It has proved (together with VDM) to be one of the most enduring formal description techniques, and has had significant industrial usage and support. Recently Z has been selected for the specification of the information model of the ODP Trader [Trader].

The development of Z has been supported through ZIP - A Unification Initiative for Z Standards, Methods and Tools. The project, [ZIP], had four main themes: standardisation of Z; methods support for Z; tools for Z and foundations of Z (for example, logic, proof rules). The project lasted for three years and finished at the end of January 1993.

There are three main reasons for extending Z to facilitate an object-oriented style. **Encapsulation** structures the specification. Data types and the operations upon them are declared together in classes. State is then local to a class as opposed to global state as in Z. **Inheritance** allows the inclusion of previously defined classes in class definitions. A hierarchy of classes and their subclasses can be developed as the Guidelines for the Definition of Managed Objects indicate. **Polymorphism** is the property that an object of a subclass can be substituted where an object of a superclass is expected.

Object-Z [Object-Z] is a specification language based on Z but with extensions to support an object-oriented specification style. Object-Z uses the concept of a class to encapsulate the descriptions of an object's state with its related operations. In addition, Object-Z provides support for inheritance, instantiation and polymorphism. Object-Z does not increase the expressive power of the Z notation, and both offer the same specification paradigm, which captures the relational aspects of state transitions within the system under study; it does, however, contain syntactic and semantic extensions to enable the object-oriented specification style to be supported explicitly.

Whilst Object-Z is not the only proposal to extend the Z language to support an object-oriented style, it is probably the most mature of the approaches; for a survey see [OOZ]. However, Object-Z is not currently in a stable form, and research is still being undertaken into the language and its semantics; this is in contrast to RAISE [RSL] which could be described as a finished product. There are clear disadvantages in using a language which is still in the process of evolving. However, by adopting a flexible approach, there is the possibility that the final version of Object-Z can be tailored to the needs of Managed Object specifications and ODP standards, [Cusack 92]. Indeed, this is the stated intent of some researchers in this area [Cusack 91].

A further factor to consider is the availability of tool support for Object-Z. RAISE has a clear advantage in this respect. Currently there is little or no tool support for Object-Z; tool support for Z exists, but the RAISE tools, coming from a single source, are better integrated.

Technical Assessment

Z specifications consist of schemas (to declare the state) and operations (which change the state). Like Z, Object-Z uses this state-based model to describe systems. This is the only model directly supported, in contrast to RAISE which offers a variety of styles to the specifier. Object-Z specifications use classes to encapsulate together the state and the operations on it. Object-Z provides direct support for expressing constraints and properties of an object's history, which makes temporal behaviour easier to describe and reason about. This can, for example, be used to express deadlock and liveness constraints.

Encapsulation, the definition of classes and objects, is achieved in Object-Z via a class definition mechanism. An Object-Z class is taken to represent a set of models; that is, a class is analogous to an ODP class type in which a class will determine the set of possible realizations that can implement it.

An object is then represented as a named member of a class. In Object-Z classes and objects have to be named, unlike RAISE where both a named and a nameless encapsulation mechanism are supported.

A visibility list in an Object-Z class nominates certain features of a class to be externally visible. In contrast RAISE uses hiding in classes to hide certain features. However, the expressive power is the same, although the specification style will obviously differ.

Object-Z supports incremental and multiple inheritance through class inclusion. The current definition of inheritance in Object-Z is compatible with that used for Managed Objects. A subclass incorporates all the features of its super-classes. As with RAISE renaming is possible for class entities upon inheritance.

There are a variety of proposed inheritance schemes that could be used in Object-Z, although no single one has dominated. Further work is needed in this area to define a notion of inheritance that directly supports ODP and Managed Object specification.

Object-Z supports ad hoc polymorphism, which is the property that an object of a subclass can be substituted when an object of a superclass is expected. Parametric polymorphism, as appears in languages such as Standard ML [SML], is not supported within Z or Object-Z; however, there is no explicit requirement from Managed Object specifications to support this.

The approaches taken by Z and RAISE with regards to subtyping are similar. RAISE supports subtyping, but defines maximal types to enable tool support for (static) type checking. There have been proposals from BT researchers to extend typing to define Object-Z classes as class types (thereby following ODP more closely), this approach would lead to subtyping. However, the issue of whether to extend Z to support subtyping appears not to have been fully resolved.

The development process of producing a new specification from an old one by adding more detail is known as refinement in Z and Object-Z [King]. There has been little work on the role of refinement in Object-Z. However, it is likely that an extension of the Z concept of refinement to Object-Z will be technically possible. In addition, using inheritance as the basis for a refinement relation is also a possibility (the subclass is the super-class with more constraints or with more detail added).

There is no explicit process algebra support for communication or concurrent aspects in either Z or Object-Z. Object-Z deals with the specification of concurrent properties by using linear temporal logic [Duke]. Temporal logic allows the intended execution sequences of objects of a class to be constrained in abstract ways. The resulting style of specification of concurrent and distributed systems is different from that using process algebras or the concurrent aspects of RAISE. Concepts needed for managed object specifications which cannot be provided by the use of linear temporal logic include dynamic aggregation of objects and sharing of one object by other objects. In addition, the duration of methods would need to be modelled via the use of an explicit attribute of a class which gives the current time and time constraints on it. Further work is needed to address these areas.

4 RAISE and the Specification Language, RSL

RAISE (Rigorous Approach to Industrial Software Engineering) is the name of an EC funded ESPRIT project (315) which ran from 1985-1990; research in this direction is currently continuing in the LaCoS (5383) and MORSE ESPRIT projects. The major results of the project include: definition of the (wide-spectrum) specification language,

RSL [RSL]; a formal (denotational) definition of RSL and a set of proof rules for reasoning about RSL specifications and designs; a methodology for program development and design in RAISE; and a set of tools to support formal development within RAISE.

RAISE has been used on a small number of pilot projects and by a number of the partners in LaCoS at a larger scale, and courses to disseminate information about various aspects of RAISE program development are offered by a number of bodies.

RSL is in a reasonably stable form, resulting as it does from a process intended to produce a standard. An early aim of the LaCoS project was to review the language, and apart from a small number of minor changes, it was deemed to need no modification. (Some work in the MORSE project is directed towards adding real-time information to the system, but this is not relevant to the specification of managed objects.)

RSL allows specification in three different paradigms: declarative (a style close to programming in Standard ML [SML], a strict functional programming language); imperative, using expressions which can cause side-effects; and concurrent, using an amalgam of CCS [Milner] and CSP [Hoare].

As might be expected, it has the advantages and disadvantages of a committee design: three programming paradigms are addressed, and both model-oriented and algebraic (property-oriented) specifications can be written.

The design appears successfully to have integrated the three programming paradigms. The language is expression-oriented, with a 'pure' functional core. On top of this are added expressions which can read or write to variables, and take input from and give output to communication channels. Sufficient checks (or imprecations) are made to ensure that side-effects and communications are restricted to appropriate parts of the language - axioms are expected not to have side-effects, for instance.

The development relation for the language contrasts with the notion of refinement familiar from VDM and Z; development is a stricter relation, requiring as it does theory extension, but on the other hand it makes modularisation of development easier to achieve, an aspect which may also carry over to test generation.

Certain aspects of language design are questionable. For example, the logical 'and' and 'or' operations are not symmetric since they are lazy in their evaluation of a second argument, a property which leads to a distressing lack of symmetry in the rules of proof for the language. In addition, the notion of concurrency differs subtly from both of those familiar from CSP and CCS, which makes intuitive understanding of its behaviour more difficult for the non-specialist user. However, without doubt it is suitable for specifying substantial systems.

Since the language is for specification and not for direct execution, it is possible for the type system to incorporate undecidable logical assertions: an object can be in a type if (and only if) it meets a particular logical property. Mechanical type checking for programming languages is essential if certain sorts of trivial error are to be found, and the same would apply to specifications. The language has a system of maximal types to which the richer types can be reduced: adherence to the maximal-type system is machine checkable. (A similar approach is used for Z.)

Technical Assessment

Classes in RSL are intended to denote sets of models, each of which may be described as objects; schemes are named classes. At its simplest, a class introduces

- a collection of type names and named types;
- a collection of variables;
- a collection of names of specified type (a signature, in other words); specifications will include variable (and channel) access descriptions;
- a collection of axioms which describe properties of the named values.

The axioms may completely specify a value, either explicitly in a declarative definition or implicitly through a set of algebraic axioms, or only specify some of its properties.

The definition of a class may be deemed to extend one or more classes, thereby giving a multiple inheritance mechanism. Inheritance is, by default, strict, but a non-strict version can be modelled by means of hiding and renaming. Classes can be defined parametrically over one another, which gives, as a special case, parametric polymorphism (as in SML and other functional languages, and in the templates of ANSI C++).

Types in the language are flexible, and not restricted to statically-checkable types. This allows, for instance, range restrictions to be type specifications.

Object creation and deletion have to be dealt with rather inelegantly using object arrays, which allow the specification of a collection of objects of unbounded size. Creation and deletion are themselves modelled by the setting of the appropriate boolean flag in an object.

RSL supports synchronous concurrency explicitly. Asynchronous communication can be modelled in standard ways.

Behavioural descriptions are possible in a number of styles. Pre- and post-conditions allow conditions to be placed on when actions take place and on their effects. Higher-level algebraic specifications allow the identification of sequences of actions which have congruent effects.

The module system and development relation allow separation of concerns within program development - it is envisaged that this may also facilitate test generation from specifications.

5 Tool Support

The tool support associated with the two languages differs in approach. The RAISE tool set is mature and powerful and could be seen as an industrial specifiers' tool set, but needs a workstation to run it. It provides proof facilities which could be an advantage when investigating automatic test generation. The tool set includes a structure-oriented editor (including a (maximal-)type checker); pretty printers generating LaTeX; translators for the constructive part of the language into Ada and C++; justification (i.e verification) tools.

The structure editors, which allow interactive construction of schemes, objects etc. are impressive. The justification editor supports interactive construction of proofs using a menu/mouse style interface. However, it is slow, and the tool is clearly not as mature as the structure editors. Support for larger-scale developments is very limited.

In contrast to RAISE, there exist a number of other sources of tools to support the specification process in Z. These include type checkers, syntax checkers and proof support tools, however, none are integrated in the same manner as the RAISE toolset. The ZIP project contains an overview of the available tools [ZIP]. ICL, for example, supply a verification environment for Z in their ProofPower system. The Formaliser specification tool, developed by Logica, is a generic tool (which is not tied to one specific language, although the bias is towards supporting Z specifications) to create and type check specifications via use of a structure editor. Unlike the RAISE toolset, these are not integrated into one system, and thus the tool support will appeal to different constituents in each case.

6 Language Standardization

Z has recently passed a work item ballot in ISO, and so will move towards standardisation through this body. There have been a variety of extensions proposed to the Z language which are claimed to be object-oriented, [OOZ]. Object-Z is one of the most mature of the object-oriented extensions to the Z language in terms of the number of applications written in the language and the international take-up of the language. However, there can be no guarantee that it will remain in the forefront or that it will be an appropriate language for standardization. It is extremely unlikely that standardization of Object-Z will begin within the next three years.

Standardising RSL is a work package in the LaCoS project, with two man years of effort devoted to it. The aim is to achieve ISO standardisation in about five years time. However, progress depends on support from other ISO National Bodies.

7 ODP and Managed Object Requirements

Both Object-Z and RAISE satisfy the general requirements made of formal description techniques supporting ODP specifications. One weakness is that Object-Z is still unstable and does not have a full semantics, although it can be translated into Z and that does have a stable semantics. Consequently there are no introductory texts nor is there a wide range of examples available.

There are more specific modelling concepts needed in ODP and Managed Object definitions. The main ODP modelling concepts include template, type, subtype, class type, polymorphism and inheritance. All of these are supported or can be supported in Object-Z.

There is work to be done on Managed Object concepts of conditional packages, atomic synchronization and allomorphic classes, and it is possible that extensions to Object-Z will be needed or desired.

Z and Object-Z offer less built-in design paradigms than RAISE. However, all the concepts provided by RAISE can also be modelled in Z and Object-Z. For example, RAISE supports concurrency by use of channels and concurrent combinators. Object-Z does not offer these, but communication can be modelled by the unification of inputs and outputs to classes and operations. Behavioural descriptions are possible through via pre- and post-conditions in a style similar to that available in RAISE.

Managed objects have already been specified in RSL, VDM, Z and Object-Z, and no overriding problems have been found [North], [SimMar], [Rudkin]. In Britain, British Telecom's (BT) Conformance Test Laboratory has done work on developing automatic test generation from process algebras, and has undertaken work on how this can be integrated into an object-oriented Z environment. In addition, there is current research (at the National Physical Laboratory (NPL) in Britain) on the development of test generation using Prolog and LOTOS [Ashford].

8 The Testing Process

Most of the current expertise in formalized conformance testing is based on the testing of communications protocols, which emphasises the procedural aspects of behaviour, in terms of the sequence of observable events.

This aspect of managed object behaviour is important, but it is also necessary to test the consistency of the management information model used to define the managed object state, and to test longer term consistency between periods of communication and between different managed objects. Doing this requires a flexible approach to testing and the combination of a variety of test techniques.

For OSI, the testing methodology and framework is defined in ISO 9646, [ISO9646]. This defines a series of test configurations, procedures and tools for test definition. The tools defined provide structure but are not truly formal, so that the scope for tool support is limited.

The number of test steps involved in a non-trivial management application will be very large. Figures of thousands or tens of thousands of test steps are typical. When operating on this scale, the cost per step of test realization must be kept very low; the process must be made as automatic as possible. It is here that the use of formal specifications for the managed objects can have major pay-backs. The current specifications use a semi-formal framework to organize information, but the heart of the behaviour specification is based on natural language, and so requires human interpretation to create the test steps.

However, the techniques for automated test generation are still being explored. Work transferred from the protocol theatre can be adopted, but is not necessarily the most effective way to create cost effective testing of all aspects of managed objects.

Formal Methods in the Testing Process

The ultimate aim of using formal methods in the testing process is to develop tools which will assist with the generation of sensible tests from formal specifications. There are currently two drawbacks to this approach.

First, fully automatic techniques generate too many tests, and hence test selection and test structure become necessary for the output to be usable. Secondly, automatic techniques do not acknowledge the relative importance of different parts of the specification.

Test generation and selection from formal specifications are active research topics in the UK; with representative work coming from both the commercial sector (eg BT) and government institutions (eg NPL).

One thread of BT's work has been to extend its LOTOS-based CO-OP work to managed objects [CusWez]. The object-oriented specifications are described by a labelled transition system, which allows general techniques, developed by Brinksma and others, to be applied.

Work at NPL has focussed on a number of areas. In aiming to generate tests for the Transport class 4 protocol [Ashford], a formalisation of test purposes as well as of the specifications themselves has shown promising results. More speculatively, there is discussion of exploiting the different description styles available in RSL to derive tests at different levels of abstraction. Related work, using the proof obligations generated during formal development to guide the search for tests is also under way.

A major manufacturer has introduced a testing methodology internally, with some degree of success. It is based on augmenting an IDL (Interface Definition Language) with pre- and post- conditions, whilst the user specifies separately which 'interesting' sets of parameters should form part of the tests. This gives some weight to the view that formal specifications of managed objects should take the form of augmented GDMO descriptions.

9 Conclusions and recommendations

The formal specification of managed objects is feasible using existing languages; the specifications produced are likely to be more precise than the existing informal or semi-formal techniques, which depend, in the last analysis, on the interpretation of English text.

However, there is considerable resistance to the use of such techniques in industry, and a lack of information about the languages and their application. An education campaign would be needed to ensure that the necessary information is made available to those involved in the specification, implementation and testing of managed objects. Due account needs to be taken of both the *de jure* and the *de facto* standardization mechanisms, and any actions taken need to be coordinated with other initiative in Europe, America and Japan.

The technical assessment of the languages (Object)-Z and RAISE indicates that either of these two languages could be used to produced specifications of managed objects. The styles would be different, reflecting the capabilities of the two languages, but the essence of the existing informal specifications could be captured. The choice of language thus rests primarily on non-technical factors such as user familiarity and degree of standardization, and on the quality of the tools available to support each language.

Z has a wide user base, and has a successful history of use. As an extension of Z, Object-Z would benefit from the position of Z in the market place. RAISE on the other hand is relatively untried; however, it is clearly powerful and offers the specifier several different design paradigms, as opposed to the single one supported by Object-Z. The most significant factor in selecting a language is its acceptability to the intended user community. From this point of view, traditional Z is the clear winner, although no formal technique is really widely established with implementors.

References

[Ashford] Automatic Test Case Generation using Prolog", S.J. Ashford, NPL Report DITC 215/95, 1993.

[Cusack 91] Object Oriented Modelling in Z For Open Distributed Systems", E Cusack, BT, 1991.

[Cusack 92] Using Z in Communications Engineering", E Cusack, BT, 1992.

[CusWez] Deriving tests for objects specified in Z", E. Cusack, C. Wezeman, in Proceedings of Z User Meeting, December 1992, Springer Verlag, 1992.

[Duke] Towards a semantics for Object-Z", David Duke and Roger Duke in VDM'90: VDM and Z, Lecture Notes in Computer Science, Springer-Verlag, Berlin, 1990.

[FormMan] Liaison to CCITT SG VII concerning the use of Formal Techniques for the specification of Managed Objects", ISO/IEC JTC1/SC21/WG4 N1644, December 1992.

[GDMO] Information Technology - Open Systems Interconnection - Structure of Management Information - Part 4: Guidelines for the Definition of Managed Objects" ISO/IEC 10165-4 (X.722).

[Hoare] Communicating Sequential Processes", C A R Hoare in Prentice Hall International Series in Computer Science, 1987.

[ISO9646] Information Technology - Open Systems Interconnection - Conformance Testing Methodology and Framework, Parts 1-5", ISO/IEC 9646.

[ISO10746] Basic Reference Model of Open Distributed Processing - Part 2: Descriptive Model, Part 3: Prescriptive Model", ISO/IEC 10746, July 1994.

[King] Z and the refinement calculus", S King in D Bjorner, C A R Hoare and H Langmaack (eds) VDM'90: VDM and Z, LNCS, Springer-Verlag, Berlin, 1990.

[Milner] Communication and Concurrency", R Milner in Prentice Hall, 1989.

[North] RSL specification of the Log Managed Object", N D North, NPL Report, 1992.

[Object-Z] Object-Z: An object oriented extension to Z", D. Carrington et. al., in S Vuong (ed), Formal Description Techniques 1989, North Holland, 1990.

[OOZ] Object Orientation in Z", S. Stepney et. al. (eds.), Springer Verlag, 1992.

[RSL] The RAISE Specification Language", The Raise Language Group, Prentice-Hall, 1992.

[Rudkin] Modelling information objects in Z", Steve Rudkin in J de Meer (ed) International Workshp on ODP, October 1991, North Holland 1992.

[SimMar] Using VDM to specify OSI managed objects", Linda Simon and Lynn S Marshall in K R Parker and G A Rose (eds), Formal Description Techniques 1991, North Holland 1992.

[SML] The Definition of Standard ML", Robin Milner, et.al., MIT Press, 1991.

[Spivey] The Z Notation, A Reference Manual", J. M. Spivey, Prentice Hall, 2nd Edition, 1992.

[Trader] Working Document on Topic 9.1 - Trader", ISO/IEC JTC1/SC21/WG7 N743, November 1992.

[ZIP] ZIP Project Final Report" in Bulletin of EATCS, 54, October 1994.

Biography

Peter Linington has been Professor of Computer Communication in the University of Kent at Canterbury since 1987. His research interests span networks and distributed systems, currently concentrating on distributed multimedia systems exploiting audio and video information. In ISO, he is currently involved in the standardization of Open Distributed Processing. He chairs the BSI panel on ODP and leads the UK delegation to the international meetings. He also chairs the internal technical review committee for the Esprit ISA project (previously ANSA).

John Derrick has been a Lecturer in Computer Science at the University of Kent since 1990. His research interests include applications of formal techniques to ODP and distributed computing. His current projects include developing techniques for the use of FDTs within ODP and formal definitions of consistency and conformance.

Simon Thompson has lectured in Computer Science at the University of Kent since 1983. His interests include functional programming, constructive type theory and the application of formal and logical methods in computing science.

55
AN APPROACH TO CONFORMANCE TESTING OF MIB IMPLEMENTATIONS

Michel Barbeau
Université de Sherbrooke
Dépt. de mathématiques et d'informatique
Sherbrooke, Québec
Canada J1K 2R1
Tel. +1-819-821-7018
E-mail:barbeau@dmi.usherb.ca

Behcet Sarikaya
University of Aizu
Computer Communications Lab.
Tsuruga, Ikki-machi, Aizu-Wakamatsu
Fukushima, Japan 965-80
Tel. +81-242-37-2559
E-mail: sarikaya@rsc.u-aizu.ac.jp

Abstract
A methodology is presented to test the conformity of managed nodes to network management standards in the SNMP framework. The first phase of the methodology consists of an object-oriented modeling of the managed node using class diagrams and SDL-92 language. The second phase takes the abstract model to systematically generate test suites. The approach is based on ISO's conformance testing methodology. Test cases are expressed in the Tree and Tabular Combined Notation (TTCN). The approach is illustrated with a recently developed Management Information Base (MIB) for the management of ATM permanent virtual links.

Keywords
Management Information Base, Simple Network Management Protocol, ASN.1, SDL-92, Class diagrams, Conformance Testing, Abstract Test Suites, TTCN.

1 INTRODUCTION

Presently there are two main frameworks for network management, namely, the OSI and Internet Engineering Task Force (IETF) frameworks. IETF has developed a simple view of network management called the Structure of Management Information (SMI) and Simple Network Management Protocol (SNMP) [9]. The approach presented in this paper has been done for IETF framework, also known as the SNMP framework.

A network in the SNMP framework is made of several managed nodes and at least one management station. Every managed node has several managed objects. Managed objects are abstractions of data processing and data communication resources, such as routing tables and counters. They represent the management view of network resources which can be physical or conceptual in nature. Managed objects and management protocol data units (PDU) constitute the management information. Management information representation in SMI is done using a subset of ASN.1 [6] with macros.

Managed objects are grouped in Management Information Bases (MIBs). MIBs are maintained by every managed node. There are several MIB models that serve different purposes and are attached to different technologies. For instance, MIB-II has been defined to manage TCP/IP networks [9] and ATOMMIB to manage permanent circuits of Asynchronous Transfer Mode (ATM) networks [1].

In this paper we develop a test design methodology for testing conformity of managed nodes to network management standards published by IETF and known as RFCs. The paper continues in Section 2 where a technique is introduced for reverse-engineering the descriptions of the managed nodes and obtaining a precise behavior model in SDL-92. In Section 3 the test design methodology is detailed. In Section 4 we discuss use of the SDL-92 specification to generate test cases. The approach is illustrated with the traffic description parameters group of the ATM MIB. Finally, Section 5 presents some concluding remarks.

2 REVERSE-ENGINEERING OF MIBS

Reverse-engineering is defined as taking something at a level of abstraction and deriving from it something at a higher level of abstraction. In the SNMP framework, MIBs are described with ASN.1, for the structural aspects, and natural language, for the behavioral aspects.

2.1 Structure of Management Information

Managed objects are data structures maintained by every managed node. In SMI, values of managed objects are of various ASN.1 types. In addition, every object has a distinct name of type `OBJECT IDENTIFIER`. For instance, managed objects in the Internet network all have the following common prefix:

```
internet OBJECT IDENTIFIER ::= { iso(1) org (3) dod(6) 1 }
```

A MIB model, or module, is defined using the following syntax:

```
ATM-MIB DEFINITIONS ::= BEGIN
IMPORTS
   MODULE-IDENTITY, OBJECT-TYPE, OBJECT-IDENTITY,
   experimental, Counter32, Integer32 FROM SNMPv2-SMI
   ...other imports
   ;
atmMIB MODULE-IDENTITY
   ...
::= { experimental 41 }
atmMIBObjects OBJECT IDENTIFIER ::= { atmMIB 1 }
   ...Definition of each group follows
END
```

Every module has a name, e.g., `ATM-MIB`. A module can import definitions from other modules, e.g., `MODULE-IDENTITY`, `OBJECT-TYPE`, `OBJECT-IDENTITY`, `experimental`, `Counter32`, and `Integer32` are imported from module `SNMPv2-SMI`. Most of the commonly used definitions have already been declared in the module `SNMPv2-SMI`. In the above, the `MODULE-IDENTITY` macro is used to define the module's identity as `experimental 41` and document its revision history. Managed objects are defined within logical groups. Each group corresponds to one aspect of the system, e.g., a protocol layer.

Managed objects within groups are defined using the `OBJECT-TYPE` macro. Macros have symbolic expansion capability such as Backus-Normal Form (BNF) rules devised for describing the syntax of programming languages. Macros have several clauses. Clause `SYNTAX` defines the data type of the object. Clause `MAX-ACCESS` specifies the level of access such as `read-create` or `not-accessible`. Clause `STATUS` serves to create versions of MIBs. Finally, clause `DESCRIPTION` introduces a textual description of the managed object.

A MIB can be seen as a collection of simple (scalar) and more complicated tabular objects. Tabular objects will be explained using ATM MIB. A tabular object called interface configuration table is defined in the ATM MIB as follows:

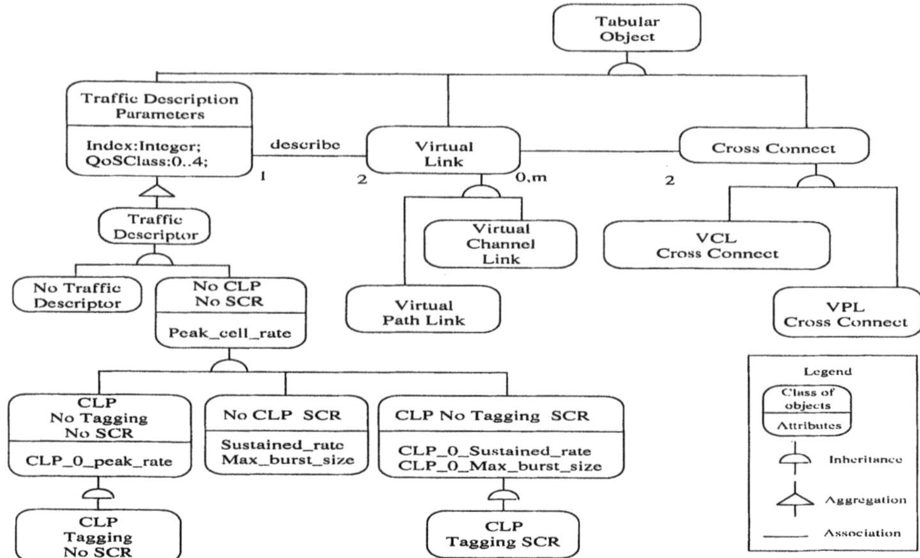

Figure 1: Class Diagram of ATM MIB

```
atmInterfaceConfTable OBJECT-TYPE
   SYNTAX SEQUENCE OF AtmInterfaceConfEntry
   MAX-ACCESS not-accessible
   STATUS current
   DESCRIPTION "a string"
::= { atmMIBObjects 2}
```

The table is defined as an unbounded list of values of type `AtmInterfaceConfEntry`, the rows of the table. `AtmInterfaceConfEntry` is defined as a record of values of various types. Such values are called columnar objects. In order to distinguish between different rows, an index (often of type `INTEGER`) is defined using the `INDEX` clause of the `OBJECT-TYPE` macro. The definition of `atmInterfaceConfTable` is completed as follows:

```
atmInterfaceConfEntry OBJECT-TYPE
   SYNTAX AtmInterfaceConfEntry
   MAX-ACCESS not-accessible
   STATUS current
   DESCRIPTION " some text"
   INDEX { ifIndex}
   ::= { atmInterfaceConfTable 1}

AtmInterfaceConfEntry ::= SEQUENCE {
   atmInterfaceMaxVpcs INTEGER,
   ... other fields}
```

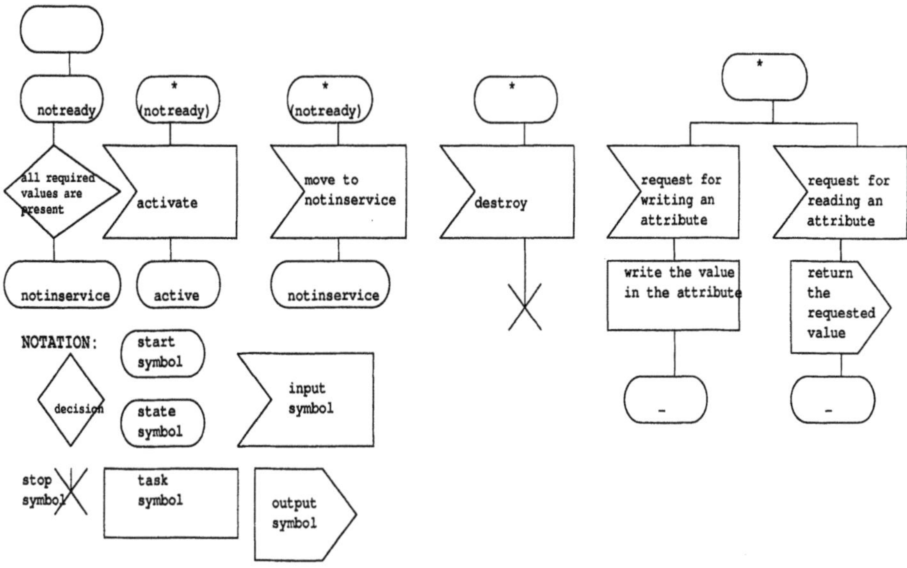

Figure 2: Tabular Object Process Diagram

2.2 Methodology

Our technique is based on class diagrams [3] and SDL-92. Class diagrams clearly show the structures to be tested and common aspects of these structures. SDL-92 specifications precisely define the behavior to be tested. Both are obtained by inspection of the ASN.1 managed object data structures, the accompanying textual description in the MIB RFCs, and SNMP protocol elements. Additional information about the system to be managed is also needed most of the time. For the example discussed in the paper, it was obtained from [5].

A class diagram shows classes of objects, subtyping relationships among classes (i.e., inheritance), containment relations (i.e., aggregation), and other associations among objects [3]. For every ASN.1 simple object type, such as `Counter32`, we define an object class. This class contains two attributes, one for storing the object value, such as `integer`, and another for storing the name of a particular object of this class, such as `snmpStatsPackets`. In addition, operations, such as `increment`, are defined to capture the semantics of the class.

Tables in the MIB are mapped to classes. Each row in a table is modeled as a class instance. Most of the fields of the table are mapped to attributes, unless they serve to subtype or establish relations between objects. A field of type `OBJECT IDENTIFIER` may serve to subtype the rows of a table. The values of the field identify the subtypes. Some of the other fields in the table may be applicable only for some subtypes. This is reflected in the class diagram as a superclass (representing the *subtyping field*) with as many subclasses as there are possible values for the field. Subtype specific fields are

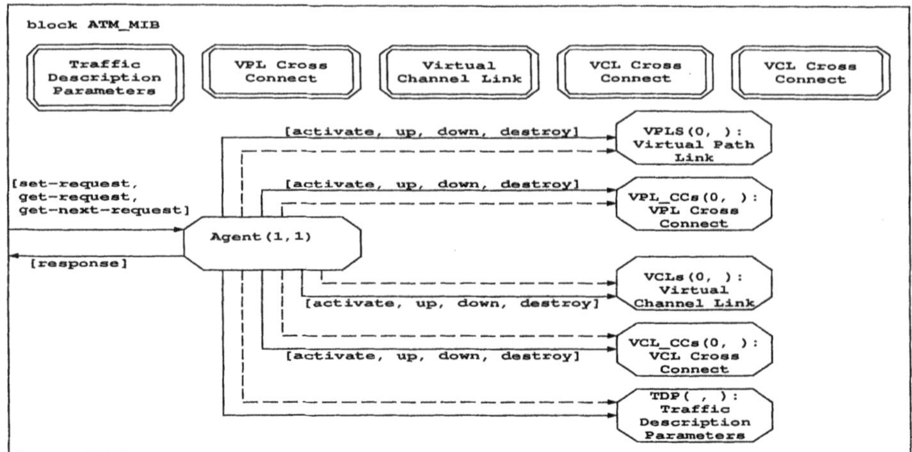

Figure 3: Block Diagram of ATM MIB

moved to the related subclasses. An aggregation relation from the superclass to the class representing the table is also created. Some fields may represent indexes in other tables. They are represented as relations among objects. Finally, superclasses are also introduced to put at on place definitions of attributes and relations common to several other classes.

In SDL-92 [4], behavior is described in terms of processes interacting with an environment. Processes consume signals and perform actions in return. The behavior of a process is modeled as an extended finite-state machine. SDL-92 allows definition of process types as well as reuse of types by inheritance. Processes can be organized into logical blocks.

Classes with behavior in the class diagram are mapped to SDL-92 process types in a block diagram. Communication and process creation relations are uncovered and also represented in the block diagram. Simple object types are static and behavior must be defined to capture the temporal dependencies between the operations, e.g., a gauge can be incremented only if its value is lower than its maximum value.

Rows of ASN.1 tables represent dynamic entities. In SMI, the states in which a row goes through during its life cycle is coded as an integer in a field of type *RowStatus*. Some of these values represent states of instances, some of them represent actions on the instances, and others represent both states and actions.

Procedures of the agent for creating and destroying rows for every kind of table must be specified in SDL-92. The procedure for a creation is initiated by a SNMP *set-request* PDU identifying a row in which the value *createAndGo* or *createAndWait* is written in its *RowStatus* field. Value *createAndGo* is used for single step creation, i.e., all the values of the fields of the row are provided in a single *set-request* PDU. Value *createAndWait* is for negotiated creation during which values of components are written one after the other allowing detailed error checking. The procedure for destroying a row is initiated by a *set-request* PDU identifying the row in which the value *destroy* is written in its *RowStatus* field.

The behavior of a process modeling a row in a table, called `Tabular Object`, is conceptualized as the following SDL-92 process type definition:

```
process type Tabular_Object
dcl value any;
signal activate, moveToNotInService, destroy;
signal read(Charstring), write(Charstring, value), result(value);
/* behavior is defined in Fig. 2 */
endprocess type
```

The behavior of a tabular object is specified in the graphical form of SDL-92 (Fig. 2). There are three states in Fig. 2: *active, notReady*, and *notInService*. Initially, an object is created and put in the *notReady* state. In the *notReady* state, as well as in other states, the agent can receive a request for writing or reading the value of a field (the two rightmost transitions in the figure). There are values of attributes that are required to be present by the agent. When all the required values are present, the object spontaneously moves to the *notInService* state (the leftmost transition). In this state the agent can put the object in the active state by sending an activate command which is modeled as a SDL signal (the second transition from left). An active object can be put in the *notInService* state by signal *moveToNotInService* (the third transition from left). From any state the object can be destroyed by sending a signal *destroy* (the fourth transition from left). The definition of `Tabular Object` can be reused by inheritance for any class of objects modeling tables.

2.3 ATM MIB Example

In an ATM network there are two kinds of virtual connections, namely, switched virtual connections and permanent virtual connections. Permanent bi-directional virtual connections are the main entities considered under this SNMP framework. There are two categories of permanent virtual connections, that is, permanent virtual channel connections (VCCs) and permanent virtual path connections (VPCs). A VCC is carried by a VPC. Conversely, a VPC can carry several VCCs. A VCC/VPC is made of virtual channel/path link's (VCL/VPL's) cross connected together.

Fig. 1 shows the main classes and relations of the ATM MIB. The class *Tabular Object* defines properties, i.e., attributes and behavior, common to traffic description parameters, virtual links, and cross connections. The class *Virtual Link* is abstract and further specialized into the subclasses *Virtual Channel Link* and *Virtual Path Link*, similarly for class *Cross Connect.* Classes *Traffic Description Parameters*, *Virtual Channel Link*, *Virtual Path Link*, *VCL Cross Connect*, and *VPL Cross Connect* originally appear as tables in Ref. [1].

Instances of the class *Cross Connect* model cross connections (CCs) between virtual links. There are two subclasses because a cross connect may be either between two VCLs or two VPLs. Note that between the class *Virtual Link* and *Cross Connect* there is the association *is connected by* linking the cross connected virtual links. Such an association is originally coded in SMI as two fields in rows storing two virtual channel identifiers (virtual path identifiers). VCLs and VPLs may be cross connected according to three topologies, i.e. point-to-point, point-to-multipoint, and multipoint-to-multipoint. A point-to-point CC associates two VCL/VPLs and is modeled as one instance of the class *Cross Connect* associated with the corresponding two instances of the class *Virtual Channel Link*/ *Virtual Path Link.* A point-to-multipoint CC associates a VCL/VPL with several other VCL/VPLs and is modeled as several instances of the class *Cross Connect.* Every instance models attachment of the single point VCL/VPL to one of the multipoint VCL/VPL. A multipoint CC is similarly modeled as several instances of the class *Cross Connect.*

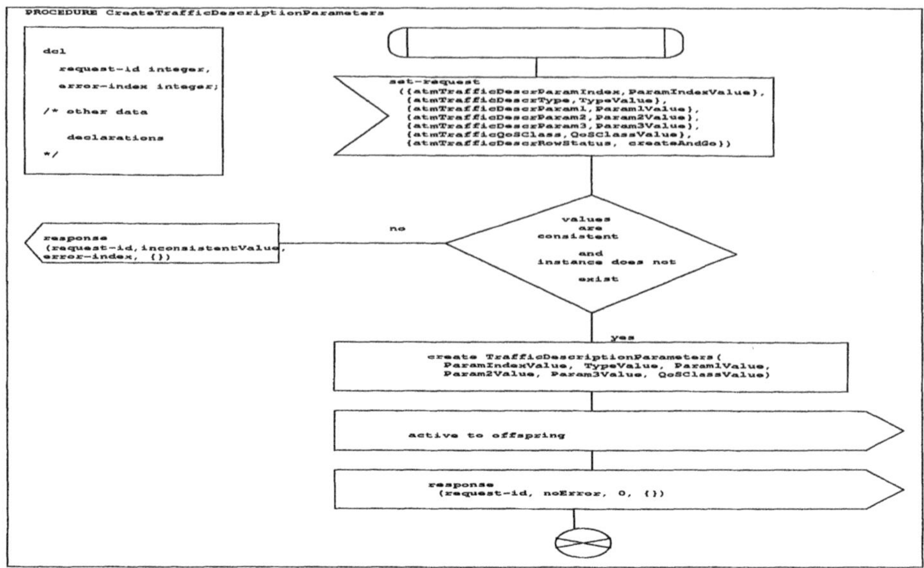

Figure 4: Traffic Description Parameters Creation Procedure

Bandwidth requirements of VCLs and VPLs are given by the users and described in terms of traffic description parameters. In Fig. 1, the class *Virtual Link* has an association to the class *Traffic Description Parameters.* The cardinality of this association is one-to-two because two sets of parameters are required to characterize the two traffic flow directions on a virtual link. An instance of class *Traffic Description Parameters* has two attributes, namely, *Index* and *QoSClass.* The attribute *Index* serves to identify the instances whereas the attribute *QoSClass* indicates the quality of service required by the connections. Class *Traffic Description Parameters* appears as a table in Ref. [1]. One of the fields in the rows of that table is of data type `OBJECT IDENTIFIER`. The values identify the seven possible ATM traffic descriptor types. This is modeled as a class with subclasses and an aggregation relation. That is, an instance of the class *Traffic Description Parameters* contains also an instance of the class *Traffic Descriptor* which have seven subclasses. In Fig. 1, acronym *CLP* stands for Cell Loss Priority and *SCR* stands for Sustained Cell Rate. In the ASN.1 representation of the MIB, attributes *Peak_cell_rate* and *CLP_0_peak_rate* are known under the names *atmTrafficDescrParam1* and *atmTrafficDescrParam2.* Attributes *(CLP_0_)Sustained_rate* and *(CLP_0_)Max_burst_size* are known under the names *atmTrafficDescrParam2* and *atmTrafficDescrParam3.*

We now discuss specification in SDL-92. The ATM MIB is encapsulated into the SDL-92 block pictured in Fig. 3. Most classes from the class diagram of Fig. 1 are mapped to SDL-92 process types. Class *Traffic Descriptor* is not mapped to a process type because it has no behavior. Hereafter, we provide a specification in SDL-92 of a procedure that must be supported by the agent for handling requests for the creation of instances of class *Traffic Description Parameters.*

Fig. 4 pictures creation of traffic description parameters of a VCL/VPL by a manager. The first action is the reception of a PDU *set-request* from the manager by the agent. The manager must provide an index (the identifier of the instance), a type (selected among the seven types that appear in Fig. 1), parameters describing the traffic characteristics and the quality of service class, and a row status value (which is *createAndGo*). The agent then decides if the traffic parameter values are consistent and makes sure that an object with the same index has not yet been created . In the affirmative, an instance of process type *Traffic Description Parameters* is created. Values of the *set-request* parameters are passed to the created process. The signal *active* is sent to the new object, a positive response PDU is returned to the manager and the procedure terminates. Otherwise, a response PDU with error status set to *inconsistentValue* is returned to the manager. The behavior of process type *Traffic Description Parameters* is as follows:

```
process type Traffic_Description_Parameters inherits from Tabular_Object
fpar
   /* unique value identifying a Traffic Description Parameters instance */
   Index INTEGER;
dcl
   QoSClass INTEGER;
endprocess type
```

Because of inheritance, instances of process type *Traffic Description Parameters* behave like instances of process type *Tabular Object.* A complete specification of the ATM MIB is described elsewhere [2].

3 ATM MIB CONFORMANCE TESTING

In this section, we shortly review conformance testing of managed objects according to the OSI conformance testing methodology. Afterwards, the methodology is applied to the design of an abstract test suite for the ATM MIB.

3.1 Conformance Testing of Managed Objects

Conformance of implementations Under Test (IUT) of protocols for OSI are tested using standardized sets of test cases called Abstract Test Suites (ATSs). ATSs are formally specified in a language called Tree and Tabular Combined Notation (TTCN) [8].

There are difficulties in applying directly the OSI methodology [7] to the conformance testing of managed nodes since it is designed for protocol testing and direct access by PDUs to the managed objects is not possible. Direct access to the agent is possible by means of SNMP PDUs. The agent in its turn has direct access to the managed objects. Therefore managed object testing can be considered as service testing. The SNMP protocol interface is defined as the Point of Control and Observation (PCO).

In the SNMP framework, the expectations placed on a given MIB are defined using the MODULE-COMPLIANCE and OBJECT-GROUP macros. Conformance information about groups of objects, referred to in a MODULE-COMPLIANCE macro, is indicated using the OBJECT-GROUP macro.

As an example, the group *atmInterfaceConfGroup* and its objects are defined as follows:

```
atmInterfaceConfGroup OBJECT-GROUP
    OBJECTS { atmInterfaceMaxVpcs, atmInterfaceMaxVccs, -- other objects }
    STATUS current
```

```
DESCRIPTION
   "a collection of objects providing configuration information
    about an ATM interface"
::= { atmMIBGroups 1 }
```

The SMI module-compliance information corresponds to the conformance statements in the OSI protocol standards. The SMI model incorporates a mechanism for the IUTs to claim their implementation capabilities. This makes it possible to select/deselect test cases, from an ATS, according to the claims of a given IUT. The SMI `AGENT-CAPABILITIES` macro is defined for this purpose. When this macro is invoked, zero or more modules are identified. For each module, the conformance groups implemented are listed. For each group, any variation is specified including the objects not implemented, not completely implemented or created only in conjunction with others. A capability statement is identified by an `OBJECT IDENTIFIER`. A management application can maintain a database of capability statements and then dynamically inspect these capability statements.

3.2 Test Methodology

In accordance with the OSI methodology, the ATS for the ATM MIB can be structured into four groups:

- Basic Interconnection Tests. This group contains a single test case whose purpose is to find out if an IUT supports SNMP. In this test, the tester sends a *get-request* PDU requesting the value of the managed object named *sysDescr*. If the tester receives a response, then the IUT passes this test and testing can be pursued.
- Capability Tests. The objective of this group is to establish whether or not a functional unit is available. If so, a representative of element of the functional unit is exercised. For MIB conformance testing, this involves checking the support of objects by the managed node.
- Valid Behavior Tests. This group contains test cases for each group defined in the ATM MIB. Valid behavior tests are designed for determining if the behavior assigned to each object has correctly been implemented. Valid behavior test design is further discussed in Section 4.
- Invalid Behavior Tests. The aim of this group is to test the responses of the IUT to syntactically or semantically invalid behaviors generated by the tester. The specification must include a description of how these exceptional cases must be treated. Syntactic encoding errors are normally captured by the ASN.1 encoding and decoding function of SNMP.
- Inopportune Behavior Tests. According to SNMP, any PDU can be sent at any time. Because of this, there will be no tests defined in this group.

4 TEST GENERATION BASED ON SDL-92

The process of obtaining a test suite from a formal specification can be viewed as the process of obtaining the behavior of an entity, called its conformance tester, exhibiting an inverse behavior. That is, inputs (outputs) to (from) the IUT are inverted to become the outputs (inputs) from (to) the tester. We call this process *behavior inversion* [10].

Test cases are selected such that all control and data flow paths in a SDL-92 specification are covered. The model of a SDL-92 specification is an extended finite-state

ASN.1 PDU Constraint Declaration	
Constraint Name:	C(p1:Integer32; p2:ErrorStatus, p3:ErrorIndex; p4:VarBindList)
PDU Type:	SNMP_PDU
Derivation Path:	
Comments:	Base Constraint for all SNMP PDUs
ASN.1 Value	
{ request_id p1, error_status p2, error_index p3, variable_bindings p4 }	

Table 1: Base Constraint

Nr	Label	Behavior description	Cref	V
1		!set_request	Set_request_base	
2		START transmissionTimer		
3	LB	?response	C(0, p2, p3, {})	
4		CANCEL transmissionTimer		
5		[p2=noError AND p3=0]		(P)
6		+POSTAMBLE		
7		[NOT (p2=noError) AND (p3=0))]		FAIL
8		?trap		
9		-> LB		
10		?TIMEOUT transmissionTimer		INCONC
11		?OTHERWISE		FAIL

Table 2: Creation of Traffic Description Parameters Test Case 1

machine. The main testing strategy is state and transition coverage. Every distinct path, consisting of one or several transitions in the extended finite-state machine, is exercised by test cases. Each test case is assigned a distinct purpose, i.e., test of a given behavior following a certain path. Parameter values, of input and output signals of the test case, are selected according to the test purpose and such that predicates of the transitions in the test case are satisfied (in order to make the test case executable). Also, as a data flow testing strategy, parameter variation and combination of parameter values is employed.

Valid behavior test generation is illustrated in this section with an example taken from the ATM MIB. The valid behavior test group contains a subgroup for each MIB group. The example, presented in the sequel, is with a group called *Traffic Descriptor Parameter.*

Several test cases are needed for testing the valid behavior of group *Traffic Descriptor Parameter.* The behavior described in Fig. 4 is used to generate test cases. In the figure, one can identify two branches. The *no* branch of the first decision node defines an invalid behavior test. Therefore, in the valid behavior test cases the *yes* branch must be used. The test case for this branch is defined in Table 2 in TTCN.

In line 1 of Table 2, a *set-request* PDU is sent by the tester to the IUT. The constraint *Set_request_base* defines the initial contents of an instance of class *Traffic Description Parameters.* Line 3 represents the expected response from the IUT, i.e., a *response* PDU. The constraint of this response is *C*, defined in Table 1, instantiated with parameter values. The *request-id* is 0 and the fourth parameter, the list of variable bindings, is empty. The received values of *error-status* and *error-index* are stored in parameters *p2* and *p3*, respectively. In line 5, if *p2* is equal to *noError* and *p3* is equal to 0, a subtree called *POSTAMBLE* is attached in line 6. Line 7 defines the condition for failing the IUT. Lines 8 to 11 define the other events that can occur instead of a *response* PDU.

The purpose of the postamble in Table 3 is to check if the *set* operation has really been performed in the MIB. Line 2 sends a *get-request* PDU to the IUT. Its constraint is *C1*, defined in Table 5. Line 4 is for handling a response to the *get-request* from the IUT. Parameters of the response define the verdict of the test case. Line 6 defines the condition for passing the test. Line 7 fails the IUT if the opposite of the condition defined on line

Nr	Label	Behavior description	Cref	V
1		POSTAMBLE		
2		!get_request	C1	
3		START transmissionTimer		
4	LB	?response	C(1, p2, p3, p4)	
5		CANCEL transmissionTimer		
6		[p2=noError AND p3=0 AND p4=TrafficNoClpNoScrBinding2]		PASS
7		[NOT (p2=noError AND p3=0 AND p4=TrafficNoClpNoScrBinding2]		FAIL
8		?trap		
9		-> LB		
10		?TIMEOUT transmissionTimer		INCONC
11		?OTHERWISE		FAIL

Table 3: Postamble for the Creation of Traffic Description Parameters Test Case

ASN.1 PDU Constraint Declaration	
Constraint Name:	Set_request_base
PDU Type:	SNMP_PDU
Derivation Path:	
Comments:	Base Constraint for all set-request PDUs for Traffic Description Parameters Creation
Constraint Value	
{ request_id 0, error_status noError, error_index 0, variable_bindings TrafficNoClpNoScrBinding }	

Table 4: Set Request Base Constraint

6 holds.

The constraint in Table 4 is the *set-request* PDU constraint.

Table 5 defines the constraint of the *get-request* PDU. In a *get-request* PDU, the variable binding list refers to an instance which has already been created, by the previous *set-request* PDU. The index value *TSP_IUT_ParIndexVal* designates the requested instance, represented as a row in a table, and the names of the requested columns are designated as *unSpecified.*

5 CONCLUDING REMARKS

We have developed a methodology for designing ATSs for testing the conformity of agents and managed objects, in managed nodes, to the MIB RFCs in the SNMP framework. In our approach, a class diagram representing classes of objects and their relations is developed. The dynamic behavior of each class is defined in SDL-92 through the concept of process type. ISO's conformance testing methodology is employed for the design of ATSs. In addition, we have identified how a MIB ATS can be structured and how some of the groups of test cases can be generated based on the SDL-92 specifications. Test cases in the ATS are specified in TTCN. An application has been made to the ATM MIB.

Use of an object-oriented specification language such as SDL-92 has several advantages. The specifications are more compact and also more readable because of non duplication of information. Test generation from these compact specifications is easier. However, the resulting test cases proved to be not so compact. This is because the test cases are designed for the instances while inheritance is on the types. The test cases need to take into account all the inherited features in the instances. Because ISO's test specification language TTCN integrates ASN.1, it was possible to specify precisely the data values in the test cases. A improvement that can be made to TTCN is the extension of constraint inheritance to ASN.1 values of type `SEQUENCE OF`, which are frequently required in MIB test cases.

Notifications (or traps) are spontaneous outputs of managed nodes. Presently, SMI conformance macros do not support notifications. More research is needed on the specification of notifications in MIBs and capture of implementation capabilities and their use in conformance test design.

Structure of ATSs need further improvements. These improvements could lead to new test groups for parameter variations and combinations. MIB integrity test cases are also left for further research.

Dependencies among MIB groups have an impact on individual test case design as well as on the overall ordering of the test cases in the ATSs. More research is needed in this direction.

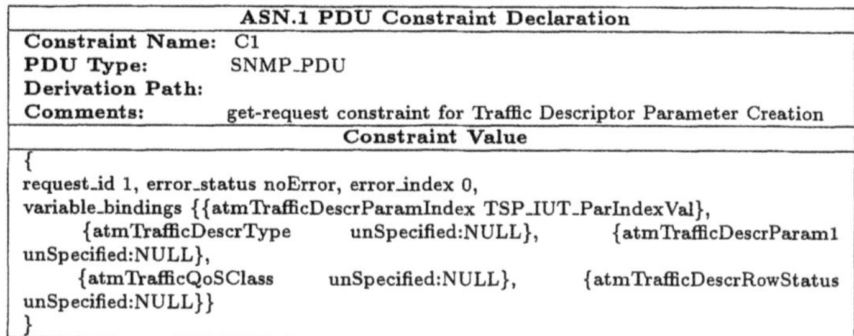

ASN.1 PDU Constraint Declaration	
Constraint Name:	C1
PDU Type:	SNMP_PDU
Derivation Path:	
Comments:	get-request constraint for Traffic Descriptor Parameter Creation
Constraint Value	

{
request_id 1, error_status noError, error_index 0,
variable_bindings {{atmTrafficDescrParamIndex TSP_IUT_ParIndexVal},
{atmTrafficDescrType unSpecified:NULL}, {atmTrafficDescrParam1 unSpecified:NULL},
{atmTrafficQoSClass unSpecified:NULL}, {atmTrafficDescrRowStatus unSpecified:NULL}}
}	

Table 5: Get-Request Constraint

References

[1] M. Ahmed and K. Tesink. *Definitions of Managed Objects for ATM Management Version 7.0*, pages 1–90. IETF, March 1994.

[2] M. Barbeau and B. Sarikaya. Formal specification of MIBs. Technical report, University of Aizu, Aizu-Wakamatsu, Fukushima, Japan, 1995.

[3] G. Booch. *Object Oriented Design with Applications.* Benjamin / Cummings, 1994.

[4] CCITT. *CCITT Specification and Description Language (SDL)*, pages 1–219. CCITT Recommendation Z.100, 1992.

[5] ATM Forum. *ATM UNI Specification, Version 3.0.* Prentice-Hall, 1993.

[6] ISO. *ISO/IEC 8824 Specification of Abstract Syntax Notation-One(ASN.1).*

[7] ISO. *ISO/IEC 9646-1: Conformance Testing Methodology and Framework- Part 1: General Concepts*, pages 1–31. ISO/IEC JTC1/SC21, 1991.

[8] ISO. *ISO/IEC 9646-3: Conformance Testing Methodology and Framework- Part 3: The Tree and Tabular Combined Notation*, pages 1–176. 1991.

[9] M.T. Rose. *The Simple Book: An Introduction to Internet Management.* Prentice Hall, Englewood Cliffs, New Jersey, second edition, 1994.

[10] B. Sarikaya. *Principles of Protocol Engineering and Conformance Testing.* Simon & Schuster, September 1993.

Michel Barbeau got his Ph.D. in Computer Science from the University of Montreal, Canada in 1991. He joined the University of Sherbrooke in 1991 and works there as a professor. His research interests include development methods for telecommunication software.

Behcet Sarikaya got his Ph. D. from McGill University, Canada in 1984. He worked at the Universities of Sherbrooke, Concordia, Montreal in Canada and Bilkent in Turkey. He joined the University of Aizu, Japan in 1993 and works there as a professor. His research interests include multimedia networking.

PART FOUR

Rightsizing in the Nineties

SECTION ONE

Plenary Session A

56

"Can we talk?"

L. Bernstein
AT&T Bell Laboratories
184 Liberty Corner Road
Warren, NJ 07059 USA
fax 908-580-4580
lbernstein@attmail.com

C.M. Yuhas
Freelance Writer
4 Marion Ave.
Short Hills, NJ 07078 USA

Abstract
The successful integration of Services, System and Network Management depends on teamwork among technicians who have trouble understanding each other. The special problems of system configuration and software reliability are addressed in the context of their implications on Services Management. A vision for the future management of complex distributed services is offered.

Keywords
Software, network management, systems management, services management, configuration

INTRODUCTION

The concept of integrated network management sounds like such a good idea. There is the illusion that such a thing exists--some people even come to symposiums to discuss it. Actually, our industry is on the brink of combining 3 types of management--system, network and services-- to achieve a totally integrated product. I've used Joan Rivers' classic line, "Can we talk?," as my title because I hoped the directness and honesty of her delivery would resonate with you. The answer to "Can we talk?" for our systems, for now, is no. Computer management systems, physical networks and service objectives use the same vocabulary to mean different things and to pursue different goals. Acknowledging the problem areas is the beginning of the solution.

Let's listen to a few of the major voices in this field. Here is Arno Penzias, Bell Labs Nobel Prize winner: "If a customer can't use it, it might as well be broken--it might as well not exist!" Clearly that is a statement of the ultimate service objective. It is the goal (albeit negatively stated) of all our work.

BITS VS BYTES

Now here is Charlotte Dennenberg, a Southern New England Telephone vice president: "The most beautiful network is one that is about to break from overuse." She is speaking from the viewpoint of telecommunications, a discipline devoted to getting bits from point A to point B reliably and efficiently. These people measure costs and manage networks to optimize bit-delivery performance. They worry about security and billing. They assume the bytes will make sense of themselves if the bits arrive intact. They juggle 4 networks: one carrying message bits, a second carrying alarms and measurements, a third to operate the network and a fourth bringing management data "outband" so that if the network fails, they can restore it or route around the failure.

Then there is Clarke Ryan, a Bell Labs vice president in network operations, who observes, "Every platform is a weak alternative to an optimal solution." His is a wide-ranging data networking perspective, focused on applications. People like Ryan see the other side. They create local area networks and establish client/server hierarchies to transfer bytes from terminal to application. They worry about balances between server and network performance. They have little enthusiasm for discussions of point-to-point switching and performance because the assumption is that of course the bits will certainly get there. In these terms, the computer system is up if one terminal can send to one printer, even if the other 999 terminals are down. To these folks, network management is "inband," with systems detecting errors for other people to isolate and correct.

Each of these viewpoints is absolutely valid, yet none is sufficient in itself. Is it any wonder that Vinton Cerf, the inventor of TCP, could remark, "Most applications have no idea what they need in network resources or how they need to be managed."? This difficulty is the main reason that large firms have not embraced distributed computing.

COMMUNICATION DIFFICULTIES

For an industry whose stock in trade is communication, we have extraordinary difficulty in talking to each other. Network managers, computer system managers and service managers each feel like Captain Picard on a Star Trek diplomatic mission to alien races.

Terms

The three management schools use the same terms to mean different things. When "disk full" is reported to a system manager, it means time to reorganize the disk use. To the network manager, it means overflow for network data. Systems managers use "fault management" to describe application anomalies. When a network manager says this, it means a box problem needs to be tracked down. Event notification, alarm distribution and logging are carried out differently.

Security

Security management presents real problems. The issues of authentication, authorization, security alarm reporting and audit trails need attention. We need to know when security is compromised. A hot topic today is the possibility of digital cash to pay for all these great system features. The problem is how to send such payment over a broadcast network without having one's credit card stolen. We cannot entirely prevent security breeches, so we had better be very good at detecting and tracking them.

Things will get even more complicated with SONET and SDH (Synchronous Digital Hierarchy) when we multiplex the control and network management data on separate channels within the same physical circuits as the messages and signals. After we understand the use of these isolated channels, we will be ready to use ATM (Asynchronous Transfer Mode) to multiplex all of these together on the same transport links, trusting decoders which will be built into routers to sort things out.

Congestion

Dr. Harry Heffes of Steven's Institute of Technology points out that traffic prediction techniques will be required to reserve bandwidth because the networks will not be able to react to surges quickly enough. Traffic jams could become horrible on the broadband networks. The "byte folks" from computer networks consider congestion management a passing cloud, while the telephony "bit folks" are consumed with avoiding it. Congestion is inadequately addressed in systems management [Vaughan94]. Messages describing component failures induced by congestion will quickly exhaust SNMP. Scalability is poor and "byte folks" need hands-on use of protocol analyzers to find and fix problems. This in itself is not bad, except for the time it takes and the custom-built arrangements required to get to the message paths so that protocol analysis can happen. This becomes a challenge for the "bit folks" to get the message streams nimbly to protocol analyzers.

Protocols

Let's examine the protocols used to convey the network management data and commands. Telephony people spend lots of effort standardizing on OSI agents and their managed objects. Client/server people use SNMP[Barbier94], but their managed objects are different. Since we have two types of network management servers and two types of networks, we will need four interfaces. Now add signaling and message networks of several varieties and watch the complexity grow. Unfortunately, we can't pick one way. The OSI base is just too expensive for simple networks and the SNMPv2 upgrade is not totally backward compatible. SNMPv2 adds security features, but they are too complicated and bulk retrieval of agent data can cause congestion.

MAJOR ISSUES

Configuration Management

Configuration management is a major issue. Reliable, cost effective and easy-to-use backup and restoral systems are needed. When it comes to software, we need to "pack it and track it" as it goes among servers, clients, managers and agents. The same holds true for data. *Data*, not *database*, management will be the issue for these complex networks. How will we trigger work to begin once we install a new feature or recover from an outage? In 1985 when client/server systems were young, we did not know we were actually doing data management. The problem we faced was to broadcast work status information from a Unisys mainframe to thirty NCR tower clients throughout the day. A work center manager who wanted to know how much work was left could ask the local client. This was a nice feature most of the day, but became critical at 3 PM when all the managers wanted to know if they had to schedule overtime to close out the day. Before we had the client/server solution, they would jam the mainframe and networks with report requests, generations and transmittals. And each wanted only that work related to their technicians. To keep the clients in step with the mainframe server, we provided an initial report whenever the client came on-line. This obviated the need for a separate record of the state of each client because we relied on the clients to customize the one comprehensive report for each work center. The server would broadcast all changes to all clients. By using this approach, we did not have to resort to complex startup and recovery procedures that cost network and server capacity. Today, better solutions elude us except for client/server systems that do not grow too fast in size or capability. Static mapping of data models or of software executables will not be good enough to handle future applications. Who will be charged with keeping this complex of systems operating sanely?

Auto-discovery of clients on a TCP/IP network has been wonderful, but this feature does not scale and has been used sparingly in telephony applications. A generalized auto-discovery feature is needed which embraces the concept that the network is the data base[Caruso1994]. It is also needed at the services level. Recently, Hewlett-Packard added centralized software to its Open View platform which manages configurations across networked systems. Instead of making multiple changes every time a user is added to the network, the administrator can use one command to configure the user's password, e-mail account and downloaded software. HP relies on a "synchronization" function to true-up the actual state of the networked applications and computers with the administrator's databases. The problem of mixing the network's physical inventory with logical data in UNIX databases is a tough one. For example, when we built a prototype to extract information from a network element and write it to a relational data base, the hardest part was getting the client protocol stack just right, especially in its interaction with the server's relational database. The database demanded versions of the protocol stack which could not be purchased for the client. The devil is in the details! Once we got the specific configuration to work, it worked well, but it is not robust to changes in the client or the server.

Software

Software may be the toughest problem to solve in building systems that manage other systems. Software has the awful propensity to fail with no warning. One manager of my acquaintance issued a memo stating, "There will be no more software bugs!" The trouble was he meant it; no joke. Even after we find and fix a bug, how do we restore the software to a known state, one where we have tested its operation? For most systems, this is impossible except with lots of custom design which is itself error-prone[Ross94].

One new idea is software rejuvenation. It is special software that gracefully terminates an application and immediately restarts it at a known, clean, internal state. It precedes failure, anticipates it and avoids it. It transforms non-stationary, random processes into stationary ones. Instead of running for a year, with all the mysteries that untried time expanses can harbor, a system is run for one day, 364 times. It is re-initialized each day, process by process, while the system continues to operate. Increasing the rejuvenation rate reduces the cost of downtime. Two years of operation have passed with no reported outages for one system set at a weekly rejuvenation interval. In one of my laboratories, a 16,000 line C program with notoriously leaky memory failed after 52 iterations. We added seven lines of rejuvenation code with the period set at 15 and the program ran flawlessly. Rejuvenation does not remove bugs, it avoids them.

SYNTHESIS

How will we bring all this together? We need to adopt a policy of "agreeing to agree before we disagree" and drop our current practice of "agreeing to disagree before we agree." Cooperation and harmony among practitioners in this field are the only hope. With customers demanding that the power of these services be unleashed, we can either make it happen together or watch others without our know-how create a new industry of service managers under our noses. Professor Ed Richter of the University of Southern California points out, "The more technically competent an organization, the harder it is to accept new technology." If we are not careful, novices will capture our industry while we squabble. We need to understand that working together improves each one's lot. There are technologies to help if we will only use them.

Photonics

We can take advantage of photonics to combine service testing and surveillance. Since photons at different frequencies do not interfere with one another as electrons do, we can send test signals at one color while customer data is being carried at another color and measure performance or monitor a system for alarms. This will lead to true non-interference testing. Len Cohen of AT&T Bell Labs Research is developing a photonic chip that will make this easy to implement[Bernstein94].

Human Factors Design

IBM seems to be leading us to friendlier systems. They will deploy standard object based technology to bind together current products and add new functions in distributed systems management encompassing network, configuration, change, performance, and storage. Their approach differs from the TMN (Telecommunications Management Network) standard by including system administration functions and by giving attention to system usability. The TMN standard is not natural for people. I have seen network managers need to combine and partition TMN functions to solve a problem. We invented "quick switch" in the 1970's to break the straight jacket of hierarchical application system design. We need something similar between TMN functions today. Will they set yet another de facto standard as we argue about the de jure ones? Bev Little commented in the April 25, 1994 issue of Forbes that "Today's systems have good facilities for humans to take over, the ones coming down the line are supposed to handle higher traffic flows, and it is not obvious that humans could intervene." Whether IBM's approach will solve this problem and be scaleable remains an open question.

Who Manages Whom

The question of whether system management is part of network management or vice versa will be hotly debated in the coming year. My view is that the network extends from my finger tip to the application executable code, so that even queue management and caching control inside the servers are right and proper network management functions. As client/server systems get deployed, savings are not as great as expected because they are much more expensive to operate and evolve than expected. The "administrator to box" ratio remains constant because of the poor systems and network management design which is slowing down progress. Planners are not faced with a "buy or build?" decision since no solution solves the existing problems and none can be built fast enough to keep up with the demand. Administrators are often faced with the problem of integrating solutions from different suppliers. People are beginning to use the INTERNET for remote management to good effect. But, this is a "send and pray" network that cannot be counted on under stress to isolate and fix problems. It may prove to be perfectly fine for detecting problems, but will it be adequate to force test messages on defined paths, or to monitor major network events, or to isolate focused overloads, or to convey commands to network elements?

State Transition Theory

A theory of state transitions for systems and network management needs to be articulated to define how the network states will be re-established after the inevitable hang or crash. With a workable theory, the state transitions will become predictable and, more importantly, testable. One set of these states will be the orderly downloading of programs and data from servers to client with appropriate recovery strategies.

ATM Data

Current network management is inadequate for the coming flood of ATM data because of incompatible network management from different suppliers[Wilson94}. SNMP cannot scale well to handle the volumes required, such as automatic topology features with huge network domains. Protocol analyzers will not be able to get to the troubled transport in time. Data base "sniffers" will be incompatible with network analysis software. With computer communications and voice communications sharing the same network, we see a threefold increase in complexity. With the emerging multimedia, we are seeing ten times this complexity. A 30:1 increase in complexity is on the horizon for our overworked system administrators. Today AT&T's Accu-Ring Service Centers provide customers with a single point of contact for SONET dual fiber ring networks [Robinson94]. These networks span customer premises, local exchange carrier central offices, competitive access providers control centers and AT&T central offices. The service center coordinates repairs and manages installation, sectionalization, and alarm monitoring. Eventually this network management service will be extended beyond communications to the actual applications on the network.

THE FUTURE

Even with all these problems, I have hope--even enthusiasm--for the future. I envision a Services Management System that detects anomalies from any source. Rather than dump a glut of raw bits, this system digests the data and transforms it to information, presenting the network manager with only what's needed to resolve the problem. Picture a manager who would sit at a display well designed for easy human use. One could see all elements of service operation from server queues to transport error seconds. The manager would extend that famous finger tip and point to any service, system or network element. Up would come graphics of decoded messages being passed or filling a buffer. The manager would take performance measurements, then instruct "service" to make changes in its operation. Finally, without stirring from the display, the manager would inject a test message into the application and trace it through the distributed system, either by allowing it to choose its path or by forcing one. This will lead to integration of boxes, links, elements, queues, utility software and data managers in a friendly, easy way.

REFERENCES

Barbier, Susan "Systems Management in the 1990s," AT&T Technical Journal, ..July/August 1994 Vol. 73 No. 4, ISSN 8756-2324 pp. 82-97

Bernstein, Lawrence "Innovative Technologies for Preventing Network Outages" AT&T ..Technical Journal, July/August 1993 Vol. 72 No. 4, ISSN 8756-2324 pp. 9-10

Caruso, Jeff(1994) , "UB Networks Embeds Network Management In The Hub," ..Network Management Systems & Strategies Vol. 6, No. 21, ISSN 1043-1217

Robinson, Michael J., Werner, Thomas S., Ward Joseph A(1994)., "Solving the Access ..Outage Problem," AT&T Technology Vol. 9 No. Fall, ISSN 0889-8979, pg. 21

Ross, Philip E.(1994), "The day the software crashed," Forbes Magazine April 25, pp. ..142-156

Vaughan, Nick(1994), "Networks and Systems Still Seeking Suitable Frameworks," ..Software Magazine October pp. 53-58

Wilson, Tim(1994), "Current Net Mgm't Inadequate for ATM," Communications Week ..November 7, No. 530 A CMP publication

BIOGRAPHIES

Lawrence Bernstein is Chief Technical Officer of the Operations Systems Business Unit at AT&T Bell Laboratories. He holds a BEE from RPI and a MEE from NYU. He has contributed to the evolution of Network Management and is a Fellow of ACM, IEEE, and Ball State. He is listed in Who's Who in America.

C. M. Yuhas is a freelance writer who has published articles on network management in IEEE Journal on Selected Areas in Communication, IEEE Network and International Journal of Systems and Network Management. She has a Bachelor's in English form Douglas and a Master's in Communications from NYU.

57

The rise of the Lean Service Provider

Keith WILLETTS, BT - Network Management Forum, UK

Global competition is making significant impacts on pricing for all communications services with consequent downward pressure on operating costs. At the same time capital and current account investment to develop new services and rebuild infrastructures for Broadband, multimedia services etc. is accelerating. This financial squeeze is causing all service providers: carriers, local operators, value added or large corporate communications divisions to seriously examine the way they do business.

The central theme of the talk will be the emergence around the world of the "lean operator" as new entrants into the marketplace with highly automated process structures and highly manageable networks emerge. Established players are also rapidly undergoing right-sizing and re-engineering programs to match new competitors. Issues such as end-to-end process automation, dealing with legacy systems and transformation of the network infrastructure (especially access networks) will be addressed.

The talk will use examples of other industries such as the deregulated airline industry and the emergence of 'lean production' by Japanese companies in the 1980's to help understand the profound restructuring that is occurring world-wide in the communications sector. A single key message will be that the established players need to change the way they operate rapidly or they will go under. Survivors will be those companies that simultaneously achieve major reductions in operating cost and major advances in customer service. This is as true for end user departments who will be replaced by outsourcing or managed services as it is the mainstream providers who will be eclipsed by lower cost, higher quality new entrants.

58

Managing Complex Systems - When Less is More

Lyman CHAPIN, Bolt Beranek and Newman, U.S.A.

In *An Introduction to Mathematics*, Alfred North Whitehead observes that "Civilization advances by extending the number of important operations which we can perform without thinking about them." That comment was made in 1911, at a time when scientific discovery and its intellectual corollaries were adding so rapidly to the store of human knowledge that Whitehead's generation would be the last to enjoy the expectation of a well-educated man that the compass of his learning include "all things that are known." Highly specialized experts dominate today's science and technology. The complexity of the systems with which we regularly interact, and the sheer quantity of information that clamors for our attention, suggest that civilization has advanced very little during the computer age.

The Internet, to take an obvious example, is surely one of the most successful large-scale distributed enterprises in history; it is used by millions of people in at least 110 countries, and is growing so rapidly that estimates of the "size of the Internet" are obsolete long before they can be published. For even its most sophisticated users, however, the Internet is a dauntingly complex system. Vint Cerf's recent assessment of the state of the Internet is telling: "It's still rocket science." The same could be said of every other large public or private network.

Engineers and other technology specialists tend to view the complexity of networks with the complacency of insiders. The prevailing engineer's viewpoint is dominated by an "engineering meritocracy" ethic that values and rewards "gurus" to whom the secrets of the network have been revealed. The tools and methodologies that are available for managing large networks reflect a corresponding lack of interest in making things simpler - the interesting problems for engineers are elsewhere. As a result, most network management systems exhibit in practice a property that Dave Oran has called the "first law of network management parameters": for every configurable component of a network management system, there are just two settings: the one that works, and all others.

The solution to this problem is <<less>> network management, not more. The last thing a network manager needs is twice as many configurable parameters to set to the wrong values, or a hundred new alerts that report irrelevant or incomprehensible events. The ideal network (from a manager's perspective) would be largely self-configuring and self-managing, requiring very little manual intervention. Unfortunately, "manageability" is not high on the list of priorities for most network engineers. A well-designed network management system can compensate for some of the consequences of a poorly-designed (from the standpoint of manageability) network, but often only by requiring the manager to exercise direct control over low-level details. The latest work on object-oriented network management models will be a step forward only if it recognizes reducing complexity as the highest priority.

SECTION TWO

Plenary Session B

59

Multimedia Information Networking in the 90's – The Evolving Information Infrastructures

Maurizio DECINA, CEFRIEL/Politecnico di Milano, ITALY

At down of the Information Age, the first objective is to provide a Global Information Infrastructure. This term describes the coming world wide interoperability of high speed networks that support a wide range of computer-based personal and professional multimedia applications. The technical foundations of the global infrastructure, the world wide Information Superhighway, are going to emerge from interactions among all players of the voice, computer data, and video information business. The global information business scenario includes a wide range of services and products:

- of the business and entertainment information industry: newspapers, books, movies, tele vision programs, advertising, on-line data services, etc.,

- of the computer industry, both hardware and software products,

- of the consumer electronics industry, both hardware and software products: television sets, CATV set-tops, video game sets, smart phones, wireless sets, etc., and finally

- of the information networking industry, where 3 major players can be identified: Telecommunications, Cable TV and Internet. The underlying technology assumption beyond all elements of the construction of the Infohighway is the staggering spread of digital technologies for processing multimedia signals and data. The presentation focuses on the 3 players in the networking arena, and gives a snapshot on their evolving network protocols, architectures and interactive multimedia services provision during this decade. In particular, the following topics are briefly reviewed:

    - current features and services of Internet, together with the ongoing work to enhance IP [Inter networking Protocol] protocol performance: addressing, routing, real time (voice & video) reservation protocols (ST II [Stream Protocol II], and RSVP [ReSerVation Protocol]), security, etc.;

- telecommunication networks evolution towards the provisioning of multimedia transport capabilities (narrowband ISDN [Integrated Services Digital Network], and broadband ATM [Asynchronous Transfer Mode]), and of signaling & control capabilities to support customer mobility, intelligent network services, and personal communications services. Recent developments to offer interactive broadband services, such as Video on Demand, on hybrid fiber-coaxial cable distribution networks.

- CATV networks evolution towards the provisioning of telephony and interactive digital broadband services. Emphasis is given to respective competitive role of these 3 participants of the networking industry during this decade, in the forthcoming process for selecting the technical pillars of the world wide platform needed to enter the Information Age.

60

Where are we going with telecommunications development and regulation in the year 2000 and beyond ?

David NEWMAN, David Newman and Associates, U.S.A.

Since the divestiture of AT&T, the political process of regulatory reform and deregulation of the telecommunications industry has swung like a pendulum from centralized federal control to decentralized state control. This dramatic change from the protection of a few major telephone companies to the allowance of competition amongst many telephone competitors, has opened the door to entrepreneurial energy and innovation. The resulting technological revolution in computer and communications technology is transforming our society. To complete this transformation will require that we meet the challenges posed by the new technologies with a pragmatic communications policy free from the distorting lens of ideology. A pro-competition policy in international communications will allow new players, greater variety in services, and more competitive pricing. To derive the most benefit from such a broad field of providers, tomorrow's communications policies must measure up favorably against real world criteria - jobs, prices, choices, international trade, and the effects of all of these on competition and innovation.

Having noted the power of competition, one must not loose sight of the reality that, in the telecommunications industry, government policy and regulation have a profound impact on technology and, in particular, on new systems. Government decisions not only shape the direction of many research and development projects, but often determine the rate of progress of new technology as well. When creating or restricting radio spectrum allocations, setting market rules, or establishing technical standards, today s policy makers effectively decide whether or not certain technologies will be able to develop and possibly succeed in the marketplace. More particularly, Federal Communications Commission (FCC) rulemakings, along with proposed legislation on the auctioning of frequency spectrum, have generated a dynamically changing regulatory environment for the communications industry in the United States; and nowhere will the rapid introduction of technological advances be more evident than in the new field of personal communications services.

At present, the FCC is allocating frequencies for personal communications services while deciding the amount of bandwidth to assign, the technical standards that should apply and, most fundamentally, who may be eligible to provide these services. The success of personal communications, however, will depend not only on government policy decisions, but also on the combined actions of engineers, business persons, economists, lawyers, and those in other disciplines. All of these interests must work together if new developments are to be devised in the laboratory and implemented successfully in the field.

From my unique position as a practicing trial lawyer and patent attorney, and a former university professor of electrical engineering and computer science, I see an ongoing interrelationship among many of the regulatory and technological issues. As we approach the twenty-first century, I see divestiture and deregulation creating a shift in demand, a shift which will be met by a change in entrepreneurial judgment as new products and services provide increased business opportunities. These opportunities will range from local personal communications networks to fiber optical transmission lines which will connect the continents. The products and services which will emerge in response to these opportunities will move us beyond the Information Age to the Age of Intellectual Property.

61

Formulating a Successful Management Strategy

Rick STURM, US West Advanced Technologies, U.S.A.

"Rightsizing" is often used as a euphemism for work force reductions. In network management, "rightsizing" is not limited to the question of how an organization can do its work with fewer people. At the same time that they are being pressured to minimize the size of their organization, managers are being asked to significantly increase the scope of their responsibilities to include such things as the management of distributed systems. This presentation will address the factors that are essential in formulating a successful strategy to respond to these conflicting demands.

SECTION THREE

Plenary Session C

62

The Paradigm Shift in Telecommunications Services and Networks

Masayoshi Ejiri
Service and Network Operations Section, Network Engineering Department
NTT Service Engineering Headquarters
1-6 Uchisaiwaicho 1-Chome Chiyoda-ku, Tokyo 100-19 Japan
Telephone: +81-3-3509-5320
Fax: +81-3-5251-7801
E-mail: ejiri@nw.hqs.cae.ntt.jp

Abstract

Liberalization of the telecommunications market has fundamentally changed the situation faced by every player involved with providing telecommunications services and networks. This paper provides an overview of the ongoing evolution of telecommunications and outlines the paradigm shift that will be necessary in such areas as service provision, network architecture and pricing.

Keywords

Customer contact, customer-defined service, customer premises equipment, information agent function, information provider, negotiations, network architecture, operation system function, service attributes, service provider, service providing structure, virtual service provider

1 INTRODUCTION

Changes in the telecommunications market are being driven by rapidly advancing technology and customer demand for increasingly sophisticated services. Successfully coping with the situation will require a paradigm shift in the way providers look at how they provide services. Currently, telecommunications services are offered through complicated interworked networks by various service providers using an array of rapidly evolving technologies. A new mechanism is needed to ensure that service providers can offer their services in a manner that looks and feels seamless from the perspective of the customer.

Customers may also want to freely choose services and then negotiate the terms of those services. To meet this demand, a negotiating mechanism for use between customers and service providers is essential. In the multimedia area, broadband services will require new pricing policies that enable providers to maximize the use of network resources and to offer lower prices to end customers.

Another essential feature of multimedia services is easy and economical access to specific information. Faced with a tremendous volume of available information, customers will want to edit this information to make it better serve their needs. An agent function for information providers becomes a key to meeting this customer demand.

To meet the needs of the new telecommunications business environment, new concepts in service, operations, management and networks must be established. These concepts cannot be built upon existing ways of thinking about telecommunications service. Instead, they require a fundamental paradigm shift in thinking toward a full realization that the multimedia era has arrived. This paper discusses the causes of the paradigm shift, the nature of the new paradigm and the crucial issues which must be dealt with under the altered circumstances.

2 REVOLUTION IN THE SERVICE PROVISION STRUCTURE

The liberalization of the telecommunications market and the resulting keen competition among providers has created a complex service environment where many service providers offer services through mutually interworked networks. With Plain Old Telephone Service (POTS), services are provided via a simple mechanism, shown in Figure 1. In this service provision structure, a carrier constructs the network and plays the role of service provider as well. Carriers deploy standardized network architecture and interfaces based on CCITT (currently ITU-T) standards. Network elements are produced in compliance with these standards. As a result, POTS service provides seamless service to customers on a global basis.

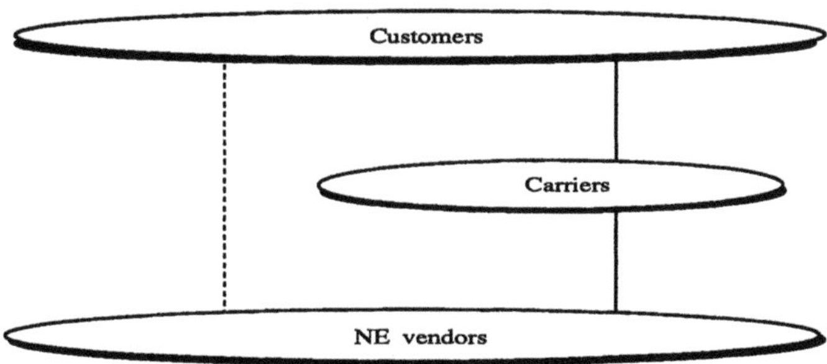

Figure 1 Traditional telephone service provision structure.

Recent advances in computer technology have brought significant improvements in Customer Premises Equipment (CPE). Furthermore, many service providers (SPs) now specialize in the full utilization of CPE capabilities with services such as Value-Added Network (VAN). The result is that several service providers now work in combination to meet customer needs (see Figure 2). Under this service provision structure, SPs and vendors tend to deploy their own network architecture and interfaces in an effort to make them de facto standards. This complicates the service provision structure and reduces customer choice, since customers may avoid otherwise desirable services to prevent an interface mismatch (Ejiri, 1994a, 1994b).

To avoid inconvenience to customers, customers, SPs and vendors need to cooperate in order to create a seamless web out of their diverse services (Network Management Forum, 1994). A virtual service provider – which integrates several service providers – could be a solution. The activities of standardizing bodies (ITU-T, ISO, etc.) and various consortia (ATM Forum, Network Management Forum, etc.) are also expected to help resolve these issues.

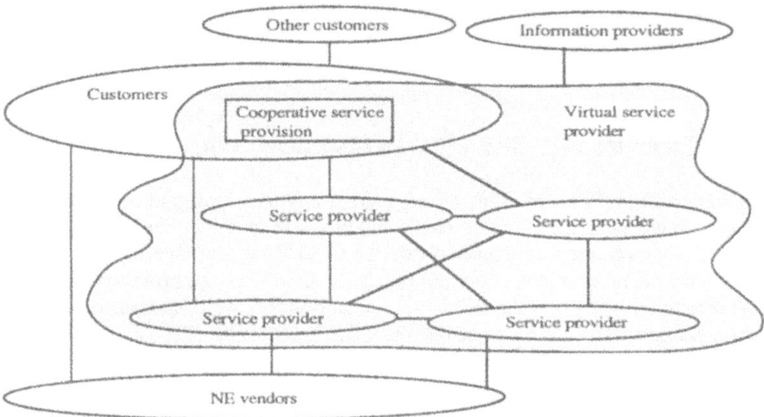

Figure 2 Future service provision structure.

3 SERVICE OFFERINGS BASED UPON CUSTOMER NEGOTIATIONS

The rapid diversification of customer demand means some customers feel restricted in their choice of SPs and vendors. Customers may also want to exert more control over service attributes and conditions. The most important issue is pricing, which is closely related to service attributes such as traffic characteristics and QOS (Quality of Service).

Due to the price of SP telecommunications, various measures have been implemented within CPE networks to minimize computer communications costs. If SPs offer lower prices along with improved CPE-SP linkage, total costs will be reduced. This will greatly stimulate the market for broadband/high-speed communications. Therefore, in computer-based multimedia networks, new concepts and functions for communication conditions negotiations are essential. Moreover, these must be consistent with the network utilization strategy between SPs and customers.

Though lower prices are urgently needed, especially for long distance broadband video communications, efforts to reduce network cost and improve bandwidth compression technology may not be enough to achieve the target price. The situation is more promising for VBR (Variable Bit Rate) video communications and high-speed computer communications. The burst-intensive nature of traffic for these services will permit network utilization strategies (based on quality variation and available time selection) that enable lower prices.

Figure 3 shows a diagram of the basic service provision structure with customer negotiations (Ejiri, 1994a). Negotiations may have static (pre-assigned) and dynamic (on demand) features. Service attributes subject to negotiation include time (point and delay), QOS, addressing and pricing. Customers negotiate with several SPs on these attributes and select their own service conditions and prices based on what's offered. SPs negotiate with customers to use their network resources at a level near 100%, obtaining maximum benefit with minimum investment.

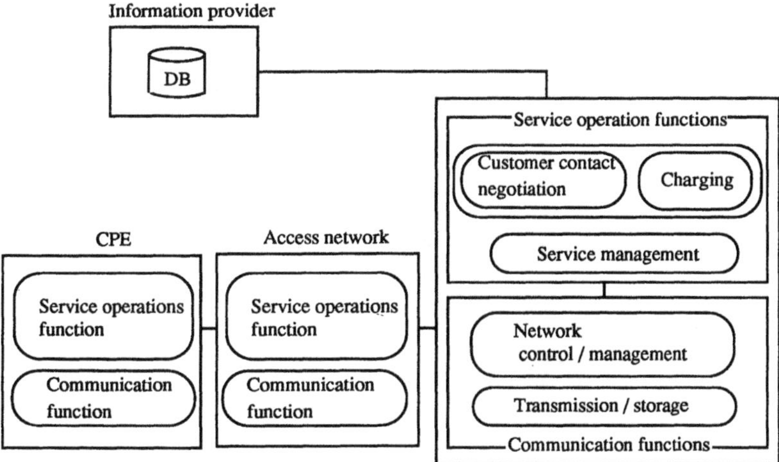

Figure 3 Basic service provision structure including customer negotiation.

In SP networks the functions indicated in Figure 4 need to be subject to negotiation. The service negotiation function manages the negotiation process based on customer demand and network resources/service status. The network resource management function allocates network resources to individual customers following requests from the service negotiation function. Various database systems tracking real-time status information on services and networks must be constructed and maintained to support the negotiation process. To ensure efficient negotiations, negotiation scenarios and intelligent functions involving the connection between CPE and SP networks will have to be developed.

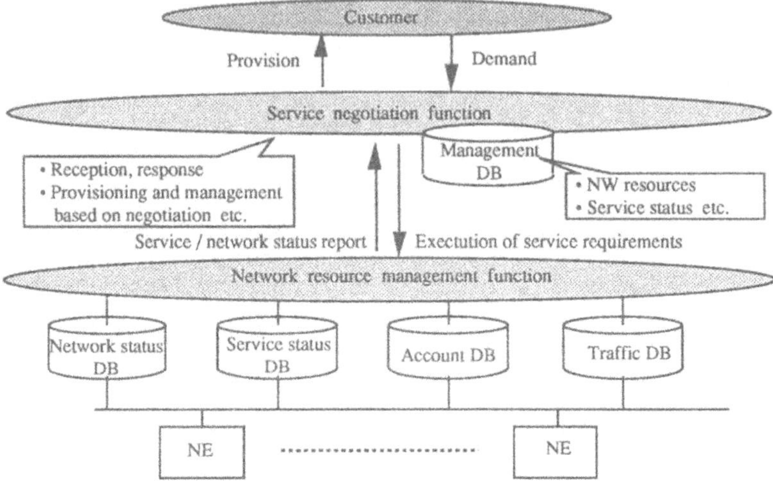

Figure 4 Function allocation for negotiations.

4 EVOLUTION OF CUSTOMER-DEFINED SERVICES

The above discussion about the negotiation of service attributes and conditions concerns only a primitive first step. Customers will want more than the ability to select services from a menu offered by SPs. They will want to change service parameters themselves, combining aspects of various services in a manner that best meets their individual needs. The trend toward customized service is currently being studied as UPT (universal personal telecommunications) by the ITU-T (ITU-T, 1994b).

Network resources will have to be rapidly and flexibly interconnected using sophisticated network operations and management functions. In order to establish the proper architecture, a layering concept should be introduced in order to identify the various functional components. This approach is similar to that used for computer networks.

In the new network architecture, both CPE and SP networks can be based on the same concept and make use of the same software packages (see Section 7). With both CPE and SP networks based on the same concept, many possibilities arise for communication among layered functions (see Figure 5). Customers and SPs can then utilize each others' functional capabilities. The reference points identified in each layer in Figure 5 are important for identifying new interfaces based on network functional architecture.

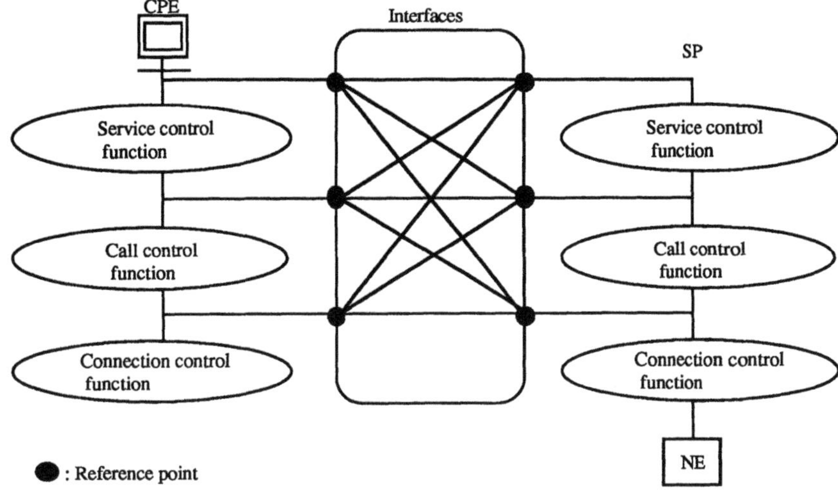

Figure 5 CPE-SP interface examples.

5 ENRICHMENT OF INFORMATION PROVIDER SERVICES

Information providers (IPs) and content providers are expected to play major roles in popularizing multimedia communications services. Since IPs use SP networks to distribute information to customers, cooperation between IPs and SPs is very important.

Attractive, useful information for customers is usually expensive to produce. A stable mechanism for drawing back costs and generating profit is essential for the success of multimedia communication services. At the same time, the mechanism should ensure that customers have easy access to the information, as well as a way to compare various features of information (quality, price, acquisition procedure, etc.) from different IPs.

Sometimes, customers have only fragmentary data about what they need and it is important that they still have a way to access the needed information, if possible. For example, a customer may know what information they need but not know which IP is best for such information. Conversely, they may know a good IP but have only the barest notion about the information they need. To satisfy various types of customer requests, an information agent function will be needed (Natarajan and Slawsky, 1992). Figure 6 shows a diagram of such a function. In this example, some possible functions are added, such as accounting for information charges and electronic mailing using an IP database as a mailbox. Using this function, customers can smoothly obtain the desired information and also add new information producing new multimedia information to be sent on. This capability will accelerate the acceptance of multimedia communications.

IPs will usually collect charges from individual customers who access the information. If SPs and IPs cooperate, another charging procedure could be implemented. Information could be free of charge to individual customers, leading to greatly increased traffic and greater income for the SP which could then be shared with the IP. This service strategy could well benefit all concerned. Customers would receive free information, while the IP and SP might both enjoy greater profits.

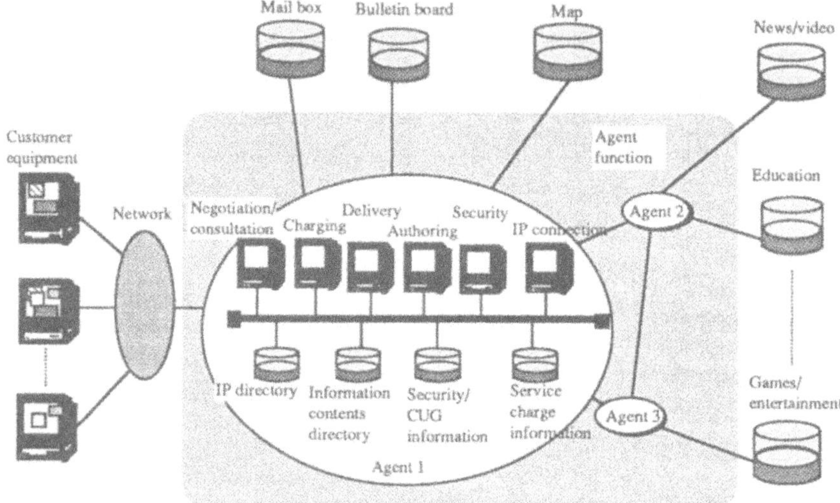

Figure 6 Agent functionality schematic.

6 IMPROVED CUSTOMER CONTACT

Negotiations between the SP and customer over service conditions will be carried out primarily through a machine-machine interface (MMI). To realize rapid, accurate and economical negotiations, the negotiation interface will have to be largely mechanized. The interface for customer contact will gradually move away from contact with humans to contact with machines (see Figure 7). The operations system function (OSF) diagrammed in Figure 7 includes negotiation and operation functions for reconfiguration, testing and related tasks.

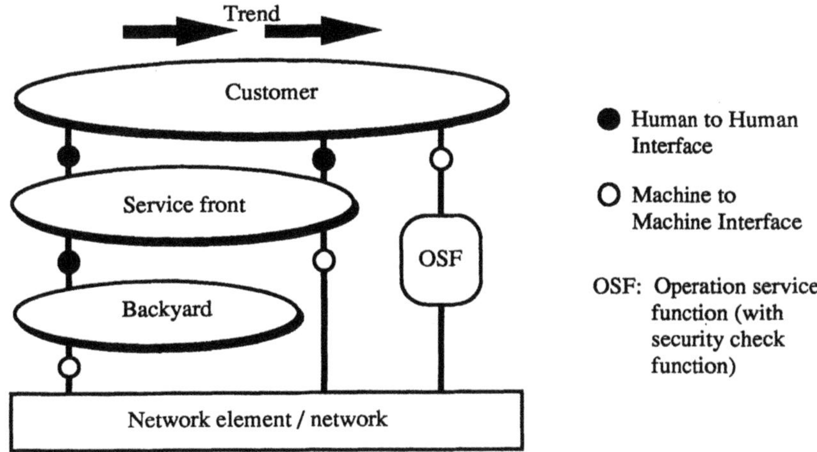

Figure 7 Operation service interfaces.

Although the mechanization of the customer contact function releases operators from their jobs on the SP side and thereby reduces operating costs, it is important to avoid the imposition of complicated procedures upon customers. To increase customer convenience, the deployment of intelligent functions in both CPE and SP networks will be necessary.

In a sophisticated, mechanized telecommunications environment, customers will want to know more about operations and management functions, as well as service capabilities (ITU-T, 1994a). In order to satisfy this demand for information, it is necessary to retain a powerful Human to Human Interface (HHI) which removes as much inconvenience as possible in communications. To provide support to operators at the HHI service front, SPs have developed sophisticated management information networking capabilities. Operators are able to obtain appropriate information quickly and efficiently when contacted by customers (ITU-T, 1993).

OSF and the customer contact point (service front) are equally important in the future service environment. HHI and MMI complement each other, offering customers the best type of contact for the information needed by customers at a particular moment.

7 NEW PARADIGM FOR NETWORK ARCHITECTURE

Many new service providers have entered into the telecommunications market, leading to a complex and rapidly changing telecommunications service provision structure. Since customers want their telecommunications services to work in a seamless fashion, it is necessary to deploy a new network architecture capable of achieving this. This, in turn, necessitates a paradigm shift in the way service providing mechanisms are conceptualized.

A diagram of the existing network is shown in Figure 8. CPE and SP networks are developed independently and are interconnected through User Network Interface (UNI). Within SP networks, the transport network and various network nodes are interconnected through Network Node Interface (NNI). ITU-T SG15 currently discusses UNI and Service Node Interface (SNI) in access networks (Matsushita, Okazaki and Yoshida, 1995).

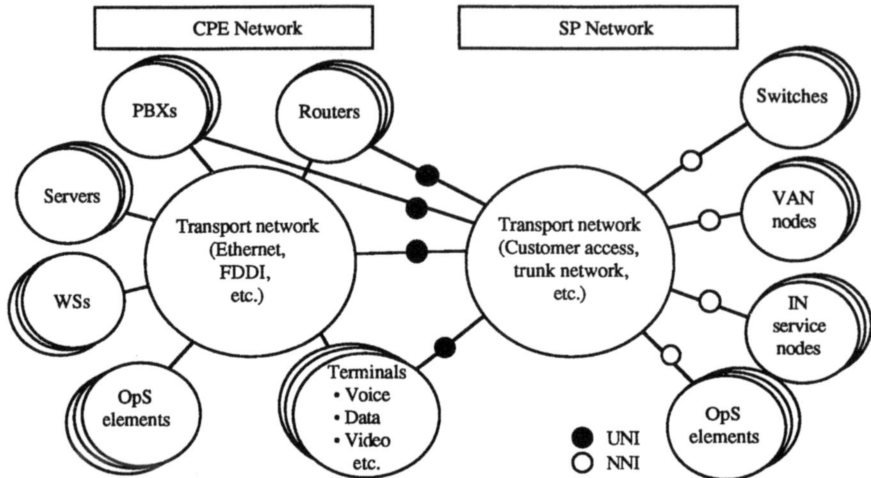

Figure 8 Existing network architecture and interfaces.

As technology advanced, the difference between CPE and SP networks narrowed. Within the CPE environment, the common architecture included a centralized information processing network based on a mainframe computer. The progress of LAN (Local Area Network) and WAN (Wide Area Network) technologies introduced distributed processing architecture into the local and world-wide environments.

In SP networks, network nodes (switching systems) used to be based on a mainframe type functionally centralized architecture, although they were distributed geographically. The need for rapid and flexible service provisioning as well as sophisticated services has forced a review of existing network architecture. The answer is a functionally distributed architecture, such as separation of service definition functions and connection functions.

Digitalization of transmission is proceeding in both CPE and SP networks. CPE networks have evolved using digital transmission technology as is the case, for example, with the Ethernet. With SP networks, digital transmission systems have been introduced into trunk networks as a first step. The digitalization process for CPE and SP networks has proceeded independently. Once access networks in SP networks are digitalized, a fence will be removed between CPE and SP networks.

CPE and SP networks are now evolving in a similar direction – toward a functionally distributed environment using similar digital processing and transmission technologies. In the emerging environment, it will be possible to use the same hardware and software for the two types of network.

Suppose the trend continues, the networks merge into functional homogeneity and one unique class of interface is established which is supported by the use of common software packages. The new network architecture shown in Figure 9 would be shared between customers and SPs. This would accelerate the smooth interworking of CPE and SP networks in the complex service offering environment illustrated in Figure 2. Customers would have the freedom to construct their networks without feeling constrained with respect to the choice of SPs or vendors when choosing services. Some customers could become SPs for other customers, thus expanding on their business opportunities.

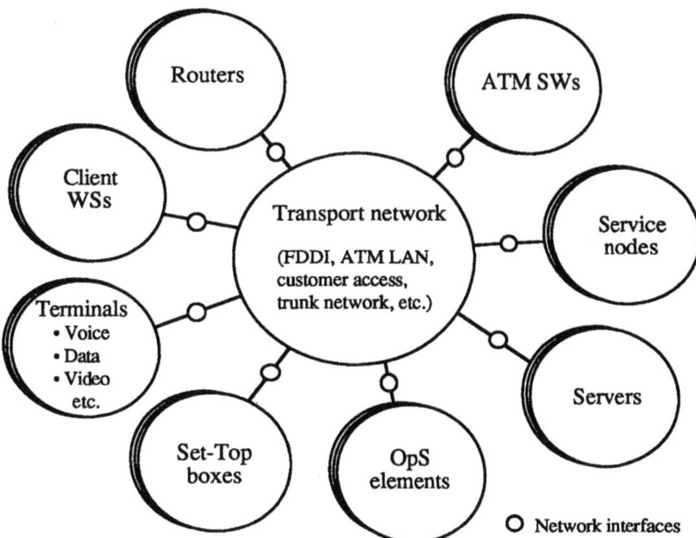

Figure 9 New network architecture and interfaces.

8 CONCLUSION

The service provision structure is becoming complex, involving a number of overlapping fields, as well as several service providers within most of these fields. Customers still want seamless service, even though they want to freely choose services from any combination of SPs and negotiate over price and service features.

To satisfy these customer demands, service providers will need to undergo a paradigm shift in the way they think about service provision. The new paradigm, from a technological standpoint, involves an agent/negotiation function as well as new network architecture integrating the service providers' networks with CPE networks in a distributed processing environment. This paradigm shift also involves a shift in pricing structure to accommodate the widespread advent of multimedia communications.

9 ACKNOWLEDGEMENTS

The author would like to express his gratitude to Messrs. Masahiko Matsushita and Noriyuki Terada for their pertinent suggestions during the preparation of this paper.

10 REFERENCES

Ejiri, M. (1994a) For whom the advancing service/network management. Keynote speech, *NOMS '94 Symposium Record,* Vol. 2, pp. 422-433.

Ejiri, M. (1994b) Advancing service operations and operations systems. *NTT Review,* Vol. 6, No. 3, pp. 31-36, 1994.

ITU-T (1993) Tele administration service. *ITU-T Draft Recommendation F.ADM.*

ITU-T (1994a) Framework recommendation on functional access networks. *ITU-T Draft Recommendation* G.9xx

ITU-T (1994b) Universal Personal Telecommunications (UPT) service description. *ITU-T Recommendation F.851.*

Maeda, M. and M. Ejiri (1994) Enhancement of service front operation, *NTT Review,* Vol. 6, No. 3, pp. 37-45.

Matsushita, M., T. Okazaki and M. Yoshida (1995) A telecommunications management integration network *IEICE TRANS. COMMUN.,* Vol. E78-B, No. 1 (January).

Natarajan, N. and G. M. Slawsky (1992) A framework architecture for multimedia information networks. *IEEE Communications Magazine,* February, pp. 97-104.

Network Management Forum (1994) Requirement capture: service management automation and re-engineering (draft), September 23.

11 BIOGRAPHY

Masayoshi Ejiri received his Bachelor's degree in Engineering from the University of Tokyo in 1967. Since joining NTT, he has worked in a number of areas, including transmission systems development and network planning and engineering. He has also directed a telephone office and managed operations systems development and telecommunications software production. He is currently in charge of strategy and system development for the Service and Network Operations Section of NTT's Network Engineering Department. Mr. Ejiri is a member of IEEE and is the General Co-Chair of IEEE/IFIP 1996 Network Operations and Management Symposium (NOMS '96).

63

An Industry Response to Comprehensive Enterprise Information Systems Management

Bill WARNER, IBM, U.S.A.

Hear Bill Warner's perspective on the systems management challenges our customers are trying to overcome; and the corresponding actions that are required by vendors who desire success in the systems management business.

Today's systems management industry is changing fast and vendors must respond as fast to the variety of needs which span the enterprise of customers large and small. This poses a tremendous challenge both for the end user and the vendors; a challenge which can be overcomed with the right plan and strategic focus on the problems our customers are trying to solve -- a focus which begins with the customers business processes and not the information technology used to achieve their success.

Bill Warner will discuss the IBM response to simplifying the management process, the openness required for technology independence, and the plan for delivering strategic new functions in the future.

64

Cooperative Management

Denis YARO, Sun Microsystems, U.S.A.

An exploration of the unification of management approaches from a variety of perspectives, including user needs, business/market realities, and technological innovation.

PART FIVE

POSTERS

65

Network Management Simulators

Anders LUNDQVIST, Nils WEINANDER, T. GRONBERG

Ericsson Infocom Consultants AB, SWEDEN

This poster session describes requirements, functions and implementation of OSI management simulation software. The described systems simulate TMN managers and agents in order to verify the management functionality of network elements and operations systems. A number of areas of use have been discovered, in addition to automated tests of Q3 interfaces.

66

On the Distributed Fault Diagnosis of Computer Networks

Henri NUSSBAUMER, Sailesh CHUTANI

Swiss Federal Institute of Technology, SWITZERLAND

We propose a general technique for the fault diagnosis of communication networks that is inspired by the theory of system-level diagnosis. This technique relies on the paradigm of comparison testing. A set of tasks, possibly implicit, is executed by the nodes in a network. The resulting agreements and disagreements in their results are used to diagnose all the faulty nodes and links with a high probability. The diagnosis algorithm proposed is applicable in a centralized as well as a distributed system. The accuracy of the diagnosis is controlled by the number of rounds of tasks performed.

67

Fault Diagnosis in Computer Networks

Martin de GROOT

University of New South Wales, AUSTRALIA

Fault diagnosis in networks of communicating devices is performed manually for all but the most common problems. Network management systems typically provide only the protocol for collecting status messages from managed nodes, and a facility for displaying these messages to the network administrator. Apart from colour coding the messages to indicate the severity, very little assistance is given to the human manager to help isolate any faults.

A system manager is generally only interested in the status messages which indicate abnormal behaviour. Such messages are commonly referred to as alarms. Chemical and electrical engineers have been interested in the possibility of automating alarms management for a long time. The most practical solution has been to build an expert system. There is a lot that computer scientists can learn from the work done in these two areas to automate this aspect of network management.

There are, however, significant differences between the task faced by process engineers and computer network managers. Although both are dealing with networks of devices, the computer network consists of more complex devices, is often much larger, and is more dynamic. While it may be feasible to build a customised expert system for, say, a blast furnace, because the process is well understood and does not change, such an ES for a computer network could never be completed before machines are upgraded or the network topology changes.

This paper is a brief discussion of the issues involved in producing an alarms management system suitable for computer networks. It will be argued that the essential problem is an extreme case of a "knowledge acquisition bottleneck". Two complementary techniques for dealing with this issue will be discussed. Firstly, we will consider a rapid knowledge base maintenance system which does not require the assistance of a knowledge engineer. Then we will briefly examine the possibility of using formal techniques to define automatic rule generation systems.

68

The Distributed Management Tree - Applying a new Concept for Managing Distributed Applications to E-mail

Vito BAGGIOLINI, Eduardo SOLANA, Jean-Francois PACCINI

Mira RAMLUCKUN, Stephane SPAHNI, Jurgens HARMS

University of Geneva, SWITZERLAND

The "Distributed Management Tree" (DMT) is a hierarchical structure designed for the management of distributed systems. The DMT has the form of an inverted tree, with nodes representing small active units for processing elements of management information. The DMT is not integrated into the system it manages but built next to it, supervising it "from the outside". The DMT has two main functionalities: (1) it extracts and refines information concerning the managed system, and (2) provides a mechanism for specifying and handling actions on the managed system. The nodes are programmed to permanently analyze the information about the managed system and to find out if it is in a normal operational state or not. If a faulty behaviour is detected, the DMT can either fix it autonomously or alert a human administrator, depending on the nature of the error. The different hierarchy levels in the tree represent the information obtained at the terminal nodes views with different levels of detail. Furthermore, they provide means to trigger complex commands and propagate them downwards, decomposing them into more elementary commands. This concept has been applied to the management of E-mail systems. A prototype has been developed for managing an important and heterogeneous fraction of the University's E-mail system.

69

A Distributed Hierarchical Management Framework for Heterogeneous WANs

Mark STOVER, Sujata BANERJEE

University of Pittsburgh, U.S.A.

The scope of network management is expanding in multiple dimensions. Local area networks (LANs) have more nodes than ever before, enterprise networks span national boundaries, wide area networks (WANs) cover the globe, and administrators want to manage their LANs all the way down to a PC's network interface and application software. In order to centrally manage these networks, the network manager faces the complexity of heterogeneous management tools and the difficulty of managing vast amount of data generated by network elements.

Many researchers have begun taking a data-centric view of network management and generally agree that a well-structured global network database is essential for effective network management. One promising paradigm identified by several researchers is to monitor and manage the network through the network management database; however, a number of issues remain-most importantly, the architecture and the data-distribution scheme of the management database. Other important issues include maintaining database consistency, minimizing network management traffic, and interoperation of multiple management standards.

Just as in the research of [1-3], we recognize the importance of a global network management database. Although the MANDATE [1] project proposed a database design that includes the distribution of some data, the focus on a central repository for structural and control data, and the lack of provision for heterogeneous interoperation of multiple management standards, presents a number of difficulties. Our research proposes a fully distributed database with the addition of a new scheme for the hierarchical distribution of network management data.

Our research is driven by these goals: to minimize the network overhead of management data, to create a flexible and scalable management framework that supports multiple management standards, and to provide continued management during network partitioning. The most important aspect of our design is the hierarchical distribution of network management data with multiple management levels. Our design relies upon a distributed database management system (DDBMS) to distribute and replicate the management data.

By distributing network management and network management data, there are a number of data-handling issues that need to be addressed. An important consideration is the data granularity at each network management level; a general rule of thumb is that as one traverses downward through the management hierarchy, the data granularity moves from coarse to fine. This granularity reflects the concerns of each level's network manager-higher-level managers will be interested in summary data while the lower-level managers are responsible for all of the data associated with every network element. Because more than one network manager may simultaneously initiate a configuration of the same managed object, there must be a concurrency control mechanism; a primary copy update mechanism is adequate to deal with these conflicts. A higher-level manager will normally have priority over a lower-level manager and has the option of preempting an operation in progress. By keeping primary copies of structural, control, and sensor data as close to the managed network element as possible, network overhead is minimized. This design permits a degree of autonomy to local management domains while ensuring that the rest of the network is aware of all management decisions that may affect them.

The DDBMS contains the management protocol traffic within the local management domains, which has two important benefits. First, local management stations can operate with multiple protocol stacks and then hide those multiple stacks from the rest of the management system. The DDBMS, with an added translation component, acts as a common agent [4] and makes the conversion between the database representation of the management data and the structure of management information (SMI) specified by the network management standards. The DDBMS will integrate and convert between existing MIBs and the database definitions. In addition, the DDBMS enables scalability because local management and database needs can be divided among many systems as required to provide adequate performance on networks of various sizes.

It is assured that networks will continue to grow in both size and complexity, and network management must evolve to accommodate this growth. Our proposed design takes a data-centric view of network management and uses the technology of distributed database management systems to provide a uniform method of managing a broad range of networks and of network elements.

References

[1] J. R. Haritsa, M. O. Ball, N. Roussopoulos, A. Datta, and J. S. Baras. MANDATE: MAnaging Networks using DAtabase TEchnology. IEEE Journal on Selected Areas in Communications, 11(9):1360-1372, December 1993.

[2] A. Dupuy, S. Sengupta, O. Wolfson, and Y. Yemini. NETMATE: A Network Management Environment. IEEE Network Magazine, pages 35-43, March 1991.

[3] O. Wolfson, S. Sengupta, and Y. Yemini. Managing Communication Networks by Monitoring Databases. IEEE Transactions on Software Engineering, 17:944-953, 1991.

[4] O. Newkerk, M. A. Nihart, and S. K. Wong. The Common Agent - A Multiprotocol Management Agent. IEEE Journal on Selected Areas in Communications, 11(9):1346-1352, December 1993

70

ISOS: Intelligent Shell Of SNMP

Jianxin LI, Benjamin J. LEON

University of Southwestern Louisiana, U.S.A.

SNMP is today's dominant network management software product. In this poster, we propose an approach to enhance the functions of SNMP through the use of an intelligent shell. The shell concept in network management is akin to that in operating systems. ISOS uses the shell script to aggregate SNMP operations. It supports the imperative features, such as sequencing, alternation, and iteration. In addition, ISOS incorporates searching and planning techniques to support query manipulation and agent-oriented programming. We claim that ISOS can relieve a network manager from tedious monitoring and controlling of the network, and it can also reduce the management traffic overload. Our prototype of ISOS is dependent on Unix shell and SNMPv2 developed by Carnegie-Mellon University.

71

A Critical Analysis of the DESSERT Information Model

Richard MEADE, Ahmed PATEL

University College Dublin, IRELAND

Declan O'SULLIVAN, Mark TIERNEY

Broadcom Eirann Research, IRELAND

This poster looks at a number of different Information Models which have been developed within various problem domains related to network management and highlights the important similarities. Furthermore it considers one particular problem domain, service provisioning, and an Information Model which was developed for it, in which the benefits of Information Modelling are particularly apparent because of the wide scope and characteristics of the domain.

Finally, we also propose a new approach to modelling networks traffic, based on this model. This will enable modelling of high level and low level details while providing a more flexible and more complete method of modelling characteristics such as network connectivity and topology. It also enables easier and more appropriate modelling of quality of service parameters.

INDEX OF CONTRIBUTORS

KEYWORD INDEX

GPSR Compliance
The European Union's (EU) General Product Safety Regulation (GPSR) is a set of rules that requires consumer products to be safe and our obligations to ensure this.

If you have any concerns about our products, you can contact us on

ProductSafety@springernature.com

In case Publisher is established outside the EU, the EU authorized representative is:

Springer Nature Customer Service Center GmbH
Europaplatz 3
69115 Heidelberg, Germany

www.ingramcontent.com/pod-product-compliance
Ingram Content Group UK Ltd.
Pitfield, Milton Keynes, MK11 3LW, UK
UKHW022316190726
13856UKWH00001B/44

* 9 7 8 1 4 7 5 7 5 5 1 6 9 *